Lecture Notes in Networks and Systems

Volume 1000

The series "Lecture Notes in Networks and Systems" publishes the latest developments in Networks and Systems—quickly, informally and with high quality. Original research reported in proceedings and post-proceedings represents the core of LNNS.

Volumes published in LNNS embrace all aspects and subfields of, as well as new challenges in, Networks and Systems.

The series contains proceedings and edited volumes in systems and networks, spanning the areas of Cyber-Physical Systems, Autonomous Systems, Sensor Networks, Control Systems, Energy Systems, Automotive Systems, Biological Systems, Vehicular Networking and Connected Vehicles, Aerospace Systems, Automation, Manufacturing, Smart Grids, Nonlinear Systems, Power Systems, Robotics, Social Systems, Economic Systems and other. Of particular value to both the contributors and the readership are the short publication timeframe and the world-wide distribution and exposure which enable both a wide and rapid dissemination of research output.

The series covers the theory, applications, and perspectives on the state of the art and future developments relevant to systems and networks, decision making, control, complex processes and related areas, as embedded in the fields of interdisciplinary and applied sciences, engineering, computer science, physics, economics, social, and life sciences, as well as the paradigms and methodologies behind them.

Indexed by SCOPUS, INSPEC, WTI Frankfurt eG, zbMATH, SCImago.

All books published in the series are submitted for consideration in Web of Science.

For proposals from Asia please contact Aninda Bose (aninda.bose@springer.com).

Xin-She Yang · Simon Sherratt · Nilanjan Dey ·
Amit Joshi
Editors

Proceedings of Ninth International Congress on Information and Communication Technology

ICICT 2024, London, Volume 5

Springer

Editors
Xin-She Yang
Middlesex University
London, UK

Simon Sherratt
University of Reading
Reading, UK

Nilanjan Dey
Techno India College of Technology
Kolkata, West Bengal, India

Amit Joshi
Global Knowledge Research Foundation
Ahmedabad, Gujarat, India

ISSN 2367-3370 ISSN 2367-3389 (electronic)
Lecture Notes in Networks and Systems
ISBN 978-981-97-3288-3 ISBN 978-981-97-3289-0 (eBook)
https://doi.org/10.1007/978-981-97-3289-0

This Springer imprint is published by the registered company Springer Nature Singapore Pte Ltd.
The registered company address is: 152 Beach Road, #21-01/04 Gateway East, Singapore 189721, Singapore

Preface

The Ninth International Congress on Information and Communication Technology will be held during 19–22 February 2024, in a hybrid mode, physical at London, UK and digital platform Zoom. ICICT 2024 was organised by Global Knowledge Research Foundation and managed by G. R. Scholastic LLP. The associated partners were Springer and Springer Nature. The conference will provide a useful and wide platform both for display of the latest research and for exchange of research results and thoughts. The participants of the conference will be from almost every part of the world, with backgrounds of either academia or industry, allowing a real multinational and multicultural exchange of experiences and ideas.

A great pool of more than 2400 papers were received for this conference from across 129 countries among which around 485 papers were accepted and will be presented physically at London and digital platform Zoom during the four days. Due to the overwhelming response, we had to drop many papers in the hierarchy of the quality. Total 70 technical sessions will be organised in parallel in 4 days along with a few keynotes and panel discussions in hybrid mode. The conference will be involved in deep discussion and issues which will be intended to solve at global levels. New technologies will be proposed, experiences will be shared, and future solutions for design infrastructure for ICT will also be discussed. The final papers will be published in ten volumes of proceedings by Springer LNNS Series. Over the years, this congress has been organised and conceptualised with collective efforts of a large number of individuals. I would like to thank each of the committee members and the reviewers for their excellent work in reviewing the papers. Grateful acknowledgements are extended to the team of Global Knowledge Research Foundation for their valuable efforts and support.

I look forward to welcoming you to the 10th edition of this ICICT Congress 2025.

Amit Joshi, Ph.D.
Organising Secretary, ICICT 2024
Director—Global Knowledge Research Foundation
Ahmedabad, India

Contents

Editors and Contributors

About the Editors

Xin-She Yang obtained his D.Phil. in Applied Mathematics from the University of Oxford and subsequently worked at the Cambridge University and the National Physical Laboratory (UK) as a Senior Research Scientist. He is currently Reader in Modeling and Optimization at Middlesex University London and Adjunct Prof. at Reykjavik University (Iceland). He is also elected Bye-Fellow at Cambridge University and the IEEE CIS Chair for the Task Force on Business Intelligence and Knowledge Management. He was included in the "2016 Thomson Reuters Highly Cited Researchers" list.

Simon Sherratt was born near Liverpool, England, in 1969. He is currently Prof. of Biosensors at the Department of Biomedical Engineering, University of Reading, UK. His main research area is signal processing and personal communications in consumer devices, focusing on wearable devices and health care. He received the 1st place IEEE Chester Sall Memorial Award in 2006, the 2nd place in 2016, and the 3rd place in 2017.

Nilanjan Dey is Assistant Prof. at the Department of Information Technology, Techno India College of Technology, India. He has authored/edited more than 75 books with Springer, Elsevier, Wiley, CRC Press and published more than 300 peer-reviewed research papers. He is Editor-in-Chief of the International *Journal of Ambient Computing and Intelligence*; Series Co-editor of *Springer Tracts in Nature-Inspired Computing* (STNIC); and Series Co-editor of *Advances in Ubiquitous Sensing Applications for Healthcare*, Elsevier.

Amit Joshi is Director of Global Knowledge Research Foundation, also Entrepreneur and Researcher who has completed his Masters' and research in the areas of cloud computing and cryptography in medical imaging. He has an experience of around ten years in academic and industry in prestigious organizations. He is Active Member of ACM, IEEE, CSI, AMIE, IACSIT-Singapore, IDES, ACEEE, NPA, and many other professional societies. He is International Chair of InterYIT at International Federation of Information Processing. He has presented and published more than 50 papers in national and international journals/conferences of IEEE and ACM. He has also edited more than 40 books which are published by Springer, ACM, and other reputed publishers. He has also organized more than 50 national and international conferences and programs in association with ACM, Springer, and IEEE to name a few across different countries including India, UK, Europe, USA, Canada, Thailand, Egypt, and many more.

Contributors

Qasem Abu Al-Haija Department of Cybersecurity, Faculty of Computer and Information Technology, Jordan University of Science and Technology, Irbid, Jordan

Sunday Achimugu Telecommunications Engineering Department, Federal University of Technology, Minna, Niger State, Nigeria

Nevena Ackovska Faculty of Computer Science and Engineering, Ss. Cyril and Methodius University, Skopje, North Macedonia

Maulana Amirul Adha Universitas Negeri Jakarta, Jakarta, Indonesia

Yasmina Alami University Hassan First of Settat, Ecole Nationale de Commerce et de Gestion, Settat, Morocco

Abdulrahman S. Alenizi The Public Authority for Applied Education and Training (PAAET), Shuwaikh Industrial, Kuwait

Khamis A. Al-Karawi Diyala University, Baqubah, Diyala, Iraq

Mohammad Alnabhan Department of Cybersecurity, Princess Sumaya University for Technology, Amman, Jordan

Abdulaziz A. Alsulami Department of Information Systems, King Abdulaziz University, Jeddah, Saudi Arabia

Badraddin Alturki Department of Information Technology, King Abdulaziz University, Jeddah, Saudi Arabia

Nova Syafira Ariyanti Universitas Negeri Malang, Malang, Indonesia

Adam Emile Aske Inland Norway University, Hamar, Norway

Junnel E. Avestro Technological Institute of the Philippines, Quezon City, Metro Manila, Philippines

Pramod Baddam Center for Information Services and HPC, Tu Dresden, Germany

Abderrahim Bajit Laboratory of Advanced Systems Engineering (ISA), National School of Applied Sciences, Ibn Tofail University, Kenitra, Morocco

Merxhan Bajrami Laigo GmbH, Friedrichshafen, Germany

Viktória Bakonyi Eötvös Loránd University, Budapest, Hungary

Israel Benjamin Luis Rivera Barreto Universidad Nacional Mayor de San Marcos, Lima, Peru

Nisrine Benhammou Ecole Nationale de Commerce et de Gestion, LR2MC, University Hassan First of Settat, Settat, Morocco

Francisco L. Benítez-Martínez COSCYBER, Fundación I+D del Software Libre (Fidesol), Granada, Spain

Habiba Bensassi Nour Ecole Nationale de Commerce et de Gestion, LR2MC, University Hassan First of Settat, Settat, Morocco

Calebe P. Bianchini Computing and Informatics Department, Mackenzie Presbyterian University, São Paulo, Brazil

Christian Ildegard Bimenyimana Université des Grands Lacs, Saint Michel, Bujumbura, Burundi

A. S. Borodulin Bauman Moscow State Technical University, Moscow, Russia

Jerry E. Borromeo Technological Institute of the Philippines, Quezon City, Metro Manila, Philippines

Khawla Bouafia Eötvös Loránd University of Budapest, Budapest, Hungary

Kumar K. Busawon Faculty of Sustainable Development and Engineering, Universite des Mascareignes, Roches Brunes, Rose-Hill, Mauritius

Robin Carlsson University of Turku, Turku, Finland

Christian Casalvieri Sapienza University of Rome, Rome, Italy

Illya Chaikovsky V. M. Glushkov Institute of Cybernetics of the NAS of Ukraine, Kyiv, Ukraine

Amália S. Chaves Departamento de Medicina Veterinária, Universidade Federal de Juiz de Fora (UFJF), Juiz de Fora, MG, Brazil

Alessandro Chiumento Twente University, Enschede, The Netherlands

Montse Cuadros Vicomtech Foundation, Basque Research and Technology Alliance (BRTA), Donostia/san-Sebastián, Spain

Gaby G. Dagher San Jose State University, San Jose, CA, USA;
Marshall University, Huntington, WV, USA;
Boise State University, Boise, ID, USA

Salih Demir Faculty of Open and Distance Education, Ankara University, Ankara, Turkey

Katherine Regina Vasquez Diaz School of Education, Faculty of Arts, Cultures and Education, University of Hull, Hull, UK

Ervin Domazet International Balkan University, Skopje, North Macedonia

Nguyen Doan Dong International School, Vietnam National University, Hanoi, Vietnam

M. N. Drakshayini School of Electronics and Communication Engineering, REVA University, Bengaluru, India

James Lucas Durham Marshall University, Huntington, WV, USA;
Boise State University, Boise, ID, USA;
San Jose State University, San Jose, CA, USA

Abdelhadi EL Allali Laboratory of Advanced Systems Engineering (ISA), National School of Applied Sciences, Ibn Tofail University, Kenitra, Morocco

Rachid EL Bouayadi Laboratory of Advanced Systems Engineering (ISA), National School of Applied Sciences, Ibn Tofail University, Kenitra, Morocco

Alaa Elkouni American University of Sharjah, Sharjah, United Arab Emirates

Nafisa Sultana Elme World University of Bangladesh, Dhaka, Bangladesh

Luis El-Moga-Sarria COSCYBER, Fundación I+D del Software Libre (Fidesol), Granada, Spain

Mohamed ElMohandes American University of Sharjah, Sharjah, United Arab Emirates

H. Essamri National School of Applied Sciences, Ibn Tofail University, Kenitra, Morocco

B. Essounaini National School of Applied Sciences, Ibn Tofail University, Kenitra, Morocco

Víctor Manuel Parasi Falcon Universidad Peruana de Ciencias Aplicadas (UPC), Monterrico, Santiago de Surco, Peru

Celia Fernández COSCYBER, Fundación I+D del Software Libre (Fidesol), Granada, Spain

Marcos P. P. Ferreira Departamento de Medicina Veterinária, Universidade Federal de Juiz de Fora (UFJF), Juiz de Fora, MG, Brazil

Almira B. Franca Departamento de Medicina Veterinária, Universidade Federal de Juiz de Fora (UFJF), Juiz de Fora, MG, Brazil

Marta Fuentes-García COSCYBER, Fundación I+D del Software Libre (Fidesol), Granada, Spain

Alessandro Gambini Sapienza University of Rome, Rome, Italy

Jess N. Garcia Technological Institute of the Philippines, Quezon City, Metro Manila, Philippines

Akash Ghosh Computing Science, Thompson Rivers University, Kamloops, B.C, Canada

Ayse Glass HafenCity University—Digital City Science, Hamburg, Germany

A. R. Glinscaya Reshetnev Siberian State University of Science and Technology, Krasnoyarsk, Russia

Priscila V. Z. C. Goliatt Departamento de Ciência da Computação, Universidade Federal de Juiz de Fora (UFJF), Juiz de Fora, MG, Brazil

Alicia Grande LIS Data Solutions, Santander, Spain

Zyrus D. Gremio Technological Institute of the Philippines, Quezon City, Metro Manila, Philippines

Mehedi Hasan World University of Bangladesh, Dhaka, Bangladesh

Paul Havinga Twente University, Enschede, The Netherlands

Joël T. Hounsou Institut de Mathématiques et de Sciences Physiques de l'Université d'Abomey-Calavi, Dangbo, Porto-Novo, Benin

Bo Huang School of Computer Science, University of Guelph, Geulph, Ontario, Canada

Adamu Hussaini Telecommunications Engineering Department, Federal University of Technology, Minna, Niger State, Nigeria

Zoltán Illés Eötvös Loránd University, Budapest, Hungary

Jamshed Iqbal School of Computer Science, Faculty of Science and Engineering, University of Hull, Hull, UK

Jonathan Jander HAW Hamburg, Hamburg, Germany

René Jäkel Center for Information Services and HPC, Tu Dresden, Germany; Center for Scalable Data Analytics and Artificial Intelligence (ScaDS.AI) Dresden/ Leipzig, Dresden University of Technology, Dresden, Germany

Salima Jazi University Hassan First of Settat, Ecole Nationale de Commerce et de Gestion, Settat, Morocco

Juharyanto Universitas Negeri Malang, Malang, Indonesia

Izzet Kale Applied DSP and VLSI Research Group (ADVRG), University of Westminster, London, UK

Firuz Kamalov Department of Electrical Engineering, Canadian University Dubai, Dubai, UAE

Murat Karakus Department of Software Engineering, Ankara University, Ankara, Turkey

Francis Kawesha University of Zambia, Lusaka, Zambia

Petr Klán Faculty of Information Technology, Czech Technical University in Prague, Prague, Czech Republic

Manjunath R. Kounte School of Electronics and Communication Engineering, REVA University, Bengaluru, India

Tom Krause HAW Hamburg, Hamburg, Germany

V. V. Kukartsev Bauman Moscow State Technical University, Moscow, Russia;
Siberian Federal University, Krasnoyarsk, Russia;
Reshetnev Siberian State University of Science and Technology, Krasnoyarsk, Russia

Pamela Kunert HAW Hamburg, Hamburg, Germany

Petre Lameski Faculty of Computer Science and Engineering, Ss. Cyril and Methodius University, Skopje, North Macedonia

Khutso Lebea University of Johannesburg, Johannesburg, South Africa

Ville Leppänen University of Turku, Turku, Finland

Indra Lesmana Universitas Negeri Malang, Malang, Indonesia

Alex Lewtschuk San Jose State University, San Jose, CA, USA;
Marshall University, Huntington, WV, USA;
Boise State University, Boise, ID, USA

Fagen Li School of Computer Science and Engineering, University of Electronic Science and Technology of China, Chengdu, People's Republic of China

Wenjun Lin Digital Healthcare Innovation Lab, School of Computer Science and Technology, Algoma University, Brampton, Ontario, Canada

David Lopez LIS Data Solutions, Santander, Spain

Mihaela Luca Institute for Computer Science, Romanian Academy, Iași, Romania

Peyton Lundquist San Jose State University, San Jose, CA, USA;
Marshall University, Huntington, WV, USA;
Boise State University, Boise, ID, USA

Luiz M. S. Maciel Departamento de Ciência da Computação, Universidade Federal de Juiz de Fora (UFJF), Juiz de Fora, MG, Brazil

Rosa M. O. F. Maciel Departamento de Ciência da Computação, Universidade Federal de Juiz de Fora (UFJF), Juiz de Fora, MG, Brazil

Radu Maftei Department of Mother and Child Medicine, Faculty of Medicine, University of Medicine and Pharmacy "Grigore T. Popa", Iași, Romania; Clinical Hospital of Obstetrics and Gynecology "Cuza Vodă", Iași, Romania; Origyn Fertility Center, Iași, Romania

Brian L. C. Maia Departamento de Ciência da Computação, Universidade Federal de Juiz de Fora (UFJF), Juiz de Fora, MG, Brazil

Maisyaroh Universitas Negeri Malang, Malang, Indonesia

Gurbanli Majid Eötvös Loránd University of Budapest, Budapest, Hungary

Youssef Makloul FSJES, L-Qualimat GRTE Laboratory, Cadi Ayyad University of Marrakech, Marrakech, Morocco

Vladyslav Malanin V. M. Glushkov Institute of Cybernetics of the NAS of Ukraine, Kyiv, Ukraine

Kochneva Maria St Petersburg University, Saint Petersburg, Russia

Daniel Matrone Computing and Informatics Department, Mackenzie Presbyterian University, São Paulo, Brazil

Darren McKie School of Computer Science, Faculty of Science and Engineering, University of Hull, Hull, UK

Daniela Mechkaroska University for Information Science and Technology "St. Paul the Apostle", Ohrid, North Macedonia

Rohan Mitra Department of Computer Science and Engineering, American University of Sharjah, Sharjah, UAE

Bálint Molnár Eötvös Loránd University of Budapest, Budapest, Hungary

Ilham Morino Laboratory of Advanced Systems Engineering (ISA), National School of Applied Sciences, Ibn Tofail University, Kenitra, Morocco

Emmanuel Samuel Mtsweni Department of Computer Science, University of South Africa, Science Campus, Florida Park, Republic of South Africa

Sadiq Abubakar Muhammed Telecommunications Engineering Department, Federal University of Technology, Minna, Niger State, Nigeria

Deivid D. Nascimento Departamento de Engenharia de Produção e Mecânica, Universidade Federal de Juiz de Fora (UFJF), Juiz de Fora, MG, Brazil

V. A. Nelyub Bauman Moscow State Technical University, Moscow, Russia

Guilherme G. Neto Coordenadoria de Engenharia de Produção, Instituto Federal do Espírito Santo (IFES), Cariacica, ES, Brazil

Aaron H. Nguyen San Jose State University, San Jose, CA, USA;
Marshall University, Huntington, WV, USA;
Boise State University, Boise, ID, USA

Gerard Niyonsaba Centre de Recherche Indépendant et d'Informatisation du Système d'Information, Centre Porta Sion, Bujumbura, Burundi

Therence Niyonsaba Institut de Statistique Appliquée de l'Université du Burundi, Nyamugerera, Bujumbura, Burundi

Järg Rainer Noennig HafenCity University—Digital City Science, Hamburg, Germany

Thierry Nsabimana Institut de Statistique Appliquée de l'Université du Burundi, Nyamugerera, Bujumbura, Burundi

Senzo Ntsele University of Johannesburg, Johannesburg, South Africa

Iryna Okhrin Center for Information Services and HPC, Tu Dresden, Germany; Center for Scalable Data Analytics and Artificial Intelligence (ScaDS.AI) Dresden/Leipzig, Dresden University of Technology, Dresden, Germany

Lorena N. C. Oliveira Departamento de Medicina Veterinária, Universidade Federal de Juiz de Fora (UFJF), Juiz de Fora, MG, Brazil

Y. Onqani National School of Applied Sciences, Ibn Tofail University, Kenitra, Morocco

Murat Osmanoglu Department of Computer Engineering, Ankara University, Ankara, Turkey

Honorato Ccalli Pacco National University of Moquegua, Pacocha, Peru

Roxanne A. Pagaduan Technological Institute of the Philippines, Quezon City, Metro Manila, Philippines

Rodrigo P. Pasquale Computing and Informatics Department, Mackenzie Presbyterian University, São Paulo, Brazil

Walace A. Paula Departamento de Ciência da Computação, Universidade Federal de Juiz de Fora (UFJF), Juiz de Fora, MG, Brazil

Jackson Phiri University of Zambia, Lusaka, Zambia

Ryan Christopher B. Pinca Technological Institute of the Philippines, Quezon City, Metro Manila, Philippines

Panu Puhtila University of Turku, Turku, Finland

Medha Pujari Western New England University, Springfield, Massachusetts, USA

Tran Duc Quynh International School, Vietnam National University, Hanoi, Vietnam

Abel van Raalte Twente University, Enschede, The Netherlands

Sampsa Rauti University of Turku, Turku, Finland

Tapu Rayhan World University of Bangladesh, Dhaka, Bangladesh

Saumya Kareem Reni Applied DSP and VLSI Research Group (ADVRG), University of Westminster, London, UK

Valeria Angelica Castillo Revelo Universidad Peruana de Ciencias Aplicadas (UPC), Monterrico, Santiago de Surco, Peru

Indira Del Rosario Mechanical Engineering, University of Saskatchewan, Saskatoon, Saskatchewan, Canada

Oana Rotaru Politehnica University of Timisoara, Timisoara, Romania; Continental Autonomous Mobility Romania, Timisoara, Romania

Ehsan Sadeghi Twente University, Enschede, The Netherlands

Ali Reza Sajun American University of Sharjah, Sharjah, United Arab Emirates

Nathaniel Salawu Telecommunications Engineering Department, Federal University of Technology, Minna, Niger State, Nigeria

Mariela Sanchez Center for Information Services and HPC, Tu Dresden, Germany

Liviu-Andrei Scutelnicu Faculty of Computer Science, "Alexandru Ioan Cuza" University of Iași, Iași, Romania;
Institute for Computer Science, Romanian Academy, Iași, Romania

Abdelkader Senouci Faculty of Engineering and Environment, Northumbria University, Newcastle, UK

Mohammed Raouf Senouci School of Computer Science and Engineering, University of Electronic Science and Technology of China, Chengdu, People's Republic of China

Ayesha Siddika World University of Bangladesh, Dhaka, Bangladesh

Ana B. G. V. Silva Departamento de Medicina Veterinária, Universidade Federal de Juiz de Fora (UFJF), Juiz de Fora, MG, Brazil

Farah Sinada Applied DSP and VLSI Research Group (ADVRG), University of Westminster, London, UK

Jeffery Bryce Molina Solano Universidad Peruana de Ciencias Aplicadas (UPC), Monterrico, Santiago de Surco, Peru

Camilla Spagnolo Free University of Bozen, Bolzano, Italy

Ulrike Steffens HAW Hamburg, Hamburg, Germany

Biljana Stojkoska Faculty of Computer Science and Engineering, Ss. Cyril and Methodius University, Skopje, North Macedonia

Dauda Usman Suleiman Telecommunications Engineering Department, Federal University of Technology, Minna, Niger State, Nigeria

Hana Sulieman Department of Mathematics and Statistics, American University of Sharjah, Sharjah, UAE

Weiqing Sun University of Toledo, Toledo, Ohio, USA

Dávid Szabó Eötvös Loránd University, Budapest, Hungary

Meisam Taheri Inland Norway University, Hamar, Norway

Kevin Tan Inland Norway University, Hamar, Norway

Vinayak Tanksale Ball State University, Muncie, IN, USA

Petrova Tatiana St Petersburg University, Saint Petersburg, Russia

Nguyen Quang Thuan International School, Vietnam National University, Hanoi, Vietnam

Marina Torres COSCYBER, Fundación I+D del Software Libre (Fidesol), Granada, Spain

Bulent Tugrul Department of Computer Engineering, Ankara University, Ankara, Turkey

Pablo Turón Vicomtech Foundation, Basque Research and Technology Alliance (BRTA), Donostia/san-Sebastián, Spain

Anabelie Villa Valdez Mindanao State University, Marawi City, Philippines

Radu Vasiu Politehnica University of Timisoara, Timisoara, Romania

Silviu Vert Politehnica University of Timisoara, Timisoara, Romania

Octávio F. Vieira Departamento de Medicina Veterinária, Universidade Federal de Juiz de Fora (UFJF), Juiz de Fora, MG, Brazil

Yvan Huaricallo Vilca Universidad Nacional Mayor de San Marcos, Lima, Peru; Universidad Tecnológica del Perú, Lima, Peru

Ghislain Vlavonou Institut de Mathématiques et de Sciences Physiques de l'Université d'Abomey-Calavi, Dangbo, Porto-Novo, Benin

Aimee Walker School of Computer Science, Faculty of Science and Engineering, University of Hull, Hull, UK

Minliang Xia Mechanical Engineering, University of Saskatchewan, Saskatoon, Saskatchewan, Canada

Yan Yan School of Computer Science, University of Guelph, Geulph, Ontario, Canada

Khaoula Zahouani FSJES, L-Qualimat GRTE Laboratory, Cadi Ayyad University of Marrakech, Marrakech, Morocco

S. A. Zamolotskii Siberian Federal University, Krasnoyarsk, Russia

Eftim Zdravevski Faculty of Computer Science and Engineering, Ss. Cyril and Methodius University, Skopje, North Macedonia

Driss Zejli Laboratory of Advanced Systems Engineering (ISA), National School of Applied Sciences, Ibn Tofail University, Kenitra, Morocco

Wenjun Zhang Biomedical Engineering, University of Saskatchewan, Saskatoon, Saskatchewan, Canada

Imran Zualkernan American University of Sharjah, Sharjah, United Arab Emirates

A New Wavelet Visual Weighting Model to Optimize Image Coding and Quality Evaluation Based on the Human Psychovisual Quality Properties

Ilham Morino, Abderrahim Bajit, Abdelhadi EL Allali, Rachid EL Bouayadi, and Driss Zejli

Abstract In this work, we suggest the development of two new image coders, namely EVIC (*Embedded Visual-based image coder*) based on the perceptual model and a designed weighting model and its optimized version VOEVIC (*Visual Optimized Embedded Visual-based image coder*) in order to obtain an improved perceptual quality compared to the standard SPIHT coder, for a given bit rate. Among other things, we propose a new VWVDP quality evaluator and its optimized version VOWVDP, and suggest the validation of objective metrics for evaluating coding quality using Human Visual System (*HVS*) quality criteria. These metrics incorporate visual weighting to eliminate any undesirable components invisible to the human visual cortex and retain only those that are perceptible in the cortical domain, and finally evaluate the differences by offering a map of visible differences VDM and a quality factor PS (*Probability Score*) for visual coding evaluation. This paper sets the objective of optimizing PAYLAOD in IoT platforms deploying computational intelligence for object detection and recognition with a view to offering green solutions to smart city applications, in areas such as health, transport, agriculture, defense and human security, for PAYLOAD size reduction while optimizing quality and reducing the cost of energy consumption and multimedia data transport times.

I. Morino (✉) · A. Bajit · A. EL Allali · R. EL Bouayadi · D. Zejli
Laboratory of Advanced Systems Engineering (ISA), National School of Applied Sciences, Ibn Tofail University, Kenitra, Morocco
e-mail: ilham.morino@uit.ac.ma

A. Bajit
e-mail: abderrahim.bajit@uit.ac.ma

A. EL Allali
e-mail: abdelhadi.elallali@uit.ac.ma

R. EL Bouayadi
e-mail: rachid.elbouayadi@uit.ac.ma

D. Zejli
e-mail: driss.zejli@uit.ac.ma

X.-S. Yang et al. (eds.), *Proceedings of Ninth International Congress on Information and Communication Technology*, Lecture Notes in Networks and Systems 1000,
https://doi.org/10.1007/978-981-97-3289-0_1

Keywords Discrete wavelet transform (DWT) · Visible difference predictor (VDP) · Visual weighting models · Probability score · Objective/subjective evaluation

1 Introduction

In their quest to improve the standard of living and quality of life of human beings, scientists have turned their attention to the development of green solutions that optimize the quality and reduce the cost of energy consumption, solve problems such as data storage, the limited memory capacity of storage machines and minimize multimedia data transfer times, all with a view to creating green, sustainable, intelligent and clean cities. The image is a powerful medium for disseminating information and communicating, and a real one in our daily lives. Its massive and intense use in many fields, such as industry, agriculture, health or even in communication via cell phones and internet broadcasting and publication, is increasing all the time, hence the need for intervention to resolve or improve the problems attached to the display, quality, storage and transfer of media data. Vision-based wavelet coding algorithms, such as the SPIHT encoder [1–3], offer very good compression results that minimize bandwidth and data transmission time over computer networks, reduce the amount of storage hardware and all the savings that go with it. An uncompressed file requires more storage capacity than a compressed one, since its size is greater than that of the compressed file.

Some algorithms have not adopted psychovisual practices and other algorithms have not used the specificities of the human visual system, such as the CSF contrast sensitivity function [4, 5], luminance masking [6] and contrast masking [7]. By working on these parameters, we can improve the quality of the restored image. However, wavelet channels are limited, so quality assessment remains the objective of the visible difference predictor VDP [8] for visual enhancement of the image compression sketch.

Our work consists of developing two encoders, namely EVIC based on the perceptual model and a designed weighting model, and its optimized version VOEVIC based on the first EVIC encoder. And since to evaluate or examine image compression techniques, we need to measure the quality of compacted images in a credible way, we are also proposing to enhance the reliability of the WVDP (*Wavelet-based Visible Difference Predictor*) evaluator [9] with a new VWVDP (*Visual Wavelet-based Visible Difference Predictor*) evaluator based on the WVDP evaluator, which in turn is based on the VDP (*Visible Difference Predictor*), and another new evaluator which is an optimized version of the first, VOWVDP (*Visual Optimized WVDP*). These two evaluators offer a PS and a VDM based on a psychomotor function. They also validate objective metrics for assessing coding quality using HVS quality criteria [10], following three assessment strategies: Subjective, Objective and Quantitative. The EVIC and VOEVIC encoders result entirely from the features of HVS, which harmonizes wavelet coefficients prior to encoding, and at the procedural and program level, these two encoders have difficulties identical to those of the standard SPIHT encoder. However, they minimize a measure of distortion perceptually and give good results in terms of application and development of specificities for coding and image quality assessment.

This paper is organized as follows: in Sect. 2, we present the block diagram of an integrated wavelet image coder based on visual properties and another block diagram of the wavelet-based Visual Weighting Model, in Sect. 3, we discuss the importance of coding quality evaluators, we present existing coding quality raters, and quality raters designed in Sect. 4, we present results obtained using visual coders developed against a reference coder and quality raters designed against an existing visible difference predictor. These new predictors produce useful SPs for measuring coding quality. And at the end, in Sect. 5, we present the conclusion of our work, along with an outlook for the future.

2 Wavelet Visually Based Image Quality Coding

The EZW [11] coding algorithm and the SPIHT coding algorithm, which offers an improvement on the EZW algorithm while retaining its features, namely its satisfactory performance and partial image restoration is achievable if the bitstream produced is suspended or truncated, oblige us to change the bitrate, leading to degraded quality. The SPIHT is based firstly on a partial organization of the information by moderately adding extra data, and secondly by amplitude through a distribution sorting algorithm, using the similarities existing at all levels of the wavelet-transformed image. An interesting feature of encoders is the notion of scalability, which enables a bitstream to be decoded at several quality levels and distinct resolutions, by a single encoding sequence. Scalability [12] is a necessity for good performance in recent compression systems.

This work presents two new encoders; the first EVIC based on the perceptual model which includes various masking effects attached to HVS perception, balancing wavelet coefficients against their visual interest, and the second is its optimized version VOEVIC. Both encoders improve the visual quality of the image for a given bit rate, compared with that obtained with SPIHT. Line 2 of Table 1 shows the result of successively applying the SPIHT (right image), EVIC (middle image) and VOEVIC (left image) encoders to the Barbara test image. The proposed encoders combine four phases (see Fig. 1): firstly, decomposition of the initial image with DWT (Discrete Wavelet Transform) [13] to establish a spatial representation of the frequency, which enables the fineness of the image parts to be specified, using a Daubechies 9/7 biorthogonal wavelet filter [14]. The mathematical characteristics of this wavelet, recommended by JPEG2000 [15], ensure perfect image recovery. In the second phase, we calculate luminance masking [16] and threshold elevation, the wavelet just the perceptible distortion JND [17] used to reduce the human observer's ability to detect quantization errors. In the third phase, CSF is applied to retain visible frequencies and eliminate all invisible frequencies in the spatial frequency (SF) sub-bands. In the fourth and final phase, we use the SPIHT encoder, an algorithm based on hierarchical tree partitioning that enables highly efficient image compression based on PSNR (maximum signal-to-noise ratio) [18], which is used to measure the quality of the initial and compressed images.

As shown in (see Fig. 1) and detailed in the following paragraphs, the weighting model used involves the calculation of luminance masking based on the pixels of the original image in the spatial domain. Next, we calculate the elevation of the contrast thresholds called contrast masking [19], which exhibits the visible thresholds necessary for the calculation of the contrast shuffle attached to the wavelet coefficients taken from the original image, the calculated luminance masking and the JND. This action promotes the masking of imperceptible contrast components and raises those that are visible by comparison with the JND. The next step consists in calculating the product of the values obtained from luminance masking and contrast masking and CSF frequency filtering in all frequency sub-bands to remove perceptible frequencies and retain imperceptible ones according to HVS, to obtain the weighted wavelet coefficients. Line 1 of Table 1 shows the result of applying this weighting model successively with the SPIHT (right image), EVIC (middle image) and VOEVIC (left image) encoders on the Barbara test image.

JND is the most narrowed dissimilarity in a defined circumstance of sensory input that is perceptible by a human being, it is a progressive function as a function of adaptation (or background) brightness, its mathematical formula (1) is as follows:

$$\mathrm{JND}(\lambda, \theta, i, j) = 1/(\mathrm{DWT}(\lambda, \theta, i, j)/V_{\mathrm{mean}})^{\alpha_{\mathrm{T}}} \tag{1}$$

λ is the decomposition level from 1 to 5, θ is the orientation of the cortical representation (horizontal, vertical and diagonal) of the HVS, DWT is the wavelet coefficients (the decomposition of the initial image in the DWT domain), V_{mean} is the contrast average and depends on image size and the α_T is JND's adjustment coefficient, which specifies the masking phenomenon.

Luminance masking is linked to the lower sensitivity of the HVS to the eccentricity inserted in the darkest and brightest bands of the image. According to Weber–Fechner's law, the smallest sensitive difference of the perceptible stimulus increases with background luminance, and this law is the primitive of luminance masking. Moreover, given the ambient lighting surrounding the screen, even noise added to very dark image areas become less visible than the same amount of noise added to mid-gray areas [19]. The overall effect of luminance masking on the sensitivity of the HVS to distortion is a U-shaped curve that defines the maximum amount of distortion tolerated for the luminance values considered. This curve is often referred to as the JND luminance masking profile; its mathematical formula (2) is as follows:

$$\alpha_L(\lambda, \theta, \mathrm{i}, \mathrm{j}) = |\mathrm{DWT}(\lambda_{\mathrm{max}}, \mathrm{LL}, \mathrm{i}, \mathrm{j})/\mathrm{V}_{\mathrm{mean}}|^{\alpha\mathrm{T}} \tag{2}$$

with $\lambda_{\mathrm{max}} = 5$, $\varepsilon = 0.6$.

Table 1 Visually weighted wavelet spectrum versus original one applied to visual coders VOEVIC (left), EVIC (middle) and original wavelet (right) and their impact on the visible errors detected by objective metrics WVDP, VWVDP and VOWVDP successively in lines 3, 4 and 5

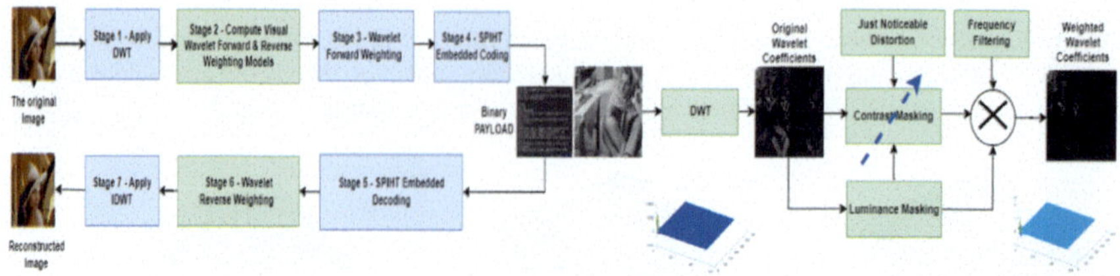

Fig. 1 Integrated wavelet image coder based on visual properties (Left) and wavelet visual-based weighting model (Right)

Contrast masking is used to examine the spatial frequency specificity of pattern perception by assessing where the signal at one spatial frequency begins in existence of other spatial frequencies its mathematical formula (3) is as follows:

$$\alpha_c(\lambda, \theta, i, j) = \max(1, (|\mathrm{DWT}(\lambda, \theta, i, j)|/(\mathrm{JND}(\lambda, \theta, i, j) * \alpha_L(\lambda, \theta, i, j))^{\varepsilon})) \quad (3)$$

The CSF provides an accurate representation of a being's ability to detect fine spatial objects, and is used to generate the curve that determines the lowest degree of contrast HSV can notice for all spatial frequencies experienced, instead of encoding the spectrum of wavelet-based image coefficients and improving the visual quality of its reconstructed variant. High spatial frequencies enable small details to be distinguished, while low frequencies match the general shape of the object.

3 Wavelet Visual-Based Image Quality

Several evaluation metrics have been used, whether mathematical metrics such as MSE (*Mean Square Error*) [20], PSNR (*Peak Signal-to-Noise Ration*) and SSIM (*Structural Similarity Index*) [21], or psychovisual metrics incorporating HVS features. These metrics are unreliable and do not provide the same error margins in the spatial domain as conventional coding, which is as inefficient as cortical coding. It is unreliable in the non-spatial domain because it generates a lot of errors. As a result, cortical quality evaluators are becoming a necessity for quality measurement. As we know, cortical filters are very demanding in terms of computer design and development. In response to this problem, we opted for coding with biorthogonal wavelet filters such as the Daubechies filter. With this in mind, we designed three evaluators, all based on the VDP [22] reference model, which is based on cortical decomposition, while our WVDP evaluators (see Fig. 2), VWVDP (see Fig. 3) and VOWVDP (see Fig. 3) are based on DWT and a weighting model (see Fig. 2) presented above.

The VDP metric excites HVS cortical references to give an index of the order of visual errors with respect to image position. One of the strongest provisions for image compression is the wavelet transform. DWT fragments the image into a narrow number of spatial frequency channels in relation to cortical fragmentation. Despite

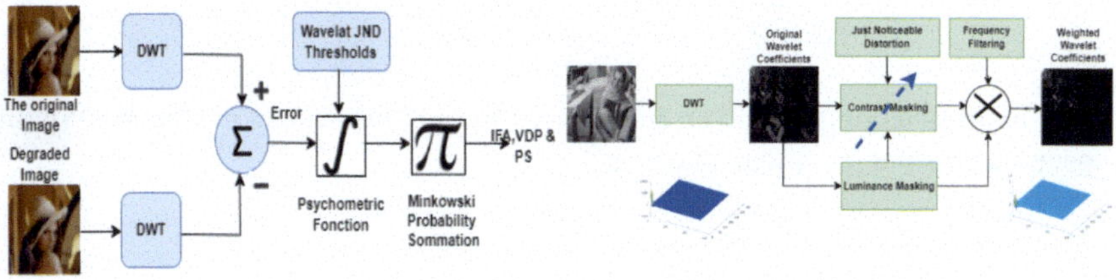

Fig. 2 Wavelet-based Visible Difference Predictor WVDP (Left) and wavelet visual-based weighting model (Right)

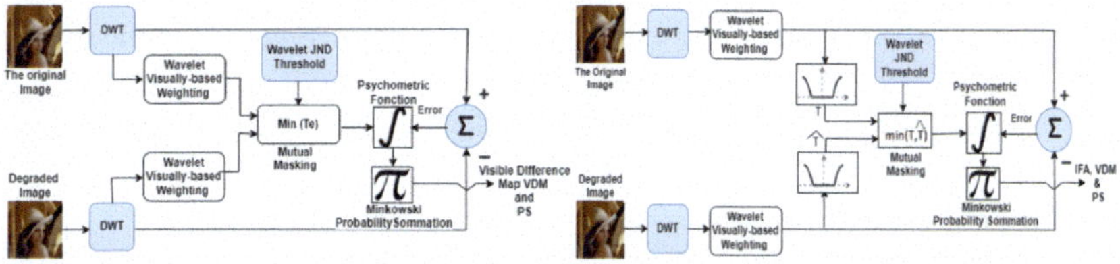

Fig. 3 Visually wavelet-based visible difference predictor WVDP (Left) and visually optimized wavelet-based visible difference predictor VOWVDP (Right)

this limitation, and to visually optimize the image compression graph, the evaluator (WVDP) aims to measure quality. It offers the FPS Probability Score quality factor using a psychometric FP function and the VDM visible difference map by summing visible errors using the Minkowski unit [23]. As shown in Fig. 2 above, both the initial and degraded images are decomposed into wavelet coefficients, wavelet discovery thresholds are performed, then the errors for all sub-bands in the DWT domain are estimated, followed by a probability summation calculation (π) based on the PF (Psychometric Function) to finally obtain the PS (Probability Score), the VDM and the IFA (Image Fidelity Assessor). Lines 3, 4 and 5 of Table 1 show the visible errors successively detected by the reference metric WVDP and its visual variants VWVDP and VOWVDP (following paragraphs), all applied to the visual encoders VOEVIC (left image) and EVIC (middle image) and their reference SPIHT (right image) on the Barbara test image.

VWVDP is an enhanced version of WVDP. It modifies the initial image and the perceptible image by removing invisible data and comparing only the fundamental data. It passes on the errors to the Minkowski summation to obtain the VDM and VPS by means of FP. The FPS plays a very important role in the development of perceptual image encoders and formulates the ability to distinguish eccentricities in wavelet channels. FP is the integral of a distribution function. It indicates the probability that the threshold has a given value. A probability map is generated at the output of the PF and the VWVDP evaluator, i.e. the probability at any pixel in the image. As a result, the probability of detection in each channel must be matched for each spatial location in the image. This is done using the following formula:

$$\mathrm{Pd}(\lambda, \theta, i, j) = 1 - \pi(1 - \mathrm{Pb}\ (\lambda, \theta, i, j))$$

The PS of detecting visible errors between a candidate image of size $M*N$ with respect the original one is calculated as follows:

$$\mathrm{PS} = 1/(M * N) \sum_{i,j}^{M,N} Pd(\lambda, \theta, i, j)^{1/Q}$$

It varies from 0 to 1, where 0 corresponds to poor quality and 1 is reached for excellent quality, given that the calculation of the probability score P for level λ, orientation θ, , and position (i, j) in the wavelet domain is expressed as follows:

$$P(\lambda, \theta, i, j) = 1 - \exp\left(-\left|\frac{D(\lambda, \theta, i, j)}{JND(\lambda, \theta, i, j)^{\alpha}}\right|^{Q}\right)$$

In which D(λ, θ, i, j) is the quantization eccentricity detection which concerns the elimination of non-redundant but consider unnecessary information, JND(λ, θ, i, j) is the minimum threshold of detecting a visible distortion, α and Q are setting factors.

In this article, we provide a new VOWVDP (see Fig. 3) evaluator based on VWVDP. It successively performs DWT wavelet transformation, JND thresholding, contrast and luminance masking, as well as mutual masking, and counts errors and PS based on FP and Minkowski summation to obtain VDM and PS by means of FP. The wavelet errors obtained from the difference between the original image and the degraded one are compared to a wavelet JND threshold; whether the errors are above or below this threshold. And this, to know whether the errors are perceptible or not.

4 Results and Discussion

In this work, we tested the EVIC encoder, based on the perceptual model and the designed weighting model (see Fig. 2), and its optimized version VOEVIC against the SPIHT reference encoder, using four test images BARBARA, MANDRILL, Image Agriculture API and BOAT. We examined their qualities using the three evaluation metrics WVDP, VWVDP and VOWVDP based on the wavelet and the designed weighting model (see Fig. 2). We used three evaluation strategies: subjective, objective and quantitative in relation to given binary budgets and a fixed observation distance.

In the first strategy, the PS probability score is acquired by examining the perceptual quality evaluation of the EVIC and VOEVIC encoder respectively against SPIHT and EVIC experienced on the test images. The acquired results are given for a fixed observation distance V and increasing bit rates varying from 64:1 to 1:1. As can be seen in Table 5, for low bit rates, many data are faintly seen in parts of the image coded by the SPIHT reference coder, while the EVIC and VOEVIC visual coders present considerably more interesting data in these parts. Also for medium bitrates, the EVIC and VOEVIC encoders exhibit outstanding quality over the whole test image, while the image encoded by SPIHT gives vague information compared with the competing visual encoders. Nevertheless, at high data rates, the differences are almost invisible for all the encoders used. And with regard to the visible errors detected, as can be seen in line 3 of Table 1, the visible errors detected by the WVDP reference evaluation metric tested on the two visual encoders VOEVIC (left image) and EVIC (middle image) and their SPIHT reference (right image) on the test image Barbara, are less interesting successively for the VOEVIC and EVIC visual encoders than for the SPIHT encoder, and similarly in lines 4 and 5, using successively the VWVDP and VOWVDP evaluation metrics for all three encoders. If we compare the errors detected with the experimental evaluation metrics, we see that the results obtained are more interesting for VOWVDP than for VWVDP, and also less significant for WVDP, successively for the three encoders VOEVIC, EVIC and SPIHT.

As for the second approach, based on the development of an objective examination, as shown in Table 6, using the WVDP objective evaluation metric (top figure) and the three encoders SPIHT, EVIC and VOEVIC on the BARBARA test image, image quality increases with increasing throughput, and the ability to detect errors is successively greater for the VOEVIC and EVIC visual encoders than for the SPIHT encoder. The difference in results between the three encoders used is not quite so remarkable, even though the two visual encoders give better results than those obtained using the SPIHT encoder. The performance obtained using the VWVDP evaluation metric (middle figure) is much more interesting than that achieved using the WVDP evaluation metric, and the result gap is very large between the two visual encoders used VOEVIC and EVIC for low bitrates, which are less than 0.5, while this gap starts to be interesting between the EVIC and SPIHT encoder as the bitrate increases and is greater than 0.2. As for the VOWVDP optimized visual evaluation metric (bottom figure), it endorses the results obtained by the VWVDP evaluation metric, and gives much better performance than the latter. As shown in the bottom figure of Table 6, the results achieved by the visual encoders are much better than those obtained by the SPIHT reference encoder. There is even a gain in quality achieved by the VOEVIC encoder compared with EVIC, and this gain is clear and significant for low and medium flow rates, which are 0.5 lower than for high flow rates. In Table 7, to measure quality, we apply the three evaluation metrics used VOWVDP, VWVDP and WVDP successively for the three encoders VOEVIC, EVIC and SPIHT on four reconstructed test images BARBARA, MANDRILL, Image Agriculture API and BOAT in relation to their original reference images and for a fixed observation distance of value $V = 4$. The results obtained in Table 6 are confirmed in Table 7 for the four test images cited above.

In the third so-called quantitative approach, as presented in Tables 2, 3 and 4, the percentage gain in quality for the VOEVIC visual encoder compared with the EVIC visual encoder is greater than zero when applying the three evaluation metrics WVDP, VWVDP and VOWVDP, and is calculated by the following formula:

Table 2 VOEVIC versus EVIC and EVIC versus SPIHT improving quality ratio based on the WVDP metric applied to a test image with respect to a fixed observation distance of value $V = 4$

Bit rate	64:1	32:1	16:1	8:1	4:1	2:1	1:1
Gain in % quality for VOEVIC/EVIC	**0.32**	**0.05**	**0.31**	**0.04**	**0.07**	**0.1**	**0.05**
Gain in % quality for EVIC/SPIHT	**32.34**	**0.18**	**0**	**0.51**	**0.50**	**0.18**	**0.17**

Table 3 VOEVIC versus EVIC and EVIC versus SPIHT improving quality ratio based on the VWVDP metric applied to a test image with respect to a fixed observation distance of value $V = 4$

Bit rate	64:1	32:1	16:1	8:1	4:1	2:1	1:1
Gain in % quality for VOEVIC/EVIC	**0.01**	**872**	**267.21**	**4.73**	**1.72**	**0.34**	**0.03**
Gain in % quality for EVIC/SPIHT	**0**	**0**	**0.5**	**17.42**	**4.26**	**1.95**	**1.01**

Table 4 VOEVIC versus EVIC and EVIC versus SPIHT improving quality ratio based on the VOWVDP metric applied to a test image with respect to a fixed observation distance of value $V = 4$

Bit rate	64:1	32:1	16:1	8:1	4:1	2:1	1:1
Gain in % quality for VOEVIC/EVIC	**215.5**	**45.13**	**16.93**	**1.25**	**0.56**	**0.13**	**0.01**
Gain in % quality for EVIC/SPIHT	**0.33**	**0.34**	**0.42**	**0.44**	**0.76**	**0.31**	**0.13**

$$(\mathrm{PS}_{\mathrm{VOEVIC}} - \mathrm{PS}_{\mathrm{EVIC}})/\mathrm{PS}_{\mathrm{EVIC}}$$

and for the EVIC visual encoder compared with the SPIHT visual encoder, applying the three evaluation metrics WVDP, VWVDP and VOWVDP, the percentage gain in quality is also greater than zero and is calculated by the following formula:

$$(\mathrm{PS}_{\mathrm{EVIC}} - \mathrm{PS}_{\mathrm{SPIHT}})/\mathrm{PS}_{\mathrm{SPIHT}}$$

All this confirms the first two approaches, subjective and objective.

5 Conclusion and Perspectives

In this work, we have presented two new visual encoders EVIC and VOEVIC and three new evaluation metrics WVDP, VWVDP and VOWDVP all of which are wavelet-based and take advantage of several weighting metrics related to the specificities of SVH, namely luminance masking, threshold elevation, CSF and contrast masking. However, images coded by our two new encoders and evaluated by our three evaluation metrics for bit rates ranging from 64:1 to 1: 1 and a fixed viewing distance, and also using three strategies—subjective, objective and quantitative—gave us very significant results, in terms of improving their quality compared with

Table 5 VOEVIC/EVIC/SPIHT image coding quality assessment applied to four test images with respect to a fixed observation distance of value $V = 4$

API_VOEVIC_Bpp128per1	API_EVIC_Bpp128per1	API_SPIHT_Bpp128per1

API_VOEVIC_Bpp32per1	API_EVIC_Bpp32per1	API_SPIHT_Bpp32per1

API_VOEVIC_Bpp8per1	API_EVIC_Bpp8per1	API_SPIHT_Bpp8per1

API_VOEVIC_Bpp2per1	API_EVIC_Bpp2per1	API_SPIHT_Bpp2per1

Table 6 Visually weighted wavelet spectrum versus original one applied to visual coders VOEVIC, EVIC and original wavelet and their impact on the visible errors detected by objective metrics WVDP (top figure), VWVDP (middle figure) and VOWVDP (bottom figure)

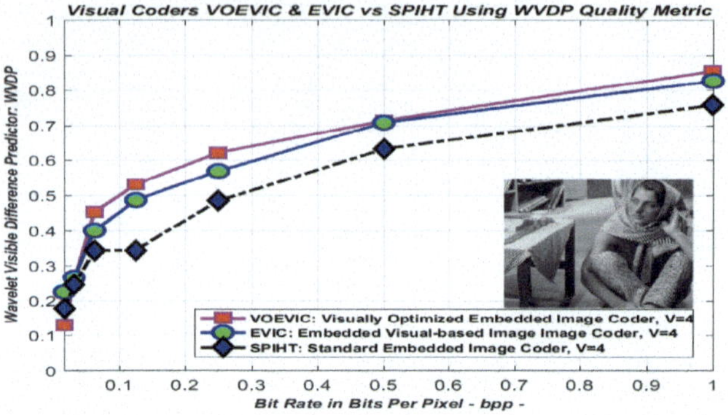

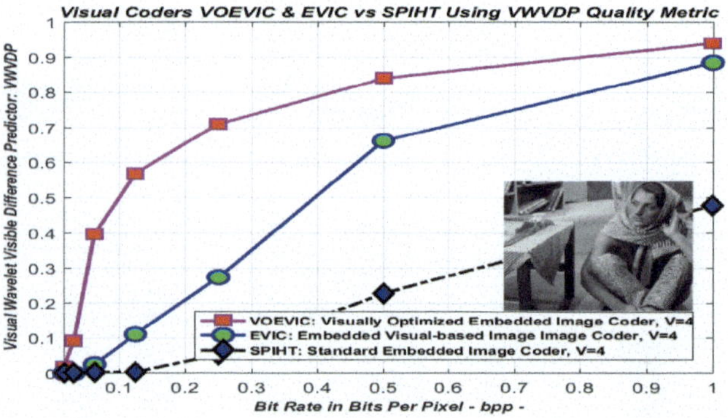

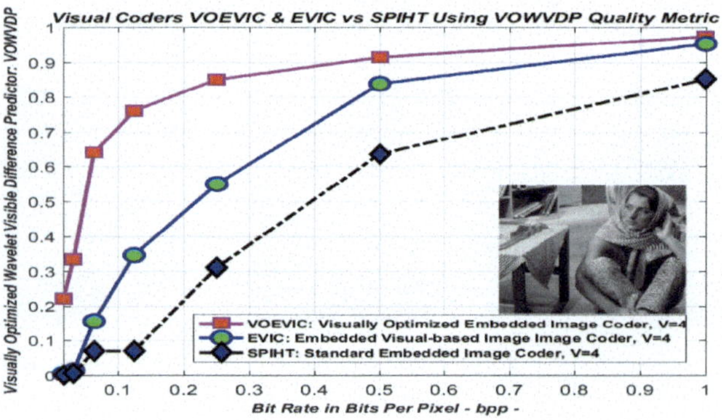

Table 7 VOWVDP/VWVDP/WVDP quality measure applied to four test images with respect to a fixed observation distance of value V = 4

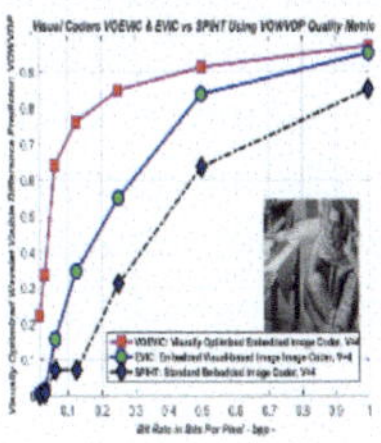	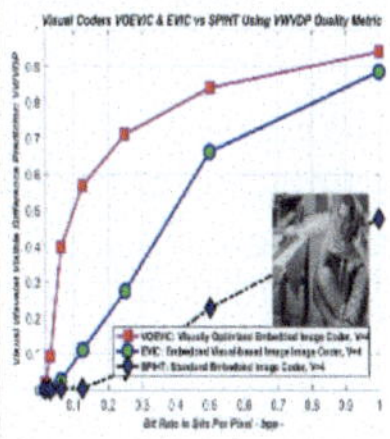	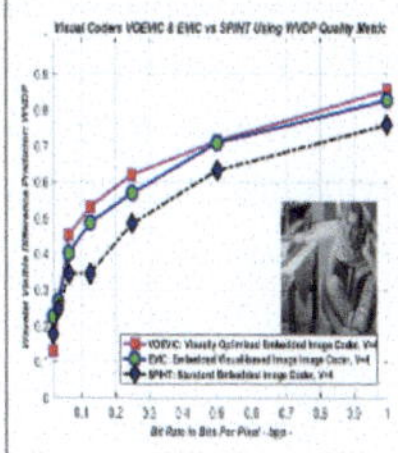
BARBARA_VOWVDP	BARBARA_VWVDP	BARBARA_WVDP
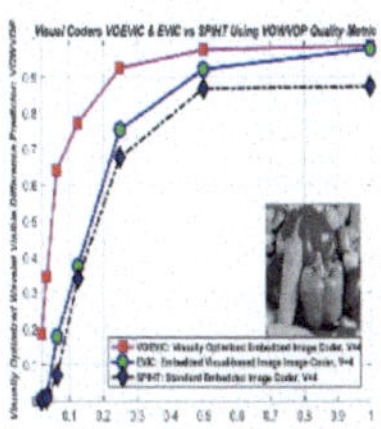	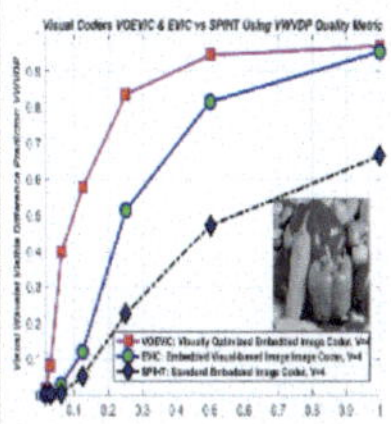	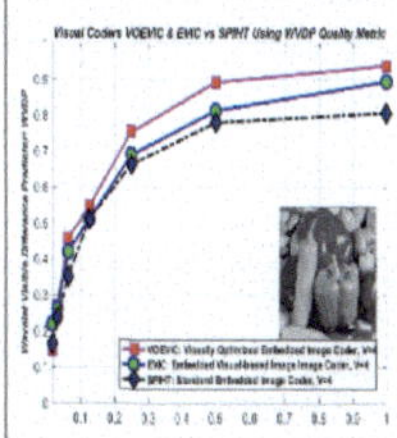
API_VOWVDP	API_VWVDP	API_WVDP
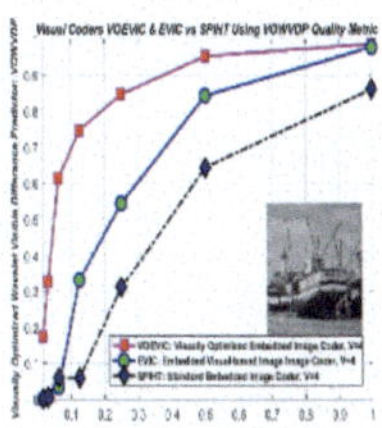	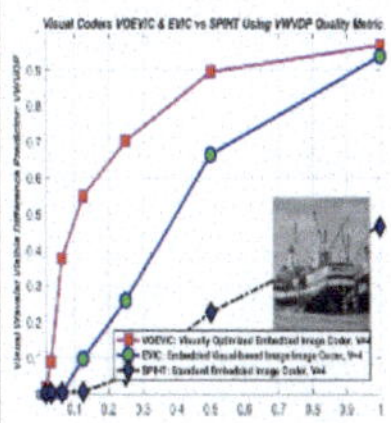	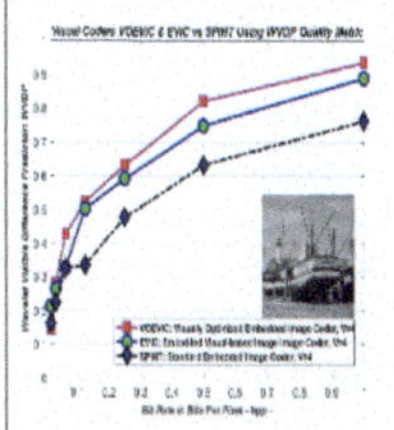
BOAT_VOWVDP	BOAT_VWVDP	BOAT_WVDP
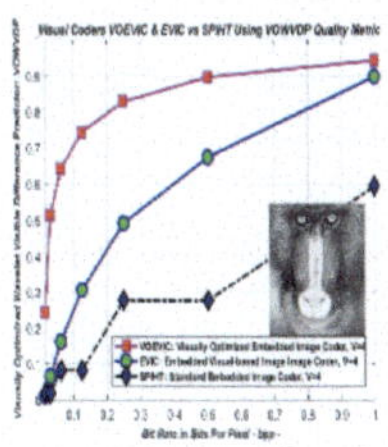	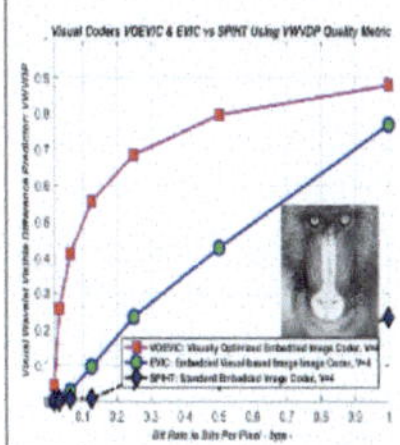	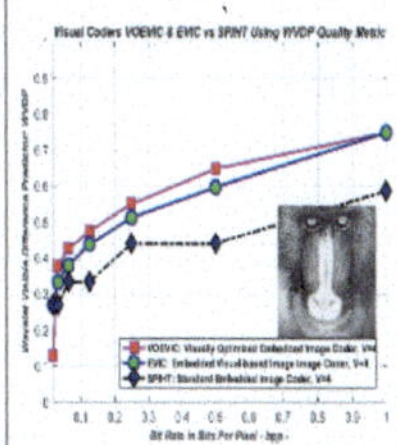
MANDRIL_VOWVDP	MANDRILL_VWVDP	MANDRILL_WVDP

the SPIHT reference coder, and also in terms of minimizing the errors detected by the quality metrics, by obtaining a PS probability score synonymous with a quality factor that quantifies the visible differences between images. Knowing that coding quality is good when the value of this factor is high, evaluation metrics are of great importance in improving coder quality and performance.

We conclude that the visual coders and quality evaluators we have developed are effective in optimizing coder performance and avoiding error generation. And note that we plan to conduct user studies using, for example, the MOS (*Mean Opinion Score*) quality factor to assess the subjective quality of images coded with EVIC and VOEVIC, and that these designed visual coders and quality assessors form part of our future work, which we intend to integrate into IoT platforms [24] to optimize PAYLOAD size and quality, reduce energy consumption costs, solve multimedia data storage problems and transport times, and all this to create green, sustainable and clean cities.

Acknowledgements On behalf of the SCAISAR Project's team, we would like to thank the Moroccan Ministry oh Higher Education, Scientific Research, and Innovation also we thank the OCP, UM6P and CNRST partners who funded this work through the SCAISAR-APRD research program on Agricultural Greenhouse of Precision.

References

1. Said A, Pearlman WA (1996) A new, fast and efficient image codec based on set partitioning in hierarchical trees. IEEE Trans Circ Syst Video Technol 6:243–250
2. Xu HL, Zhong SH (2009) Image compression algorithm of SPIHT based on block-tree. J Hunan Human Inst Eng 19(1):58–61
3. Guo Y, Yu M, Lin X, Liang Y (2009) Binary-tree SPIHT coding for true bidimensional wavelet decomposition. In: Second International conference on intelligent networks and intelligent systems
4. Itoh N, Okamoto A, Sagawa K, Mitani S, Yosida T (2009) Considerations of contrast sensitivity function of low vision. In IEEE 13th International symposium on consumer electronics, Japan. https://doi.org/10.1109/ISCE.2009.5156941
5. Xu Z, Zhuang Y, Hou F, Chan L, Feng L, Ye Q, He Y, Zhou Y, Jia Y, Yuan J, Lu Z, Li J (2022) Assessing the contrast sensitivity function in myopic parafovea: a quick contrast sensitivity functions study. Front Neurosci. https://doi.org/10.3389/fnins.2022.971009
6. Nercessian S, Panetta K, Agaian S (2013) Non-linear direct multi-scale image enhancement based on the luminance and contrast masking characteristics of the human visual system. IEEE Trans Image Process 22(9)
7. Hua Y, Liu L, Zhao Q (2015) Medical image quality assessment via contrast masking. In: 8th International congress on image and signal processing (CISP). Shenyang, China
8. Mantiuk R, Myszkowski K, Seidel H (2004) Visible difference predicator for high dynamic range images. In: IEEE International conference on systems, man and cybernetics. The Hague, Netherlands. https://doi.org/10.1109/ICSMC.2004.1400750
9. Bajit A, Nahid M, Tamtaoui A, Bouyakhf EH (2007) A perceptually optimized wavelet embedded ZeroTree image coder. Int J Comput Inf Eng 1(6)
10. Wang B, Wang Z, Liao Y, Lin X (2008) HVS-based structural similarity for image quality assessment. In: 9th International conference on signal processing, 26–29 October (2008)
11. Venkata Sainath Gupta T, Naveen Ch, Satpute VR, Gandhi AS (2014) Image security using chaos and EZW compression. In: Students conference on engineering and systems. India
12. Domariski M, Madkowiak S, Blaszak L, Luczak A (2002) Fine granularity in multi-loop hybrid coders. In: International conference on image processing. Rochester, NY, USA
13. N. Aina, A. Razak, R. Aripin, M. T. Ismail: Denoising Malaysian time series data: a comparison using discrete and stationary wavelet transforms. In International conference on science and social research (CSSR 2010). Malaysia, pp 412–415

14. Vonesch C, Blu T, Unser M (2007) Generalized Daubechies wavelet families. IEEE Trans Signal Process 55(9)
15. Kim B, Lee J, Yoon K (2014) A parallel implementation of JPEG2000 encoder on multi-GPU system. In: 16th International conference on advanced communication technology. Pyeongchang, South Korea
16. Bajit A, Najid M, Tamtaoui A, Lassioui A (2019) A perceptually optimized embedded image coder and quality assessor based both on visual tools. Adv Sci Technol Eng Syst J 4(2)
17. Wu J, Shi G, Lin W, Liu A, Qi F (2013) Just noticeable difference estimation for images with free-energy principle. IEEE Trans Multimedia 15(7)
18. Keleş O, Yılmaz M, Tekalp A, Korkmaz C, Doğan Z (2021) On the computation of PSNR for a set of images or video. In: Picture coding symposium (PCS). Bristol, United Kingdom. https://doi.org/10.1109/PCS50896.2021.9477470
19. Nahid M, Bajit A, Zyane A, Lahby M (2014) Wavelet visible difference measurement based on human visual system criteria for image quality assessment. J Theor Appl Inf Technol 66(3)
20. Pu Y, Wang W, Xu Q (2012) Image change detection based on the minimum mean square error. In: Fifth International joint conference on computational sciences and optimization. Harbin, China
21. Fu W, Sheng Y, Xiong D (2017) Optimal reconstruction design for JPEG-coded image using structural similarity index. In: 12th IEEE conference on industrial electronics and applications (ICIEA). Siem Reap, Cambodia. https://doi.org/10.1109/ICIEA.2017.8282837
22. Bajit A, Nahid M, Tamtaoui A, Benbrahim M (2020) A psychovisual optimization of wavelet foveation-based image coding and quality assessment based on human quality criterions. Adv Sci Technol Eng Syst 5(2):225–234
23. Martino V, Tralli G (2020) On the Minkowski formula for hypersurfaces in complex space forms. In: International mathematics research notices, pp 1–27
24. Morino I, Beloualid S, EL Allali A, Bajit A (2023) Deploying computational intelligence AI and psychovisual quality optimization tools in embedded image coding to enhance the MIoT PAYLOAD features. In: International symposium on automatic control and emerging technologies (ACET). Kenitra, Morocco

Optimizing the IoT PAYLOAD Encryption Watermarking-Oriented Applying Computational Intelligence and Visual Quality Coding to Improve the Vehicular Speed Controller Platform's Features

Y. Onqani, Abderrahim Bajit, B. Essounaini, Abdelhadi EL Allali, H. Essamri, and Driss Zejli

Abstract In this article, we'll outline a road transportation Internet of Things platform that enables real-time road flow data collection, adaptation, and encryption to enhance user experience, communicate in excellent quality at a bit rate appropriate for cloud storage, processing, and decision-making. The recommended approach entails compiling all the crucial road data, which include: the speed of movement, the geolocation of the vehicle, the vehicle picture, the detection of the license plate, the detection and recognition of the driver's face, as well as suspicious situations like telephone use, strange driving, failure to obey traffic signals, and so on. A relative QR code is produced by this solution. With the help of an advanced object recognition technology, our solution creates a relative QR code that contains all of the vehicle's data and ensures that it is appropriate for the car being recognized. The recommended approach demonstrates how psychovisual optimization tools affect watermarking coding quality in comparison with reference yields by SPIHT standard coder, how computational intelligence predicts lost data, and seeks to optimize platform data features in terms of CoAP payload time transportation duration, ECC encryption bandwidth storage and space occupation, and communication energy consumption to finally provide a high-quality and time/energy lossless watermarked content to enhance the security of the QR code during its transmission. To achieve these goals, the proposed system incorporates real-time data collection from the various nodes in the IoT network topology, data compression, synchronized handshaking between the cloud and the IoT nodes, and missing data prediction in the event of node failure to accomplish these aims.

Y. Onqani (✉) · A. Bajit · B. Essounaini · A. EL Allali · H. Essamri · D. Zejli
National School of Applied Sciences, Ibn Tofail University, Kenitra, Morocco
e-mail: yousseufonqani@gmail.com

X.-S. Yang et al. (eds.), *Proceedings of Ninth International Congress on Information and Communication Technology*, Lecture Notes in Networks and Systems 1000,
https://doi.org/10.1007/978-981-97-3289-0_2

Keywords CoAP · IoT · ECC lightweight encryption · Psychovisual optimizations · Tools wavelet quality watermarking · Computational intelligence · Vehicular object detection

1 Introduction

One of the main issues in our connected world is road safety. The health of people, social stability, and the economy as a whole are all severely affected by traffic accidents. Excessive vehicle speed continued to be a common and dangerous problem among the many reasons causing these catastrophes. Every year, lives are placed in danger due to the multiple risks involved with driving at high speeds. To decrease accidents and protect lives, it is crucial to have sophisticated and intelligent speed control systems. The mainstays of conventional techniques for regulating vehicle speed are the application of speed limits set by authorities and the use of speed radars. These radars record the speed of a vehicle at a single location on the road [1], but they don't offer continuous monitoring or the capacity to find suspicious vehicles that abruptly alter their behavior or lanes. Other possibilities include devices that can identify vehicles via their license plates but are not specifically built for regulating speed [2]. Furthermore, while GPS monitoring devices for vehicles can offer speed data, they are not designed for quick response when speed limits are violated. The main drawbacks of current speed control technologies are their inability to quickly foresee possible problems, effectively identify vehicles exhibiting unpredictable lane changes or reckless driving, and effectively transmit information to the appropriate authorities [3]. These flaws highlight the urgent need for the creation of more sophisticated and intelligent speed control systems. Our novel strategy entails creating a special QR code for each image including suspicious automobiles in order to guarantee data privacy [4]. This technique minimizes energy use while capturing all pertinent data about the vehicles in the image to maximize resource usage. When we discuss "information related to suspicious vehicles," we're discussing a lot of important information, like GPS coordinates, unique vehicle identifiers (like license plate IDs), speed information, the location of the suspected vehicle in the image, and other essential variables. These flaws highlight the urgent need for the development of advanced and intelligent speed control systems. Our novel strategy entails creating a special QR code for each image including suspicious automobiles in order to guarantee data privacy. This technique minimizes energy use while capturing all pertinent data about the vehicles in the image to maximize resource usage. When we discuss "information related to suspicious vehicles," we're discussing a lot of important information, like GPS coordinates, unique vehicle identifiers (like license plate IDs), speed information, the location of the suspected vehicle in the image, and other essential variables.

Our vehicle control of speed platform is a creative and all-encompassing solution that uses IoT and AI for proactive detection, real-time monitoring, and improved safety. This clever strategy shows the transformative power of technology in boosting road safety and human life while optimizing the use of energy resources. It also lowers the danger of accidents on our roads. Our capacity to quickly and accurately identify suspect vehicles, notably by examining license plates for improved road safety, is greatly aided by AI [5]. To provide readers with an in-depth understanding of our IoT-based vehicle speed control solution, this paper is divided into numerous thoughtfully constructed sections. Titled "Road Safety Platform Architecture," goes into more depth about our ground-breaking approach. The technology, hardware, and software used to support the connectivity, management, and analysis of data from connected items are the basis of the IoT platform we have created in the third section, titled "Vehicle Recognition and Image Watermarking Integration," we pivot our focus toward the seamless integration of artificial intelligence (AI) within our road safety platform, while also delving into the intricacies of watermarking integration.

2 Road Safety Platform Architecture

In general, an IoT platform is a collection of technology, hardware, and software components created to make it possible to connect, manage, and process data from connected things. Sensors for data collection, microcontrollers, or CPUs for local processing, communication modules for data transmission, IoT protocols like MQTT, CoAP, and 6LowPAN, specifically created to optimize IoT communications, and IoT servers/clouds for data storage, management [6], and analysis are the fundamental components of an IoT platform. IoT protocols are designed to use less energy, less bandwidth, and have lower latency than traditional protocols like HTTP, making them appropriate for IoT devices with limited resources. IoT can use a range of topologies for its network configurations, including the star model for close-by devices linked to a central hub, the tree model for nodes far from the hub, the mesh model for strong and complex connectivity, and the three-tier cluster model, which combines all three models (Fig. 1).

An IoT platform has certain restrictions that are essential to guaranteeing its effective operation. First and foremost, execution time is crucial because IoT devices need to digest data fast in order to make decisions in real time [7]. Additionally, as memory resources are frequently scarce in IoT devices, memory occupancy must be optimized. To increase gadget longevity, power consumption must be managed effectively. The performance of an IoT platform can be significantly impacted by the selection of a communication protocol and topology that are improper. A poor decision could cause delays, high memory utilization, or ineffective power use. The size of the payload, or the quantity of data transmitted in each message, can also have an impact on how much memory and energy the IoT platform uses. Therefore, proper payload size control is crucial. Finally, the performance of the IoT platform might be impacted by the absence of artificial intelligence [8] and the impact of data prediction at the cloud level. An IoT platform with computational intelligence at the node and cloud levels, visual coding, a secure blockchain [9], ECC encryption, synchronized exchange between the cloud and nodes, and data sharing via the CoAP

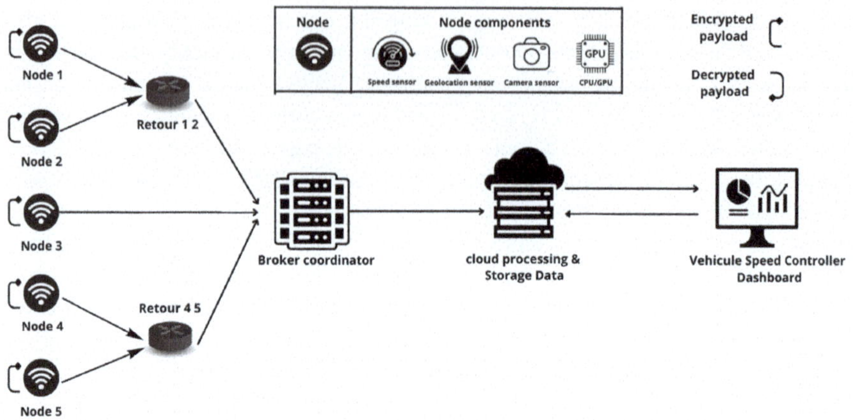

Fig. 1 Proposed IoT transport architecture

protocol have been offered as a solution to these limitations. The embedded system in this solution is at the cloud and IoT node levels. An embedded system with an I2C master and many I2C slaves, as shown in Fig. 2, has been developed at the IoT node level to gather, process, and transmit real-time data from the slaves to the master. The master collects the data sent by the slaves, packs it into a small file, encrypts it, and then sends the encrypted file to the cloud using an IoT communication protocol. This strategy aims to address the aforementioned limits while improving the IoT platform's performance.

The master commands the various slaves to perform four tasks in real time at the node level: gathering data samples, processing them locally, alerting them locally via LEDs, and communicating them to the master. This communication occurs synchronously between the master and slaves. With the exception of the image recorded by the camera, which will be aesthetically optimized independently by different image processing, all of this information is kept in an image file created by the QR code generator [10] that will be visually optimized by additional image processing. To create the watermarking coding that ensures data integrity [11], the image will be combined with the QR code. The following section will go into great detail about this idea. The synchronous communication principle utilized at

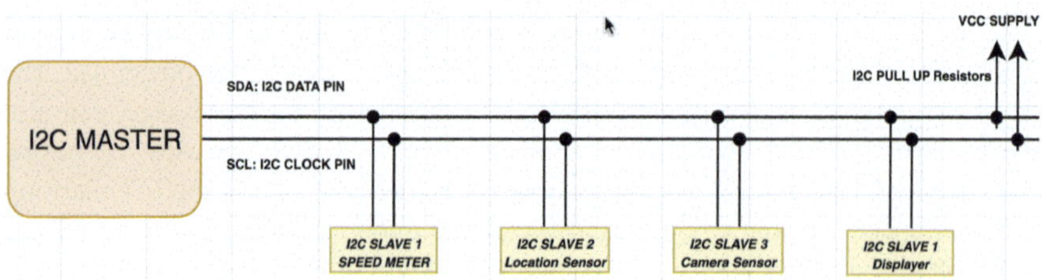

Fig. 2. 2C master and slave communication hardware configuration

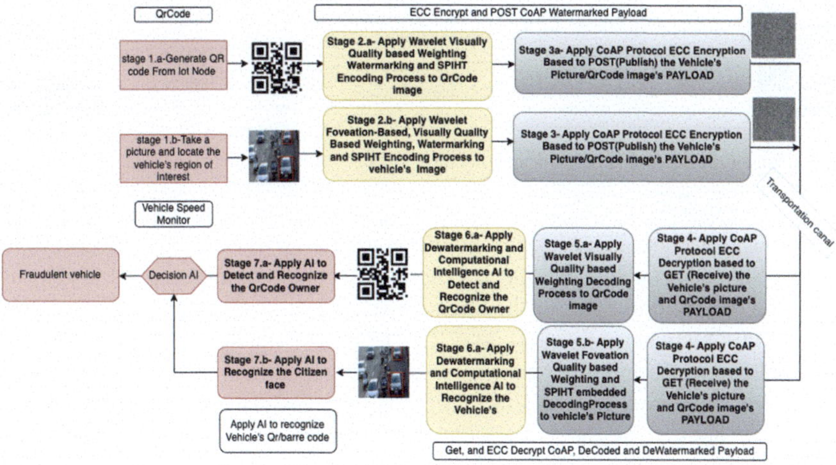

Fig. 3 Visual encoding/watermarked/ECC encryption/decryption/decoding/dewatermarked the image of the transport

the embedded system level (IoT node) has been implemented at the level of the entire IoT system between the cloud and the nodes in the IoT platform in accordance with the exchange shown in Fig. 3.

3 AI Vehicle Recognition and Image Watermarking Integration

Artificial intelligence (AI) has a crucial role by enabling quick and accurate analysis of visual data obtained from deployed IoT cameras on the ground. Our IA model was created specifically for this challenging task. Our IA model begins to operate when the system's speed sensor snaps a picture of a scene with questionable vehicles in it. It examines the image in search of license plate and vehicle plaques using the principles of the network of convolutional neurons (CNN) and data from a variety of vehicle databases and license plate [12], as well as specific vehicle models. Our model was learned by using backpropagation, a complicated technique that adjusts its parameters automatically to reduce the errors between predictions and actual objects found in the dataset [13]. This allows our IA model to learn to recognize the unique characteristics of vehicles and registration plates.

In Fig. 4 our algorithm locates the vehicles in the image, recognizes them, and then carefully examines the image to differentiate the vehicles from the license plates' alphanumeric characters. Activation functions that support this approach, such as the ReLU function, allow the model to learn complex patterns in the visual data. The margin of error reduces as a result of the model's iterative learning process, which improves the accuracy of vehicle identification and license plate recognition.

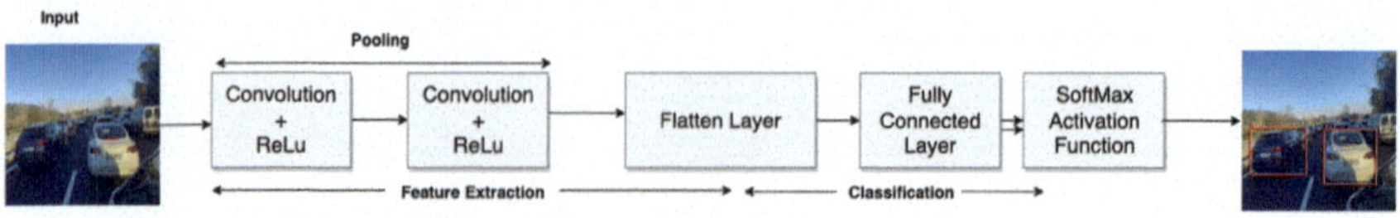

Fig. 4 Convolutional neural network architecture

Computational intelligence is at the heart of our approach, with the ability to determine areas of interest in collected multimedia data. Thanks to a detailed analysis of the quality properties of the human visual system (HVS) [14], we optimize multimedia content using foveal coding. This technique enables us to identify essential visual elements, minimizing payload size while maintaining outstanding visual quality. A concrete example of this process is the management of an IoT camera. The image captured by the camera undergoes special processing to optimize its multimedia content. This processing relies on psychovisual coding based on human visual quality criteria. It aims to adapt the image's frequency content by weighing it with a visual model that filters out or eliminates any components deemed unnecessary or undesirable. This approach helps to reduce the size of PAYLOAD [15], thus improving the overall performance of the IoT platform. For the vehicular image, we'll apply optimal coding based on the same visual weighting principle based on regions of interest, i.e., foveal coding. As a result, only visually relevant information in these regions of interest alone will be retained, and other components (outside the regions of interest) will be eliminated, thus optimizing coding. For QR code, we will apply non-foveal visual weighting. As a result, only visually relevant information will be maintained and the rest will be degraded.

In Fig. 5 at stage 2, watermarking involves the embedding of imperceptible information, known as a watermark, within the content [16, 17]. This process is applied to ensure the integrity, authenticity, and ownership of the media. In this IoT platform for radar systems with vehicle detection, watermarking is utilized to securely embed a QR code containing crucial information into the image of a suspect vehicle. The embedded watermark should be imperceptible to human observers, ensuring that it doesn't degrade the quality or appearance of the original image.

In stage 3 of the process, an additional layer of security is implemented by applying ECC encryption to both the foveally coded image of the vehicle and the watermarked image. ECC is a cryptographic technique that leverages the mathematics of elliptic curves over finite fields to provide secure encryption [18]. When applied to the images, ECC ensures that the visual and encoded information is encrypted in a way that is computationally secure and resistant to unauthorized access. Specifically, ECC uses the properties of elliptic curves to generate cryptographic keys, making it challenging for adversaries to decipher the encrypted content without the corresponding decryption key [18].

The adoption of the constrained application protocol (CoAP) holds several advantages [19]. Given the resource-constrained nature of IoT devices involved in traffic surveillance, CoAP emerges as a fitting communication protocol. CoAP ensures

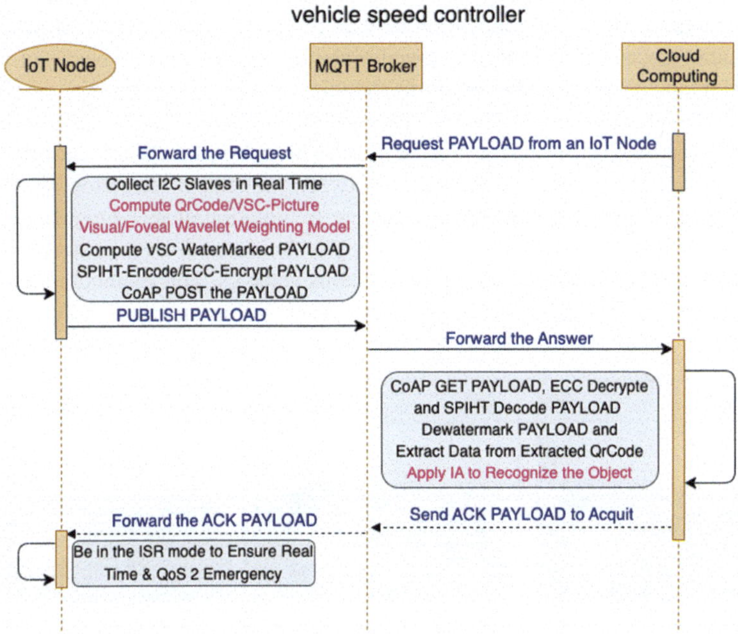

Fig. 5 Transport IoT platform cloud nodes synchronized and ECC encrypted handshaking

efficient data exchange, a critical factor in an IoT ecosystem where devices such as sensors and cameras may have limited processing capabilities and bandwidth.

This communication protocol is advantageous in our system where devices may need to conserve energy. Devices can stay in a low-power state and wake up only when necessary for communication.

4 Results and Discussion

In the implemented architecture, a parallel processing approach is adopted for enhanced efficiency. The node tasked with vehicle detection seamlessly integrates multiple data streams simultaneously. Initially, artificial intelligence algorithms are applied for general vehicle detection. On one pathway, a speed sensor captures real-time data, precisely measuring the speed of vehicles within the radar's range. Concurrently, another pathway handles positional information, pinpointing the location of each vehicle at the exact moment of detection.

The IoT system operates in parallel, and once the IoT module, equipped with its speed sensor, identifies a speed limit violation, the process is seamlessly handed over to the camera. In this subsequent parallel process, the camera captures the image of the vehicle in violation. This staged parallelism optimizes computational resources, ensuring that the initial AI application focuses on broad vehicle detection, while the subsequent process, triggered by a speed violation, engages the camera and AI for detailed image retrieval of the offending vehicle.

In Fig. 6, vehicles detected by the AI are outlined in blue, while those identified as exceeding the speed limit are highlighted in red. This color-coded distinction provides a clear and intuitive way to differentiate between the general detection of vehicles and the specific subset that is considered to be in violation of speed regulations.

In our system, we employ a foveal coding technique, which is centered around areas of interest to extract potentially suspicious vehicles. This method allows us to focus computational resources on specific regions within an image, enhancing the

Fig. 6 Detection of vehicles in blue and suspicious vehicles in red

efficiency of vehicle detection. Once the region of interest, containing the vehicle, is isolated (highlighted in red), we apply a deep learning algorithm to determine the license plate details as shown in Fig. 7. This algorithm, trained through deep neural networks, is adept at recognizing and extracting intricate patterns, enabling accurate identification of license plates even in complex visual scenarios. The combination of foveal coding and deep learning ensures a targeted and robust approach to license plate recognition, contributing to the overall effectiveness of our system in accurately identifying vehicles of interest.

Fig. 7 License plate identification and QR code generation

The system consolidates critical data, including the vehicle's speed, spatial coordinates, location, and license plate details. This comprehensive set of information is then encoded into a QR code. The QR code serves as a compact and efficient means of encapsulating the multifaceted details associated with the detected vehicle infringement. With a single QR code, the system establishes a unified data package for the detected violation. This QR code not only streamlines data management but also facilitates a secure and rapid transfer of information within our IoT platform.

In the subsequent step, a watermarking process is applied to the vehicle's image, involving the insertion of the QR code. Watermarking is a technique used to embed imperceptible and robust information directly into digital media, such as images, to ensure its integrity and authenticity. In this specific case, the QR code, representing essential details, is strategically embedded as a watermark within the image of the vehicle. The purpose of this watermarking process is to conceal the QR code in a manner that is imperceptible to the human eye but can be reliably extracted later. It acts as a hidden identifier, linking the visual data of the vehicle with the encoded information in the QR code. This dual-layered approach ensures that even if the visual content is altered or manipulated, the essential data remains intact and recoverable through the concealed QR code.

In Fig. 8, the illustration reveals a tripartite composition. On the rightmost side, we observe the QR code, encapsulating crucial information such as the vehicle's speed, spatial coordinates, location, and license plate details. Positioned in the middle is the foveally coded image of the vehicle. On the far left, the figure displays the watermarked image, which ingeniously combines the visual content of the suspicious vehicle with the concealed QR code.

The application of watermarking in this context not only enhances the security of the information but also provides a means to trace back the encoded details to their visual source. This process contributes to the robustness and reliability of the system in maintaining the integrity and authenticity of the data throughout its lifecycle, from detection to subsequent analysis.

In Fig. 9 (right), we compare the execution time of transmitting a watermarked image coded by visually based watermarking with the execution time of sending the same image coded by SSEWIC in the transport sector. The results demonstrate that visually based watermarking offers significant advantages in terms of transmission time, indicating its efficiency in reducing time overhead during data transfer. In this Fig. 9 (left) compares the execution time for encoding and decoding a transport image using the EFIC and SPIHT encoders. The illustration clearly indicates that EFIC is notably faster in both encoding and decoding processes, underscoring its efficiency as a more suitable encoder for this specific application in the transport sector.

In Fig. 10 (right), we present two charts displaying the visual weighting and its important impact in the coding quality on images related to vehicular speed controller systems-related QR code image. The left chart contrasts visual encoding with SPIHT applied to a vehicular speed controller QR code image, while the right one examines the visual coder foveally oriented alongside SPIHT applied to a vehicular speed controller picture. The visual encoding segment in the left chart reflects the image's interesting quality with respect to its conqueror SPIHT, allowing viewers

Fig. 8 Watermarked images

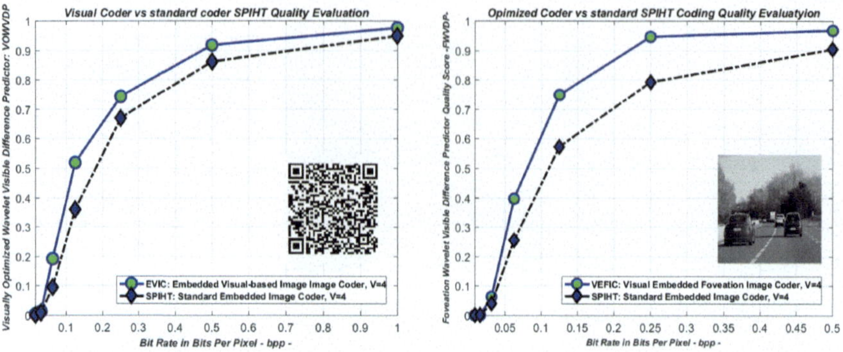

Fig. 9 Visual encoding versus SPIHT encoding applied to vehicular speed controller QR code image (Left), foveation encoding versus SPIHT encoding applied to vehicular speed controller picture (Right)

to perceive the net quality difference for a given binary budget and a fixed observation conditions. Turning to the right chart, we see the visual foveally oriented quality coding with respect to its conqueror SPIHT, which depicts the vehicular speed controller picture after undergoing its spectrum a visual regulation managing a wavelet-based weighting quality model to be applied to the original one aiming to filter or eliminate any undesirable content. This method mimics the human visual system quality properties by encoding the image's regions of interest with higher resolution and the peripheral ones with reduced detail, often employed to optimize the image content. This approach, similar to its use in the left chart, aims to compactly represent the image data, potentially reducing storage and processing requirements. These visual representations facilitate a clear comparison of the effects of various encoding methods on vehicular speed controller images, offering insights into their potential benefits, such as data efficiency, image quality, and suitability for specific application scenarios.

Figure 10 displays a comparison of transportation times for the following payloads: The VSC image and its watermarked version employ roughly two approaches, namely encoding and decoding. Both images produce resulting files in PNG format, with different compression ratios ranging from 64:1 (64 bits per pixel) for very low data rates, up to 1:1 (1 bit per pixel) for high binary rates.

The second approach, which is the subject of our contribution, involves encoding the VSC image and its watermarked version. The latter moderately follows visual weighting models and sends the resulting binary files to the cloud in an encrypted format using the IoT communication protocol. Upon arrival at the cloud, the cloud performs the reverse operations: ECC decryption, SPIHT decoding, dewatermarking, and reverse weighting, finally reconstructing the VSC and its watermarked version by angle. This approach helps reduce the transmission rate and energy consumption compared to the first approach.

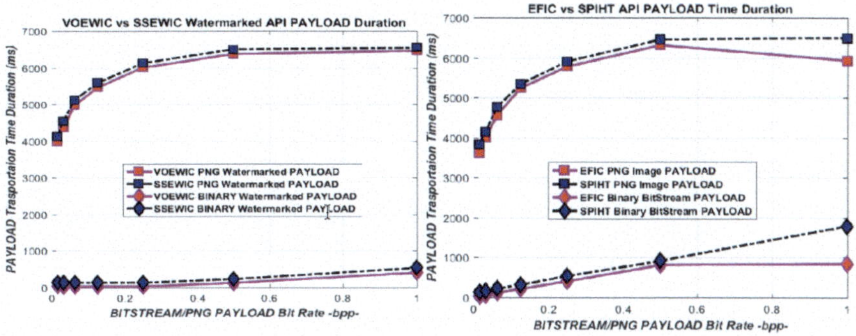

Fig. 10 Standard watermarking versus its visual version (Left), and standard coding SPIHT versus its visual version (Right) payload time duration applied to vehicular speed controller picture with respect to scalable bit rate (128:1, 64:1, 32:1, 16:1, 8:1, 4:1, 2:1) and a fixed observation distance of V = 4

5 Conclusion

This article focused on a cutting-edge IoT platform created specifically for the transportation industry to handle its specific difficulties and security needs. High-resolution cameras, dynamic QR code generators, and QR code readers are all included in our IoT nodes, which provide safe data exchange. This platform successfully addresses the complicated problems relating to the identification of vehicles and license plates, improving the safety and effectiveness of transportation operations. We've implemented cutting-edge watermarking methods to strengthen data security in the transportation industry. Additionally, in order to address issues with data storage, execution time, memory consumption, and bandwidth utilization, we investigated and compared a number of encoding techniques, including SPIHT, and EFIC. Our goal is to identify the most effective encoding technique in order to maximize resource usage and simplify data transfer in the transport domain. In addition, our platform depends on reliable communication protocols, especially CoAP [20] and ECC, to guarantee the secure transfer of photos, whether or not they contain digital watermarks [21]. While utilizing artificial intelligence for decision-making processes, the sent data is encoded, encrypted, decoded, and dewatermarked. Our technology stands out for its combination of AI-driven license plate and vehicle recognition, which boosts security and effectiveness. This AI-powered technology makes a substantial contribution to the identification of vehicles and license plates, enhancing the efficiency and safety of transportation operations as a whole. Our transport IoT platform's results show significant improvements in resource optimization, lower energy use, and sustained image quality, all while maintaining a constant viewing experience. The results of the study's quantitative, subjective, and objective evaluations support the value of the solution we produced. In conclusion, our transport IoT platform offers a significant step in ensuring the security, effectiveness, and dependability of data transfer within the transportation industry. We have improved security and decision-making in this crucial arena, making it safer and more effective, by integrating AI vehicles and license plate recognition.

References

1. Fitzpatrick C et al (2016) The application of continuous speed data for setting rational speed limits and improving roadway safety. Saf Sci 85:171–178. https://doi.org/10.1016/j.ssci.2016.01.020
2. Shobayo O et al (2020) Development of smart plate number recognition system for fast cars with web application. Appl Comput Intell Soft Comput 2020:1–7. https://doi.org/10.1155/2020/8535861
3. Zhou H et al (2021) Vehicle speed preview control with road curvature information for safety and comfort promotion. Proc Inst Mech Eng Part D: J Automob Eng 235(6):1527–1538. https://doi.org/10.1177/0954407020972851
4. Wei Y et al (2017) Optical image encryption using QR code and multilevel fingerprints in gyrator transform domains. Opt Commun 403:62–67. https://doi.org/10.1016/j.optcom.2017.06.087.
5. Onim Md.SH et al (2022) BLPnet: a new DNN model and Bengali OCR engine for automatic licence plate recognition. Array 15:100244. https://doi.org/10.1016/j.array.2022.100244
6. Patel C, Doshi N (2020) A novel MQTT security framework in generic IoT model. Procedia Comput Sci 171:1399–1408. https://doi.org/10.1016/j.procs.2020.04.150
7. Chawla S et al (2021) Smart education: a proposed IoT based interoperable architecture to make real time decisions in higher education. Rev Geintec 11(4):5643–5658. https://doi.org/10.47059/revistageintec.v11i4.2589
8. Campolo C et al (2021) Virtualizing AI at the distributed edge towards intelligent IoT applications. J Sens Actuator Netw 10(1):13. https://doi.org/10.3390/jsan10010013
9. Alrubei S et al (2021) The use of blockchain to support distributed AI implementation in IoT systems. IEEE Internet Things J 9(16):14790–14802. https://doi.org/10.1109/jiot.2021.3064176
10. Roh B-H et al (2023) Mixed reality-enabled multilateral collaboration application platform with AI and IoT convergence. In 2023 IEEE International conference on consumer electronics (ICCE). https://doi.org/10.1109/icce56470.2023.10043498
11. Qi W et al (2021) Generic reversible visible watermarking via regularized graph Fourier transform coding. IEEE Trans Image Process 31:691–705. https://doi.org/10.1109/tip.2021.3134466
12. Tom RJ et al (2022) Car license plate detection and recognition using modified U-Net deep learning model. In: 2022 8th International conference on smart structures and systems (ICSSS). https://doi.org/10.1109/icsss54381.2022.9782176
13. Xie X et al (2021) Oriented R-CNN for object detection. In: 2021 IEEE/CVF International conference on computer vision (ICCV). https://doi.org/10.1109/iccv48922.2021.00350
14. Al-Azawi M et al (2015) Human attention-based regions of interest extraction using computational intelligence. https://doi.org/10.1109/ieeegcc.2015.7060025
15. Alduais NAM et al (2018) APRS: adaptive real-time payload data reduction scheme for IoT/WSN sensor board with multivariate sensors. Int J Sens Netw 28(4):211–211. https://doi.org/10.1504/ijsnet.2018.096458
16. Hua G et al (2020) Informed histogram-based watermarking. IEEE Signal Process Lett 27:236–240. https://doi.org/10.1109/lsp.2020.2965331
17. Singh P et al (2022) Region-based hybrid medical image watermarking scheme for robust and secured transmission in IoMT. IEEE Access 10:8974–8993. https://doi.org/10.1109/access.2022.3143801

18. Shantha A et al (2019) Analysis and implementation of ECC algorithm in lightweight device. https://doi.org/10.1109/iccsp.2019.8697990
19. Ukil A et al (2014) Lightweight security scheme for IoT applications using CoAP. Int J Pervasive Comput Commun 10(4):372–392. https://doi.org/10.1108/ijpcc-01-2014-0002
20. Jung J-H et al (2020) CoAP-based streaming control for IoT applications. Electronics 9(8):1320. https://doi.org/10.3390/electronics9081320
21. Hasan HR et al (2017) Secure lightweight ECC-based protocol for multi-agent IoT systems. https://doi.org/10.1109/wimob.2017.8115788

Influence of Ethical Behavior on Job Satisfaction of Members of Information System Project Team

Emmanuel Samuel Mtsweni

Abstract The information system projects (ISPs) are experiencing a higher failure rate because ISP team members are having poor individual performance which result in poor ISP team performance. ISP team is failing to become high performing ISP teams. The poor ISP team performance results in ISP teams failing to meet their main objective of ensuring the success of ISPs. The poor performance results from poor job satisfaction of the members which makes the members of ISP team to have poor collaboration, teamwork, well-being, burnout, and less work engagement. The main purpose of the study was to evaluate the influence of ethical behavior on the job satisfaction of the members of ISP team. The researcher conducted a quantitative study using survey approach. The quantitative data was collected from the members of the ISP team using online structured questionnaire. The researcher performed quantitative data analysis using Smart PLS 4.0. The researcher performed partial least square-structured equation model (PLS-SEM) with path analysis technique. The results indicates ethical leadership, ethical climate, and ethical culture improves the job satisfaction of the members of ISP teams. The job satisfaction of the members of the ISP teams can be improved through ethical climate, ethical culture and ethical climate. The ethical leadership makes ISP team members to have improved ethical climate and ethical culture, and ethical climate is the driver for good ethical culture within ISP teams and among the individual members which result in ISP teams having good ethical behaviour.

Keywords Information systems · Information system projects · Job satisfaction · Ethical leadership · Ethical climate · Ethical culture

E. S. Mtsweni (✉)
Department of Computer Science, University of South Africa, Science Campus, Florida Park, Republic of South Africa
e-mail: mtswees@unisa.ac.za

X.-S. Yang et al. (eds.), *Proceedings of Ninth International Congress on Information and Communication Technology*, Lecture Notes in Networks and Systems 1000,
https://doi.org/10.1007/978-981-97-3289-0_3

1 Introduction

Information system projects (ISPs) are human-oriented in nature and humans formulate ISP teams. ISP teams are the ones that perform all the tasks to ensure the success of ISP projects by relying on their skills, expertise and knowledge [1, 2]. The individual members are experiencing a high level of burnout, stress, and low job satisfaction which makes them to be less productive, and become high-performing members of ISP teams. The less productive members do not perform their tasks to the required quality standard and deliver their tasks late which results in the failure of ISPs [3]. The failure of ISPs is currently at an alarming rate to an extent that is estimated that one out of eight ISPs result in the success and the remaining seven ISPs result in failure. The high failure rate of ISPs resulted in different organizations not realizing any benefits from commissioned ISPs in the form of return on investment and organization further remained with same challenge led to the commissioning of ISPs which limits them from having a competitive edge over their competitors [4].

The satisfied members of ISP teams, especially with their jobs can forge good and quality relationships with other members of their respective ISP teams which makes them have good collaboration and teamwork with other members [5]. The good quality relationships, teamwork, and collaboration make the members share knowledge, support other members, and encourage teamwork which makes the entire ISP team to achieve their main objectives which includes completing the tasks with accurate quality standard and on time and ensuring the success of ISP [6]. Satisfied members with their jobs are engaged in their work and perform the work to the required quality standard which makes them to be productive members [5]. The engaged members are ready and prepared to perform the most challenging tasks and put extra effort, dedication, and time to ensure that they deliver the quality work and meet their deadlines to ensure the ISP team achieve their objectives [7].

The satisfied members with their jobs can manage their burnout, fatigue, and stress which is currently at a high rate and limits the members from being productive. The burnout, fatigue, and stress make the members not to be able to deliver the tasks on time, with the required quality standard and make them not to be productive which includes not making them to be high performing members [8]. It is important for the ISP team to consist of the members who have good job satisfaction because they formulate a high performing ISP team whereby the members do not require any supervision when it comes to them performing their tasks and ISP teams can themselves. The job satisfaction is influenced by the different elements, and it is important for the senior management and team leaders of ISP teams to ensure the members are satisfied with their jobs for the members to be high performing members of ISP teams [9].

The handling, leadership, management, interactions, and the working environment of members develop and improve their job satisfaction [10]. The main purpose of this study was to evaluate the influence of ethical behavior on the job satisfaction of the members of ISP teams. This study will create an understanding of how job satisfaction can be developed among the members and the role of ethical behavior in developing job satisfaction which includes the management and handling the members of ISPs teams. The researcher started by outlining the theoretical background, followed by the research design and methodology, the data analysis performed and outlining the conclusion which includes the results of the research study.

2 Theoretical Background

2.1 Job Satisfaction

Job satisfaction is the emotional state of the member as the result of their job experiences [11]. The job satisfaction of the member is basically based on the overall emotional feelings, and the opinions of the member on their work and roles within ISP teams which includes their work conditions and the status that they have within ISP teams [12]. Job satisfaction has powerful consequences on the members which makes the members have positive work attitude and improved performance which makes members to be high performing [13]. Job satisfaction is further explained with the reliance of theory of equity where equity theory refers to the balance of the member's input and output in the work workplace. The input is in the form of time, efforts, loyalty, flexibility, skills, and trust in senior management to name few. The output is in the form of salary, job security, appraisals, reputation, achievements, and the benefits which is inclined to the treatment of the members [14].

The members who have good job satisfaction have good professional identity [15]. Professional identity makes the members have good attitude and devotion to their profession including their roles within IS teams. Professional identity is reflected in the members' desire to continue working within the IS projects or ISP teams, and their higher degree in liking being the part of the ISP team [16]. Professional identity becomes the motivation for the members to dedicate themselves to their work, which makes them work harder and performs the most challenging tasks when there is a need for such tasks to be performed. Professional identity makes the individual members to be productive and become more committed to their roles within ISP teams [15].

Job satisfied members have good work engagement and resilience to their work [17]. Work engagement makes the members to have positive and affective state of motivation in relation to their work which include work related well-being which is characterized by absorption, dedication, and vigor. The vigor, dedication, and absorption make the members manage burnout and their poor well-being, especially psychology well-being. This makes them to be effective and efficient which results in them being productive [18]. The resilience of members gives them an ability to recover from negative events and have positive emotions most of the time which makes them to deliver quality work. Resilience makes the members become stronger, more confident, and more productive by experiencing the stressful and challenging events and can overcome negative events. These events become the motivation for them to work harder to ensure that they perform their tasks to the required quality standard[19]. Resilience makes members alleviate stress and being able to overcome stress which is a norm within ISPs [18].

The members that are satisfied with their jobs collaborate, have good relationships and act in an ethical manner at work [19]. Collaboration enables the members to share knowledge and work as a unit with the aim of achieving the main objective of ISP teams of ensuring success of ISPs [5]. The members with job satisfaction are motivated to perform their work to the required quality standard. Motivation makes members put extra effort and time when performing tasks to meet their deadlines. The ISPs have immense pressure because the time allocated to them is not sufficient which brings lot of pressure to the members to deliver tasks on time with that unrealistic time [19].

2.2 *Ethical Culture*

Ethical culture is the subset of organizational culture. Ethical culture represents the multidimensional interplay among formal and informal systems that govern the behavior of members which focus on promoting the ethical behavior and eliminating unethical behavior within ISP teams and organizations [20]. The formal system of ethical culture consists of leadership, authority, structure, and policies while informal systems consist of one behavior and perceived norms. The informal and formal systems of ethical culture are influenced heavily by the senior management and team leaders within ISP teams through formal policies, authority, and structures that guide the ethical behavior among the members of ISP teams. The senior management and team leaders influence members by being role players or role models and leading by example of accepted ethical behavior including ethical decision-making [21–23]. Ethical culture shares two important dimensions with organizational culture, namely values and beliefs and cross-fertilization among different levels [24]. Ethical culture is concerned with the promotion of fairness and justice with the aim of members following procedures and to avoid any punishment for being unethical [25]. Ethical culture gave so much attention to the outcomes which is ethical behavior [26] and ethical reasoning [27].

Ethical culture is all about the fairness and justice within ISP teams and organizations, creating the open collaborations and allowing all members to participate in decisions on what is an accepted ethical behavior [28, 29]. Fairness makes members feel appreciated in ISP teams with the work they do within ISP teams, and their work because they are being treated in a fair manner and with integrity, which brings justice. This makes the members enjoy their job which results in them having positive emotions and attitude toward their jobs [22].

Open collaboration allows members to support each other by sharing knowledge, expertise, and working as a single unit, which makes them to have teamwork, and ensures that every member is competitive and achieves their objectives, which includes delivering quality work and on time. Collaboration makes members who struggle to be in a good position to get the required support on time, which positions them to be able to perform their tasks and deliver them on time. This makes them feel good about their work and looking forward to performing more tasks because they feel good and can perform well, and they are able to deliver their tasks on time which makes them productive [28]. Ethical culture promotes open communication and makes members participate in decision-making on acceptable ethical behavior. This creates transparency and makes the members to be proud and enjoy their jobs. It gives the members access to accurate and sufficient information for them to be productive [29].

H_1: Ethical culture has positive and significant relationship with job satisfaction.

2.3 Ethical Climate

Ethical climate is the shared perceptions by the members on what is accepted and regarded as correct ethical behavior and how ethical situations should be managed within ISP teams [29]. There are two dimensions of ethical climate namely ethical decision-making and the ones that drives ethical decision-making. The first dimension is based on the ethical decision-making, namely egoism, benevolence, and principles. The second dimension which drives ethical decision-making is based on individual, local, and cosmopolitan [24]. The ethical climate is based on the consensus that is held by the ISP team on what is right or wrong behavior among members or within the ISP team [30].

Ethical climate exists at psychological level and team level or organizational level. Psychological ethical climate is based on ethical atmosphere experienced by the members. The psychological ethical climate is the member's assessment of their own ISP team ethical climate which is affected by ISP team's ethical climate and distinctive characteristics in the member's perception about their respective team climate. Teamethical climate is the ethical atmosphere that is perceived by the consensus with all the members of ISP team. Team ethical climate is the team level assessment of the team ethical climate or organizational ethical climate which is determined by the team's ethical climate or organizational climate [31].

Ethical climate of the ISP team takes into consideration the ISP team's procedures, policies, and practices with moral consequences [32]. Members who consider their ISP team to be ethical in the way they do things are likely to be treated fairly. Fairness indicates the high level of ethical climate within ISP teams which increases the engagement by members in the job roles. Members who experience high level of ethical climate reciprocate ISP team ethical climate with a positive attitude to the jobs as compared to those with low perceive of ethical climate [31]. The elements of ethical climates of caring and rules make the members to be happy about their jobs which increases their job satisfaction [21].

Ethical climate makes members work in an efficient and effective manner because of how their work environment is handled and managed which becomes enabling environment for the members [23]. Ethical climate guides how members interact which take into considerations morals, integrity, ethical decision-making, and communication to name few. This reduces stress, fatigue, and burnout because ethical climate makes members to be productive and encourages good treatment of the individual members within ISP teams [22]. Ethical climate has a high potential in making members have good and positive ethical behavior which result in good attitude toward their jobs [23].

H_2: ethical climate increases the job satisfaction of members of ISP team.

H_3: Ethical climates increases the ethical culture within ISP team.

2.4 Ethical Leadership

Ethical leadership is the demonstration of the normative acceptable behavior based on the personal actions and interpersonal relationships and promotes the acceptable ethical behavior by open communication and ethical decision-making [33]. Ethical leadership consists of two dimensions, namely moral person and moral manager. Moral person dimension is related to the qualities such as honesty, trustworthiness, and fairness, especially in the interactions, relationships, and approaches with the members. Moral dimension, a leader that has moral principles is regarded to have moral principles in their personal life and professional life which makes them gain the reputation of honesty, caring, integrity, and fairness. Moral manager dimension focuses on the endeavors of the ethical leader for driving and promoting acceptable ethical behavior. Moral leader displays ethical behavior and ethical standards to members [34]. Ethical leaders rely on reward and punishment to promote ethical behavior within ISP teams, and they are the role models by "walk the walk" and "talk the walk" approaches or mechanisms [32].

Ethical leadership credibility exerts meaningful influence on members in the form of influencing them to do what is beneficial to ISP team, especially their actions which includes their behaviors [35]. This is done with the purpose of ensuring the ISP team achieves its goals and objectives, which includes ensuring the success of ISPs. Ethical leaders promote effective interactions, communication, relationships, and collaboration among members [29]. Effective interactions, collaboration, relationships, and communication make members to share knowledge, to work as a unit, to support each other and have access to sufficient information and knowledge required for the members to complete their tasks with correct functionality which is based on quality and within allocated time. Effective interactions, communication, and collaboration make members to be high performing members because receive support which makes them to have good attitude toward their jobs [6].

Ethical leaders provide motivation, capacity, freedom, and knowledge which result in good work engagement because members become committed, enjoy their work, are prepared to go extra mile, and put extra effort to ensure that they meet their deadlines which makes them productive [29]. Motivation drives the members to gain more passion, dedication, and resistance with the purpose of ensuring that they perform their tasks properly and work hard to meet their deadlines which improves their productivity [17]. ISP are knowledge intensive in nature and members use their knowledge to perform tasks and ethical leadership enables the ISP team to overcome the knowledge hiding. The role of ethical leadership in providing motivation, capacity, freedom, and knowledge makes the members enjoy their work which leads to good job satisfaction [6].

Ethical leaders gain trust by ensuring transparency within the respective ISP teams. Trust makes the members to have assurance that leaders and senior management have the best interest for the members by doing things in a correct manner. Leaders will always make decisions in the members' best interest and taking into consideration the members [6]. Transparency within ISP teams is ensured through open communication and involvement of members in decision-making. The open communication makes members voice their concerns and discuss uncomfortable issues in relation to what is acceptable and right ethical behavior. The process of involvement of members in decision-making makes the members understand and influence the decisions that are made within the ISP teams which makes them to be part of the decision. The members feel appreciated and accepted, which makes them work harder and ensures that the ISP team to be high performing ISP team. The members are most likely to love and appreciate their workplace and their work and have positive attitude toward the jobs [36].

H_4: Ethical leadership increases job satisfaction of the members of ISP team.

H_5: Ethical leadership increases the ethical culture within ISP teams.

H_6: Ethical leadership increases the ethical climate within ISP teams.

3 Research Design and Methodology

Researcher conducted quantitative research study using survey approach which was influenced by nature and the purpose of the study of evaluating the influence of ethical behavior on the job satisfaction of the ISP team members. This includes validation of hypothesis [37]. Quantitative data was collected using a structured online questionnaire and was collected from the members of ISP teams that were based in the Republic of South Africa. The members needed to have minimum of two years working within ISP teams when this study was conducted. The profile of the sample is depicted in Table 1.

The questionnaire was developed using the existing literature. The researcher relied on the concepts that were used heavily by different researchers. The constructs were measured using Likert Scale items which were measured using 7-point Likert Scale whereby 1 represents strongly disagree while 7 represented strongly agree. A

Table 1 Sample profile of respondents

Category	Options	Number	Percentages (%)
Gender	Female	64	19.6
	Male	263	80.4
Number of projects	1–20 projects	135	41.3
	21–40 projects	107	32.7
	41–60 projects	63	19.3
	More than 60 projects	22	6.7
Experiences	1–5 years	101	30.9
	6–10 years	118	36.1
	11–15 years	36	11
	16–20 years	46	14.1
	More than 20 years	26	8
Roles	Business analysts	37	11.3
	Data analysts	24	7.3
	Developers	200	61.2
	Project managers	35	10.7
	Quality assurance	12	3.7
	Senior manager	19	5.8

total of 327 responses were collected which met the minimum required responses required for the researcher to perform types of analysis to achieve the objective and validating the hypothesis of the research study [38]. The details of constructs that were used in this research study were summarized in Table 2.

The researcher adhered to the ethical requirement in this research study. The researcher applied for the ethical clearance from the School Ethics Board and the ethical certificate was granted prior to the data collection process. The participation in this study was voluntary and the anonymity of the respondents was ensured through using synonyms. The respondents needed to give consent prior completing the questionnaire.

4 Data Analysis

Quantitative data analysis was conducted. Partial least square-structured equation model (PLS-SEM) was performed using path analysis technique was used in this study. Smart PLS 4,0 was used to conduct quantitative data analysis.

Table 2 Construct definitions and items

Construct	Definition	Items	Key reference
Ethical leadership	Ethical leadership is the type of leadership that is characterized by fairness, integrity, respect, and transparency in the way in which the leaders are conducting themselves and the way they are doing things	10	[33]
Ethical culture	Ethical culture is the culture within ISP teams that creates and promotes ethical behavior and makes the ISP teams have acceptable behavior. The ethical culture provides clarity on the expected issues of ethical behavior within ISP teams	5	[39]
Ethical climate	Ethical climate is the ethical standards that are based on the set of criteria within ISP team and provides norms that guide the ISP teams in decision-making and the behavior within ISP teams that needs to meet the ethical standards	3	[23]
Job satisfaction	Job satisfaction is the emotional state of the member of ISP team as the result of their job experiences	3	[9]

4.1 Validity and Reliability

The researcher needed to ensure validity and reliability in this research study. The researcher started by using the existing literature to ensure validity and reliability of the constructs. The researcher further ensured all the Likert Scale items had a factor loading that had a minimum value of 0.5 and above [38]. All the Likert Scale items met this requirement in this study. The reliability was ensured using Cronbach's Alpha (α), Composite Reliability (RHO_A), and Composite Reliability (RHO_C). The α needed to have a minimum value of 0.5 for the construct to be considered having good reliability [40]. All the constructs met the minimum value of α and were considered acceptable. The RHO_A and RHO_C needed to have a minimum value of 0.7 for the construct to be considered to have acceptable reliability [38]. All the constructs had the RHO_A and RHO_C that met the minimum values that are required for the construct to be considered having good reliability. All the constructs had good reliability.

Validity was measured using average variance extracted (AVE) and Heterotrait-Monotrait ratio correlation. AVE was used to measure convergent validity and discriminant validity was measured using Heterotrait-Monotrait ratio correlation. The construct needs to have AVE value that is above 0.5 for good convergent validity [38]. All the constructs had the AVE values that were more than 0.5 which indicated that all the constructs passed the convergent validity in this research study (Table 3).

Discriminant validity was measured using Heterotrait-Monotrait ratio correlation. Heterotrait-Monotrait ratio correlations needed to be below 0.85 for the constructs to have good discriminant validity [40]. All the constructs had the Heterotrait-Monotrait ratio correlation that was below 0.85. Therefore, all the constructs had good discriminant validity test in this research study (Table 4).

Table 3 Measurement Model

Construct	Items	Loading	Alpha	RHO_A	RHO_C	AVE
Ethical leadership	ETL_1	0.517	0.928	0.952	0.940	0.614
	ETL_2	0.743				
	ETL_3	0.602				
	ETL_4	0.897				
	ETL_5	0.765				
	ETL_6	0.820				
	ETL_7	0.898				
	ETL_8	0.878				
	ETL_9	0.878				
	ETL_10	0.517				
Ethical culture	ETC_1	0.825	0.787	0.807	0.855	0.545
	ETC_2	0.748				
	ETC_3	0.774				
	ETC_4	0.533				
	ETC_5	0.776				
Ethical climate	EC_1	0.930	0.904	0.918	0.939	0.838
	EC_2	0.892				
	EC_3	0.924				
Job satisfaction	JS_1	0.898	0.906	0.912	0.941	0.842
	JS_2	0.946				
	JS_3	0.908				

Table 4 Heterotrait-Monotrait ratio matrix

	Ethical Climate	Ethical Culture	Ethical Leadership	Job Satisfaction
Ethical climate				
Ethical culture	0.577			
Ethical leadership	0.437	0.658		
Job satisfaction	0.727	0.674	0.608	

4.2 *Significance of Structural Model*

The structural model was assessed using R^2 and Q^2. The R^2 is statistical variance indicates how the endogenous variable is explained by the exogenous variable [40]. Ethical climate had R^2 value of 0.185 which indicated ethical leadership had 18.5% variance which was low. Ethical culture had R^2 value of 0.442 which indicated ethical leadership and ethical climate had a 44.2% variance which was moderate.

Table 5 Structural model assessment (R^2 and Q^2)

Endogenous variable	R^2	Q^2
Ethical climate	0.185	0.173
Ethical culture	0.442	0.359
Job satisfaction	0.577	0.333

Job satisfaction had R^2 value of 0.577 which indicated ethical leadership, ethical climate, and ethical culture had 57.7% variance which was moderate.

The Q^2 is predictive relevance which measures the predictive relevance of the model. Q^2 is used to establish predictive relevance endogenous variables. Q^2 value that is above zero indicates the predictive relevance and the constructs are well-constructed [40]. Ethical climate had Q^2 value of 0.173 which was predicted and was ethical leadership. Ethical culture had Q^2 value of 0.359 which was predicted by ethical climate and ethical leadership. Job satisfaction had Q^2 value of 0.333 which was predicted by ethical leadership, ethical climate, and ethical culture. Therefore, the predictive relevance was at existence because all the Q^2 values were above zero (Table 5).

Effect of size (f^2) was further evaluated to measure the strength of endogenous variables in predicting the exogenous variable [40]. The $f^2 \geq 0.02$ is considered small, $f^2 \geq 0.15$ is considered medium, and $f^2 \geq 0.35$ is considered large [38]. Ethical climate had f^2 value of 0.113 which was small and there is no effect, while ethical climate had f^2 value of 0.377 on job satisfaction, which was considered large. Effect of ethical climate on job satisfaction cannot be ignored. The ethical culture had the f^2 value of 0.041 which was small. Ethical leadership had f^2 value of 0.228 on ethical climate which was medium, and its influence cannot be ignored. Ethical leadership further had f^2 value of 0.359 on ethical culture and the influence cannot be ignored. Lastly ethical leadership had f^2 value of 0.108 on job satisfaction, which was small.

Model fit was further evaluated, which was measured using SRMR, Chi-Square, and NFI. The SRMR needed to be ≤ 0.008, while Chi-Square needed to be greater than 0.01 and NFI needed to be greater than 0.90 [40]. The SRMR was 0.005, while Chi-Square was 0.301 and NFI was 0.91 which indicated the model fit was of a good standard in this research study (Table 6).

Table 6 Model fit

Mode	Estimated model	Fit criteria
SRMR	0.005	≤ 0.008
Chi-Square	0.301	> 0.01
NFI	0.91	> 0.90

4.3 Hypothesis Testing

Validation and testing of relationships among different constructs was performed to test and validate the hypotheses of the research study. The relationships were evaluated using the path coefficient (β), t-values, and p-values. The β represents the strength and sign of relationship, especially the effect of exogenous variable(s) to an endogenous variable. The t-values and p-values are used to test the significance of the relationship between exogenous variables and endogenous variables. The t-value needs to be greater than 1.64 while p-value needs to less than 0.05 for the relationship to be considered significant [40].

The first relationship was between ethical culture and job satisfaction, and the relationship was significant and positive ($\beta = 0.175, t = 3.085, p = 0.002$). Therefore, H_1 was supported. The relationship between ethical climate and job satisfaction was positive and significant ($\beta = 0.467$, $t = 9.421$, $p = 0.000$). Therefore, H_2 was supported. The relationship between ethical climate and ethical culture was positive and significant ($\beta = 0.279$, $t = 9.212$, $p = 0.000$). Therefore, H_3 was supported. The relationship between ethical leadership and job satisfaction was positive and significant ($\beta = 0.276, t = 3.416, 0.001$). Therefore, H_4 was supported. Relationship between ethical leadership and ethical culture was positive and significant (β = 0.496, t = 12.916, p = 0.000). Therefore, H_5 was supported. The last relationship in this research study was between ethical leadership and ethical climate which was significant and positive ($\beta = 0.431, t = 5.827, p = 0.000$). Therefore, H_6 was supported. This indicates all the relationships were supported (Table 7).

Table 7 Summary of structural model (hypotheses outcomes)

Endogenous Variable	β	T	Correlation	P-Value	Outcome
Ethical Leadership → Job Satisfaction	0.276	3.416	0.585	0.001	Supported
Ethical Culture → Job Satisfaction	0.175	3.085	0.575	0.002	Supported
Ethical Climate → Job Satisfaction	0.467	9.421	0.672	0.000	Supported
Ethical Leadership → Ethical Culture	0.496	12.916	0.616	0.000	Supported
Ethical Leadership → Ethical Climate	0.431	5.827	0.431	0.000	Supported
Ethical Climate → Ethical Culture	0.279	9.212	0.492	0.000	Supported

5 Discussion of Results

The good ethical behavior by ISP team leaders, senior management, and members makes other members enjoy and have greater interest in their work environment including their work. The ethical behavior creates an atmosphere where the members can trust each other including the senior management and team leaders, and treat each other with integrity, respect, and motivates them to give it all to their work to deliver tasks on time. Ethical behavior makes the treatment and management of the members to be of a good and acceptable standard which makes them efficient, effective, and productive which results good individual performance. The members who are efficient, effective and productive feel good about their work and working environment which makes them yield positive emotions towards their work and their work environment that make them to be happy with their jobs and result in good job satisfaction.

Leadership plays an important role in shaping, molding, providing direction, motivation, and driving how members perceive their jobs and their work environment [22]. Ethical leadership provides a superior way of doing things including molding thebehavior by promoting ethical behavior, encouraging open communication, ensuring transparency, treating members with integrity, respect, moral way, and enabling ethical decision-making within ISP teams [23, 30]. Ethical leadership makes senior management and team leader with ISP teams to manage the members with integrity, respecting and interact with each other with integrity, and manage them in a way that is satisfactory and enables them to enjoy their roles which makes them productive. This includes members enjoying their workplace, job roles, and the tasks that they are performing. This makes the individual members get joy and happiness for doing their work within ISP teams which results in good job satisfaction.

The working environment that embeds ethics, ethical culture, and ethical climate encourages the members to have good ethical behavior which motivates them in doing what is right all the time and acting in an ethical manner all the time. This is enforced using policies, rules, procedures, norms, and the leaders being the driving force of ensuring ethical behavior in ISP teams is implemented and practiced by members. The members are encouraged and forced to have good ethical behavior results in creating work environment that has work ethic. The work ethic makes the members work in the environment that is supportive, treats members well, has trusts, and provides all necessary required skills and resources that members need to be competent in their respective roles and allocation of work and duties in a fair manner which makes the members to be productive.

The ethical leadership creates a working environment that embeds ethical culture and ethical climate and promotes good management and treatment of individual members in a fair manner. The ethical leaders are the driving force of ensuring the implementation of ethical culture and ethical climate in a working environment, especially within ISP teamsand rely on different mechanisms and approaches which includes them being instrumental. The ethical leaders rely on their capability of doing things in a correct manner and ethical manner in ensuring that they become exemplary to the members when it comes to what is acceptable ethical behavior. Therefore, the ethical leaders become the instruments of implementing and enforcing the ethical culture and ethical climate by the way, they do things within ISP teams.

6 Conclusion

This research study indicated that ethical behavior makes the members have good and improved job satisfaction. The ethical behavior is one of the driving forces in ensuring the members develop and have job satisfaction. The elements that make the members have good job satisfaction are ethical leadership, ethical culture, and ethical climate. Therefore, the ethical behavior, especially ethical culture, ethical climate, and ethical leadership are not only for enhancing the relationships, efficiency, effectiveness and productivity of the members but also enhancing the job satisfaction of the members which triggers on how the members perceive and feel about their jobs which result in enhanced job satisfaction.

In theoretical perspective, this study incorporates the elements of ethical behavior to the job satisfaction of the members especially the role and influence of ethical behavior in driving the job satisfaction of the members. The job satisfaction is improved by doing things right and ethical with the members, especially on how they are being treated and managed, which rely oninfusing honesty, transparency, respect, morals and integrity. The ethical behavior makes the members yield positive emotions and feel good about what they are doing for living including their jobs, especially meeting their performance standards which is the indication for productivity. This includes them perceiving the job roles as important and fulfilling. The enabling working environment as the result of ethical behavior makes the members to be high performing members which provides motivation for them to feel good about their jobs and have high performing ISP teams.

The ethical behavior provides a practical approach that makes the members have improved and enhanced job satisfaction. This is done by influencing different elements such as emotions and how members perceive their jobs that give the members improved job satisfaction. The members behave in an ethical manner especially acceptable ethical behavior makes it easier for other members to have good job satisfaction which indicates they are creating the environment which is fulfilling for other members for them to feel good about their jobs and the members end up having good job satisfaction. The members that work in an ethical environment or environment that prioritize ethics and ethical behavior have good relationships and collaborate with other members which makes them to have access to wide variety of support such having access to sufficient and quality tacit knowledge that enable them to be high performing members.

The ISP teams and their respective organizations need to have ethics programs and code of ethics within the respective ISP teams and organizations to promote ethical leadership, ethical climate, and ethical culture. The ethics programs and code of ethics enable the management to create transparency on ethics and promote ethical behavior among the members of ISP teams. The senior management, team leaders and members of ISP teams need to make the ought and sign ethics pledge that they will abide to ethics or ethical norms all the time for them to have good ethical behavior and make ethical decisions irrespective of the situation. By signing ethics pledge by members, team leaders and senior management within ISP teams provide a good position to enforce the members, senior management and team leaders to adhere to good ethical standards, practices, values, and behaviors all the time which improve their capabilities of being ethical leaders. This will include them treating the members with integrity, respect, honesty, and transparency which results in them being trusted by the members of ISP team. This further results in the ISP team having good ethical culture because the leaders are the source and instrument of developing and maintaining ethical culture within ISP teams. This will further enhance the two elements of ethical leaders, namely moral person and moral leader which result in the development of ethical leaders.

Ethical leaders need to ensure they abide to the values and norms of the ethical leaders. The values include fairness, honesty, moral, trustworthiness, open communication, and transparency [34]. Ethical leaders in doing create the enabling working environment for themembers to be competitive and enjoy performing their duties. The members develop good and positive attitude toward their work and work environment which makes them have improved job satisfaction. An environment that infuses ethical culture and climate equips and facilitates members to be in good position to perform their tasks to the required quality standards which makes them to be productive and meet the deadlines. Then the working environment becomes the one that is enabling for productivity.

This study focused on ISP teams whereby the emphasis was about improving the job satisfaction among the members ISP teams. The study relied on the ethical leadership, ethical climate, and ethical culture which indicates other element such as work ethic was not included in this study which needs to be explored. There is a need to understand the influence of the ethical work environment on the performance, collaboration, job satisfaction, trust, and relationships of the members. In conclusion, the ethical behavior makes the members have improved and good job satisfaction.

References

1. Sharma R, Singh JN (2022) Systematic literature review and comparative centred software development. In: 4th International conference on advances in computing, communication control and networking. IEEE, pp 2239–2246
2. Guveyi E, Aktas MS, Kalipsiz O (2020) Human factor on software quality: a systematic literature review. In: 20th International conference in computational sciences and its applications. Springer International Publishing, pp 918–930
3. Venkatesh V, Thong JY, Chan FK, Hoehle H, Spohrer K (2020) How agile software development reduce work exhaustion: insights on role perceptions and organizational skills. Inf Syst J 30(4):733–761

4. Liu M, Balamurugan S, Seetharam TG (2021) Impact of stress on software developers by moderating the relationship through emotional intelligence in a work environment. Aggression Violent Behav 9
5. Nayak S, Jena D, Patnaik S (2021) Mediation framework connecting knowledge contract, psychological contract, employee retention, and employee satisfaction: an empirical study. Int J Eng Bus Manage 13
6. Bhatti SH, Kiyani SK, Dust SB, Zakariya R (2021) The impact of ethical on project success: the mediating role of trust and knowledge sharing. Int J Manag Proj Bus 14(4):982–998
7. Motlhanke SD, Naong MN (2021) The impact of emotional intelligence on behavioural factors during transition: a case of free state technical and vocational education and training colleges. SA J Hum Resour Manage 19
8. Agarwal RN (2015) Stress, job satisfaction and job commitment's relation with attrition with special reference to Indian IT sector. Manage Innov Competitive Advantage
9. Tripp JF, Riemenschneider C, Thatcher JB (2016) Job satisfaction in agile development as work redesign. J Assoc Inf Syst 17(4)
10. Rezvani A, Chang A, Wiewiora A, Ashkanasy NM, Jordan PJ, Zolin R (2016) Manager emotional intelligence and project success: the mediating role of job satisfaction and trust. Int J Project Manage 34(7):1112–1122
11. McAllister CP, Harris JN, Hochwarter WA, Perrewé PL, Ferris GR (2017) Got resource? A multi-sample constructive replication of perceived resource availability's role in work passion-job outcomes relationships. J Bus Psychol 32:147–164
12. Troesch LM, Bauer CE (2017) Second career teachers: Job satisfaction, job stress and the role of self-efficacy. Teach Teach Educ 67:389–398
13. Zhai Q, Lindorff M, Cooper B (2013) Workplace quanxi: its dispositional antecedents and mediating role in the affectivity-job satisfaction relationship. J Bus Ethics
14. Adams JS, Freedman S (1976) Equity theory revisited: comments and annotated bibliography. Adv Exp Soc Psychol 1(9):43–90
15. Chen H, Liu F, Pang L, Liu F, Fang T, Wen Y, Chen S, Xie Z, Zhang X, Zhao Y, Gu X (2020) Are you tired of working amid the pandemic? The role of professional identity and job satisfaction against job burnout. Int J Environ Res Public Health 17(24)
16. Mahmoudi-Gahrouei V, Tavakoli M, Hamman D (2016) Understanding what is possible across a career: professional identity development beyond transition to teaching. Asia Pac Educ Rev 17:581–597
17. Riyanto S, Herlissha N (2020) Job satisfaction management, work motivation and employee engagement to improve employee performance. Int J Bus Manage Invention (IJBMI) 9(7):11–16
18. Lu L, Lu AC, Gursoy D, Neale NR (2016) Work engagement, job satisfaction, and turnover intentions: a comparison between supervisors and line-level employees. Int J Contemp Hosp Manag 28(4):737–761
19. Kamalanabhan TJ, Sai LP, Mayuri D (2009) Employee engagement and job satisfaction in the information technology industry. Psychol Rep 105(3):759–770
20. Treviño LK, Butterfield KD, McCabe DL (1998) The ethical context in organizations: Influences on employee attitudes and behaviours. Bus Ethics Q 8(3):447–476
21. Mayer DM (2014) A review of the literature on ethical climate and ethical culture. The Oxford handbook of organizational climate and culture
22. Schaubroeck JM, Hannah ST, Avolio BJ, Kozlowski SW, Lord RG, Treviño LK, Dimotakis N, Peng AC (2012) Embedding ethical leadership within and across organizational levels. Acad Manag J 55(5):1053–1078
23. Lu CS, Lin CC (2014) The effects of ethical leadership and ethical climate on employee ethical behaviour in the international port context. J Bus Ethics 124:209–223
24. Schein EH (2010) Organizational culture and leadership. Wiley
25. Zhang J, Chiu R, Wei LQ (2009) On whistleblowing judgement and intention: the roles of positive mood and organizational ethical culture. J Manag Psychol 24(7):627–649

26. Ruiz-Palomino P, Martinez-Cañas R (2014) Ethical culture, ethical intent, and organizational citizenship behavior: the moderating and mediating roles of person-organization fit. J Bus Ethics 120:95–108
27. Sweeney B, Arnold D, Pierce B (2010) The impact of perceived ethical culture of the firm and demographic variables on auditors' ethical evaluation and intention to act decisions. J Bus Ethics 93:531–551
28. Ardichvili A, Mitchell JA, Jondle D (2009) Characteristics of ethical business cultures. J Bus Ethics 85:445–451
29. Engelbrecht AS, Heine G, Mahembe B (2017) Integrity, ethical leadership, trust and work engagement. Leadersh Organ Dev J 38(3):368–379
30. Aryati AS, Sudiro A, Hadiwidjaja D, Noemijati N (2018) The influence of ethical leadership to deviant workplace behavior mediated by ethical climate and organizational commitment. Int J Law Manage 60(2):233–249
31. Wang YD, Hsieh HH (2012) Towards a better understanding of the link between ethical climate and job satisfaction: a multilevel analysis. J Bus Ethics 105:535–545
32. Treviño LK, Den Nieuwenboer NA, Kreiner GE, Bishop DG (2014) Legimating the legitimate: a grounded theory study of legitimacy work among ethics and compliance officers. Organ Behav Hum Decis Process 123(2):186–205
33. Brown ME, Treviño LK, Harrison DA (2005) Ethical leadership: a social learning perspective for construct development and testing. Organ Behav Hum Decis Process 97(2):117–134
34. Zhu W, Zhang X, He H, Wang G, Zhang X (2019) Ethical leadership with both "moral person" and "moral manager" aspects: scale development and cross-cultural validation. J Bus Ethics 158:547–565
35. Piccolo RF, Greenbaum R, Hartog DN, Folger R (2010) The relationship between ethical leadership and core job characteristics. J Organ Behav 32(2):259–278
36. Hendrikz K, Engelbrecht AS (2019) The principled leadership scale: an integration of value-based leadership. SA J Ind Psychol 45(1)
37. Oates BJ, Griffiths M, McLean R (2022) Researching information systems and computing. Sage
38. Hair Jr JF, Hult GT, Ringle CM, Sarstedt M, Danks NP, Ray S (2021) Partial least squares structural equation modelling (PLS-SEM) using R: a workbook. Springer Nature
39. Cortes-Mejia S, Cortes AF, Herrmann P (2021) Sharing strategic decisions: CEO humility, TMT decentralization, and ethical culture. J Bus Ethics 1–20
40. Ringle C, Da Silva D, Bido D (2015) Structural equation modeling with the Smart PLS. Braz J Mark 13(2):56–73

To Be or Not To Be—After AI Revolution

Viktória Bakonyi, Zoltán Illés, and Dávid Szabó

Abstract During the last decades, IT solutions were growing continuously in almost all parts of life, so the labor market needed more and more programmers and other IT professionals it seemed to be a perfect career. At the end of 2022 freely available, artificial intelligence (AI) solutions were published, which among others are able to handle programming tasks, give suggestions, create codes, and debug programs. This new possibility raised two important questions. One about informatics education. How can we use up this AI tool in education, and how can we decide what they really know, or they simply use ChatGPT during tests? The other question is more serious, if AI can solve implementation tasks, does it mean the end of IT careers in the near future for a lot of young people? We asked the opinion of our students, who will be programmers studying either in Hungarian or in English training. One eye weeps the other laughs. Some of them already used it in non-ethical way during their studies but the higher their skills the less they fear about not to have a real IT career. This is in line with our view that the more in-depth knowledge one has, the less one should fear the effects of the AI revolution.

Keywords ChatGPT · IT professional · Education

V. Bakonyi (✉) · Z. Illés · D. Szabó
Eötvös Loránd University, Pázmány s 1/c 1117, Budapest, Hungary
e-mail: hbv@inf.elte.hu

Z. Illés
e-mail: illes@inf.elte.hu

D. Szabó
e-mail: sasasoft@inf.elte.hu

X.-S. Yang et al. (eds.), *Proceedings of Ninth International Congress on Information and Communication Technology*, Lecture Notes in Networks and Systems 1000,
https://doi.org/10.1007/978-981-97-3289-0_4

1 Introduction

We are living in a digitalized world which permeates all aspects of life. The online environment around us is highly strengthened according to the effect of Covid-19 emergency [1]. The quickly growing automatization sounded the alarm bell in the labor market. It was said that a lot of jobs, which can be made by robots will disappear soon. It seemed that IT area is not among the losers, because automatization and digitalization need programmers and other IT professionals.

Previously, we can use tools which produces part of a program code from UML, e.g., enterprise architect. It was said that the goal is to implement applications without human work, but it might be thought as the product of the far future. From years ago, we faced with customer service chatbots from whom we can discuss something in the filtered topic.

From the end of 2022 AI-based tools, ChatGPT and the like were published, and anybody may use it freely. It caused a "revolution," the penetration of trying this tool was quick and huge [2]. These are using artificial intelligence that can act as a virtual partner with whom you can discuss any topic in the same way you would with a human. Moreover, they can create new essays, (picture or music) in any given topic—which shocked the field of education immediately [3]. Who knows after this, who did the homework or a test, the student, or the AI tool. (There are applications, which detect it [4–6].) Some university professors tried them, and they passed their tests. It is a real fear from the teachers that lazy students try to neglect work having such possibility [7]. Some of the schools banned its usage, some of them created ethical rules [8]. It is clear this tool can be used for an excellent personal help for learning [9–11]. They are used already in a lot of places in the industry, e.g., in advertising industry.

They can support you (even the free version) in the field of informatics as well, giving ideas in implementation, helping in debugging, explaining the code, or even writing a code after describing the task. And here comes the double problem of IT area. If this auto-coding tools are excellent than the students may spare the hard work and get their diploma easier but what comes later? Will the industry need so many programmers if they work can be substituted by AI tools? This way the future career of a lot of them is in danger, though remains enough work for more skilled, complex-thinking, and creative persons. Therefore, we asked their opinion and their experiences about AI tools, their usage, and their possible fear from losing their future career.

2 AI Tools—Student Survey 2023 May

Generally, we were interested whether they know about it, how they like it and for what they use it and what they think about their future career. We have both Hungarian and English trainings so we can compare their attitudes.

2.1 Survey Environments

We used Google Forms to keep anonymity. We asked all students attending or courses in spring semester from operating system (fourth semester) and computer systems (first cross-semester) and some secondary school students to fill the forms. Filling the form was not compulsory. Therefore, the data collected from them does not represent the opinion of everybody. Meanwhile, it means that we can get only trends.

Hungarian Version

Available at: https://forms.gle/sPiyphgYR97eQSvL8

Filled by 233 students:

- Secondary school: 49
- First year: 13 (40)
- Non-first year: 171 (241).

English Version

Available at: https://forms.gle/jz29enbuv5nTRZTT9

Filled by 34 students:

- Secondary school: –
- First year: 10 (20)
- Non-first year: 24 (60).

2.2 Awareness, Liking, Confidence, and Frequency of Usage

Awareness and Liking

First question "Have you tried the new ChatGPT or other voice-based AI options?".

We do not have too many answers from first year students (in a cross semester, mainly they failed the subject in autumn). In a good secondary school about half of them heard about it. As it was waited, programming students are interested in this new possibility and most of them already tried it.

Almost the same percentage of awareness we can see in the case of Hungarian and English training; see Table 1.

They may choose between the answers: Tried, liked; Tried, not liked; Not tried, interested; Not tried, not interested. See Table 2 and Fig. 1.

Table 1 Awareness of new AI tools among students in 2023 May

	Trial-HU (%)	Trial En (%)
Secondary school	59.18	–
First year student	100.00	100.00
Non-first year	91.81	91.67

Table 2 Awareness and likeness of new AI tools among students in 2023 May

	Tried, liked (%)	Tried, not liked (%)	Not tried, interested (%)	Not tried, not interested (%)
Secondary school HU	44.90	14.29	18.37	22.45
Secondary school EN	0.00	0.00	0.00	0.00
First year student HU	76.92	23.08	0.00	0.00
First year student EN	80.00	10.00	0.00	10.00
Non-first year HU	78.36	13.45	4.68	3.51
Non-first year EN	95.83	4.17	0.00	0.00

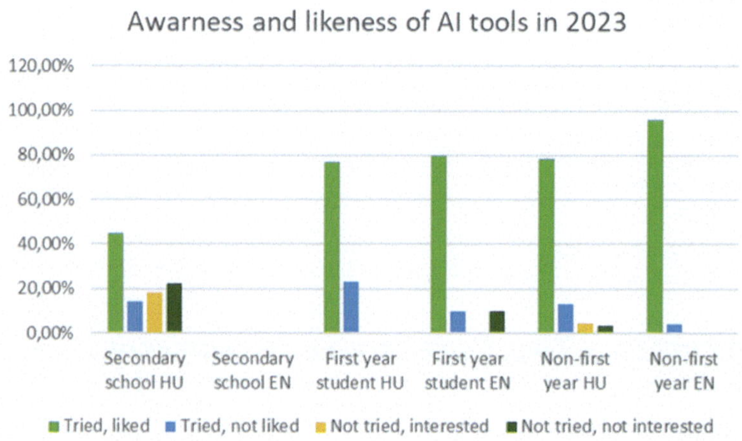

Fig. 1 Awareness and likeness of new AI tools among students in 20,223, May

Almost everybody tried it (more than 90%), but Hungarian students were more critical toward the tools. English students do not learn on their mother tongue, so they need more help—they are more grateful to have any special help. Moreover, they arrive from different countries and have different pre-knowledge and that is why learning is a bigger challenge for them.

Frequency of AI Tools Usage

Let us see the details. How frequently do they use it? They can grade it between 1 and 5 (1 means never).

It is seen at Fig. 2 that English students use AI more frequently than Hungarians. Remember, they liked it better also!

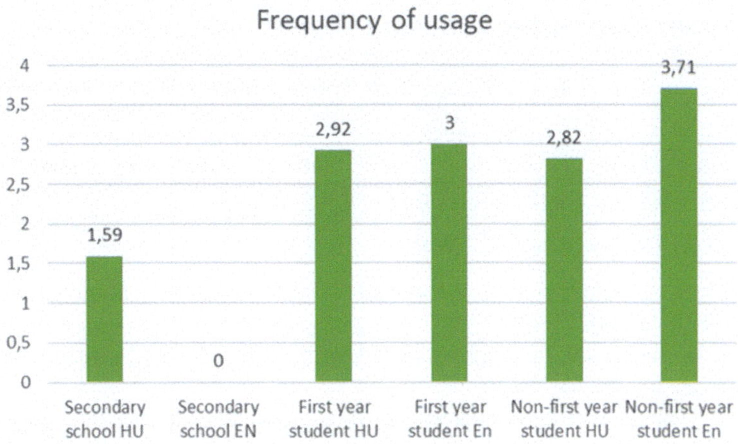

Fig. 2 How often they use AI tools (graded between 1 and 5)

Confidence

The next question asks about confidence. How confident are you that the answer is correct? They could give a grade between 1 and 5. (1 means that somebody does not believe AI correctness at all) See Fig. 3.

Quite interesting result, that the more practiced students are less believe the confidence. At this moment, we must agree with them, AI code implementation is not perfect. In a lot of times, first semester tasks are solved perfectly, but sometimes it makes ridiculous, simple errors. (Naturally we made some testing of the system with our previous exams and homework.) To tell the truth to find an error sometimes is more difficult than to implement a code by your own.

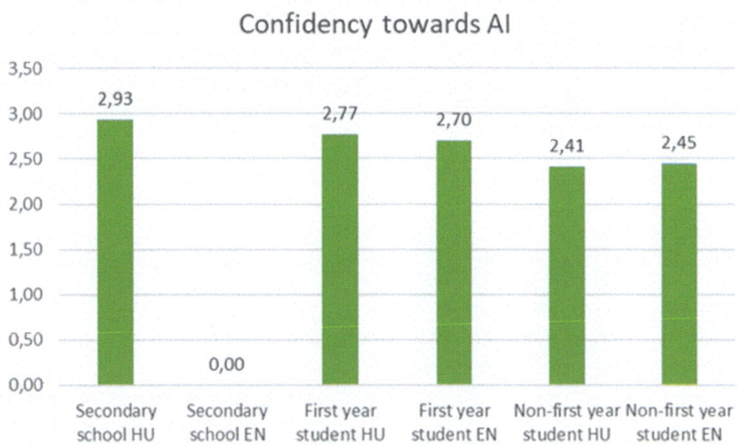

Fig. 3 Do they believe in the correctness of AI answers?

2.3 *Future Career, Professional Usage*

We asked about their plans too. Do they fear about losing their future career possibilities? According to the spread of AI?

Our question was: Do you fear that in the future ChatGPT or voice, image-based AI tools will take over some of the programming tasks?

Fear that AI Do Programmers' Work

They had to grade their answers between 1 and 5 (1 means they do not fear about this possibility at all).

In Fig. 4, we can see, that more than 80% of them fear from AI spread somehow (gave a grade bigger than 1).

Grade 3 means that they think about AI as potential rival, but they are not in panic. What is interesting that non-first grade English students gave 5 at the highest percentage?

Why they are so unsure of their future? Contrary Hungarian non-first year students are more confidents. In other groups, we are not able to see such great difference—though it is true, the number of the participants is not high.

We have seen in Fig. 3 that the average confidence against AI is less in the case of non-first year students. So, we have to examine in what circumstances they use AI.

Do they use it only for fun or for professional purposes too**?**

Usage of AI

The question was: What did you use it for, if you used it?

- Talking, testing, having fun with machine intelligence.
- To solve everyday problems (e.g., writing a letter, finding out about a topic (where the cow buzzards live).
- Solve professional and/or school tasks (e.g., programming, comprehension, composition).
- Voice-based generation.
- Image-based cloning, image, motion picture generation.
- Other.

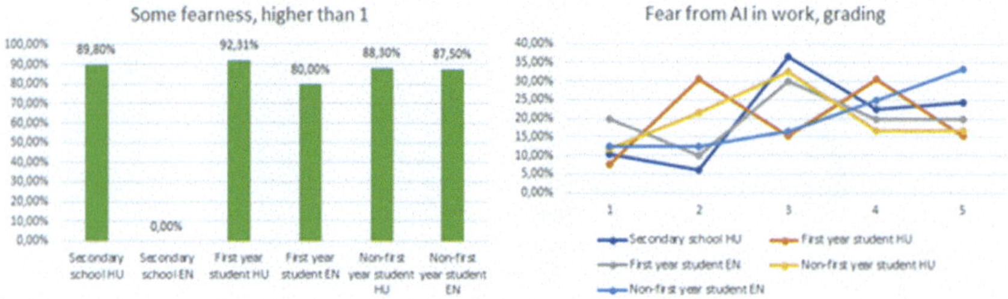

Fig. 4 Fear about their future

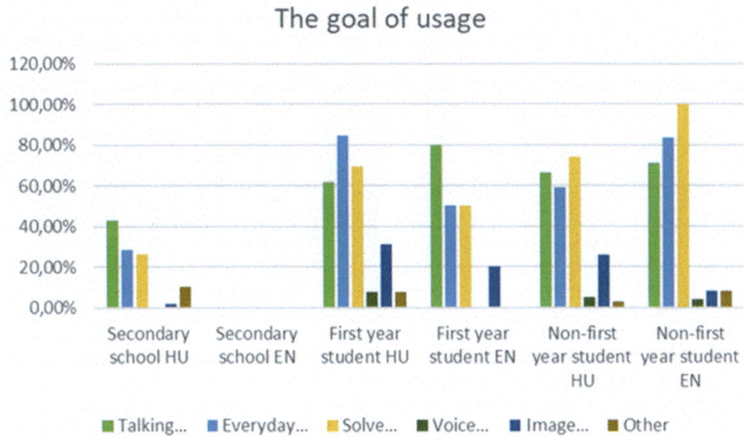

Fig. 5 Goal of usage

We can see in Fig. 5 that all the examined groups used AI for fun, for everyday problems, and for professional purposes too! Voice, image, and other usage are not in their focus yet.

Professional usage is in our main interest naturally. More than 50% of the students use it for (school) work too. The highest rates are in the case of non-first year students (73%, 100%), especially English students. Remember to Fig. 3 which showed that the more practiced the student, the less they believe in the infallibility of AI! We must understand why they still use it and for what in their work?

Professional Work

Our next question was: For what do you use it in school or other professional tasks? The possible answers were:

- To interpret the task/concepts.
- To theoretically build the solution algorithm (what steps are needed).
- For specific tasks, implementation.
- Existing concrete solution, code interpretation.
- To prepare documentation, summarize a topic.
- Other.

There was nobody, who checked "other."

In Fig. 6, the reader can see that all of the examined groups use AI (mainly ChatGPT) for all of the given tasks, but there are differences naturally. First year foreign students use it in a very high percentage for creating the documentation it is understandable because they have to create it not in their mother tongue. Each group uses AI for understanding better the task description, and the idea of the topic, so it simply helps them in the learning process.

The same we can say about the interpretation of code, if they do not understand a demo program it is like a personal assistant to help them to understand the solutions.

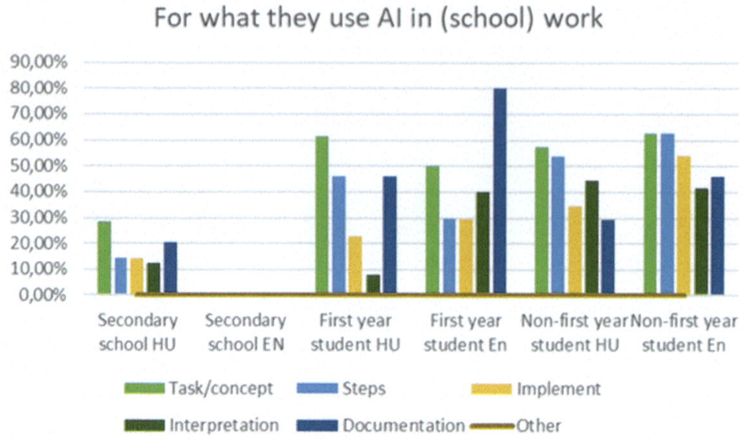

Fig. 6 For what do they use AI in schoolwork?

In the case of first year students, we understand that they feel themselves lost what algorithm (steps) they have to use but we were surprised that non-first year students also use this kind of help (53%, 62%). In their case the usage of different types of work is quite balanced—they used AI for everything.

The biggest surprise is, that a lot of students use AI for implementation too, meanwhile they do not trust in it! More than 50% of English non-first year students use ChatGPT for implementing a given task (54%).

Create, Interpret, Document—Three Types of Professional Work

For a better overview, we have also created a diagram that distinguishes between three types of use: creation (algorithm or code generation), interpretation (task, concept, or concrete code), and documentation see in Fig. 7. (Though documentation can be a task for deeper understanding the main focuses.)

Professors are happy if the students have a personal assistant to explain everything which is not clear, but we are a bit sad, if we see they want to use it for creating something instead of them.

We tried a lot of tasks from first year subjects and non-first year tasks, and sometimes the AI gives a misleading solution, where finding the error is often more difficult than writing the code yourself. Why non-first year students use AI for creation, for implementation?

2.4 *Cheating*

We had another question: Have you ever used it in any accountancy? (exam, test, etc.) They can grade it between 1 and 5 (1 means never).

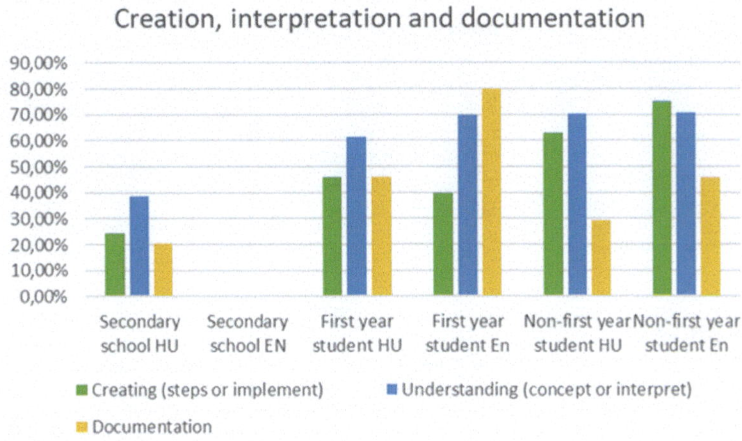

Fig. 7 Main usage types in professional work

At Fig. 8, you can see that during the tests, exams how many of them used at least once the AI tool. (Naturally, it was against the rules to use any external help.) It is another question how frequently they use it.

The reader may notice that the frequency is not too high (it is below 3 in each group), but it is growing after the first year especially in the case of English students. At first sight, we can think that those, who frequently use AI tools during tests have got a bigger chance to be unsure about their future, but it is not true. See in Table 3 that in the case of Hungarian non-first year students—whose number is the biggest in this survey and who are close to get their diploma, the average fear of losing their career is at very similar in the case of never and in the case of frequently cheating fellows. Naturally the survey is based on their words and not sure that everything is true in the answers (especially when they are speaking about cheating).

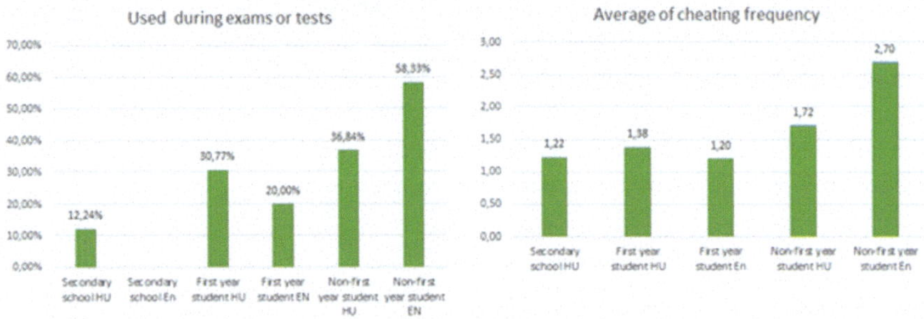

Fig. 8 Cheating at least once, and average of cheating frequency (graded between 1 and 5)

Table 3 Average fear from future, based on cheating grade

Frequency of cheating	1	2	3	4	5
	3.28	4.25	4.00	0.00	5.00
First year student HU	3.78	1.67	2.00	0.00	0.00
First year student En	2.88	4.00	0.00	0.00	0.00
Non-first year student HU	2.89	3.52	3.13	3.33	3.14
Non-first year student En	4.11	4.00	3.50	0.00	2.86

3 Summary

After analyzing these data, we can see some trends. Students naturally use the new AI tools for all purposes, for fun, and for professional work too. They do not confident absolutely to AI, but they still use it quite frequently.

The type of professional usage may give a very good personal help to them to learn better and to help foreign students to learn easier not in their native tongue. We made three types of professional usage, ask AI to create something (code or algorithm), to explain something (idea or code), and to do documentation. These are on different levels of thinking about a problem.

First year students try to ask for help in what they need the best, higher semester students use it in all of their different types of work. Fortunately, they do not lose their hope toward their career. Therefore, they are working hard to achieve their goals. They see that AI is not a perfect tool, human creativity is not forgettable—now, for a while.

So, we also think what most of our surveyed students think. Exactly now we can still answer, reflect, and respond to the familiar pollemma in the title: To be! It is worth to be a student who works hard to be an expert of informatics and to be a professor, who is to keep on teaching with using new tools, like AI too.

The only problem what universities must solve is to ensure fairness during exams. We can try different solutions, but going back to oral face-to-face exams from theory is a sure but very hard, tiring solution. The task is given for the professors of informatics trainings, where a lot of exams are organized on computers.

In the education today, there is no general solution, but in our experience, before each computer exam professors need to find the wording for which there is currently no correct AI answer—now, for a while! What about tomorrow? We'll see!

References

1. Bakonyi V, Illés Z (2022) Education challenges after Covid-19. In: 2022 20th International conference on emerging eLearning technologies and applications (ICETA). Stary Smokovec, Slovakia, pp 34–39. https://doi.org/10.1109/ICETA57911.2022.9974620

2. Mauran C, ChatGPT monthly traffic has dropped for the first time. Available at https://mashable.com/article/chatgpt-monthly-traffic-decline
3. Bennett D (2023) ChatGPT is an OK law student. Can it be an OK lawyer? Blomberg Newslett. Available at https://rb.gy/fw2xp
4. AI detector GPTZero available at https://gptzero.me/
5. AI detector, ZeroGPT available at https://www.zerogpt.com/
6. Tangerman V, There's a problem with that app that detects GPT-written text: it's not very accurate. Available at https://futurism.com/gptzero-accuracy
7. Blitchok A, How are colleges and universities responding to ChatGPT? Available at https://www.universities.com/news/how-are-colleges-and-universities-responding-to-chatgpt
8. Johnson A, ChatGPT in schools: here's where it's banned—and how it could potentially help students. Forbes. Available at https://www.forbes.com/sites/ariannajohnson/2023/01/18/chatgpt-in-schools-heres-where-its-banned-and-how-it-could-potentially-help-students/?sh=2a9ee8f86e2c
9. Fütterer T et al, ChatGPT in education: global reactions to AI innovations. Preprint. available at https://www.researchgate.net/publication/370658302_ChatGPT_in_Education_Global_Reactions_to_AI_Innovations. https://doi.org/10.21203/rs.3.rs-2840105/v1
10. Khaddage F, Flintoff K (2023) Goodbye to structured learning ChatGPT in education, is it a threat or an opportunity? In: Conference: Society for information technology & teacher education international conference at New Orleans. Available at https://www.researchgate.net/publication/370838637_Goodbye_to_Structured_Learning_ChatGPT_in_Education_is_it_a_Threat_or_an_Opportunity
11. Golgiyaz S (2023) ChatGPT in computer software education. In: Conference: ICHEAS, 4th International conference on health, engineering and applied sciences at Dubai. Available at https://www.researchgate.net/publication/371762576_ChatGPT_in_Computer_Software_Education

Use of Analytical Methods in Variant Forecasting Tasks

V. V. Kukartsev, A. S. Borodulin, S. A. Zamolotskii, V. A. Nelyub, and A. R. Glinscaya

Abstract The article presents an overview of modern data analytics methods and techniques used in forecasting tasks and considers key aspects of using data analytics in the context of the hospitality industry. The hospitality industry is particularly dependent on accurate forecasts, as effective management of occupancy, pricing, and guest service is critical to the successful operation of hotels. The application of data analytic techniques to forecast various phenomena and events and to assess the quality of forecasts is emphasized. The application of such techniques can reduce costs, increase revenue, and improve overall guest satisfaction.

Keywords Labor market · Socio-economic development · Sustainable economic development · Economy · Society

1 Introduction

The article considers the actual aspects of using analytical methods in the context of variant forecasting in various spheres, especially in the field of labor market and socio-economic development. Ensuring sustainable economic development and the impact of this process on the economy and society as a whole is very important. The application of forecasting is examined by applying it to a hotel reservation dataset.

Hotel booking through the Internet is becoming more and more popular [1–5]. But it is possible to cancel a reservation due to various circumstances. Reservation

V. V. Kukartsev · A. S. Borodulin · V. A. Nelyub
Bauman Moscow State Technical University, 105005 Moscow, Russia

V. V. Kukartsev · S. A. Zamolotskii
Siberian Federal University, 660041 Krasnoyarsk, Russia

V. V. Kukartsev · A. R. Glinscaya (✉)
Reshetnev Siberian State University of Science and Technology, 660037 Krasnoyarsk, Russia
e-mail: anna_glinskaja@rambler.ru

X.-S. Yang et al. (eds.), *Proceedings of Ninth International Congress on Information and Communication Technology*, Lecture Notes in Networks and Systems 1000,
https://doi.org/10.1007/978-981-97-3289-0_5

cancelation predictions are aimed at understanding the expected demand which will help to ultimately increase the profit [6, 7].

This paper discusses several data analysis techniques to solve the forecasting problem:

- K-nearest neighbors method [8];
- The method of support vectors with a linear kernel [9];
- Support vector method with RBF kernel [10];
- Decision tree [11];
- Random forest method [12];
- AdaBoost [13];
- Naive Bayesian classifier [14];
- Linear discriminant analysis [15];
- Quadratic discriminant analysis [16];
- Perceptron [17].

The models were compared, and as a result, the model with the highest accuracy was selected.

In recent years, online hotel booking has become increasingly popular due to the convenience it provides to consumers [18–20]. Consumers can directly browse the various options available on different travel websites or hotel chains, compare prices and amenities offered by competitors, and ultimately choose the right provider with ease [21–23]. Typically, these transactions are made between intermediary companies that represent multiple properties rather than directly contacting the hotels themselves [24].

Canceling a hotel reservation usually involves informing the hotel or the booking platform that you will not be staying at the hotel as planned [25]. Possible reasons for canceling a hotel reservation may be unforeseen circumstances requiring a long trip to another location [26]. Other examples include double bookings caused by human error or system failures [27]. Sometimes customers realize that their chosen destination no longer suits them, especially couples getting married abroad, newlyweds, families vacationing together, and business delegates attending conferences [28]. Finally, personal preferences such as wanting more privacy than originally advertised, food quality being lower than expected, or simply realizing that it becomes unsafe once they arrive in the area.

Cancelation forecasts aim to understand the expected demand, which helps set rates and marketing strategies to attract the desired customer base [29]. To calculate accurate forecasts, data analysis is used to compare historical records with current rates to identify patterns that may affect future results. Several data analysis techniques for solving the forecasting problem are discussed next [30].

2 Materials and Method

The research develops a model that consists of perceived value, trust in a third-party online booking site, and trust in hotels. Models that are a combination of historical and advanced booking predictions are also proposed. Logistic regression is applied to find out which is more important, low price or convenience. A big data clustering method is developed to forecast the hotel accommodation market as a time series. The factors influencing the decision to click on a link in online agency search engines are studied. The effect of price discrimination on the perceived fairness of booking intentions is also studied. The problem of ranking a hotel in a list is solved—the corresponding formula is derived. Methods using which hotels can promote their product in social media are outlined. With the help of data analysis, a web-oriented recommendation system for online hotel booking is also created. The method of content analysis is identified as one of the most promising ways to assess guest satisfaction with hotel services.

The principal component method is used to visualize the multidimensional array. After the data needs to be classified into two classes.

The "Hotel Reservations Dataset" containing the following input features was taken as the data under study:

- no_of_adults—number of adults is an integer from the interval [0, 4], mathematical expectation: 1.84, standard deviation: 0.52;
- no_of_children—number of children—integer from the interval [0, 10], mathematical expectation: 0.11, standard deviation: 0.40;
- no_of_weekend_nights—number of weekend nights (Saturday or Sunday)—integer from the interval [0, 7], mathematical expectation: 0.81, standard deviation: 0.87;
- no_of_week_nights—number of nights per week (Monday through Friday)—integer from the interval [0, 17], mathematical expectation: 2.20, standard deviation: 1.41;
- required_car_parking_space—parking space requirement—binary value {0, 1}, expectation: 0.03, standard deviation: 0.17;
- lead_time—number of days between booking date and arrival date—integer from the interval [0, 443], mathematical expectation: 85.23, standard deviation: 85.93;
- arrival_year—year of arrival—integer {2017, 2018}, mathematical expectation: 2017.82, standard deviation: 0.38;
- arrival_month—month of arrival—integer from the interval [1, 12], mathematical expectation: 7.42, standard deviation: 3.07;
- arrival_date—arrival date—integer from the interval [0, 31], mathematical expectation: 15.60, standard deviation: 8.74;
- repeated_guest—whether the client is a repeat guest—binary value {0, 1}, expectation: 0.03, standard deviation: 0.16;
- no_of_previous_cancellations—number of previous canceled bookings—integer from the interval [0, 13], expectation: 0.02, standard deviation: 0.37;

- no_of_previous_bookings_not_canceled—number of previous uncanceled bookings—integer from the interval [0, 58], expectation: 0.15, standard deviation: 1.75;
- avg_price_per_room—average price per day—a rational number from the interval [0.0, 540.0], mathematical expectation: 103.42, standard deviation: 35.09;
- no_of_special_requests—number of special requests—integer from the interval [0, 5], mathematical expectation: 0.62, standard deviation: 0.79;
- type_of_meal_plan—meal plan type—categorical {Meal Plan 1, Meal Plan 2, Meal Plan 3, Not Selected};
- room_type_reserved—room type—categorical {Room_Type 1, Room_Type 2, Room_Type 3, Room_Type 4, Room_Type 5, Room_Type 6, Room_Type 7};
- market_segment_type—type of market segment—categorical {Online, Offline, Corporate, Complementary, Aviation}.

The output variable booking_status denotes booking cancelations and is a categorical value {Canceled, Not_Canceled}.

The dataset includes 36,275 rows: 24,390 with the value of the output variable Not_Canceled, and 11,885 with the value of the output variable Canceled.

3 Results

Categorical attributes were converted to numeric attributes using LabelEncoder.

To visualize the data, the method of principal components was chosen—displaying the combination of features in a two-dimensional space. In this case, the sample shifts it so that the mean values are equal to 0 (Fig. 1).

In order to determine the significance of the factors, correlation analysis was performed. The correlation heat map is presented in Fig. 2.

To visualize the classification process, a fragment of the decision tree shown in Fig. 3 is given.

Accuracy was chosen as the comparison parameter. The comparison results are shown in Table 1.

As can be seen from the table, the most efficient model is the support vector method with RBF kernel.

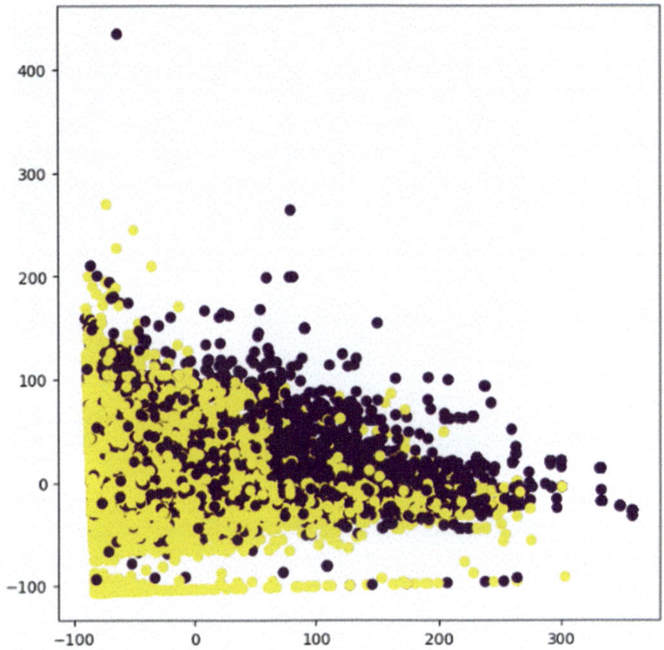

Fig. 1 Visualization by the method of principal components

4 Conclusion

Thus, the online hotel booking process is a relatively simple task with convenient accessibility. But there are cancelations due to unexpected events or dissatisfaction, these situations usually occur due to miscommunication or technological errors. Overall satisfaction ratings show strong user approval reflecting successful completion of bookings with high satisfaction levels. In addition, the implementation of advanced technology integration has improved user interaction, further streamlining processes while reducing the need for physical presence in the field.

As a result, several classifiers are proposed that can be used to cancel reservations, which can increase the profit of the hospitality industry. A comparison of the models is made, and it is found that the support vector method with RBF kernel is the best for this task.

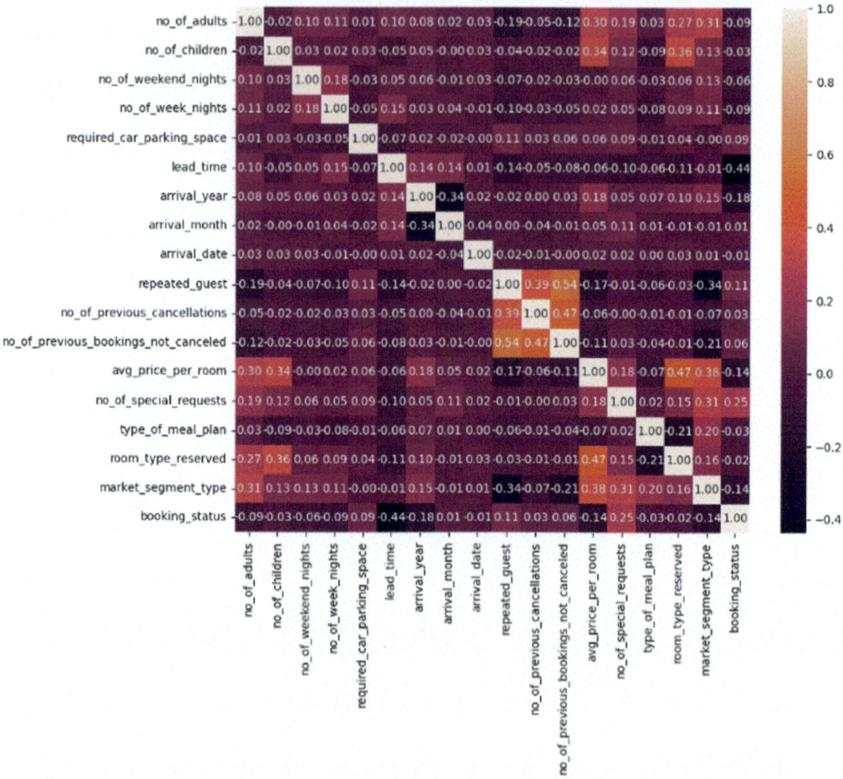

Fig. 2 Correlation heat map

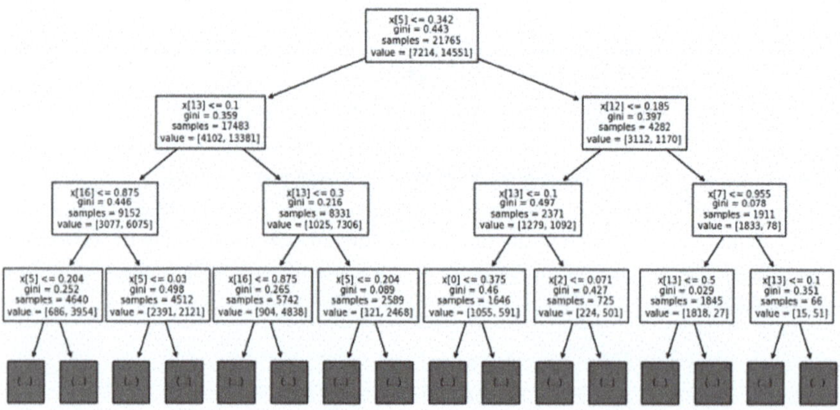

Fig. 3 Fragment of a decision-making tree

Table 1 Comparison of model accuracies

Model	Accuracy
K-nearest neighbors method	0.849
Support vector method with linear kernel	0.796
Support vector method with RBF kernel	0.850
Decision tree	0.844
Random forest method	0.685
AdaBoost	0.828
Naive Bayesian classifier	0.429
Linear discriminant analysis	0.803
Quadratic discriminant analysis	0.449
Perceptron	0.817

References

1. Rassokhin A et al (2022) Different types of basalt fibers for disperse reinforcing of fine-grained concrete. Mag Civ Eng 109(1):10913
2. Shutaleva A et al (2022) Migration potential of students and development of human capital. Educ Sci 12(5):324
3. Efremenkov EA et al (2021) Research on the possibility of lowering the manufacturing accuracy of cycloid transmission wheels with intermediate rolling elements and a free cage. Appl Sci 12(1):5
4. Gutarevich VO et al (2023) Reducing oscillations in suspension of mine monorail track. Appl Sci 13(8):4671
5. Malozyomov BV et al (2023) Overview of methods for enhanced oil recovery from conventional and unconventional reservoirs. Energies 16(13):4907
6. Strateichuk DM et al (2023) Morphological features of polycrystalline CdS1−xSex films obtained by screen-printing method. Crystals 13(5):825
7. Malozyomov BV et al (2023) Study of supercapacitors built in the start-up system of the main diesel locomotive. Energies 16(9):3909
8. Malozyomov BV et al (2023) Substantiation of drilling parameters for undermined drainage boreholes for increasing methane production from unconventional coal-gas collectors. Energies 16(11):4276
9. Masich IS, Tyncheko VS, Nelyub VA, Bukhtoyarov VV, Kurashkin SO, Borodulin AS (2022) Paired patterns in logical analysis of data for decision support in recognition. Computation 10(10):185
10. Masich IS, Tynchenko VS, Nelyub VA, Bukhtoyarov VV, Kurashkin SO, Gantimurov AP, Borodulin AS (2022) Prediction of critical filling of a storage area network by machine learning methods. Electronics 11(24):4150
11. Barantsov IA et al (2023) Classification of acoustic influences registered with phase-sensitive OTDR using pattern recognition methods. Sensors 23(2):582
12. Bosikov II, Martyushev NV, Klyuev RV, Savchenko IA, Kukartsev VV, Kukartsev VA, Tynchenko YA (2023) Modeling and complex analysis of the topology parameters of ventilation networks when ensuring fire safety while developing coal and gas deposits. Fire 6(3):95
13. Mikhalev AS et al (2022) The Orb-weaving spider algorithm for training of recurrent neural networks. Symmetry 14(10):2036
14. Moiseeva K et al (2023) The impact of coal generation on the ecology of city areas. In: 2023 22nd International symposium INFOTEH-JAHORINA (INFOTEH). IEEE, pp 1–6

15. Kukartsev V et al (2022) Analysis of data in solving the problem of reducing the accident rate through the use of special means on public roads. In: 2022 IEEE International IOT, electronics and mechatronics conference (IEMTRONICS). IEEE, pp 1–4
16. Kireev T et al (2022) Analysis of the influence of factors on flight delays in the United States using the construction of a mathematical model and regression analysis. In: 2022 IEEE International IOT, electronics and mechatronics conference (IEMTRONICS). IEEE, pp 1–5
17. Kukartsev V et al (2022) Prototype technology decision support system for the EBW process. In: Proceedings of the computational methods in systems and software. Springer International Publishing, Cham, pp 456–466
18. Kukartsev V et al (2022) Methods and tools for developing an organization development strategy. In: 2022 IEEE International IOT, electronics and mechatronics conference (IEMTRONICS). IEEE, pp 1–8
19. Malozyomov BV (2023) Improvement of hybrid electrode material synthesis for energy accumulators based on carbon nanotubes and porous structures. Micromachines 14(7):1288
20. Bukhtoyarov VV et al (2023) A study on a probabilistic method for designing artificial neural networks for the formation of intelligent technology assemblies with high variability. Electronics 12(1):215
21. Shutaleva A et al (2021) Environmental behavior of youth and sustainable development. Sustainability 14(1):250
22. Kosenko EA, Nelyub VA (2022) Evaluation of the stress–strain state of a polymer composition material with a hybrid matrix. Polym Sci Ser D 15(2):240
23. Nelyub VA, Komarov IA (2022) Technology of treatment of carbon fibers under electromagnetic influences of various origins to produce high-strength carbon fiber reinforced plastics. Russ Metall 13(9):1696
24. Nelyub VA, Fedorov SY, Malysheva GV, Berlin AA (2022) Properties of carbon fibers after applying metal coatings on them by magnetron sputtering technology. Fibre Chem 53(4):252–257
25. Nelyub VA (2021) The effect of copper and zinc coatings on the properties of carbon fibers and composites based on them. Polym Sci Ser D 142(14):260–264
26. Nelyub VA, Fedorov SY, Malysheva GV (2021) The study of the structure and properties of elementary carbon fibers with metal coatings. Inorg Mater Appl Res 124(12):1037–1041
27. Potapenko I et al (2021) Analysis of the structure of Germany's energy sector with self-organizing Kohonen maps. In: International conference on business information systems. Springer International Publishing, Cham, pp 5–13
28. Kukartsev VV et al (2023) Using digital twins to create an inventory management system. E3S Web of Conf
29. Gladkov AA et al (2023) Development of an automation system for personnel monitoring and control of ordered products. E3S Web Conf
30. Borodulin AS et al (2023) Using machine learning algorithms to solve data classification problems using multi-attribute dataset. E3S Web Conf

Toward a Classification of Factors Influencing eWOM in Online Shopping Through a Systematic Literature Review

Nisrine Benhammou, Khaoula Zahouani, Habiba Bensassi Nour, and Youssef Makloul

Abstract The rise of social media and the Internet has drastically changed the way consumers interact with brands. They have become more active consumers who are also more likely to provide feedback and recommendations. This interaction is referred to as eWOM, and it is one of the most vital sources of information for online shopping. The purpose of our study is to review the literature on the various factors that influence eWOM through a systematic literature review of current research between 2010 and 2022. It aims to identify the multiple antecedents and attributes of this interaction and to classify them based on their nature into two main categories, namely the attributes relating to the individual including those related to the communicator, to the receiver, and those related to the interaction between the two. The second category includes the attributes relating to the information. Our study attempts to provide a conceptual model project proposal based on our results of classification of antecedents and attributes that influence eWOM in online shopping for future potential corporate/academic research on this subject, as well as contribute to the advancement of our research about the online shopping experience and its impact on e-reputation.

Keywords eWOM · Online shopping antecedent · Attribute · Factor · Systematic literature review

N. Benhammou (✉) · H. Bensassi Nour
Ecole Nationale de Commerce et de Gestion, LR2MC, University Hassan First of Settat, Settat, Morocco
e-mail: benhammou.nisrine@gmail.com

K. Zahouani · Y. Makloul
FSJES, L-Qualimat GRTE Laboratory, Cadi Ayyad University of Marrakech, Marrakech, Morocco

X.-S. Yang et al. (eds.), *Proceedings of Ninth International Congress on Information and Communication Technology*, Lecture Notes in Networks and Systems 1000,
https://doi.org/10.1007/978-981-97-3289-0_6

1 Introduction

In an era dominated by technology, online shopping has become a pivotal aspect of global e-commerce, favored for its efficiency and convenience, providing expansive access to product information across various channels like blogs and social media platforms [1, 2]. The Internet's role in consumer communication has led to the emergence of electronic word-of-mouth (eWOM), differentiating from traditional word-of-mouth by its rapid dissemination and significant influence on consumer behavior [2, 3]. Recognizing eWOM's potential, this study conducts a systematic literature review spanning twelve years, focusing on the antecedents and attributes that affect eWOM in online shopping [4]. Aimed at enriching existing literature, it seeks to identify and categorize these factors into two principal groups, to understand their impact on electronic buyer opinions. The paper is structured to first define eWOM, followed by outlining the research methodology used to extract and analyze the pertinent antecedents and attributes. This comprehensive review culminates in the proposition of a conceptual model for practical application and suggests directions for future research in the domain [3].

2 Electronic Word-of-Mouth (eWOM)

The rise of the Internet has catalyzed the shift from traditional word-of-mouth (WOM) to electronic word-of-mouth (eWOM), fundamentally altering how consumers share and access opinions about products and services [5, 6]. eWOM encompasses feedback from current, former, or potential customers distributed broadly across online platforms such as social media, blogs, and review sites [3, 7]. Unlike WOM, which is characterized by face-to-face exchanges, eWOM operates asynchronously, allowing opinions to persist online and reach wide audiences, often anonymously [8]. This digital form of communication has a profound impact on consumer behavior, heavily influencing purchasing decisions and shaping consumer attitudes and perceptions due to its perceived credibility over company-provided information [9, 10]. Studies by ACNielsen and EMarketer have confirmed the consumer preference for peer reviews, establishing eWOM's trustworthiness [11]. Moreover, research by TW et al. links eWOM, especially customer-to-customer interactions, to the building of loyalty intentions, signifying its importance in the marketplace [12]. eWOM has thus become a critical communication tool in the e-commerce landscape [13].

3 Methodology and Results

A systematic literature review, as outlined by Denyer and Tranfield [14], is a foundational step in management research, setting the stage for empirical studies or offering an in-depth survey of a particular academic field. This methodology involves a thorough strategy to identify relevant studies, critically assess and synthesize them, and then report the findings comprehensively. Such reviews serve dual purposes: They enrich the academic community by summarizing and evaluating existing knowledge and aid practitioners by providing actionable insights, examples, and practice-oriented recommendations. The review process adheres to Denyer and Tranfield's structured steps, ensuring a rigorous and systematic approach to literature analysis [14].

- **Step 1**: Research question formulation
- **Step 2**: Locating studies related to the research question
- **Step 3**: Studies selection and evaluation
- **Step 4**: Analysis and synthesis
- **Step 5**: Reporting and using the results.

Step 1—Research Question Formulation

As with any research, the main decision in preparing a review is to establish its purpose [15]. By formulating clear questions as suggested by Hedges and Cooper [16], research is then clearer and more consistent throughout its stages and process. Thus, for our systematic literature review, our research question is formulated as follows: What are the antecedents and attributes that influence eWOM in online shopping?

Step 2—Locating Studies Related to the Research Question

As argued by Denyer and Tranfield [14], systematic literature reviews aim to seek, select and appraise as much research as possible relevant to the formulated research question(s), using methods such as searches in databases, in specialized bibliographies or manual searches of specialized journals. For our systematic literature review, we explored during our research the Scopus and Web of Science databases, as well as several publishing houses including Emerald, Elsevier, Springer, ScienceDirect, Taylor & Francis, Safe, for which we used keywords such as eWOM, antecedents of eWOM, determinants of eWOM, eWOM factors, eWOM influencers, eWOM in online shopping. Once downloaded, all papers were imported into the ZOTERO software with the elimination of duplicate ones.

Step 3—Studies Selection and Evaluation

In this phase, Denyer and Tranfield [14] recommend that for the sake of process clarity, systematic literature reviews should adopt clear criteria for determining the suitability of each identified study. This ensures that each study genuinely addresses the question posed by the review. The rationale behind including or excluding sources is then documented. These rationales effectively form the selection benchmarks. In our systematic literature review, we established criteria for what should be included or excluded based on:

- **The paper's date of publication**: We have chosen to examine and evaluate the papers published over the period of the last twelve years, i.e., from 2010 to 2022 to be in line with recent publications and results on this subject, given the speed of technological advance and scientific research as well.
- **The papers' language**: We selected the papers written in French and English given our mastery of both languages.
- **The papers' subject**: In accordance with our research question announced for our systematic literature review, we have retained the papers relating to the antecedents and attributes of the eWOM.
- **The papers' field of studies**: Still in accordance with our research question, we have opted for papers relating to online shopping.

Thus, our inclusion and exclusion criteria are presented in Table 1.

In the previous step, we downloaded and imported into the ZOTERO software a total of 50 papers that we examined, and to which we applied the inclusion and exclusion criteria above mentioned. At the end of this stage, 36 papers were eliminated and 14 were retained for the next stage of analysis.

Step 4—Analysis and Synthesis

After gathering the appropriate papers from relevant sources according to our selection criteria, the next step is data analysis and synthesis. As part of the research question formulated for our systematic literature review, all the antecedents and attributes of eWOM and their conceptual models have been identified and synthesized. Then, the antecedents and attributes of eWOM that have been frequently mentioned in the literature were taken into consideration. Before the step of classifying the antecedents and attributes of eWOM, we examined all their definitions in the literature; thus, antecedents and attributes that have the same definition have been grouped under the same antecedent/attribute. Then, we proceeded to the categorization in two categories, namely the attributes relating to the individual and those relating to information. The first includes three subcategories including the attributes

Table 1 Inclusion and exclusion criteria

Exclusion criteria	Inclusion criteria
• Papers published before 2010 • Papers written in languages other than French and English • Papers related to aspects of eWOM other than antecedents and attributes • Papers relating to fields of study other than online shopping	• Articles dealing with the conceptual models of the antecedents and attributes of the electronic WOB • Articles written in French and English • Fields of study: e-commerce • Articles published between 2010 and 2022

of the communicator, those of the receiver, and finally the attributes resulting from the interaction between the communicator and the receiver.

Attributes of the Individual

- Communicator attributes: Source credibility, Satisfaction.
- Receiver attributes: Trust, Internet user experience.
- Attributes of the interaction between communicator and receiver: Tie strength, Homophily, The degree of influence.

Attributes of the Information

- The valence of information
- The volume of information
- The quality of information
- The quality of the message.

Step 5—Reporting and Using the Results

Figure 1 represents a conceptual model project proposal of antecedents and attributes influencing eWOM in online shopping based on the previous classification for potential future avenues of research on this subject and for further exploration.

Source Credibility. It hinges on the perceived expertise, trustworthiness, and product experience of the information provider [17]. Credible sources have been found to enhance persuasion and foster positive product attitudes [18]. These three dimensions: expertise (knowledge), trustworthiness (receiver's trust), and experience (familiarity with the product) form the core of source credibility [19, 20].

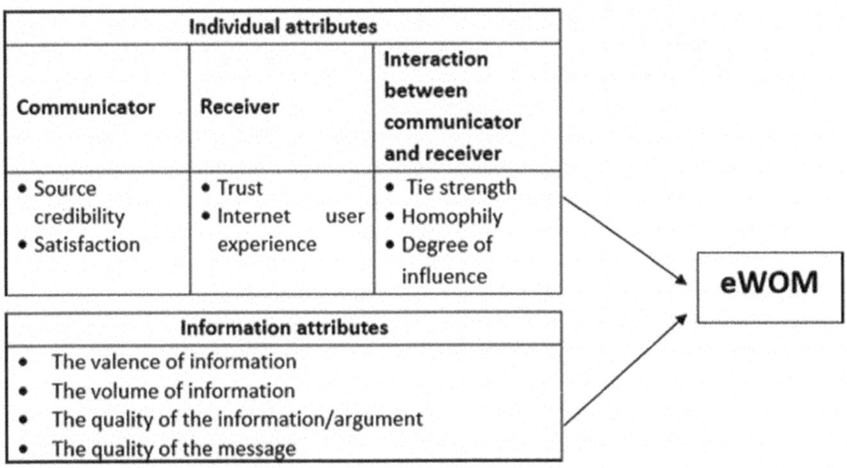

Fig. 1 Conceptual model project proposal of eWOM

Satisfaction. De Matos and Rossi [21] define satisfaction as the positive experience resulting from the use of the brand which in turn develops repurchase intentions. Oliver [22] developed the expectation-disconfirmation theory which showed that consumers always try to match two dimensions relating to brand performance namely expected performance and perceived performance. If these two dimensions are in correspondence, the consumer then becomes satisfied with this brand [21].

Trust. A cornerstone concept in marketing research within both tangible and digital marketplaces is the readiness to rely on an exchange partner [23]. It gains prominence in eWOM contexts, where the perceived trustworthiness of online messages fosters its development [21]. In virtual environments, trust facilitates the sharing and adoption of information [24], and is especially pertinent within sophisticated technological platforms and the evolving sphere of social media [25]. Mangold and Faulds noted that social media is often trusted more than traditional brand messaging, and Billal et al. [26] contend that this trust can alleviate concerns about online purchases.

Internet User Experience. The rise of the Internet and online shopping has connected consumers more than ever, yet their online activity varies with Internet literacy. Frequent Internet users, particularly in Singapore, show a higher tendency to shop online [27], and those versed in online reviews are inclined to trust them [28]. Frias et al. highlight that Internet novices may find online information searches challenging and less discerning due to minimal media experience [29]. Experienced users, aware of the mixed quality of online information, may view eWOM with skepticism, which could lessen its influence on them compared to newcomers [22, 30].

Tie Strength. As defined by Mittal et al. [31], tie strength quantifies the robustness of connections between individuals in social networks. It assesses user engagement and the depth of interpersonal bonds [32], incorporating aspects of familiarity, emotional support, similarity, and cooperation [33]. Relationships with strong ties signify close, supportive bonds [34, 35], whereas weak ties are associated with less frequent contact or unfamiliar individuals [34].

Homophily. The tendency of individuals to associate with others who share similar characteristics, such as beliefs and demographics, facilitates interpersonal communication and enhances the trust and influence of electronic word-of-mouth (eWOM) when communicators and receivers share attributes [36, 37]. Additionally, consumer decision-making is significantly shaped by interpersonal influence, which is split into normative (conforming to expectations) and informational (acquiring information from knowledgeable peers), both influencing consumer attitudes and engagement with eWOM and social networks [38, 39].

The Valence of Information. The valence, as defined by Frijda [40], refers to an event or object's intrinsic attractiveness or aversion. Based on information integration theory, consumers merge different opinions into an overarching evaluation, representing the eWOM's perceived valence [41]. Unfavorable information tends to have a greater influence than positive data [42, 43]. Mizerski [44] noted that negative product details resulted in a more pronounced product impact, with such products often perceived as low quality [42, 45]. In contrast, positive information is seen as more ambiguous [45].

The Volume of Information. The volume of information, denoting the number of opinions or reviews, affects consumer behavior [46]. As per Lopez et al. [30], eWOM volume enhances product notoriety. A large number of reviews indicates many consumers have experienced the product, providing a reliable impression of its outcomes [13, 47, 48]. Consequently, review volume can hint at product market performance, influencing inferred product quality [49].

The Quality of Information/Argument. Argument quality, also mentioned in the literature as information quality refers to the persuasiveness of arguments within an informative message [50]. Cheung et al. [51] see it as the receiver's perception of the argument's convincing nature. Within social media, argument quality influences the attitude of message recipients [52]. Valid online reviews lead to positive brand attitudes [53], enhancing brand equity [54]. In contrast, invalid content generates negative perceptions [51]. Delone and McLean [55] stress the importance of accuracy, relevance, and timeliness in high-quality reviews, asserting these factors bolster the review's persuasiveness.

The Quality of the Message. According to Slater and Rouner [56], message quality encompasses the stylistic attributes of the content. Castellano and Dutot [57] emphasized that for effective communication, a message should be straightforward, accurate, and pertinent to the context of the interaction between the sender and recipient. Sundar [58] outlined five essential criteria for a quality message: clarity, coherence, completeness, conciseness, and good writing. The perception of the message by the receiver can be influenced based on these criteria, as noted by Castellano and Dutot [57].

4 Conclusion, Limits, and Implications

Through our systematic literature review, we were able to synthesize the attributes and factors influencing eWOM that have been mentioned in the marketing literature, into two categories: the first relating to the individual namely the communicator, the receiver and the interaction between the two; the second category relates to information. We have proposed through our study conceptual model project proposal based on our results of categorizing antecedents and attributes that influence eWOM for potential future avenues of research in this regard, as well as for exploration as part of our advancement of our research work on the online shopping experience and its impact on e-reputation. This research has certain limitations that should be noted. The first limitation is the specific context of the study. We have identified the attributes and factors that influence the eWOM in an online shopping context without specifying whether they are products or services, and which online shopping field it is. We can wonder if these results can be generalized in the field of clothing, cosmetics, technology tools, e-learning platforms, etc. A second limitation lies in the country of the study. The attributes and factors that influence eWOM may not be similar between two different countries. This precision could provide more specific and insightful results.

The findings of this study have several practical implications: They offer a framework for businesses to understand the dynamics of eWOM, allowing them to craft strategies that align with the identified factors. Practitioners can use the conceptual model to assess their current eWOM strategies and identify areas for improvement.

By recognizing the dual categories of influence (individual and information attributes), businesses can tailor their communications to enhance their e-reputation effectively.

The study can inform customer relationship management by highlighting the importance of individual-level interactions and personalized communication.

References

1. Eneizan B, Alsaad A, Alkhawaldeh A, Enaizan O (2005) E-wom, trust, usefulness, ease of use, and online shopping via websites: the moderating role of online shopping experience, vol. 13, p 12
2. Goldsmith RE, Horowitz D (2006) Measuring motivations for online opinion seeking. J Interact Advert 6(2):3–14
3. Hennig-Thurau T, Gwinner KP, Walsh G, Gremler DD (2004) Electronic word-of-mouth via consumer-opinion platforms: what motivates consumers to articulate themselves on the internet? J Interact Mark 18(1):38–52
4. Bruhn M, Schoenmueller V, Schafer DB (2012) Are social media replacing traditional media in terms of brand equity creation? Manag Res Rev 35(9):770–790
5. Sen S, Lerman D (2007) Why are you telling me this? An examination into negative consumers reviews on the web. J Interact Mark 21(4):76–94
6. Buttner BO, Goritz AS (2008) Perceived trustworthiness of online shops. J Consum Behav 7(1):35–50
7. Kaplan M, Haenlein M (2010) Users of the world, unite! The challenges and opportunities of social media. Bus Horizons 53(1):59–68
8. Qiu L, Pang J, Li KH (2012) Effects of conflicting aggregated rating on eWOM review credibility and diagnosticity: the moderating role of review valence. Decis Support Syst 54(1):631–643
9. Cheung CM, Lee MK, Rabjohn N (2008) The impact of electronic word-of-mouth: the adoption of online opinions in online customer communities. Internet Res 18(3):229–247
10. Jalilvand MR, Samiei N (2012) The effect of electronic word of mouth on brand image and purchase intention: an empirical study in the automobile industry in Iran. Mark Intell Plan 30(4):460–476
11. Nielsen AC (2008) Trends in online shopping: a global Nielsenreport. The Nielsen Company, Vienna. Available at: http://at.nielsen.com/site/documents/GlobalOnlineShoppingReportFeb08.pdf. Accessed 5 May 2014
12. Gruen TW, Osmonbekov T, Czaplewski AJ (2006) EWOM: the impact of customer-to-customer online know- how exchange on customer value and loyalty. J Bus Res 59(4):449–456
13. Chatterjee P (2001) Online reviews: do consumers use them? Adv Consum Res 28:129–133
14. Denyer D, Tranfield D (2009) Producing a systematic review. In: Buchanan D, Bryman A (eds) The Sage handbook of organizational research methods. Sage, London, pp 671–689

15. Light RJ, Pillemer DB (1984) Summing up: the science of reviewing research. Harvard University Press
16. Hedges LV, Cooper H (1994) Research synthesis as a scientific enterprise. The handbook of research synthesis, pp 285–299
17. Chaiken S (1980) Heuristic versus systematic information processing and the use of source versus message cues in persuasion. J Pers Soc Psychol 39(5):752–766
18. Khong KW, Wu YL (2013) Measuring the impact of celebrity endorsement on consumer behavioural intentions: a study of Malaysian consumers. Int J Sports Mark Spons 14(3):157–177
19. Wu PC, Wang Y-C (2011) The influences of electronic word-of-mouth message appeal and message source credibility on brand attitude. Asia Pac J Mark Logist 23(4):448–472
20. Martin WC, Lueg JE (2013) Modeling word-of-mouth usage. J Bus Res 66(1):801–808
21. De Matos CA, Rossi CAV (2008) Word-of-mouth communications in marketing: a meta-analytic review of the antecedents and moderators. J Acad Mark Sci 36(4):578–596
22. Dellarocas (2006) Strategic manipulation of internet opinion forums: implications for consumers and firms. Manage Sci 52(10):1577–1593
23. Moorman C, Deshpande R, Zaltman G (1993) Factors affecting trust in market research relationships. J Mark 57(21):81–102
24. Ridings CM, Gefen D, Arinze B (2002) Some antecedents and effects of trust in virtual communities. J Strat Inf Syst 11(3 & 4):271295
25. Dutot V (2014) Adoption of social media using technology acceptance model: the generational effect. Int J Technol Human Interact 10(4):18–35. https://doi.org/10.4018/ijthi.2014100102
26. Billal M, Jianqiu Z, Dukhaykh S, Fan M, Trunk A (2021). Understanding the effects of EWOM antecedents on online purchase intention in China. Information 12(5):192. https://doi.org/10.3390/info12050192
27. Liao Z, Cheung TM (2001) Internet-based e-shopping and consumer attitudes: an empirical study. Inf Manage 38(5):299–306
28. Park C, Wang Y, Yao Y, Kang YR (2011) Factors influencing EWOM effects: using experience, credibility, and susceptibility. Int J Soc Sci Hum 74–79. https://doi.org/10.7763/IJSSH.2011.V1.13
29. Chevalier, Kicka M (2006) Web designers and web users: influence of the ergonomic quality of the web site on theinformation search. Int J Hum-Comput Stud 64(10):1031–1048
30. López M, Sicilia M (2014) Determinants of E-WOM influence: the role of consumers' internet experience. J Theor Appl Electron Commer Res 9(1):7–8. https://doi.org/10.4067/S0718-18762014000100004.
31. Mittal V, Huppertz JW, Khare A (2008) Customer complaining: the role of tie strength and information control. J Retail 84(2):195–204
32. Money RB, Gilly MC, Graham JL (1998) Explorations of national culture and word of mouth referral behavior in the purchase of industrial services in the US and Japan. J Mark 62(4):76–87
33. Granovetter MS (1973) The strength of weak ties. Am J Sociol 78(6):1360–1380
34. Steffes EM, Burgee LE (2009) Social ties and online word of mouth. Internet Res 19:42–59. https://doi.org/10.1108/10662240910927812
35. Pigg KE, Crank LD (2004) Building community social capital: the potential and promise of information and communications technologies. J Commun Inf 1(1):58–73
36. Festinger L (1957) A theory of cognitive dissonance. Stanford University Press, Oxford, UK
37. Sweeney J, Soutar G, Mazzarol T (2012) Factors enhancing word-of-mouth influence: positive and negative service-related messages. Eur J Mark 48(1–2):336–359
38. D'Rozario D, Choudhury PK (2000) Effect of assimilation on consumer susceptibility to interpersonal influence. J Consum Mark 17(4):290–307
39. Deutsch M, Gerard HB (1955) A study of normative and informational influence upon individual judgment. J Abnorm Soc Psychol 51(3):629–636
40. Frijda NH (1986) The emotions: studies in emotion and social interaction. Maison de Sciences de l'Homme, Paris

41. Anderson NH (1976) Equity judgments as information integration. J Pers Soc Psychol 33(3):291
42. Skowronski JJ, Carlston DE (1987) Social judgment and social memory: the role of cue diagnosticity in negativity, positivity, and extremity biases. J Pers Soc Psychol 52(4):689–699. https://doi.org/10.1037/0022-3514.52.4.689
43. Chiou J-S, Cheng C (2003) Should a company have message boards on its web sites? J Interact Mark 17(3):50–61. https://doi.org/10.1002/dir.10059
44. Mizerski RW (1982) An attribution explanation of the disproportionate influence of unfavorable information. J Consum Res 9(3):301–310. https://doi.org/10.1086/208925
45. Herr PM, Kardes FR, Kim J (1991) Effects of word-of mouth and product attribute information on persuasion: an accessibility-diagnosticity perspective. J Consum Res 17(4):454–462. https://doi.org/10.1086/208570N.A
46. Liu Y (2006) Word-of-mouth for movies: its dynamics and impact on box office revenue. J Mark 70(3):74–89
47. Park C, Lee TM (2009) Information direction, website reputation and eWOM effect: a moderating role of product type. J Bus Res 62(1):61–67
48. van Birgelen M, Roderik R, Jörg H (2010) Explaining consumer intentions to adopt online consumer reviews. In:Beckmann SC, Ringberg T, Ritter T (eds) Proceedings of the 39th European marketing academy conference (EMAC). European Marketing Academy (EMAC), Copenhagen, p 194
49. Chevalier JA, Mayzlin D (2006) The effect of word of mouth on sales: online book reviews. J Mark Res 43(3):345–354
50. Bhattacherjee A, Sanford C (2006) Influence processes for information technology acceptance: an elaboration likelihood model. MIS Q 30(4):805–825
51. Cheung MY, Luo C, Sia CL, Chen H (2009) Credibility of electronic word-of-mouth: informational and normative determinants of on-line consumer recommendations. Int J Electron Commer 13(4):9–38
52. Cacioppo JT, Petty RE, Morris KJ (1983) Effects of need for cognition on message evaluation, recall, and persuasion. J Pers Soc Psychol 45(3):805–818
53. Sijoria C, Mukherjee S, Datta B (2018) Impact of the antecedents of EWOM on CBBE. Mark Intell Plann 36(5):528–542. https://doi.org/10.1108/MIP-10-2017-0221
54. Kim AJ, Ko E (2012) Do social media marketing activities enhance consumer equity? An empirical study of luxury fashion brand. J Bus Res 65(10):1480–1486
55. Delone WH, McLean ER (2003) The DeLone and McLean model of information systems success: a ten-year update. J Manag Inf Syst 19(4):9–30
56. Slater MD, Rouner D (1996) How message evaluation and source attributes may influence credibility assessment and belief change. J Mass Commun Q 73(4):974–991. https://doi.org/10.1177/107769909607300415
57. Castellano S, Dutot V (2017) Investigating the influence of eWord-of-mouth on e-reputation. Int Stud Manage Organ 47(1):42–60. https://doi.org/10.1080/00208825.2017.1241088
58. Sundar SS (1998) Effect of source attribution on perception of online news stories. J Mass Commun Q 75(1):55–68. https://doi.org/10.1177/107769909807500108

Away from Theory: A Cloud Storage Application with Public Key Searchable Encryption Feature

Mohammed Raouf Senouci, Abdelkader Senouci, Fagen Li, and Kumar K. Busawon

Abstract Navigating the complex realm of secure cloud storage, our application offers a uniquely robust and user-friendly data storage solution, adeptly bridging AES-256 encryption and public key searchable encryption to safeguard user data while maintaining accessible retrieval. Specifically, the Searchable Encryption submodule leverages the unaltered scheme proposed by Senouci et al. In J Syst Archit 119:102271 [1], foundational in enabling secure, efficient data searchability without necessitating decryption. Furthermore, detailed access control and comprehensive audit logging empower users with precise control and transparent oversight of their stored data, enhancing data security and user trust. With an architecture that synergizes stringent security protocols and user-centric design, the application presents a paradigm shift in secure cloud storage, achieving a harmonious balance between impenetrable data security and user accessibility, thereby setting a new benchmark in secure data storage and retrieval in cloud environments.

Keywords Searchable encryption · Cloud computing · Data outsourcing · Sequence diagram · Flowchart diagram · Privacy

M. R. Senouci · F. Li
School of Computer Science and Engineering, University of Electronic Science and Technology of China, Chengdu 611731, People's Republic of China
e-mail: senoucimedraouf@gmail.com

F. Li
e-mail: fagenli@uestc.edu.cn

A. Senouci (✉)
Faculty of Engineering and Environment, Northumbria University, Newcastle, UK
e-mail: senouciaek@gmail.com

K. K. Busawon
Faculty of Sustainable Development and Engineering, Universite des Mascareignes, Concorde Avenue, Roches Brunes, Rose-Hill, Mauritius
e-mail: kbusawon@udm.ac.mu

X.-S. Yang et al. (eds.), *Proceedings of Ninth International Congress on Information and Communication Technology*, Lecture Notes in Networks and Systems 1000,
https://doi.org/10.1007/978-981-97-3289-0_7

1 Introduction

In the realm of secure cloud storage and encrypted data retrieval, various schemes have been proposed and analyzed in terms of security, efficiency, and usability [1]. Sun et al. [2] explore privacy-preserving keyword search over encrypted data, emphasizing the importance of protecting both the privacy of the keyword and the queried data. An interesting approach to enable public key encryption with conjunctive field keyword search is discussed by Park et al. [3], where the objective is to ensure data retrievability with keyword-specific queries.

The work by Baek et al. [4] revisits public key encryption with keyword search, providing a renewed perspective on search functionality over encrypted data. This work is especially relevant when considering the risks and challenges associated with maintaining the confidentiality of the data while still enabling keyword-specific search functionality. Byun et al. [5] highlight potential vulnerabilities in keyword search schemes over encrypted data by exploring off-line keyword guessing attacks, underlining the constant security considerations that must be adhered to when dealing with encrypted search capabilities.

Zeng et al. [6] present a lattice-based authorized searchable encryption model, which proposes a specialized keyword approach, facilitating a unique dimension to the exploration of searchable encryption mechanisms and broadening the scope of methodologies in the field. Another progressive approach by Lu et al. [7] showcases a verifiable conjunctive dynamic searchable symmetric encryption that ensures forward and backward privacy, revealing the continual evolution of searchable encryption by balancing verifiability and privacy. Wang et al. [8] introduce a secure and efficient ranked keyword search over outsourced cloud data, showcasing the practicality of enabling users to retrieve their data in a structured manner while preserving the security and integrity of the encrypted data.

Our work harnesses the robust and secure foundation laid by Senouci et al. [1] by implementing their efficient and secure certificateless searchable encryption scheme, safeguarding against keyword guessing attacks, directly within our SearchableEncryption submodule without significant alterations. This foundation allows our system to retain a proven, secure base, enabling us to focus on additional functionalities and improvements in the secure and efficient retrieval of encrypted data in a cloud storage environment.

1.1 Contribution

Motivated by the discernible gap between user-friendly interfaces and staunch data security in existing cloud storage applications, our approach amalgamates AES-256 encryption with public key searchable encryption, crafting a secure yet accessible data storage and retrieval environment. Furthermore, by embedding detailed access control and audit logging, users are not only assured of their data's security but also endowed with transparent control and oversight over their stored data. Our contributions hence lie in orchestrating complex, secure operations into a coherent, accessible application, affording users both peace of mind and ease of use.

1.2 Organization

This paper delineates the architecture, operation, and benefits of our cloud storage application. Section 2 provides an overview of the application and its features. Subsequent sections (i.e., Sects. 3 to 5) delve into the distinct modules and their respective functionalities, elucidating the secure, seamless operations of the application, with Sect. 6 shedding light on the interaction between these modules. The paper concludes in Sect. 7, reflecting on key advantages and exploring prospective future developments.

2 Application Overview

The designed cloud storage application emerges as a secure haven for sensitive data storage, combining user-friendly interfaces with top-tier security mechanisms. The architecture is meticulously formulated, embedding features that address various aspects of secure data storage, retrieval, and management in the cloud environment.

2.1 Comprehensive Feature Set

With a core focus on stringent security and user accessibility, the application incorporates:

- Robust Encryption: Leveraging AES-256 encryption ensures that all data is securely encrypted before it is stored in the cloud, providing a strong defense against potential breaches.
- Searchable Encryption: Implementing public key searchable encryption facilitates users to securely search their encrypted data without the necessity for decryption, safeguarding data while maintaining usability.
- Precision Access Control: Users maintain full control over their data, setting permissions on individual files and folders, thus dictating who has access and to what extent.
- Transparent Audit Logging: By logging all user activities, the application provides a transparent audit trail, enabling users to see who accessed their data and when, thereby enhancing accountability.

2.2 *Submodule Interplay*

The application is composed of various submodules that collectively form a cohesive, secure data storage solution:

- Core Module: Orchestrating the functionality between submodules, the Core module manages data flows and ensures smooth interoperability among all other submodules, being the central decision-making entity.
- Auth Submodule: Managing user registration and authentication, this submodule safeguards access to application functionalities, ensuring only authorized users can access and manipulate data.
- FileManager Submodule: Overseeing file management operations, including encryption and decryption of files during upload and download, respectively, it ensures data at rest is always encrypted.
- SearchableEncryption Submodule: Responsible for enabling encrypted data search functionality, this submodule generates encrypted indexes to allow secure, seamless file searches without requiring decryption.

Every submodule performs its distinct function yet collaboratively they yield a powerful, user-friendly, and highly secure cloud storage solution, ensuring that user data is always protected, accessible, and under their control.

3 Auth Submodule

Ensuring a strong user authentication mechanism, the Auth submodule is pivotal in safeguarding user accounts and associated data. The system allows users to register and log in, pivotal for personalized data storage and retrieval. During registration, it validates the uniqueness of usernames via the Core module, ensuring no duplicity. When logging in, it validates user credentials, ensuring only legitimate access.

In terms of security, especially during the login phase, if either the username or password (or both) are incorrect, a generic message such as "Invalid Credentials" is returned. This design choice aims to mitigate vulnerabilities against enumerative attacks and subsequent brute-force attacks by providing minimal hints to an adversary attempting unauthorized access.

The following sequence diagram presented in Fig. 1 depicts the process flow during user registration and login. In this diagram, the interaction starts with the user entering credentials. The Auth submodule validates initial input format and forwards data to the Core module, which interfaces with the database to validate user credentials. Whether granting access or denying with an "Check Status" message, the Auth module provides feedback to the user.

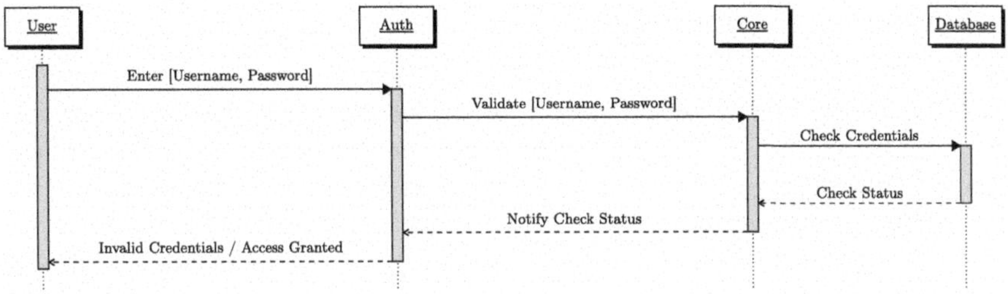

Fig. 1 Sequence diagram for the Auth submodule

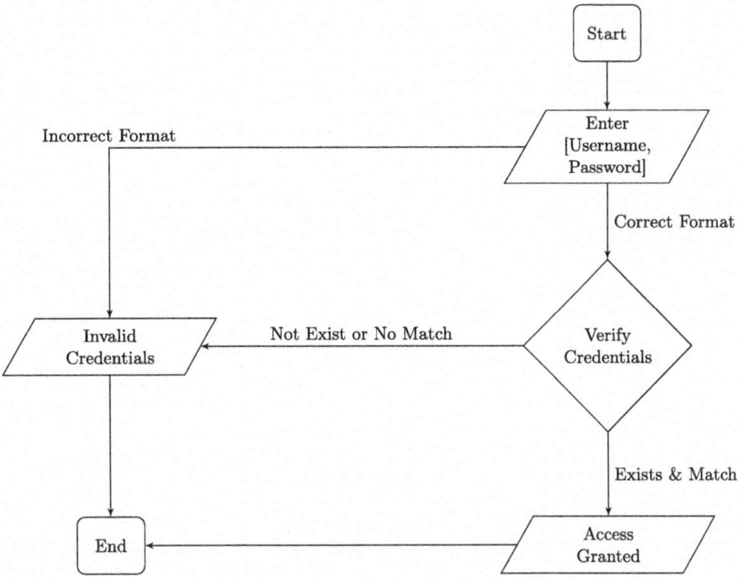

Fig. 2 Flowchart diagram for the Auth submodule

Similarly, the flowchart diagram presented in Fig. 2 illustrates the decision-making processes in the Auth submodule. The flowchart unveils the logical flow from user input through credential verification, leading to either access being granted or a generic error message, safeguarding against revealing specific credential errors.

The Auth submodule meticulously ensures the safeguarding of user credentials and data, utilizing a minimal-feedback approach during login to shield against potential adversarial attacks, thereby maintaining a secure, user-friendly environment.

4 FileManager Submodule

The FileManager submodule is integral for safeguarding data during storage and retrieval, employing AES-256 encryption to encrypt files before cloud storage, and decrypting files during download. The submodule collaborates with the Core module to ensure that the encrypted files are securely stored and retrieved from the database. Secure file management is paramount, as it not only ensures confidentiality but also plays a vital role in ensuring that the user's data is protected against unauthorized access and potential threats.

The sequence diagram shown in Fig. 3 represents the interaction flow during file uploads and downloads. Uploaded files pass through encryption in the FileManager

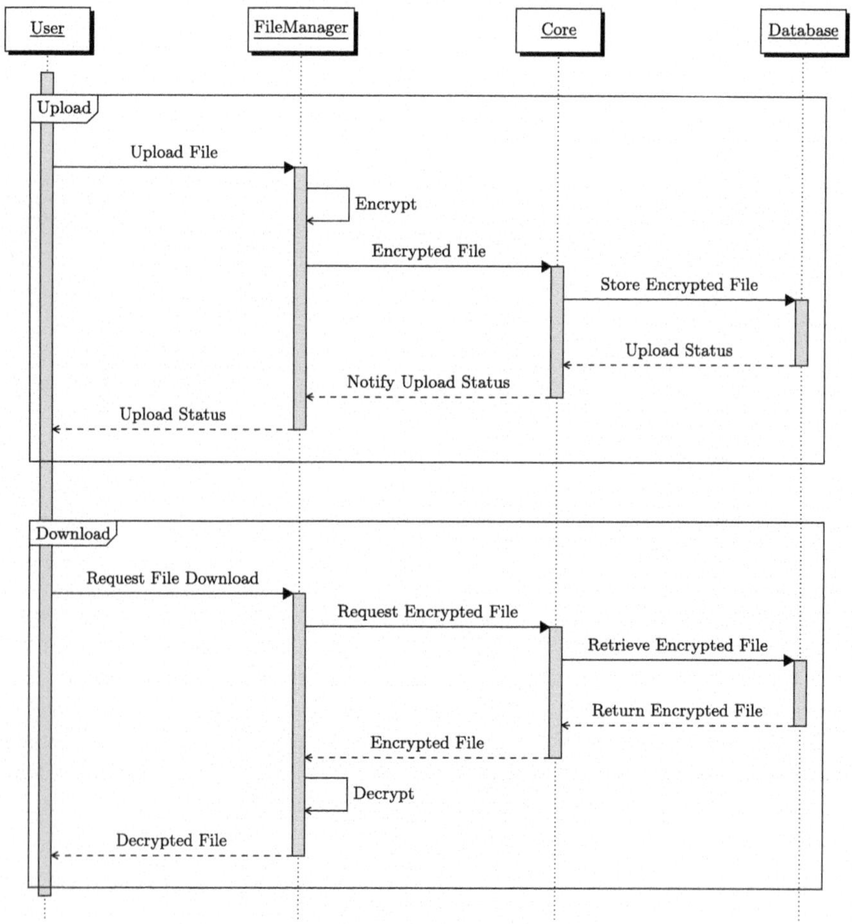

Fig. 3 Sequence diagram for the FileManager submodule

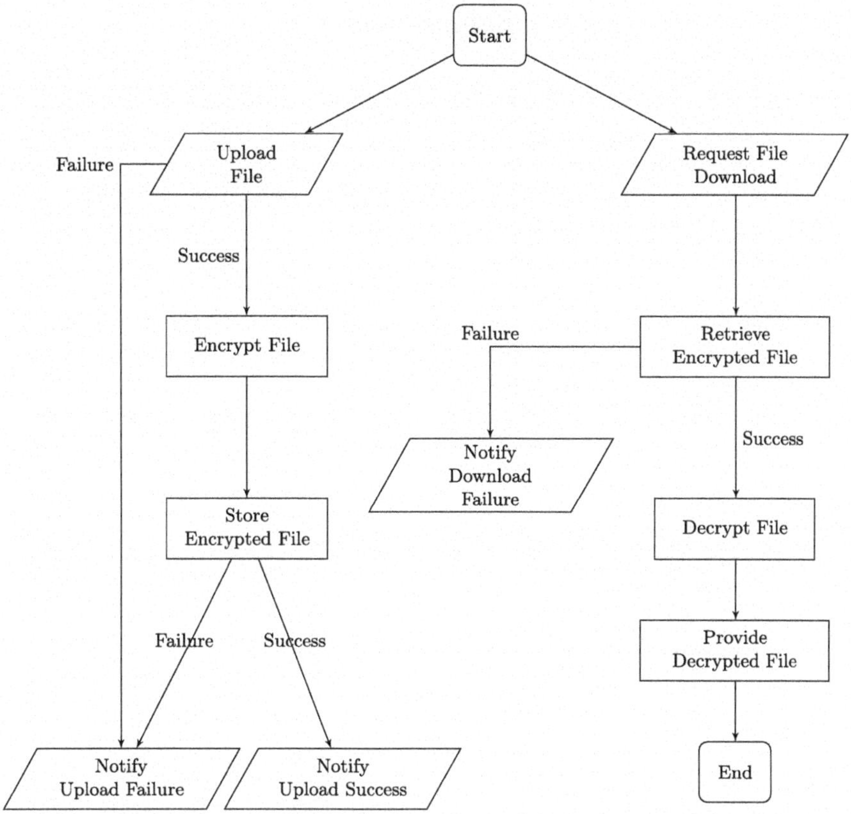

Fig. 4 Flowchart diagram for the FileManager submodule

submodule before storage in the database. Conversely, file download requests initiate encrypted file retrieval, followed by decryption in the FileManager submodule, and ultimately, the file is provided to the user.

In addition, the flowchart diagram presented in Fig. 4 illustrate whether the user is uploading or downloading a file, success, and failure pathways are outlined, providing visual insight into the logical progression through the FileManager submodule's operations.

Through rigorous AES-256 encryption and decryption, the FileManager submodule ensures the secure handling of user files, thereby preserving data confidentiality and integrity while ensuring that the user experience remains seamless and secure.

5 SearchableEncryption Submodule

The SearchableEncryption submodule embodies a revolutionary approach to encrypted data handling, enabling users to perform keyword searches on their encrypted files without necessitating decryption. By creating an encrypted index upon file upload, users can execute searches using keywords, maintaining user accessibility alongside uncompromised security.

The following sequence diagram presented in Fig. 5 depicts the interaction sequence within the system during file upload and search. This sequence diagram portrays the role of the SearchableEncryption submodule in managing file uploads and searches. Encrypted indexes are created and stored during uploads, and during searches, a trapdoor query enables secure, encrypted file retrieval without decryption.

Note that we utilize the searchable encryption scheme proposed by Senouci et al. [1], which is integral to our approach for securely indexing and searching encrypted

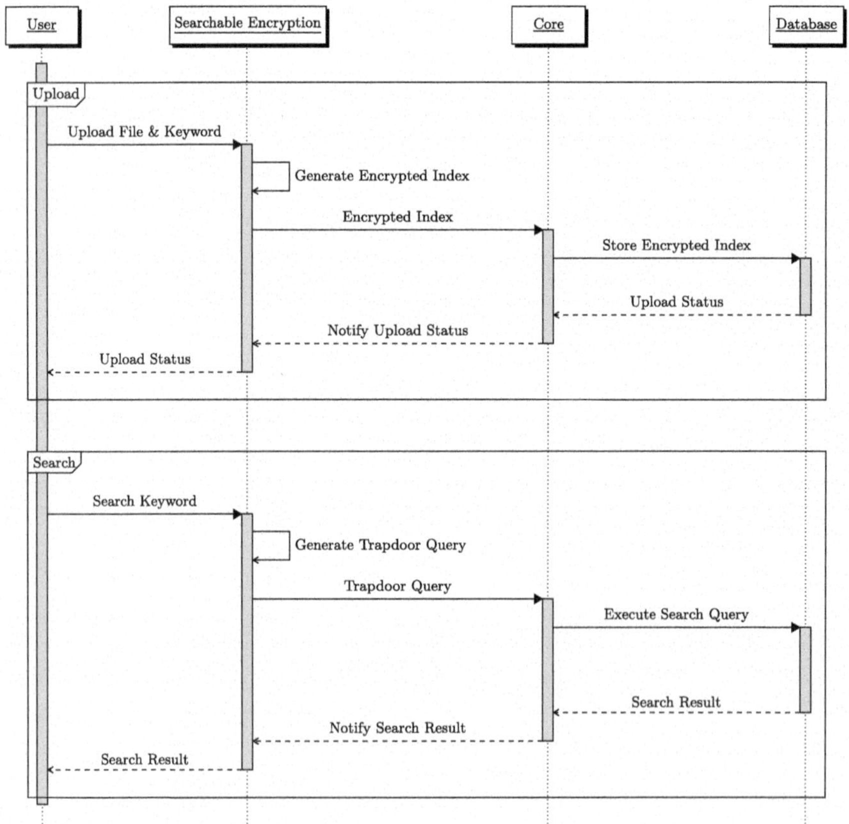

Fig. 5 Sequence diagram for the SearchableEncryption submodule

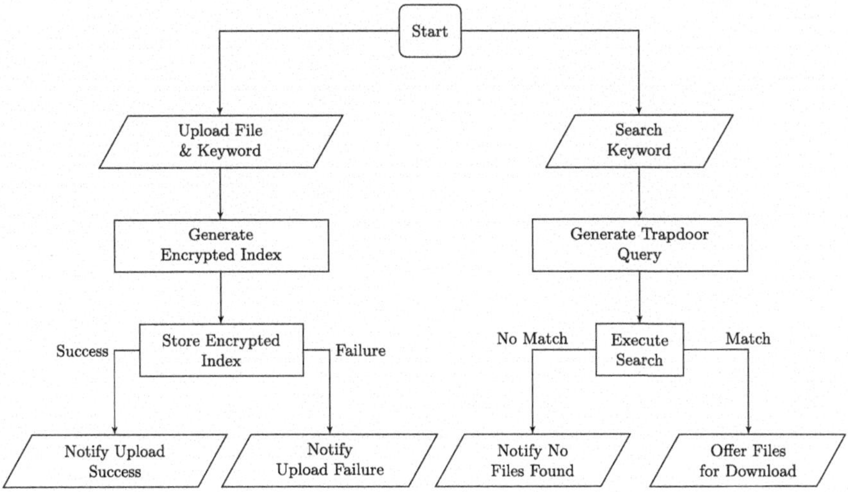

Fig. 6 Flowchart diagram for the SearchableEncryption submodule

data. This scheme has been implemented without alterations, ensuring that our module benefits from its proven robustness and efficacy in the realm of secure, searchable encryption.

Similarly, the flowchart diagram presented in Fig. 6 exhibit the logic flow within the SearchableEncryption submodule, crucial during file uploads and searches. In the upload phase, the file and keyword are processed to generate an encrypted index, which is then stored. For searches, keywords create a trapdoor query that aids in securely retrieving relevant encrypted files, either presenting files for download or notifying of no matches.

Through the innovative use of searchable encryption, the SearchableEncryption submodule navigates the tightrope between user accessibility and stringent data security, providing a secure yet user-friendly environment within the cloud storage application.

6 Interaction Among Submodules

Each submodule within the cloud storage application plays a pivotal role, yet it is through their collective interaction, managed by the Core module, that secure, seamless operation is realized. Ensuring a user-centric, secure experience, operational procedures like upload, download, and search are meticulously designed to be restricted to authenticated users, thereby enhancing data security and integrity.

The flowchart diagram presented in Fig. 7 articulates the logic flow and decision points within this multilateral interaction. In the flowchart, the logged-in users can

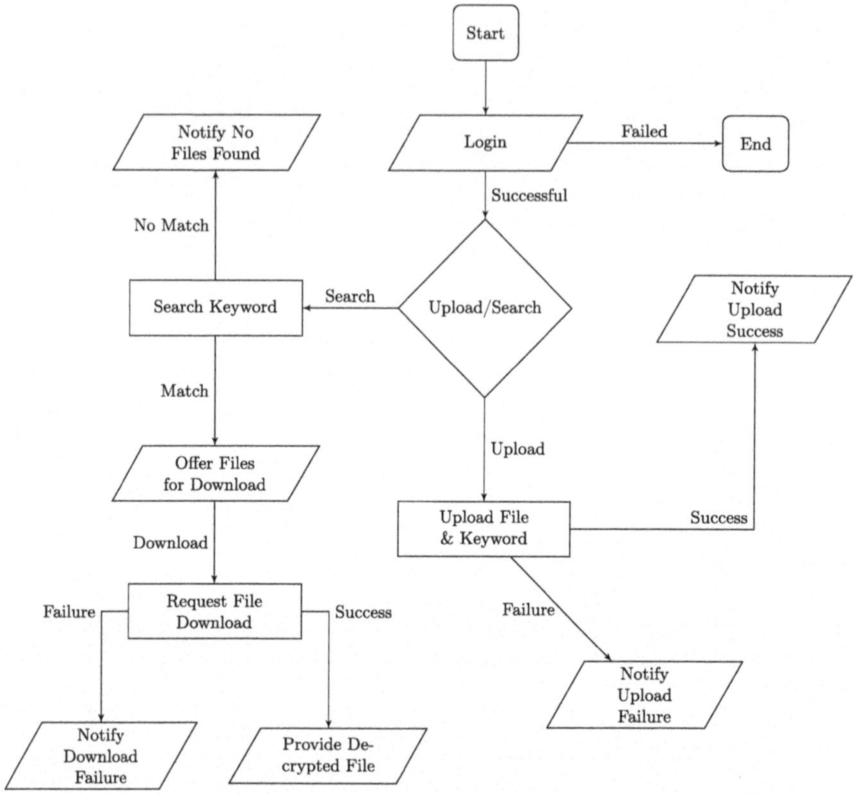

Fig. 7 Flowchart diagram for the interaction among submodules

select between uploading, downloading, and searching. Each process, albeit distinct, is executed seamlessly, leveraging the harmonious interaction between the submodules.

In addition, the sequence diagram shown in Fig. 8 serves to illuminate the intricate interactions across modules, particularly during file uploads. In this sequence diagram, authenticated users commence file uploads, which entails a collaboration between the FileManager and SearchableEncryption submodules, under the orchestration of the Core module, to securely store encrypted files and their corresponding indexes.

Through safeguarding crucial operations behind an authentication requirement and ensuring a secure, symbiotic interaction between submodules, the system balances robust security with a smooth user experience, proving a user-friendly yet impenetrable cloud storage solution.

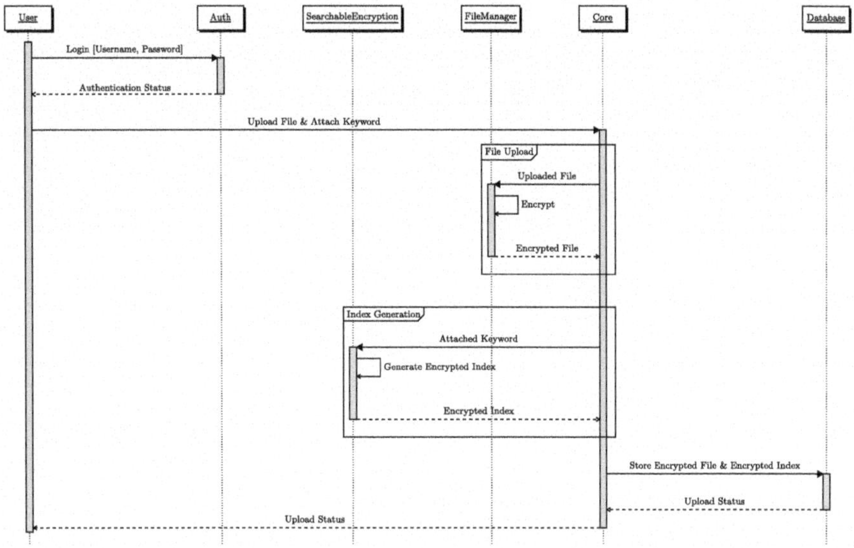

Fig. 8 Sequence diagram for the interaction among submodules

7 Conclusion

In an era where data is simultaneously an asset and a vulnerability, our cloud storage application stands out as a bastion of data protection. By interweaving cutting-edge technologies into a user-centric design, it offers both unparalleled security and effortless usability. As we tread into the future, we are committed to incorporating groundbreaking technologies like blockchain and quantum cryptography to offer an even more fortified and advanced storage solution.

References

1. Senouci MR, Benkhaddra I, Senouci A, Li F (2021) An efficient and secure certificateless searchable encryption scheme against keyword guessing attacks. J Syst Archit 119:102271. https://doi.org/10.1016/j.sysarc.2021.102271
2. Sun W, Lou W, Hou YT, Li H (2014) Privacy-preserving keyword search over encrypted data in cloud computing. Secure Cloud Comput 189–212
3. Park DJ, Kim K, Lee PJ (2005) Public key encryption with conjunctive field keyword search. Inf Secur Appl 73–86
4. Baek J, Safavi-Naini R (2008) Susilo W (2008) Public key encryption with keyword search revisited. Comput Sci Appl ICCSA 5072:1249–1259
5. Byun JW, Rhee HS, Park H-A, Lee DH (2006) Off-line keyword guessing attacks on recent keyword search schemes over encrypted data. Secure Data Manage 4165:75–83
6. Zeng F, Xu C (2015) A novel model for lattice-based authorized searchable encryption with special keyword. Math Probl Eng 2015:1–7. https://doi.org/10.1155/2015/314621

7. Lu H, Chen J, Ning J, Zhang K (2022) Verifiable conjunctive dynamic searchable symmetric encryption with forward and backward privacy. Comput J bxac084. https://doi.org/10.1093/comjnl/bxac084
8. Wang C, Cao N, Ren K, Lou W (2012) Enabling secure and efficient ranked keyword search over outsourced cloud data. IEEE Trans Parallel Distrib Syst 23:1467–1479. https://doi.org/10.1109/TPDS.2011.282

Developing a Serious Game for Acute Pain Detection by Utilizing Virtual Reality and Brain-Computer Interfaces

Meisam Taheri, Adam Emile Aske, and Kevin Tan

Abstract Brain-computer interface (BCI) technology can be used to measure pain. Brain activity is recorded and analyzed using BCI devices. The use of brain-computer interface (BCI) technology for pain measurement is an area of ongoing research and development. While there is no single definitive proof at this time, there is a growing body of evidence and research that supports the idea that BCI technology can provide objective and precise measurements of the pain. BCI enables researchers to observe and evaluate pain-related neural activity real-time. This could lead to the identification of neural patterns linked with pain, which could then be used to develop personalized therapy strategies for each patient. Furthermore, we could combine this technology with VR to make a simulation that can monitor the patient's pain experience and assist with further therapy. This will provide a safe and controlled environment for patients to better understand their pain by allowing patients to interact simulation elements using their brain activity and get real-time feedback. Overall, the combination of brain-computer interface and virtual reality technology has immense potential to revolutionize the way we detect and ease pain. The aim is to develop a simulation that provides patients with an immersive and interactive environment, allowing them to engage in activities that can help manage their pain, such as meditation, relaxation exercises, and cognitive-behavioral therapy. In this study, authors worked on a method to show how pain can be detected and visualized and used in the simulation.

Keywords BCI · VR · Pain · Acute pain · Game · Simulation

M. Taheri (✉) · A. E. Aske · K. Tan
Inland Norway University, Hamar, Norway
e-mail: meisam.taheri@inn.no

X.-S. Yang et al. (eds.), *Proceedings of Ninth International Congress on Information and Communication Technology*, Lecture Notes in Networks and Systems 1000,
https://doi.org/10.1007/978-981-97-3289-0_8

1 Introduction

Brain-computer interface (BCI) represents a technology with considerable promise, particularly in its capacity to assist individuals confronting physiological and physical challenges [1]. Additionally, BCI has the potential to serve as a means for pain measurement through the recording of electrical brain activity. When a person experiences pain, various parts of the brain become activated, which can be detected by measuring the electrical signals generated by the brain using electrodes placed on the scalp. The EEG signal is typically analyzed in the frequency domain, which involves breaking down the signal into different frequency components using mathematical techniques such as Fourier analysis. Different frequency components are associated with different types of brain activity, such as alpha, beta, delta, and theta waves. Research has shown that different patterns of EEG activity are associated with distinct types of pain, such as acute pain and chronic pain. For example, acute pain is associated with an increase in high-frequency beta waves, whereas chronic pain is associated with changes in the low-frequency alpha and theta waves [2]. EEG can also be used to measure the effects of pain medications and other pain management strategies on brain activity, which can provide insights into how these treatments work, how they can be optimized, and how effective the simulation and trainings are. Pain is a complex sensation affected by several factors. Therefore, the objective assessment of pain is a complex challenge. Research is being done into pain assessment, and modern technologies and methodologies are driving the field forward. By employing modern artificial intelligence (AI), new paradigms and methodologies can be developed to accomplish objective pain assessment from brain activity. Fourier analysis is a mathematical technique commonly used to analyze EEG signals to reduce the complexity in analyzing the data. This technique involves breaking down a complex signal into its component frequency parts, which can help identify patterns of neural activity associated with pain.

2 Background

2.1 Brain-Computer Interfaces

A brain-computer interface is a system which can record brain activity and turn it into commands for computers to execute. BCIs can restore independence to disabled people by translating brain activities to control signals for computers and external devices [3]. Other uses for BCIs include epilepsy detection and intervention, car drowsiness detection, and intervention and emotion tracking [4].

Signal Acquisition. Brain activity can be recorded with several methods, such as EEG, electrocorticography (ECoG), functional magnetic resonance imaging (fMRI), and positron emission tomography (PET). Recording methods are categorized as invasive and noninvasive. An invasive recording requires surgery, often inserting signal capturing devices placed to the brain's surface. However, invasive recording methods are associated with higher risk than their noninvasive counterparts due to surgery. Additionally, the signal from invasive devices degrades over time. Noninvasive recording methods like ECoG are the most used compared to invasive recording method. Compared to different methods of signal acquisition, EEG is the most used method for recording brain activity [4].

Signal Processing. Once brain activity has been recorded as digital signals, the signals must be processed as the signal commonly contains noises. Noise includes signal artifacts such as eye blinks, heartbeats, and movement; environmental factors can also induce noise in the signals. This stage identifies the noise and extracts it from the signal, revealing the brain activity. These unwanted signal artifacts are removed through mathematical techniques, such as notching and filtering. Different recording methods yield different types of noise.

Feature Classification. This classification is needed to analyze the signal and distinguish features and artifacts corresponding to the user's intent. This is achieved through several methods, such as AI algorithms and statistical and visual analyses of the signals. Algorithms are calibrated on a set of signals with known intents. After calibration, the algorithms can classify brain activity into intent.

Control. Once the brain activity has been translated into the user's intent, this stage turns the intent into a command for a computer to execute. This can involve controlling external devices, such as robots and communication tools. This stage usually includes providing feedback to the user based on their brain activity. By providing feedback, the user is informed about the performance of the BCI and can more effectively change their brain activity to improve it.

Electroencephalography. EEG records electrical activity in the brain. This is done by placing electrodes on the scalp which measure the electrical potential in the brain. Intracranial EEG is also used, where electrodes are placed on or inside the brain. Due to the scalp and skull, scalp-based EEG results in a lower signal-to-noise ratio than intracranial methods. For scalp-based EEG, electrodes are placed according to the "10–20" system where the location and lateralization of the electrode determines each electrode's label [5]. Brain regions are marked with different colors. The blue section represents the frontal lobe, yellow represents the parietal lobe, red represents the occipital lobe, and the green section represents temporal lobes. This system ensures consistency across disciplines using EEG.

2.2 Serious Games

One of the issues addressed with conventional rehabilitation is that therapy sessions might be being tedious due to the repetition of activities and simplicity of the tasks. Gamifying the tasks can motivate, engage, and foster adherence to treatment in patients [6]. Therapists can also benefit from automated and customizable sessions for individual patients. Francesco discussed in their article that serious games' biggest strength is their effectiveness in increasing motivation and engagement. A game's ability to provide real-time feedback about performance is a crucial factor [7]. Therefore, our focus in this study is to design and develop a serious game that is customizable and engaging enough to keep patients interested in doing the therapy session. This will allow patients with less mobility to have a system that they can work with at home or any place they prefer to.

2.3 Fourier Analysis

Acute pain is often associated with an increase in high-frequency beta waves, whereas chronic pain is associated with changes in the low-frequency alpha and theta waves. By analyzing the power spectrum of the EEG signal, researchers can identify these patterns of neural activity and use them to measure pain. The mathematical formula used for Fourier analysis is called the Fourier transform (FT). There are different types of Fourier transforms, such as the discrete Fourier transform (DFT) and the fast Fourier transform (FFT), but they all use similar principles. The Fourier transform converts a signal from the time domain to the frequency domain, by decomposing the signal into a sum of sine and cosine waves with different frequencies [8].

2.4 Pain

Pain is a complex and subjective experience that is an integral part of the human body's central nervous system (CNS) [9]. It is a crucial mechanism that notifies the body of potential harm and further damage. Pain is a multi-dimensional experience that involves both sensory and emotional components. Pain management is pivotal for many patient groups across the world. The consequences of unrelieved pain are associated with increased risk of chronic pain and reduced quality of life for patients [10, 11]. Most pain assessments are conducted using self-reported Visual Analog Scales (VAS) and questionnaires. However, these methods lack objectivity. Additionally, it is not suitable for many patient groups, for example, nonverbal and cognitively impaired patients. Pain management is primarily done using opioids [11]. This poses several issues, such as side effects and addiction. And therefore, opioids are not suitable for every patient; however multimodal analgesia is appropriate for everyone [10]. Objective pain assessment is a key factor in administering appropriate analgesia for patients. This raises the need for objective pain assessment methodologies and technologies.

3 Methodology

3.1 Pain Detection

The objective of pain detection is to objectively assess pain from EEG signals. This is implemented through AI, specifically a deep learning (DL) algorithm. DL algorithms can make predictions and classifications by identifying patterns in data. Typically, DL algorithms require large datasets for effective training and classification. However, for this article, access to substantial data containing scalp EEG recordings during acute pain sensation was unavailable. To address this limitation, a novel dataset was created by introducing pain-correlated signal artifacts into EEG recordings where no actual pain was experienced. Pain artifacts were defined from studying articles: [12–27]. A pain artifact is defined as:

$$A = \{b, c, e, d, t\} \tag{1}$$

where (b) represents the affected band ranges. (c) represents the channels affected. (e) signifies the effect. (d) represents the delay. (t) represents the duration. The EEG recordings were obtained from one of the authors during daily activity. The EEG data was sampled into smaller sections. For half of the samples, a pain artifact, timing, and intensity level ranging from 1 to 10 were randomly selected. The remaining samples are not imposed with pain. The chosen artifact, timing, and intensity are saved as labels for the AI algorithm.

The set of all the samples is defined as:

$$S_n = \{S_1, S_2, \ldots, S_N\} \tag{2}$$

where N is the total amount of samples.

The label for each sample is defined as:

$$L = \{(A_1, T_1, I_1), (A_2, T_2, T_3), \ldots, (A_N, T_N, I_N)\} \tag{3}$$

where L is the set of all samples with imposed pain artifacts and N is the total amount of samples. Each label contains the artifact, timing, and intensity.

The signal is imposed with the pain artifact by attenuating frequencies by the intensity at specific times, and its formula is:

$$K_n = \mathrm{STFT}(S_n[A_c])[A_b] + A_e * I_n * \mathrm{rect}(t, T + A_d, T + A_d + A_t) \tag{4}$$

where *S_n* is the original EEG sample. *A_c* returns the channels defined by the artifact. STFT() is the short-time Fourier transform function resulting in a time–frequency domain signal. From the time–frequency domain, the frequency amplitudes over the whole-time dimension at *A_b* frequencies are selected. t represents the time. The frequencies are attenuated by the effect A_e multiplied by the intensity I. This attenuation is only applied while the rectangular function returns 1, and the rectangular function is defined as:

$$\mathrm{rect}(t, a, b) = \begin{cases} 1, \text{ if } a \leq t \leq b \\ 0, \text{ otherwise} \end{cases} \tag{5}$$

The reconstructed signal is defined as:

$$F_n = \mathrm{ISTFT}(K_n) \tag{6}$$

where ISTFT () is the inverse short-time Fourier transformation function and S_n is the reconstructed signal in the time domain. The signal is reconstructed to the time domain via the inverse short-time Fourier transform.

Figure 1 illustrates how a pain artifact affects the signal. The left signal shows the EEG recording at a single electrode. The right signal shows the reconstructed signal after a pain artifact has been imposed on it using the methodology explained in this section. At the 3 s mark, certain frequencies exhibit an increase, resulting in an overall enchantment in signal strength.

Figure 2 provides spectrograms before and after the introduction of a pain artifact. The modified spectrogram demonstrates heightened activity in the 5–30 Hz range, lasting for 1 s, beginning at the 3 s mark.

The method described for imposing pain artifacts on EEG signal was implemented with custom Python scripts and files. To get real EEG data, EEG was recorded by the authors for 2 h during regular daily activity while wearing an EEG recording device. Electrodes FP1, FP2, AF3, AF4, F3, F4, F7, F8, FC5, FC6, T3, T4, T5, T6, P3, and P4 were used and recorded at 125 Hz. The 2 h of EEG data was split into 5 s segments and saved in files accessible to the Python scripts. To generate a sample with imposed pain artifact, a Python script loads a sample from the real EEG data. Then it chooses a pain artifact and uses the algorithm explained earlier to impose pain artifacts on it.

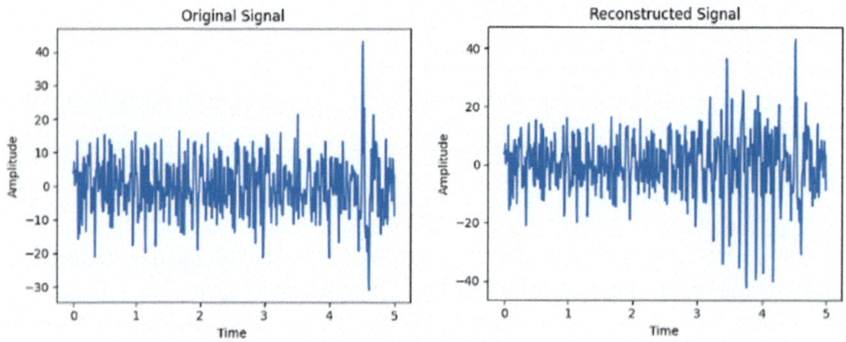

Fig. 1 Example of a time series imposed with a pain artifact

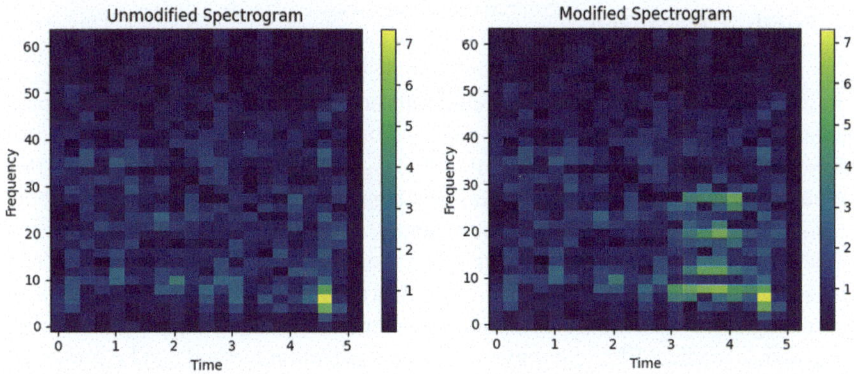

Fig. 2 Example of a spectrogram imposed with a pain artifact

A convolutional neural network (CNN) model is used for classification and is implemented with the TensorFlow API in custom Python scripts. The CNN model is comprised of layers and starts with an input layer. The EEG samples were recorded with 16 channels at 125 Hz for 5 s. Thus, the shape of the input layer is 16 by 625. Following the input layer, 3 convolutional layers are present, each followed by an activation layer and a pooling layer. Following the third convolution, a flattened layer and a fully connected layer complete the CNN model. The model is compiled with a binary cross-entropy loss function and Adam optimizer. The CNN model trains to classify each sample into either no pain or pain, hence a binary loss function. The goal being that the CNN could learn to classify real pain from no pain because it has trained to classify fake pain from no pain.

4 Results and Discussion

4.1 Results from Pain Detection CNN

EEG samples with real pain sensations were obtained by one of the authors of this article. EEG was recorded in 5 s samples, and pain was induced by the author submerging a hand in icy water. This was done 20 times. Three CNN models were trained on EEG samples, respectively, with 500, 1000, and 10,000 sample sizes. Half the samples contained generated pain artifacts, and the rest were regular EEG samples without pain. Every model was able to distinguish between no pain and pain with higher than 95% accuracy. All the models were fed to EEG samples with real pain. However, no correlation between real pain sensation and the output of the CNN algorithm was observed. This means that the CNN model did not find correlation between the imposed pain artifacts to real pain artifacts in EEG signals. Further improvements can be made with a CNN model created for EEG data.

5 Further Research

The algorithm for pain detection yielded unsuccessful classification of real pain signals. The CNN algorithm was trained on thousands of samples and could accurately classify the imposed pain from the EEG signals without imposed pain, but this did not translate to classification of real pain signals. Several factors may contribute to this. The chosen pain artifacts, for example, reduction in alpha waves over the frontal cortex, may not correlate with acute pain and therefore calibrate the CNN with incorrect information. The real EEG data may not have been recorded with high enough resolution. Only 16 channels were recorded, a relatively small number compared to medical EEG equipment. The pain detection method can be improved in several ways for increasing the likelihood of success. For this study, only time-domain signals were analyzed by the CNN algorithm. More insight can be gained by including frequency and time–frequency domain signals into the analysis. The EEG was recorded with a lack of electrodes placed at the central regions of the scalp. This results in electrodes FCz and Cz not being included in the EEG recordings. Both electrodes have potential for pain detection [28]. With higher resolution EEG and more electrodes, more information could be extracted from the brain activity. The EEG data with real pain sensations was recorded 20 times, each for 5 s. More EEG recordings with real pain sensations are needed to improve testability of the CNN model.

6 Conclusion

Research was conducted to define signal artifacts that correlate with subjective pain sensations. EEG data where no pain sensation occurred was imposed with these artifacts to create a novel dataset. A CNN algorithm was trained on EEG data imposed with signal artifacts and tested with EEG data where real pain occurred. The objective was to objectively assess pain by analyzing brain activity. Despite not resulting in successful classification of pain, this method can be further improved in pursuit of objective pain assessment using noninvasive EEG.

References

1. Taheri M, Kalnikaite D (2022) A study of how Virtual Reality and Brain Computer Interface can manipulate the brain. In: The 5th international conference on software engineering and information management (ICSIM)
2. Pinheiro ESDS et al (2016) Electroencephalographic patterns in chronic pain: a systematic review of the literature. PloS One 11(2):e0149085. https://doi.org/10.1371/journal.pone.0149085
3. Engel A, Moll C, Fried I, Ojemann G (2005) Invasive recordings from the human brain: clinical insights and beyond. Nat Rev Neurosci 6:35–47

4. JJ Shih, DJ Krusienski, JR Wolpaw (2012) Brain-computer interfaces in medicine. Mayo Clin Proc 87(3):268–279
5. Plass-Oude Bos D (2006) EEG-based emotion recognition. Influence Visual Auditory Stimuli 56:1–17
6. Bruno Bonnechère SVSJ (2019) Chapter 39—Rehabilitation. In: DHM and posturography. Academic Press, pp 541–547
7. Bellotti F, Kapralos B, Lee K, Moreno-Ger P, Berta R (2013) Assessment in and of serious games: an overview. Adv Hum-Comp Int
8. Wikipedia (Online). Available: https://en.wikipedia.org/wiki/Fourier_transform. Accessed 1 Apr 2023
9. Raja SN, Carr DB, Cohen M, Finnerup NB, Flor H, Gibson S, Keefe FJ, Mogil JS, Ringkamp M, Sluka KA, Song XJ (2022) The revised international association for the study of pain definition of pain: concepts, challenges, and compromises. Pain 161(9):1976–1982
10. Hyland SJ, Wetshtein AM, Grable SJ, Jackson MP (2022) Acute pain management pearls: a focused review for the hospital clinician. Healthcare (Basel)
11. Sinatra R (2010) Causes and consequences of inadequate management of acute pain. Pain Med 11(12):1859–1871
12. Kong Y, Feng L, Li H, Cui H, Xie X, Xu S, Hu Y (2021) Low back pain assessment based on alpha oscillation changes in spontaneous electroencephalogram. Neural Plast Neuropathic Pain
13. Bunk SF, Lautenbacher S, Rüsseler J, Müller K, Schultz J, Kunz M (2018) Does EEG activity during painful stimulation mirror more closely the noxious stimulus intensity or the subjective pain sensation? Somatosens Mot Res 3–4:192–198
14. Vijayakumar V (2018) Automated detection and quantification of pain using electroencephalography. Retrieved from the University of Minnesota Digital Conservancy
15. Panavaranan P, Wongsawat Y (2013) EEG-based pain estimation via fuzzy logic and polynomial kernel support vector machine. In: 6th Biomedical engineering international conference
16. Reyes-Galaviz R, Mendoza Montoya O, Antelis J (2022) Detection of pain caused by a thermal stimulus using EEG and machine learning. In: Pattern recognition
17. Chen D, Zhang H, Kavitha PT, Loy FL, Ng SH, Wang C, Phua KS, Tjan SY, Yang SY, Guan C (2022) Scalp EEG-based pain detection using convolutional neural network. IEEE Trans Neural Syst Rehabil Eng 274–285
18. Koyama S, LeBlanc BW, Smith KA, Roach C, Levitt J, Edhi MM, Michishita M, Komatsu T, Mashita O, Tanikawa A, Yoshikawa S (2018) An electroencephalography bioassay for preclinical testing of analgesic efficacy. Sci Rep 8
19. Sun G, Wen Z, Ok D, Doan L, Wang J, Chen ZS (2021) Detecting acute pain signals from human EEG. J Neurosci Methods 347
20. Yanagisawa T, Fukuma R, Seymour B, Tanaka M, Hosomi K, Yamashita O, Kishima H, Kamitani Y, Saitoh Y (2020) BCI training to move a virtual hand reduces phantom limb pain. Neurology 95(4):417–426
21. Ploner M, Sorg C, Gross J (2017) Brain rhythms of pain. Trends Cogn Sci 21(2):100–110
22. Zis P, Liampas A, Artemiadis A, Tsalamandris G, Neophytou P, Unwin Z, Kimiskidis VK, Hadjigeorgiou GM, Varrassi G, Zhao Y, Sarrigiannis PG (2022) EEG recordings as biomarkers of pain perception: where do we stand and where to go? Pain Ther 369–380
23. Tayeb Z, Bose R, Dragomir A, Osborn LE, Thakor NV, Cheng G et al (2020) Decoding of pain perception using EEG signals for a real-time reflex system in prostheses: a case study. Sci Rep
24. Jingjing L, Jianwei Z, Kan Z, Jijian Z (2022) Predictive value of EEG-derived pain threshold index for acute postoperative pain in children. Front Pediatr 10
25. Prichep LS, John ER, Howard B, Merkin H, Hiesiger EM (2011) Evaluation of the pain matrix using EEG source localization: a feasibility study. Pain Med 12(8):1241–1248
26. Savignac C, Ocay DD, Mahdid Y, Blain-Moraes S, Ferland CE (2022) Clinical use of electroencephalography in the assessment of acute thermal pain: a narrative review based on articles from 2009 to 2019. Clin EEG Neurosci 124–132

27. Bunk SF, Lautenbacher S, Rüsseler J, Müller K, Schultz J, Kunz M (2018) Does EEG activity during painful stimulation mirror more closely the noxious stimulus intensity or the subjective pain sensation? Somatosens Mot Res 35(3–4):192–198
28. Urits I, Seifert D, Seats A, Giacomazzi S, Kipp M, Orhurhu V, Kaye AD, Viswanath O et al (2019) Treatment strategies and effective management of phantom limb–associated pain. Curr Pain Headache Rep 23(64)
29. Pinheiro ES, Queirós FC, Montoya P, Santos CL, Nascimento MA, Ito CH, Silva M, Nunes Santos DB, Benevides S, Miranda JG, Sá KN (2016) Electroencephalographic patterns in chronic pain: a systematic review of the literature. PLoS One 25(11):2
30. Schulz E, Zherdin A, Tiemann L, Plant C, Ploner M. Decoding (2012) Decoding an individual's sensitivity to pain from the multivariate analysis of EEG data. Cereb Cortex 22(5):1118–1123

Hidden Knowledge Extraction for Association Rule Mining Technique Based on the Apriori and Frequent Pattern Growth Algorithms: A Case Study

Thierry Nsabimana, Therence Niyonsaba, Christian Ildegard Bimenyimana, Ghislain Vlavonou, Joël T. Hounsou, and Gerard Niyonsaba

Abstract Association rule mining technique is a branch of data mining area. It's purpose is to extract the hidden knowledge from databases D with transactions T. Due to the huge amount of data from multiple sources continuously being captured and stored in databases D, many industries, enterprises, to name just a few, are more interesting in exploring association rules from their transactional databases in order to increase their profits. In this paper, the simulation study is investigated based on the Apriori and Frequent Pattern Growth (FP-G) algorithms which are popular utilized in the discovery of the relevant hidden frequent items from transactional databases. This study focuses on the design and the implementation of the appropriate data structure using linked lists of array. The simulation of the algorithms is evaluated using a supermarket dataset named as retail market basket data so as to analyze the performance characteristics of the implemented algorithms, on a 8GB Random Access Memory (RAM) laptop computer. Furthermore, the investigation study uses the designed useful data structure and the features of the algorithms during the extraction of frequent itemsets on supermarket dataset, which are then used for generating the association rules, and performance costs estimated for different values of association rule length, minimum support, and time-consuming performance.

Keywords Data mining · Apriori · FP-Growth · Frequent itemsets · Association rule · Support

---

T. Nsabimana (✉) · T. Niyonsaba
Institut de Statistique Appliquée de l'Université du Burundi, Nyamugerera, Bujumbura, Burundi
e-mail: thierry.nsabimana@ub.edu.bi

C. I. Bimenyimana
Université des Grands Lacs, Saint Michel, Bujumbura, Burundi

G. Vlavonou · J. T. Hounsou
Institut de Mathématiques et de Sciences Physiques de l'Université d'Abomey-Calavi, Dangbo, Porto-Novo, Benin

G. Niyonsaba
Centre de Recherche Indépendant et d'Informatisation du Système d'Information, Centre Porta Sion, Bujumbura, Burundi

X.-S. Yang et al. (eds.), *Proceedings of Ninth International Congress on Information and Communication Technology*, Lecture Notes in Networks and Systems 1000,
https://doi.org/10.1007/978-981-97-3289-0_9

1 General Introduction

The purpose of data mining is to extract the relevant knowledge in transactional database to be used for decision making in an organization. It is widely considered as an attractive tool since it is applied in public and private sectors. With the growth of data being collected, stored, and shared from different sources, government and non-government organizations are relying on the use of data mining tool to identify useful information from their storage databases (i.e., employee database, sales database, fire insurance database, etc.). This way of extracting patterns from structured and unstructured data, data mining applies multiple mining approaches and algorithms including clustering, classification, association rule mining, prediction, and outlier analysis. In this study, we only emphasize on the association rule mining technique using the Apriori and FP-G algorithms.

The Apriori algorithm is one the data mining algorithms used to produce rules based on the association rule mining approach. The property of this algorithm stated that all non-empty subgroup of a Frequent Pattern is also frequent. This property aims to minimize significantly the process of searching candidate k-itemsets that at the end become infrequent [1].

Association rule mining technique is a data mining technique that is used for knowledge extraction in significant volume of data. The results of this technique are presented as an implication of the form, if $X \subset T$, $Y \subset T$, $X \cap Y = \emptyset$, $X \bigcup Y$ and $P(Y/X)$ [1], where X is at the antecedent and Y at the consequent of the rules. It is designed and implemented for the purpose of discovering the useful frequent itemset in transactional database. Thus, there are two performance metric analysis in the association rule mining approach named as support and confidence, which are used to identify the frequent itemset that meet some minimum support and confidence thresholds.

The FP-Tree is one of data mining algorithms used for mining frequent itemsets from database with transaction T. It is also used to reduce the problem of generating too many candidates k-itemsets at the become infrequent.

This study focuses on the design and the implementation of the appropriate data structure using linked lists of array. The simulation of the algorithms is evaluated using supermarket dataset, to analyze the performance characteristics of the implemented algorithms on a 8 GB RAM laptop computer. Furthermore, the investigation study uses the data structure materials and algorithmic features during the frequent itemsets generation on market basket dataset, which are then used for generating the association rules, and performance costs estimated for different values of association rule length, minimum support, and time-consuming performance. The main objectives are to:

- Determine the performance characteristics of the Apriori and FP-Growth algorithms on a laptop computer capacity of 8 GB of RAM.
- Design and implement the Apriori and FP-G algorithms using useful data structure and retail market basket dataset, for generating frequent itemsets, association rules, and estimated performance costs.

- Compare the performance of the Apriori and FP-G algorithms, and determine the time complexity usage for mining association rules.

This paper is organized as follows. Section 2 discusses the problem statement formulation and proposed solution. Section 3 states the related works. Section 4 presents the research methodology. Section 5 presents the implementation of the Apriori algorithm, and useful data structure used for exploring the association rules. Section 6 presents the achieved results and discussion. The final section presents the conclusion and perspective.

2 Problem Statement Formulation

2.1 Problem Associated with Apriori Algorithm

With new technology, the amount of data is being collected, stored, and shared by industries, organizations, and research institutes. According to the journal Forbes and IDC [2, 3], the projection shows that the significant volume of that digital data will increase from 33 ZB in 2018 to 175 ZB in 2025. Storage capacity and analysis tools are major concerns in the era of Big Data. Due to these issues, the most of these data are dormant, not analyzed, and not explored to serve the decision support. In the process of finding frequent itemset and association rules from database D with transactions T, Apriori algorithm has some challenges:

1. It scans many time a database D with transactions T.
2. It generates multiple candidates including frequent itemset as well as infrequent itemset.
3. Database with transactions T increases running time of the model and visits.
4. Determining low minimum support and confidence threshold increases the generation number of frequent itemset and scans. More space for storage is required
5. Determining high min support and confid threshold misses useful candidates.

2.2 Proposed Solution of the Problem

To identify the association rules from database with transaction T, we need to:

1. Design and implement the algorithms on the appropriate data structure.
2. Data structures named as an array of linked lists are explored to manage candidate k-itemsets to minimize significantly memory wastage.
3. Apply different possible minimum support and confidence threshold to analyze the performance of the algorithms.
4. Divide a database D with transactions T into multiple number of instances.
5. Employ hash function technique to minimize the memory wastage.

3 Related Works

3.1 Apriori Algorithm Related Work

Data mining technique uses multiple mining techniques named as (1) association rule mining, classification, clustering, and prediction. Association rule mining is data mining technique that uses to discover a hidden interesting association in a large database *D* with transaction *T*. Gama and Widnyani [4] optimized the performance of Apriori algorithm to get frequent itemset and to form association rules in a short time. The modifications they did in their study are obtained by using techniques of combination reduction and iteration limitation. Using in medical field, the improved Apriori algorithm, Zhang and Ma showed the we could dig out association rule models about the properties and nature of the disease from a medical database, which can assist doctors in medical diagnosis. They showed that it is a worthy research direction that using data mining method to process and analyze the data of disease prevention and drug treatment [5]. In their research [6], Wang and Gao aim at the performance bottleneck of traditional Apriori algorithm when the dataset is slightly large, by parallelization and improves the Apriori algorithm based on Map Reduce mode. The local frequent item sets on each sub-node in the cluster were calculated, then all the local frequent item sets were merged into the global candidate item sets, and finally, the frequent item sets that meet the conditions were filtered according to the minimum support threshold [6]. The Apriori Data Mining algorithm has been used to calculate frequency patterns and to optimize the medical device inventory management [7]. In this work [8], Jiao Yabing improved Apriori algorithm in data mining association rules. The Apriori acquires more memory space for candidate generation process, Ingle and Suryavanshi improved the Apriori algorithm by reducing the acquired memory space for generation process. The developed pruning strategy by them decrease the scans required to generate candidate item sets [9].

3.2 Frequent Pattern Growth Related Work

Nurmayanti et al. implemented the Apriori algorithm and the Frequent Pattern Growth (FP-Growth) algorithms to extract and analyze the association rules from sales transaction data. Their experimental results showed that 10 top rules were generated by the Apriori algorithm, whereas 4 top rules were generated by the FP-Growth algorithm [10]. In the study conducted by Zhang et al., they used the improved FP-Growth algorithm to eliminate the different impact of different teachers on the test paper scoring standards and they coded the scores of each student using it to the correlation between the failed courses [11].

4 Research Methodology

The proposed research methodology is shown in Fig. 1.

4.1 Data Description

The dataset used to test the implemented both algorithms is retail market basket data produced by the Author Tom Brijs, such file contains a transaction. In that file, each line has its own transaction identifier labeled as TID. Each transaction identifier represents the sold article in the corresponding transaction. The dataset contains 88162 transactions and 16470 items [12]. In our experiment, we have utilized 121 transactions and 647 items because of memory capacity.

4.2 Frequent Pattern Growth Algorithm

Han et al. invented FP-G algorithm [13]. Their purpose is to prevent generating more k-candidates that have no chance to become frequent. There are more than one phases required to build the FP-Tree algorithm:

- Phase 1: For the first time, the algorithm scans the database with transactions T, to discover frequent items based on their support count frequencies. The set of frequent items is sorted in the order of descending support count.
- Phase 2: At the top of the tree, the root labeled as null is created.
- Phase 3: For the second time, the algorithm scans again the database and sorts the frequent items based on the support count. Finally, the branch of transaction is built. Table 1 depicts the transactional database of items used for building FP-Tree using a min support threshold equal to two (2).

The first transaction $T_{100} : I_2, I_1, I_5$ is ordered based on the descending support count metric. At this point, we have three nodes, I_2, I_1, I_5 . I_2 is linked to the root known as (null), I_1 is linked to I_2, and I_5 is linked to I_1. First transaction finishes.

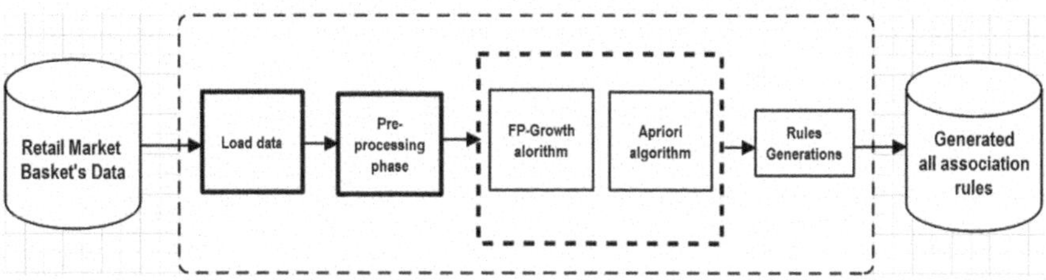

Fig. 1 Steps of research methodology

Table 1 Database with transactions TID [14]

TID	List of items Ids
T100	I1, I2, I5
T200	I2, I4
T300	I2, I3
T400	I1, I2, I4
T500	I1, I3
T600	I2, I3
T700	I1, I3
T800	I1, I2, I3, I5
T900	I1, I2, I3

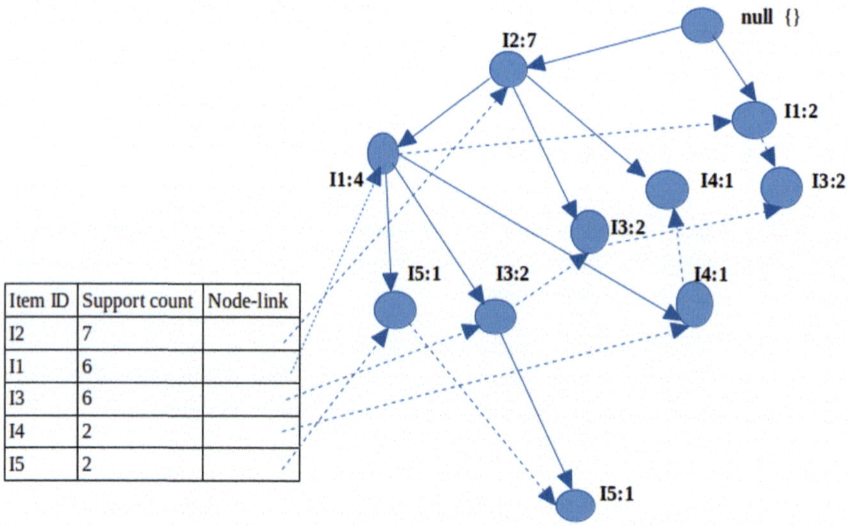

Fig. 2 Frequent Pattern Tree construction [1]

The second transaction T_{200} : I_2, I_4 is ordered based on the descending support count metric. At this point, we have two nodes, I_2, I_4 . I_2 is connected to the root of the tree, I_4 is connected to I_2. second transaction finishes.

The construction process going on until it reaches the final transaction. The construction of the Frequent Pattern Tree is shown in Fig. 2.

4.3 Apriori Algorithm Pseudo-Code

The proposed pseudo-code algorithm is based on the research work done by Zhao and Sourav [13]. The steps of the implementation of the algorithm are shown in Algorithm 1.

Algorithm 1 : Apriori Algorithm Pseudo-Code

Input :Supermarket data set, Min Support, Min Confid Thresholds
Outcome :Set of frequent k-itemsets, Set of association rules
Procedure :
lg1 = Larg1-itemsets;//Initializing large1-itemsets
for (size=2 ; $lg_{size-1} \neq \emptyset$; size++) //instruction for counting frequent size-itemsets

C_{size}= gen_$cand$_apriori(lg_{size-1}); //New candidates generated from large items

for (Database D with Transaction T)//database comprises transaction

C_{ts}=subgroup(C_{size},T);//Candidates comprise in Transaction
for Candidates contain in C_{ts}

Counter=support_count(c)+1

lg_size= all set of frequent size-itemsets satisfy the criteria of greater than or equal Min Support
$L1_f = \cup_{lg\_size}$; /* all frequent itemsets */
R_{gen}=Gen_Rul($L1_f$,Min Confid) /*Generating set of association rules*/

4.4 Performance Metrics Used for the Analysis of the Rules

In this study, two performance measures used to identify the association rules were demonstrated. The support count of each frequent itemset is achieved by computing the formula 1. Now the frequent k-itemsets are determined. The next stage is to find out all set of associations rules that satisfy the minimum confidence threshold, by computing the formula 2.

$$\text{Supportcout}(A, B) = \frac{\text{All support value of A and B}}{|N|}, \tag{1}$$

where N is the total number of transactions containing in dataset.

$$\text{Confidence}(A \Longrightarrow B) = \frac{\text{Support}(A \cup B)}{\text{Support}(A)}. \tag{2}$$

Table 2 Original database [1]

TID	List of items
T1	1, 11, 5
T2	1, 11, 39, 41
T3	2, 6, 10
T4	2, 7, 39, 32
T5	1, 2, 30, 40
T6	1, 9, 12, 30
T7	4, 35, 39, 20, 25

5 Implementation of the Algorithms

5.1 *Useful Data Structure Identification*

In this study, useful data structures for managing data known as linked lists of an array on a hash function were used. Furthermore, a hash function technique was utilized to identify the real array index where an itemset was kept in. Table 2 depicts 7 transactions of items. This algorithm allows to find out the size of the frequent k-itemsets and association rules. The first step helps to generate candidate 1-itemset. The algorithm scans the database D with transaction T by counting their support values and keep them in an integer array. All generated candidate 1-itemset is given in Table 3a. The second step allows to generate candidate itemsets of size 2. The generation of the size of two is computed by combining 2 items in transactions database. The combination of 2 items requires the extra memory which is the inconvenient. The generation of candidate 2-itemsets is given in Table 3b. Due to this issue, hash function is used to compute the hash value of an itemset. The hash values of the 2-itemsets is given in Table 3c. The process continues until the generation of all rules.

5.2 *Analysis of the Memory Requirements of the Algorithms*

The simulation experiments were performed on LENOVO Think Book E 5430 Laptop with Intel(R) Core(TM) i5-43000 M CPU @ 2.60 GHz and 8.00 GB of RAM. Running on W10 with HDD 500 GB. The used dataset comprises in total 88162 transactions and 16470 items. Due to the limitation of the computer memory, we applied a part of this dataset file comprising 121 transactions and 647 items. Different steps of generating candidate t-itemsets are shown below:

(a) Let P be the total transactions in database, $P = 647$.
(b) Equation (3) is the computation of the combination of t-itemsets.

$$\binom{p}{t} \ or \ C_t^p = \frac{p!}{t!(p-t)!}. \quad (3)$$

Table 3 The generation of candidate 1 and 2-itemsets and hash values of the 2-itemsets

(a) Candidate 1-itemset

Array index	Count	Array index	Count	Array index	Count
0	0	15	0	30	2
1	4	16	0	31	0
2	3	17	0	32	1
3	0	18	0	33	0
4	1	19	0	34	0
5	1	20	1	35	1
6	1	21	0	36	0
7	1	22	0	37	0
8	0	23	0	38	0
9	1	24	0	39	3
10	1	25	1	40	1
11	2	26	0	41	1
12	1	27	0	42	
13	0	28	0	43	
14	0	29	0	44	

(b) Candidate 2-itemset

Array index	2-itemsets	Count	2-itemsets	Count	2-itemsets	Count
0	1,2	1	2,3	0	3,4	0
1	1,3	0	2,4	0	3,5	0
2	1,4	0	2,5	0	…	
3	1,5	0	2,6	1		
4	1,6	0	2,7	1		
5	1,7	0	…	…		
…	…	…	2,9	0		
…	1,9	1	2,10	1		
	1,10	0	2,11	0		
	1,11	2	2,12	0		
	1,12	1	…	…		
	…	…	2,20	0		
	1,20	0	…	…		
	…	…	2,25	…		
	1,25	0	…	…		
	…	…	2,30	1		
	1,30	2	…	…		
	…	…	2,32	1		
	1,32	0	…	…		
	…	…	2,35	0		
	1,35	0	…	…		
	…	…	2,39	1		
	1,39	1	…	…		
	1,40	1	2,40	1		
	1,41	1	2,41	0		

(c) Hash value of the selected 2-itemsets

Itemset	Hash value
1,2	1+2 mod 33 = 3
1,9	1+9 mod 33 =10
1,11	1+11 mod 33 = 12
1,12	1+12 mod 33= 13
1,30	1+30 mod 33 = 31
1,39	1+39 mod 33 = 7
1,40	1+40 mod 33 = 8
1,41	1+41 mod 33 = 9
2,6	2+6 mod 33 = 8
2,7	2+7 mod 33 = 9
2,10	2+12 mod 33 = 12
2,30	2+30 mod 33 = 32
2,32	2+32 mod 33 = 1
2,39	2+39 mod 33 = 8
2,40	2+40 mod 33 = 9

(c) Every t-itemset contain in transaction T is labeled as an integer that needs 4 bytes of storage. We obtain the Eq. (4) for every candid t-itemsets.

$$t \times 4 = 4t \text{ bytes.} \tag{4}$$

Data structure known as Pointer was applied to form linked lists. For now, 4 bytes are then required. Equation (5) states that every t-itemsets node in data structure named as linked lists requires at the minimum

$$4t + 4 = 4(t + 1)\text{bytes.} \tag{5}$$

(d) Equation (6) shows the limitation memory demands for generating candidate t-itemsets, by multiplying 4(t+1) bytes with the result achieved in step b

$$C_p^t \times 4(t + 1) = 4(t + 1) \times C_p^t\text{bytes.} \tag{6}$$

(e) For every search space, we identify and keep all set frequent t-itemsets as well as candidate t-itemsets

$$t \times 4(t+1) \times C_t^p = 4t(t+1) \times C_p^t \text{bytes}. \quad (7)$$

(f) Additional data structures to the storage are mandatory needed. For an array, memory demands for each candidate t-itemsets= $t \times 4$= 4k bytes.
(g) The availability of Random Access Memory in our laptop computer of 8 GB RAM: **8GB-0.27 GB= 7.73 GB**

We calculate the combinations of the candidates 2-itemsets, by applying Equation (7). We get $4t(t+1) \times C_p^t = 4 \times 2(2+1) \times C_2^p = 8(2+1) \times C_2^p$bytes. For $\binom{p}{2}$ *or* $C_2^p = \frac{p!}{2!(p-2)!} = \frac{647!}{2!(647-2)!} = \frac{646 \times 647}{1 \times 2} = 208981$ Composition of 2-items in database D with transactions T. By replacing its value, we obtain C_2^p=24× 208981= 0.0046711 GB. We then compare the achieved outcomes with the availability of Random Access Memory we find in g, it proves that 7.73 GB is greater than 0.0046711 Of RAM. In addition, we calculate the combinations of candidate 3-itemsets, formula (7) achieved in step e) is used. We find $4t(t+1) \times C_p^t = 4 \times 3(3+1) \times C_3^p = 48 \times C_3^p$bytes. For $\binom{p}{3}$ *or* $C_3^p = \frac{p!}{3!(p-3)!} = \frac{647!}{3!(647-3)!} = \frac{645 \times 646 \times 647}{1 \times 2 \times 3} = 44930915$ composition of 3-items in database D with transactions. By replacing its value, we achieved C_3^p= 48× 44930915=2.0085684 GB. We then compare the achieved outcomes with the availability of Random Access Memory we find in g, it proves that 7.73 GB is greater than 2.0085684 GB of RAM. For combination of 2-itemsets and 3-itemsets, it demonstrates that no issues of memory capacity in the generation of 2- and 3-itemsets. We need now to show the behavior of the combinations of 4-itemsets. In this point, formula (7) we achieved in e is used. We obtain $4t(t+1) \times C_p^t = 4 \times 4(4+1) \times C_4^p = 80 \times C_4^p$bytes. For $\binom{p}{4}$ *or* $C_4^p = \frac{p!}{4!(p-4)!} = \frac{647!}{4!(647-4)!} = \frac{644 \times 645 \times 646 \times 647}{1 \times 2 \times 3 \times 4} = 7233877315$ composition of 4-items in database D with transactions T. By replacing its value, we achieved C_4^p= 80× 7233877315=538.965859636387 GB.

By comparing the outcomes we achieved with the availability of the RAM in g, it proves that 7.73 GB is less than 538.9 GB for 8GB. For 4-itemsets, our study proves that the memory capacity will be low for 8GB RAM.

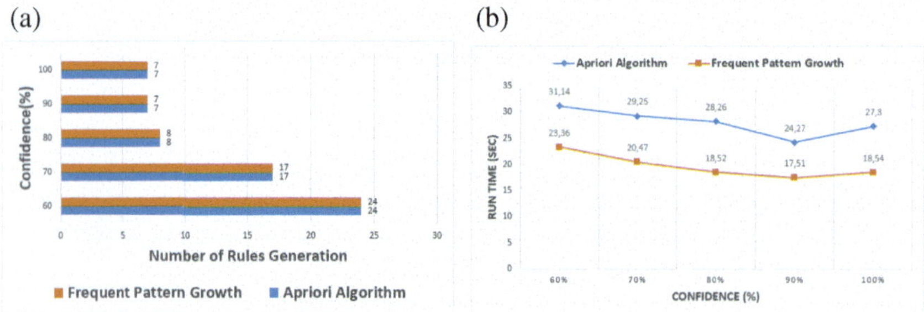

Fig. 3 **a** Number of association rules generated by Frequent Pattern Growth algorithm and Apriori algorithm using multiple confidence, **b** Comparison of running time between Frequent Pattern Growth algorithm and Apriori algorithm using multiple confidence

6 Experimental Results and Discussion

This study emphasized on the use of the Apriori and FP-Growth algorithms on retail market basket data. This is performed using the programming environment and programming language known as anaconda-python and mlxtend library. In this experiment, we use 121 out of 88162 transactions and 647 out of 16470 items. Figure 3a shows different generated association rules when used multiple minimum confidence 60%, 70%, 80%, 90%, and 100% with minimum support 4%. It can be seen that whenever increase the confidence, the number of rules generation is decreased for both algorithms. Figure 3b shows the comparison of the running time of the model using different confidence. It can be seen that Frequent Pattern Growth algorithm has better performance than Apriori algorithm in running time of the model.

7 Conclusions

The purpose of this study was first to design and implement the model using the Apriori and FP-Growth algorithms based on the appropriate data structure. The second was to find out the frequent k-itemsets and using them to generate the whole association rules on market basket data. The third was to compare the two algorithms in terms of running time, support, and confidence of the generated rules. Regarding to the outcomes, FP-Growth algorithm has better performance than Apriori algorithm in running time of the model and whenever increase the confidence the number of rules is decreased for both algorithms.

References

1. Nsabimana T (2015) Association rule mining using the apriori algorithm: a simulation study using retail market basket data. Doctoral dissertation, University of Buea, Cameroon
2. Reinsel D, Gantz J, Rydning J (2018) The digitization of the world from edge to core. Seagate
3. Naha B, Banerjee S, Mondal S, A survey on cloud-denial of service. Am J Adv Comput. https://doi.org/10.15864/ajac.1205
4. Gama AWO, Widnyani NI (2020) Simple modification for an Apriori algorithm with combination reduction and iteration limitation technique. Knowl Eng Data Sci (KEDS) 3(2):89–98
5. Zhang W, Ma D (2014) Medical diagnosis data mining based on improved Apriori algorithm. J Netw 9(5)
6. Wang H-B, Gao Y-J (2021) Research on parallelization of Apriori algorithm in association rule mining. Procedia Comput Sci 183:641–647
7. Avizenna MH (2022) J Appl Data Sci 3(4):143–151
8. Yabing J (2013) Research of an improved Apriori algorithm in data mining association rules. Int J Comput Commun Eng 2(1)
9. Minal G. Ingle NY, Suryavanshi (2015) Association rule mining using improved Apriori algorithm. Int J Comput Appl (0975 - 8887) 112(4)

10. Nurmayanti WP, Sastriana HM, Rahim A, Gazali M, Hirzi RH, Ramdani Z, Malthuf M (2021) Market basket analysis with Apriori algorithm and frequent pattern growth (Fp-Growth) on outdoor product sales data. Int J Educ Res Soc Sci (IJERSC) 2(1):132–139
11. Zhang J, You C, Huang J, Li S, Wen Y (2020) Research on application of frequent pattern growth algorithm in academic early warning. In: Proceedings of the 2020 8th International conference on information and education technology, pp 116-121
12. Brijs T (2003) Retail market basket data set. In: Workshop on frequent itemset mining implementations (FIMI'03)
13. Zhao Q, Bhowmick SS (2003) Association rule mining: a survey. Nanyang Technological University, Singapore
14. Han J, Kamber M (2006) Mining frequent patterns, associations, and correlations. Data mining: concepts and techniques, 2nd edn. San Francisco, pp 227–283

Integration of Distributed Intrusion Detection Systems in IoT Infrastructure

Christian Ildegard Bimenyimana, Thierry Nsabimana, Ghislain Vlavonou, Joël T. Hounsou, and Gerard Niyonsaba

Abstract Internet of Things (IoT) is an attractive innovation technology which utilized to interrelate all IoT devices worldwide with the help of the best high speed Internet. Nowadays, industries' data are being transferred over heterogeneous IoT devices (e.g., smart house, smart watch, sensors, and so forth) connected with each other via power Internet. Nevertheless, this emerging innovation technology is under risks due to the lack of safe security. IoT network requires strong security to detect malicious behavior accurately. Intrusion Detection System (IDS) is a smart tool/software used to monitor input/output network traffic and generate an alarm message whenever malicious activities occurred. In this study, we build a smart IDS based on machine learning (ML) approaches using data mining tool known as Knime and Anaconda-Python to classify intrusions in IoT network. The evaluation of the model is conducted based on two tools for efficiently predicting and classifying real-world network intrusions from Coburg Intrusion Detection Dataset (CIDDS-001) for better decision making.

Keywords CIDDS-001 · Machine learning · IoT · IDS · Knime

C. I. Bimenyimana (✉)
Université des Grands Lacs, Saint Michel, Bujumbura, Burundi
e-mail: barundure@gmail.com

T. Nsabimana
Institut de Statistique Appliquée de l'Université du Burundi, Nyamugerera, Bujumbura, Burundi

G. Vlavonou · J. T. Hounsou
Institut de Mathématiques et de Sciences Physiques de l'Université d'Abomey-Calavi, Dangbo, Porto-Novo, Benin

G. Niyonsaba
Centre de Recherche Indépendant et d'Informatisation du Système d'Information, Centre Porta Sion, Bujumbura, Burundi

X.-S. Yang et al. (eds.), *Proceedings of Ninth International Congress on Information and Communication Technology*, Lecture Notes in Networks and Systems 1000,
https://doi.org/10.1007/978-981-97-3289-0_10

1 Introduction

With new advanced technology, Internet of Things (IoT) is nowadays considered as an attractive innovation technology since it increases exponentially in industries, institutions, transport sector, etc. Due to the exponential growth of the IoT infrastructures in our society, a critical question upraised concerns about their security [1]. IoT devices need to be prepared, installed, and configured before deployment in buildings environment either public sector or private sector by considering the three security measures like integrity, availability, and confidentiality of computer resources. It could be better enough the IoT users to be trained in how the IoT infrastructures designed, communicated to each other, and kept them in safe. The outgoing and incoming data via IoT devices must be transferred and stored in secured environment. The smart IDSs based on machine learning are highly recommended to reduce the damage that could be occurred in certain moment (Fig. 1).

In this study, we apply machine learning algorithms such as KNN, Decision Tree (DT), Random Forest (RF) and Linear Regression (LR), and data mining Knime tool for efficient attack classification from CIDDS-001. For the performance analysis of

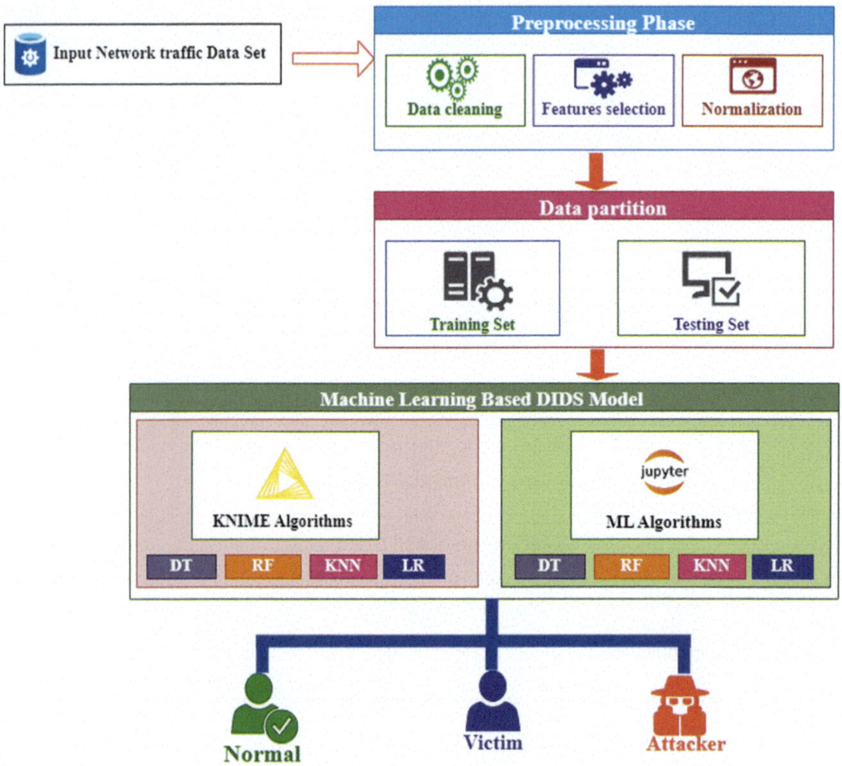

Fig. 1 Proposed security framework for the intrusion detection in the IoT devices

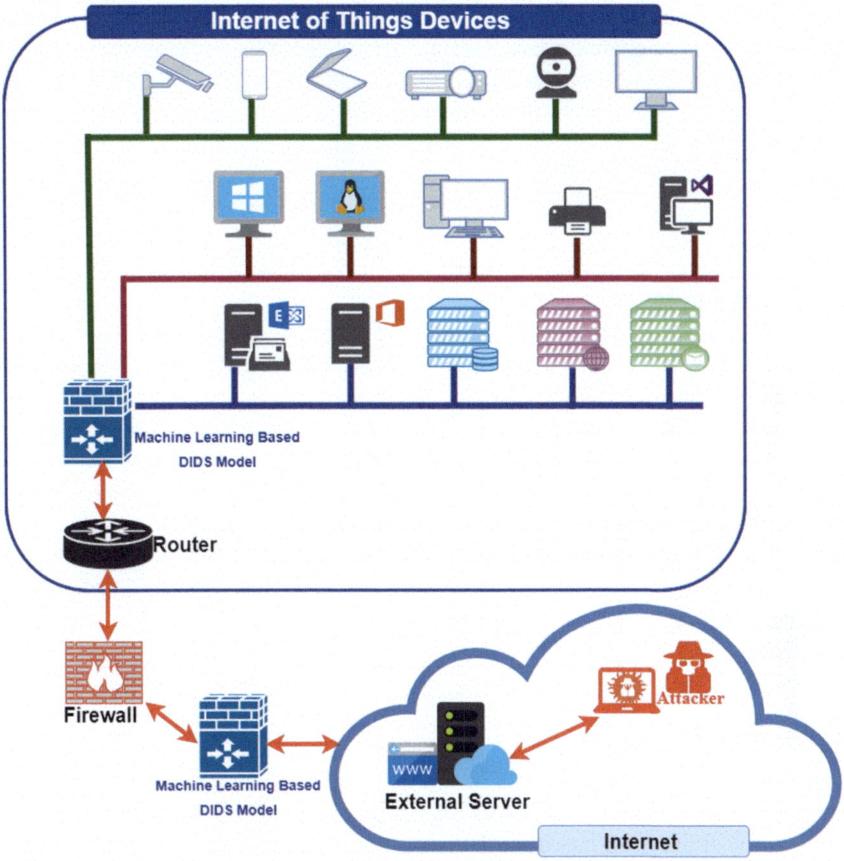

Fig. 2 Proposed deployment model of a distributed intrusion system

our model, precision, recall, and accuracy were used. The proposed framework using ML-based DIDS and Knime have been evaluated using CIDDS-002 dataset. This research proposed also a deployment model-based Distributed Intrusion Detection System (DIDS). The proposed DIDS model controls the network traffic outside of the IoT infrastructure and inside of the IoT as can be seen in Fig. 2.

2 Machine Learning Background

Machine learning, ML in short form, is a part of Artificial Intelligence (IA) which is used to automate mathematical models for predicting the desired outcomes based on the input features without being explicitly programmed. AI technologies are actual applied by the researchers, engineers, DevOps to solve a real-world problems. For understanding better the concept of ML, let's take an example of the real-world problem predicting network attack. Given a dataset D consists of real-world attacks with normal traffic. Base of the size of the dataset, we need to predict the attack traffic. To do that, we need to make function that maps the input variable x to the output variable y:

$$f(x) \to y, \tag{1}$$

where x $(x_1, x_2, x_3, x_4, x_n)$ is the input features, i.e., Flow ID, Source IP, Source Port, Destination IP, Destination Port, etc., y $(y_1, y_2, y_3, y_4, y_n\ )$ is the expected output features based on the input dataset, i.e., DDoS, Botnet, Dos Slow HTTP Test, WebAttack:SQL, etc. In ML, there are two main learning methods, namely supervised and unsupervised learning methods. Supervised learning method can be understood as learning with labeled data, where the expected output is recognized, and there is common relationship between the input features of x and the output features of y, while unsupervised learning method can be defined as learning without labeled data, where we don't recognize the output features, and there is no common relationship between the input features of x and the output features of y. In our paper, we consider ML as signature and an anomaly-based detection approach, where supervised learning technique is applied.

3 Related Works

Different researchers in the domain of IoT security challenges have proposed multi-form security framework-based machine learning by combining with open sources data mining tools labeled as Weka, Knime, Rapid Mine, etc. The authors Basha et al. [2] suggested a new avenue to evaluate the performance of their model built on machine learning algorithms like Decision Tree and Naive Bayes, and using Knime data mining tool. The implementation of their model was conducted with UC Irvine machine learning repository. According to their achieved results, reported that the performance of the model is 99.465% of accuracy rate. The authors [3] utilized intrusion CIDDS-001 dataset which was split into the present subsets: 70% for training set and 30% for testing test. They used only one machine learning algorithm labeled as Decision Tree to classify normal and harmful data. Their proposed model achieved with the highest accuracy of 100%. The authors [4] proposed a model built on four machine learning algorithms such as Shallow Neural Network (SNN), Decision Tree (DT), Bagging Tree (BT), K-Nearest Neighbor (kNN), and Support Vector Machine (SVM). The evaluation of the their model is performed based on the IoTID20 dataset, which was split into 70% training set, 25% testing set, and 5% of validation set. The performance evaluation reported an accuracy of 100%. The authors Alsulami et al. [5] proposed machine learning model to classify network intrusions in an IoT infrastructure. Their implemented model is differentiated between normal and abnormal activities.

4 Methodology

In this study, we proposed a Distributed Intrusion Detection Systems in IoT infrastructure using four machine learning algorithms as KNN, Decision Tree (DT), Random Forest (RF), and Linear Regression (LR), and data mining Knime tool for efficient attack classification from CIDDS-001 [6]. The work proposed also a deployment model-based Distributed Intrusion Detection System (DIDS). The proposed DIDS model controls the network traffic outside of the IoT infrastructure and inside of the IoT as can be seen in Fig. 2. The learning process of our model has been performed in different stages, and each stage will be explained in the upcoming section.

4.1 Proposed Security Framework for the Intrusion Detection in the IoT Infrastructure

Figure 1 depicts the security framework of the proposed distributed intrusion detection model. The design of the proposed model consists of four stages such as the input network traffic, preprocessing phase, data partition, and classifiers.

1. Preprocessing Phase: We receive a huge amount of data from different sources with their particular forms named as structured/SQL, unstructured/NoSQL, document, videos, images, and audio. Tools are required to manage them like MongoDB, Neo4J, Cassandra, Hadoop Spark. In this study, data are loaded, transformed using auto-encoder, cleaned and normalized, and fed it to ML model for learning process. Some data captured from IoT devices consist of a lot of features, and some of them are irrelevant which may lead poor performance of the model. The feature selection step is more required so as to choose the best features which will improve the performances in terms of accuracy, running time of the model.
2. Data partition, in this study, data is splitted into two sub sets such as training set for learning the model and testing set for evaluating the performance of the model.
3. Machine learning-based DIDS model. After data partition in different subsets, the data is trained using various classifiers for a comparative analysis. The classifiers used for this analysis are the following KNN, Decision Tree (DT), Random Forest (RF), Linear Regression (LR), and data mining Knime tool. They are all the supervised machine learning approaches and data mining techniques.

4.2 Proposed Deployment Model-Based Distributed Intrusion Detection System (DIDS)

The proposed deployment model named as distributed Intrusion Detection System could control safely the surity of the IoT device in outside of the IoT devices and after the router insider the IoT devices, as depicted in Fig. 2. The proposed machine learning model-based DIDS is evaluated using CICIDS-002.

5 Experimental Setup

The experiment was performed on HP Laptop with Intel(R) core(TM) i5-1035G1 CPU@1.00 GHz, 1.19 GHz and 12 GB. The operating system was type 64-bit, using window 11 Home and HDD 1To. The different algorithms are implemented using Python programming language and Anaconda-Navigator environment.

5.1 Performance Analysis Measures

Four performance measures such as precision and accuracy were used to evaluate the performance of the proposed ML-based DIDS model for the IoT Devices.

5.2 Evaluation

The achieved results using four classifiers will be analyzed, evaluated, and compared based on the performance analysis in terms of metrics.

The experiment 1 is the comparison of different classifiers in terms of accuracy using Knime data mining tool. We can analyze the highest accuracy for two classifiers such as Random Forest and Decision Tree as depicted in Fig. 3. The lowest accuracy value has been observed using Linear Regression algorithm.

The experiment 2 is the comparison of different classifiers in terms of accuracy using Anaconda-Python environment. The achieved result of accuracy value is depicted in Fig. 4. The highest accuracy is observed using Linear Regression, K-Nearest Neighbor, and Decision Tree algorithms. The lowest accuracy has been obtained using Random Forest.

The experiment 3 compares different classifiers in terms of precision using the Knime tool. In this case, we are aiming for the algorithm with the highest precision compared with the other algorithms used in the present work. The four algorithms were used on the same dataset, and according to the obtained results after experimentation (as shown in Fig. 5), we realize that Decision Tree algorithm and Random Forest algorithm come up with 100% detection rate, whereas Linear Regression algorithm ends up with 90% detection.

In experiment 4, we provide the results of the four machine learning algorithms. According to the results obtained, we found that three algorithms, namely Decision Tree algorithm, Linear Regression algorithm, and K-Nearest Neighbor (KNN) achieved an excellent classification estimated at 100%, while Random Forest Algo-

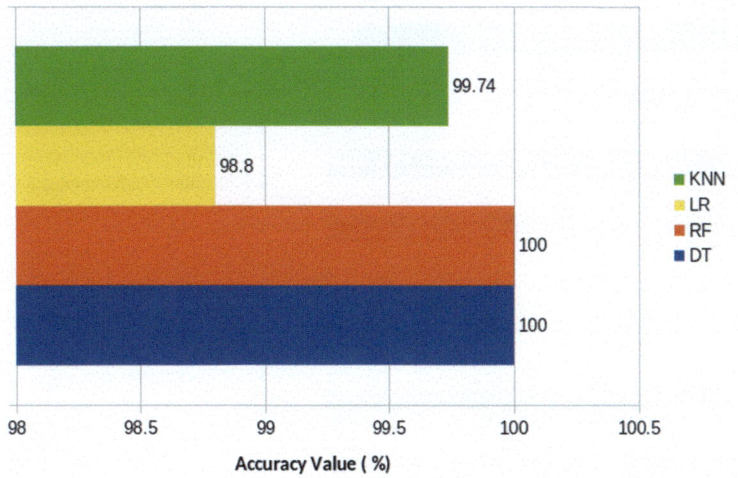

Fig. 3 Accuracy comparison between the algorithms using Knime tool

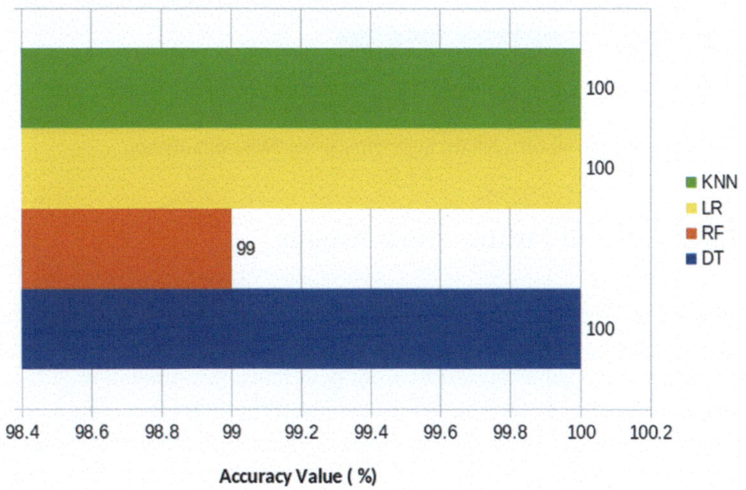

Fig. 4 Accuracy comparison between the algorithms using Anaconda-Python

rithm achieved a classification of 90%, which is relatively low comparatively to the other algorithms. Figure 6 shows the details of the classification performed by each algorithm.

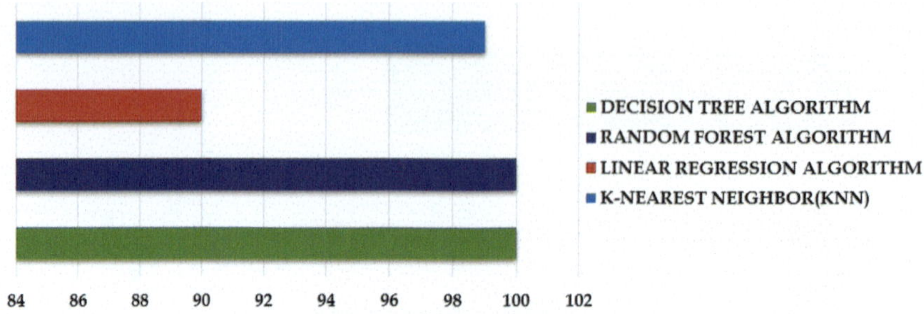

Fig. 5 Accuracy comparison between the algorithms using Anaconda-Python

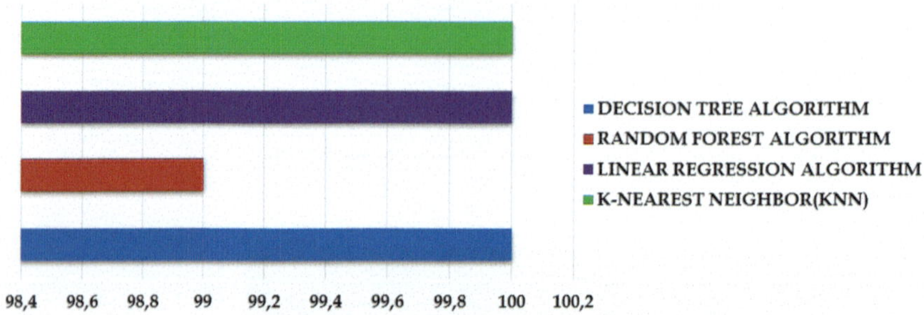

Fig. 6 Accuracy comparison between the algorithms using Anaconda-Python

6 Conclusion and Future Work Avenue

In this study, we designed and implemented the Distributed Intrusion Detection System for IoT infrastructure using machine learning algorithms and data mining tool. According to the achieved results from both tools, the simulation of the algorithms using Anaconda-Python has been achieved the highest accuracy value of 100% using Linear Regression, K-Nearest Neighbor, and Decision Tree algorithms than the results obtained in multiple algorithms using data mining Knime tool. In addition, in terms of precision metric, the obtained results using the Knime tool and machine learning algorithms, we can assume that machine learning end up with the most wonderful goal in attack type classification with three algorithms out of four used algorithms. These algorithms are Decision Tree, Linear Regression, and K-Nearest Neighbor within a classification of 100% and Random Forest achieved with a classification of 90%. In the Knime datamining tool, Decision Tree algorithm and Random Forest algorithm achieved a classification of 100%, while two others, K-Nearest Neighbor and Linear Regression Algorithm, achieved a classification of 99% and 90%, respectively. Regarding to the results achieved the proposed model, our model could be integrated in IoT devices so as to mitigate the inbound and outbound network attacks. In the future work, we plan to conduct the same experiment in online experiment using deep learning algorithms and Snort-IDS.

References

1. Bharati S, Podder P (2022) Machine and deep learning for IoT security and privacy: applications, challenges, and future directions. Security Commun Netw 2022:1–41
2. Basha SM, Rajput DS, Poluru RK, Bhushan SB, Khalandar Basha SA (2018) Evaluating the performance of supervised classification models: decision tree and Naïve Bayes using KNIME. Int J Eng Technol
3. AL-Jawazneh A, Salameh JB (2022) A model for cloud intrusion detection system using feature selection and decision tree algorithms. Int J Sci Technol Res 10(03)
4. Alsulami AA, Al-Haija QA, AT, Alqahtani A (2021) An intrusion detection and classification system for IoT traffic with improved data engineering. Appl Sci 12336
5. Alsulami AA, Abu Al-Haija Q, Tayeb A, Alqahtani A (2022) An intrusion detection and classification system for IoT traffic with improved data engineering. Appl Sci 12(23):12336
6. https://github.com/markusring/CIDDS

Predictive Analysis Model to Improve the Control of Eating Disorders in Adolescents in Metropolitan Lima Based on Machine Learning

Jeffery Bryce Molina Solano, Valeria Angelica Castillo Revelo, and Víctor Manuel Parasi Falcon

Abstract *Objective* Eating disorders in adolescents have been a recurring issue within our society, significantly impacting their mental and physical well-being. This study aims to implement a predictive analysis model based on machine learning and the EDI-3 evaluation method to forecast eating disorders (ED) in adolescents. *Method* Using the machine learning tool with the Random Forest algorithm and the Eating Disorder Inventory-3 (EDI-3) assessment method, a dataset of information was gathered from a sample of 500 adolescent students. This dataset will contribute to the training of the predictive model, enabling it to identify patterns of eating behaviors in end-users. *Results* Through data cleaning of the final dataset, 70% of the information was utilized for training the predictive model, and 30% was allocated for subsequent validation. This approach yielded an effectiveness rate of 92.27% upon completion of the training. Furthermore, in order to validate these results, Cronbach's alpha coefficient was employed, resulting in a score of 0.70, indicative of a satisfactory level of reliability. *Discussion* The results obtained strongly support one of the main objectives of the study conducted, as it significantly surpasses the 85% accuracy threshold. Our results suggest that eating behavior patterns can be crucial factors in making predictions that enable the early identification of positive cases.

Keywords Eating disorders · Machine learning · Random Forest · Inventory of eating disorders

J. B. M. Solano (✉) · V. A. C. Revelo · V. M. P. Falcon
Universidad Peruana de Ciencias Aplicadas (UPC), Prolongación Primavera 2390, Monterrico, Santiago de Surco, Peru
e-mail: u201916712@upc.edu.pe

V. A. C. Revelo
e-mail: u201911916@upc.edu.pe

V. M. P. Falcon
e-mail: pcsivpar@upc.edu.pe

X.-S. Yang et al. (eds.), *Proceedings of Ninth International Congress on Information and Communication Technology*, Lecture Notes in Networks and Systems 1000,
https://doi.org/10.1007/978-981-97-3289-0_11

1 Introduction

Presently, eating disorders (EDs) among adolescents are emerging as a recurring concern within our societal framework. These instances have exhibited a notable upward trajectory in recent years, thus compelling the imperative implementation of corresponding measures. Lacking judicious control over this predicament, adolescents are poised to contend with prospective social and personal repercussions, including but not limited to depression, impairment of the digestive system, social isolation, etc. [1].

Regarding the contemporary pace of life, it is discerned that psychological, familial, and societal factors are the underlying catalysts behind inadequate dietary practices. These causal factors may be exacerbated or ameliorated contingent upon the consumption of appropriate sustenance. During the phase of adolescence, it is observable that young individuals tend to juxtapose their physical attributes with those of their admired peers due to diminished self-esteem. This significantly affects the dietary habits of adolescents, leading to a drastic change in their diet to conform to social stereotypes, which can risk their mental and physical health [1].

As per data furnished by the Ministry of Health (MINSA), eating disorders are prevalent in 6% of adolescents residing in Lima and Callao (7.1% among females and 4.9% among males). Furthermore, a breakdown of this data reveals that adolescent females exhibit a prevalence of 1.6% in bulimic behaviors, while males display a figure of 0.3%. Conclusively, the compilation of information regarding anorexia nervosa indicates that 0.7% of female adolescents grapple with this mental affliction [2].

In accordance with the comprehensive data collated by the Ministry of Health (MINSA), it is noteworthy that during the COVID-19 pandemic, Peru documented an 11.5% incidence of eating disorders within the entire populace, with adolescents exhibiting a prevalence ranging from 7 to 11%. This discerning outcome underscores the gravity of eating disorders (EDs) as a pressing societal concern [3].

1.1 Importance of the Problem

Eating disorders arise from numerous factors: psychological, familial, and social, which encourage the development of eating disorders in young adolescents, causing them physical harm such as muscle weakness and heart failure, mental distress like depression, self-harming behaviors, suicidal thoughts, and in the worst cases, death. On the other hand, eating disorders lead to poor academic performance and development in adolescents, affecting them personally and professionally in the long term [1].

2 Contribution

A predictive analysis model for eating disorders is proposed for teenagers residing in Metropolitan Lima, employing two assessment methods for data collection: Eating Disorder Inventory 3 (EDI-3) for physical and attitudinal diagnosis. Furthermore, machine learning will be used as the data analysis tool. The research is focused on enhancing control and diagnosis to contribute to the treatment of eating disorders.

2.1 Related Works

In [4], the authors introduce a questionnaire that assesses social media use, focusing on time, followed users, and comparison frequency. The goal is to identify eating disorder presence or development, using scales from the "Eating Disorder Inventory" (EDI—BD and EDI—DT). They promoted their study through social media and clinic pages, targeting those aged 15–35 with social media accounts. The study consisted of three phases: gathering general information, analyzing social media usage (frequency, hours, body comparisons, and video posts), and evaluating body image using the EDI-2 survey and EDI—DT and EDI—BD scales. The findings indicate that social media use can influence eating disorders, primarily through body image dissatisfaction when comparing oneself to admired users. Uncontrolled social media usage in adolescents and young adults can heighten body dissatisfaction and the pursuit of thinness as an ideal.

In [5], the authors introduce the Eating Attitudes Test-26 (EAT-26) as a tool for identifying the risk of developing an eating disorder. This instrument comprises 26 questions with six response options. A score exceeding 20 at the end of the assessment indicates the potential risk of an eating disorder. After evaluating 700 adolescent females and utilizing Cronbach's alpha coefficient, the study found that the EAT-26 questionnaire achieved a strong reliability score of 0.92. Importantly, this assessment revealed that approximately 13.9% of the total users assessed were at risk of developing an eating disorder.

In [6], the authors introduce a machine learning (ML) model employing decision tree classification to analyze eating behaviors in Binge Eating Disorder (BED) patients. They recruited 1341 participants, 38% of whom exhibited recurrent binge eating behaviors in the past month. Data was collected through online questionnaires with participants' consent, and the decision tree algorithm was used to analyze and classify the data. To prevent overfitting, the data was partitioned for training, model construction, and test sample selection. The results demonstrated that the model effectively identified and classified 70% of BED patients based on their binge eating behavior type: non-recurrent and recurrent.

In article [7], the author proposes an analysis of the issue to identify the outcomes of the placebo effect in patients with eating disorders or anxiety-related illnesses. First, data from different patients who received a placebo was collected, and patients gave their consent at the time of enrollment in the research project. After obtaining the data, it was approved by the Institutional Review Board. Second, various algorithms were used to implement the placebo response prediction model. The algorithms used included support vector, decision tree, logistic regression, KNN, GNB, and AdaBoost. Finally, the "XGBoost" algorithm was used for data reduction and final accuracy. At the end of the research and experimentation process, it was found that the Binge Eating Frequency (BEF) was 56%, indicating a significant response to the placebo. Additionally, 30% of the patients who participated in the experimental stage reduced their binge eating after the 6-week duration of the project.

In article [8], the authors propose an ML-based model to determine the predominant polarity (PP) of a patient with symptoms of bipolar disorder (BD). First, patients with BD were recruited and selected from the outpatient clinic of the Bipolar Disorder Research Program. Second, relevant information for the study was collected through a questionnaire. Third, the information was analyzed using an ML algorithm, and patients with PP were classified using the "Random Forest" (RF) algorithm. Finally, the results were validated using cross-validation. The study findings revealed that the predictive algorithm achieved a 64.8% accuracy rate in its predictions and can effectively discern and categorize individuals with Bipolar Disorder (BD).

3 Predictive Analysis Model of Eating Disorders in Adolescents

In this section, the main concepts of the proposed project will be presented, with the aim of providing a comprehensive overview of the methodologies and technologies that support the research.

Adolescence [9]: It is the period when childhood ends and the transition to adulthood begins, typically starting at the age of 12 and ending at 18 years old, depending on the individual's adaptive and functional value. Adolescence consists of three stages: early adolescence (occurring between 12 and 13 years old, marked by the onset of sexual hormones in adolescents' bodies), middle adolescence (occurring between 14 and 16 years old, characterized by psychological changes), late adolescence (occurring from 17 years old until the end of adolescence, as individuals begin to accept their bodies and define their identities).

Eating Disorders [10]: Eating disorders are mental illnesses caused by eating patterns that reduce necessary nutrition or involve poor dietary habits. These behaviors are linked to emotional, physical, and psychological thoughts, leading to insecurities in individuals when it comes to their physical well-being. Some of the most common eating disorders include anorexia, bulimia, binge eating disorder, obesity, among others.

Machine Learning [11]: Machine learning is a technology that learns through the collection and analysis of data. This technology also provides practical solutions for everyday life through continuous learning, such as voice recognition, autonomous driving, among others.

EDI-3 [12]: The Eating Disorder Inventory-3 (EDI-3) assesses the behaviors and traits of individuals with eating disorders. It is designed for individuals aged 12 and older who have eating disorders. This questionnaire consists of 12 scales, including concerns about thinness, bulimia, body dissatisfaction, low self-esteem, personal alienation, interpersonal insecurity, among others.

Predictive Analysis Model [13]: A predictive analysis model is a set of techniques and strategies that, along with automation tools, data collection, and predictive algorithms, can produce results with a high degree of accuracy to extrapolate hidden trends.

Cloud Computing [14]: Cloud computing involves multiple access to applications, development environments, software resources, or IT infrastructure, providing optimized processes. It also offers improvements in server control and information security.

Bulimia [15]: Bulimia is a type of eating disorder characterized by excessive food intake followed by feelings of guilt or general discomfort, which drive individuals to induce vomiting, engage in excessive physical exercise, or take laxatives repeatedly in their daily lives.

Anorexia [16]: Anorexia is a type of eating disorder that involves abstaining from food consumption to decrease body weight, which affects one's health due to the loss of defenses in the body. It also develops a fear of consuming food, further impacting body weight and leading to anxiety and stress problems.

Psychological Patterns [17]: Psychological patterns are behaviors performed at specific times due to routines generated by an individual's experiences. These types of attitudes can either positively or negatively affect the individual depending on the final outcomes.

Body Dissatisfaction [18]: Body dissatisfaction in individuals is based on the rejection of their current body image. For this reason, individuals make drastic decisions that affect them both physically and mentally, potentially leading to the development of an eating disorder.

4 Method

4.1 Participants

The project's participants will consist of 500 secondary school adolescents, aged between 13 and 18 years, as the preferred selection criteria for the project.

4.2 Procedure

Physical questionnaires were conducted to collect information for model validation. Participants answered the survey questions (EDI-3). The responses will be entered into the model's database to assess the likelihood of developing any eating disorder among the participants.

4.3 Predictive Model

A predictive analysis model for ED is proposed for teenagers residing in Metropolitan Lima using the psychological assessment method and data collection EDI-3 for physical, attitudinal, and psychological diagnosis. Additionally, for data analysis, the ML tool and Azure cloud services will be used. The research is focused on developing a predictive analysis model that allows for the early identification of eating disorders in teenagers. As shown in Fig. 1, the presented proposal consists of seven phases: in the first phase, the user accesses the predictive analysis web application; in the second phase, the questionnaires for the EDI-3 methods aimed at teenagers are found. In phase 3, the questionnaire responses are stored in the cloud; subsequently, the responses are analyzed using ML to provide a predictive analysis of eating disorders. In phase 5, the system stores the responses in the patient's history within the application, and then in phase 6, the results of the tests performed are visualized. Finally, in phase 7, specialist support is provided for the final diagnosis after analyzing the results of the predictive analysis model.

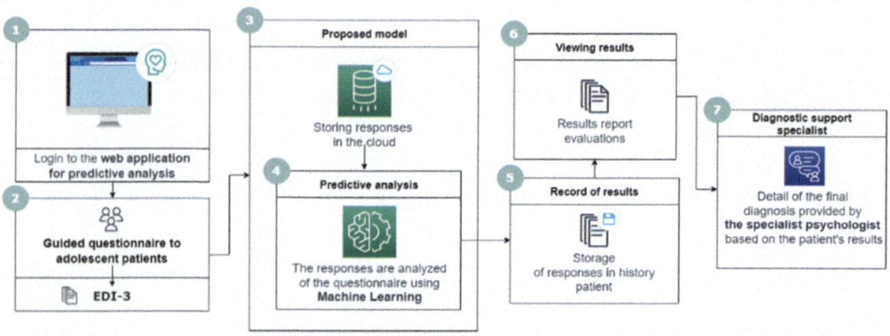

Fig. 1 Phases of the predictive model of eating disorders are shown to obtain the results that show the probability of developing an eating disorder

4.4 Architecture

The project architecture integrates interconnected layers, facilitating the integration of a predictive analysis model for eating disorders into the medical care process. It features web-based prediction software supporting medical care, incorporating security measures, session logging, and a core machine learning module utilizing the "Random Forest" algorithm for data processing. Key modules were developed in Python, HTML, and CSS. The entire infrastructure is hosted on a Microsoft Azure cloud server. In sum, this architecture harmonizes medical care, predictive analysis, and technological infrastructure to enhance the diagnosis of eating disorders through web-based software on the Microsoft Azure platform.

4.5 Main Contribution

The primary contribution of this research lies in the use of machine learning techniques and evaluation tools to accurately estimate the probability of developing an eating disorder and identify the behavioral patterns exhibited by adolescents. This optimizes the process by leveraging cloud services, which facilitates the analysis of physical and mental patient characteristics. The result is an immediate pre-diagnosis that supports specialized psychologists in the field.

5 Experimentation

Data Collection. For the data collection, a survey with questions based on the EDI-3 questionnaire was provided to a total of 500 school students aged between 13 and 18 years old. The obtained information was utilized for training the predictive model for eating disorders and general psychological habits.

Data Preprocessing. This phase involved cleaning atypical data from the collected surveys, such as false patterns or incomplete responses. Additionally, categorical variables were identified, and the information's characteristics were normalized.

Model Training. Using Random Forest and machine learning techniques, the model was configured in Azure machine learning. Firstly, a virtual machine with the following specifications was created: Virtual Machine Tier (Dedicated), Virtual Machine Size (Standard), and Maximum Number of Nodes (Four Nodes).

Then, the collected data file was loaded for model training.

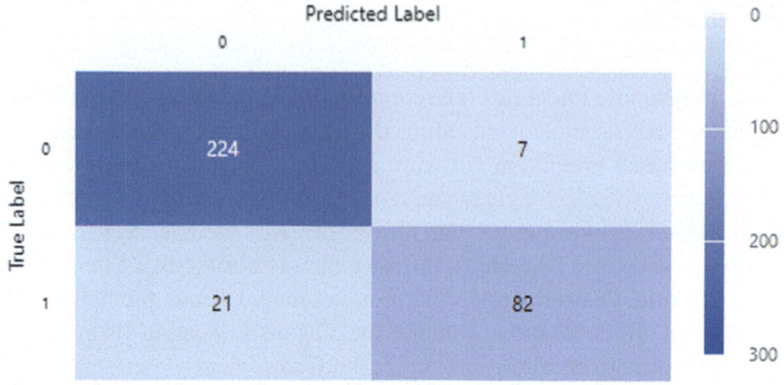

Fig. 2 Results of the predictive model are displayed, showing the quantities of false positives and false negatives obtained. The results obtained in the confusion matrix indicate the variables of true positives, false positives, true negatives, and false negatives that were subsequently detected after the training was conducted. Additionally, it provided a final view on the percentage of accuracy that was achieved

6 Results

During the predictive model training, various predictive analysis algorithms were compared to assess the performance of each, evaluating accuracy results and resulting confusion matrices. In this way, the "Random Forest" algorithm was chosen, using the "MaxAbsScaler" function, achieving a final accuracy of 92.27% after training with the surveys conducted, surpassing the results of the other algorithms. Furthermore, the confusion matrix of the result identified a minimum of false positives and false negatives, as shown in Fig. 2.

7 Validation

Cronbach's alpha coefficient obtained, with a value of 0.70, reflects satisfactory reliability and internal consistency in the responses collected in the survey used to assess the patient. This result suggests that the formulated questions are moderately related to each other and exhibit acceptable and appropriate consistency (Table 1).

In Table 2, correlations between the variables of the Eating Disorder Inventory are presented. In general terms, these correlations are moderate, but some are more

Table 1 Summary item statistics

	Average	Minimum	Maximum	Range	Max /min	Variance
Element variances	3.221	1.178	8.361	7.183	7.095	1.571

Note: The variances of the summarised items are shown

Table 2 Correlations are significant at the 0.01 level (two-sided)

		DT	B	BD
DT	Pearson correlation	1	0.355*	0.184**
	Sig. (bilateral)		< 0.001	< 0.001
	N	478	478	478
B	Pearson correlation	0.355**	1	0.217**
	Sig. (bilateral)	< 0.001		< 0.001
	N	478	478	478
BD	Pearson correlation	0.184**	0.217*	1
	Sig. (bilateral)	< 0.001	< 0.001	
	N	478	478	478

Note: The results of the correlations are shown where moderate results are presented, improving the reliability of the predictive model. ** Correlation is significant at the 0.01 level (bilateral)

significant than others. For example, thinness obsession (DT) showed a higher correlation with bulimia (B) ($rw = 0.355$) than with body dissatisfaction (BD) ($rw = 0.184$). On the other hand, bulimia (B) showed a lower correlation compared to body dissatisfaction (BD) ($rw = 0.217$).

8 Discussion

In the conducted study, data for training the predictive analysis model was obtained through surveys administered to 500 school-going adolescents. Following the requisite data cleaning process, 70% of the information was used for training, and 30% for performance validation. Using the Random Forest algorithm, an impressive 92.27% effectiveness was achieved. Additionally, Cronbach's coefficient of 0.70 was obtained, indicating a reliable and satisfactory result.

The results obtained substantiate the initial project objectives, which aimed for an accuracy rate exceeding 85% upon completion of the training of the predictive analysis model for eating disorders. Favorable rates of true positives were identified in the confusion matrix, further validating the effective outcomes of the machine learning model.

9 Conclusion

In this study, various analyses were conducted to interpret the thoughts and behaviors of adolescents as traits that could help identify eating disorders (EDs). Through training a machine learning model with a dataset generated from the EDI-3 inventory, effective and optimized results were obtained. Additionally, validations using Cronbach's alpha coefficient tool yielded a final score of 0.7, indicating a reliable assessment. Meetings with expert specialists in the psychological field provided alignment with performance and results.

It is crucial to note that the process of comparing prediction algorithms was of paramount importance, as it involved evaluating the context in which data collection and validation were to take place. Consequently, the options focused on the XGBoost, decision tree, and Random Forest algorithms, with the latter offering a higher effectiveness percentage and a more reliable confusion matrix.

The study suggests a room for enhancing the ED dataset with more information through collaboration with academic institutions and extended data collection. This approach could lead to improved predictive model metrics, potentially achieving an effectiveness rate above 92.27%.

References

1. Apaza M A (2022) Factores relacionados a trastornos alimenticios en adolescentes de una institución educativa-Arequipa 2022 [Trabajo de suficiencia profesional para optar el Título Profesional de Médica Cirujana]. Universidad Nacional San Agustin
2. Ministerio de Salud (2017) Documento técnico: situación de salud de los adolescentes y jóvenes en el Perú. Recuperado de http://bvs.minsa.gob.pe/local/MINSA/4143.pdf
3. Chumpitaz M (2020). ¿Por qué la anorexia, la bulimia y los"atracones" de comida se agudizan durante la pandemia? El Comercio. https://elcomercio.pe/somos/historias/coronavirus-peru-por-que-la-anorexia-labulimia-y-los-atracones-de-comida-se-agudizan-durante-la-pandemia-noticia/
4. Jiotsa B, Naccache B, Duval M, Rocher B, Grall-Bronnec M (2021) Social Media use and body image disorders: association between frequency of comparing one's own physical appearance to that of people being followed on social media and body dissatisfaction and drive for thinness. Int J Environ Res Public Health 18(6):2880. https://doi.org/10.3390/ijerph18062880
5. Lazo Montoya Y, Quenaya A, Mayta-Tristán P (2015) Influencia de los medios de comunicación y el riesgo de padecer trastornos de la conducta alimentaria en escolares mujeres en Lima, Perú. Archivos Argentinos de Pediatría 113(6). https://doi.org/10.5546/aap.2015.519
6. Linardon J, Messer M, Helms ER, McLean C, Incerti L, Fuller-Tyszkiewicz M (2020) Interactions between different eating patterns on recurrent binge-eating behavior: a machine learning approach. Int J Eat Disord 53(4):533–540. https://doi.org/10.1002/eat.23232
7. Goyal RK, Kalaria SN, McElroy SL, Gopalakrishnan M (2022) An exploratory machine learning approach to identify placebo responders in pharmacological binge eating disorder trials. Clin Transl Sci. https://doi.org/10.1111/cts.13406
8. Belizario GO, Junior RGB, Salvini R, Lafer B, Dias RDS (2019) Predominant polarity classification and associated clinical variables in bipolar disorder: a machine learning approach. J Affect Disord 245:279–282

9. Santillano Cárdenas I (2009) La adolescencia: añejos debates y contemporáneas realidades. Ultima Década 17(31). https://doi.org/10.4067/S0718-22362009000200004
10. Tanner AB (2023) Unique considerations for the medical care of restrictive eating disorders in children and young adolescents. J Eat Disord 11(1):33. https://doi.org/10.1186/s40337-023-00759-2
11. Holzinger A (2017) Introduction to machine learning and knowledge extraction (MAKE). Mach Learn Knowl Extr 1(1):1–20. https://doi.org/10.3390/MAKE1010001
12. Momeñe J, Estévez A, Griffiths MD, Macía P, Herrero M, Olave L, Iruarrizaga I (2022) Eating disorders and intimate partner violence: the influence of fear of loneliness and social withdrawal. Nutrients 14(13):2611. https://doi.org/10.3390/nu14132611
13. Morris EL, Taylor SF, Kang J (2022) On predictability of individual functional connectivity networks from clinical characteristics. Hum Brain Mapp 43(17):5250–5265. https://doi.org/10.1002/hbm.26000
14. Schilder CMT, Sternheim LC, Aarts E, Elburg AA, Danner UN (2021) Relationships between educational achievement, intelligence, and perfectionism in adolescents with eating disorders. Int J Eat Disord 54(5):794–801. https://doi.org/10.1002/eat.23482
15. Camacho C, Boratyn GM, Joukov V, Vera Alvarez R, Madden TL (2023) ElasticBLAST: accelerating sequence search via cloud computing. BMC Bioinf 24(1):117. https://doi.org/10.1186/s12859-023-05245-9
16. Fuertes Ortiz J, Pérez R, Hernández A (2010) Trastornos de la conducta alimentaria. Anorexia y bulimia. Offarm 29(2):72–77. https://www.elsevier.es/es-revista-offarm-4-articulo-trastornos-conducta-alimentaria-anorexia-bulimia-X0212047X10489696
17. Carpinelli L, Watzlawik M (2023) Anorexia Nervosa in adolescence: parental narratives explore causes and responsibilities. Int J Environ Res Public Health 20(5):4075. https://doi.org/10.3390/ijerph20054075
18. Izydorczyk B, Sitnik-Warchulska K, Lizińczyk S, Lipiarz A (2019) Psychological predictors of unhealthy eating attitudes in young adults. Front Psychol 10. https://doi.org/10.3389/fpsyg.2019.00590
19. Casullo M, Pérez M (2003) Inventario para la evaluación de conductas alimentarias. Universidad de Buenos Aires U B.A
20. Levinson CA, Vanzhula IA, Christian C (2019) Development and validation of the eating disorder fear questionnaire and interview: preliminary investigation of eating disorder fears. Eat Behav 35:101320. https://doi.org/10.1016/j.eatbeh.2019.101320

A Predictive Model to Support Decision-Making for the Accreditation of Learning Programs Using Data Mining and Machine Learning

Francis Kawesha and Jackson Phiri

Abstract Terabytes of data are produced by higher education institutions every year, and this data is crucial for determining how countries will develop. There are significant amounts of this educational data in a variety of relatively recent formats. We suggest a model for gathering, securing, and analyzing this substantial amount of data. A quality assurance body sets standards, and the analysis of the data is used to assess the institution in relation to those standards for the accreditation of higher education programs. Therefore, the model supports the decision-making process in accreditation evaluation. The research in this paper takes academic staff appropriateness and adequacy into consideration when proposing a model for data mining and machine learning-based accreditation criteria prediction.

Keywords Educational data · Knowledge discovery · Accreditation process · Higher learning institutions (HEIs) · Learning programs · Quality assurance · Predictive analytics · Machine learning · Data mining

1 Introduction

Accreditation is a vital tool in higher education, ensuring the quality and effectiveness of learning programs by setting standards based on the institution's operations and curriculum, thereby enabling the delivery of the best possible education [1]. Accreditation measures curriculum quality, attracts top students, and aids informed decision-making. External agencies assess higher education institutions' eligibility and meet required standards [2, 3]. The traditional accreditation process, prone to human error and time-consuming, requires the development of predictive models

F. Kawesha (✉) · J. Phiri
University of Zambia, 10101 Lusaka, Zambia
e-mail: francis.kawesha@cs.unza.zm

J. Phiri
e-mail: jackson.phiri@cs.unza.zm

X.-S. Yang et al. (eds.), *Proceedings of Ninth International Congress on Information and Communication Technology*, Lecture Notes in Networks and Systems 1000,
https://doi.org/10.1007/978-981-97-3289-0_12

for decision-making support [4]. These models can analyze factors and data related to higher education institutions and their learning programs, such as student performance, faculty qualifications, curriculum design, resource allocation, facilities, and program characteristics like curriculum design and industry standards [5, 6].

Implementing a predictive model in the accreditation process can provide several benefits [7]. First, it can improve the efficiency of the accreditation process by automating data collection and analysis, reducing the time and effort required for manual evaluation [8]. Second, a predictive model can enhance the accuracy and objectivity of decision-making by removing any potential bias or subjectivity associated with human judgment [9]. Additionally, a predictive model can provide early warning indicators and identify areas of improvement for institutions that may be at risk of not meeting accreditation standards [10]. Furthermore, integrating a predictive model into the accreditation process can facilitate continuous quality improvement in higher education [11].

Predictive models use machine learning and deep learning algorithms to analyze historical data to predict student performance trends and improve learning programs. This helps accrediting bodies identify at-risk institutions and provide valuable insights for decision-making in accreditation processes [12]. Further, predictive models analyze quiz scores and program performance to identify programs with consistently below-average scores, enabling investigation into instruction quality and resource allocation for accrediting bodies. AI models are increasingly used in data analytics for forecasting, such as predicting course abandonment probabilities and predicting academic achievement based on learning behavior [13]. AI models, for example, were used to develop an early warning system for identifying students who have a high chance of leaving school during their first year [13]. However, in order to forecast student engagement during online learning, machine learning models were used to evaluate behavioral data from students [13]. Predictive models help personalize learning experiences by analyzing student data, identifying effective activities based on progress and comprehension, enhancing engagement and motivation, and predicting university program intake [14]. In this study, we apply machine learning algorithms to cluster data and identify the data related to the learning program under submission and the standards set in the accreditation criteria by the Higher Education Authority.

The structure of this paper is as follows: We discussed related works and a review of the literature in Sect. 2. The paper examines the goals of the study in Sect. 3. The study's findings are covered in Sect. 4 of the paper. The conclusion and areas recommended for additional research were presented in Sect. 5 of the paper.

2 Literature Review

This section of the paper considers literature that focuses on similar research work.

2.1 Accreditation Process with ABET

Academic accreditation is a quality assurance process where an educational institution is evaluated against specific standards set by an accreditation authority, often prioritizing academic programs or overall university accreditation to influence stakeholders' decisions [15, 16]. The Accreditation Board for Engineering and Technology (ABET) is used throughout the study as a running example [16]. ABET evaluates academic programs using eight criteria, according to Scales et al. [16, 17]: students, program objectives, results, curriculum, faculty, facilities, institutional support, and improvement. Qualified professors, staff credentials, counseling, and faculty involvement are required [16]. Departments or colleges applying for ABET accreditation must submit a self-study report demonstrating completion of all requirements. The process involves four parts: creating the report, submitting it to ABET, and reviewing it [16]. ABET visits the institution in person and reviews the report. The institution receives an exit report in the fourth phase that assesses the program's preparedness [16]. The program's accreditation and the subsequent accreditation period are then determined by ABET [16].

2.2 Big Data in University Teaching

Higher education institutions produce data that meets the criteria for big data [18]. Numerous IT systems, such as those for student registration, finances, human resources, and institutional research, are managed by universities [18]. These systems produce enormous amounts of data each semester, including unstructured and semi-structured data from students, staff, and professors, including internet traffic, activities, and sensor data, in addition to structured data [18].

Therefore, using big data techniques to analyze data from higher education institutions is preferable to using conventional techniques [19]. Researchers and businesses can examine vast datasets using this big data analysis to find important trends [19]. Additionally, a university initiative evaluates over 600,000 student records from 3 million course transactions using big data [19]. This helps identify common factors affecting academic program completion and retention, as well as at-risk students, enhancing institutions' capacity to identify areas in need of development [20].

2.3 A Framework for Automation to Evaluate Academic Accreditation Requirements in Saudi Arabian Universities

Recently, governments all over the globe have turned their focus to higher education institutions in an effort to maintain the quality of education, enhance learning outcomes, and establish and promote the establishment of social and economic competencies within the country [21]. With an emphasis on quality policies, institutional mission practices, beliefs, and stakeholder expectations and aspirations, higher education frequently experiences continuous growth [21].

At the institutional and program levels, Saudi Arabian universities are subject to a number of protocols and accreditation requirements [21]. The Commission collaborates with agencies to schedule evaluations, enabling quality assurance and self-studies, with each organization conducting a comprehensive self-study every five years. Self-appraisal scales provided by the National Academic Accreditation and Assessment Commission (NCAAA) are the models utilized as the foundation for self-studies [21].

2.4 An Integrated Perspective on Data Analytics in Higher Education

Data analytics in higher education is transforming how institutions operate, manage resources, and support students [22]. Data analytics aids in generating institutional research reports, ensuring institutions meet accreditation requirements, and demonstrating accountability to stakeholders [22]. Terms like learning analytics (LA), academic analytics (AA), and educational data mining (EDM) have been developed as a result of the study's done in educational data analytics in higher education [22] (Table 1).

Table 1 Types of analytics [22]

Analytics	Analysis level/object analysis	Beneficiaries
Learning analytics (LA)	Course: social networks, discourse analysis, conceptual building, and intelligent curricula	Student, Academic Staff
	Departmental: predictive modeling, patterns of success, and/or failure	Student, Academic Staff
Academic analytics (AA)	Institutional: student profiles, achievements of scholars, information flow	Sponsors, managers, and marketers
	Provincial: contrasts between various systems	Sponsors and managers
	National and international	Government, Regulatory Agencies

2.5 *An Updated Survey on Learning Analytics and Educational Data Mining*

Data mining in education was the subject of a previous survey that was revised and refined and published in this journal in 2013 [23]. Academic analytics, institutional analytics, teaching analytics, and educational data science are some of the terms that are related to this text, which examines the application of learning analytics and educational data mining in education and highlights developments in the field over the previous ten years [23]. Institutional analytics (IA) gathers, analyzes, and visualizes data about academic program activities including courses and degree programs, research, student fee income, course evaluation, resource allocation, and management in order to create institutional insight [23].

3 Methodology

Our knowledge discovery framework, which is built on a data science model, is presented in this next section.

The theory proposed by Barneveld, Arnold, and Campbell [24], states that academic analytics focuses on management at the institutional and faculty levels, where educational data is essential to operational and financial decision-making [24]. Furthermore, academic analytics can handle system comparisons to the benefit of educational donors, administrators, authorities, and governments, claim Siemens and Long [25]. The model aids academics and staff in accreditation by enhancing data collection, analysis, confidence in reports, and decision-making through educational big data tools, ensuring accurate measurement and analysis of results.

In this study, we proposed the knowledge discovery in educational data for accreditation processes at the Higher Education Authority, to take into consideration staff alignment, course burden, and staff moonlighting in the accreditation criterion. The model is a component of the HEA-IMIS, that is utilized by the Higher Education Authority (HEA) to automate the accreditation process. As illustrated in Fig. 1, the HEA-IMIS gathers data about institutions, learning programs, educational initiatives, teaching methods, academic staff, various institutional documentation, etc. Additionally, we proposed the educational data mining with academic analytics (EDM/AA) support system model, which combines the information and data management components mentioned above [26].

The following accreditation criteria, staff alignment, course workload, and staff moonlighting are taken into consideration when implementing the suggested model in this study. The following section provides more information about the implementation.

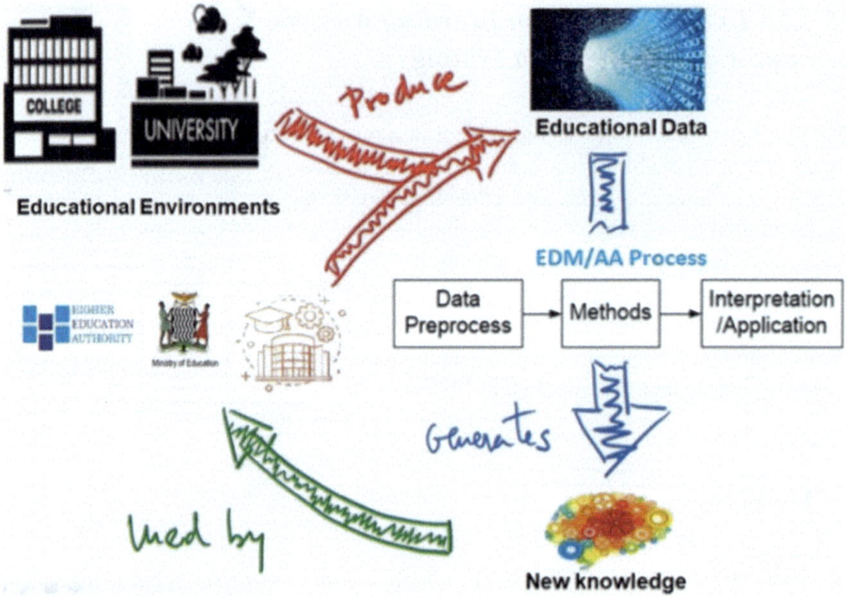

Fig. 1 Proposed knowledge discovery framework

3.1 Data Preprocess

At this phase, decisions are made regarding the data's storage location, its intended use, and whether the information gathered is relevant to the objective. The process of feature selection entails reducing the number of variables that are utilized to forecast a specific result. The objective is to make the model easier to read, to simplify it, to make algorithms more efficient at computing, and to prevent overfitting. The educational data provided by the Higher Education Institution (HEI) shall include but not be limited to bio-data, academic data, qualifications, work and academic experience, specialization, assigned courses, among other data. The information needs for the model are further defined for the data collection such as staff workload, staff alignment, and staff moonlighting to ensure the objective is achieved.

3.2 Methods

Model Design and Implementation

The study evaluated a number of techniques, including the General Linear Model, Logistic Regression, Fast Large Margin, Deep Learning, Decision Tree, Random Forests, Gradient Boosted Trees, and Support Vector Machine, to predict the appropriateness and adequacy of academic staff. Prediction accuracy was measured using the Confusion Matrix to determine the most appropriate technique to employ. The DM procedure aimed to create predictions using a predictive model and characterize behaviors using a descriptive model. Predictive models used known outcomes, while descriptive models used identified patterns in data to guide decision-making.

Statistical techniques like Logistic Regression and time series can be used when analyzing the reasons for success or failure [26]. However, when forecasting is the primary goal, methods like neural networks [27], Support Vector Machines [28], Decision Trees [29], and Random Forests [30] are more effective and produce more accurate results. Statistical methods are employed to develop models which accurately forecast the final values, while supervised optimization issues enable machine learning techniques to automatically match input data with target values. The confusion matrix indicates model effectiveness, but no classifier performs well for prediction outcomes. Therefore, it's crucial to identify the most effective classifiers for the studied data. [31].

3.3 Web Design and Implementation

A synthetic dataset and a reproduction of the Higher Education Authority's database structure are created. The Authority data is utilized as a testbed for the framework's application. MySQL is used to store data in the form of tables. MySQL Representational State Transfer (REST) Application Programming Interface (API) data services are used to publish data in JavaScript Object Notation (JSON) format. PHP is utilized to create an online interface for the integrated framework. SQL queries are employed to store and retrieve data. The proposed integrated framework is put into practice using the processes listed below:

1. Create a dataset using the data source, first.
2. Perform a dataset preprocessing.
3. Data conversion to a MySQL database.
4. Set up a RESTful API responsible for extracting processed data
5. Produce information in JSON format.
6. Create data extraction services for the integrated framework.
7. Define the information needs based on the accreditation process criteria, for this study the categories considered were staff academic alignment, academic staff course workload, and academic staff moonlighting.
8. Present documentation in accordance with the requirements for the applicable accreditation.

3.4 Interpretation/Application

The experimental phase used RapidMiner machine learning, a user-friendly data mining tool suitable for both experienced and novice data scientists. Widgets with

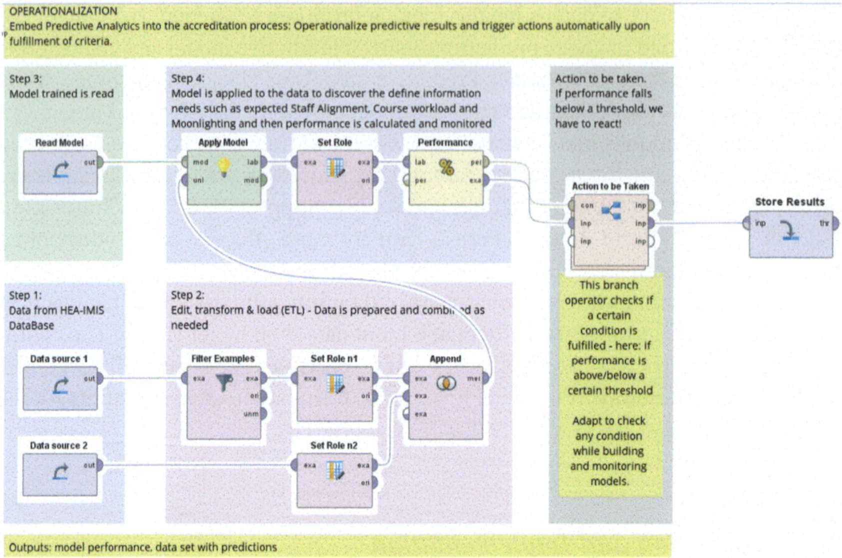

Fig. 2 Flowchart for operationalization of the predictive analytics—AA process

specialized functions like data retrieval, preprocessing, visualization, modeling, or evaluation were stacked into workflows to perform data analysis. Figure 2 illustrates how various components of a workflow were combined to create comprehensive data analysis charts (Fig. 3).

Further, a script in Python through RapidMiner software was implemented as part of the workflow, following the steps below:

1. Load your dataset containing the features and the accreditation status (target variable).
2. Define the features and target variable.
3. Split the dataset into training and testing sets.
4. Perform text vectorization using TF-IDF to convert textual features into numerical representations.
5. Initialize and train a Gradient Boosting (XGBoost) model and a Random Forest model.
6. Make predictions using both models and evaluate their performance using accuracy and a classification report.

Program Code

```
# Import necessary libraries
import pandas as pd
from sklearn.model_selection import train_test_split
from sklearn.feature_extraction.text import TfidfVectorizer
from xgboost import XGBClassifier
```

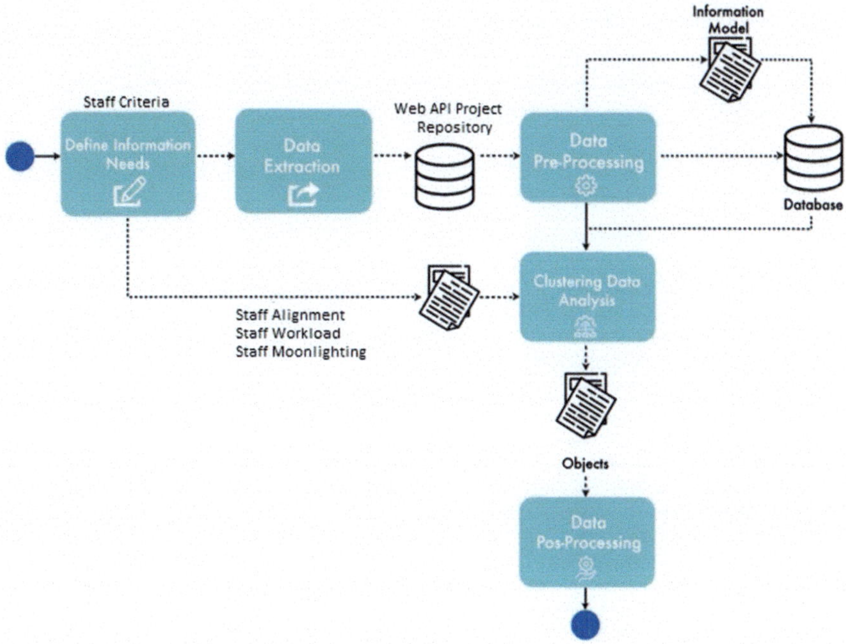

Fig. 3 Flowchart for implementation of web design

```
from sklearn.ensemble import RandomForestClassifier
from sklearn.metrics import classification_report, accuracy_score

# Load your dataset (CSV, Excel, or other formats)
data = pd.read_csv("learning_programs_data.csv")

# Define features and target variable
features = [
   "Staff Alignment
   "Staff Workload",
   "Staff Moonlighting",
]
target = "Accredited" # Assuming "Accred-
ited" is your binary accreditation status (1 for accred-
ited, 0 for not accredited)

# Split the dataset into training and testing sets
X = data[features]
y = data[target]
X_train, X_test, y_train, y_test = train_test_split(X, y, test_
size=0.2, random_state=42)

# Text vectorization using TF-IDF
tfidf_vectorizer = TfidfVectorizer(max_features=1000, stop_
words="english")
X_train_tfidf = tfidf_vectorizer.fit_transform(X_train)
```

```
X_test_tfidf = tfidf_vectorizer.transform(X_test)

# Initialize and train a Gradient Boosting (XGBoost) model
xgb_model = XGBClassifier()
xgb_model.fit(X_train_tfidf, y_train)

# Make predictions using XGBoost
y_pred_xgb = xgb_model.predict(X_test_tfidf)

# Evaluate the Random Forest model
accuracy_rf = accuracy_score(y_test, y_pred_rf)
report_rf = classification_report(y_test, y_pred_rf)

# Initialize and train a Random Forest model
rf_model = RandomForestClassifier()
rf_model.fit(X_train_tfidf, y_train)

# Make predictions using Random Forest
y_pred_rf = rf_model.predict(X_test_tfidf)

# Evaluate the Random Forest model
accuracy_rf = accuracy_score(y_test, y_pred_rf)
report_rf = classification_report(y_test, y_pred_rf)

# Print the evaluation results for both models
print("XGBoost Model:")
print("Accuracy:", accuracy_xgb)
print("\nClassification Report:\n", report_xgb)

print("\nRandom Forest Model:")
print("Accuracy:", accuracy_rf)
print("\nClassification Report:\n", report_rf)
```

Staff Academic Alignment

The International Standard Classification of Education (ISCED) framework is used to ensure that academic staff members are assigned to courses according to their specialization and qualifications from accredited higher education institutions [32]. The ISCED is a framework for putting together, gathering, and evaluating comparable statistics on education across national borders [32].

Course workload

HEA analyzes course workload based on prerequisites for teaching complete courses, determining required contact hours based on trackable workload. Full courses require a minimum of 10 contact hours and a maximum of 12 h per week. The model uses parameters like good, moderately high, or very high.

Academic Staff Moonlighting

The study examines staff moonlighting by analyzing academic staff's bio information, workloads, and matching them with higher education institutions and learning programs, comparing matches based on defined information. According to Amini-Philips [33, 34], many professors experience illness and burnout as a result of extra work or side jobs. Moonlighting in teaching can lead to stress, decreased effectiveness of instruction, and increased preoccupation among lecturers [34, 35]. Scholars who put in a lot of overtime at a second job are more likely to struggle in their first teaching role [34, 35]. As stated by Hobbs and Stutz [34, 36], academic staff believe moonlighting negatively impacts instruction quality; this model helps accrediting process evaluators understand teaching job numbers.

4 Results

The dataset for this study was extracted from MySQL Database using PL/SQL [37]. The extracted raw data consisted of 1814 observations and 18 features [37]. In order to investigate the raw data and extract valuable insights from the dataset, we used a variety of EDA techniques. We deleted the category characteristics and reduced the feature list to nine by using several Python functions, while also taking into mind the crucial features required for clustering.

4.1 Model Evaluation

As shown in Table 2, Random Forest, an ensemble learning method, demonstrated strong performance with an accuracy of 95%. It maintained a high precision rate of 90%, indicating a low rate of false positives. The AUC of 0.961 suggests a good ability to discriminate between classes, and the F1-score of 95% reflects a balanced trade-off between precision and recall. Gradient Boosting, another ensemble learning approach, exhibited similar impressive results. It achieved an accuracy of 95% and maintained a precision of 90%. The AUC of 0.961 suggests strong overall predictive performance, and the F1-score of 95% indicates a favorable balance between precision and recall.

Random Forest and Gradient Boosted Trees algorithms have demonstrated high-quality performance on the given task, making them strong candidates for practical applications in similar scenarios.

4.2 Confusion Matrix

The confusion matrix results provide insights into the models' performance, specifically in terms of recall for both classes ("True Yes" and "True No"). The Random Forest achieved a perfect 100% recall for the "True Yes" class and a high 95% recall for the "True No" class. It exhibited zero false positives and false negatives, indicating excellent performance. The Gradient Boosting model also excelled with 100% recall for the "True Yes" class but had a slightly lower recall (90.62%) for the "True No"

Table 2 Summary of model evaluation results

Model	Accuracy	Classification Error	AUC	Precision	Recall	F-Measure	Sensitivity	Specificity
Generalized Linear Model	68%	32%	0.759	77%	39%	51%	39%	91%
Logistic Regression	56%	44%	0.43		0%		0%	100%
Fast Large Margin	70%	30%	0.817	89.30%	38%	53%	38%	97%
Deep Learning	79%	21%	0.905	90%	60%	72%	60%	95%
Decision Tree	56%	44%	0.5		0%		0%	100%
Random Forest	95%	4.60%	0.965	90%	100%	95%	100%	92%
Gradient Boosted Trees	95%	4.60%	0.961	90%	100%	95%	100%	92%
Support Vector Machine	56%	44%	0.5		0%		0%	100%

Table 3 Random Forest confusion matrix results

	True no	True yes	Class precision
Pred. no	263	0	100.00%
Pred. yes	24	232	90.62%
Class recall	91.64%	100.00%	

Table 4 Gradient Boosted tree confusion matrix results

	True no	True yes	Class precision
Pred. no	263	0	100.00%
Pred. yes	24	232	90.62%
Class recall	91.64%	100.00%	

class. It did not produce any false positives but did have a small percentage (9.38%) of false negatives in the "True No" class.

These results shown in Tables 3 and 4, demonstrate the models' ability to correctly identify instances of criteria required for the provided information needs: Staff Alignment, Course workload, and Moonlighting, with the Random Forest achieving a slightly higher recall for the "True No" class. The choice of the best model may depend on the specific requirements and constraints of the task, such as the tolerance for false negatives in the "True No" class. In the instance of True Yes, the learning program satisfied the criteria required for the provided information needs: Staff Alignment, Course workload, and Moonlighting.

4.3 ROC Curve

The ROC curve was utilized to assess the model's performance at different classification levels, plotting the true positive rate or recall against the false positive rate, thereby evaluating its effectiveness. The ROC curve that our model produced is displayed (Fig. 4).

The ROC curve for the Random Forest algorithm achieved an AUC of 0.961. This high AUC value indicates the model's strong ability to satisfy the criteria required for the provided information needs, with a significant area under the curve. The ROC curve for the Gradient Boosting algorithm also yielded an AUC of 0.961, demonstrating its excellent discriminative power in the binary classification task. Random Forest and Gradient Boosting algorithms indicate that both models have strong discriminative performance. They exhibit similar capabilities in terms of distinguishing between positive and negative instances, as evidenced by the high AUC score. These results suggest that both Random Forest and Gradient Boosting models are well-suited for the tasks of decision support for the accreditation process.

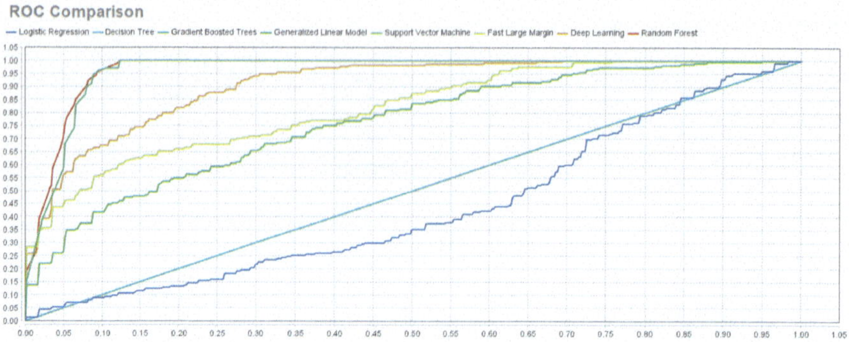

Fig. 4 ROC comparison

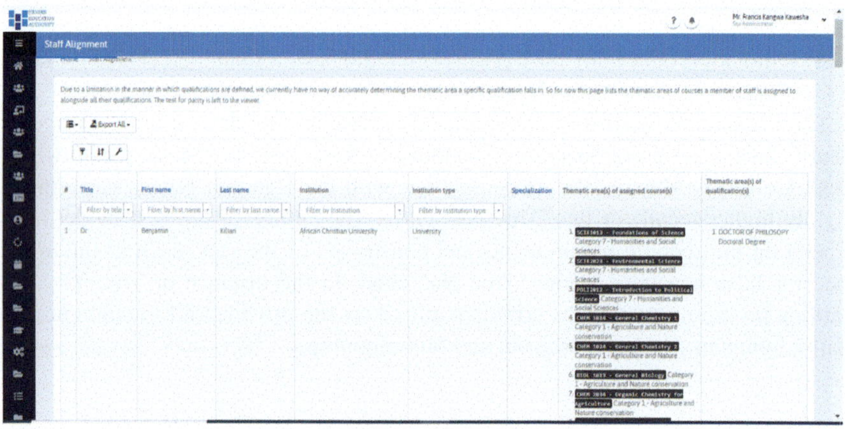

Fig. 5 Web visualization of academic staff alignment results

4.4 Web Implementation Results

Figure 5 shows the web implementation results for staff alignment data, allowing visualization of academic staff thematic areas, qualifications, specializations, and their appropriateness for courses, guiding evaluators, and submitters on the appropriateness of academic staff.

5 Discussion and Conclusion

The research successfully demonstrated the feasibility of using data mining and machine learning techniques to predict the accreditation outcomes of learning programs. This is a valuable application of technology in the educational sector.

It analyzes data to determine if an academic learning program meets accreditation requirements from the Higher Education Authority. The selection of Random Forest and Gradient Boosting as the ideal algorithms for the task is a crucial finding in the research. These algorithms are well-suited for complex classification tasks, offering high predictive accuracy and robustness against overfitting. Random Forest is known for its ability to handle high-dimensional data and feature selection, while Gradient Boosting excels in boosting the predictive performance by iteratively improving the model's accuracy. The research demonstrated that both Random Forest and Gradient Boosting models outperformed other machine learning algorithms in predicting the outcomes of accreditation for learning programs.

While Random Forest and Gradient Boosting are powerful algorithms, their effectiveness is highly dependent on the quality and quantity of the training data. Garbage in, garbage out—data quality remains a critical consideration. The complexity of these algorithms may present challenges in terms of interpretability. Ensuring that decision-makers can understand and trust the model's output is crucial.

Further, this paper suggests that qualifications, specialization areas, course allocations, workloads, and side jobs are crucial predictors for academic personnel's suitability in the accreditation process at the Higher Education Authority. In light of this, the researcher suggests that future studies take into account the additional HEIs that the HEA currently oversees as a result of the Amendment to the Higher Education Act No. 4. of 2013. Further, the methods can be developed for all the other accreditation criteria such as introduction, rationale, teaching and learning plan, curriculum, internal quality assurance, financial resources and regulations and should expand the generalizability of the models to different accreditation systems, and continually monitoring and improving the accuracy and fairness of these data-driven decision support systems in the context of educational accreditation.

Acknowledgements We thank the higher education institutions in Zambia that took part in the study as well as the Higher Education Authority in Zambia's management for their assistance with the research.

References

1. Antwi J et al (2021) Global accreditation practices for accelerated medically trained clinicians: a view of five countries. Hum Resour Health 19(1). https://doi.org/10.1186/s12960-021-00646-4
2. Braithwaite J et al (2021) Strengthening organizational performance through accreditation research-a framework for twelve interrelated studies: the ACCREDIT project study protocol. BMC Res Notes 4(1). https://doi.org/10.1186/1756-0500-4-390
3. Ahmad MI, Widianto E, Sari RP, Dewi CK, Bendriyanti RP (2021) Online accreditation assessment with SISPENA: survey on PKBM assessors in East Java. Adv Soc Sci Educ Humanit Res. https://doi.org/10.2991/assehr.k.211210.043
4. Drumm S, Moriarty F, Rouse MJ, Croke D, Bradley C (2020) The development of an accreditation framework for continuing education activities for pharmacists. Pharmacy 8(2):75. https://doi.org/10.3390/pharmacy8020075

5. Al-Jaghoub S, Al-Adwan A, Al-Yaseen H, Al-Soud A, Areiqat A (2019) Challenges of improving effectiveness and efficiency of the higher educational system in developing countries. Prob Perspect Manage 17(1):19–31. https://doi.org/10.21511/ppm.17(1).2019.03
6. Ibrahim F, Susanto H, Haghi PK, Setiana D (2020) Shifting paradigm of education landscape in time of the COVID-19 pandemic: revealing of a digital education management information system. Appl Syst Innovation 3(4):49. https://doi.org/10.3390/asi3040049
7. Vasilyeva ER, Nurutdinova A (2018) The academic model of managing integration processes: study case of the multicultural educational space. In: SHS web of conferences, Jan 2018. https://doi.org/10.1051/shsconf/20185001223
8. Tawafak RM, Romli A, Malik SI, Shakir M, Alfarsi GM (2019) A systematic review of personalized learning: comparison between e-learning and learning by coursework program in Oman. Int J Emerg Technol Learning (IJET) 14(09):93
9. Valdivia A, Sánchez-Monedero J, Casillas J (2021) How fair can we go in machine learning? Assessing the boundaries of accuracy and fairness. Int J Intell Syst 36(4):1619–1643. https://doi.org/10.1002/int.22354
10. Markovic-Petrovic G, Vukovic M, Jovic-Vranes A (2018) The impact of accreditation on health care quality in hospitals. Vojnosanit Pregl 75(8):803–808. https://doi.org/10.2298/vsp160728390m
11. Kőmíves PM, Pilishegyi P, Novák N, Nagy AS, Körösparti P (2019) The role of the higher education in the development of the agriculture. Int J Inf Educ Technol 9(9):607–612
12. Crasta LC, Shailashri VT (2023) A systematic review on the employability prediction model for the management students. Int J Case Stud Bus IT Educ 1–25
13. Truong T-C, Diep QB (2023) Technological spotlights of digital transformation in tertiary education. IEEE Access 11:40954–40966
14. Nor NAM, Mohamed A, Mutalib S (2020) Prevalence of hypertension: predictive analytics review. IAES Int J Artif Intell (IJ-AI) 9(4):576
15. Council for Higher Education Accreditation (CHEA) (2010) The Value of Accreditation, Washington, DC
16. Hussain M, Al-Mourad M, Mathew S, Hussein A (2016) Mining educational data for academic accreditation: aligning assessment with outcomes. Glob J Flex Syst Manag 18(1):51–60
17. Scales K, Owen C, Shiohare S, Leonard M (1998) Preparing for program accreditation review under ABET Engineering Criteria 2000: choosing outcome indicators. J Eng Educ 87(3):207–210
18. Gubbi J, Buyya R, Marusic S, Palaniswami M (2013) Internet of Things (IoT): a vision, architectural elements, and future directions. Futur Gener Comput Syst 29(7):1645–1660. https://doi.org/10.1016/j.future.2013.01.01
19. Reed DA, Dongarra J (2015) Exascale computing and big data. Commun ACM 58(7):56–68
20. Ali Aljarallah N, Kumar Dutta A (2022) Developing a quality automation framework to assess specifications for academic accreditation in Saudi Arabian universities. TEM J 667–674. https://doi.org/10.18421/tem112-21
21. Nguyen A, Gardner L, Sheridan D (2020) Data analytics in higher education: an integrated view. J Inf Syst Educ 31(1):61
22. Romero C, Ventura S (2020) Educational data mining and learning analytics: an updated survey. WIREs Data Min Knowl Disc 10(3). https://doi.org/10.1002/widm.1355
23. Van Barneveld A, Arnold KE, Campbell JP (2012) Analytics in higher education: Establishing a common language. Educause Learn Initiative 1(1):1–11
24. Siemens G, Long P (2011) Penetrating the fog: analytics in learning and education. Educause Rev 46(5):30
25. Arias Ortiz E, Dehon C (2013) Roads to success in the Belgian French community's higher education system: predictors of dropout and degree completion at the Université Libre de Bruxelles. Res High Educ 54:693–723
26. Yağci M (2022) Educational data mining: prediction of students' academic performance using machine learning algorithms. Smart Learn Environ 9(1)

27. Huang S, Fang N (2013) Predicting student academic performance in an engineering dynamics course: a comparison of four types of predictive mathematical models. Comput Educ 61:133–145
28. Thammasiri D, Delen D, Meesad P, Kasap N (2014) A critical assessment of imbalanced class distribution problem: the case of predicting freshmen student attrition. Expert Syst Appl 41(2):321–330
29. Delen D (2010) A comparative analysis of machine learning techniques for student retention management. Decis Support Syst 49(4):498–506
30. Asif R, Merceron A, Ali SA, Haider NG (2017) Analyzing undergraduate students' performance using educational data mining. Comput Educ 113:177–194
31. UNESCO Institute for Statistics (2012) International standard classification of education: ISCED 2011. Comp Soc Res 30
32. Amini-Philips C (2019) Moonlighting activities and lecturers' wellbeing in Nigerian universities. Adv Soc Sci Res J
33. Sakyi EK, Agomor KS (2020) Moonlighting in Ghana's higher education institutions: exploring lecturers' experiences at the Ghana Institute of Management and Public Administration (GIMPA). J Appl Res High Edu 13(1):180–194
34. Champion S (2010) Increased accountability', teachers' effort and moonlighting. University Graduate School of Business, Stanford
35. Living 'paycheck to paycheck' these teachers moonlight to make ends meet. Dallas News (2016). https://www.dallasnews.com/news/education/2016/09/21/living-paycheck-to-paycheck-these-teachers-moonlight-to-make-ends-meet/. Accessed 28 May 2023
36. Kasimba P (2020) Quality assurance in Zambian higher education: a new dawn. Int J Sci Res (IJSR) 9–10
37. Higher education authority, data from: higher education authority—integrated management information system. Digital Repository. Accessed 9 Jan 2023

Design of a Blockchain-Based Organ Donation and Transplantation Management Framework

Murat Osmanoglu, Murat Karakus, Bulent Tugrul, and Salih Demir

Abstract Several requirements and obstacles are involved in organ donation and transplantation, including registration, matching donors with recipients, organ removal, organ delivery, and transplantation. Several legal, clinical, ethical, and technical barriers prevent the desired results. Moreover, immutability and transparency are two key requirements in such a system to maintain sustainability and trust among potential donors and patients. Therefore, a comprehensive organ donation and transplantation framework is essential to ensure a fair, efficient, and confident process. A blockchain-based organ donation and transplantation framework, with important qualities such as immutability, transparency, decentralization, and anonymity, would solve the above-mentioned challenges and concerns of present organ donation and transplantation systems. These characteristics can facilitate trust, security, and privacy in a blockchain-based donation and transplantation system, restoring trust among stakeholders. Although blockchain technology has the capability to provide an efficient and trustworthy remedy to the aforementioned challenges, it is critical to thoroughly and diligently identify and analyze all of the modules and requirements of such a framework to construct a trustable organ donation and transplantation management framework for the key stakeholders. In this regard, we envision the modules and requirements design of a blockchain-based organ donation and transplantation system using the SysML modeling language. The suggested design ensures immutability, decentralization, transparency, trust, security, and privacy among system actors.

M. Osmanoglu (✉) · B. Tugrul
Department of Computer Engineering, Ankara University, Ankara, Turkey
e-mail: mosmanoglu@ankara.edu.tr

B. Tugrul
e-mail: btugrul@eng.ankara.edu.tr

M. Karakus
Department of Software Engineering, Ankara University, Ankara, Turkey
e-mail: mrtkarakus@ankara.edu.tr

S. Demir
Faculty of Open and Distance Education, Ankara University, Ankara, Turkey
e-mail: sdemir@ankara.edu.tr

X.-S. Yang et al. (eds.), *Proceedings of Ninth International Congress on Information and Communication Technology*, Lecture Notes in Networks and Systems 1000,
https://doi.org/10.1007/978-981-97-3289-0_13

Keywords Blockchain · Organ donation · Organ transplantation · SysML

1 Introduction

Organ transplantation has been one of the most significant developments in modern medicine. It is the only chance of survival for patients in the final stages of their disease. Despite advances in the transplant process and increased awareness among people, hundreds of people die each year from the failure of a vital organ while waiting for a suitable organ [20]. As a result of the adverse conduct experienced in some countries in recent years, it has been difficult for new initiatives to increase organ donation to meet their objectives. Organ donation numbers are falling rapidly due to the abuse of such authorities. Therefore, establishing trust between organ donors and patients is crucial [7]. In addition, current practices in health systems in many countries lack transparency. The lack of transparency regarding the organ transplantation process and criteria leads to an increase in public suspicion about the fairness and equity of the process. Transparency in transplantation waiting lists should allow patients to follow the procedures and activities more closely while encouraging others to trust the system and consider organ donation. In this context, a blockchain-based organ donation and transplantation framework would remedy the problems and concerns of the current organ donation and transplantation schemes employing its key features, such as immutability, transparency, decentralization, and anonymity. These features can enable trust, security, and privacy in a blockchain-based donation and transplantation framework, restoring confidence among all parties.

Blockchain is an innovative technology that ensures the transparency and accountability of data [18]. Health and finance data are considered sensitive data within the scope of many countries. A particular law or legislation protects the storage and sharing of such data [23]. On the other hand, a blockchain requires all data to be stored on the blockchain so other participants can see what data has been stored by others in the network. Medical data must be stored with cryptologically secure methods to prevent a breach of data privacy [9, 12]. Performing analyses or organ-matching functions on encrypted texts require the development of new algorithms. However, there are solutions where sensitive patient data is not stored in the chain, and only hashes are stored [13]. In addition to complying with the law, storing only hash values in the chain will not lead to security flaws. Consequently, blockchain-based solutions require evaluating certain features to ensure compliance with legal requirements [8, 16].

Although blockchain technology offers an effective and reliable potential solution to the above-mentioned problems, it is vital to carefully and meticulously define and analyze all the components and requirements of such a framework to establish a reliable organ donation and transplantation management framework for the relevant parties [1, 2, 10, 15]. *SysML* is a graphical modeling language that describes the essential features of a system design, such as structure, behavior, and requirements, using various diagrams and constructs [21]. *SysML* orchestrates the analysis, specification, design, validation, and verification of complex systems such as hardware, software, data, personnel, procedures, and facilities. There are nine types of diagrams in *SysML* [5]. Block definition, package, internal block, parametric, and requirement diagrams specify the structural aspects of the system. In addition, behavioral aspects are presented in use case, sequence, state machine, and activity diagrams [6]. By using these diagrams, the components of the system can be viewed from different perspectives.

In this study, we present the design of a blockchain-based organ donation and transplantation framework concerning components and requirements using *SysML* modeling language. The proposed design provides immutability, decentralization, and transparency and enables trust, security, and privacy among the participants in the framework. In the rest of the paper, we give an overview of blockchain technology to make the paper self-contained in Sect. 2. We give a high-level design of components needed in the proposed system in Sect. 3. In Sect. 4, we discuss the requirements necessary to maintain the proposed management framework before concluding in Sect. 5.

2 Overview of Blockchain Technology

An early motivation for the development of blockchain was to address the problem of double spending in electronic cash systems. It also eliminated the need for a trusted third party to store transaction records, which was needed in the traditional financial system [11, 14]. Blockchain applications have expanded beyond cryptocurrency since its announcement. However, it is likely to remain the most widespread use case for a while. The popularity of blockchain is due to its features, such as security, anonymity, consensus, decentralization, immutability, and transparency. Records stored on a blockchain are individually encrypted to ensure security and anonymity. A blockchain network ensures that all participants in the network agree on the validity of each record, thereby ensuring the fairness of the results and the trustworthiness of the network [24].

A blockchain usually stores records of transactions performed by valid users within a system. When a transaction occurs, it is stored in distributed databases by following certain steps: (i) a block representing a certain number of transactions is generated and distributed to each participating node in the network, (ii) the nodes independently verify the transactions by running a consensus algorithm, (iii) some nodes receive a *reward* for their activities for the consensus algorithm, and (iv) the newly generated block is appended to the end of the existing blockchain.

As a peer-to-peer network, blockchain eliminates the need for a trusted third party. Thus, it solves the well-known single-point-of-failure problem associated with centralized systems controlled and managed by third parties. Often, in systems with multiple equal participants, establishing trust is a challenge to reach an agreement on some solutions between all participants. Blockchains run on a well-defined consensus algorithm to ensure trust and eliminate the need for external authorities.

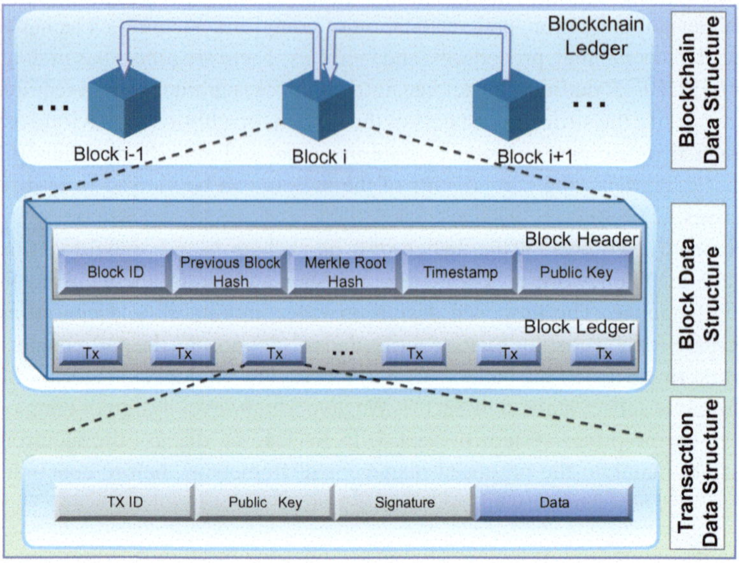

Fig. 1 A pictorial representation of general blockchain, block, and transaction data structures

Nodes communicate and make decisions in accordance with a distributed consensus protocol. Consensus mechanisms affect many aspects of blockchain performance, including flexibility, scalability, and durability [19].

Various consensus protocols provide different properties depending on the use case and requirements [3, 22]. One of the most common consensus mechanisms used in negotiations is proof-of-work (PoW). The PoW protocol requires nodes to solve a complex mathematical puzzle to validate a new block. In addition, this makes it hard for malicious actors to attack the network since it requires a lot of computing power. Another well-known consensus method is proof-of-stake (PoS), which does not require mining and determines block publishing by participant share [17].

A high-level representation of blockchain, block, and transaction data structures is shown in Fig. 1. Whenever a block is published on a blockchain, each user checks it to ensure no invalid transactions are contained. Additionally, the cryptographic hash of the previous block is verified and included in the current block. Therefore, all blocks are validated up to the first block, called Genesis. All previous blocks are now immutable and cannot be modified, so the validation process is no longer required. As a result, the blockchain and the data it stores are secure, and integrity is ensured.

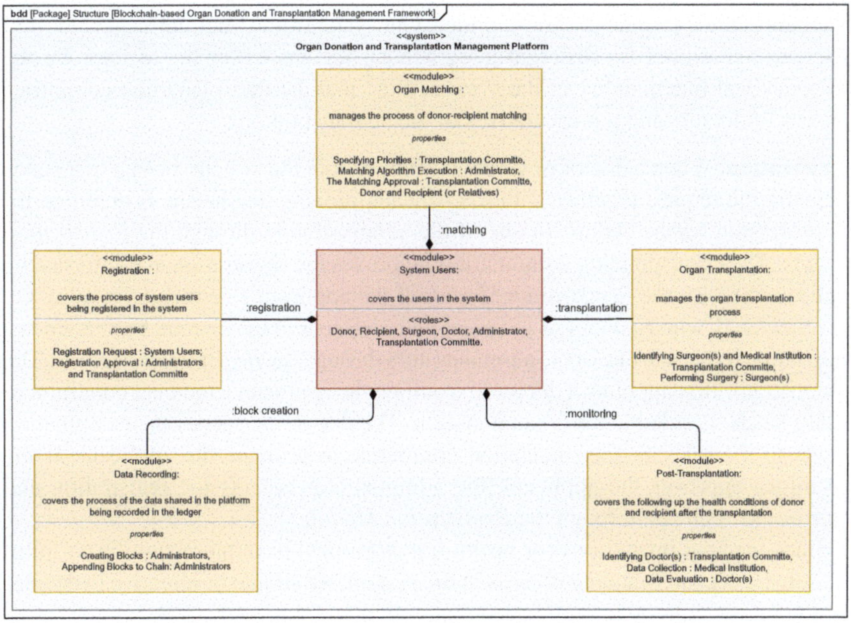

Fig. 2 An illustration of organ donation and transplantation management framework

3 Organ Donation and Transplantation Management Framework

This section gives a high-level design of components needed in the proposed blockchain-based organ donation system using the *SysML* modeling language. In our framework, each component is represented by a module, and arrows represent relationships among different modules. As shown in Fig. 2, our framework consists of six modules: *Registration, Users, Organ Matching, Organ Transplantation, Post-Transplantation,* and *Data Recording*.

Organ donation, in particular, comprises life-critical systems, and users in this system may adversely affect the quality of health care through their decisions; even they may put human life in danger. Thus, users in our system must be health professionals with well-established identities. In this sense, users must get a certificate from a relevant medical institution or governmental organization showing that they are qualified for the roles they claim to register in the system. Our model has seven different types of users: donor, recipient, surgeon, doctor, administrator, and transplantation committee.

Administrators are the officers who represent the medical institutions (or health organizations) related to the organ donation system, such as hospitals, health directorates, and health associates. Administrators manage the registration process and maintain the platform in our system. They are also responsible for designating the members of the transplantation committee. Our system allows administrators to form different committees for different geographical regions or organs to improve the efficiency and functionality of the system. Note that the transplantation committee consists of doctors and representatives of medical institutions.

Registration. When a doctor or surgeon wants to join the system, he/she first sends a registration request together with a certificate, proving that he/she is qualified for the profession he/she claims, to one of the administrators through the *Registration Module*. The corresponding administrator approves the registration and informs the applicant through the *Registration Module* if the applicant's certificate is valid. On the other hand, when a donor or recipient demands to join the system, he/she sends a registration request to one of the administrators through the *Registration Module*. The administrator first appoints a doctor to examine the applicant's medical condition to ensure he/she is suitable for organ donation. The doctor then sends the examination results to the relevant transplantation committee to evaluate the applicant. If the committee approves the applicant, the administrator accepts the registration and informs the applicant through the *Registration Module*.

Since medical institutions (or health organizations) have more significant influence on the decision-making process in the system, we should be attentive to designing the registration process of a new institution to the system. When a new institution wants to enroll in the system, it first submits a registration request to one of the administrators via the *Registration Module*. Unlike the registration process of other users, the administrator calls for the other administrators' opinions on the applicant's admission by initiating a voting scheme through the *Registration Module*. If most administrators in the voting pool accept the registration, the corresponding administrator approves it and informs the applicant through the *Registration Module*.

Organ Matching (or Donor-Recipient Matching). After a transplantation committee is formed, it specifies the priorities the matching algorithm utilizes through the *Organ-Matching Module*. When a patient (or recipient) or donor is registered in the system with examination records indicating the patient or donor is eligible for organ donation, the administrators execute the matching algorithm on the records to find a fitting match based on the priorities specified by the corresponding transplantation committee through the *Organ-Matching Module*. If the algorithm finds a match, the required information about the matching donor and recipient will be sent to the corresponding transplantation committee to review the match through *Organ-Matching Module*. If the committee approves the match, the system will ask for the consent of the matching donor and recipient. In the organ-matching process, when the committee, the donor, or the recipient does not approve the match, the system re-executes the matching algorithm to find another match. Otherwise, it finalizes the organ-matching process for both the donor and recipient through *Organ-Matching Module*.

Organ Transplantation. After the organ-matching process is completed with a matching donor and recipient, the transplantation committee specifies a surgeon qualified to perform the corresponding surgery and a medical institution suited for the implementation of the surgery via *Organ Transplantation Module*. Once the surgeon and the medical institution are informed about the surgery, they execute the surgery and report the surgery records to the platform through *Organ Transplantation Module*.

Post-Transplantation. After the transplantation, the transplantation committee specifies some eligible doctors to monitor the donor and recipient's health conditions. In addition to being used to examine both the donor's and recipient's health, data collected through this process also allows the priorities utilized by the matching algorithm to be updated, thus improving the matching algorithm.

Data Recording. In our system, time is considered as divided into time segments called slots. In each time slot, a leader is selected among the administrators registered in the system through *Data Recording Module*, according to the consensus mechanism the administrators have agreed on. Each leader creates a new block consisting of all the data shared on the platform during its time slot and broadcasts it to other users. After being voted by the administrators through a consensus mechanism, the users on the platform add the new block to the chain.

4 System Requirements

This section analyzes the requirements to properly and securely maintain a blockchain-based organ donation system. The relationship between system requirements and modules is demonstrated in Fig. 3.

As stated in Sect. 3, the organ donation system comprises life-critical processes. It is, therefore, required that users of this system be health professionals. During registration, they must provide evidence of their qualifications for the roles they claim. During registration, qualified health staff should examine the donor and recipient to determine if they are suitable for organ donation. To manage a reliable donor-recipient matching process, the relevant health professionals should verify the examination results of both the donor and recipient. On the other hand, revealing the examination results to unintended parties, especially criminals involved in organ trafficking, may put both the donor's and recipient's lives in danger. To avoid such cases, the data of both donors and recipients should be securely shared and recorded in the system.

One of the most critical organ donation and transplantation system components is the donor-recipient matching procedure [4]. This component should be carefully shaped for a more reliable and effective system. To manage the organ-matching procedure correctly and efficiently, the priorities utilized by the algorithm should be specified and provided to the platform by the relevant health experts. Also, the matching procedure should be operated transparently so that all related parties are willing to participate in the process. Besides not leading to severe complications for donor and recipient, the results of the matching algorithm should be validated by qualified health professionals.

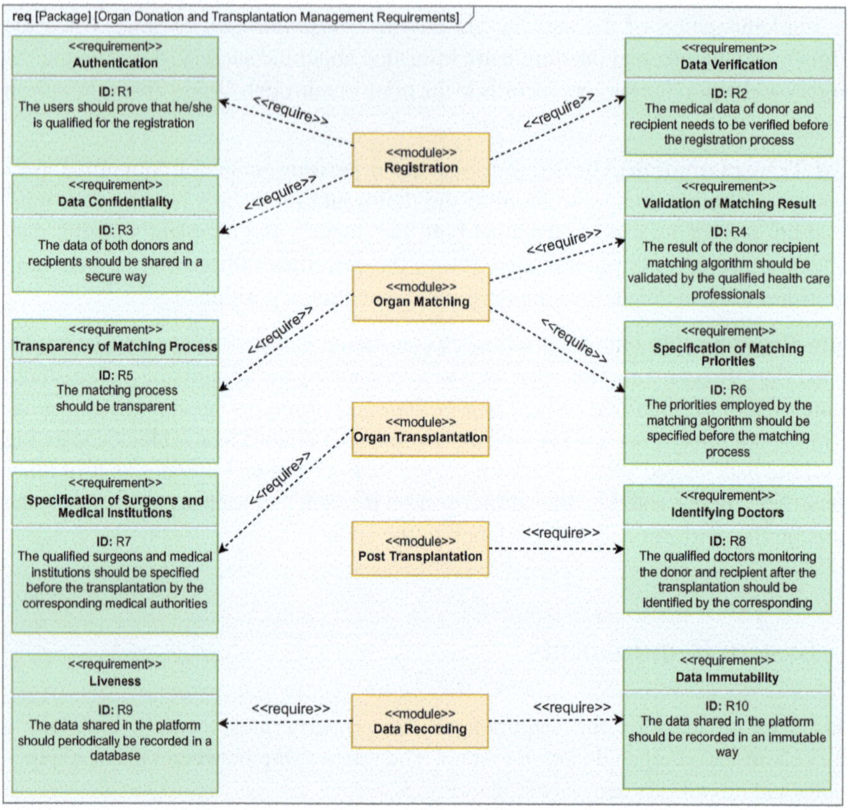

Fig. 3 Requirements to properly and securely maintain a blockchain-based organ donation and transplantation management system

After the organ-matching procedure is finalized with a matching donor and recipient, the corresponding medical authorities must specify some qualified surgeons and medical institutions where the surgery is executed. Besides, to monitor the donor and recipient after transplantation and to collect data that will be used to improve the matching procedure, some eligible health personnel should be specified by the relevant authorities.

Finally, the data shared on the platform should periodically be recorded in a database to keep the system functional. Also, the data should be stored in an immutable way to avoid some malicious users altering it to manipulate the matching process.

5 Conclusions

This study presented a trustworthy and transparent framework for blockchain-based organ donation and transplantation management. Organ donation and transplantation organizations face significant challenges, which has led to the development of this framework, which utilizes blockchain technology's inherent strengths to address these challenges. We designed the framework to provide a safe and efficient platform for all organ donation and transplantation stakeholders that minimizes the risk of data manipulation, fraud, and errors while enhancing data integrity and transparency. The framework's necessary components and their relations to each other were demonstrated using SysML diagrams.

A successful implementation of such a framework, however, requires overcoming various technical, regulatory, and ethical challenges. Creating standardized data formats, ensuring interoperability with existing healthcare systems, and conforming to privacy regulations such as GDPR and HIPAA are necessary. Furthermore, addressing ethical concerns about consent management and data access is crucial to ensuring the framework's success and adoption. Finally, this framework represents an important step toward a more efficient, secure, and transparent organ transplantation system. Further research, pilot implementations, and collaboration among healthcare organizations, regulatory bodies, and technology experts will be essential to bring this vision to fruition.

References

1. Anselmo A, Materazzo M, Di Lorenzo N, Sensi B, Riccetti C, Lonardo MT, Pellicciaro M, D'Amico F, Siragusa L, Tisone G (2023) Implementation of blockchain technology could increase equity and transparency in organ transplantation: a narrative review of an emergent tool. Transp Int 36:10800
2. Attaran M (2022) Blockchain technology in healthcare: challenges and opportunities. Int J Healthc Manage 15(1):70–83
3. Bodkhe U, Mehta D, Tanwar S, Bhattacharya P, Singh PK, Hong WC (2020) A survey on decentralized consensus mechanisms for cyber physical systems. IEEE Access 8:54371–54401
4. Briceño J, Ayllón MD, Ciria R (2020) Machine-learning algorithms for predicting results in liver transplantation: the problem of donor-recipient matching. Curr Opin Organ Transp 25(4):406–411
5. Delligatti L (2013) SysML distilled: a brief guide to the systems modeling language. Addison-Wesley
6. Elbüz A, Osmanoglu M, Tanriöver Ö (2024) A SysML-based approach for designing an ideal blockchain-based data trading platform. Gazi Üniversitesi Mühendislik Mimarlik Fakültesi Dergisi 39(1):509–520
7. Etheredge HR (2021) Assessing global organ donation policies: opt-in vs opt-out. Risk Manage Healthc Policy 14:1985–1998
8. Feng Q, He D, Zeadally S, Khan MK, Kumar N (2019) A survey on privacy protection in blockchain system. J Netw Comput Appl 126:45–58
9. Ghazal TM, Hasan MK, Abdullah SNHS, Bakar KAA, Al Hamadi H (2022) Private blockchain-based encryption framework using computational intelligence approach. Egyptian Inform J 23(4):69–75

10. Hawashin D, Jayaraman R, Salah K, Yaqoob I, Simsekler MCE, Ellahham S (2022) Blockchain-based management for organ donation and transplantation. IEEE Access 10:59013–59025
11. Judmayer A, Stifter N, Krombholz K, Weippl E (2017) Blocks and chains: introduction to bitcoin, cryptocurrencies, and their consensus mechanisms. Synth Lect Inform Secur Priv Trust 9(1):1–123
12. Kang P, Yang W, Zheng J (2022) Blockchain private file storage-sharing method based on ipfs. Sensors 22(14):5100
13. Lima C (2018) Blockchain-gdpr privacy by design how decentralized blockchain internet will comply with gdpr data privacy. IEEE Blockchain Group. https://blockchain.ieee.org/images/files/pdf/blockchain-gdpr-privacy-by-design.pdf
14. Nakamoto S (2008) Bitcoin: a peer-to-peer electronic cash system. Decentr Bus Rev
15. Niyigena C, Seol S, Lenskiy A (2020) Survey on organ allocation algorithms and blockchain-based systems for organ donation and transplantation. In: International conference on Information and Communication Technology Convergence (ICTC), pp 173–178
16. Osmanoglu M, Selcuk AA (2022) Privacy in blockchain systems. Turkish J Electr Eng Comput Sci 30(2):344–360
17. Singh A, Kumar G, Saha R, Conti M, Alazab M, Thomas R (2022) A survey and taxonomy of consensus protocols for blockchains. J Syst Archit 127:102503
18. Tariq U, Ibrahim A, Ahmad T, Bouteraa Y, Elmogy A (2019) Blockchain in internet-of-things: a necessity framework for security, reliability, transparency, immutability and liability. IET Commun 13(19):3187–3192
19. Turner SW, Karakus M, Guler E, Uludag S (2023) A promising integration of sdn and blockchain for iot networks: a survey. IEEE Access 11:29800–29822
20. Vanholder R, Domínguez-Gil B, Busic M, Cortez-Pinto H, Craig JC, Jager KJ, Mahillo B, Stel VS, Valentin MO, Zoccali C et al (2021) Organ donation and transplantation: a multi-stakeholder call to action. Nat Rev Nephrol 17(8):554–568
21. Wolny S, Mazak A, Carpella C, Geist V, Wimmer M (2020) Thirteen years of sysml: a systematic mapping study. Softw Syst Model 19(1):111–169
22. Xu J, Wang C, Jia X (2023) A survey of blockchain consensus protocols. ACM Comput Surv 55(13s)
23. Yaqoob I, Salah K, Jayaraman R, Al-Hammadi Y (2022) Blockchain for healthcare data management: opportunities, challenges, and future recommendations. Neural Comput Appl 34(14):11475–11490
24. Zheng Z, Xie S, Dai HN, Chen X, Wang H (2018) Blockchain challenges and opportunities: a survey. Int J Web Grid Serv 14(4):352–375

Audio Style Transfer Using a Modified Lightweight TimbreTron Pipeline

Alaa Elkouni, Mohamed ElMohandes, Imran Zualkernan, and Ali Reza Sajun

Abstract Neural networks have gained extensive usage in various tasks such as classification, object detection, and style transfer, particularly with images and text. Audio style transfer has made progress in several areas, but there is still room for development. This paper presents a lightweight pipeline to solve the audio style transfer problem, modifying the TimbreTron style transfer such that the reduced computational demand allows deployment on edge hardware. Notably, it is difficult to apply common methodologies like accuracy, recall, or precision monitoring while reviewing the findings. Instead, evaluation depends heavily on personal preferences as it is based on subjective auditory perception. The technique that was used to overcome the aforementioned difficulties is examined in the paper. In a survey that was done, most respondents, 54.5%, claimed that the audio quality was identical when switching from Classical to Jazz and back to Classical. Additionally, 36.4% of the participants said that the audio similarity was slightly noticeable.

Keywords TimbreTron · Audio style transfer · CycleGANs · Classical · Jazz · Mel-spectrogram

A. Elkouni · M. ElMohandes · I. Zualkernan · A. R. Sajun (✉)
American University of Sharjah, Sharjah, United Arab Emirates
e-mail: b00068908@aus.edu

A. Elkouni
e-mail: g00084191@aus.edu

M. ElMohandes
e-mail: b00083108@aus.edu

I. Zualkernan
e-mail: izualkernan@aus.edu

X.-S. Yang et al. (eds.), *Proceedings of Ninth International Congress on Information and Communication Technology*, Lecture Notes in Networks and Systems 1000,
https://doi.org/10.1007/978-981-97-3289-0_14

1 Introduction

A developing area of research, audio style transfer demands more improvement and development. The importance of addressing this problem stems from its wide range of significant applications. Nowadays, as seen in Copet et al., audio can be generated using text and the results are impressive [1]. Although the concept of style transfer is not new, there have been significant advancements in the field of image style transfer. Notably, using neural networks has made it easier for images to change and take on the properties of various visual aspects. For example, an image of a person can be changed to a cartoon character or transform a landscape to a painting as done by Gatys et al. where they used convolutional neural networks (CNNs) [2]. The use of advanced architectures, notably generative adversarial networks (GANs), allowed for an improvement in the results for image style transfer as shown in the work of Liao and Huang [3]. For this project, the domain of choice has been audio, with a focus on developing a pipeline to achieve audio style transfer. This issue of audio style transfer is significant because it has several real-world applications in the industries of audio engineering, entertainment, and music creation. For instance, audio style transfer can be used to remix previously released audio recordings or to produce brand new, distinctive sounds for musical compositions. Additionally, it can be used to enhance audio recordings. While the issue is of high importance, there are numerous challenges facing those aiming to achieve audio style transfer. Firstly, the lack of big, high-quality datasets for training GAN models for audio style transfer is a major difficulty as it hinders researchers from being able to leverage the vast successes seen by GANs. Despite the fact that some datasets, like the NSynth dataset, are accessible, they are modest in comparison to the datasets used for visual style transfer. Furthermore, audio data is frequently vulnerable to differences in the recording environment, the quality of the microphone, and other elements that might make it more challenging to achieve consistent results. Lastly, there are difficulties caused by the arbitrary nature of audio style. What one individual finds appealing in terms of fashion may not be what another person enjoys. Using GANs to transfer audio styles is a difficult but significant problem that calls for specialized methods and datasets.

Engel et al. have used WaveNet autoencoders to perform audio synthesis [4]. Mital has proposed using approaches like Ulyanov's approach to obtain audio style transfer as well as methods to optimize time-domain audio signal [5]. Other approaches use CNNs and CycleGANs to perform audio style transfer, for example, Barry and Kim used CNNs for musical audio style transfer using multiple time–frequency representations to combine content and style components [6]. The original audio is altered to match a desired style while preserving the content. Huzaifah and Wyse proposed the use of 1D CNN in audio style transfer [7].

In the field of voice conversion, Pasini proposed MelGAN-VC, an approach that enables the conversion of audio signals of arbitrary length from a source voice to a target voice [8]. It also involves computing spectrograms from waveform data and subsequently performing domain translation using a GAN architecture. The proposed MelGAN-VC architecture addresses the challenges of audio style transfer and demonstrates promising results. Singing style transfer has also been explored using GANs. Wu et al. presented a method for singing style transfer based on GANs, which enables the transformation of the singing voice from one style to another [9]. Although their focus was on singing voice, their approach highlights the potential of GANs in capturing and transferring style characteristics in audio. Zhang et al. introduced a method for learning latent representations that enable style control and transfer in speech using unsupervised learning [10]. They proposed the use of a variational autoencoder. Although their focus was on speech synthesis, their work sheds light on the importance of latent representations in capturing and manipulating style-related aspects of audio. This concept can be extended to audio style transfer tasks, where the aim is to transfer the musical timbre from one style to another. Furthermore, Kameoka et al. proposed StarGAN-VC, a non-parallel many-to-many voice conversion method that has real-time implementations [11]. Although the focus of these papers was on voice conversion, their approach provides insights into the use of GANs for style transfer tasks in audio domains.

For this project, we utilized a modified version of the TimbreTron pipeline, originally proposed by Huang et al., for musical timbre transfer [12]. The TimbreTron pipeline demonstrates the potential of combining generative adversarial networks GAN with the CQT representation of the audio to capture and manipulate timbre-related aspects of music. Our modification builds upon this pipeline and focuses on addressing the challenges specific to audio style transfer.

We use the musical timbre which is the quality of the musical sound, and it is distinct from its pitch and intensity to perform this task. Therefore, our project seeks to contribute to this field by examining a different strategy for audio style transfer, testing it, and investigating how it could be used in music and audio engineering.

2 Methodology

In this section, an overview of the methodology employed in this study for audio style transfer using CycleGAN and a modified version of the TimbreTron pipeline is discussed. The methodology encompasses key steps including data selection, data cleaning and feature engineering, our proposed pipeline, and neural network architecture selection. We describe the process of selecting the appropriate data, focusing on the Classical and Jazz genres from the publicly available GTZAN Dataset. Data cleaning and feature engineering techniques are then applied to prepare the audio data for input into the GAN model. The selection of the TimbreTron pipeline, built upon CycleGAN and modifying it, is discussed, highlighting its significance in capturing and manipulating timbre-related aspects of music and how the modified pipeline performs. Finally in Sect. 3, an outline of the validation methodology adopted, which includes a sample input and video demonstration, is used to assess the effectiveness of the style transfer process. By presenting these methodological details, we establish a robust foundation for conducting audio style transfer experiments.

2.1 Data Selection

Data selection is critical in training a GAN for audio style transfer. However, due to limited access to relevant data and the time-intensive nature of GAN training, certain adjustments were necessary to streamline the process and achieve the desired outcomes. Huang et al. used their own dataset that contained the MIDI audio format to perform the timbre style transfer [12].

In this study, a publicly available GTZAN Dataset was used, accessible through Kaggle. This comprehensive dataset comprises ten distinct music genres, including blues, Classical, Country, Disco, Hip-Hop, Jazz, Metal, Pop, Reggae, and Rock. To simplify the task and mitigate complexities, the focus was narrowed down to the Classical and Jazz genres, which were specifically chosen to limit the amount of speech components. Each genre consisted of a collection of 100 audio clips.

During the dataset preprocessing phase, one corrupted audio clip was identified within the Jazz genre and subsequently removed to ensure data integrity. By leveraging the GTZAN Dataset and strategically selecting the Classical and Jazz genres, this research aims to tackle the challenges of audio style transfer while considering data availability, limited speech available in the data, and dataset suitability for the intended purpose. These measures contribute to a more focused and streamlined approach in training the GAN model for audio style transfer.

2.2 Data Cleaning and Feature Engineering

The process of data cleaning started by partitioning the data into 3-s clips to not only augment the dataset but also facilitate efficient training of the model. Shorter clips are advantageous as they can be more effectively processed by the model compared to longer clips. To prepare the audio data for input into the GAN for audio style transfer, a variety of techniques can be used. These techniques encompass:

- Time-domain representation: This method involves capturing periodic measurements of the sound wave's amplitude to represent audio signals in the time domain.
- Fourier analysis: By decomposing the signal into its constituent frequencies, audio signals can be represented in the frequency domain, enabling further analysis.
- Mel-spectrogram: Deep learning models commonly employ Mel-spectrograms as a standard representation for audio data. This technique involves performing a STFT on an audio signal and applying a filter bank that delineates frequency components over time.
- CQT: Similar to the Fourier transform as it also transforms the audio from a time domain to a frequency domain with the frequency values geometrically spaced [12].

Table 1 Default parameters used in the function from the Librosa library to convert the audio to Mel-spectrograms

Parameters	Value
N_fft	2048
Hop_length	512
Window length	None
Window	Hann
Centering the frame	True
Pan_mode	Reflect
Power for magnitude estimation	2.0
Number of Mel bands	128

- Waveform images: By plotting the waveform over time or utilizing a spectrogram representation, audio signals can be transformed into visually interpretable 2D images.

Ensuring consistent audio quality across all samples was a primary challenge in our preprocessing phase. We leveraged the Librosa library to convert the audio files, which were predominantly in the .wav format, into numpy arrays. This conversion not only facilitated efficient data manipulation but also allowed for the extraction of the audio's sample rate. Subsequently, we used dedicated functions from the Librosa library to transform the audio into Mel-spectrograms. Throughout this process, we consistently adhered to standardized parameters, such as the window length, the number of Mel bands, centering the frame, and power for magnitude estimation. Table 1 enumerates these parameters along with their respective values, ensuring that a uniform quality and format were maintained across the dataset.

Following the execution of this function, the Mel-spectrograms were prepared for integration into CycleGAN. To obtain audible outputs, an additional function was applied, which took the output from the CycleGAN and converted it back into the.wav format. This function essentially reversed the process of the initial function, utilizing the same parameters.

2.3 *Our Proposed Pipeline*

Figures 1 and 2 show a schematic overview of the whole data processing pipeline for both the original and suggested pipelines. According to Huang et al., there may be a significant computational cost associated with training the Conditional WaveNet models [12]. They applied CQT and decided to employ the Conditional WaveNet model. A CycleGAN was then fed the changed data, which was in the CQT format. According to Huang et al., recovering the original waveforms proved to be a challenging job without a clear technique, so the audio waveforms were created using the Conditional WaveNet Synthesizer. The authors also mention that WaveNet models are expensive in terms of the waveform generation [12].

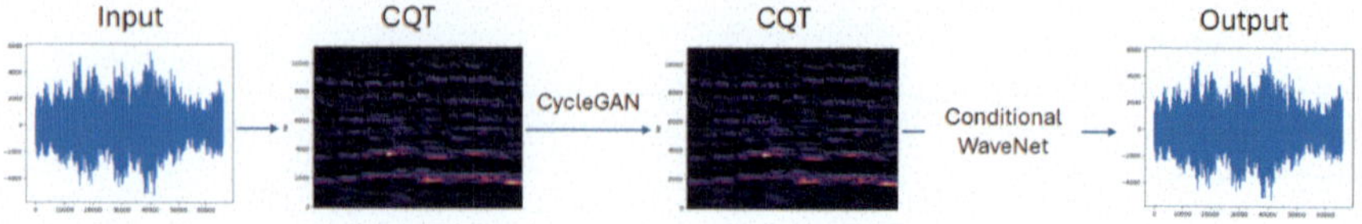

Fig. 1 Original TimbreTron pipeline

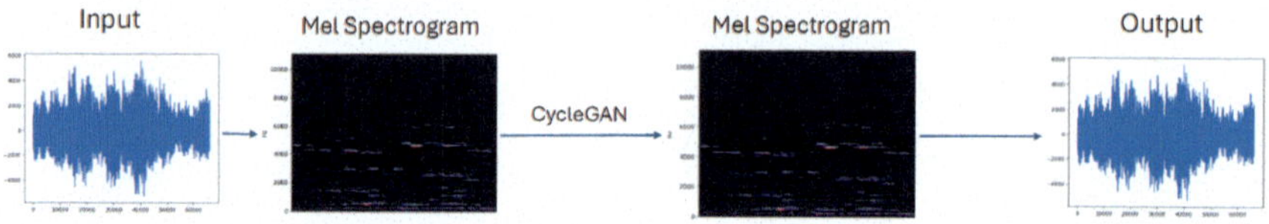

Fig. 2 Our proposed pipeline

By retrieving the waveform from the Mel-spectrogram using the Librosa library, we were able to skip the use of the Conditional WaveNet model for waveform synthesis, which reduced the number of parameters required for training compared to the original TimbreTron pipeline. Therefore, to represent our waveform and feed it into the CycleGAN model, we employed a Mel-spectrogram. This reduction in the number of parameters in our suggested model is therefore beneficial for applications which require immediate processing. Therefore, we can deploy our lightweight pipeline on edge devices for real-time applications where resources are scarce because there are far less parameters in it than in the original pipeline.

2.4 Architecture

CycleGAN is the neural network architecture that has been selected to be used to solve the present problem. It consistently produced better results, according to a thorough study of the material that has already been published because it can handle unpaired data. CycleGAN was used specifically in this study for style translation [13]. Originally conceptualized for image translation tasks, CycleGAN has since proven its versatility across diverse domains, extending its utility to audio style translation, among others. CycleGAN, which was initially developed for picture translation, has demonstrated its application in a variety of fields, including audio style translation.

In our simulation, the audio recordings' Mel-spectrograms are processed by the CycleGAN framework, which creates an audio file with the relevant transmitted style based on the chosen generator. With two generators present, it is possible to switch between Classical and Jazz styles, with each generator facilitating the opposite change.

The use of the CycleGAN architecture in this research project shows how effective it is at transferring audio styles in the desired way. Our model successfully facilitates the translation between Classical and Jazz styles and back to Classical and the other way around by using the special characteristics of CycleGAN and the Mel-spectrogram format, broadening the potential applications of this technology in the area of audio processing and beyond.

3 Discussion and Results

3.1 Validation Methodology

In order to validate the efficacy of our proposed pipeline, testing was performed on a sample original input and the result was analyzed through a video demonstration. Validating the performance of a model in audio style transfer typically necessitates the availability of reference data for comparing the generated outputs. However, obtaining ground truth data can be challenging, particularly when dealing with subjective terms like artistic style. Audio styles are highly interpretative and can vary based on individual preferences and creative objectives. As a consequence, establishing a definitive ground truth for assessing the precision of style transfer becomes a complex task.

While validation remains an essential aspect of many deep learning projects, it is acknowledged that certain scenarios, such as audio style transfer with CycleGAN, may present difficulties or render traditional validation methods less applicable. In such cases, it becomes imperative to prioritize the exploration of style transfer possibilities, emphasizing the creative element and artistic objectives. Encouraging creativity in the audio output becomes paramount, acknowledging the subjective nature of audio styles and fostering an environment conducive to artistic expression.

To gain insights into the effectiveness of our conversion, we employed a validation approach involving the utilization of the output from the Classic to Jazz transfer as input for the Jazz to Classic transfer. Figure 3 explains the pipeline that was used to validate the model's performance in being able to transfer from Classical to Jazz and back to Classical. This is one of the properties of CycleGAN which is to make sure we can reconstruct the input back using the other generator.

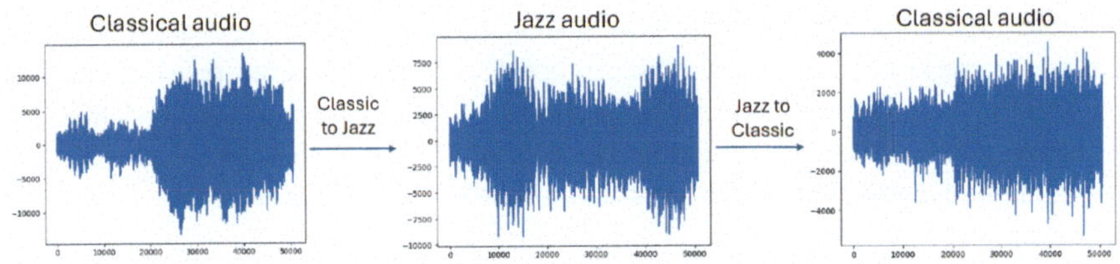

Fig. 3 Validation methodology of transforming Classical to Jazz and back to Classical

Remarkably, as we can see in Fig. 3 this process yielded a highly similar Classical audio output to the original Classical audio, providing evidence of the successful conversion achieved by our lightweight approach. The Classical audio output has some noise and attenuation when compared to the Classical audio input; however, the overall pattern is the same. This validation can even be shown by the results from the second question in the survey which is discussed in the results section. This validation method demonstrates the effectiveness of our style transfer and reinforces the fidelity of the generated outputs.

3.2 Results

To evaluate the effectiveness of the CycleGAN model for audio style transfer, we primarily focused on three types of losses: cycle consistency loss, adversarial loss, and identity loss. These losses are crucial to preserving the original identity of the audio files and shed light on the dynamics of the model's training and convergence. In contrast, traditional performance analysis criteria such as accuracy, precision, and recall were not employed. Given the unique nature of the problem, which lacks these conventional measurements, using them would make it challenging to accurately describe and analyze the model. The YouTube video shows how the losses of the model changeover time by keeping track of these losses while training. Audio demos were added to the video on YouTube in addition to losses. Viewers can hear the audio samples created by our CycleGAN model by playing them throughout the video. They are able to examine the audio outputs using their own individual perceptions and judgments due to this qualitative assessment. It offers a chance to evaluate the generated audio's overall quality and the level of detail with which the required artistic styles are captured. Traditional performance measurements might not be there, but by focusing on losses and utilizing subjective evaluation through audio demonstrations, you can gain valuable information about the potential and effectiveness of the CycleGAN model for audio style transfer. This method emphasizes the value of artistic expression and the imaginative exploration of many audio styles while acknowledging the particularity and subjectivity of the work.

A survey was conducted that consisted of two main questions. The survey had a total of 11 respondents that were given the audio files to listen to and answer the questions based on that. The first question was about how well the Jazz audio sounds after the conversion from the Classical audio to the Jazz audio. While the second question was about how well the model was able to generate the Classical audio that is expected to be the same as the original Classical audio after passing into the Classical to Jazz generator and then back to Classical using the Jazz to Classical generator. Tables 2 and 3 summarize these results. This method of validation was also done by Huang et al. where they had a human study to know if they had good results or not [12].

Our proposed pipeline emphasizes maintaining audio quality, as highlighted by our survey results, where most respondents found the audio quality consistent.

Table 2 Survey question 1

Question 1: On a scale of 1 to 5 on how well the model was able to perform the style transfer from classic to Jazz					
Scale	1	2	3	4	5
–	0%	27.3%	18.2%	36.4%	18.2%

Table 3 Survey question 2

Question 2: Did the two Classic files sound the same before and after the conversion back to Classic?		
Yes	Somewhat	No
54.5%	36.4%	9.1%

Furthermore, we recognize the importance of data quality and have taken measures to inspect and eliminate potential noise from our dataset. As for artifacts introduced during the style transfer, it could be because of some speech components available in the data, this is an area that needs further exploration and will be addressed in future improvement to our pipeline.

For a comprehensive visual summary and a demonstration of our code execution, please refer to the accompanying video presentation available at: [https://www.youtube.com/watch?v=48HstmZGSKQ].

4 Future Work and Conclusion

4.1 Future Work

The project has demonstrated that the modified lightweight TimbreTron pipeline, a CycleGAN-based system, has the capacity to transfer audio style, allowing the transformation of audio signals from one style or genre to another. However, restrictions on the dataset, the length of training as well as some speech components in some audio files may have hindered the model's performance. The following techniques can be investigated to enhance the outcomes:

- Enhanced Dataset: A diversified dataset with varied genres, instruments, and increasing speech samples can enrich the model's understanding of style intricacies and allow for more versatile audio transfers.
- Optimized Training: Extending training duration and fine-tuning hyperparameters can help the model discern intricate style correlations better.
- Feedback: The model's performance can be improved repeatedly by gathering user feedback on the transferred audio from musicians and audio professionals.

- Ensemble Models and Transfer Learning: Merging results from diverse CycleGAN models or leveraging existing models can speed up the learning process and elevate performance.
- Incorporating Speech Style: By training CycleGAN with varied speech samples, the model's applicability could extend to transferring speech styles alongside musical ones. This adaptation could facilitate advanced voice modifications to emulate specific dialects. However, this approach is challenging, requiring extensive datasets and extended training.
- Real-time and Interactive Applications: There is potential to further investigate interactive real-time applications. This exploration could benefit live performances and on-the-fly style changes, yet it would need to address inherent challenges such as processing speed, latency, and audio fidelity.

As we look ahead, the challenge remains not just in increasing the quantity of data but ensuring its quality, ensuring that our model's transformations are both accurate and artistically resonant. Building upon our foundational work, there's vast potential to reshape and expand the horizons of audio style transfer.

5 Conclusion

This paper provides a modified lightweight TimbreTron pipeline for audio style transfer. The study focuses on modifying audio signals' timbre to produce the required genre. The modified TimbreTron pipeline, based on CycleGAN, was chosen for its capacity to capture timbre-related characteristics. Data cleaning techniques are used. Due to the arbitrary nature of audio style, an evaluation using video is used and can be accessed using the link provided in Sect. 3.2. A survey was also conducted to support our results. The dataset can be expanded, hyperparameters can be changed, ensemble models and transfer learning can be used, and speech style transfer can be investigated. The model's performance can be improved by gathering feedback from audio professionals, and its scope can be extended to real-time applications.

Acknowledgements Some of the code such as the CycleGAN model and the Dataloader, used in this paper are accessible at the following public GitHub repository: [https://github.com/edivadiranatnom/Machine-Learning-Project/tree/master].

References

1. Copet J, Kreuk F, Gat I, Remez T, Kant D, Synnaeve G, Adi Y, Défossez A (2023) Simple and controllable music generation. http://arxiv.org/abs/2306.05284
2. Gatys LA, Ecker AS, Bethge M (2016) Image style transfer using convolutional neural networks. In: 2016 IEEE conference on computer vision and pattern recognition (CVPR). IEEE, Las Vegas, NV, USA, pp 2414–2423

3. Liao Y, Huang Y (2022) Deep learning-based application of image style transfer. Math Probl Eng 2022:1–10. https://doi.org/10.1155/2022/1693892
4. Engel J, Resnick C, Roberts A, Dieleman S, Norouzi M, Eck D, Simonyan K (2017) Neural audio synthesis of musical notes with WaveNet Autoencoders. In: Proceedings of the 34th international conference on machine learning. PMLR, pp 1068–1077
5. Mital PK (2017) Time domain neural audio style transfer, http://arxiv.org/abs/1711.11160
6. Barry S, Kim Y (2018) "Style" Transfer for musical audio using multiple time-frequency representations
7. Huzaifah M, Wyse L (2020) Applying visual domain style transfer and texture synthesis techniques to audio—insights and challenges. Neural Comput and Applic 32:1051–1065. https://doi.org/10.1007/s00521-019-04053-8
8. Pasini M (2019) MelGAN-VC: voice conversion and audio style transfer on arbitrarily long samples using Spectrograms. http://arxiv.org/abs/1910.03713
9. Wu C-W, Liu J-Y, Yang Y-H, Jang J-SR (2018) Singing style transfer using cycle-consistent boundary equilibrium generative adversarial networks. http://arxiv.org/abs/1807.02254
10. Zhang Y-J, Pan S, He L, Ling Z-H (2019) Learning latent representations for style control and transfer in end-to-end speech synthesis. http://arxiv.org/abs/1812.04342
11. Kameoka H, Kaneko T, Tanaka K, Hojo N (2018) StarGAN-VC: non-parallel many-to-many voice conversion with star generative adversarial networks. http://arxiv.org/abs/1806.02169
12. Huang S, Li Q, Anil C, Bao X, Oore S, Grosse RB (2019) TimbreTron: a WaveNet(CycleGAN(CQT(Audio))) pipeline for musical timbre transfer. http://arxiv.org/abs/1811.09620
13. Zhu J-Y, Park T, Isola P, Efros AA (2020) Unpaired image-to-image translation using cycle-consistent adversarial networks. http://arxiv.org/abs/1703.10593

Deep Dive into Invoice Intelligence: A Benchmark Study of Leading Models for Automated Invoice Data Extraction

Merxhan Bajrami, Nevena Ackovska, Biljana Stojkoska, Petre Lameski, and Eftim Zdravevski

Abstract Invoice data extraction is crucial in contemporary business environments, streamlining the processing and organization of voluminous invoice data for efficient management and decision-making. This paper presents a thorough benchmark study, evaluating the performance of five state-of-the-art models—LayoutLM, LiLT, Donut, Yolov8x, and Yolov5x—in the automated extraction of data from invoices. Leveraging transfer learning, each model is fine-tuned and assessed based on a private dataset comprising diverse invoice types. The evaluation metrics encompass precision, recall, F1-score, and accuracy, providing a comprehensive performance analysis. Results indicate that Yolov8x outperforms its counterparts with a commendable accuracy of 0.913 and an F1-score of 0.928, whereas LiLT trails with an accuracy of 0.537. Furthermore, the paper sheds light on the real-world deployment performance of these models, considering factors such as GPU usage, inference time on GPU and CPU, loading time, and loading GPU usage. LayoutLM demonstrates balanced performance with moderate resource consumption, while other models exhibit varied efficiencies and resource utilizations. Through this investigation, the paper furnishes invaluable insights into the strengths and limitations of each model, offering guidance for practitioners in selecting the optimal model tailored to specific invoice data extraction tasks and operational constraints. The findings of this study

M. Bajrami (✉)
Laigo GmbH, Eckenerstraße 65, 88046 Friedrichshafen, Germany
e-mail: merdjan.bajrami@laigo.ai; bajrami.merdzan@students.finki.ukim.mk

N. Ackovska · B. Stojkoska · P. Lameski · E. Zdravevski
Faculty of Computer Science and Engineering, Ss. Cyril and Methodius University, Skopje 1001, North Macedonia
e-mail: nevena.ackovska@finki.ukim.mk

B. Stojkoska
e-mail: biljana.stojkoska@finki.ukim.mk

P. Lameski
e-mail: petre.lameski@finki.ukim.mk

E. Zdravevski
e-mail: eftim.zdravevski@finki.ukim.mk

X.-S. Yang et al. (eds.), *Proceedings of Ninth International Congress on Information and Communication Technology*, Lecture Notes in Networks and Systems 1000,
https://doi.org/10.1007/978-981-97-3289-0_15

underscore the importance of discerning model selection and performance evaluation for invoice data extraction applications, serving as a robust reference for future research and practical implementations in the field.

Keywords Invoice data extraction · Layout analysis · OCR · Intelligent document processing · LayoutLM · Yolo

1 Introduction

With the advent of digitalization, traditional financial ecosystems are evolving. The voluminous influx of data mandates the need for efficient, automated extraction systems, especially in the realm of German accounting where invoice formats are manifold [1]. This paper bridges this exigency by rigorously evaluating the efficacy of five leading models, providing a compass for future applications in this domain. Within this context, the task involves the automated extraction of pre-defined key-fields of interest from invoice documents, acting as a foundational step for various applications crucial to business operations, such as automated accounting, proactive financial forecasting, and monitoring.

Industry experience indicates that a substantial segment of data managed by businesses is inherently unstructured, with invoices standing out as the most common examples of such data. As the operational scale of businesses expands, so does the volume of invoices they generate and need to process. This increase renders traditional manual processing methods not only outdated but also markedly inefficient and susceptible to human errors. The adoption of automated invoice data extraction techniques, therefore, represents a pivotal advance in the field, serving to enhance the efficiency of data processing workflows significantly while concurrently minimizing error rates. As a well-identified and common problem in accounting companies in Germany [1], our study undertakes a detailed examination and comparison of the performance metrics of five state-of-the-art models, namely LayoutLM [2], Donut [3], LiLT [4], Yolov8x [5], and Yolov5x [6]-in the task of invoice data extraction.

One of the standout features of this research is the novel approach adopted in evaluating and comparing these diverse models against a common benchmark tailored for the task of invoice data extraction. While earlier studies have typically opted for the assessment of these models either in isolation [7] or within disparate contexts, to the best of our knowledge, the present study constitutes the first scholarly endeavor offering a direct, head-to-head comparison of these models, focusing explicitly on their performance in invoice data extraction. This methodological choice facilitates a nuanced understanding of the strengths and weaknesses inherent to each model, providing valuable insights that are expected to inform and optimize their deployment in real-world scenarios, particularly those related to the extraction of data from invoices.

The structure of the paper is organized as follows: Sect. 2 introduces the reader to the existing body of literature and the state of research in the field of invoice data extraction. Section 3 details the methodology underpinning the study, with a focus on the selection criteria for the models and the dataset designated for evaluation purposes. Additionally has a subsection that provides an in-depth account of the implementation process, covering aspects such as fine-tuning of the models and optimization of hyperparameters. A detailed analysis and interpretation of the results obtained are presented in Sect. 4, followed by Sect. 5, which concludes the paper by summarizing the key findings and their implications. Section 6 outlines the prospective directions for future research in this domain, highlighting potential areas of interest and avenues for further investigation. Finally, Sect. 7 is about the acknowledgments for the company that gave the data and developers that put their model's open source.

2 Related Work

Information extraction from documents begins with OCR frameworks like Doctr, EasyOCR, and Tesseract, essential for detecting and recognizing text in business documents [7]. With the growing need for automated data extraction due to the surge in e-commerce activities, conventional OCR and text analysis methods have proven impractical and expensive [8]. While advanced methodologies like pattern matching offer alternatives, their accuracy is often limited without specified patterns, although recurrent and graph neural networks help overcome these limitations efficiently [8].

The semi-structured nature of invoices makes information extraction intricate, as data is contextually located [9]. While many extraction systems rely on heuristic methods and expert-curated trigger words, they require laborious manual annotations for new templates [9]. Neural network approaches, although data-intensive, eliminate the need for manual curation, providing efficiency [9].

Leon's thesis [10] highlighted the effectiveness of deep learning methods for extracting semi-structured data from invoices, with NLP approaches outperforming others in terms of overall efficiency and accuracy [10]. The study indicates the promising potential of NLP methodologies for invoice information extraction, underscoring the need for further research and exploration in this domain.

Yao et al. [11] addressed issues in manual invoice processing, such as inefficiency, errors, and security risks, by proposing a deep learning-based invoice recognition system. The system enhances low-contrast images through preprocessing techniques and then applies a combination of YOLOv3 and CRNN models for accurate and efficient invoice detection and recognition. This approach minimizes errors and resource losses associated with traditional methods, as verified by experimental results [11].

In the realm of information extraction from business documents, Chua and Duffy [12] addressed the challenge of extracting structured information without relying on detailed annotations. They introduced a novel approach named Deep Conditional Probabilistic Context Free Grammars (DeepCPCFG), which is capable of parsing complex two-dimensional documents. The system employs recursive neural networks to create an end-to-end mechanism designed to identify the most probable

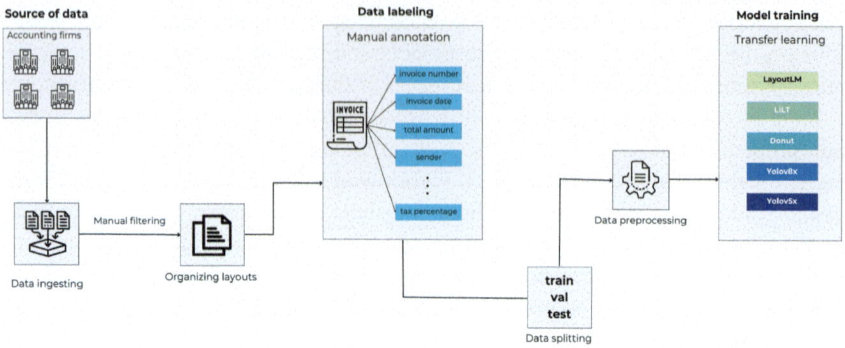

Fig. 1 Visual representation of the research workflow

parse, representing the structured information to be extracted. Remarkably, the proposed system is trained end-to-end with scanned documents serving as input and relational records as labels, which are extracted from existing databases. This approach eliminates the need for manual document annotations, thereby reducing the associated labor and costs. Applied to the extraction of information from scanned invoices, the DeepCPCFG demonstrated state-of-the-art results, even without the use of hand annotations.

3 Methodology

The research methodology, illustrated in Fig. 1, outlines the comprehensive process followed from initial data collection to the final stages of model training.

3.1 Data Acquisition

The data used during this research was collected from four accounting firms in Friedrichshafen. While the names of the firms remain confidential to preserve anonymity, it's noteworthy that each firm contributed a disparate number of files, ranging from 500 to 1000.

The data acquired from these firms underwent a careful initial review process where unique invoice layouts were identified. An invoice layout is termed 'unique' if it originates from different clients and exhibits distinct structures, notably in the location of key fields and the presence of other metadata. This process distilled the bulk data down to 250 unique layouts.

For each unique layout identified, ten sample invoices were selected. This selection aimed to encapsulate the diversity and variability within each layout, yielding a rich, representative dataset consisting of 2500 files in total.

Labeling was performed manually using a commercially available tool. The process involved drawing bounding boxes around each key field, essential for the Named Entity Recognition (NER) approach, as referenced in [13]. The key fields annotated were as follows: INVOICE DATE, INVOICE NUMBER, INVOICE RECEIVER, INVOICE RECEIVER ADDRESS, INVOICE SENDER, INVOICE SENDER ADDRESS, INVOICE SENDER BIC, INVOICE SENDER IBAN, INVOICE SENDER TAX NUMBER, INVOICE TAX PERCENTAGE, INVOICE TOTAL AMOUNT BRUTTO, and INVOICE TOTAL AMOUNT NETTO.

Following the labeling process, the dataset was split into training, validation, and test sets. The training set comprised 1750 files, while the validation and test sets consisted of 375 files each. Each set featured entirely distinct invoice layouts, thereby safeguarding against data leakage during the model evaluation phases.

Prior to model training, the dataset underwent a series of image preprocessing steps. Noise reduction was applied to minimize background interference, while rescaling was employed to normalize image dimensions across the dataset. Text Normalization converted textual content within images to a consistent format, facilitating smoother model training. Augmentation techniques were also utilized to synthetically expand the dataset, fostering robust model learning, and improved generalization.

3.2 Model Training

Model training was initiated with standardized hyperparameters across all models, followed by a series of experiments conducted during the fine-tuning phase. Details pertaining to the experiments and fine-tuning process are elaborated in the ensuing Experiments section.

Figure 2 shows three different invoice styles, protecting private data. Each invoice has a unique setup and design, highlighting the different ways invoices can look. The variety in these examples shows why it's challenging to create a one-size-fits-all solution for automatic data pulling and why advanced systems are needed to handle different invoice types. Due to the confidential nature of the data contained within these invoices, precautions have been taken to anonymize and secure the data to prevent the disclosure of sensitive information.

3.3 Models Overview

LayoutLM [2]: Ingeniously integrates textual and layout information, addressing the challenges in document image understanding. It is chosen for its excellent perfor-

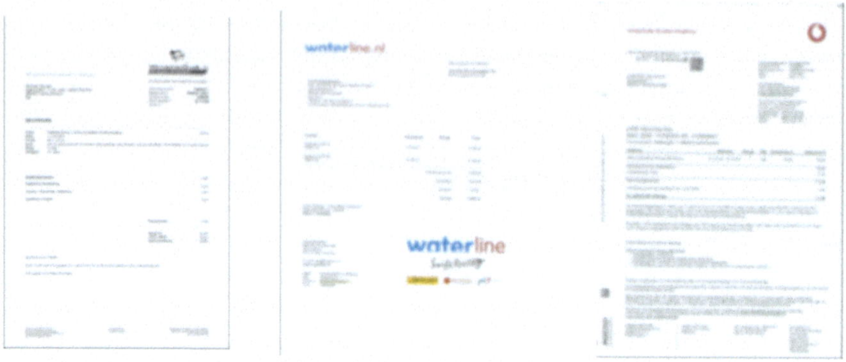

Fig. 2 Visual representation of diverse invoice layouts

mance in understanding scanned documents and extracting information seamlessly, thanks to its deep understanding of both text and layout features.

Language-independent Layout Transformer (LiLT) [3]: LiLT offers a language-agnostic approach to structured document understanding. It's selected for its ability to be pre-trained on one language and fine-tuned for others, making it versatile and highly effective for invoices in different languages without compromising performance.

Donut [4]: A simplistic yet powerful OCR-free Visual Document Understanding (VDU) model, Donut is crucial for tasks requiring reading text and understanding document layouts. Its selection stems from its ability to efficiently and accurately perform on various VDU tasks without relying on traditional OCR, making it a cost-effective and flexible choice.

Yolov8x [5]: Although primarily an object detection model, Yolov8x offers high speed and accuracy. It is chosen for its ability to quickly locate and identify text blocks on invoices, facilitating efficient text extraction and reading, which is foundational for the invoice data extraction process.

Yolov5x [5]: Similar to Yolov8x but with a focus on being lightweight, Yolov5x is selected for its speed and efficiency in detecting text within invoice images. It is especially valuable in scenarios where computational resources are limited but high performance is non-negotiable.

3.4 Experiments

For each model, transfer learning techniques were applied using their respective pre-trained versions available, initializing with weights trained on extensive datasets. This approach provides a foundational understanding of text and structure that is further fine-tuned to adapt to the specificities of invoice data extraction.

The fine-tuning process for each model was conducted using a cross-entropy loss function and the Adam optimizer, iterating for a maximum of 100 epochs. To avoid overfitting and identify optimal model configurations, early stopping technique was utilized, monitoring performance on the validation set, and for this, we used Optuna[14]. This strategy not only ensured efficient training but also minimized the risk of overfitting, leading to robust and generalizable models capable of adeptly handling diverse invoice data extraction tasks.

During the preprocessing stage, the Tesseract [15] OCR engine was employed to convert the images of invoices into machine-encoded text. Tesseract is an open-source OCR engine that supports various languages and can recognize different text layouts. Its robustness and accuracy in text recognition make it a valuable tool for extracting text data from invoice images, thereby serving as a crucial step in the data preparation process for model training. Moreover, when bounding boxes were identified in the invoice images, Tesseract OCR was utilized to extract the text within these regions accurately. This approach facilitated the precise extraction and subsequent analysis of relevant information from the invoices for the training of our models.

For each experiment of training, we have used the 70–10–10 splitting ratio for the dataset. Through a meticulous evaluation process, the study offers a comparison of each model's strengths and limitations, providing insights into their respective performances and applications in invoice data extraction.

4 Results

In this section, we delve into the comprehensive analysis and interpretation of the results garnered from the experiment, focusing on the performance of each model across various metrics.

4.1 Overall Performance Analysis

The results obtained from the models are summarized in Fig. 3.

Precision LayoutLM demonstrates a good precision at 0.925, making it highly reliable for positive identifications in the data extraction from invoices. But Yolov8x is the highest with 0.942 precision and this indicates that when they identify a positive, it is correct 0.942 of the time. Yolov5x, with a precision of 0.815, shows the lowest precision among the considered models, suggesting that it might generate more false positives than other models.

Recall Yolox8x exhibits the highest recall at 0.901, ensuring it identifies the majority of actual positive cases, making it valuable for applications where missing a positive identification would be costly. In contrast, LiLT has the lowest recall at 0.581, suggesting it might miss a substantial number of positive cases.

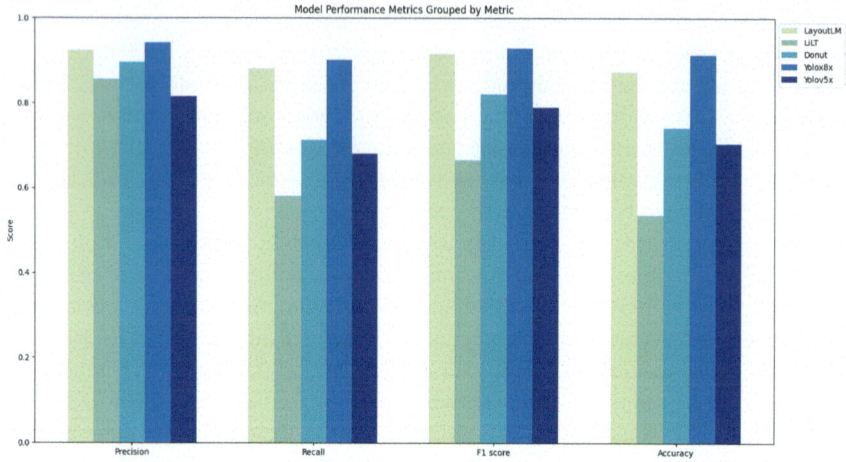

Fig. 3 Results of the models

F1-score Yolox8x also tops the list in terms of the F1-score, with 0.928, which indicates a balanced performance between precision and recall. This score makes Yolox8x a strong, all-around model for data extraction from invoices. LiLT, again, has the lowest F1-score at 0.666. This low score reveals that the model's performance is not well-balanced between precision and recall, which may limit its effectiveness in various applications.

Accuracy With an accuracy of 0.913, Yolox8x proves to be the most accurate model among the ones considered for the task of invoice data extraction. This suggests that in general, Yolox8x's predictions (both positive and negative) are mostly correct. LiLT, with 0.537 accuracy, is the least accurate model. This low accuracy indicates that its predictions are less reliable, and it often misclassifies the data.

4.2 Performance on Key Fields

Analyzing the performance on extracting specific key fields from invoices provides deeper insights into the models' capabilities. Each key field represents different challenges, requiring the models to navigate and accurately extract varied information types from the invoices. From the performance scores across key fields (see Fig. 4), it is evident that each model has distinct strengths and weaknesses.

4.2.1 Model-Based Results Interpretation

From the heatmap at Fig. 4, results we can state the following things:

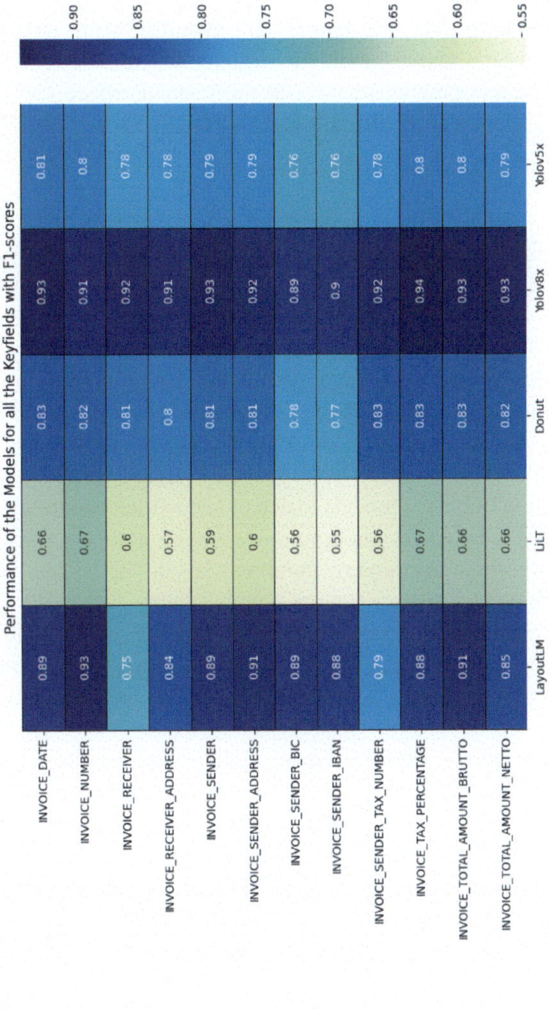

Fig. 4 Heatmap displaying model results for each key field

1. Yolov8x's Superiority: Yolov8x consistently outperforms the other models across most metrics and key fields, underscoring its robustness and effectiveness in invoice data extraction tasks. It stands out as the most versatile and robust choice for this dataset, making it particularly valuable for practitioners in the field.

2. LiLT's Limitations: LiLT demonstrates consistent underperformance across almost all fields. This underperformance may suggest a need for more extensive fine-tuning or may indicate that its architecture is less suited to the nuanced task of invoice information extraction.

3. Consistent Performance of LayoutLM: LayoutLM offers steady and reliable performance, making it a viable candidate for diverse invoice extraction tasks. It is particularly adept at recognizing fields with more standardized formats or unique identifiers, such as INVOICE NUMBER and INVOICE TAX PERCENTAGE.

4.2.2 Key Field Analysis

Based on the investigations for each keyfield and each model performance per keyfield, we can see the following things:

1. Easiest Keyfield: INVOICE DATE is among the easier key fields for models to recognize accurately. This ease of recognition likely results from the standardized formats in which dates are represented on invoices, providing models with clear patterns to identify and extract.

2. Challenging Keyfields: INVOICE SENDER BIC and INVOICE SENDER IBAN are challenging fields for models to detect and extract accurately, as indicated by their lower F1-scores across multiple models. The complexity in these fields' structures, especially the alphanumeric combinations in IBANs, poses a significant challenge to accurate extraction.

3. Uniform Performance: Some models, like Donut, exhibit relatively uniform performance across the key fields. This uniformity suggests a general approach to information extraction that does not significantly excel or falter with any specific type of data.

4. Standardized versus Variable Data: Fields containing highly standardized information (e.g., INVOICE DATE, INVOICE TAX PERCENTAGE) are generally easier for models to recognize. In contrast, fields with variable structures or those that blend with surrounding text are more challenging to extract accurately. Address-related fields, like INVOICE RECEIVER ADDRESS and INVOICE SENDER ADDRESS, illustrate this challenge due to their variability in length, format, and structure.

4.3 Resource Consumption Results

Resource consumption is a pivotal factor when deploying models, especially in a production environment where efficiency and speed are paramount [16]. In addition

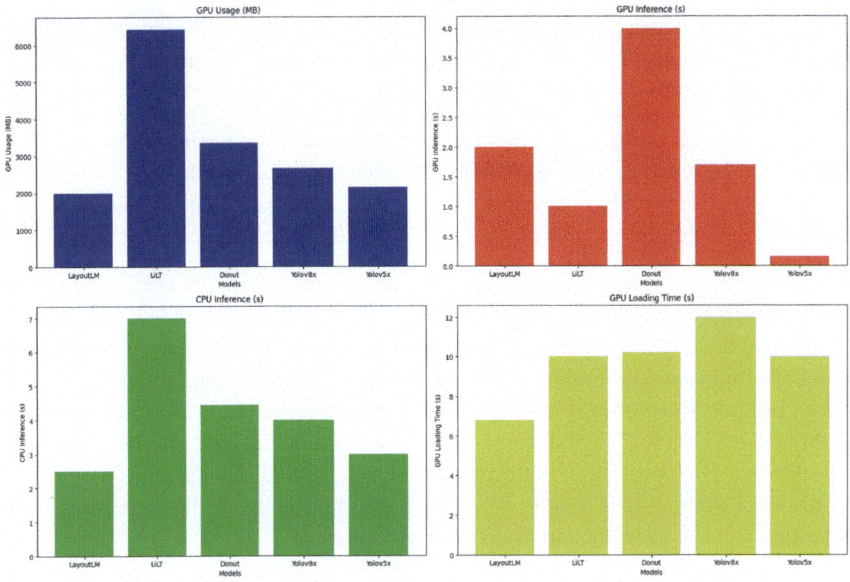

Fig. 5 Real-world analysis performance

to their accuracy and efficiency in extracting data, it is critical to evaluate the models based on their resource consumption and performance on hardware. This subsection provides an in-depth analysis of each model's performance when deployed on a test server equipped with an NVIDIA GTX 3070–12 GB GPU. For the tests, a single Torchserve worker was utilized, and evaluations were performed under CPU and RAM conditions, considering overhead storage necessary for data processing.

Analyzing the results from Fig. 5, we can see the following things:

GPU Usage: LiLT demands significantly more GPU memory (6433 MB), while Yolov5x is the most economical, requiring only 2153 MB. Efficient GPU usage is crucial for ensuring that the model can run smoothly, particularly in environments where multiple processes are executed simultaneously.

Loading GPU Usage: Again, LiLT consumes more resources (3221 MB) during the loading phase, while LayoutLM is more moderate, utilizing 1249 MB of GPU memory.

Inference Time: Inference time directly impacts the user experience and the system's overall efficiency.

GPU Inference Time: Yolov5x stands out with the fastest GPU inference time (0.16 s), significantly outpacing the other models. In contrast, Donut takes considerably more time (4 s), which may affect its usability in real-time or time-sensitive applications.

CPU Inference Time: While all models exhibit reasonable CPU inference times, it's noteworthy that LiLT takes the longest (7 s), whereas LayoutLM and Yolov5x are relatively faster.

Loading Time: Loading time is essential for applications requiring frequent model loading and unloading. Yolov8x has the longest GPU loading time at 12 s, while LayoutLM is faster with a loading time of 6.8 s. These variations in loading times can be crucial in dynamic environments where models are intermittently loaded and unloaded.

Implications: Understanding these hardware performance metrics is vital for selecting a model that not only meets accuracy requirements but also aligns with the hardware constraints and performance expectations of the deployment environment [17].

For Limited Resources: For environments with limited GPU memory, Yolov5x might be a preferable choice due to its low consumption and fast inference time.

Balancing Performance: If the deployment environment can afford more substantial resource usage, LiLT, despite its higher consumption, might be considered if its extraction accuracy significantly benefits the application.

Real-time Applications: For real-time applications where inference time is crucial [18], Yolov5x's rapid GPU inference offers a distinct advantage.

5 Conclusion

This study embarked on a comprehensive exploration and analysis of the performance of five distinguished deep learning models—LayoutLM, LiLT, Donut, Yolov8x, and Yolov5x—in the context of automated invoice data extraction. Through rigorous experimentation and careful analysis of each model's precision, recall, F1-score, and accuracy in extracting specific key fields from invoices, we have unearthed crucial insights and understandings pivotal to the realm of invoice intelligence.

Yolov8x emerged as the frontrunner, demonstrating superior performance consistently across various metrics and key fields, cementing its status as a robust and effective tool for invoice data extraction tasks. Its versatility and robustness make it an invaluable asset for practitioners seeking efficient solutions for invoice data extraction, offering a harmonious blend of precision and recall.

In contrast, LiLT faced challenges, trailing in performance across most key fields. This observation signals the imperative need for further refinement and fine-tuning to enhance its efficacy and reliability in the invoice data extraction landscape. The underperformance of LiLT may be addressed through more granular fine-tuning or modifications to its architecture to better align with the requirements of invoice data extraction tasks.

LayoutLM showcased consistent and reliable performance, standing as a dependable candidate for a wide array of invoice extraction tasks. Its proficiency in recognizing standardized formats and unique identifiers was particularly noteworthy, illuminating its underlying strengths in pattern recognition and the discernment between commonly occurring text and distinctive identifiers.

Our analysis also delved into the ease or challenge posed by different key fields in invoices. Fields with standardized formats, such as INVOICE DATE and INVOICE TAX PERCENTAGE, were generally more accessible for models to recognize and extract accurately. Conversely, fields with variable structures or those easily blending with surrounding text, exemplified by INVOICE SENDER BIC and INVOICE SENDER IBAN, presented more formidable challenges to the models.

Moreover, the study observed that some models exhibited uniform performance across various key fields, pointing toward a generalized approach to information extraction that neither excels nor falters significantly with any specific data type. This balanced performance profile may be advantageous in certain applications where consistency is prioritized.

Our work also was concentrated on measuring the performance of these models in real-world application. For this, we did testings and observations in a controlled environment in our server side with different parameters. The findings for this are that Yolov5x is the lightest and best to use for real-time predictions and not only, due to the fast response and memory consumptions. On the other hand, LayoutLM and Yolov8x have similar numbers for all parameters except that Yolov8x takes significantly more time for GPU inference. An interesting thing we can see that LiLT takes double the time from other models when running on CPU, whereas Donut when tested with GPU takes slightly more time than the other models.

In conclusion, this benchmark study contributes novel and invaluable insights to the ongoing dialogue and research surrounding automated invoice data extraction using deep learning. The findings here not only serve as initial guiding practitioners and researchers to make informed decisions on model selection but also as a foundation for future explorations and improvements in the vibrant field of invoice intelligence. The continuous evolution of deep learning models promises further advancements and breakthroughs, and with meticulous research, the horizon of possibilities in intelligent invoice data extraction is boundless.

6 Future Work

Future endeavors in this field of research will aim to extend the breadth and depth of understanding regarding invoice data extraction. One significant direction of this extension involves bringing other state-of-the-art models into the comparative framework established in this study. By doing so, a broader panorama of performance metrics and capabilities of different models can be evaluated and analyzed, shedding light on their respective strengths and weaknesses in various tasks related to invoice data extraction.

Furthermore, our future work is directed toward curating and incorporating a more diverse and challenging dataset into the testing paradigm. By introducing documents of varied complexities and types, the resulting dataset will not only be richer but will also pose a wider array of challenges representative of real-world scenarios. This approach will facilitate a more stringent and comprehensive assessment of the models, allowing for a discerning understanding of their adaptability and efficiency in handling diverse invoice data extraction tasks in practice.

Lastly, these concerted efforts in future work aim to provide a deeper understanding of the landscape of invoice data extraction. With a more robust and nuanced knowledge base, the development and implementation of practical, efficient, and accurate invoice data extraction solutions can be expedited, catering to the multifaceted needs of various industries and applications effectively. The insights and advancements garnered through these future endeavors will undeniably contribute to propelling the field forward, ultimately facilitating the creation of tools and systems that are not only theoretically sound but also practically invaluable.

Acknowledgements We would like to acknowledge the support of Laigo GmbH company for providing the datasets and computing resources used in this study. We also would like to thank the developers of LayoutLM, LiLT, Donut, Yolo, and Tesseract for making their models publicly available.

References

1. Pargmann J, Riebenbauer E, Flick-Holtsch D, Berding F (2023) Digitalisation in accounting: a systematic literature review of activities and implications for competences. Empirical Res Vocat Educat Training 15(1):1
2. Xu Y, Li M, Cui L, Huang S, Wei F, Zhou M (2019) Layoutlm: pre-training of text and layout for document image understanding. arXiv preprint arXiv:1912.13318. revised Jun 2020 [Online]
3. Kim G, Hong T, Yim M et al. (2021) Ocr-free document understanding transformer. arXiv preprint arXiv:2111.15664. revised Oct 2022 [Online]
4. Wang J, Jin L, Ding K (2022) Lilt: a simple yet effective language-independent layout transformer for structured document understanding. arXiv preprint arXiv:2202.13669. [Online]
5. Redmon J, Divvala S, Girshick R, Farhadi A (2015) You only look once: unified, real-time object detection. arXiv preprint arXiv:1506.02640. revised May 2016 [Online]
6. Bajrami M, Zdravevski E, Lameski P, Stojkoska B (2023) A comprehensive analysis of layoutlm and donut for document classification. In: Proceedings of the 20th International conference on informatics and information technologies—CIIT 2023. Available: http://hdl.handle.net/20.500.12188/27397
7. Geletka M et al. (2022) Information extraction from business documents: a case study. In: Proceedings of recent advances in Slavonic natural language processing, RASLAN 2022, pp 35–46. https://nlp.fi.muni.cz/raslan/2022/paper18.pdf
8. Varma O et al. (2022) Technical invoice data extraction system: state of the art, research challenges and countermeasures. J Conf Name
9. Hamdi A et al. (2021) Information extraction from invoices. Technical report, University de La Rochelle, L3i, La Rochelle, France. https://hal.science/hal-03418385/document
10. Leon D (2021) Extracting information from pdf invoices using deep learning. Technical report, Stockholm, Sweden. https://www.diva-portal.org/smash/get/diva2:1608779/FULLTEXT01.pdf
11. Yao X, Sun H, Li S, Lu W (2022) Invoice detection and recognition system based on deep learning

12. Chua FC, Duffy NP (2023) Deepcpcfg: Deep learning and context free grammars for end-to-end information extraction, Palo Alto, CA, USA, Ernst & Young (EY) AI Lab
13. Pakhale K (2023) Comprehensive overview of named entity recognition: models, domain-specific applications and challenges
14. Akiba T, Sano S, Yanase T, Ohta T, Koyama M (2019) Optuna: a next-generation hyperparameter optimization framework. In: Proceedings of the 25th ACM SIGKDD international conference on knowledge discovery and data mining, pp 2623– 2631. https://doi.org/10.1145/3292500.3330701
15. Google: Tesseract OCR (2021) Google. [Online]. https://github.com/tesseract-ocr/tesseract
16. García-Martín E, Rodrigues CF, Riley G, Grahn H (2019) Estimation of energy consumption in machine learning
17. Sze V, Chen Y-H, Emer J, Suleiman A, Zhang Z (2016) Hardware for machine learning: challenges and opportunities (invited paper)
18. Sarker IH (2021) Machine learning: algorithms, real-world applications and research directions. SN Comput Sci 2(160)

Combined ECG Analysis Approach Based on Detection of Deviation from the Personal or Population Norm Using Wearable Devices

Vladyslav Malanin and Illya Chaikovsky

Abstract This paper presents a strategy for developing a real-time electrocardiogram (ECG) monitoring and analysis system using wearable devices, emphasizing personalization and continuous cardiac health management. The proposed methodology introduces a two-stage approach: generating a personal ECG norm model and establishing a robust monitoring architecture. The process involves systematic ECG data collection, pre-processing to eliminate noise, training a personalized model, and periodically updating this model with new data to establish a baseline for individual cardiac activity. The conceptual architecture assures consistent ECG monitoring, immediate data processing, and comprehensive analysis to detect any abnormalities in the ECG data, thereby providing timely and reliable ECG assessments. Expected outcomes include accurate detection of cardiac events, immediate intervention capabilities, and continuous monitoring. The implementation presents several challenges such as maintaining data quality and model accuracy, ensuring user compliance, safeguarding data privacy, and navigating the technical limitations of wearable devices. Solutions like employing advanced and adaptive algorithms, providing user training, employing robust data encryption methods, and developing optimized algorithms are discussed to mitigate these challenges. The paper concludes by identifying future research and development directions, including improving model efficacy, enhancing user interaction, bolstering data security, and refining algorithms, positioning the personalized, real-time ECG monitoring system as a crucial element in individualized healthcare and management of cardiac health.

Keywords Electrocardiogram (ECG) · Artificial intelligence (AI) · Wearable devices · Healthcare technology · Cardiac health management · Signal processing · Population norm · Real-time monitoring · Automated ECG analysis

---

V. Malanin (✉) · I. Chaikovsky
V. M. Glushkov Institute of Cybernetics of the NAS of Ukraine, Kyiv, Ukraine
e-mail: vladislav.malanin@gmail.com

X.-S. Yang et al. (eds.), *Proceedings of Ninth International Congress on Information and Communication Technology*, Lecture Notes in Networks and Systems 1000,
https://doi.org/10.1007/978-981-97-3289-0_16

1 Introduction

1.1 Background

The integration of artificial intelligence (AI) in deciphering electrocardiogram (ECG) signals has shown promising results in cardiac diagnostics. AI has the potential to enhance the detection and monitoring of cardiac anomalies and pathologies by analyzing the complex rhythmic patterns and subtle anomalies present in ECGs. Additionally, AI models for asynchronous ECG signals obtained from smartwatches have been investigated for the automated detection of acute myocardial infarction [1].

1.2 Problem Statement

The traditional methodologies for interpreting electrocardiogram (ECG) data have limitations, especially in real-time analysis and managing large-scale data from continuous monitoring. However, the proliferation of wearable devices has introduced new challenges in maintaining accuracy during real-time analysis and managing the voluminous data derived from continuous monitoring. Additionally, patients with persistent diseases like diabetes or chronic heart conditions present further complexities in ECG analysis, requiring an analytical approach that can navigate through the nuanced complexities of their ECG data with precision. To address these challenges, there is a need to develop ECG analysis approaches that are proficient in general ECG data interpretation and capable of individualized model training based on the unique ECG profiles of patients with chronic conditions. The development of ECG analysis approaches that can handle real-time analysis, manage large-scale data from wearable devices, and provide individualized model training for patients with chronic conditions is crucial for precision and reliable cardiac diagnostics and management.

1.3 Objectives

This research aims to dive into the complex world of ECG data interpretation using a combined ECG analysis approach based on detection of deviation from the personal or population norm using wearable devices with a focus on providing a practical solution for real-time, precise, and patient-specific analysis results, especially for individuals with chronic health conditions. The central objectives of this research are:

1. **Enhance Accuracy**: Increase the precision of identifying and interpreting ECG data, particularly variations and deviations from personal or population norms, with a special focus on handling subtle, clinically relevant changes in the data from diverse and disease-affected populations.
2. **Persistent Real-time Analysis**: Suggest an approach capable of conducting persistent real-time ECG analysis, ensuring prompt detection and practical insights into cardiac events and anomalies, while managing the data accuracy, limited computing powers, and reliability issues commonly encountered with wearable devices.
3. **Adaptability and Scalability**: Create an architecture of a process that remains relevant and applicable to both current and emerging ECG-related technologies, with built-in capabilities to adapt and scale in response to new data, technological advancements, and changes in healthcare practices.
4. **Personal Model Training**: Suggest a system capable of personal model training while maintaining its actuality to ensure that the AI adapts to individual cardiac data over time.

The above objectives guide the research toward developing a solution that bridges the gap between general ECG analysis and personalized, persistent real-time cardiac healthcare, with enhanced functionality for wearable devices, and effectiveness for population, especially individuals with chronic conditions.

2 Literature Review

2.1 *AI Analysis of ECG to Detect Deviation from the Population Norm*

Artificial intelligence (AI) has made significant advancements in the analysis of electrocardiogram (ECG) data, particularly in detecting deviations from established population norms. Machine learning techniques have been utilized to develop models that can recognize a range of cardiac events and irregularities by comparing against typical patterns [2]. However, the detection of subtle deviations, especially those indicative of early or latent conditions, remains a challenge. Advanced AI techniques, such as deep convolutional neural networks, have shown promise in detecting subtle signals and patterns in ECGs. These AI-enabled algorithms have the potential to predict diseases that were previously unpredictable with ECGs [3]. Moreover, AI-augmented electrocardiogram detection has been used to identify left ventricular systolic dysfunction in the general population [4]. In conclusion, AI analysis of ECG data has shown great potential in detecting deviations from population norms [5].

2.2 Real-Time ECG Monitoring via Wearable Technology

Wearable devices provide immediate insights into cardiac health, but they also pose challenges in terms of accuracy and management of continuously streaming data. To make significant progress in this field, it is crucial to develop solutions for real-time data processing and analytics while ensuring the reliability of data derived from wearables. One approach to tackle these challenges is the utilization of low-power high-data-transmission multi-lead ECG acquisition sensor systems [6]. The performance of wearable ECG devices can also be enhanced by using capacitive coupling electrodes which measure impedance [7]. Additionally, the accuracy of ECG parameters from wearable single-lead ECG systems for 24-h monitoring has been analyzed and compared, including parameters related to heart rate, heart rate variability, and the detection of premature atrial and ventricular contractions [8]. Overall, the use of wearable technology for real-time ECG monitoring holds great potential for proactive detection and management of potential cardiac issues. However, it is crucial to address challenges related to data accuracy and management.

2.3 Multi-step or Combined Analysis for ECG Data Interpretation

The landscape of ECG analysis involves a combination of non-AI and mixed methodologies. Combination of AI and non-AI approaches to analyze ECG lacks coverage in research papers, where non-AI methods are usually applied for pre-processing of datasets for either training ML models or feeding the ECG data to already trained models for results. Non-AI methods rely on predefined algorithms and heuristic strategies, utilizing established cardiac models and clinical guidelines. These methods employ techniques such as filtering, transforming, and threshold-based detection to interpret ECG data and classify cardiac events. They offer stability and reliability based on established medical knowledge. Various methods and algorithms have been proposed for the classification of ECG signals, including support vector machine (SVM) based methods [9]. Some of the algorithms used for ECG classification include CWA, direct ECG characteristic, Fourier transform, power spectrum analysis, complexity measure, TLS, adaptive Kalman filter, TLS parameter, and fuzzy ARTMAP neural network classification. ECG denoising and quality assessment are probably the most common non-AI applications for modern ECG signal processing. Several techniques have been proposed to address these challenges. One approach is based on empirical mode decomposition (EMD) and source separation [10]. This method utilizes independent component analysis (ICA) and source separation to denoise ECG signals. Another method combines empirical mode decomposition, sample entropy, and an improved threshold function [11]. This approach evaluates the effectiveness of the denoising method under different input signal-to-noise ratios (SNR) and compares it with previous techniques. A novel approach for ECG denoising is proposed based on transformation matrix for non-decimated wavelet transform (WT) and wavelet/total variation (WATV) denoising [12]. This method aims to reduce noise in ECG signals using a combination of wavelet transform and total variation denoising. Another approach combines variational mode decomposition (VMD) with non-local means and wavelet domain filtering for ECG denoising [13]. VMD is used to decompose the ECG signal into different modes, which are then filtered using non-local means and wavelet domain filtering. Principal component analysis (PCA) has been employed for ECG signal denoising [14]. This technique utilizes PCA to reduce noise in ECG signals.

In summary, the combination of non-AI and AI methodologies in ECG analysis offers the potential to improve diagnostic precision and adaptability. However, in most cases non-AI methods are utilized for pre-processing or quality assessment of ECG data rather than signal analysis.

2.4 *Individual Model Training Coupled with Wearable Technology*

The synthesis of individual model training with data from wearable technology is a promising approach in personalized healthcare [15]. This methodology involves tailoring AI models to the unique cardiac signatures of individual patients, which can improve the diagnostic and predictive precision of the system. By integrating wearable technology data with individual model training, healthcare solutions can be customized to each patient's specific cardiac profile and conditions, potentially setting a future standard in personalized healthcare provision.

One of the studies introduces a novel framework for patient-specific ECG beat classification, utilizing a specialized architecture involving two parallel-connected long short-term memory (LSTM)-based recurrent neural networks, trained on initial 5-min segments of individual patient ECG records and validated against subsequent data, with promising results in ventricular and supra-ventricular ectopic beat classification when compared to other state-of-the-art techniques [16]. Notably, the developed model demonstrates superior performance metrics and proves efficient and lightweight, indicating its potential utility in continuous, automated ECG analysis in clinical settings.

3 Methodology

The suggested methodology presents a two-component strategy for ECG analysis: creating a personal ECG norm model (Fig. 1) and establishing a consistent monitoring architecture (Fig. 2).

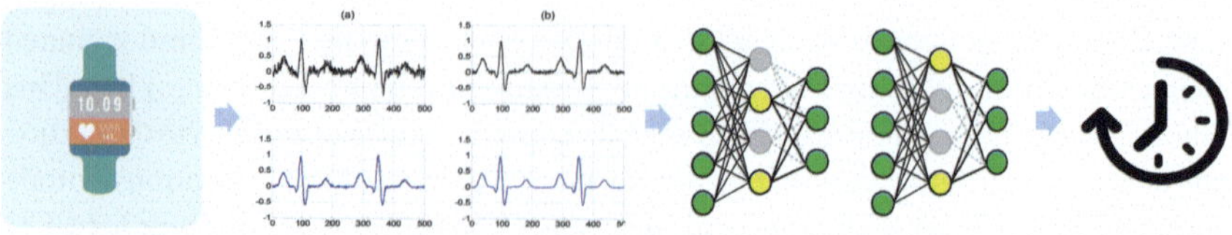

Fig. 1 Diagram of suggested steps for a personal ECG norm model generation and reinforcement using an ECG dataset collected by a wearable device

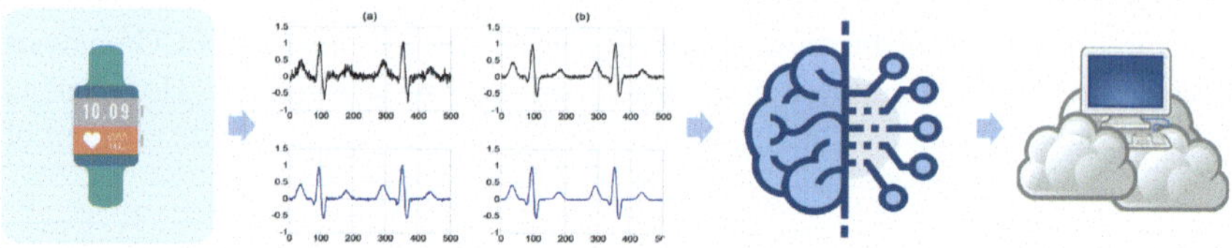

Fig. 2 Diagram of suggested steps for an architecture of persistent monitoring process with ECG analysis

3.1 Personal ECG Norm Model Generation Process

- **ECG Dataset Collection Using Wearable Devices**. The first step is collecting ECG data using wearable devices to capture a wide range of an individual's cardiac activity across various conditions. It is important to collect enough qualified data for model training, and thus a significant amount of time may be needed with different types of activities of an individual undergoing a data collection. The volumes of recorded ECG signals required for a robust model training may differ significantly depending on the capabilities of a recording device.
- **ECG Denoising and Pre-processing**. Next, algorithms are applied to the collected ECG data to remove noise and artifacts, assessing quality of the gathered data, yielding clean data essential for building a precise and accurate personal model. If there is not enough qualified data after pre-processing, the first step must be repeated.
- **Personal Norm Model Training**. Using the refined ECG data, a personalized model is trained to establish a baseline of an individual's regular cardiac activity, against which future data can be compared to spot anomalies. During the initial stages of ECG data collection, the population norm model can temporarily replace the personal one to start the monitoring straight away.
- **Periodic Reinforcement of the Model with Recently Collected Personal Datasets**. The model is suggested to get periodically reinforced with new data or retrained entirely to ensure it continues to accurately reflect an individual's cardiac status, incorporating any new patterns or changes over time. Reinforcement of the model is supposed to happen on a short-term basis (i.e., daily) to reflect slow aging cardiac changes. Retraining, on the contrary, should be performed when individual's cardiac activity or its signal processing undergoes significant

changes (medical conditions, hardware replacement, major software upgrades, etc.).

3.2 Concept Architecture

- **ECG Real-Time Monitoring**. The architecture begins with the real-time monitoring of ECG via wearable devices, continuously recording cardiac activity to deliver a steady data stream.
- **ECG Signal Denoising and Pre-processing**. Concurrently, the raw ECG data is immediately denoised and pre-processed on the same device to ensure data quality for real-time analysis.
- **Persisting Analysis on a Wearable Device to Detect Deviation**. The pre-processed data undergoes a lightweight on-the-fly analysis on the wearable device to identify any deviations from the personal or population norm, crucial for the immediate detection of abnormal health conditions and further analysis.
- **External Analysis of Abnormal ECG**. When deviations are identified, the ECG data is transferred for a deeper external analysis where more complex algorithms conduct a detailed analysis of the abnormal ECG, enabling accurate diagnosis and management. Potentially, with growing computing capabilities of wearable devices, the entire cycle may be done within the closed ecosystem with no data transfer to external devices.

4 Expected Results and Requirements

4.1 Expected Results

Deploying a personalized, real-time ECG monitoring and analysis system through a structured model training process and an integrated concept architecture is anticipated to deliver several outcomes:

- **Immediate Identification**: The system should provide prompt detection of deviations or anomalies in an individual's ECG, as compared to their personal or population norm.
- **Enhanced Accuracy**: By leveraging personal models, which are regularly updated with new data, the system is expected to demonstrate high accuracy in identifying actual cardiac events and minimizing false alarms.
- **Timely Intervention**: Early detection of potential cardiac events allows for swift medical intervention, which can be crucial in managing and potentially mitigating adverse cardiac events.

- **Continuous Monitoring**: The usage of wearable devices ensures potentially uninterrupted monitoring of an individual's cardiac activity, providing comprehensive data for ongoing health management.

4.2 Requirements

In order to achieve the aforementioned expected results, several requirements need to be fulfilled in both the model training process and the overall concept architecture:

- **Data Quality**: The ECG data collected must be of high quality, necessitating reliable wearable devices and effective denoising and pre-processing algorithms.
- **Robust Models**: The personal norm models must be robust and adaptive, capable of accurately reflecting an individual's typical cardiac activity and identifying deviations.
- **Computational Efficiency**: The algorithms deployed must be computationally efficient to ensure prompt analysis without unreasonable device overloading or draining device battery.
- **Secure Data Transmission**: Ensure safe and secure transmission of ECG data during external analysis to protect patient privacy and comply with healthcare data regulations.
- **User-Friendly Interface and Accessibility**: The wearable device and any associated software should have a user-friendly interface to facilitate ease of correct use. The system should be accessible and usable by individuals with varied levels of technological proficiency, and appropriate training should be available.

5 Challenges and Solutions

Creating and implementing a personalized, real-time ECG monitoring and analysis system presents various challenges, both from technical and user-adoption perspectives. Addressing these challenges necessitates the identification and application of strategic solutions.

5.1 Challenges

- **Data Integrity**: Ensuring consistent collection of high-quality, reliable ECG data from wearable devices amidst varied external conditions.
- **Model Accuracy**: Maintaining high accuracy in the personal models despite the inherent variability in individual ECG data.

- **User Compliance**: Achieving widespread user adoption and compliance, given the technological and usability barriers that may exist for certain user demographics.
- **Security and Privacy**: Protecting the sensitive health data during collection, transmission, and storage phases, abiding by healthcare data regulations.
- **Technical Limitations**: Wearable devices may have limitations in terms of computational power, battery life, and data storage capacity, which could impact the efficacy of real-time analysis and continuous monitoring.

5.2 Solutions

- **Advanced Algorithms**: Employing advanced denoising and pre-processing algorithms to ensure that the ECG data used in model training and real-time analysis is of the highest quality.
- **Adaptive Model Training**: Utilizing adaptive algorithms that continually refine the personalized model with new data, ensuring it remains accurate and reflective of the individual's current cardiac health status.
- **User Training and Support**: Implementing comprehensive user training programs and providing ongoing support to facilitate user adoption and ensure correct usage of the wearable devices. The initial setup of a wearable device may be conducted by trained medical personnel upon medical check-ups or specific healthcare programs.
- **Data Encryption or Local Computing**: Employing robust encryption methods to secure data during transmission and ensuring secure storage solutions that are compliant with healthcare data protection regulations. Nevertheless, healthcare data regulations may vary, thus more attention to local laws is required to avoid legal and ethical challenges. For cases, where medical data transmission and/or external processing is restricted, the local computing directly on a wearable device can be used as a compromise.
- **Optimized Algorithms**: Developing computationally efficient and lightweight algorithms that ensure effective real-time analysis while minimizing the computational load on wearable devices to conserve battery life and optimize data storage.

Addressing these challenges with effective and practical solutions is crucial to ensuring the feasibility, reliability, and success of the personalized ECG monitoring and analysis system, thereby allowing it to deliver on its promise of providing precise and timely insights into individual cardiac health.

6 Conclusion and Future Directions

Building a personalized ECG monitoring and analysis system involves careful planning and execution across several stages, including data collection, model training, and the application of a comprehensive architecture to ensure real-time monitoring and accurate anomaly detection. This endeavor poses multiple challenges like maintaining data quality, ensuring model precision, user adherence, and securing health data.

The methodologies discussed propose a pathway toward a future where ECG monitoring can actively contribute to real-time health management. The suggested conceptual framework aligns with the emerging focus on personalized healthcare.

Looking ahead, specific areas warrant additional research and development:

- **Model Effectiveness**: Additional studies to determine the sustained accuracy and reliability of personal models in accurately depicting and forecasting individual cardiac incidents.
- **User Interaction**: Ongoing improvements to user interfaces and user experience to ensure the systems are accessible and user-friendly across various user groups.
- **Data Security**: More research to enhance the safety and privacy of data during all stages of the process, especially considering new cybersecurity threats.
- **Algorithm Enhancement**: Developing new algorithms that can further improve the accuracy and efficiency of real-time ECG analysis, especially on wearable devices with limited computational capabilities and lower ECG signal quality.

In conclusion, the development of integrated, real-time, and personalized ECG monitoring systems is an ongoing journey, connecting technological advancements and comprehensive healthcare solutions. The aim is to make these systems not just supplementary tools but essential elements in guiding personalized healthcare, offering a blend of technological capabilities and clinical expertise. Through continued research, technological progression, and a consistent emphasis on user-centered design and data security, we step toward a future where personalized, real-time ECG monitoring becomes a seamless, trustworthy, and essential aspect of daily health management.

References

1. Han C, Song Y, Lim H, Tae Y, Jang J, Lee B et al (2021) Automated detection of acute myocardial infarction using asynchronous electrocardiogram signals—preview of implementing artificial intelligence with multichannel electrocardiographs obtained from smartwatches: retrospective study. J Med Internet Res 23(9):e31129. https://doi.org/10.2196/31129
2. Shameer K, Johnson K, Glicksberg B, Dudley J, Sengupta P (2018) Machine learning in cardiovascular medicine: are we there yet? Heart 104(14):1156–1164. https://doi.org/10.1136/heartjnl-2017-311198

3. Han C, Kang K, Kim T, Uhm J, Park J, Jung I et al (2022) Artificial intelligence-enabled ECG algorithm for the prediction of coronary artery calcification. Front Cardiovascular Med 9. https://doi.org/10.3389/fcvm.2022.849223
4. Medina-Inojosa KJ, Noseworthy P, Rodeheffer R, Lopez-Jimenez F, Attia Z et al (2021) Artificial intelligence–augmented electrocardiogram detection of left ventricular systolic dysfunction in the general population Mayo Clinic Proc 96(10):2576–2586. https://doi.org/10.1016/j.mayocp.2021.02.029
5. Chaikovsky AP (2023) Advances in the analysis of electrocardiogram in context of mass screening: technological trends and application of AI anomaly detection. In: Qaisar SM, Nisar H, Subasi A (eds) Advances in non-invasive biomedical signal sensing and processing with machine learning. Springer, Cham. https://doi.org/10.1007/978-3-031-23239-8_5
6. Wang L, Zhang W, Guan M, Jiang S, Fan M, Abu P et al (2019) A low-power high-data-transmission multi-lead ECG acquisition sensor system. Sensors 19(22):4996. https://doi.org/10.3390/s19224996
7. Sakuma DA, Wang J (2016) Performance of human body communication-based wearable ECG with capacitive coupling electrodes. Healthcare Technol Lett 3(3):222–225. https://doi.org/10.1049/htl.2016.0023
8. Cai Z, Li J, Zhang X, Shen Q, Murray A, Liu C (2019) How accurate are ECG parameters from wearable single-lead ECG system for 24-hours monitoring. https://doi.org/10.22489/cinc.2019.187
9. Tang X, Mo Z (2007) Electrocardiogram classification method based on SVM. https://doi.org/10.2991/iske.2007.282
10. Talbi M, Abid S, Cherif A (2015) EMD-based ECG denoising using source separation. J Mech Med Biol 15(05):1550082. https://doi.org/10.1142/s0219519415500827
11. Zhang D, Wang S, Li F, Tian S, Wang J, Ding X et al (2020) An efficient ECG denoising method based on empirical mode decomposition, sample entropy, and improved threshold function. Wirel Commun Mob Comput 2020:1–11. https://doi.org/10.1155/2020/8811962
12. Rezgui BN, Talbi M (2022) ECG denoising based on transformation matrix for non-decimated wavelet transform and wavelet/total variation (WATV) denoising. https://doi.org/10.21203/rs.3.rs-2164025/v1
13. Singh P, Pradhan G (2018) Variational mode decomposition based ecg denoising using non-local means and wavelet domain filtering. Australas Phys Eng Sci Med 41(4):891–904. https://doi.org/10.1007/s13246-018-0685-0
14. Talbi M (2015) ECG denoising based on PCA and using R peaks detection. https://doi.org/10.5220/0004998201190125
15. Chaikovsky MB, Hodakovsky V, Budnyk V (2019) Method for assessing the cardiac-vascular system based on gender-age and individual standards of ECG and HRV. Ukraine Patent and Trademark Office, Ukraine Patent No. u201911161. https://base.uipv.org/searchINV/search.php?action=viewdetails&IdClaim=268272
16. Boda S, Mahadevappa M, Dutta PK (2023) An automated patient-specific ECG beat classification using LSTM-based recurrent neural networks. Biomed Signal Process Control 84:104756. https://doi.org/10.1016/j.bspc.2023.104756

Atenea Lab: A Softwarized IoT Laboratory for Research and Knowledge Transfer

Marta Fuentes-García, Celia Fernández, Marina Torres, Luis El-Moga-Sarria, and Francisco L. Benítez-Martínez

Abstract The Fourth Industrial Revolution (so-called Industry 4.0) has had a fast growth. A lot of companies are investing more and more money in digital transformation. One of the main trends is using sensors to optimize and enhance industrial processes, as a part of the Industrial Internet of Things (IIoT). Sometimes it would be interesting having an intermediate platform to perform testing prior to deploy the complete IoT ecosystem. This is particularly important in two cases: (i) small and medium enterprises (SMEs), to reduce the risks of errors and minimize the impact of their investment, and (ii) R+D+i, where it is needed to evaluate solutions and perform experiments in a controlled way. Atenea Lab is a software-based platform that allows simulating a wide range of realistic IoT scenarios. It permits testing solutions, performing experiments, and evolving the technology readiness level (TRL) of prototypes reducing costs and making it easier to knowledge transfer.

Keywords IoT · Laboratory · Research · Testing · Software · Cybersecurity

1 Introduction

Industry 4.0 has become a reality during the last years. More and more companies embrace the revolution and are willing to apply new technologies into their production process cycle. Even some small and medium enterprises (SMEs) have started to adopt new solutions such as the Internet of Things (IoT) to optimize and improve their products. There is a wide range of areas in which IoT is being applied (e.g. agriculture, manufacturing, or health care). In the particular case of Spain (more specifically in Andalusia), one of the main bases to reactivate the economy after the crisis that we are suffering points to the optimization of farming techniques [1]. For this reason, deploying sensors in cultivated fields and using the IoT as a monitoring

WWW home page: https://www.fidesol.org/.

M. Fuentes-García · C. Fernández (✉) · M. Torres · L. El-Moga-Sarria · F. L. Benítez-Martínez
COSCYBER, Fundación I+D del Software Libre (Fidesol), Granada, Spain
e-mail: cfernandez@fidesol.org

X.-S. Yang et al. (eds.), *Proceedings of Ninth International Congress on Information and Communication Technology*, Lecture Notes in Networks and Systems 1000,
https://doi.org/10.1007/978-981-97-3289-0_17

mechanism might imply a huge advance in the agriculture sector [2, 3]. This is only one example about how the Fourth Industrial Revolution is reaching to each sector in our society.

However, sometimes the investment in the design and deployment of an IoT system cannot be identified at a first glance as a potential benefit in a medium-long time. In addition, it is needed to take into account the need of experts to make valuable the collected data as well as interpret them properly.

On the other hand, this revolution has been accelerated mainly due to the situations derivated from the COVID and the war in Ukraine. Furthermore, the rapid evolution of artificial intelligence (AI) techniques has promoted that, sometimes, cybersecurity does not walk at the same pace as the technology neither that the cyber criminal does. This affects particularly to the IoT context, due to devices have often restricted resources and their dynamic nature. Most of the existing solutions were designed for traditional networks and need to be adapted to current environments [4]. In addition, these solutions are usually expensive and unaffordable for SMEs.

1.1 Motivation

There are a number of variate projects in Fidesol, some of them are IoT-related. In an IoT project, it is interesting to have the possibility of carrying out a first deployment that helps us to make an idea on how it is going to work. We believe this should be something similar to a sketch or mockup. On the other hand, EGIDA is a special kind of project where we have had different sub-projects and prototypes that we needed to build. These projects and prototypes could be deployed in different sectors (e.g., industry or food). Actually, EGIDA is a National Excellence Network whose goal is to do a step up in relation to security and privacy. A wide range of security problems and challenges are covered in EGIDA, and some of them are IoT-related. This encouraged us to think about an IoT laboratory to be able to simulate our use cases and try our solutions. These solutions are the result of a previous research in the corresponding topic.

1.2 Proposal and Key Points

We propose to create an IoT laboratory that allows us to develop different IoT environments. The main goal is to be able to simulate different devices and communication protocols to reproduce real-life scenarios in a controlled way. Thanks to this laboratory, we will be able to:

- Design and test IoT solutions for our clients, reducing the developing costs.
- Design and test cybersecurity solutions for IoT scenarios in a controlled and isolated way, ranging from encryption to anomaly detection.

- Promote knowledge transfer from basic to applied research and, then, to real world and companies.
- Generate realistic datasets for researching, for example, in machine learning (ML) and artificial intelligence (AI) algorithms.

1.3 Contribution

Thus, the goals and needs of the IoT laboratory which we call ***Atenea Lab*** are:

- To be a **bridge** between basic research (theoretical-practical experiments) and prototypes (some of them deployed in the real world).
- To provide an **abstraction** on the devices that only takes into account some of their relevant features, such as power consumption, battery life, or behavior. It is not needed to simulate hardware or specific components (e.g. brands or specific processors).
- To have a kind of **Lego**®[1] where we can build whatever we need under the knowledge that it is **realistic**. In addition, with this approach, we will have **flexibility** and **versatility** to build our scenarios or use cases in an easy manner.
- To be able to **implement our solutions** once we have simulated the use case. We will be able to launch our experiments (previously designed) and evaluate their performance.

The rest of the paper is organized as follows. In Sect. 2, we review the state of the art and related work to our proposal. In Sect. 3, we present the Atenea Lab, describing the concept, definition, and high-level architecture. In Sect. 4, we present a first prototype to illustrate the use of the Atenea Lab. Finally, in Sect. 5 we present the main conclusion of this paper and future lines of work.

2 State of the Art

The use of IoT devices has exponentially grown during the last years. This has been particularly significant for those industries that aim to optimize their production and monitor their activities. Prior to carry out the deployment of an IoT ecosystem, it would be advisable for companies to test the proposed design. To do that, there are different options. One of these options is hiring external companies, which usually are limited by the IO-link standard [5]. Another alternative is testing the sensors in a controlled physical environment, under simulated conditions. We can find some examples of this in IoT Lab,[2] that was developed as a result of an European project,

---

[1] https://bit.ly/3YX6u4r, https://en.wikipedia.org/wiki/Lego.

[2] https://iotlab.com/es/.

Urbana IoT Lab,[3] or *Laboratorio de internet de las cosas (IoT).*[4] However, this type of solution is usually an expensive and complex process. For this reason, there have been other proposals that aim to simulate real conditions using software rather than physical environments [6–9].

For example, IoT device simulator allows connecting a huge number of devices and provides a flexible and scalable solution simulating the IoT architecture using software.[5] It is needed to pay for using this solution according to the amount of hours per day of running the simulation and the amount of devices to be simulated. Other solutions are focused on data simulation, to generate both sensor values and emulate possible problems during the connection.[6] OMNeT++ is an IoT simulation software for Massive IoT. It was developed using C++ and following a component-based architecture. One of the main uses of this solution is to study network issues [6]. FLoRaSaT is an OMNeT++-based solution that is used to simulate IoT environments using the standard LoRaWAN Low-Power Wide Area (LPWAN) protocol [7]. Both solutions are restricted to specific communication protocols and the scenario allows a reduced amount of devices [6, 7]. Other authors propose solutions that aim to generate huge amount of 5G data [8] or that simulate the code that will be run over the actual layers of the IoT architecture [9]. The former can generate a number of IoT devices and allows interacting with the real world. The latter is focused on low-level features, such as communications network details. Finally, VIoLET is an environment for IoT emulation at a large scale that is built using virtual machines [10]. Most of the aforementioned solutions have at least one of the following limitations: lack of a (usable) Graphical User Interface (GUI), high costs (especially for researching purposes or for SMEs), or limited possibilities for researching.

3 Our Proposal: Atenea Lab

There are different architectures for IoT environments [11–14]. This work is aligned with the latest, which proposes the following layers (from bottom to top), and it is represented in Fig. 1:

- **Physical Layer** (e.g., sensors and gateways). This is also known by other authors as "perception" layer.
- **Communications Layer** (e.g., data transport and data format unification).
- **Management Layer** (e.g., to allow devices join to the environment or to store values in a database).
- **Application and Analytics Layer** (e.g., data visualization, analysis, and reporting).

[3] https://urbanasmart.com/iot-lab/.
[4] https://bit.ly/3Nzsf7U.
[5] https://go.aws/3XdfdzO.
[6] https://bit.ly/3NALApi.

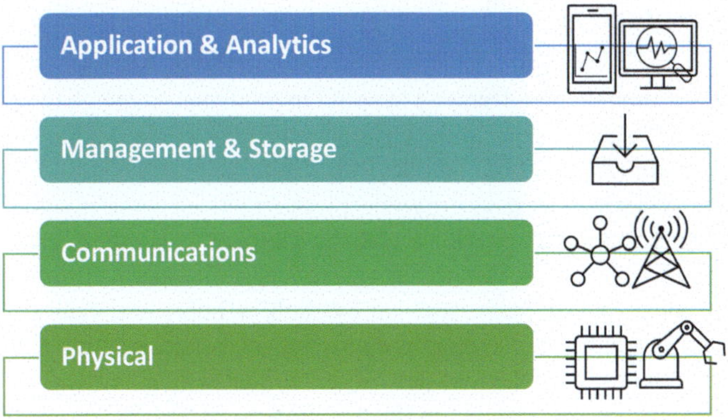

Fig. 1 IoT layers covered by Atenea Lab

Building an IoT laboratory is not an easy procedure, even more if we want to have an abstraction that allows us to build any use case. For this reason, we propose an incremental design and development of the lab. To do that, we started with a simplified use case. Like this, we identified which are the minimum elements that are needed to build a complete scenario. An IoT environment is composed of devices, which are connected by a communication protocol that allows them sending their values to a processing center. In addition, each IoT device should have at least one sensor, which is in charge of detect and measure the value of interest. The minimum IoT scenario could be a device with a sensor (e.g. a temperature sensor), which is sending the measured values to an end device (e.g. a server or smartphone) for its processing. To send the values, it needed to implement a communication protocol. Finally, the values should be stored in a persistent medium.

According to this first use case, we have designed and developed the first release of Atenea Lab. This functional prototype has the following features:

- Physical layer. It allows adding devices, which are composed of at least one sensor and also have a behavior simulation function assigned.
- Communications layer. It has the MQTT protocol implemented, which is one of the most extended communications protocols in IoT environments.
- Management and Storage layer. It has a Mongo DB instance running for the persistent storage of the received data.
- Application and Analytics layer. It implements an intuitive, usable, user-centered GUI that allows building and editing IoT scenarios, as well as visualizing the simulation results after running such scenarios.

The following paragraphs aim to describe the way in which we have built the Atenea Lab.

3.1 Architecture

Atenea Lab was thought and designed keeping in mind one single goal: reaching an abstraction of IoT devices that can be customized and/or configured to simulate the behavior and main resources of such devices. To do that, we proposed following a **Lego**® approach in which we have different "pieces" that, not being an actual/true representation of the real world, provide a realistic scenario. In addition, these "pieces" allow us to build different use cases, ranging from agrifood to manufacturing (IIoT), but including other possibilities, such as e-health or home automation. Having a set of protocols available to use together with the devices, like a "glue" that allows the communication among devices and edge computers (e.g., servers or personal computers), is another key element in this solution.

The architecture of the solution has been designed and developed following a component-based approach to provide independence among the layers. It is based on containers, where each element is executed isolated from the remaining. This allows the system to grow horizontally and deploy complex IoT networks in an easy way. Furthermore, this provides independence, flexibility, and scalability of the laboratory, making it possible to include different communication protocols as well as new databases in the future easily.

Thus, the architecture of the solution allows users to build a wide variety of simulated IoT environments, where their main structure is outlined as follows:

- **Environment**. An environment is a scenario where the user can define the executions along a defined time. This configuration includes a description, the communication protocol employed, and the type of database. In this release, it is only possible to select the MQTT protocol and the Mongo DB database.
- **Device**. Belonging to an environment, one or more devices can be configured with certain properties, such as name, description, and the list of attached sensors.
- **Sensor**. A sensor is defined as an entity with several properties, including a descriptive name and description, data types, units, values, and the data generation interval. A behavior generation function is also assigned to each sensor, which sends the generated data by using very small binary packages, to simulate the real-time conditions.

By creating and combining the outlined items, a great variety of scenarios can be simulated. To define a scenario, at least one of each element is required.

3.2 User Operation

Atenea Lab has been designed following a user-centered approach. As a result, we propose an interactive dashboard that allows users to configure a custom IoT scenario in a simple and easy way. First of all, it is needed to create a scenario, including name, description, communication protocol, and database. Once the environment is created, a set of devices and sensors can be easily created by an interactive wizard. During the process, the user is asked to provide a name and description for each device and sensor. Sensors need to complete additional information, including type of the measurement, units, rate of measurements, and type of the generation functions. In this release of the Atenea Lab, these functions are:

- **Constant**. The same value is returned along the execution.
- **Linear function**. An initial value and an increment are required.
- **Random values**. A completely random value in a defined interval is generated in each measurement.
- **Trigonometrical-based values**. A value is computed by using trigonometrical functions (sine, cosine, exponential sine, exponential cosine). This type of function is more useful to simulate more complex behaviors, such as temperature degradation or fuel consumption.

After all devices and sensors are properly configured, the environment can be started. The system automatically deploys a set of containers in the infrastructure and starts data generation. After the defined time in the environment, the execution is automatically stopped. It can also be manually stopped and/or paused.

In general, the flow of an execution after deploying an environment is the following: (i) for each sensor in a device, an execution thread is created, where periodically (following a defined rate), a value is read by following the mathematical function defined; (ii) each measurement is sent using the selected communication protocol, which unpacks the data and sends the information to the database via API. The gathered data can be subsequently analyzed by querying this database. A practical demo can be found in https://bit.ly/AteneaLab.

4 Atenea Lab in Practice

4.1 Test and Limitations

The Atenea Lab has been tested under different stress conditions (*e.g.*, increasing the sending frequency, defining a high number of sensors and devices) to verify its correctness and performance. Below can be found some of the main limitations that we have found:

- **Data loss**. When a high frequency of sending data is defined (1 s), a medium number of devices (9 devices), and a high number of sensors per device is defined (18 sensors/device), it is possible that some data are lost. This is due to the broker cannot manage such amount of data at a high pace. However, this might not be realistic, since most times it is not needed to send information so frequently neither having such amount of sensors.

– **Band with reduction**. When there is a high number of sensors and devices defined and the frequency of sending data is also high, the reception of the messages by the subscribers might be delayed. In any case, the received messages are stored in the database.

4.2 An Agrifood Prototype

Primary sector is one of the most relevant in Andalusia. One of the most important areas within this sector is agriculture. After years of crisis, one of the bases for the reactivation of this sector is to optimize the cultivation techniques and embrace the digital transformation. This implies the application of IoT as a monitoring mean, which has provided lots of benefits to those enterprises that have implemented it.

In the context of the EGIDA project,[7] and in collaboration with the University of Granada, we have developed a prototype for anomaly detection in network traffic for the agrifood sector. The goals of this prototype were: testing the Atenea Lab in a realistic use case in the context of agrifood, and applying for the first time MSNM sensor (MSNM-S)[8] in an IoT-based scenario. To do this, we built a simple use case in Atenea Lab with two devices: one for the environment and another for the soil. For this prototype, we configured the environment with MQTT as the communication protocol and MongoDB for the database. Each device had a temperature and a humidity sensor. We had one MQTT publisher for each device that sent the generated data for the sensors to the MQTT broker. A subscriber read the collected data and stored them into the database. We ran the scenario to verify that it was working properly. After that, we deployed four instances of the MSNM-S: one on each device, another in the MQTT broker, and the last one in the database. Then, we started the simulation of the complete scenario, including the MSNM-S and collecting data under normal operation conditions. Finally, we launched some controlled attacks to try the capability of anomaly detection of the MSNM-S in this new context. The attacks were properly detected, yielding more than satisfactory results. More information and results of these experiments can be found in a paper that will be sent to the Special Sessions of this Forum.

5 Conclusions and Future Work

In this work, we present Atenea Lab, an IoT laboratory that allows building multiple scenarios in a simple way. This platform has been designed and built with the purpose of reducing the gap between academia and enterprises (particularly, SMEs), helping

---

[7] https://egidacybersecurity.com/.

[8] https://github.com/nesg-ugr/msnm-sensor.

to knowledge transfer and digital transformation. This is due to Atenea Lab helps to perform research experiments in realistic environments, obtain IoT datasets, and evolve IoT prototypes until TRL 5 or 6.

This first release of Atenea Lab is a functional prototype that allows: (i) building a complete IoT scenario (sensors, devices, communication protocol and database), (ii) simulating a complete IoT scenario (sensor behavior can be simulated, communication, and storage are working properly), and (iii) interacting with Atenea Lab in a graphical and intuitive manner (thanks to its user-centered design). The stress testings performed have unveiled some limitations that can be easily solved. In addition, a first use case has been implemented using the Atenea Lab. This is a cybersecurity prototype for anomaly detection. The prototype has been tested under different attacks, being all of them properly detected.

As a future work, we plan to evolve the Atenea Lab at least in the following points: (i) allow registering different users, (ii) make the laboratory accessible externally from Fidesol, (iii) define and try more complex use cases, (iv) perform additional stress testing to define better limitations, and (v) evaluate usability and user experience with different groups of users.

Acknowledgements This work was funded by Cervera Centros Tecnológicos from the Centro Español para el Desarrollo de Tecnología (CDTI) in the context of the EGIDA project (CER-20191012). In addition, authors of this paper would like to acknowledge and recognize the work of all our intranet at Fidesol, who have made this project possible.

References

1. PWC (2020) How fake news has exploited COVID-19. https://cutt.ly/6geEVST, [Online, Accessed on 07 Oct 2020]
2. Ray PP (2017) In: Internet of things for smart agriculture: technologies, practices and future direction. vol 9. IOS Press, pp 395–420
3. Tzounis A, Katsoulas N, Bartzanas T, Kittas C (2017) Internet of things in agriculture, recent advances and future challenges. Biosyst Eng 164:31–48
4. Fuentes-García M, Camacho J, Maciá-Fernández G (2021) Present and future of network security monitoring. IEEE Access 9:112744–112760
5. IFM (2021) Moneo: la plataforma IIoT para industria y fabricación"
6. Josifović K, Boljević S, Ninković V, Turćinović N (2019) Simulating massive IOT environmental monitoring scenario using omnet++. In: 2019 27th Telecommunications forum (TELFOR), pp 1–4
7. Fraire JA, Madoery P, Mesbah MA, Iova O, Valois F (2022) Simulating lora-based direct-to-satellite IOT networks with florasat. In: 2022 IEEE 23rd International symposium on a World of wireless, mobile and multimedia networks (WoWMoM), pp 464–470
8. Dayalan UK, Fezeu RAK, Salo TJ, Zhang ZL (2022) Kaala. In: Proceedings of the ACM SIGCOMM workshop on networked sensing systems for a sustainable society, ACM
9. Barriga JA, Clemente PJ (2022) Designing and simulating IoT environments by using a model-driven approach. In: 2022 17th Iberian conference on information systems and technologies (CISTI). IEEE
10. Baheti S, Badiger S, Simmhan Y (2020) VIoLET. 5(3):1–39

11. Mrabet H, Belguith S, Alhomoud A, Jemai A (2020) A survey of IoT security based on a layered architecture of sensing and data analysis. vol 20(13)
12. Altexsoft (2020) In: IoT architecture: the pathway from physical signals to business decisions
13. Simmons A (2020) Internet of things (IoT) architecture: layers explained
14. Agar R (2023) IoT architecture guide. Major and additional layers of IoT system

Exploring the Antecedents of Purchase Intention for Organic and Healthy Foods: A Systematic Literature Review

Yasmina Alami and Salima Jazi

Abstract Nowadays, we are witnessing the development of health products' market such as health foods which represent new growth drivers for companies; well-being becoming the new concern of consumers. In this regard, the notion of food is intimately linked to the notion of health. Several studies show that a healthy diet contributes to maintaining the proper functioning of the body and promotes well-being in general. The expression "prevention is better than cure" has never been more relevant. Indeed, consumers of all ages are increasingly concerned about their health. This trend is encouraging them to learn more about organic products and seek out healthier foods. Our study represents a systematic literature review of current research between 2010 and 2023 related to antecedents and attributes influencing purchase intention for organic products and healthy foods. The aim of this study is, firstly, to synthesize the different antecedents and attributes that influence the purchase intention of organic and healthy food products, and secondly, to propose a conceptual model founded on the Theory of Planned Behavior (TPB) Ajzen's in Org Behav Human Dec Proc 50(2):179–211 (1991), which states that behavior is the result of reasoned decisions. This theory identifies three factors influencing intention, namely attitude toward behavior, subjective norms, and perceived behavioral control. We propose through our study a conceptual model based on the TPB model in exploring the antecedents of purchase intention, in addition to incorporating other antecedents identified by existing research in the literature.

Keywords Purchase intention · Healthy food · Organic food · Systematic literature review

Y. Alami · S. Jazi (✉)
University Hassan First of Settat, Ecole Nationale de Commerce et de Gestion, LR2MC Settat, Morocco
e-mail: jazisalima@gmail.com

X.-S. Yang et al. (eds.), *Proceedings of Ninth International Congress on Information and Communication Technology*, Lecture Notes in Networks and Systems 1000,
https://doi.org/10.1007/978-981-97-3289-0_18

1 Introduction

During the last decades, consumers have become worried about the quality of the food on their plates: its origin, composition, etc. A new food trend has emerged, that of the consumption of organic healthy food. Thus, we notice important changes in consumer behavior, which are increasingly demanding and seeking information before purchasing a specific food product. These changes impact the consumer's purchase intention, which consequently influences their final behavior. Our study represents a systematic literature review of current research between 2010 and 2023 related to antecedents and attributes influencing purchase intention with the goal of complementing and expanding the current literature. The objective of this study is to synthesize the different antecedents and attributes that influence purchase intention in order to propose a conceptual model. Thus, our main research question is: ***Which attributes and antecedents influence purchase intention of organic and healthy food products?*** Our paper is structured as follows. The first section defines our main theoretical concept being purchase intention. In the second section, we present our detailed research methodology with a proposed conceptual model based on the Theory of Planned Behavior [1] for potential future research on this topic.

2 Purchase Intention

Purchase intention refers to the set of motivations stimulated by products and incites the subjective probability of purchasing these products [2, 3]. It is a multifaceted process that is intricately linked to consumer behavior, perceptions, and attitudes, making it an effective tool for forecasting the buying process [4]. As per Morwitz and Schmittlein [5], purchase intentions can be a measure to guess the behavior of customers buying decision.

3 Organic Food

Several authors have highlighted the pro-environmental and health benefits of organic foods. Many researchers have emphasized the positive aspects of organic food [6–9]. Fotopoulos and Krystallis [10] have identified consumers' perceptions that organic food is not only healthier but also more flavorful and as safe as conventional options. Lea and Worsley [11] have noted that organic food is generally considered healthier when compared to conventional food products. Moreover, consumers tend to perceive organic food as safer due to the absence of pesticides, fertilizers, chemicals, preservatives, hormones, and antibiotics in their production [12–14]. Health stands as a primary motivator for consumers when selecting organic food products. This choice is influenced by concerns about health problems. Hence, some consumers opt for organic food as a protective measure [15–19].

4 Healthy Food

As per Hansen and Thomsen [20], the literature review reveals three different definitions of healthy eating.

4.1 Healthy/Unhealthy Eating Definition

Consumers of food products often frame healthy eating as a delicate equilibrium between the consumption of healthy and unhealthy food [21, 22]. This perspective allows for the concept that consuming less healthy items can be offset by increased consumption of fruits and vegetables [20]. This phenomenon is further supported by the concept of licensing, which advises that customers are more disposed to buy unhealthy foods after adding healthy items to their shopping carts [23].

4.2 Mind/Body Healthy Eating Definition

This second definition emphasizes an equilibrium between physical and mental well-being [20]. Customers following this definition believe in the intimate connection between the body and mind [24] and consider emotional well-being an integral part of healthy eating [25, 26].

4.3 Healthy Eating Guidelines

Health authorities promote an official definition of healthy food, referred to as healthy eating guidelines [20]. Consumers who adhere to this definition may perceive a risk to their health if they do not follow these guidelines [20]. Moreover, these consumers are inclined to articulate the connection between health and food using scientific and nutritionally informed language [27, 28].

5 Methodology and Results

Denyer and Tranfield [29] assert that in the realm of management and business studies, a systematic literature review typically serves as the initial phase of the scientific research process. It is commonly presented either as a concise introduction to an empirical study or as a comprehensive examination of the existing body of literature. They define this process as a meticulous approach involving the review and scrutiny of extant literature. It entails employing a comprehensive strategy to identify and locate research within the existing literature, followed by a critical evaluation and analysis of the selected literature. The final step involves synthesizing the findings and presenting a detailed account of the review. Importantly, systematic literature reviews serve not only to inform the research community but also to provide valuable guidance to practitioners through the provision of information, examples, and practical recommendations. As such, for our systematic literature review, we will follow the methodology proposed by Denyer and Tranfield [29], which is structured around the following steps:

- **Step 1**: Preparation of the research question
- **Step 2**: Search for studies and works related to the research question
- **Step 3**: Selection and evaluation of studies
- **Step 4**: Analysis and synthesis
- **Step 5**: Results and reporting.

5.1 Step 1—Preparation of the Research Question

As pointed out by Light and Pillemar [30], the key decision in preparing a review is to establish its purpose. By formulating clear questions as suggested by Cooper and Hedges [31], the research is then more clear and consistent throughout its stages and process. Thus for our systematic literature review, our research question is formulated as follows: ***Which attributes and antecedents influence purchase intention of organic and healthy food products?***

5.2 Step 2—Search for Studies and Works Related to the Research Question

As proposed by Denyer and Tranfield [29], systematic literature reviews aim to search, select, and evaluate as much as possible research relevant to the research question(s) formulated, using methods such as database searches, specialized bibliographies, or manual searches of specialized journals and newspapers. For our systematic literature review, we used the Google Scholar search engine, which is a service of the Google search engine that allows the search of scientific articles and publications.

Table 1 Inclusion and exclusion criteria

Exclusion criteria	Inclusion criteria
• Articles published before 2010 • Articles written in languages other than English and French • Articles dealing with other aspects of purchase intent other than antecedents and attributes	• Articles discussing conceptual models of antecedents and attributes of purchase intention • Articles written in French and English • Articles published between 2010 and 2023

We explored several databases during our search, including Springer, ScienceDirect, and Elsevier; for which we used keywords such as *purchase intention, determinants of Purchase intention, Purchase intention AND Organic food, Purchase intention AND healthy food.* Once downloaded, all articles were imported into the ZOTERO software with the elimination of duplicate articles.

5.3 Step 3—Selection and Evaluation of Studies

As recommended by Denyer and Tranfield [29], this step in the systematic literature review process is essential to maintain transparency. It involves the utilization of explicit selection criteria to evaluate the relevance of each identified study. The objective is to determine if each study effectively addresses the research question posed in the review. During this phase, all decisions regarding the inclusion or exclusion of sources of information are meticulously documented. The rationale for inclusion and exclusion forms the basis of these selection criteria. In our systematic literature review, we applied a specific set of inclusion and exclusion criteria, as outlined in Table 1.

In the previous step, we downloaded and imported a total of 135 articles into ZOTERO, which we reviewed and to which we applied the inclusion and exclusion criteria mentioned above. At the end of this step, 102 articles were eliminated and 33 were retained for the next step of analysis.

5.4 Step 4—Analysis and Synthesis

After gathering the appropriate articles from relevant sources according to our selection criteria, the next step is the analysis and synthesis of the data. Within the framework of the research question formulated for our systematic literature review. The most frequent variables, which are attitude, subjective norms, and perceived behavioral control, represent the determinants of intention according to Azjen's Theory of Planned Behavior [1]. In addition to the variables of the Theory of Planned Behavior, we have grouped the most frequent antecedents of purchase intention, we have examined all their definitions in the literature and then we have assembled them under the same antecedent/attribute. This will allow us to develop our proposed conceptual model:

Product quality: Perceived quality, Perceived product quality, Product quality, Food safety, Food safety concern, Nutritional content.
Environmental concern: Environmental concern, Perceived environmental responsibility, Environmental Knowledge.
Price consciousness: Perceived price, Price consciousness.
Health consciousness: Health, Health behavior, Health value, Health Consciousness, Health orientation, Consumer health concern, Healthy food knowledge, Knowledge about health, Health awareness.

5.5 Step 5—Results and Reporting

Figure 1 represents a proposed conceptual model of the antecedents and attributes that influence purchase intention for potential future research on this topic.

For the design of our model, we have taken inspiration from the Theory of Planned Behavior model and added four other variables raised in our study, which were the most frequent ones, namely: environmental concern, health consciousness, price consciousness, and product quality.

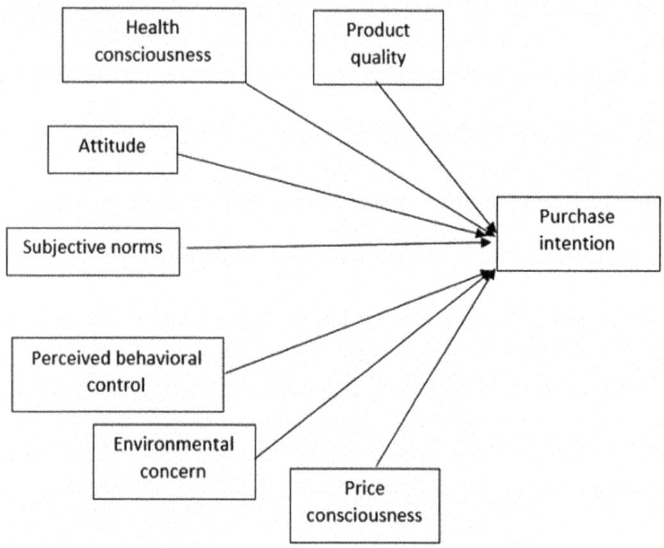

Fig.1 Conceptual model project proposal of antecedents of purchase intention

Theory of Planned Behavior (TPB). Ajzen's Theory of Planned Behavior (TPB) [1] is a significant social cognitive model that explains variations in volitional behavior and is an extension of Fishbein and Ajzen's Theory of Reasoned Action (TRA) [2]. TPB introduces the concept of Perceived Behavioral Control (PBC), addressing criticisms of TRA, especially in analyzing behaviors with limited volitional control [32, 33]. PBC directly influences both intention and behavior, leading to the model's name change from TRA to TPB [1, 34]. According to Ajzen [1], TPB emphasizes that intention is the primary determinant of an individual's behavior, with higher intention associated with more vigilant behavior monitoring. Additionally, Hidayat et al. [35] suggest that TPB is applicable in studying consumer switching behavior related to healthy food products.

Attitude. In TPB, attitude is defined as "the evaluative effect of positive or negative feelings that individuals have toward performing a particular behavior" [36]. It indicates how individuals assess a specific subject [1]. In the context of organic food consumption, attitude significantly influences purchase intention, with studies demonstrating a positive relationship [37–40]. Studies in the field of organic food have additionally shown that one's attitude can serve as a strong predictor of behavioral intention [41–43].

Subjective norms. As described by Fishbein and Ajzen [36], involve an individual's perception of others' approval or disapproval of a specific behavior, reflecting perceived social pressure [1]. Multiple studies have revealed a strong connection between customers' intention to buy organic food and their subjective norms [43, 44]. Reference groups such as family, friends, and colleagues significantly influence the purchase and consumption of organic food [45].

Perceived Behavioral Control. One of the fundamental concepts in TPB is the perceived behavioral control, and it signifies the extent to which an individual believes they can effectively carry out a specific behavior [1, 6, 46]. Ajzen [46] recommended that perceived behavioral control should be understood as one's perception of control over the execution of a behavior. In the context of food purchase, perceived behavioral control encompasses the consumer's ability to overcome disincentives to purchase organic food products and the availability of resources that can facilitate this behavior [45].

Product Quality. Perceived product quality pertains to the perceived capacity of a product to provide satisfaction compared to available alternatives [47]. It is becoming gradually important in the context of the fast consumption of organic food [48, 49]. Consumers often buy organic food because they perceive it to have superior or excellent quality compared to conventionally grown counterparts [50]. Consequently, product quality acts as a powerful motivator, changing positive attitudes into real buying actions [51]. Seven food attributes, including appearance, freshness, naturalness, absence of substitutes, nutritional value, safety, and taste, can be surveyed to measure the quality of organic food [52].

Environmental concern. Representing individuals' awareness and support for addressing environmental issues [53], relates to their apprehensions about environmental quality and national well-being [54]. It strongly connects to commitment for environmental protection [55]. In the context of organic food purchases, environmental concern significantly drives purchase intention [56], actual purchases [57], and willingness to pay a price premium [58, 59]. It also leads to eco-friendly behaviors like food waste reduction [60] and recycling [61], making environmentally conscious consumers willing to pay a premium for ethically oriented products [62, 63].

Price consciousness. Price consciousness plays an important role in purchase decisions, mainly for buyers who frequent discount grocery retailers [64]. Given that organic food items are usually more expensive than their non-organic counterparts [65], high prices can deter both purchase intention [66] and consumption [67]. The price difference between organic and conventional products varies depending on factors such as country, product type, and processing degree [68]. For example, in products like fruits and vegetables, price differences can range from 50 to 100%, occasionally reaching up to 250% [69, 70]. Price premiums for organic products typically fall between 15 and 50% though there is variation, particularly with dairy products [71, 72].

Health consciousness. According to Jayanti and Burns [73], health consciousness refers to the extent to which individuals integrate health considerations into their daily routines. It's a key driver in selecting organic products [74, 75] and influences willingness to pay more for them [75, 76]. Consumers often perceive organic food as offering superior health benefits [77]. Health-conscious individuals are more likely to prioritize their health, adopting measures to enhance it, such as consuming nutritious food and engaging in regular exercise [78]. They are also more inclined to invest in their well-being by purchasing healthy products [79] and are influenced by health labels when making buying decisions [80].

6 Conclusion and Limits

Through our systematic literature review, we listed the attributes and antecedents that influence purchase intention that we've found in the marketing literature. We then proposed through our study a conceptual model based on the Theory of Planned Behavior [1] and antecedents that influence purchase intention, for potential future research on this topic, as well as for our research progress. There are some limitations to this research that should be noted. The first limitation is the specific context of the study. We have identified the attributes and factors that influence purchase intention without specifying whether we can generalize them also for online purchase intention. We can question whether these results can be generalized to other organic products, i.e., cosmetics, personal care products, cleaning products, etc. Another limitation is about the country of the study. The attributes and factors that influence purchase intention may change from one country to another.

References

1. Ajzen I (1991) The theory of planned behavior. Org Behav Human Dec Proc 50(2):179–211. https://doi.org/10.1016/0749-5978(91)90020-T
2. Fishbein M, Ajzen I (1975) Belief, attitude, intention and behaviour: an introduction to theory and research
3. Conner M, Armitage CJ (1998) Extending the theory of planned behavior: a review and avenues for further research. J Appl Soc Psychol 28(15):1429–1464. https://doi.org/10.1111/j.1559-1816.1998.tb01685.x
4. Mirabi V, Akbariyeh H, Tahmasebifard H (2015) A study of factors affecting on customers purchase intention case study: the agencies of bono brand tile in Tehran. J Multidisciplinary Eng Sci Technol 2(1):267–273
5. Morwitz VG, Schmittlein D (1992) Using segmentation to improve sales forecasts based on purchase intent: which "intenders" actually buy? J Market Res 29(4):391–405. https://doi.org/10.2307/3172706
6. Chen MF (2007) Consumer attitudes and purchase intentions in relation to organic foods in Taiwan: moderating effects of food-related personality traits. Food Qual Pref 18(7):1008–1021. https://doi.org/10.1016/j.foodqual.2007.04.004
7. Chiu TS, Ortiz J, Chih WH, Pang LC, Huang JJ (2019) Antecedents of consumers' citizenship behaviour towards organic foods. J Consum Behav 18(4):332–349. https://doi.org/10.1002/cb.1774
8. McCarthy B, Murphy L (2013) Who's buying organic food and why? Political consumerism, demographic characteristics and motivations of consumers in North Queensland. Tourism Manage Stud 9(1):72–79
9. Roitner-Schobesberger B, Darnhofer I, Somsook S, Vogl CR (2008) Consumer perceptions of organic foods in Bangkok Thailand. Food Pol 33(2):112–121. https://doi.org/10.1016/j.foodpol.2007.09.004
10. Fotopoulos C, Krystallis A (2002) Purchasing motives and profile of greek consumer a countrywide survey. British Food J 104(9):730–765. https://doi.org/10.1108/00070700210443110
11. Lea E, Worsley T (2005) Australians' organic food beliefs, demographics and values. British Food J 107(11):855–869. https://doi.org/10.1108/00070700510629797
12. Jolly D (1991) Determinants of organic horticultural products consumption based on a sample of California consumers. Acta Horticulture 295:141–148. https://doi.org/10.17660/ActaHortic.1991.295.18
13. Scheifferstein H, Ophius P (1998) Health-related determinants of organic food consumption in the Netherlands. Food Qual Pref 9(3):119–133. https://doi.org/10.1016/S0950-3293(97)00044-X
14. Shaw R, McDonagh P, Prothero A, Shutz C, Stanton J (2007) Who are organic food consumers? A compilation and review of why people purchase organic food. J Cons Behav 94–110. https://doi.org/10.1002/cb.210
15. Makatouni A (2002) What motivates consumers to buy organic food in the UK? Results from a qualitative study. British Food J British Food J 104(3/4/5):345–352. https://doi.org/10.1108/00070700210425769
16. Zanoli R, Naspetti S (2002) Consumer motivations in the purchase of organic food: a means-end approach. British Food J 104(8):643–653. https://doi.org/10.1108/00070700210425930
17. Kotler P, Armstrong G, Cunninghan P (2005) Principles of marketing. Pearson Prentice-Hall, Toronto
18. O'Donovan PM (2002) Irish consumer preference for organic meat. British Food J 104(3/4/5):353–370. https://doi.org/10.1108/00070700210425778
19. Hougthon E, Ritson C (2007) Food consumers and organic agriculture. In Cooper JN (eds) Handbook of organic food safety and quality. Woodhead Publishing limited, Cambridge, pp 74–94. https://doi.org/10.1533/9781845693411.1.74

20. Hansen T, et Thomsen T (2018) The influence of consumers' interest in healthy eating, definitions of healthy eating, and personal values on perceived dietary quality». Food Policy 80:55–67. https://doi.org/10.1016/j.foodpol.2018.09.002
21. Luomala HT, Paasovaara R, Lehtola K (2006) Exploring consumers' health meaning categories: towards a health consumption meaning model. J Consum Behav 5(3):269–279. https://doi.org/10.1002/cb.178
22. Thompson CJ, Troester M (2002) Consumer value systems in the age of postmodern fragmentation: the case of the natural health microculture. J Consumer Res 28(4):550–571. https://doi.org/10.1086/338213
23. Hui SK, Bradlow ET, Fader PS (2009) Testing behavioral hypotheses using an integrated model of grocery store shopping path and purchase behavior. J Consumer Res 36(3):478–493. https://doi.org/10.1086/599046
24. Forstmann M, Burgmer P, Mussweiler T (2012) The mind is willing, but the flesh is weak: the effects of mind-body dualism on health behavior. Psychol Sci 23(10):1239–1245. https://doi.org/10.1177/0956797612442392
25. Geeroms N, Verbeke W, Van Kenhove P (2008) Health advertising to promote fruit and vegetable intake: application of health-related motive segmentation. Food Qual Prefer 19(5):481–497. https://doi.org/10.1016/j.foodqual.2008.02.004
26. Geeroms N, Verbeke W, Van Kenhove P (2008) Consumers' health-related motive orientations and ready meal consumption behavior. Appetite 51(3):704–712. https://doi.org/10.1016/j.appet.2008.06.011
27. Coveney J (2005) A qualitative study exploring socio-economic differences in parental lay knowledge of food and health: implications for public health nutrition. Pub HealthNutr 8:290–297. https://doi.org/10.1079/PHN2004682
28. Delormier T, Frohlich KL, Potvin L (2009) Food and eating as social practice—understanding eating patterns as social phenomena and implications for public health. Sociol Health Illn 31:215–228. https://doi.org/10.1111/j.1467-9566.2008.01128.x
29. Denyer D, Tranfield D (2009) Producing a systematic review. In: Buchanan D, Bryman A (eds) The Sage Handbook of organizational research methods. Sage, London, pp 671–689
30. Light RJ, Pillemer DB (1984) Summing up: the science of reviewing research. Harvard University Press
31. Hedges LV, Cooper H (1994) Research synthesis as a scientific enterprise. The Handbook of Research Synthesis, pp 285–299
32. Liao CJ, Chen L, Yen CY (2007) Theory of planning behaviour (TPB) and customer satisfaction in the continued use of e-service: an integrated model. Comput Hum Behav 23:2804–2822. https://doi.org/10.1016/j.chb.2006.05.006
33. Burton R (2004) Reconceptualising the 'behavioural approach' in agricultural studies: a socio-psychological perspective. J Rural Stud 20(3):359–371. https://doi.org/10.1016/j.jrurstud.2003.12.001
34. Ajzen I (1985) From intentions to actions: a theory of planned behaviour. In: Kuhl J, Beckmann J (eds) Action- control: from cognition to behaviour. springer, Heidelberg, pp 11–39. https://doi.org/10.1007/978-3-642-69746-3_2
35. Hidayat A, Wijaya T, Ishak A, RejekiEkasasi S, Zalzalah GG (2021) Model of the consumer switching behavior related to healthy food products. Sustainability 13(6):3555. https://doi.org/10.3390/su13063555
36. Fishbein M, Ajzen I (1977) Belief, attitude, intention, and behavior: an introduction to theory and research. Addison-Wesley, Reading, MA
37. Enax L, Krajbich I, Weber B (2016) Salient nutrition labels increase the integration of health attributes in food decision-making. Judgment Decision Making 11(5):460–471. https://doi.org/10.1017/S1930297500004563
38. Boobalan K, Nawaz NRMH, Gajenderan V (2021) Influence of altruistic motives on organic food purchase: theory of planned behavior. Sustainability 13(11):6023. https://doi.org/10.3390/su13116023

39. Wang ES-T, Tsai B-K (2014) Consumer response to retail performance of organic food retailers. Br Food J 116(2):212–227. https://doi.org/10.1108/BFJ-05-2012-0123
40. Teng C-C, Wang Y-M (2015) Decisional factors driving organic food consumption: generation of consumer purchase intentions. Br Food J 117(3):1066–1081. https://doi.org/10.1108/BFJ-12-2013-0361
41. Armitage CJ, Conner M (2001) Efficacy of the theory of planned behaviour: a meta-analytic review. Br J Soc Psychol 40(4):471–499. https://doi.org/10.1348/014466601164939
42. Asif M, Xuhui W, Nasiri A, Ayyub S (2018) Determinant factors influencing organic food purchase intention and the moderating role of awareness: a comparativeanalysis. Food Qual Prefer 63:144–150. https://doi.org/10.1016/j.foodqual.2017.08.006
43. Tarkiainen A, Sundqvist S (2005) Subjective norms, attitudes and intentions of Finnish consumers in buying organic food. British Food J 107(11):808–822. https://doi.org/10.1108/00070700510629760
44. Al-Swidi A, Mohammed RafiulHuque S, Haroon Hafeez M, Noor MohdShariff M (2014) The role of subjective norms in theory of planned behavior in the context of organic food consumption. British Food J 116(10):1561–1580. https://doi.org/10.1108/BFJ-05-2013-0105
45. Scalco A, Noventa S, Sartori R, Ceschi A (2017) Predicting organic food consumption: a meta-analytic structural equation model based on the theory of planned behaviour. Appetite 112:235–248. https://doi.org/10.1016/j.appet.2017.02.007
46. Ajzen I (2002) Perceived behavioral control, self-efficacy, locus of control, and the theory of planned behavior. J Appl Soc Psychol 32(4):665–683. https://doi.org/10.1111/j.1559-1816.2002.tb00236.x
47. Monroe KB, Krishnan R (1985) The effect of price on subjective product evaluations, In Jacoby J. et Olson J.C.
48. Magnusson MK, Arvola A, Hursti U-KK, Aberg L, Sjödén P-O (2003) Choice of organic foods is related to perceived consequences for human health and to environmentally friendly behaviour. Appetite 40(2):109–117. https://doi.org/10.1016/S0195-6663(03)00002-3
49. Prakash G, Singh PK, Yadav R (2018) Application of consumer style inventory (CSI) to predict young Indian consumer's intention to purchase organic food products. Food Qual Prefer 68:90–97. https://doi.org/10.1016/j.foodqual.2018.01.015
50. Vindigni G, Janssen MA, Jager W (2002) Organic food consumption: a multi-theoretical framework of consumer decision making. British Food J 104(8):624–642. https://doi.org/10.1108/00070700210425949
51. Joshi Y, Rahman Z (2015) Factors affecting green purchase behaviour and future research directions. Int Strategic Manage Rev 3(1–2):128–143. https://doi.org/10.1016/j.ism.2015.04.001
52. Dolezalova H, Picha K, Navratil J, Vesela M, Švec R (2016) Perception of quality in decision making regarding purchase of organic food. Food Safety Manage 17(153):86–91
53. Dunlap R, Jones R (2002) Environmental concern: conceptual and measurement issues. In Dunlap R, Michelson W (eds) Handbook of environmental sociology. Greenwood, London
54. Bickart BA, Ruth JA (2012) Green eco-seals and advertising persuasion. J Advert 41(4):53–69. https://doi.org/10.2139/ssrn.2036977
55. Bamberg S (2003) How does environmental concern influence specific environmentally related behaviors? A new answer to an old question. J Environ Psychol 23:21–32. https://doi.org/10.1016/S0272-4944(02)00078-6
56. Yin S, Wu L, Du L, Chen M (2010) Consumers' purchase intention of organic food in China. J Sci Food Agric 90:1361–1367. https://doi.org/10.1002/jsfa.3936
57. Albayrak T, Caber M, Moutinho L, Herstein R (2011) The influence of skepticism on green purchase behavior. Int J Bus Soc Sci 2
58. Sriwaranun Y, Gan C, Lee M, Cohen DA (2015) Consumers' willingness to pay for organic products in Thailand. Int J Soc Econ 42:480–510. https://doi.org/10.1108/IJSE-09-2013-0204
59. Krystallis A, Fotopoulos C, Zotos Y (2006) Organic Consumers' profile and their willingness to pay (WTP) for selected organic food products in Greece. J Int Consum Mark 19:81–106. https://doi.org/10.1300/J046v19n01_05

60. Kilbourne W, Pickett G (2008) How materialism affects environmental beliefs, concern, and environmentally responsible behavior. J Bus Res 61:885–893. https://doi.org/10.1016/j.jbusres.2007.09.016
61. Ureña F, Bernabéu R, Olmeda M (2008) Women, men and organic food: differences in their attitudes and willingness to pay. A Spanish case study. Int J Consum Stud 32:18–26. https://doi.org/10.1111/j.1470-6431.2007.00637.x
62. Trudel R, Cotte J (2008) Does being ethical pay? Wall Str J 12
63. Caruana R (2007) A sociological perspective of consumption morality. J Consum Behav 6(5):287–304. https://doi.org/10.1002/cb.222
64. Steenhuis IHM, Waterlander WE, de Mul A (2011) Consumer food choices: the role of price and pricing strategies. Public Health Nutr 14:2220–2226. https://doi.org/10.1017/S1368980011001637
65. Gschwandtner A (2018) The organic food premium: a local assessment in the UK. Int J Econ Bus 25:313–338. https://doi.org/10.1080/13571516.2017.1389842
66. Zakowska-Biemans S (2011) Polish consumer food choices and beliefs about organic food. Br Food J 113:122–137. https://doi.org/10.1108/00070701111097385
67. Bruschi V, Shershneva K, Dolgopolova I, Canavari M, Teuber R (2015) Consumer perception of organic food in emerging markets: evidence from saint Petersburg Russia. Agribusiness 31(3):414–432. https://doi.org/10.1002/agr.21414
68. Michelsen J, Hamm U, Wynen E, Roth E (1999) The European market for o growth and development. Alemania 7
69. Thomposon G (1999) Consumer demand for organic foods. Am Soc Horticultural Science 95th Conference. E. U.
70. Glaser LK, Thompson GD (1999) Demand for organic and conventional frozen vegetables. In: Agricultural and Applied Economics Association (AAEA) Conferences.1999 Annual meeting, August 8-11, Nashville, TN.https://doi.org/10.22004/ag.econ.21583
71. Menghi A (1997) Consumer response to ecological milk in Sweden. Department of Economic SLU Master Thesis
72. Glaser LT (2000) Demand for organic and conventional beverage milk. Ann Meet de la Western Agric Assoc. E.U. https://doi.org/10.22004/ag.econ.36346
73. Jayanti RK, Burns AC (1998) The antecedents of preventive health care behavior: an empirical study. J Acad Mark Sci 26(1):6–15. https://doi.org/10.1177/0092070398261002
74. Hsu S-Y, Chang C-C, Lin T (2016) An analysis of purchase intentions toward organic food on health consciousness and food safety with/under structural equation modeling. Br Food J 118:200–216. https://doi.org/10.1108/BFJ-11-2014-0376
75. Kriwy P, Mecking R-A (2011) Health and environmental consciousness, costs of behaviour and the purchase of organic food. Int J Consum Stud 36:30–37. https://doi.org/10.1111/j.1470-6431.2011.01004.x
76. Didier T, Lucie S (2008) Measuring consumer's willingness to pay for organic and Fair Trade products. Int J Consum Stud 32(5):479–490. https://doi.org/10.1111/j.1470-6431.2008.00714.x
77. Wier M, Jensen KO, Andersen LM, Millock K (2008) The character of demand in mature organic food markets: great Britain and Denmark compared. Food Policy 33(5):406–421. https://doi.org/10.1016/j.foodpol.2008.01.002
78. Michaelidou N, Hassan LM (2008) The role of health consciousness, food safety concern and ethical identity on attitudes and intentions towards organic food. Int J Consumer Stud 32(2). https://doi.org/10.1111/j.1470-6431.2007.00619.x
79. Teng C-C, Lu C-H (2016) Organic food consumption in Taiwan: motives, involvement, and purchase intention under the moderating role of uncertainty. Appetite, 105:95–105. https://doi.org/10.1016/j.appet.2016.05.006
80. Mai R, Hoffmann S (2015) How to combat the unhealthy= tasty intuition: the influencing role of health consciousness. J Public Policy Mark 34(1):63–83. https://doi.org/10.1509/jppm.14.006

Communicating Universe

Petr Klán

Abstract The paper presents a picture of the communicating Universe based on a combination of the worlds of physics and mathematical communication (information) theory. It gives a new view of whole physics, which connects the description of micro- and macro-worlds and brings some new knowledge from the point of communication in the Universe. It is shown that the mathematical tool for communication is deeply rooted in the true nature of physical reality.

Keywords Physics · Communication · Shannon information · Odds · Probability · Ratio · Linking quantum and relativity

1 Introduction

We communicate not only information in our lives. Additionally, it includes ideas, knowledge, experiences, advances, emotions, feelings, etc. C.E. Shannon published the first systematic paper on this subject called A Mathematical Theory of Communication in 1948. The cornerstone of this theory lies in the use of probability. The mathematical quantity, which is applied not only to problems of everyday lives, but also serves to modern scientific thought. This mathematical quantity began to be studied only in the advent of scientific thought. Today, mathematical probability is applied not only to problems of everyday life but also widely in modern science. In this paper, we use probability for the communication of information.

In the last chapter of [1], M. Brooks ponders the true nature of reality and proposes using a language of communication and information. The same chapter (close to the end) introduces the words of physicist J.A. Wheeler from 1990 that in time we will learn to understand all of physics and describe it in the language of information. The author of this paper has reviewed a significant amount of literature but has not found any relevant references to build on directly. Consequently, the author begins with an

P. Klán (✉)
Faculty of Information Technology, Czech Technical University in Prague, Prague, Czech Republic
e-mail: petr.klan@fit.cvut.cz

X.-S. Yang et al. (eds.), *Proceedings of Ninth International Congress on Information and Communication Technology*, Lecture Notes in Networks and Systems 1000,
https://doi.org/10.1007/978-981-97-3289-0_19

elementary but crucial consideration of information based on communication [5]. The paper proposes to use Shannon information to connect the micro- and macro-worlds. The paper also proposes a way to transfer micro–world concepts to the macro–world level, and vice versa, by using Shannon information.

Many books on physics have been published [6]. When learning important formulae describing both micro- and macro-worlds with all their implications of rigor and magic, it is evident that they are full of mathematical ratios, from the definition of speed and the law of gravity to relativity and quantum theory. This awakens the idea that these may be just the conditions that underlie the expression of communication and information in the observable Universe. However, how to realize that mathematical ratios and communication of information go hand in hand is not yet generally known. The work [3] of the author solves the situation for time ratios only.

If we have some relationship in the form of a mathematical ratio, then we look for a simple way to express it in the language of mathematical communication of information. In other words, we need to get some bridge connecting the world of current physics with the more imaginary world of mathematical communication. This will help to find how in the communicating Universe information is transferred while maintaining its description by mathematical tools. The paper proposes to use for this purpose odds for success known from the probability theory, which are specific ratios associated with subjective probabilities [2]. These probabilities are then fundamental for determining mathematical information.

The mentioned procedure allows us to transform ratios into information quantities obtained from a communication. The same algorithm can be followed when using ratios known from derivatives (from the context of calculus). It results in a new view of physical relationships as sources of mathematical information. This makes the Universe as a subject of communication theory originated by C.E. Shannon. Notice that mathematical entropies do not figure in the considerations (while respecting their importance) since they represent nothing more than the averaging of information from our point of view.

The paper is organized as follows. Section 2 summarizes the concept of mathematical or Shannon information. Section 3 introduces odds and probabilities. Section 4 proposes simple transformation rules for transiting from the purely physical world to the new information world. Section 5 shows some results illustrating the properties of information-based relations. Conclusions are given in Sect. 6.

2 Quantifying Shannon Information

Claude Shannon, the author of the mathematical theory of communication, stated that it is possible to measure information based on the content of a message received through communication, regardless of its length. The concept of mathematical or Shannon information is argued in [5, 7]. The following text briefly summarizes the essentials.

Things that are highly likely, unlikely, or somewhere between happen in the Universe. All the time. The essence of measuring amount of information that such things bring in the form of a "message" is simple. The more surprising the message's content is, the quantitatively more information it includes. For example, new days repeat each day, and Venus cycles every 584 days. Thus, the Venus cycle is associated with less probability because it occurs more rarely. Therefore, it is more surprising. A message about it brings more information than about a new day. Messages with a low probability content give a lot of information. In other words, the amount of information is inversely proportional to the probability.

If denote the amount of information as I and the probability as p, then the Shannon information I is defined by

$$I = 1/p.$$

The information, therefore, forms a hyperbolic geometry. The less probability is (e.g., for rare events), the more information will provide. If events occur with a probability of $1/2$, then they give information $I = 2$. Note that it corresponds to a non-normed bit or the most uncertain situation. Figure 1 graphically presents selected values of I depending on probabilities p summarized in Table 1.

Note that a base 2 logarithm is used for some practical reasons of communication: $I = \log_2(1/p) = \log_2 1 - \log_2 p = -\log_2 p$. The latter is measured in bits. The

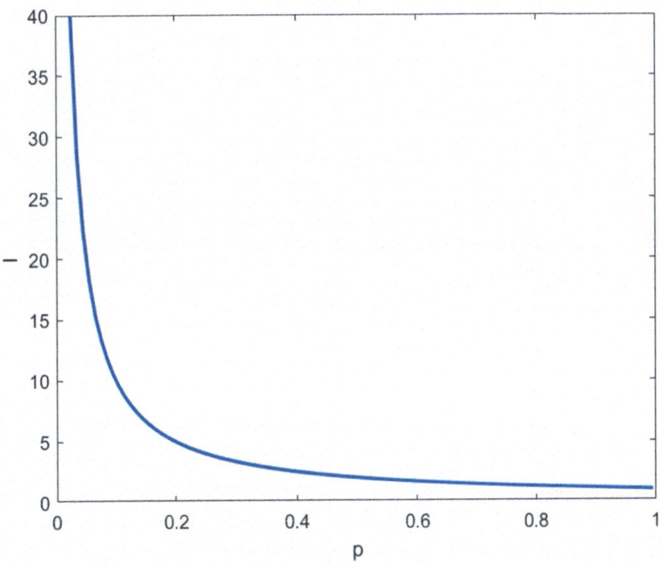

Fig. 1 Probability p versus information I

Table 1 Some values of information

p	1	1/2	1/4	1/8	1/16	1/64
I	1	2	4	8	16	64

expressions $I = 1/p$ and $I = -\log_2 p$ differ only in the scaling factor. We use the first simpler one for purposes of this paper because it much better fulfills its needs.

The $I = 1/p$ represents an ingenious way to obtain the information in the quantitative way when we know the probability. The following section, therefore, focuses on how to find these probabilities.

3 Probabilities and Odds

Sometimes in the real world, we use that the odds for success given by simple number ratios are indicative of our probability estimates. For example, suppose that we propose the ratio of 2 to 1 odds that we win a computer game. This means that we think the probability of our winning is $2/3$. Note more generally that such a concept of odds for success considers the ratio of the number of favorable possibilities to the number of unfavorable ones.

It is customary to quote typical odds as ratios of positive integers having no common factors and the larger number comes first [2]. However, odds can be quoted more generally as ratios of positive real numbers a and b as used in this paper. Let us look at the relation between odds and probabilities.

As apparent, they are closely related. In more analytic detail, if the odds for success are a to b, where a and b are positive numbers, and p is the probability of success, then a and b are in the same ratio as p and $1 - p$, or in other words the ratio of the probability of success to the probability of failure $a/b = p/(1 - p)$. The solution of this equation for p yields

$$p = \frac{a}{a+b}.$$

With this method, it is possible to convert odds to probabilities. Thus, the relationship we have just established between odds and probabilities can be applied to calculate the amount of information

$$I = \frac{1}{p} = 1 + \frac{b}{a}.$$

In the above example, the odds for success are 2 to 1, the probability of success is $p = 2/(2+1) = 2/3$, and $I = 1/p = 1 + 1/2 = 1.5$. However, in this way, it is possible to look at more complicated ratios of odds, for example, π to 1. The probability of success is $\pi/(\pi + 1) = 0.7585$ which is associated with $I = 1 + 1/\pi = 1.3183$. On the other hand, the odd for success determined as 1 to π gives higher information $I = 1 + \pi = 4.1415$.

The above relation for I states that a large number of b in ratio (or an imaginary pie divided by a large number of pieces) yields a proportionally significant amount of information. It gives a base relationship that assigns a way of calculating information to each well defined mathematical ratio of positive real numbers.

Both probabilities and information in this section are interpreted subjectively. We feel (as persons) that a to b are fair odds for success, assigning to such a success the subjective probability $p = a/(a + b)$ or information $I = 1 + b/a$. Now, transfer these ideas to physical quantity ratios.

4 Communicating Physics

Let's imagine the entire physics as a big person. She (or he) has plenty of odds for success in the form of ratios a/b. Assign to such success the subjective probabilities $p = a/(a + b)$ or mathematical information $I = 1 + b/a$. The word subjective is justified here because physics is inductive and covers the observable universe. However, there is unclear how much of the universe is involved. As introduced before, the entire physics as the big person feels that ratios used in physics are fair odds due to the Universe.

They are, for example, ratios defining average speed s/t (total distance/time), average acceleration $v/t = s/t^2$ (total speed/time) or Newton's law of gravitation $F = Gm_1m_2/r^2$, where m_1 and m_2 are the masses of the objects, r is their separation, and G is the gravitational constant. The information I is therefore equal to:

1. For speed: $1 + \frac{t}{s}$.
2. For acceleration: $1 + \frac{t}{v} = 1 + \frac{t^2}{s}$.
3. For law of gravitation: $1 + \frac{r^2}{Gm_1m_2}$.

Thus, in this interpretation, increasing time (in the first case, linearly) and increasing time or distance (in the second and third case, quadratically) increase the information. Conversely, increasing total distance or masses has opposite linear effects. Note that some known physical time ratios are studied in [3].

Instantaneous velocity is given as differential ratio ds/dt and instantaneous acceleration by d^2s/dt^2. Information of this velocity kind is given similarly to $1 + dt/ds$ and similarly for acceleration $1 + dt^2t/d^2s$. It results in the conclusion that speed and acceleration act against the information trend. The situation is interesting for any variables a and b ratios, where the derivative becomes $d(a/b) = (b\,da - a\,db)/b^2$, and information given by the denominator then increases quadratically.

In the atom micro–worlds, the physical properties are frequently covered by subjective probabilities directly. For example, the quantum world is based on wave functions Ψ. In general, they are complex numbers without physical interpretation. However, their magnitudes $|\Psi|^2$ always are real numbers that directly give probabilities. The information is therefore given as $1/|\Psi|^2$. Such an example can be the radial probability of a hydrogen atom depending on the radius of the atom r in the form $p(r) = (4/a^3)r^2e^{-2r/a}$, where $a = 5.291772\,10^{-11}$. By definition, the information I is taken directly as $1/p$.

Just this possibility gives all of physics to one picture. All the physics (micro- and macro-world) is thus full of information and communication. This information can be measured, compared, and examined for properties. For all physics, it is therefore possible to use the phrase communicating physics. It is illustrated in the next section.

5 One Picture of Micro- and Macro-world

The preceding considerations and examples show that probability and information can view all physics as one unified picture. Doing it in a logically consistent way means it can be built upon. Let's illustrate the united information view of completely different cases: the macro-world and the micro-world.

Assume the aforementioned gravitational law and the radial probability of the hydrogen atom. Within a single image Fig. 2, it is possible to plot the information *I* depending on the separation *r* and the radius *r*, both in a proportional scale. It can be seen that from the beginning, the character of information growth is practically identical and starts to differ somewhere in the middle of the scale. The lower curve corresponds to the law of gravity, and the upper curve to the hydrogen atom.

Further investigation of the informational world of physics will probably be about learning the properties and discovering the informational regularities of the macro- and micro-worlds. It is possible that some information dependencies will surprisingly have similar information flows.

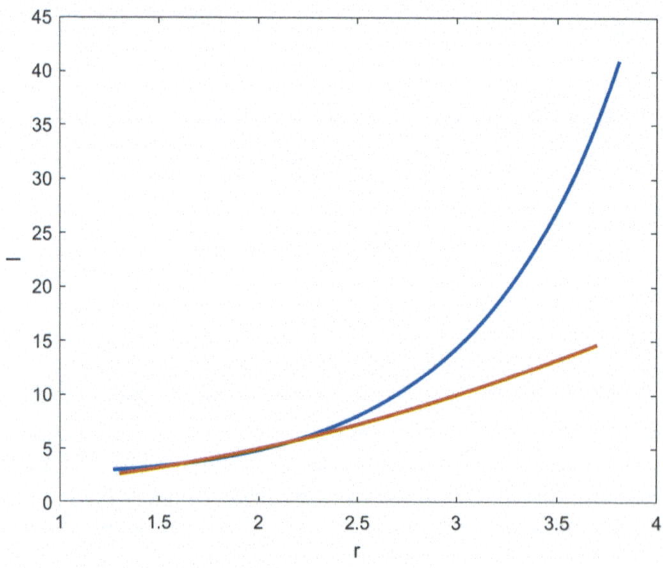

Fig. 2 Information *I* versus separation *r* (bellow) and radius *r* (above)

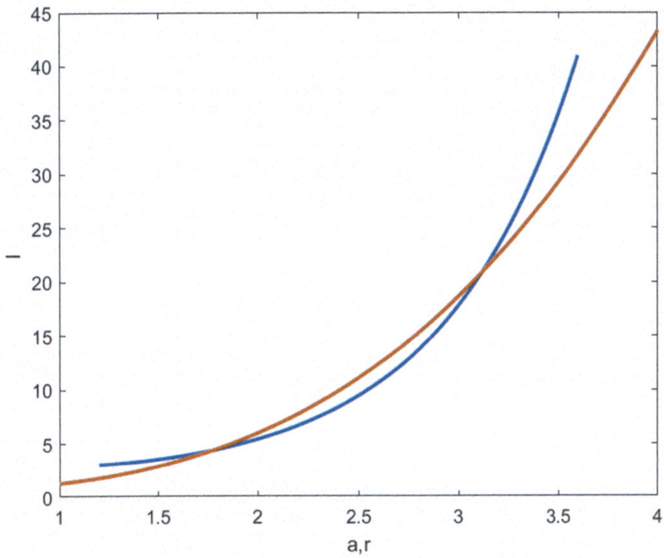

Fig. 3 Information I versus half-axe a and radius r (above)

Kepler's third law and the radial probability of the hydrogen atom can also be compared from the perspective of information communication. Kepler's third law states that the ratios T^2/a^3 and T'^2/a'^3 are the same for planets P and P', where T and T' denote the periods (in planetary years) of the planets and a and a' denote the half-axes (farthest from the centers of orbits). In that case, $I = 1 + a^3/T^2$ $(T = 1)$. The information I depending on the a is shown in Fig. 3 (flatter curve) and proportionally compared with the radial probability of the hydrogen atom. Planets with further orbits have longer planetary years and higher information content. This is logical, because events that occur less frequently are generally more surprising.

Again, it can be seen that in an area, both curves, one from the macro-world and one from the micro-world, have a similar character in a single picture. This confirms the words of physicist J.A. Wheeler mentioned at the beginning of this paper —to understand and describe physics in the language of information. The author of this paper adds that the understanding of the information language that is proposed is only at dawn.

For example, some information dependencies will need to be explained, such as the following. Do we live in a privileged world with the fifty–fifty chance $I = 2$, as the following example shows?

Special relativity uses two different time dilation factors t and t'. The ratio of these factors, according to special relativity, is given by the coefficient $t/t' = 1/\sqrt{1 - v^2/c^2}$, where v is the relative speed and c is the speed of light. Therefore, if we define $I = 1 + t'/t = 1 + \sqrt{1 - v^2/c^2}$, then the information content as a function of speed is shown in Fig. 4. Speeds close to 0 represent states with one bit of information. As the speed increases, the information decreases toward a minimum.

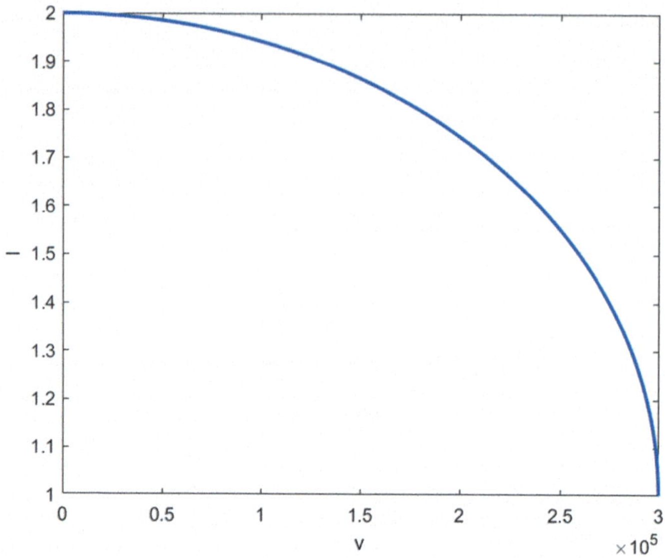

Fig. 4 Information *I* versus speed *v* in special relativity

6 Conclusion

The paper proposes an application of the Shannon information as a linking element in communicating Universe. At the same time, it also designs how to transfer concepts from the micro-world to the macro-world and vice versa.

Would it improve our understanding of the Universe? Do we describe the entire Universe by communication of information? As introduced, some physicists ask about the true nature of reality. This paper uses the Shannon information to answer. The true nature of reality lies in communicating information. In a sense, the mathematical communication concept gives all of the Universe into one communication-based picture. We live in the communicating Universe, giving one picture of the entire physics, as illustrated in Fig. 5.

On the left is the micro-world-hydrogen probability density visualization, and on the right is the macro-world-the Sun. The worlds are connected by communication of the Shannon information.

Based on the results of this paper, several conclusions and questions can be formulated:

1. Any increase by one in the ratio of two physical quantities provides information.
2. Shannon's information provides a common framework for describing the micro- and macro-worlds.
3. Physical relations where time or distance occur in the denominator carry more information if time or distance increases. Therefore, is the reason for expanding space to increase information?

Fig. 5 One picture physics. Left: hydrogen probability density visualization. Right: the Sun. Realized in virtual metaverse Resonite

4. By examining the physical formulas containing the ratio, one can find that a significant proportion (if not most) includes distances or times in the denominator. So they increase the information. Is this why spacetime plays such a role in physics?
5. If speed in the special relativity is equal to zero, information is equal to the most uncertain situation with probability of $1/2$. By other words, speeds close to 0 relates to one bit. Speed of light eliminates the information. Therefore, are we living in communicating Universe with the luxurious greatest possible degree of uncertainty?
6. If in an one physical ratio distance and time are present, it seems that time is prioritized in communicating information. So is in Universe time more important than space?

R. Penrose in his The road to reality [4] claims that there is something like a triangle of reality with vertices: the mental world, the mathematical world and the physical world. The mathematical world is subordinate to the mental world and the physical world to the mathematical one. On the other hand, the mental world is subordinated to the physical world, thus closing the whole cycle. Shannon's information originated in the mental world, is formulated in the mathematical world, and describes communication in the physical world. The latter then provides information to the mental one. In this case, the communicating universe is in perfect harmony with R. Penrose's triangle of reality cycle.

References

1. Brooks M (2010) In: The big questions: physics. Quercus Publishing, UK edition
2. Freund JE (1993) In: Introduction to probability. Dover Publications
3. Klán P (2023) Information: from Communication to physical principles. In: Arai K (eds) Advances in information and communication. FICC 2023. Lecture notes in networks and systems, vol 651. Springer, Cham, pp 187–194. https://doi.org/10.1007/978-3-031-28076-4_16
4. Penrose R (2007) In: The road to reality: a complete guide to the laws of the Universe. Vintage (2007)
5. Pierce JR (1980) In: An introduction to information theory: symbols signals and noise. Dover Publications
6. Walker J (2018) In: Fundamentals of physics. 11th edn. Wiley
7. Wheeler JA (1990) Information, physics, quantum: the search for links. In: Zurek WH (ed) Complexity, entropy, and the physics of information, Addison-Wesley, Redwood City, pp 354–368. https://philpapers.org/archive/WHEIPQ.pdf

An Overview on Diagnosis of Endometriosis Disease Based on Machine Learning Methods

Liviu-Andrei Scutelnicu, Radu Maftei, and Mihaela Luca

Abstract Endometriosis is a complex gynecological condition characterized by the presence of endometrial-like tissue outside the uterus. Diagnosis and management of endometriosis are challenging due to diverse clinical manifestations and lack of definitive diagnostic methods. Medical imaging, particularly ultrasound and magnetic resonance imaging (MRI), has shown potential in visualizing and assessing endometriosis lesions. In this research paper, we review recent advances in machine learning algorithms applied to endometriosis diagnosis, focusing on the importance of training datasets in improving diagnostic accuracy. We analyze the contributions of various studies and discuss the challenges and future directions in this field.

Keywords Endometriosis · Machine learning · Training datasets · Accuracy

L.-A. Scutelnicu (✉)
Faculty of Computer Science, "Alexandru Ioan Cuza" University of Iași, General Berthelot Street, No. 16, 700483 Iași, Romania
e-mail: andrei.scutelnicu@uaic.ro

L.-A. Scutelnicu · M. Luca
Institute for Computer Science, Romanian Academy, Iași Branch, T. Codrescu Street, No. 2, 700481 Iași, Romania

R. Maftei
Department of Mother and Child Medicine, Faculty of Medicine, University of Medicine and Pharmacy "Grigore T. Popa", University Street, No. 16, 700115 Iași, Romania

Clinical Hospital of Obstetrics and Gynecology "Cuza Vodă", Cuza Vodă Street, No. 34, 700038 Iași, Romania

Origyn Fertility Center, Palace Street, No. 3C, 700032 Iași, Romania

X.-S. Yang et al. (eds.), *Proceedings of Ninth International Congress on Information and Communication Technology*, Lecture Notes in Networks and Systems 1000,
https://doi.org/10.1007/978-981-97-3289-0_20

1 Introduction

Endometriosis,[1] a chronic gynecological condition, affects millions of individuals worldwide and poses significant challenges to both patients and healthcare providers. Characterized by the abnormal growth of endometrial tissue outside the uterus, this condition often leads to debilitating symptoms, including pelvic pain, infertility, and menstrual irregularities. Despite its prevalence and impact, endometriosis remains notoriously difficult to diagnose accurately.

Accurate and timely diagnosis of endometriosis is crucial for effective management and improved patient outcomes [1]. Unfortunately, the complex and heterogeneous nature of the disease, coupled with the limitations of traditional diagnostic methods, has led to diagnostic delays and suboptimal patient care. The need for a reliable and non-invasive diagnostic approach has spurred research into the integration of medical image analysis and machine learning.

The diagnosis of endometriosis is fraught with challenges [2]. The variability in disease presentation, ranging from superficial lesions to deep infiltrating nodules, complicates both clinical evaluation and imaging interpretation. Additionally, the lack of specific biomarkers and the overlap of symptoms with other conditions further hinder accurate diagnosis. Conventional diagnostic methods, such as clinical examination and laparoscopy, are invasive, time-consuming, and operator-dependent.

Medical imaging has emerged as a promising avenue for enhancing endometriosis diagnosis [3]. Techniques such as ultrasound [4] and magnetic resonance imaging (MRI) offer non-invasive visualization of pelvic structures and endometriotic lesions. However, the vast amount of image data generated presents a challenge in itself. Enter medical image analysis, a field that leverages machine learning algorithms to automatically extract meaningful insights from these images, enabling accurate lesion detection, characterization, and classification [5].

This paper delves into the intersection of endometriosis and medical image analysis, focusing on the application of machine learning techniques for accurate and non-invasive diagnosis. Chapters 2 to 9 explore key aspects of this intersection, from the clinical manifestations of endometriosis to the integration of machine learning models into clinical practice. By the end, readers will have gained a comprehensive understanding of the potential, challenges, and future prospects of machine learning-based endometriosis diagnosis.

The primary objectives of this paper are as follows:

- to provide an in-depth exploration of the clinical manifestations and challenges associated with endometriosis diagnosis;
- to elucidate the role of medical imaging modalities, such as ultrasound and MRI, in visualizing endometriosis lesions;
- to discuss the significance of constructing a comprehensive training dataset for machine learning-based endometriosis diagnosis;

[1] https://www.who.int/news-room/fact-sheets/detail/endometriosis.

- to detail the process of feature extraction and selection, enabling the transformation of raw medical images into informative attributes;
- to comprehensively explore various machine learning algorithms, including supervised learning and deep learning, for accurate endometriosis diagnosis;
- to elucidate the methodologies of model training, validation, and evaluation, ensuring robustness and clinical relevance;
- to discuss the challenges and considerations in integrating machine learning models into clinical workflows for real-world impact;
- to outline future directions, ethical implications, and the broader implications of machine learning in medical image analysis, using endometriosis as a case study.

This paper is relevant to a multidisciplinary audience, including clinicians, researchers, data scientists, and policymakers. It serves as a comprehensive resource for those interested in leveraging machine learning and medical image analysis to improve the accuracy and efficiency of endometriosis diagnosis. By providing insights into the challenges, methodologies, and future prospects of this innovative approach, the paper aims to drive collaboration and advance the field of medical AI in the context of endometriosis.

2 Endometriosis and Medical Imaging

Endometriosis, characterized by the presence of endometrial tissue outside the uterine cavity, is a multifaceted gynecological disorder with a spectrum of clinical manifestations. The condition is often associated with pelvic pain, dysmenorrhea, dyspareunia, dysuria, dyschezia, and depression in advanced cases of the disease with major pain and infertility [6]. While the exact etiology of endometriosis remains elusive, several theories, including retrograde menstruation and immune dysfunction, have been proposed. The wide spectrum of clinical presentations, ranging from superficial peritoneal implants to deeply infiltrating lesions and ovarian endometriomas, poses significant diagnostic challenges. It's important to mention adenomyosis as well, which can be considered an independent pathology but equally debilitating, causing pain, reproductive failures, and heavy bleeding. In this context, it's worth exploring the potential utility of elastography and uterine mapping. Using 3D scanning focused on the uterus [7, 8], the differences in ultrasound emission characteristics can be leveraged to distinguish a healthy myometrium from one infiltrated by adenomyosis.

Accurate and timely diagnosis of endometriosis is crucial to initiate appropriate management strategies and alleviate patients' suffering. Medical imaging plays a pivotal role in visualizing endometriosis lesions, enabling clinicians to make informed decisions [9]. Two primary imaging modalities, ultrasound and magnetic resonance imaging (MRI), have gained prominence in endometriosis diagnostics.

Ultrasound, a widely accessible and cost-effective imaging modality, provides real-time visualization of pelvic structures. Transvaginal ultrasound (TVUS) offers high-resolution imaging of the pelvis, enabling the identification of endometriotic lesions, ovarian cysts, and other associated findings [10]. Color Doppler ultrasonography aids in evaluating blood flow patterns within endometriosis lesions, contributing to their characterization. Doppler imaging can distinguish between active and non-active endometriosis lesions based on vascularity, further refining diagnostic accuracy.

MRI, renowned for its superior soft tissue contrast and multi-planar imaging capabilities, offers detailed insights into endometriosis lesions [11]. T1-weighted and T2-weighted sequences provide valuable anatomical information, allowing precise localization and size estimation of lesions. Diffusion-weighted imaging (DWI) and dynamic contrast-enhanced MRI (DCE-MRI) offer functional data by assessing tissue cellularity and blood perfusion, respectively [12]. Such advanced imaging techniques contribute to improved lesion characterization and staging.

Medical imaging modalities play a pivotal role in endometriosis diagnosis, offering non-invasive visualization and evaluation of lesions [13]. Ultrasound's real-time imaging, portability, and lack of ionizing radiation make it a valuable tool, especially in young patients and for serial follow-up. MRI, on the other hand, provides unparalleled soft tissue contrast and multi-planar capabilities, facilitating the accurate depiction of complex endometriosis lesions.

However, each modality has its limitations. Ultrasound's operator-dependent nature and challenges in visualizing deep-seated lesions may hinder its sensitivity in certain cases. MRI's availability, cost, and potential for overdiagnosis due to its high sensitivity to minor lesions warrant careful consideration.

The synergy between medical imaging and machine learning holds transformative potential in the realm of endometriosis diagnosis. By harnessing the information-rich data provided by ultrasound and MRI, machine learning algorithms can be trained to identify subtle patterns, differentiate between normal and abnormal tissues, and assist in accurate lesion classification. Chapter 3 delves into the foundation of this synergy-the construction of a comprehensive and diverse training dataset.

3 Constructing an Informative Training Dataset

The success of machine learning models hinges upon the quality and diversity of the training dataset [14]. In the context of endometriosis diagnosis, constructing a meticulously curated dataset is paramount. Such a dataset serves as the foundation for training models to distinguish between normal pelvic structures and endometriosis-related abnormalities.

An informative training dataset encompasses a diverse range of endometriosis cases, capturing various stages, lesion types, and anatomical locations. Inclusion of cases that represent the spectrum of disease manifestations ensures the model's ability to generalize to real-world scenarios. This diversity enables the algorithm to recognize subtle patterns associated with different endometriosis presentations.

Accurate annotations by domain experts are crucial to provide ground truth labels for the training dataset. Expert clinicians with expertise in endometriosis diagnosis meticulously identify and annotate lesions within medical images. These annotations contribute to the dataset's reliability and serve as the reference standard for training the machine learning model [15].

Prior to model training, thorough data preprocessing is essential. Image normalization, noise reduction, and resolution standardization enhance dataset consistency and reduce variability. Preprocessing steps ensure that images are in a suitable format for model training, promoting uniform feature extraction and subsequent analysis.

Careful attention must be paid to potential bias within the dataset. Bias, whether due to age, ethnicity, or other factors, can influence the model's performance and generalizability. Ensuring a representative dataset by actively mitigating bias is essential [16]. Additionally, ethical considerations related to patient consent, data privacy, and data sharing must be addressed when curating and utilizing the dataset.

Dataset augmentation techniques further enhance the dataset's diversity. Augmentation involves applying transformations to existing images, such as rotation, flipping, or adding noise, to simulate variations in imaging conditions and patient demographics [17]. This approach enriches the dataset and enables the model to better adapt to novel data during real-world application.

To capture a comprehensive view of endometriosis lesions, the training dataset can benefit from multi-modal imaging integration. Combining data from ultrasound and MRI modalities provides a more holistic representation of the disease. Multimodal datasets facilitate the development of models that leverage complementary information from different imaging sources [18].

Curating a training dataset is an iterative process. As new data becomes available and clinical insights evolve, the dataset should be continually refined and expanded. Regular updates ensure that the model remains relevant and effective in capturing the evolving landscape of endometriosis presentations.

Challenges in dataset curation include acquiring a sufficient volume of annotated data, ensuring data quality and reliability, and addressing potential biases. Future directions involve the establishment of standardized datasets for benchmarking, collaborative efforts among researchers to share curated data, and the development of guidelines for ethical data usage.

4 Feature Extraction and Selection

Feature extraction is a critical step in translating raw medical image data into quantifiable and informative attributes. In the context of endometriosis diagnosis, extracting relevant features from ultrasound and MRI images is essential for training accurate machine learning models.

Endometriosis lesions exhibit diverse shapes, including irregular contours, cystic formations, and nodular structures. Shape descriptors, such as area, perimeter, eccentricity, and circularity, capture geometric characteristics that distinguish normal anatomical structures from abnormal lesions. These descriptors provide the model with essential clues for differentiation [19].

Texture features quantify variations in pixel intensities within an image region. Endometriosis lesions often display distinct textures, such as fine granularity, irregular patterns, or heterogeneous signal intensities [20]. Texture analysis techniques, including Gray-Level Co-Occurrence Matrix (GLCM) and Local Binary Pattern (LBP), extract texture features that enhance the model's ability to discriminate between different tissue types [21].

Endometriosis lesions' spatial relationships with neighboring structures offer valuable diagnostic insights. Proximity to organs, borders, and relative positions within the pelvis can aid in lesion characterization. Spatial attributes, such as distance maps and neighborhood relations, contribute to the model's understanding of lesion context.

While medical images may contain numerous features, not all contribute equally to accurate diagnosis. Feature selection techniques enhance model efficiency by identifying and prioritizing the most informative attributes. By reducing dimensionality and noise, feature selection optimizes the model's ability to discern relevant patterns.

Filter methods assess feature relevance independently of the machine learning algorithm. Statistical tests, such as chi-squared or mutual information, quantify the relationship between each feature and the target class. Features with the highest discriminatory power are retained, enhancing the model's discriminative capabilities [22].

Wrapper methods evaluate feature subsets by training and testing the machine learning model iteratively. Techniques like recursive feature elimination (RFE) and forward/backward selection systematically assess feature combinations, optimizing model performance. Wrapper methods consider feature interactions, leading to improved diagnostic accuracy.

Embedded methods incorporate feature selection within the model training process. Regularization techniques, such as L1 (Lasso) regularization, encourage sparse feature sets by penalizing irrelevant attributes. Embedded methods automatically adjust feature weights during model optimization, resulting in more efficient and focused models.

The rise of deep learning revolutionizes feature extraction by enabling automatic and hierarchical pattern recognition. Convolutional Neural Networks (CNNs) excel at capturing intricate image features through multiple layers of convolution and pooling. End-to-end learning, where CNNs learn both feature extraction and classification, eliminates the need for handcrafted features [23].

Transfer learning leverages pre-trained CNN architectures, fine-tuning them for endometriosis diagnosis. By utilizing models pre-trained on large image datasets, transfer learning capitalizes on learned image features, even when the target dataset is smaller. This approach accelerates model development and enhances generalization.

Integrating features from multiple imaging modalities enhances the model's diagnostic accuracy. Ultrasound and MRI offer complementary information, and their fusion enables a more comprehensive analysis of endometriosis lesions. Feature-level fusion combines attributes extracted from different modalities, providing a holistic view of the disease.

Feature extraction and selection are not without challenges. Complex lesions may require sophisticated feature extraction techniques, and noisy or low-resolution images can impact feature quality. The choice of features and selection methods should be guided by clinical insights and domain expertise. Ensuring that selected features are clinically meaningful and interpretable remains a priority.

5 Machine Learning Algorithms for Endometriosis Analysis

Supervised learning algorithms form the cornerstone of endometriosis analysis, enabling accurate classification and diagnosis based on annotated training data. Within this category, several algorithms have demonstrated effectiveness in handling medical image data for endometriosis detection.

SVMs are widely adopted for their ability to classify complex and nonlinear data. In the context of endometriosis analysis, SVMs learn to distinguish between normal pelvic structures and endometriosis lesions by identifying optimal hyperplanes in the feature space [24]. SVMs leverage kernel functions to project data into higher-dimensional spaces, enabling efficient separation of classes.

Random Forests are ensemble methods that aggregate multiple decision trees to enhance classification accuracy. RFs partition the feature space and make decisions based on the majority vote of constituent trees. Their ability to capture complex feature interactions and mitigate overfitting is particularly advantageous for endometriosis classification.

The advent of deep learning, particularly CNNs, has revolutionized medical image analysis. CNNs excel at capturing hierarchical patterns and spatial relationships within images, making them well-suited for endometriosis diagnosis [23].

CNNs consist of multiple convolutional and pooling layers that progressively learn and extract intricate features. These networks automatically learn relevant features from raw medical image data, eliminating the need for handcrafted feature extraction. Pre-trained CNN architectures, such as VGG, ResNet, and Inception, provide a foundation for endometriosis analysis and can be fine-tuned for enhanced performance.

Transfer learning leverages pre-trained CNN models to expedite model development and improve generalization. Fine-tuning enables the adaptation of pre-trained models to endometriosis-specific features. Data augmentation techniques, such as rotation, scaling, and flipping, increase the training dataset's diversity and enhance model robustness.

Ensemble methods combine multiple algorithms to achieve superior predictive performance compared to individual models.

Voting classifiers aggregate predictions from multiple algorithms, employing majority voting to make final decisions. By leveraging the strengths of different algorithms, voting classifiers enhance endometriosis classification accuracy and mitigate individual model biases.

Stacking and blending involve training a meta-model that combines predictions from diverse base models. These techniques capture higher-level feature interactions and offer enhanced generalization. Stacking, in particular, uses a secondary model to learn how to best combine the base models' predictions.

While these machine learning algorithms show promise in endometriosis analysis, challenges persist. Class imbalance, particularly in rare endometriosis subtypes, can impact model performance. Interpretability of deep learning models remains a concern, requiring efforts to develop methods for model explanation and visualization. Careful hyperparameter tuning and model evaluation are essential to ensure optimal performance and generalization.

6 Model Training, Validation, and Evaluation

Model training involves feeding the curated dataset into the chosen machine learning algorithm to enable it to learn and generalize from the data. The dataset is typically split into training and validation subsets. During training, the algorithm iteratively adjusts its internal parameters to minimize the difference between predicted and actual outcomes.

Hyperparameters, settings that control the model's behavior, play a crucial role in training. Parameters such as learning rates, regularization strengths, and kernel sizes impact the model's convergence and performance. Hyperparameter tuning, often performed using techniques like grid search or random search, optimizes model behavior for optimal accuracy and generalization.

Cross-validation is employed to assess model robustness and prevent overfitting. Common cross-validation techniques, such as k-fold cross-validation, divide the dataset into k subsets. The model is trained on k-1 subsets and validated on the remaining subset in each iteration. Cross-validation averages performance metrics across iterations, providing a more accurate estimate of the model's real-world performance.

Accurate evaluation of the model's diagnostic capabilities is essential. Performance metrics gauge the model's ability to distinguish between normal and endometriosis-affected pelvic structures. Common metrics include [25, 26]

- accuracy quantifies the proportion of correctly classified instances. While valuable, accuracy alone may be insufficient when dealing with class-imbalanced datasets, where the majority class dominates;
- sensitivity (true positive rate) measures the model's ability to correctly identify endometriosis cases, while specificity (true negative rate) gauges its accuracy in

recognizing normal cases. Balancing sensitivity and specificity are crucial to avoid false positives and negatives;

- the AUC-ROC curve depicts the trade-off between sensitivity and specificity at different decision thresholds. A higher AUC value indicates improved discrimination power, with an AUC of 0.5 denoting random chance and an AUC of 1 representing perfect discrimination.

Overfitting occurs when the model memorizes the training data and performs poorly on unseen data. Regularization techniques, such as L1 (Lasso) and L2 (Ridge) regularization, penalize large model weights, preventing over-reliance on individual features [27]. Data augmentation, dropout layers, and early stopping during training contribute to improved model generalization.

Machine learning model's clinical relevance lies in their ability to provide meaningful insights to healthcare providers. Integration of these models into clinical workflows empowers clinicians with automated and objective diagnostic support. Model outputs, such as heat maps highlighting potential endometriosis lesions, aid in diagnosis and treatment planning.

Model training, validation, and evaluation entail challenges such as class imbalance, dataset bias, and interpretability of deep learning models. In the realm of endometriosis diagnosis, the class imbalance dilemma is pronounced due to the relative rarity of this condition in comparison to the larger pool of negative cases. This poses challenges such as model bias and skewed evaluation metrics. To mitigate these issues, various strategies come into play. Oversampling techniques generate synthetic data to amplify the minority class, while weighted loss functions emphasize the correct classification of the underrepresented group during training. Careful choice of evaluation metrics, including precision, recall, and F1-score, allows for a more comprehensive assessment of model performance, ensuring that no endometriosis case is overlooked, and false positives are minimized. These approaches collectively address the intricacies of class imbalance, enhancing the diagnostic accuracy and equity of machine learning models in the context of endometriosis diagnosis.

Addressing these challenges involves collaborative efforts between clinicians, machine learning experts, and regulatory bodies. Ongoing research is directed toward enhancing model interpretability, fine-tuning for specific endometriosis subtypes, and optimizing clinical integration.

7 Clinical Integration and Real-World Impact

Integrating machine learning models into clinical workflows represents a transformative shift in endometriosis diagnosis. The successful translation of models from research environments to real-world clinical settings requires careful consideration and collaboration between clinicians, researchers, and technology experts.

Machine learning models serve as valuable clinical decision support tools. Clinicians leverage model outputs, such as lesion probability maps or diagnostic confidence scores, to aid in their diagnostic assessments. The automated nature of these tools enhances diagnostic accuracy, potentially reducing variability between different clinicians.

Beyond diagnosis, machine learning models contribute to treatment planning and patient management [28]. Accurate lesion characterization enables tailored treatment strategies, optimizing patient outcomes. Models can predict disease progression, guide surgical interventions, and assist in post-operative monitoring.

Clinical integration of machine learning models presents unique challenges. Regulatory approvals, compliance with medical standards, and data privacy considerations must be addressed. Models' robustness to diverse patient populations, imaging protocols, and equipment variations requires comprehensive validation and adaptation.

Successful clinical integration necessitates close collaboration between clinicians and data scientists. Clinicians provide domain expertise, ensuring that models align with clinical requirements and guidelines. Model interpretability techniques, such as saliency maps and feature visualization, enable clinicians to understand and trust model decisions.

Patient-centric care remains paramount. Transparency in model predictions, informed consent for data usage, and clear communication of model limitations uphold ethical principles. Ensuring that machine learning models augment, rather than replace, clinical judgment maintains the human touch in patient care.

The integration of machine learning into clinical practice has the potential to revolutionize endometriosis diagnosis and patient management. Timely and accurate diagnosis translates to improved quality of life for patients. Additionally, the experience gained from endometriosis diagnosis can serve as a blueprint for similar applications in other medical domains.

The clinical integration of machine learning models in gynecology and endometriosis diagnosis demands meticulous adherence to regulatory hurdles, encompassing FDA approval or CE marking, stringent data privacy compliance, and ethical considerations. Harmonizing with DICOM and ISO standards, alongside collaboration with medical societies, ensures that these innovative tools align with clinical guidelines and seamlessly integrate into existing healthcare systems, bolstering their credibility and clinical utility.

Clinicians and data scientists engage in a collaborative partnership, where clinicians contribute their deep domain knowledge in gynecology and endometriosis diagnosis, ensuring the tool's clinical relevance. Data scientists harness this expertise to develop robust machine learning models that can analyze complex medical images with precision. This synergy between medical practitioners and AI specialists paves the way for a groundbreaking diagnostic solution, poised to transform endometriosis diagnosis, offering earlier detection and improved patient outcomes.

Collaborative research initiatives between medical professionals and technology experts drive continuous advancements. Feedback loops and iterative improvements refine machine learning models, enhancing their clinical utility and ensuring that they remain aligned with evolving clinical needs.

8 Future Directions and Challenges

The future of endometriosis diagnosis lies in harnessing the synergy between multiple imaging modalities. Integrating data from ultrasound, MRI and potentially other emerging techniques, such as elastography or hyperspectral imaging, enhances the richness of information available for analysis. Multi-modal fusion facilitates comprehensive lesion characterization and provides a more holistic view of the disease.

The interpretability of machine learning models remains a challenge for clinical adoption. Advances in explainable AI aim to demystify model decisions, enabling clinicians to understand the rationale behind predictions. Techniques such as feature attribution maps and attention mechanisms shed light on the model's decision-making process, fostering trust and facilitating model integration into clinical practice.

In the realm of explainable AI, a multitude of innovative techniques are emerging to illuminate the inner workings of machine learning models, offering clinicians and patients valuable insights. From feature importance attribution and rule-based models to attention mechanisms and counterfactual explanations, these approaches bridge the gap between black-box algorithms and transparent decision-making, empowering stakeholders to better understand and trust AI-driven diagnoses and predictions in healthcare.

Endometriosis encompasses a spectrum of presentations, each with distinct clinical implications. Tailoring machine learning models to recognize specific endometriosis subtypes, such as deep infiltrating lesions or ovarian endometriomas, holds potential for more precise diagnosis and personalized treatment strategies.

Collaborative efforts to create standardized datasets and benchmarking frameworks are crucial for evaluating and comparing the performance of machine learning models. Establishing common datasets with well-defined annotations enables fair model comparisons and facilitates advancements in endometriosis diagnosis across different research groups.

The translation of machine learning models into clinical practice necessitates adherence to medical regulations and standards. Gaining regulatory approvals, such as FDA clearance, requires rigorous validation, safety assessments and compliance with medical device guidelines. Addressing these considerations paves the way for widespread clinical adoption.

As machine learning algorithms become integral to medical practice, ethical considerations become paramount. Ensuring patient privacy, obtaining informed consent for data usage, and mitigating biases are essential to uphold ethical standards. Open discussions among stakeholders, including patients, clinicians, and technology experts, guide responsible AI deployment.

Overcoming the challenges and realizing the full potential of machine learning in endometriosis diagnosis necessitates continued collaboration between medical professionals, researchers, data scientists, and policymakers. Interdisciplinary research fosters innovation, addresses complex challenges, and drives the field forward.

The future of machine learning-based endometriosis diagnosis is promising. With advancements in imaging technology, algorithmic sophistication, and clinical integration, the accuracy, efficiency, and accessibility of diagnosis will significantly improve. Collaborative efforts and a patient-centered approach will drive the field's evolution, culminating in enhanced patient care and improved outcomes.

9 Conclusion

The integration of machine learning into medical image analysis represents a paradigm shift in the diagnosis of endometriosis. By leveraging the synergistic power of medical imaging and advanced algorithms, clinicians gain access to accurate, objective, and timely diagnostic insights. The journey from raw medical images to predictive models holds the promise of revolutionizing patient care and outcomes.

In preparing the training dataset, we developed a web interface, which aims to annotate ultrasound images, both visually and at a metadata level (specific to the pathology), the interface described in Scutelnicu et al. [29].

Machine learning-based endometriosis diagnosis offers the potential for earlier detection and improved patient outcomes. Timely identification of endometriosis lesions, coupled with accurate staging and characterization, empowers clinicians to initiate appropriate interventions promptly. Reduced diagnostic delays can alleviate patient suffering and lead to more effective treatment strategies.

The individualized nature of endometriosis is well-suited to the capabilities of machine learning models. Tailored treatment plans, guided by accurate lesion characterization and disease progression predictions, hold the promise of optimizing patient care. Personalized interventions based on patient-specific data contribute to more effective management and better quality of life.

The journey from algorithm development to clinical integration requires collaboration and iterative refinement. Bridging the gap between research and clinical practice entails interdisciplinary cooperation, rigorous validation, and adherence to ethical and regulatory standards. This collaborative effort ensures that machine learning models make a meaningful impact on patient care.

The lessons learned from machine learning-based endometriosis diagnosis serve as a blueprint for the broader field of medical AI. The challenges faced, such as dataset curation, model validation and clinical integration, mirror those encountered in other medical domains. As the field advances, the principles and insights gained from endometriosis analysis pave the way for similar applications in diverse healthcare contexts.

The evolution of machine learning-based endometriosis diagnosis is an ongoing journey marked by continuous research, innovation, and collaboration. The integration of advanced imaging modalities, refined algorithms, and comprehensive clinical workflows will further enhance the accuracy and clinical utility of these models. The road ahead holds the promise of improved diagnostic accuracy, better patient outcomes, and a brighter future for individuals affected by endometriosis.

References

1. Allaire C, Bedaiwy MA, Yong PJ (2023) Diagnosis and management of endometriosis. Can Med Assoc J (CMAJ) 195(10):E363–E371. https://doi.org/10.1503/cmaj.220637
2. Agarwal SK, Chapron C, Giudice LC, Laufer MR, Leyland N, Missmer SA, Singh SS, Taylor HS (2019) Clinical diagnosis of endometriosis: a call to action. Am J Obstet Gynecol 220(4):354.e1-354.e12. https://doi.org/10.1016/j.ajog.2018.12.039
3. Chiorean DM, Mitranovici MI, Toru HS, Cotoi TC, Tomuț AN, Turdean SG, Cotoi OS (2023) New insights into genetics of endometriosis—a comprehensive literature review. Diagnostics 13:2265. https://doi.org/10.3390/diagnostics13132265
4. Liu Y, Zhou Q, Peng B, Jiang J, Fang L, Weng W, Wang W, Wang S, Zhu X (2022) Automatic measurement of endometrial thickness from transvaginal ultrasound images. Front Bioeng Biotechnol 10(853845):29. https://doi.org/10.3389/fbioe.2022.853845
5. Javaid M, Haleem A, Singh RP, Suman R, Rab S (2022) Significance of machine learning in healthcare: features, pillars and applications. Int J Intell Netw 3:58–73. ISSN 2666-6030. https://doi.org/10.1016/j.ijin.2022.05.002
6. Missmer SA, Tu FF, Agarwal SK, Chapron C, Soliman AM, Chiuve S, Eichner S, Flores-Caldera I, Horne AW, Kimball AB, Laufer MR, Leyland N, Singh SS, Taylor HS, As-Sanie SS (2021) Impact of endometriosis on life-course potential: a narrative review. Int J Gen Med 14:9–25. https://doi.org/10.2147/IJGM.S261139
7. Hanafi M (2013) Ultrasound diagnosis of adenomyosis, leiomyoma, or combined with histopathological correlation. J Human Reprod Sci 6(3):189–193. https://doi.org/10.4103/0974-1208.121421
8. Wang X, Bao N, Xin X, Tan J, Li H, Zhou S, Liu H (2022) Automatic evaluation of endometrial receptivity in three-dimensional transvaginal ultrasound images based on 3D U-Net segmentation. Quan Imaging Med Surg [Online] 12(8):4095–4108. https://doi.org/10.21037/qims-21-1155
9. Pascoal E, Wessels JM, Aas-Eng MK, Abrao MS, Condous G, Jurkovic D, Espada M, Exacoustos C, Ferrero S, Guerriero S, Hudelist G, Malzoni M, Reid S, Tang S, Tomassetti C, Singh SS, van den Bosch T, Leonardi M (2022) Strengths and limitations of diagnostic tools for endometriosis and relevance in diagnostic test accuracy research. Ultrasound Obstetr Gynecol 60. https://doi.org/10.1002/uog.24892
10. Daniilidis A, Grigoriadis G, Dalakoura D, D'Alterio MN, Angioni S, Roman H (2022) Transvaginal ultrasound in the diagnosis and assessment of endometriosis-an overview: how, why, and when. Diagnostics (Basel, Switzerland) 12(12):2912. https://doi.org/10.3390/diagnostics12122912
11. Bianek-Bodzak A, Szurowska E, Sawicki S, Liro M (2013) The importance and perspective of magnetic resonance imaging in the evaluation of endometriosis. Biomed Res Int 2013:436589. https://doi.org/10.1155/2013/436589
12. Kinh Do R, Reyngold M, Paudyal R, Oh JH, Konar AS, LoCastro E, Goodman KA, Shukla-Dave A (2020) Diffusion-weighted and dynamic contrast-enhanced MRI derived imaging metrics for stereotactic body radiotherapy of pancreatic ductal adenocarcinoma: preliminary

findings. Tomography (Ann Arbor, Mich.) 6(2):261–271. https://doi.org/10.18383/j.tom.2020.00015
13. Nisenblat V, Bossuyt PM, Farquhar C, Johnson N, Hull ML (2016) Imaging modalities for the non-invasive diagnosis of endometriosis. Cochrane Database Syst Rev 2(2):CD009591. https://doi.org/10.1002/14651858.CD009591.pub2
14. Zhong K, Feng G, Shen L, Luo J (2018) Deep learning for steganalysis based on filter diversity selection. Sci China Inf Sci 61:1–3. https://doi.org/10.1007/s11432-018-9640-7
15. Sarker IH (2021) Machine learning: algorithms, real-world applications and research directions. SN Comput Sci 2:160. https://doi.org/10.1007/s42979-021-00592-x
16. Shahbazi N, Lin Y, Asudeh A, Jagadish HV (2023) Representation bias in data: a survey on identification and resolution techniques. ACM Comput Surv 55(13s):39. Article 293. https://doi.org/10.1145/3588433
17. Mumuni A, Mumuni F (2022) Data augmentation: a comprehensive survey of modern approaches. Array 16:100258. ISSN 2590-0056. https://doi.org/10.1016/j.array.2022.100258
18. Viswanath SE, Tiwari P, Lee G, Madabhushi A (2017) Alzheimer's disease neuroimaging initiative: dimensionality reduction-based fusion approaches for imaging and non-imaging biomedical data: concepts, workflow, and use-cases. BMC Med Imaging 17(1):2. https://doi.org/10.1186/s12880-016-0172-6
19. Ravishankar TN, Jadhav HM, Kumar NS, Ambala S, Pillai M (2023) A deep learning approach for ovarian cysts detection and classification (OCD-FCNN) using fuzzy convolutional neural network. Meas Sens 27:100797. ISSN 2665-9174. https://doi.org/10.1016/j.measen.2023.100797
20. Ştefan RA, Ştefan PA, Mihu CM, Csutak C, Melincovici CS, Crivii CB, Maluțan AM, Hîțu L, Lebovici A (2021) Ultrasonography in the differentiation of endometriomas from hemorrhagic ovarian cysts: the role of texture analysis. J Pers Med 11(7):611. https://doi.org/10.3390/jpm11070611
21. Wan S, Lee HC, Huang X, Xu T, Xu T, Zeng X, Zhang Z, Sheikine Y, Connolly JL, Fujimoto JG, Zhou C (2017) Integrated local binary pattern texture features for classification of breast tissue imaged by optical coherence microscopy. Med Image Anal 38:104–116. ISSN 1361-8415. https://doi.org/10.1016/j.media.2017.03.002
22. Sun X, Liu Y, Li J, Zhu J, Liu X, Chen H (2012) Using cooperative game theory to optimize the feature selection problem. Neurocomputing 97:86–93. ISSN 0925-2312. https://doi.org/10.1016/j.neucom.2012.05.001
23. Yamashita R, Nishio M, Do RKG et al (2018) Convolutional neural networks: an overview and application in radiology. Insights Imaging 9:611–629. https://doi.org/10.1007/s13244-018-0639-9
24. Parlatan U, Inanc MT, Ozgor BY, Oral E, Bastu E, Unlu MB, Basar G (2019) Raman spectroscopy as a non-invasive diagnostic technique for endometriosis. Sci Rep 9(1):19795. https://doi.org/10.1038/s41598-019-56308-y
25. Abrao MS, Andres MP, Miller CE, Gingold JA, Rius M, Neto JS, Carmona F (2021) AAGL 2021 endometriosis classification: an anatomy-based surgical complexity score. J Minim Invasive Gynecol. https://doi.org/10.1016/j.jmig.2021.09.709
26. Akter S, Xu D, Nagel SC, Bromfield JJ, Pelch K, Wilshire GB, Joshi T (2019) Machine learning classifiers for endometriosis using transcriptomics and methylomics data. Front Genet 10(766):4. https://doi.org/10.3389/fgene.2019.00766
27. Hastie T (2020) Ridge regularization: an essential concept in data science. Technometrics 62(4):426–433. https://doi.org/10.1080/00401706.2020.1791959
28. Sivajohan B, Elgendi M, Menon C, Allaire C, Yong P, Bedaiwy MA (2022) Clinical use of artificial intelligence in endometriosis: a scoping review. NPJ Digit Med 5(1):109. https://doi.org/10.1038/s41746-022-00638-1
29. Scutelnicu LA, Maftei R, Bălteanu A (2023) Image annotation platform for training dataset in endometriosis automatic detection. In: International conference on computer vision and pattern recognition (ICCVPR-23), Dubai, United Arab Emirates, November 2023, accepted for publication in Scopus, Web of Science

Blockchain-Integrated Chatbot

Daniela Mechkaroska and Ervin Domazet

Abstract Artificial intelligence and blockchain are two transformational technologies that have radically converged in blockchain-integrated chatbots. The revolutionary combination of blockchain and chatbot capabilities is explored in depth in this paper, shining light on its potential to change user interactions, enhance security, and promote trust across a range of industries. Data security, transparency, and accountability issues that arise during chatbot interactions are addressed through blockchain integration. Blockchain reduces the risks related to errors, privacy invasion, and unethical acts by documenting interactions on an immutable and transparent ledger. The multiple advantages of blockchain-integrated chatbots are explored in this study, including enhanced data privacy, transparency, open transaction tracking, and automated dispute resolution. Real-world use cases also highlight the revolutionary potential of this confluence, notably in the context of food delivery services. This paper demonstrates how blockchain-integrated chatbots have the potential to transform the way companies interact with their customers, ensuring secure and reliable interactions in the digital age, through in-depth study and case studies.

Keywords Artificial intelligence · Blockchain · Chatbot · Food delivery service

1 Introduction

Chatbots are artificial communication systems that are gaining popularity, but not all of their security issues have been satisfactorily addressed. Chatbots are used by people to help with a variety of tasks, including shopping, bank communication,

D. Mechkaroska (✉)
University for Information Science and Technology "St. Paul the Apostle", Ohrid, North Macedonia
e-mail: daniela.meckaroska@uist.edu.mk

E. Domazet
International Balkan University, Skopje, North Macedonia
e-mail: ervin.domazet@ibu.edu.mk

X.-S. Yang et al. (eds.), *Proceedings of Ninth International Congress on Information and Communication Technology*, Lecture Notes in Networks and Systems 1000,
https://doi.org/10.1007/978-981-97-3289-0_21

meal delivery, healthcare, and automobiles. However, it adds a new security risk and generates significant security issues that must be resolved. Modern chatbots use machine learning (ML) and natural language processing (NLP) instead of rule-based algorithms. These methods absorb information from a discussion, which may include private information [1].

Attackers have become smarter as new technologies have emerged. Hackers may issue an invitation to a legitimate bot to start a chat. If the bot accepts the invitation, believing it is from another legitimate bot, and proceeds with the interaction, the legitimate bot may provide its company's confidential data to the counterfeit bot, becoming a victim [2].

The importance of safe and open communication channels between companies and customers has never been more essential as the global digital ecosystem develops. Traditional interactions, which frequently include middlemen and are susceptible to data breaches, have created a demand for innovative solutions that can ensure data privacy, stop fraud, and inspire user confidence. The potential of chatbots that are coupled with blockchains becomes apparent in this situation. A new paradigm that reimagines how transactions, orders, and interactions are carried out in the digital age is created by combining the capabilities of chatbots, which enable intuitive and personalized discussions, with the inherent security, and immutability of blockchain.

Blockchain is an effective method of protecting consumer data and enables secure digital transactions that cannot be altered. This means that only those parties participating in a transaction have access to client data, which is safeguarded. Additionally, digital identities and client authentication can be created using blockchain technology, guaranteeing the security and safety of the data [3].

This research paper explains the advantages and, more importantly, the disadvantages of conventional artificial intelligence (AI) chatbots in customer service. It then discusses the dynamics of blockchain-integrated chatbots, highlighting their potential advantages and looking at a variety of use cases, particularly in the context of food delivery services, while exposing the complexity of their integration. We want to shed light on how these two cutting-edge technologies, blockchain and chatbot, may work together to change consumer experiences, expedite company processes, and build the foundation of trust that is so important in today's data-driven world.

2 Chatbots in Customer Service—Advantages and Drawbacks

AI chatbots are software programs that replicate human discussions with clients by using NLP and AI. They can conduct basic duties like scheduling appointments, processing payments, or changing account information in addition to providing information and responding to frequently asked inquiries. Without the need for human agents, AI chatbots can be linked with a variety of channels, including websites, mobile apps, social media, and messaging apps, to deliver customer care around-the-clock [4].

They are programmed based on NLP techniques to communicate in real time, in order to advise, support or simply talk to the interlocutor. The implementation of AI in e-services is no longer a rarity, so chatbots are also becoming more familiar and popular in customer service, and are most often assigned to sales and support functions in an area of e-commerce [5].

Despite their benefits, AI chatbots have certain risks and cons as well. For instance, they might not be able to respond effectively or empathetically because they lack the human touch and empathy needed to comprehend the customer's emotions, tone, or intent. Additionally, they might not be able to handle difficult, confusing, or delicate circumstances like grievances, disagreements, or crises without escalating the matter to a human agent, which could harm consumer loyalty, satisfaction, and confidence. Moreover, they could give incorrect or irrelevant information, misinterpret the customer's question, or give the same response more than once, all of which can harm the chatbot's credibility and the company's reputation. Last but not least, AI chatbots may bring up certain ethical and legal issues, such as how to safeguard consumer privacy and data, how to guarantee the chatbot's responsibility and transparency, and how to adhere to the necessary norms and legislation [4, 6]. Various technologies can help improve the skills of chatbots and solve the problems that surround them. In this paper, we will talk about the benefits of blockchain technology in existing chatbots and provide a theoretical framework of a blockchain-integrated chatbot.

2.1 Food Delivery Chatbot

An AI-powered virtual assistant created specifically to make ordering and delivering meals to consumers easier is called a food delivery chatbot. Users can use it as an interactive platform that offers a single conversational interface to browse menus, personalize orders, make payments, track delivery, and get real-time information. A meal delivery chatbot participates in human-like interactions using NLP and ML, improving the customer experience and streamlining the entire food ordering process.

E-commerce chatbots have grown dramatically in popularity during the previous years. Working as advisors "on the side" or as the first line of sales, these chatbots are responsible for thousands of product sales. The food industry, particularly the home delivery service sector, has undoubtedly embraced this trend, acquiring all of the positive elements of this type of revolution [7]:

- Your chatbot can assist numerous consumers at the same time. You're not going to miss any more phone orders because clients find a busy line and give up.
- If your restaurant also serves seated customers, no one should answer a phone call and be distracted from regular service.
- Your clients write orders clearly and check them several times. No more erroneous orders due to misunderstandings or poorly written notes.

- Instead of calling your restaurant many times, your clients can amend orders at any time, even after they have been placed.
- Reaching out to prior clients would be far easier than distributing paper leaflets that no one reads.

2.2 Building Our Food Delivery Chatbot

Here, we propose a cutting-edge chatbot system for meal delivery that revolutionizes the ordering process by seamlessly fusing AI with food services.

The idea of this workflow model is used from different food delivery services [7–9].

1. **Greeting and Menu Presentation**: The chatbot welcomes users and presents the options on the restaurant menu.
2. **Order Selection**: From the menu, users select the items, quantity, and choices they want.
3. **Customization**: Chatbot takes care of dietary needs, allergies, and special requests.
4. **Review of Order**: Customers affirm their personalized order.
5. **Payment Processing**: It is done by a chatbot in a secure manner.
6. **Order Confirmation**: Users receive an order summary and an expected delivery date upon receiving an order confirmation.
7. **Order Completion**: The restaurant gets ready to deliver the order.
8. **Delivery Tracking**: A chatbot offers up-to-the-minute delivery information.
9. **Delivery Confirmation**: Users are notified when their orders are delivered, according to delivery confirmation.
10. **Feedback Gathering**: The chatbot asks people for reviews and ratings.

The incorporation of an AI-driven chatbot in meal delivery services is an example of how food services may be able to improve user engagement, personalization, and convenience.

Although there are many advantages to using a typical food delivery chatbot, it's necessary to be aware of any difficulties that can occur when it is put into use. Here are some typical things to think about [10–12].

- Concerns about **privacy** and **security** are raised when user data is gathered and stored for order processing. It is essential to preserve user information and ensure compliance with data protection laws.
- It may be challenging for the chatbot to effectively handle some **complex orders** because they may include several goods, specific instructions, and delivery instructions.
- It may be difficult for the chatbot to respond in a clear and useful manner since different users may have different **communication preferences** and **language styles**. There is still no substitute for the organic flow of human dialogue in terms

of bots. The majority of chatbots can't change their language to sound like humans, therefore they frequently struggle to recognize slang, misspellings, and irony.
- **The lack of emotion** or personalization in a bot's response to a consumer is a major turnoff for many customers. Customers expect empathy when they are frustrated with a brand or company, especially when there is a problem. An unsatisfactory customer experience results from the bot's inability to improvise and lack of human touch when a discussion deviates from its intended course.
- To ensure that chatbots are telling your customers the right things and properly portraying your business, they need to be **optimized frequently**. The bot will need to be updated when user needs and corporate goals shift, which necessitates looking back at previous bot talks to uncover common queries our bot encounters.

3 Integration of Blockchain Technology in Chatbot

The rise of chatbots has transformed business-customer interactions thanks to conversational commerce. However, issues with trust, user privacy, and data security continue. A solution to these problems is provided by blockchain, which is recognized for its strong security features and decentralized architecture. Blockchain technology integration with chatbots offers a number of possible advantages and approaches to various problems in the area of conversational commerce [13, 14].

Let's imagine a chatbot-based payment assistance application. Most consumers would first be hesitant to give chatbots their sensitive personal or financial information, like their credit card number. Customers can be reassured that their personal information has been utilized carefully by looking at the successful transaction history if you employ a chatbot powered by blockchain technology because the transaction history will be recorded on a distributed database that can be read by anyone [15].

3.1 Theoretical Framework of a Blockchain-Integrated Chatbot

Depending on unique needs and goals, one should choose between permissioned and permissionless blockchain. Hybrid Blockchains contain sensitive data and private transactions on a network that is only accessible to authorized members, such as consortiums or companies. However, less sensitive information will be kept on the public chain, promoting inclusion and transparency.

In a food delivery system, a hybrid approach may be used, in which some system components—such as order processing and payment management—might be handled on a permissioned blockchain for security and privacy, while others—such as tracking deliveries and user reviews—might be recorded on a permissionless blockchain for openness and accessibility.

The idea of the workflow model given in the previous section will be adapted for the blockchain-based food delivery chatbot:

1. **Greeting and Menu Presentation**: The chatbot extends a warm greeting to users and displays the available menu items.
2. **Order Selection**: From the menu, users select the items, quantity, and choices they want. When a consumer uses the chatbot to place an order, the chatbot gathers all the relevant data, including the customer's address, the food items they want, and their payment information.
3. **Customization**: Chatbot takes care of dietary needs, allergies, and special requests.
4. **Review of Order**: Customers affirm their personalized order.
5. **Smart Contract for Order Creation**: The chatbot starts a smart contract created expressly for this order. The delivery address, delivery time, and order total are all predetermined criteria and conditions that are included in this smart contract.
6. **Payment Processing**: The chatbot processes payment securely through blockchain-based transactions.
7. **Smart Contract for Payment Confirmation**: The payment is verified and recorded on the blockchain using a smart contract. The smart contract has the ability to automatically validate payments made by customers. It examines the legitimacy and execution of the payment. Otherwise, unless the payment is verified, the contract won't move forward.
8. **Order Confirmation**: Users receive an order summary and an expected delivery date upon receiving an order confirmation.
9. **Order Completion**: The restaurant gets ready to deliver the order.
10. **Delivery Coordination**: The blockchain is used by the chatbot to track delivery information and cooperate with delivery services. The smart contract can allocate the order to a delivery driver as soon as the payment has been confirmed. This assignment may be based on a number of variables, including workload, distance from the delivery point, or a pre-defined algorithm.
11. **Delivery Tracking**: Through a blockchain-tracked status, the chatbot gives real-time delivery updates.
12. **Order Delivered**: Users receive their order.
13. **Smart Contract for Delivery Confirmation**: On the blockchain, a smart contract verifies that the delivery was successful. To confirm the delivery, the delivery team can use a mobile app or talk with the chatbot. This confirmation can trigger the smart contract to release payment to the delivery person.
14. **Dispute Resolution**: The smart contract may deal with any issues, such as when the food isn't delivered as promised. It may have predetermined guidelines for refunds or compensation, or, if necessary, it may demand a manual review by a human operator.
15. **Customer Feedback on Blockchain**: For transparency, user evaluations and ratings are stored on the blockchain. The chatbot might ask the client to provide a review or feedback after the delivery. The chatbot may give rewards like

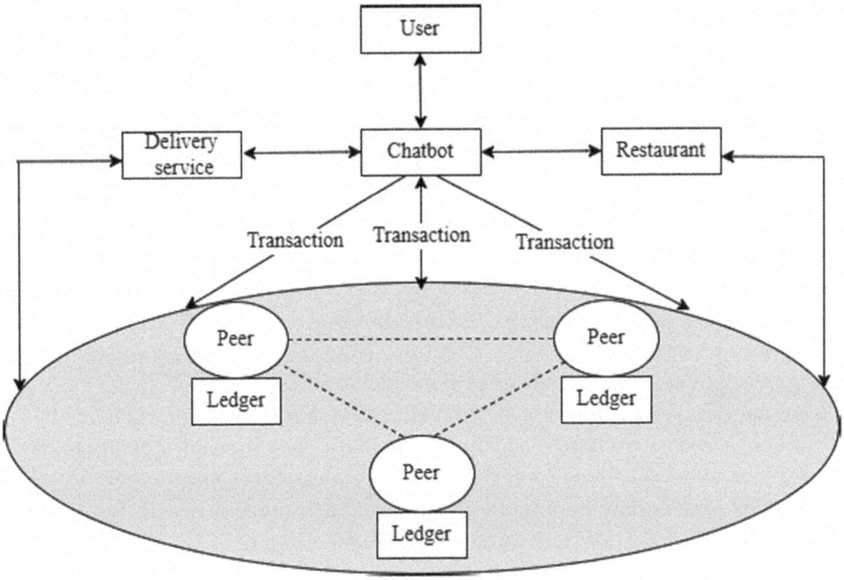

Fig. 1 Flowchart of a blockchain-based food delivery chatbot

discounts or loyalty points for favorable comments based on specified rules in the smart contract.

16. **Immutable Record**: The blockchain maintains an immutable record of each of these exchanges and transactions. As a result, the full order history, from placement to delivery, is safely preserved and tamper-proof.

The flowchart of theoretical framework of a blockchain-integrated food delivery chatbot is given in Fig. 1.

Users who have been given access to the hybrid blockchain platform can participate in all of its functions, including looking at transactions and adding or modifying transactions. Users can rest easy knowing that their identity will always be kept private and not disclosed. In this method, the **user's privacy** is protected.

4 Discussion—Advantages and Disadvantages of a Blockchain-Integrated Chatbot

The importance of blockchain-integrated chatbots in determining the future of the food delivery sector is growing as it develops. They offer a number of advantages, including greater security, transparency, and a wider range of use cases, if we compare it to the traditional food delivery chatbot. It does, however, have several downsides and difficulties relating to complexity and expense.

A normal food delivery chatbot system can encounter a number of problems that could be solved by blockchain technology. The following is how blockchain might address some of the previously identified problems:

- On the blockchain, smart contracts are self-executing contracts that are written. When the requirements are satisfied, they are started by pre-defined events and run automatically. Many manual operations might potentially be automated by smart contracts, lowering the likelihood of mistakes and fraud. By posting all transactions on the blockchain, they may also improve accountability and transparency [16]. By recording significant interactions on an immutable ledger, smart contracts bring an extra degree of **security**, **transparency**, and trust to the blockchain-integrated process. This increases user trust in the ordering process and makes the experience of receiving meals more dependable and seamless [17].
- **Language barriers** may not be immediately addressed by blockchain, but it can enhance data exchange and interoperability, which enables chatbots to be combined with translation tools to better serve a variety of users.
- **Complex orders** may be reliably documented and processed with fewer errors because to blockchain's transparent and traceable ledger.
- **Secure payment processing** via the chatbot can be facilitated by blockchain's cryptography protocols. This reduces the possibility of fraud and guarantees the security of payment information [18].
- The chatbot can use blockchain to give users with verifiable information, fostering a sense of **trust** in the information presented. Blockchain has the potential to **authenticate** the origin and authenticity of information [19].
- **Reducing costs and fees for restaurants, providing lower prices to customers, real-time tracking of orders and drivers, reduced waiting time** are made possible by the synthesis of Blockchain and chatbot [20].

While the advantages of blockchain-integrated chatbots are generally recognized, it is as important to shed light on a less-explored element of this merger—the potential drawbacks and problems [21, 22].

- **Complex Development**: Using blockchain technology complicates the chatbot creation process. Blockchain ideas must be understood by developers, and their code must be adjusted properly.
- **Costs Increased**: Because blockchain knowledge and infrastructure are required, developing and maintaining a blockchain-integrated chatbot can be more expensive.
- **Blockchain Scalability**: As the number of transactions increases, many blockchain networks, particularly public ones like Bitcoin and Ethereum, encounter scalability concerns. This can cause delays in the processing of chatbot conversations.
- **Compliance with Regulations**: Integrating blockchain may introduce regulatory and compliance challenges, particularly regarding data protection and consumer protection regulations.

- **Maintenance**: Blockchain networks necessitate ongoing upkeep, updates, and monitoring. Failure to update the blockchain can result in malfunctions or concerns with chatbot compatibility.

5 Conclusion

In this research, we investigated the incorporation of blockchain technology into chatbots, with a special emphasis on its applicability in the food delivery industry. Our analysis uncovered a number of benefits and drawbacks linked with this integration, providing light on both the intriguing potential and the coming obstacles.

The integration of blockchain technology into chatbot systems was at the heart of our research. We proposed a conceptual framework for how this integration could improve transparency, security, and user confidence. Our analysis and debate were built on this paradigm.

While traditional food delivery chatbots and their blockchain-integrated chatbots are set to play crucial roles in the changing food delivery landscape, it's the blockchain integration that has the potential to be a transformative force, profoundly influencing the future of food delivery services. Although effective and user-friendly, traditional chatbots may need to change to keep up with new technologies and shifting consumer tastes.

The expectations and predictions are that traditional and blockchain-integrated chatbots will coexist in the food delivery industry as it continues to develop, each serving a different market segment and meeting a different set of needs. The decision between them will be influenced by a number of elements, such as the necessary level of security and transparency, regulatory compliance, and the readiness to invest in blockchain technology. In the end, the delicate balancing act between creativity and pragmatism will determine the future of food delivery services, with both types of chatbots assisting in the industry's continual development and evolution.

Future study can investigate the integration of chatbots with IoT devices in the food delivery process as the IoT ecosystem continues to grow. Utilizing IoT data for order tracking, delivery planning, and real-time updates may fall under this category. Moreover, we plan to implement a conversational flow that allows customers to browse menu items, place orders, provide delivery information, and track their orders while designing the user interface for a food delivery chatbot with Xenioo integrated with blockchain.

References

1. Hasal M, Nowaková J, Saghair KA, Abdulla H, Snášel V, Ogiela L (2021) Chatbots: security, privacy, data protection, and social aspects. https://doi.org/10.1002/cpe.6426
2. Bbntimes. https://www.bbntimes.com/technology/how-blockchain-can-improve-chatbots. 15 Dec 2021

3. Blockchain—8 ways blockchain is revolutionising the restaurant industry. https://business.getonbloc.com/blockchain-in-the-restaurant-industry/#:~:text=The%20use%20of%20blockchain%20technology,inefficiencies%20in%20their%20supply%20chain
4. Linkedin. https://www.linkedin.com/advice/3/what-benefits-challenges-using-ai-chatbots-customer. 01 Mar 2023
5. Ashfaq M, Yun J, Yu S, Correia LSMI (2020) Chatbot: modeling the determinants of users'satisfaction and continuance intention of AI-powered service agents. Telemat Inf 54:101473. ISSN 0736-5853. https://doi.org/10.1016/j.tele.2020.101473
6. Rapp A, Curti L, Boldi A (2021) The human side of human-chatbot interaction: a systematic literature review of tenyears of research on text-based chatbots. Int J Hum-Comput Stud 151:102630. ISSN 1071-5819. https://doi.org/10.1016/j.ijhcs.2021.102630
7. Gupshup. https://www.gupshup.io/resources/blog/6-advantages-of-restaurant-chatbots#:~:text=The%20chatbot%20can%20display%20menus,short%2C%20a%20thriving%20food%20business! https://marutitech.medium.com/how-to-order-food-conveniently-through-chatbot-12f2cfa82f59. 14 Jul 2021
8. Blink. https://www.blinkco.io/blog/purpose-of-the-online-ordering-system/. 25 Jul 2023
9. MarutiTech. https://marutitech.medium.com/how-to-order-food-conveniently-through-chatbot-12f2cfa82f59. 14 Apr 2017
10. ChatDesk. https://www.chatdesk.com/blog/pros-and-cons-of-chatbots. 08 Dec 2022
11. IAPP. https://iapp.org/news/a/chatbots-ai-and-the-future-of-privacy/. 31 Mar 2023
12. Hasal M, Nowaková J, Ahmed Saghair K, Abdulla H, Snášel V, Ogiela L (2021) Chatbots: security, privacy, data protection, and social aspects. Concurr Comput: Pract Exp 33(19):e6426
13. Zavolokina L, Zani N, Schwabe G (2023) Designing for trust in blockchain platforms. IEEE Trans Eng Manage 70(3):849–863. https://doi.org/10.1109/TEM.2020.3015359
14. Bhuiyan MSI, Razzak A, Ferdous MS, Chowdhury MJM, Hoque MA, Tarkoma S, BONIK (2020) A blockchain empowered chatbot for financial transactions. In: IEEE 19th International conference on trust, security and privacy in computing and communications (TrustCom). Guangzhou, China, pp 1079–1088. https://doi.org/10.1109/TrustCom50675.2020.00143
15. AI Network. https://medium.com/ai-network/everyones-blockchain-chatbot-with-blockchain-feat-ai-network-blockchain-997c62e0c94f. 29 Sept 2021
16. Wonderment. https://www.linkedin.com/pulse/blockchain-transparency-accountability-wonderment-apps/. 02 Mar 2023
17. Cozzio C, Viglia G, Lemarie L, Cerutti S (2023) Toward an integration of blockchain technology in the food supply chain. J Bus Res 162:113909. ISSN 0148-2963. https://doi.org/10.1016/j.jbusres
18. AnalyticSteps. https://www.analyticssteps.com/blogs/role-blockchain-technology-secured-payments. 15 Feb 2020
19. Batista D, Kim H, Lemieux VL, Stancic H, Unnithan C (2021) Blockchains and provenance: how a technical system for tracing origins, ownership and authenticity can transform social trust. In: Building decentralized trust: multidisciplinary perspectives on the design of blockchains and distributed ledgers, pp 111–128
20. https://medium.com/nerd-for-tech/decentralizing-food-delivery-app-with-blockchain-integration-5982c4ff57c2. 02 Mar 2023
21. Chainlink. Blockchain scalability: execution, storage, and consensus. https://chain.link/education-hub/blockchain-scalability. 09 Aug 2023
22. Paradigm. Understanding blockchain latency and throughput. https://www.paradigm.xyz/2022/07/consensus-throughput. 22 Jul 2022

Switching from DOORS to DOORS Next Generation. Identifying the Main Usability Issues. A Case Study

Oana Rotaru, Silviu Vert, and Radu Vasiu

Abstract Automotive engineering requires a scientific approach to requirement management because it is a challenging and dynamic field. This paper discusses the importance of usability in requirements management tools, such as IBM Rational DOORS and DOORS Next Generation, commonly used in the automotive engineering. It presents a usability evaluation of these tools, emphasizing key usability issues. The analysis is anchored in a case study examining an automotive company's transition from DOORS to DOORS Next Generation, with a focus on the experiences of requirements and quality engineers, but also interns.

Keywords RM tool—requirements management tool · RE-requirements engineering · SW development · SW RM-SW requirement management · IBM rational DOORS · DOORS next generation · DOORS NG · Usability evaluation · User testing

1 Introduction

The automotive industry is a fast-paced environment. Cutting-edge technology and sustainable practices are leading the way ahead in the automobile industry, ushering in an era of electrification, driverless cars, and a renewed focus on environmental consciousness [1].

Requirements engineering and management are an important step when it comes to an automotive project and it implies the process of establishing, outlining, verifying, and managing the requirements and constraints of a project or product, all of these being crucial for determining the functional and technical qualities of a vehicle

O. Rotaru (✉) · S. Vert · R. Vasiu
Politehnica University of Timisoara, Vasile Parvan No.2, Timisoara, Romania
e-mail: oana.rotaru@student.upt.ro

O. Rotaru
Continental Autonomous Mobility Romania, Siemens No.1, Timisoara, Romania

X.-S. Yang et al. (eds.), *Proceedings of Ninth International Congress on Information and Communication Technology*, Lecture Notes in Networks and Systems 1000,
https://doi.org/10.1007/978-981-97-3289-0_22

and for ensuring traceability and consistency throughout the development process [2].

Requirements are built based on the existing factors of the environment and the purpose needed to be achieved. The requirements evolve during a project, becoming gradually detailed, solid and sophisticated [3].

As long as the complexity grows and the requirements become more solid and sophisticated, the requirement management activities must be accompanied by a systematic tool capable to support the complexity [3]. The requirement management (RM) tools are used by the system requirements engineers and managers, but also by the developers, software requirement engineers, requirement managers, and test engineers. As highlighted in [3], some of the main basic tasks usually done by the users of an RM tool include editing artifacts, importing/exporting artifacts, traceability creation, review of requirements or generating reports. Two of these kinds of tools are DOORS and DOORS Next Generation (DOORS NG), which will be analyzed in the following section.

To accomplish these tasks and more others, to fulfill the stakeholders needs, the RM tools must be usable. According to Nielsen [4], usability consists of multiple components and the main attributes of it are the learnability, efficiency, memorability, errors and, satisfaction, all these attributes being explained in the following section.

IBM Rational Doors (9.7) and DOORS NG are the two RM tools which we chose to evaluate from the user experience perspective. The motivation behind it is triggered by the transition from DOORS to DOORS NG, taking part within one of the departments from an automotive company.

A part of this study was also presented in the paper [5], which is submitted for publication, a paper focused on the evaluation of the collaboration features of RM tools, such as DOORS NG.

For this study, the main research questions we aim to answer to are:

RQ1. Which are the primary usability challenges identified by the mixed participant group when using DOORS NG?

RQ2. How do users with prior experience in DOORS perceive DOORS NG? Are there any similarities between these two?

RQ3. From the perspective of novice users, which platform-DOORS or DOORS NG-achieves a better SUS score?

2 Theoretical Background

2.1 Requirements Engineering Process

The most data-intensive and media-rich activities in software development projects are requirements engineering (RE) tasks, including stakeholder identification, needs analysis, requirement communication and change management. Early RE activities involve decision-making, often with uncertainties about cost, schedule, user needs, system functionality, and converting business objectives to technical specifications [2].

A requirements engineering phase implies four activities: elicitation, specification, validation and verification management [6].

The elicitation is the process of learning about all the key components of the project. The customer will supply specifics about their requirements and vital background data.

The requirements specification/documentation phase implies the system requirements that are elicited in the early stage of the RE process. This stage also implies the reviews activity with the verification and validation teams [7].

The next phase is the verification and validation phase, where the verification part ensures that the software is built based on the customers' needs and the validation side ensures that the software is implementing the right functions [6].

RM is an ongoing activity that occurs concurrently with the three aforementioned stages. It is essential because requirements are dynamic entities, being created, used, updated and deleted throughout their development lifecycle. Therefore, prioritizing requirements management is crucial from the project's outset rather than waiting for complexity to arise [8].

A proper illustration sustaining the entire requirements engineering process is presented in Fig. 1, where one simplified state chart of a requirements lifecycle is presented.

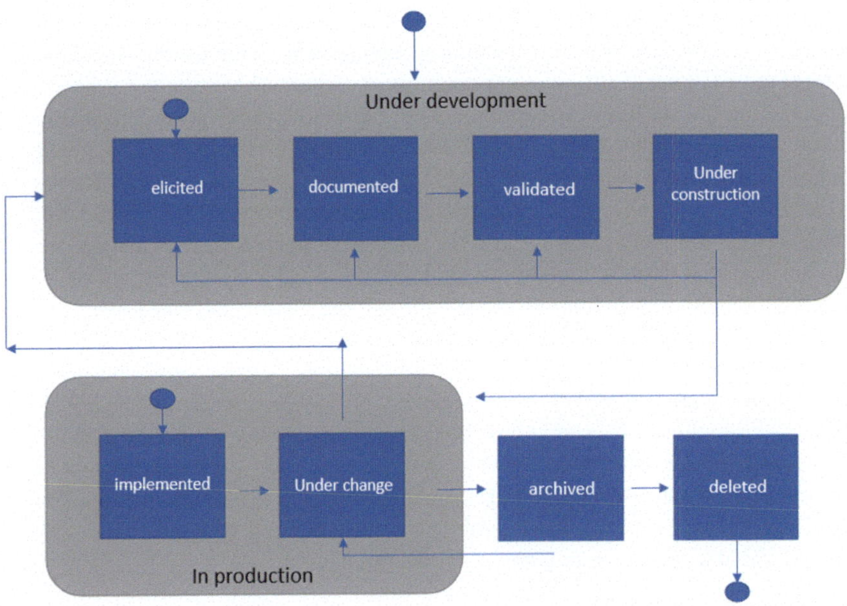

Fig. 1 State chart of a requirements lifecycle [8]

2.2 Requirements Engineering and Management Tools

Requirements are built based on the existing factors of the environment and the purpose needed to be achieved. The requirements evolve during a project, becoming gradually detailed, solid, and sophisticated [3].

The specifications for a system are already a model of the system that must be created—it is typically too complicated for humans to grasp everything at once. To manage the complexity, requirements management (RM) accompanied by a systematic tool must be built [3].

According to Hoffman et al. [3], to sustain these, an RM tool must provide: acquisition, specification, grouping and attribution support for the requirements, support on deriving the requirements into detailed levels as customer, system, component, design levels, support for test cases, tracking options for statuses and support on collaborative work.

The RM tools are especially used by the requirement engineers and managers, but also by the developers and test engineers.

The requirements engineering process, facilitated by RM tools, is integrated within the V-model, a software development process depicting the connection between development phases and testing. According to [9], the V-model divides the process into two sections: the initial half encompasses requirements analysis, function design, and software development, while the latter half encompasses verification, validation and release, as depicted in Fig. 2 [10].

As summed up in [3, 11], some of the capabilities of an RM tool in order to facilitate the whole RE process are: organization of requirements with attributes and possibility of reuse (standard attributes or possibility of user created attributes), traceability support (traceability links to requirements or other artifacts), variant management, remote working (the tools should work even if it doesn't have Internet connection), multiple views can be created and managed by the users, performance (large amounts of data), collaboration features (essentials for technical reviews activities and discussions), also collaborative working on the same development tasks, easily adapting into business processes (compatibility with pre-existing tools), exporting/importing data functionality and support, documentation of the history and baselining (partial database of the artifacts states at a given point in time).

All of these capabilities contribute to the creation, management, and use of a good quality requirement [12]. As explained in [11], a good requirement should be: atomic—it should fulfill a single stakeholder need or a quality attribute; attainable—the needs must be achievable; consistent/free of conflict—any requirement shall neither conflict with any other requirement nor with legal regulations and applicable standards, comprehensible and unambiguous, testable and of course, traceable—derived and able to satisfy an upper-level requirement.

DOORS and DOORS NG

DOORS is a tool developed in 1991 by Quality Systems and Software Ltd (QSS), attained by Telelogic in 2000 and by IBM company in 2008, as it was renamed IBM

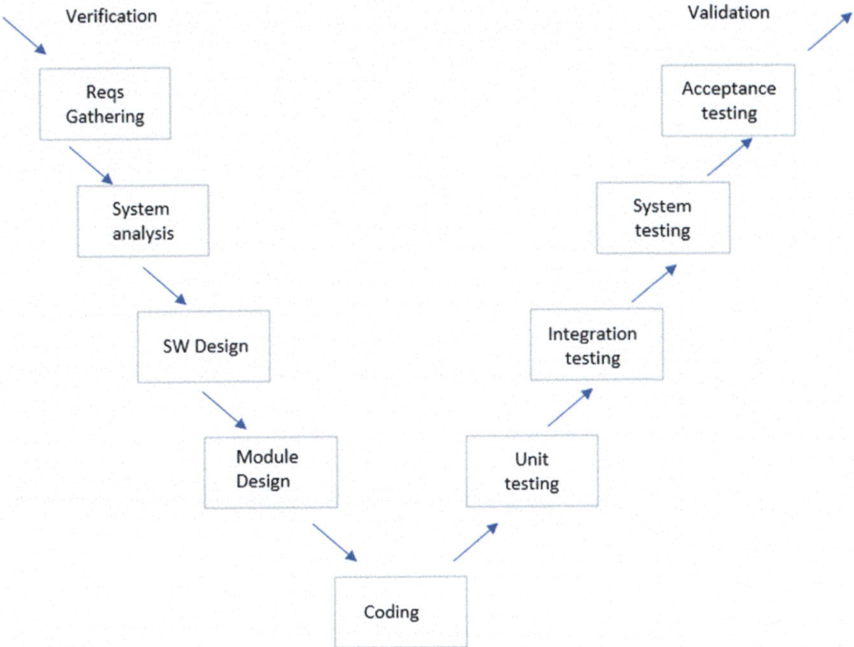

Fig. 2 V-model lifecycle for the automotive SW development [10]

Rational DOORS. Later in 2019, IBM Rational DOORS turned into IBM Engineering Requirements Management DOORS [13].

As highlighted in [13], some of the features DOORS has are: offering support on elicitation, documentation, analyzing, traceability, import/export from MS Word, Excel, PowerPoint, Outlook, Adobe FrameMaker, the ability to offer linkage between requirements, test cases, test results, freedom of attributes creation and reusability of documents, baselining, multiple views.

As also mentioned in [14], DOORS offers the possibility of using descriptive modules as an alternative to the plain test importer to import text files into the tool. These descriptive modules can be created using DXL [15]

When it comes to managing and composing requirements, DOORS offers an interface which allows the users to organize data in projects and folders, create modules and create and generate reports [14].

IBM Engineering Requirements Management DOORS NG is an RM tool having the purpose to store, organize, link and share product requirements within a project. The tool uses a web client and the Jazz platform to achieve the definition, management and reporting on requirements [16].

The main tasks that can be done by requirements engineers, managers or SW development teams using DOORS NG include: editing artifacts, importing/exporting artifacts from documents, creating diagrams and wireframes, creation of links between the artifacts or external links, creation of traceability, review of artifacts, baselining, generating reports, creation of traceability with work items and also, a new feature, creation and management of streams and changed sets [16].

2.3 Usability

Usability measures how easily a specific user in a specific context can use a product in order to accomplish a specific goal in an efficient and satisfying way [17].

According to Nielsen [4], usability consists of multiple components and the main attributes of it are the learnability, efficiency, memorability, errors and satisfaction.

The learnability represents the easiness of accomplishing basic tasks by a user, at the first encounter, while memorability refers to the easiness of using the same product after a period of not using it. The efficiency refers to the speed of performing tasks after the users learned the product and the "errors" attribute refers to the number of errors the users do and their severity. Also, the satisfaction is the overall pleasure of the user when he/she uses the interface [18].

There are many methods to run usability evaluations for different digital products.

One of the most used method is the user testing, which involves real users testing the product, with tasks performed in various scenarios recorded for later analysis [19]. Metrics such as task completion time, completion rate and error types are documented. Typically conducted in a specialized laboratory with specific equipment, the sessions involve observers, a facilitator and a moderator [20].

An additional technique used in the usability evaluation is the Think-Aloud Protocol. This approach encourages participants to verbalize their thoughts while navigating through the tasks they are handed during the user testing session. During this time, the researchers document the verbalized insights for further analysis [21].

The focus group method is another commonly used usability evaluation method. It was proven as an effective method for assembling individuals with comparable backgrounds or experiences to deliberate on a specific subject of interest. The participants group is steered by a moderator, who raises topics for discourse and engages a dynamic conversation among the group members [22].

The System Usability Scale Questionnaire, or SUS, is also an instrument used for the usability evaluation and it is the most frequently used for the usability measurements. The survey consists of 10 statement questions and the users are asked to rate their perception of agreement or disagreement on each of them. SUS is a standardized questionnaire but it is easy to customize it according to the particular tested product [23].

The product reaction cards is another useful method when it comes to usability measurement. It can be done either physically, with cards to be picked by the participants, or electronically, with the help of a Word or Excel file. The participants are asked to choose a certain number of cards expressing their own experience with the tested product [24].

3 Usability Evaluation of the RM Tools

3.1 *The Evaluation Methodology*

As previously mentioned, the tools evaluated were IBM Rational DOORS 9.7 and DOORS NG 7.0.2 (current used versions in the company), two RM tools frequently used in the automotive sector. For confidentiality reasons, two playground areas were created by the authors for the user testing session.

Participants

For our study, we engaged a total of 10 participants. Five of the participants only had basic knowledge of requirements engineering, with no previous use of DOORS or DOORS NG tools. The other five participants were totally new in DOORS NG, but proficient in IBM Rational DOORS.

The first purpose of the evaluation was to attain the user testing results for DOORS NG, where all the mixed participants are active: experienced with a similar tool and totally beginners. In this case, we acquired the main usability issues found and we also obtained the SUS postquestionnaire score and the choices of product reactions cards.

The second purpose of the evaluation was to obtain separate results for DOORS NG user testing, where the participants are only the half of them, the ones already experienced with DOORS. Here, after the user testing, we also applied the focus group method in order to understand the participants' expectations before the user testing session and their overall feedback on the new RM tool, in comparison with the familiar tool.

The third aim of the usability evaluation was to acquire another set of results for the user testing of the other half of the participants, the beginners, who have never used an RM tool before. Here, we gathered separate SUS scores for both platforms, the chosen reaction cards and the main usability issues found for both platforms.

The demographic profile of the usability evaluation participants is presented in Table 1.

Procedure

The main usability evaluation method used in our study was the user testing. It was organized remotely, via Microsoft Teams, with each participant at a time. Figure 3 illustrates one of the users, during the user testing session.

First, the participants filled out a prequestionnaire with questions on details of their profession/position in the company, their previous experience with requirement management tools and their interest in this kind of platforms. The participants with previous experience in DOORS were questioned about their favorite features/pleasant experience of the well-known tool and their expectations upon the new platform, which was about to be tested, DOORS NG.

The participants received 9 tasks to perform in DOORS and DOORS NG. The tasks were the same for both platforms.

Table 1 Demographic profile of the user testing participants

Criteria	Variation	Number
Gender	F	5
	M	5
Age	18–26	5
	26–35	2
	35–50	3
Profession/position in the company	Student/Intern	5
	Requirements engineer and manager	3
	Quality engineer	2
Experience with DOORS	First time	5
	Proficient	5
Experience with DOORS NG	First time	10

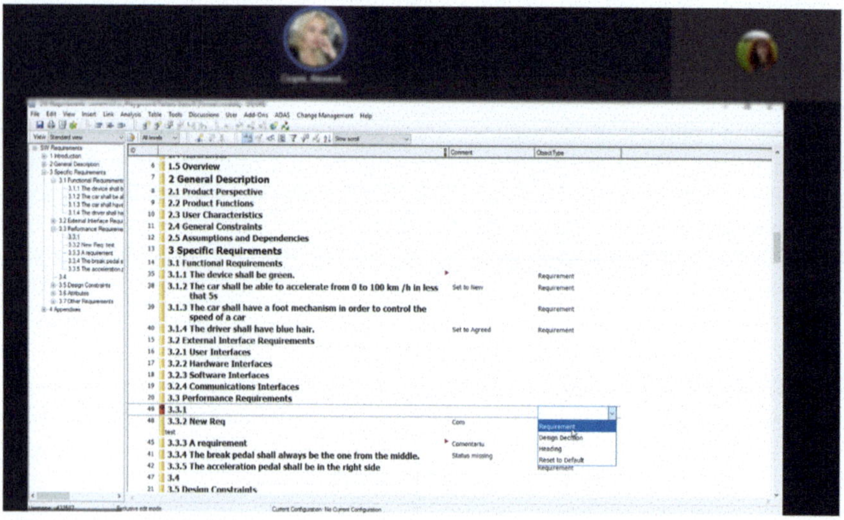

Fig. 3 Participant during user testing session

While solving the tasks, one observer recorded the time needed to finish them. Two observers took notes on the participants' comments, which were encouraged to think aloud while doing the tasks [21].

The tasks handed to the participants were tasks concerning basic routine activities that are usually done in requirements management tools. Some usual activities are: creating views, filtering requirements after different attributes values, adding new requirements artifacts, navigating through the requirements document, traceability linkages with upper-level requirements and linkage with test cases, changing attributes or baselining the documents.

After the completion of the tasks, we applied the focus group method on half of the participants, those with previous experience in DOORS, who have only performed tasks for the new tool. The aim of the focus group session was to gather the participants honest feedback on the new platform tested, on the new user interface in comparison with the old platform. We also aimed to get feedback on the complexity of DOORS NG and on the differences or similarities between the RM tools.

Also, after finishing the user testing session, the participants were asked to complete the SUS questionnaire and choose the product reaction cards, based on their immediate experience.

3.2 *The Evaluation Results*

DOORS NG: Evaluation Results from All Participants

As mentioned in the previous subsection, the authors gathered results from the evaluation of DOORS NG with 10 mixed participants, who never used the tool, half of them with already gained experience in a similar platform.

The main usability issues found after the user testing concern activities like: not finding how to apply filters for 20% of participants; and not being able to navigate through the requirements document with a navigation sidebar—the module explorer is a widget that should be installed in order to use it and it's not visible aside the document. This problem was encountered by 100% of participants.

Another usability challenge identified was the difficulty in linking or visualizing a test case to a requirement artifact, an issue encountered by 80% of the participants. The participants attributed this difficulty to the multitude of link types, which can be overwhelming and confusing upon initial interaction with the platform. Additionally, the participants emphasized the necessity of acquiring technical knowledge to successfully create linkages for the requirement artifacts within the new tool.

Following the user testing, we identified several valued features of the platform, such as the multiple methods available for adding comments, adding attributes in the table view, and modifying attributes. Approximately 40% of the participants were able to discern three distinct methods for executing these actions, primarily utilizing icons and detailed sidebars.

An additional feature that was favorably received, as indicated by 40% of the participants, is the extensive range of standard actions that can be performed through right-clicking on the artifacts, actions like editing the attributes, adding comments, adding different types of links, inserting new artifacts and copying/removing artifacts.

In Table 2, we present a concentrated situation of the average time spent on each task and the average rates of completing the tasks by the participants. To sum up, the time spent on the actual user testing for one participant is an average of 26 min, while an average of the completion rate is 74.2%.

Table 2 Average time and completion rates per task-all participants

Task	Average time taken (min)	Average completion rate (%)
1—Changing views	0.5	100
2—Displaying comments	2	100
3—Applying filters	2	83
4—Document navigation-module explorer	4	10
5—Adding requirements artifacts	3	95
6—Linkage between 2 requirements	4	60
7—Editing attributes for a new artifact	2	100
8—Linkage to a test case	5	30
9—Making modifications visible for anyone	3	90

After the user testing, the participants completed the SUS questionnaire and after that, they were asked to choose the reaction cards based on their recent experience with DOORS NG.

The resulted SUS score was 58, which is considered to be Graded D, with the adjective rating Poor [25]. This suggests that there is significant room for improving the usability.

It's worth mentioning that the interpretation of SUS scores can be somewhat subjective and can depend on the context. For example, an SUS score of 58 might be seen as acceptable for a complex system like DOORS NG, while the same score might be seen as poor for a system that is intended to be simple and easy to use.

Based on the product reaction cards selected by participants, the most frequently chosen descriptors were "Complex", "Confusing", "Difficult" and "Stressful". However, some participants also selected more positive descriptors, including "Accessible", "Familiar" and "Attractive". These terms provide a mixed assessment of the product, indicating areas of both strength and potential improvement.

DOORS NG: Usability Findings from Participants with Prior Experience in DOORS

As one of the purposes of the study was to achieve separate feedback from the participants with previous experience in a similar RM tool, DOORS, after the actual user testing, the focus group method was applied on this group of participants.

Regarding the user interface, 40% of the participants considered that the tool has a more attractive interface than DOORS, while 60% of them considered that the user interface can become overwhelming, with lots of icons and sidebars which are not very helpful in the end.

All participants expressed dissatisfaction with the newly introduced streaming feature, considering it a source of confusion. This was particularly evident when compared with the DOORS platform, which incorporates only the baseline concept, thereby offering a more straightforward user experience.

Some similarities between the tools were also emphasized, like the way of adding comments in the table view of the requirements, or the way of creating filters or different views.

A significant concern was raised to the module explorer, which is not immediately visible, but rather hidden behind an installable widget. This issue was accompanied by a preference expressed by the participants: instead of having a sidebar of views, they suggested that the module explorer should be prominently displayed, with the views ideally compressed under a dropdown menu.

Four participants asserted that prior experience with DOORS facilitated their navigation through DOORS NG, attributing this to familiar features and an already established mindset. Antithetically, one participant believed that having experience with a similar tool creates a basis for comparison, which could lead to confusion and overthinking simple actions.

The System Usability Scale was calculated separately for this group of participants, yielding a resultant score of 48. This suggests that the participants found the system to be quite challenging to use.

Additionally, the product reaction cards selected by this group of participants predominantly featured descriptors like "Complex", "Difficult" and "Confusing", indicating areas of potential usability improvement. However, the term "Useful" was also chosen, suggesting that despite the complexity and difficulty perceived by the participants, they also found value in the system's functionality.

DOORS NG and DOORS: Evaluation Results from New Users

Although a portion of these results is presented in one previous subsection regarding the evaluation of DOORS NG, this section will encompass the derived results corresponding to the participant with no prior experience in RM tools.

The main usability issue for this group using DOORS NG was the inability to find the module explorer for document navigation, a problem faced by all participants, who also struggled to filter content by chapter number. Additionally, four out of five participants struggled to link requirements to test cases, and three couldn't make modifications visible to all users, highlighting areas for usability improvement in DOORS NG.

Despite the challenges mentioned before, the participants identified three alternative scenarios for adding traceability linkages and two additional scenarios for creating new artifacts when these were required.

In the context of DOORS, the primary usability issues were related to linking to test cases, a difficulty experienced by two out of five participants. In addition, one participant encountered problems when attempting to add new requirements or to establish linkages between two requirements artifacts.

Figures 4 and 5 emphasize the module explorer feature as a left sidebar, present in DOORS, an appreciated asset by the participants in the context of document navigation and the left sidebar as a list of views from DOORS NG.

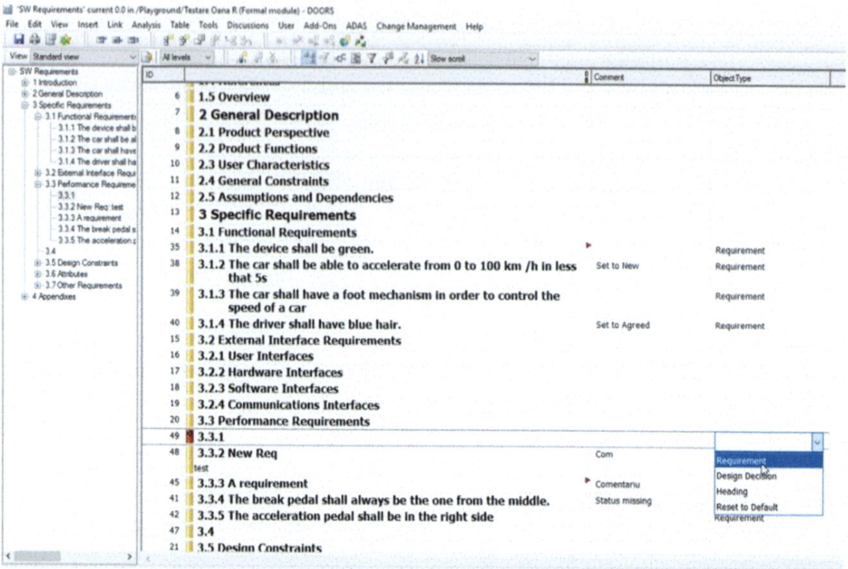

Fig. 4 Module explorer in DOORS

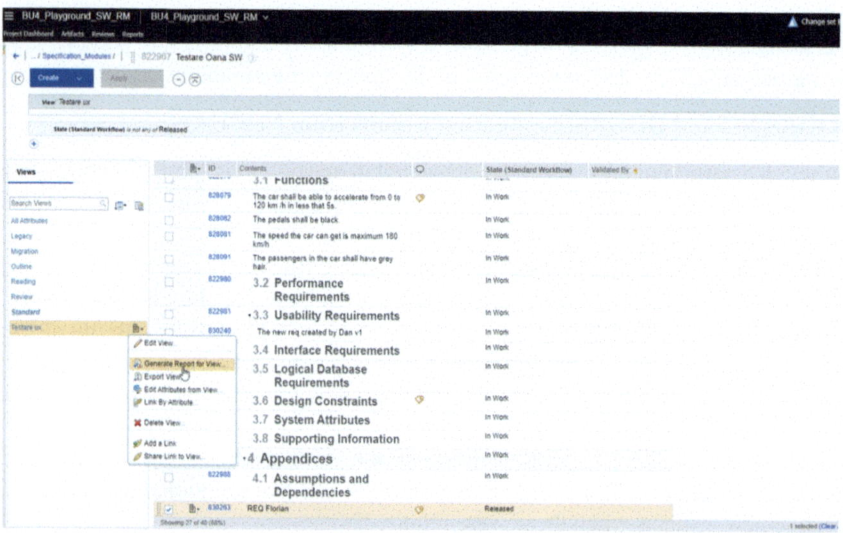

Fig. 5 Left views sidebar in DOORS NG

Table 3 Average time taken and completion rates for the new users of DOORS NG (DNG) and DOORS (D)

Tasks	Average time taken (min)		Average completion rate (%)	
	DNG	D	DNG	D
1—Changing views	1	0.5	100	100
2—Displaying comments	1	2	100	100
3—Applying filters	2	0.33	60	100
4—Document navigation-module explorer	1	1	0	80
5—Adding requirements artifacts	2	2	90	90
6—Linkage between requirements	2	2	80	80
7—Editing attributes for a new artifact	2	2	90	100
8—Linkage to a test case	3	2	20	60
9—Making modifications visible for anyone	2	0.5	40	100

Table 3, presented subsequently, sets side by side the average time taken for each task with the average completion rates for this specific group of five participants, providing a task analysis of task performance.

The average total time required to complete the tasks was 16 min for DOORS NG and 13 min for DOORS.

System Usability Scale (SUS) scores were computed for both platforms as part of this user testing segment. DOORS NG achieved a score of 68, translating to an adjective rating of "Okay" and a grade of C. Otherwise, DOORS attained a score of 70, corresponding to a "Good" rating and a grade of B.

The reaction cards most frequently selected by participants for DOORS NG included "Accessible", "Complex" and "Rigid". For DOORS, the most chosen descriptors were "Easy to use", "Old", "Stable" and "Clear".

In the final analysis, participants found the tools to be quite similar. While DOORS NG was perceived as having a more user-friendly interface, DOORS was considered more stable and was appreciated for providing user hints during various actions.

4 Conclusion

The scope of this paper was to identify the main challenges encountered during the transition from a requirements engineering and management tool to another, in this case, from IBM Rational DOORS to DOORS NG, inside an automotive company department. The study was realized with the help of 10 participants, with various experience and knowledge, who actively participated in user testing sessions and three sets of results were extracted and interpreted.

The first set of results answers to the first research question, concerning the main usability challenges encountered for DOORS NG, for all the participants.

The primary usability issues identified from the user testing include difficulties in applying filters, navigating through the requirements document due to the hidden module explorer widget, and linking or visualizing a test case to a requirement artifact due to the overwhelming variety of link types. Despite these challenges, participants appreciated the platform's multiple methods for adding comments, modifying attributes, and the extensive range of standard actions accessible through right-clicking on artifacts. These findings highlight both the areas of improvement and strengths of the platform.

The second set of results addressed the second research question, regarding the impact of prior experience with DOORS on the first encounters in DOORS NG. Here we applied the focus group method on this group of participants, after the user testing session.

We can conclude that users with prior experience in DOORS had mixed feedbacks on DOORS NG. While 40% found the interface more attractive, 60% found it overwhelming due to numerous icons and sidebars. All participants noticed and found the new streaming feature confusing compared to DOORS' simpler baseline concept. The similarities between the tools, such as adding comments and creating filters, were appreciated. The hidden module explorer was a significant concern, with a preference for a more visible placement. Four participants felt their DOORS experience aided their navigation in the new tool, while one felt it led to confusion. This group SUS score was 48, indicating serious challenges, while the reaction cards highlighted descriptors like "Complex", "Difficult", but also "Useful", indicating room for improvement but also recognized functionality.

The third research question addressed comparing results on the SUS scores of both platforms, tested with novice users.

Along with the usability issues faced, such as struggling with finding the module explorer, filtering content by chapter numbers, or inability to submit the modifications, the SUS score revealed for DOORS NG was 68, graded C.

In DOORS' case, the usability issues encountered were linking to test cases, linking between requirements, or adding new requirements. Here, the SUS score was 70, graded B, slightly better for usability than DOORS NG. The overall feedback from the novice users was that the RM tools are quite similar, with DOORS NG having a more attractive user interface, but with DOORS having an airy yet stable user interface, with user hints provided during certain actions.

In conclusion, as part of a future research in this field, we aim to expand the study to other RM tools. The focus will be on specific features like requirement traceability, collaboration and communication, reporting and visualization, in order to highlight patterns of usability issues for this type of tools. This extended study will lead to a deeper understanding of usability challenges and will provide valuable insights for the requirements management tools field.

Acknowledgements We thank all the participants in the study, employees of Continental Autonomous Mobility Timisoara, who provided important insights over the study, through the user testing sessions and focus group discussions.

References

1. Frąckiewicz M (2023) Autonomous driving: unveiling its impact on the automotive industry. TS2 SPACE. Accessed 13 Jun 2023. [Online]. Available: https://ts2.space/en/autonomous-driving-unveiling-its-impact-on-the-automotive-industry/
2. Shakeri Z, Noaeen M, Ruhe G (2016) Requirements engineering visualization: a systematic literature review. https://doi.org/10.1109/RE.2016.61
3. Hoffmann M, Kuhn N, Weber M, Bittner M (2004) Requirements management tools. pp 308. https://doi.org/10.1109/ICRE.2004.1335687
4. Nielsen J (201) Chapter 1—What Is Usability?. In: Wilson C (ed) User experience re-mastered. Boston, Morgan Kaufmann, pp 3–22. https://doi.org/10.1016/B978-0-12-375114-0.00004-9
5. Rotaru O, Vert S, Vasiu R (2023) Usability evaluation of requirements collaboration features in RM tools. In: 2023 International conference on engineering and emerging technologies, Istanbul, Turkey, IEEE, Oct 2023. [Online]. Available: https://ieeexplore.ieee.org/xpl/conhome/1825364/all-proceedings, manuscript submitted for publication
6. "Requirements Engineering" Jama Software. Accessed 13 Jun 2023. [Online]. Available: https://www.jamasoftware.com/requirements-management-guide/requirements-gathering-and-management-processes/requirements-engineering/
7. Franch X, Palomares C, Quer C, Chatzipetrou P, Gorschek T (2023) The state-of-practice in requirements specification: an extended interview study at 12 companies. Requir Eng. https://doi.org/10.1007/s00766-023-00399-7
8. Glinz M, van Loenhoud H, Staal S, Bühne S (2020) Handbook for the CPRE foundation level according to the IREB standard. version 1.0.0, November 2020
9. "Applying the V-Model in Automotive Software Development" Accessed 14 Jun 2023. [Online]. Available: https://www.einfochips.com/blog/v-model-in-automotive-software-development/
10. Tierno A, Santos M, Arruda B, Rosa J (2016) Open issues for the automotive software testing. pp 8. https://doi.org/10.1109/INDUSCON.2016.7874609
11. Carrillo de Gea JM, Ebert C, Hosni M, Vizcaíno A, Nicolás J, Fernández-Alemán JL (2021) Requirements engineering tools: an evaluation. IEEE Softw 38(3):17–24. https://doi.org/10.1109/MS.2021.3058394
12. "Characteristics of Good Requirements | Enterprise Architect User Guide" Accessed 14 Jun 2023. [Online]. Available: https://sparxsystems.com/enterprise_architect_user_guide/14.0/model_domains/characteristics_of_good_requirements.html
13. Oduko OO (2023) Comparison of requirement management software. ERA. Accessed 20 Jun 2023. [Online]. Available: https://era.library.ualberta.ca/items/42bc4612-17ef-4f70-b900-a08e8104413e
14. "IBM Documentation" Accessed 21 Jun 2023. [Online]. Available: https://www.ibm.com/docs/en/engineering-lifecycle-management-suite/lifecycle-management/7.0.0?topic=next-comparison-doors-doors
15. "DXL Reference Manual". Available : https://www.ibm.comdocs/en/SSYQBZ_9.5.0/com.ibm.doors.requirements.doc/topics/dxl_reference_manual.pdf
16. "IBM Documentation" (2023) Accessed 21 Jun 2023. [Online]. Available: https://www.ibm.com/docs/en/engineering-lifecycle-management-suite/doors/9.7.2?topic=access-viewing-adding-comments-discussions
17. "What is Usability?," The Interaction Design Foundation (2023) Accessed 26 Jun 2023. [Online]. Available: https://www.interaction-design.org/literature/topics/usability
18. WL in R-BU Experience (2023) Usability 101: introduction to usability. Nielsen Norman Group. Accessed 26 Jun 2023. [Online]. Available: https://www.nngroup.com/articles/usability-101-introduction-to-usability/
19. https://www.facebook.com/torsten.tromm . User testing: definition, how to, tools, methods | Userpeek.com. Accessed 06 Jul 2023. [Online]. Available: https://userpeek.com/blog/user-experience-testing/

20. Bastien JMC (2010) Usability testing: a review of some methodological and technical aspects of the method. Int J Med Inf 79(4):e18–e23. https://doi.org/10.1016/j.ijmedinf.2008.12.004
21. Mcdonald S, Edwards H, Zhao T (2012) Exploring think-alouds in usability testing: an international survey. IEEE Trans Prof Commun 55:2–19. https://doi.org/10.1109/TPC.2011.2182569
22. Mishra L (2016) Focus group discussion in qualitative research. TechnoLearn Int J Educ Technol 6(1):1. https://doi.org/10.5958/2249-5223.2016.00001.2
23. Rotaru O, Vert S, Vasiu R, Andone D (2020) Standardised questionnaires in usability evaluation. Applying Standardised Usability Questionnaires in Digital Products Evaluation. pp 39–48. https://doi.org/10.1007/978-3-030-59506-7_4
24. Product Reaction Card—an overview | ScienceDirect Topics. Accessed 21 Jul 2021. [Online]. Available: https://www.sciencedirect.com/topics/computer-science/product-reaction-card
25. Will T (2023) Measuring and interpreting system usability scale (SUS). UIUX Trend. Accessed 07 Jul 2023. [Online]. Available: https://uiuxtrend.com/measuring-system-usability-scale-sus/

A Study on Image Enhancement and Noise Filtering of Kato-Katz Images

Farah Sinada, Saumya Kareem Reni, and Izzet Kale

Abstract Neglected tropical diseases affect more than 24% of the world's poorest populations. Some NTDs are diagnosed using kato-katz images; the kato-katz technique is the most commonly used diagnostic technique for soil-transmitted helminthiasis. The technique involves microscopic examination of sieved stool samples and tallying the parasitic eggs present. Kato-katz images present a unique and complex problem when it comes to image enhancement for feature detection. The foreground objects of interest (the parasite eggs) have features that match with the rest of the foreground objects in the image, making it difficult to distinguish and differentiate them. Hence sensible noise filtering and feature enhancement strategies are required for successful object identification and diagnosis. This paper will describe the results of studies on noise filtering and image enhancement of kato-katz images. Due to the 'noisy' and 'noise-like' nature of the images and the large number of objects in the images, typical noise filtering methods produce poor results. To our best knowledge, this is a pioneering work looking at image enhancement and noise filtering of kato-katz images. This work builds upon previous research by the authors concerning kato-katz images.

Keywords Kato-katz images · Image feature extraction · Image enhancement · Noise filtering · Neglected tropical diseases

F. Sinada (✉) · S. K. Reni · I. Kale
Applied DSP and VLSI Research Group (ADVRG), University of Westminster, London W1W 6UW, UK
e-mail: f.sinada@westminster.ac.uk

S. K. Reni
e-mail: s.reni@westminster.ac.uk

I. Kale
e-mail: kalei@westminster.ac.uk

X.-S. Yang et al. (eds.), *Proceedings of Ninth International Congress on Information and Communication Technology*, Lecture Notes in Networks and Systems 1000,
https://doi.org/10.1007/978-981-97-3289-0_23

1 Kato-Katz Images

1.1 Image Features

Neglected tropical diseases (NTDs) are infections caused by parasitic worms that affect 24% of the world's population, in tropical and sub-tropical areas [1]. They have a severe negative impact on a patient's life, causing malnutrition, stunted growth, and affect work performance [2]. The kato-katz technique is the most commonly used technique in the field for diagnosing soil-transmitted helminthic infections due to it being low-cost and not resource-intensive [2]. Kato-katz images are collected using the kato-katz technique, which involves minimal processing of a stool sample that is then examined under a microscope [3]. As such, the images are unstained and vary greatly in quality. These images are usually:

- of a heterogenous nature,
- have spurious intensity patterns,
- have objects that have similar characteristics to the parasitic eggs,
- and the eggs lack a continuous distinctive edge.

Figure 1 shows a kato-katz image with parasite eggs.

Prior to the application of any object identification techniques, the kato-katz images undergo some preprocessing in order to enhance the features of the foreground objects of interest, which in this case are the parasite eggs.

A major obstacle encountered in each study conducted as part of this research thus far has been the spurious intensity patterns coupled with the fine details that must be preserved. It is also difficult to distinguish between the background and foreground regions of the image because of the heterogenous nature of the regions. As can be seen in Fig. 1, even though the parasite eggs have somewhat defined edges, their morphology differs in each instance. Similarly, the features of the objects' interior greatly match and overlap with the features exterior to these objects. These disparities and overlapping regions have the greatest effect in terms of the noisy features in the images and the segmentation of the regions of interest. Hence managing them by

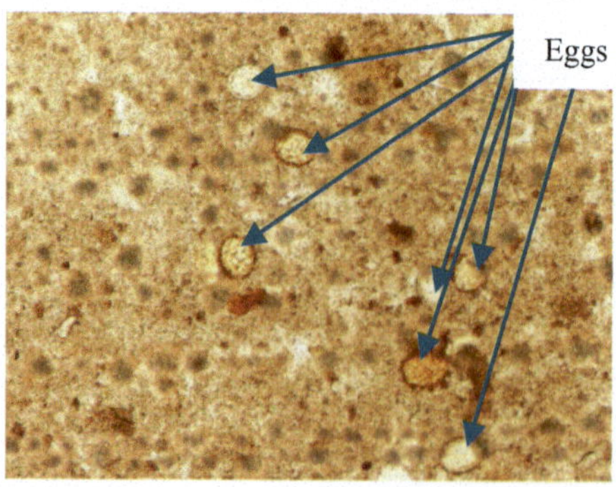

Fig. 1 Kato-katz image with 7 parasitic eggs, marked with arrows

filtering the noise features will greatly enhance the diagnostic probability of the model. In summary, there are two goals for noise filtering of kato-katz images:

- To decrease the 'noisiness' of the background and turn it into a uniform, well-illuminated background,
- To enhance/make the eggs more prominent by improving the readability of the eggs' borders.

The following section describes the studies conducted on noise filtering as part of NTD diagnosis in kato-katz images. The aim was to find methods that reduced or eliminated the unwanted patterns, creating cleaner images which would allow conventional segmentation algorithms to be applied successfully. The research has looked at pixel-based approaches and approaches that look at entire regions of an image. It should be noted this is the first work of its kind that has been carried out on the effectiveness of techniques, such as filters, on kato-katz images, or images from NTDs.

2 Rank Order Filtering

As mentioned in the previous section, the spurious intensity patterns in the images are the main contributors that decrease the efficacy of results achieved from segmentation methods on kato-katz images [4, 5]. As the intensity patterns are not confined to a set area within the image, a method must be obtained that can identify the patterns pertaining to the objects of interests (OOI).

Rank order filtering is a method by which the impulses in a signal can be ordered according to rank [6]. This method can be used to isolate the frequency of the noise, thereby facilitating its removal. The objective of using the filter on the kato-katz images was to enhance selective regions such as parasite egg borders and remove or smoothen the other regions. The target image/region is set and the size of the window, N, and the percentile, P are chosen manually [7]. These filters are effective in image processing as it allows sample padding to avoid any edge effects.

The *p*th percentile denotes how the pixel values in the neighbourhood get ordered. For kato-katz images, each region is chosen as a matrix and a neighbourhood is defined using defined window sizes. The filter then calculates the *p*th percentile of the intensities around each pixel within the matrix. The following results were from an experiment with a window size ranging from 2 to 8. Beyond 8 the window was too large and small but important details, such as the parasite egg border, were lost. The percentiles tested were: 5, 20, 35, 50, 65, 80, and 95. At 50% the function operates as a median filter and close to 0 or 100 an approximate lower/higher signal envelope is calculated. Figure 2 displays the results achieved on one image using the above window sizes and percentiles.

As can be seen from Fig. 2 for smaller window sizes ($W = 2$ or 3) the filter has little to no effect in enhancing the object features. Similarly, when $N = 8$, the window

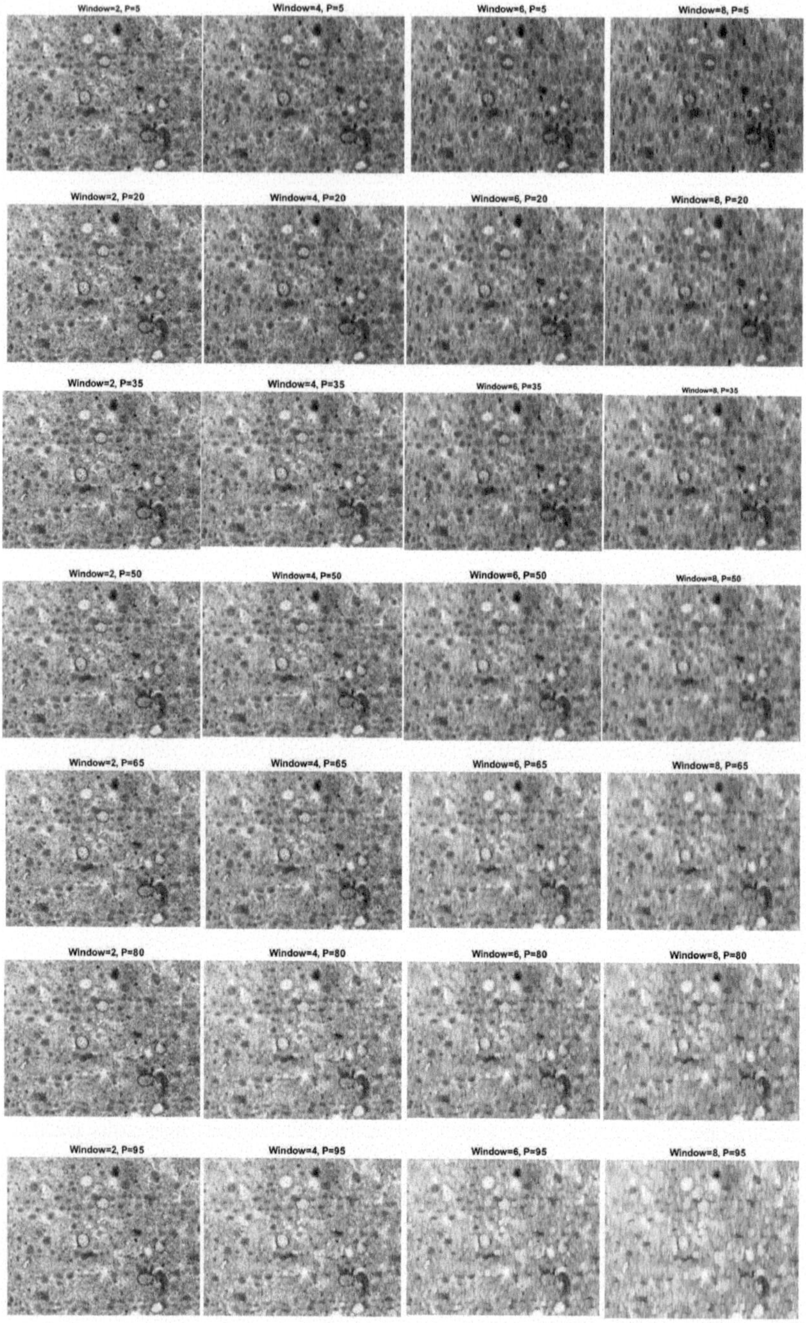

Fig. 2 Rank order filtering results on one image. Each column is a window size, from left to right $W = 2$—8. Each row is a percentile, top to bottom, $N = 5, 20, 35, 50, 65, 80, 95$

is too large and has a detrimental effect, eliminating cell borders. Hence an effective window size observed was within the range of 4–6.

With regards to the percentile chosen, the results achieved were comparable to those achieved using erosion, a morphological filtering operation, with a disc/rectangle/diamond shaped structuring element [5]. Any effective smoothing of the intensity patterns or the removal of unwanted objects is also accompanied by the erasure of cell borders or complete disappearance of the cells themselves, as seen in the images at the 95th percentile, Fig. 2. Each iteration of this file generated 28 images, from which the most suitable image with the most successful noise suppression must be chosen. This is a manual procedure as a window size and percentile must be attuned for each image, which is not viable in a large dataset. Similarly, as with any digital filter, as the order increases so does the complexity. The issue of global applicability of the method to the dataset is therefore pertinent.

3 Gabor Filters

As seen in the previous section rank order filters are not ideal in the noise filtering and feature enhancement procedure in kato-katz images. Hence another approach that was investigated as part of this research was texture analysis. Gabor filters are linear filters used in texture analysis. They can be used to identify the magnitude and phase or direction, of a signal within an image [8]. The Gabor filter when applied to the kato-katz images was expected to analyze specific frequency contents and their orientation in a localized region in the image. Many wavelengths and orientations were tested as seen in Fig. 3 which shows the Gabor filter applied to a kato-katz image using a wavelength of 4, orientation of 90°. The filter was applied to both the image as a whole and to a sub-section of the image containing the OOIs. As displayed in Fig. 3, the texture analysis of kato-katz images does not pinpoint a clear angle or magnitude of origin. No discernible magnitude or phase, that is unique to either the cells or to the background noise, were determined when used on a full image as well as an image sub-section. Although employing different techniques, the results achieved were similar to other image processing techniques such as morphological erosion that employ different angles of orientation, wherein they also have a detrimental effect on all the OOIs in the image.

4 Other Noise Filtering Techniques

As seen from the experiments involving the rank order and Gabor filters, removing noise and noisy objects results in the elimination of the parasite eggs in the images in tandem with the noisy background elements. In this section, a common noise removal method is detailed, wherein noise is first added to an image and using noise removal techniques, subsequently eliminated along with the noise that was already present

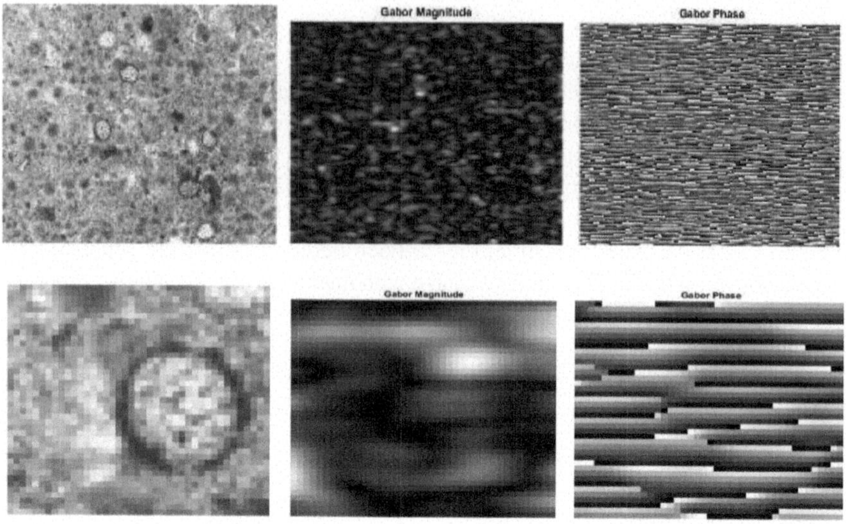

Fig. 3 Gabor filter on a kato-katz image, using a wavelength of 4, orientation of 90°. Top row full image, bottom row, image sub-section. First column image, second column Gabor magnitude, third column Gabor phase

within the images. This study, conducted primarily on a single image, tested whether these noise removal techniques would be useful in clearing the spurious intensity patterns that exist in these images.

Five different noise addition and removal techniques were tested. Firstly, noise was added to the test images. The types of noise tested were: Salt and Pepper, Gaussian, local variance, Poisson, and Speckle as shown in Fig. 4. The added noise was then filtered out using a Wiener filter. Expected results would be the removal of the added noise plus any additional noise already present in the image, creating a smoothing effect. However, the removal of the added noise had very little effect on the overall image, with almost no change in the heterogenous intensities present. These tests were also conducted on the image sub-sections and yielded the same results.

The Gaussian noise filter improves the illumination of the image as seen in Fig. 4; however, the illumination effect is different for each image and the OOIs, due to the large disparity in image quality. In this particular image, the egg cell walls (object boundaries) deteriorate, which has implications for future detection methods. Local variance creates the same type of noise as the Gaussian filter without the illumination effect. After removal of the local variance noise with the Wiener filter, the image quality degrades. With the Poisson filter, the difference after noise removal is negligible. Speckle noise, once removed, also worsens image quality, and increases the graininess of the image. The noise filtering technique that had the poorest results was the Salt and Pepper filter; the image, with the superimposed Salt and Pepper noise, after passing through the Wiener filter, which smoothened the image, creating an image with worse quality than at the start of the process.

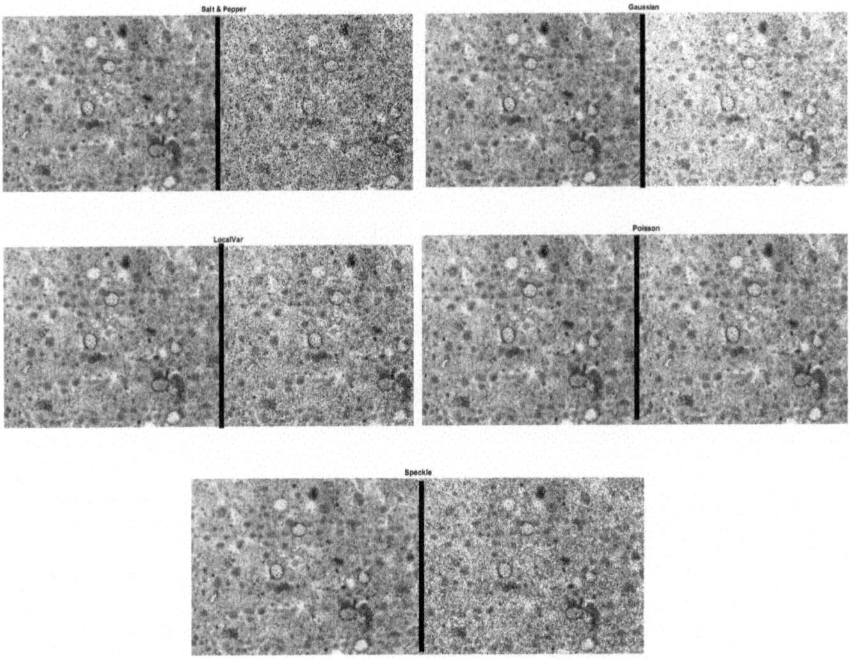

Fig. 4 Each section has, original image left of divider, with noise added right of divider. Top left to right: Salt and Pepper, Gaussian, local variance, Poisson, and Speckle

A control test was done using the Wiener filter on its own, the results of which are shown in Fig. 5. A control image shows the effect of using the Wiener filter on the image directly, without the addition of any noise. The result is similar to a weak Gaussian or blurring filter. Filtering of the other five noise techniques should result in an evening of the non-uniform illumination, accompanied by more pixel uniformity in each area. That is to say, a decrease in the extreme bright and dark areas within the image. The final results were grainier images where the objects in the image were indiscriminately broken up as shown below.

As stated, the results were also tested on image sub-sections containing cells, Figs. 6 and 7. The same results were achieved as with the full images, with noise removal techniques at best having little to no effect, and at worst, decreasing image quality. The most successful filters, such as the Weiner on its own and the Poisson filters created a slight smoothing effect. The other filters increased the graininess of the images and lowered the connectivity of the border areas.

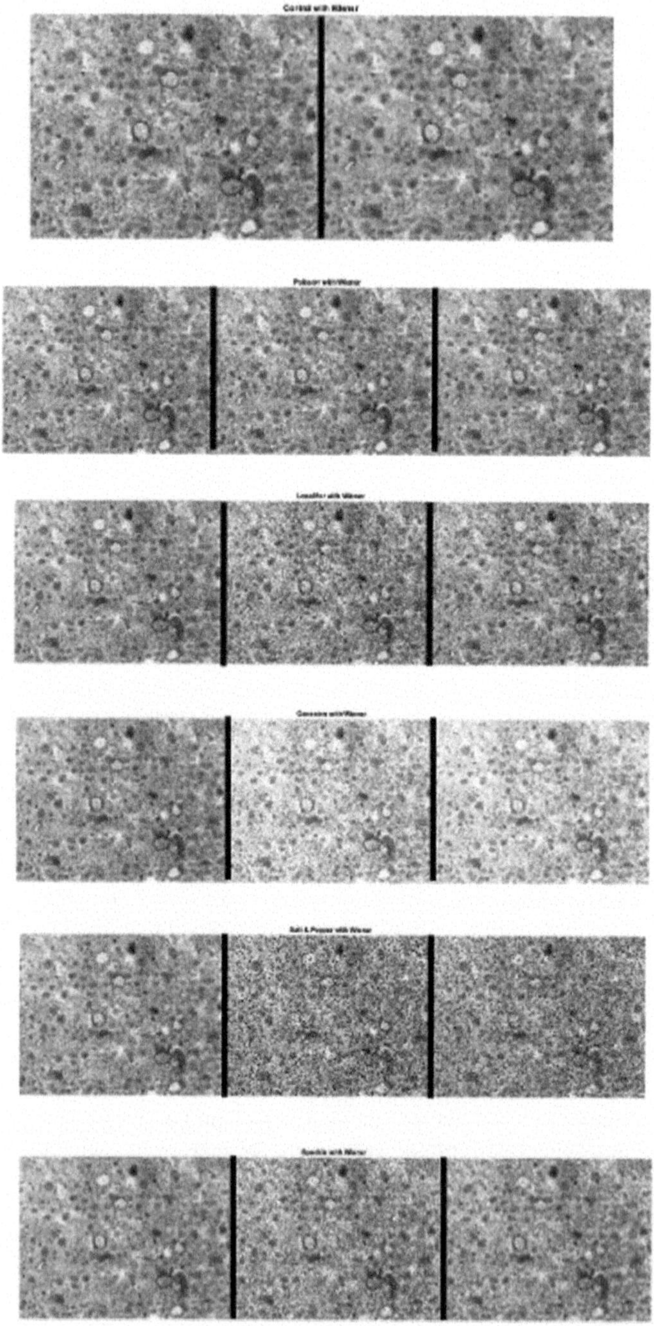

Fig. 5 Three image montages in each section, left—original 'control' image, middle—with noise added, right—noise removed. Starting top row to bottom: image with: control—Wiener filter only, Poisson, local variance, Gaussian, Salt and Pepper, and Speckle

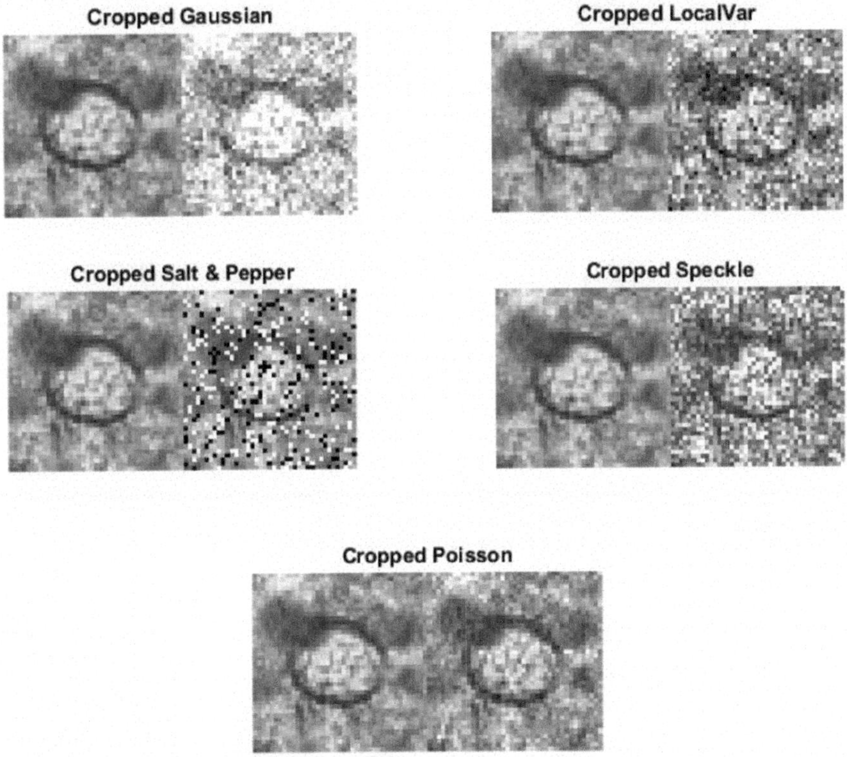

Fig. 6 Each section has, original image left, with noise added right. Top left to right: Gaussian, local variance, Poisson, Salt & Pepper, and Speckle

5 Noise Evaluation

As evident from the experiments described in the previous sections, applying noise filtering techniques not only deteriorates the image quality but also removes important features pertaining to object identification. This was primarily due to the overlay between noise features and OOIs. Hence, another experiment was conducted to better understand the noise and intensity patterns in the images. For this, a series of 'control' images were created. The parasite eggs were removed from the image, creating what will be from now on referred to as an 'empty' image. This is displayed in the Fig. 8.

One of the pervasive issues with noise removal in these images is that there is a collection of different noise types in the images and a great deal of 'randomness', as in the lack of a discernible pattern, in the unwanted noisy regions. Using the empty images, the aim of this experiment was to find a pattern or find prominent characteristics that are descriptive of the noise, which can lead to the development of an algorithm to either minimize it, improving matching efforts, or, allow for

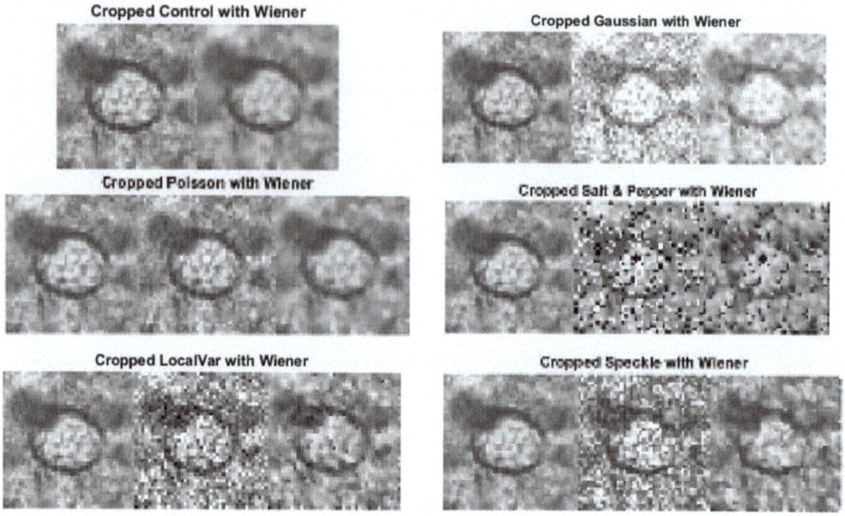

Fig. 7 Three image montages in each section, left—original image, middle—with noise added, right—noise removed. Top left to bottom right image with: control—Wiener filter only, Poisson, local variance, Gaussian, Salt & Pepper, and Speckle

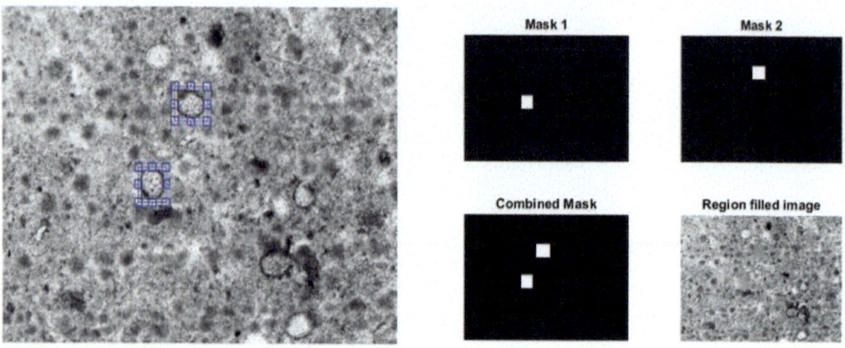

Fig. 8 Left, two eggs highlighted for masking. Right, binarized masks and final result with masked regions

a deductive approach where the 'random' areas are searched for and non-random areas are classified as the regions of interest (ROI), as previously stated.

To create the images, masking was used. Two cells are chosen, and a rectangle drawn around them to create a mask, as seen in Fig. 8. The masks created an empty region that was then filled using inward interpolation, which is an estimation technique that uses the surrounding data points, to remove the object in question. It is important to note that the masked region did not create a smooth transition, but instead the area masked was clearly outlined, as seen in Fig. 8. Using these images,

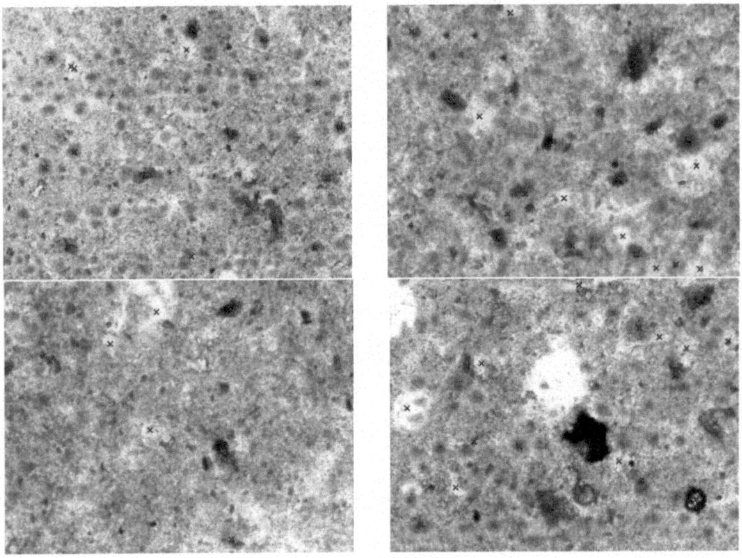

Fig. 9 Locations of ROIs marked with red crosses on 'empty' images

the studies conducted previously could be conducted again to find data points for comparison.

The noise removal techniques were then applied, with their effect on the overall image quality noted.

Performing feature extraction on the empty images and extracting the locations of the ROIs gave interesting results as shown in Fig. 9. These control images and the information received on the ROIs can be used for further processing and feature extraction. There were many regions that were picked up owing to the fact that they have very similar region characteristics to the parasite eggs. By studying these characteristics, the aim is to find a threshold or set of characteristics that can be used to differentiate these areas from the parasite eggs. The following section explains a statistical inference analysis carried out on the images after the masking process.

6 Entropy

Closely related to the subject of noise is entropy. Entropy is a statistical measure of randomness that can be used to characterize the texture of the input image or a smaller region of the cell or noise. If there is true randomness or entropy within each image, perhaps that can be measured and used as a possible method for the elimination of false positive results, or the unwanted noise.

First, the entropy of the images was collected using two different methods: an entropy filter and a neighbourhood approach to calculate the entropy in a 9×9 area, calculating the entropy value around the centre pixel. The neighbourhood approach result had the overall average calculated. Both had the same results of 7.0735. It is also important to note that when the entropy was compared with the image and the empty version, the results were almost identical, as seen below, Table 1. The small margins are again present, with the differences of around 1/100th, too small to be useful as a separator.

The entropy of an image sub-section that contained an egg and one that did not was compared, Table 2. Looking at an image sub-section instead of the image allows us to have a much tighter comparison of an area that contains an egg and one that does not. The values from the sub-sections were averaged for each image. As can be seen, there is a slightly larger margin, however it is still too small to be useful in segmentation.

Finally, the entropy of the cell border and interior areas was also calculated as before, but for each sub-section, in an effort to find a differentiator between the inside and outside of the egg. As described above, structuring elements were used to extract the border and cell interior pixel intensity values, Table 3.

There is a definite difference between the area within the cell and the area outside it. However as can be seen, the value is the same whether the image contains an egg

Table 1 Entropy on full images

Image	Full image	Empty image
1	7.0810	7.0326
2	7.0855	7.0284
3	7.0598	7.0146
4	7.3019	7.2616

Table 2 Entropy on image sub-sections

Image	True positive entropy mean	False positive entropy mean
1	7.001554	6.944483
2	6.877727	6.772731
3	6.752983	6.918311
4	7.092132	7.059375

Table 3 Entropy on image sub-sections, comparing cell border with interior

Image	True positive entropy mean			False positive entropy mean		
	Cell border	Outer area	Inner area	Cell border	Outer area	Inner area
1	0.526413	0.567683	0.152586	0.526413	0.529307	0.152586
2	0.526413	0.570518	0.152586	0.511067	0.625906	0.152586
3	0.526413	0.610139	0.152586	0.526413	0.568276	0.152586
4	0.526413	0.53381	0.152586	0.526413	0.55718	0.152586

or not. There also does not seem to be a difference between the cell border and the outer area. As the cell border is needed to define the egg, it is important to find a metric that can identify it. Further studies are required to work on the effectiveness of entropy on masked images.

7 Conclusion

This paper described various noise reduction and feature enhancement algorithms performed on kato-katz images and their results. Due to the unique nature of kato-katz images, processing and feature extraction is complex compared to traditional pathological image analysis. This is so far the first study conducted of this nature and hence the findings are not only significant to kato-katz images but also relevant to similar pathological image analysis. Among the four methodologies applied, the control image extraction, followed by masking and entropy calculation proved to be most effective in distinguishing the OOIs from their background/neighbourhood but only when applied to singular images; the results were not applicable to a set of images. The experiments were to an extent, futile despite the intricacy of the image background and foreground elements. This study is currently progressing towards a hybrid approach combining the outcomes of multiple techniques explained in this paper to identify the central tendencies of intensities and thereby distinguishing the regions. Even though this can increase the computational complexity, the diagnostic probabilities can be improved to a great extent.

Acknowledgements Thanks to the Parasitology division of University College London Hospital (UCLH), and Public Health England for providing us with the images and appropriate guidance and advice on the clinical aspects of the research. We thank the Funds for Women Graduates (FfWG) for grants to support living expenses.

References

1. World Health Organisation (WHO) (2021) Soil-transmitted helminth infections. Accessed 02 May 2021. [Online]. Available: https://www.who.int/news-room/fact-sheets/detail/soil-transmitted-helminth-infections
2. Feasey N, Wansbrough-Jones M, Mabey DCW, Solomon AW (2010) Neglected tropical diseases. Br Med Bull 93(1):179–200. https://doi.org/10.1093/bmb/ldp046
3. Gray DJ, Ross AG, Li YS, McManus DP (2011) Diagnosis and management of schistosomiasis. BMJ 342(7807). https://doi.org/10.1136/bmj.d2651
4. Sinada F, Reni SK, Kale I (2020) Image analysis and feature extraction of Kato-Katz images for neglected tropical diseases diagnosis. In: IEEE 63rd Midwest symposium on circuits and systems (MWSCAS), Springfield, MA, IEEE, Aug 2020. https://doi.org/10.1109/MWSCAS48704.2020.9184535

5. Sinada F, Reni S, Kale I, Chiodini P, Shrivastava J (2019) A reflective analysis of image processing operations on Kato-Katz images for the pathological diagnosis of neglected tropical diseases. In: 7th E-Health and bioengineering conference (EHB), Iasi, IEEE, Nov 2019. https://doi.org/10.1109/EHB47216.2019.8970011
6. Scikit-image team, 'Rank filters—skimage 0.21.0 documentation'. Scikit-Image. Accessed 11 Aug 2023. [Online]. Available: https://scikit-image.org/docs/stable/auto_examples/applications/plot_rank_filters.html
7. Salarian A (2008) 'Rank-Order Filter'. MATLAB central file exchange. Accessed 15 Mar 2022. [Online]. Available: https://uk.mathworks.com/matlabcentral/fileexchange/22111-rank-order-filter
8. Jain AK, Ratha NK, Lakshmanan S (1997) Object detection using gabor filters. Pattern Recognit 30(2):295–309. https://doi.org/10.1016/S0031-3203(96)00068-4

Synthetic Data for Feature Selection

Firuz Kamalov, Rohan Mitra, and Hana Sulieman

Abstract This study aims to perform a comparative analysis of popular feature selection algorithms in the regime of imbalanced data using synthetic datasets. The use of synthetic data has several advantages over the traditional, real-life data including the direct evaluation of the algorithms without the classifier and metrics bias. As a result, we obtain a more accurate evaluation of the algorithms than existing approaches. The results—based on eight different algorithms and eight different datasets—show that sequential forward selection and entropy-based selection achieve the highest precision in the regime of imbalanced data.

Keywords Feature selection · Imbalanced data · Synthetic data · Comparative study

1 Introduction

Feature selection is an important component of many machine learning pipelines. The correct identification of relevant features can drastically improve the performance of the learning algorithm. As a result, a significant amount of research has been directed towards developing effective feature selection algorithms. In particular, several attempts have been made to determine the relevant features in the regime

Supported by the American University of Sharjah research grant: FRG22-E-S84.

F. Kamalov (✉)
Department of Electrical Engineering, Canadian University Dubai, Dubai, UAE
e-mail: firuz@cud.ac.ae

R. Mitra
Department of Computer Science and Engineering, American University of Sharjah, Sharjah, UAE
e-mail: b00085023@alumni.aus.edu

H. Sulieman
Department of Mathematics and Statistics, American University of Sharjah, Sharjah, UAE
e-mail: hsulieman@aus.edu

X.-S. Yang et al. (eds.), *Proceedings of Ninth International Congress on Information and Communication Technology*, Lecture Notes in Networks and Systems 1000,
https://doi.org/10.1007/978-981-97-3289-0_24

of imbalanced data, where the distribution of the class labels is skewed [1–3]. Conventionally, feature selection algorithms are evaluated on real-life data which is not suitable in the context of imbalanced data. This research aims to utilize a hitherto unused approach of analyzing feature selection algorithms based on synthetic data.

Imbalanced data creates a unique challenge for learning algorithms. Since most algorithms aim to optimize the overall performance on the dataset, it leads to bias against the minority class labels. Feature selection algorithms are also susceptible to this bias. Filter methods select the best features based on a similarity measure between individual features and the target variable. Thus, the majority class labels exert the most influence in the selection process. Similarly, embedded methods rely on the results of a learning model which are also biased for imbalanced data. To deal with imbalanced data, several approaches have been proposed including sampling, cost-sensitive learning, and others.

In the existing literature, feature selection algorithms for imbalanced data are evaluated using real-life data. This approach has several inherent disadvantages. First, the use of real-life data relies on the accuracy of the classifier used in feature evaluation. The choice of the classifier can create a distortion in feature evaluation—irrelevant features may have a good performance with certain deep neural networks. Second, the traditional metrics used for real-life data such as classification accuracy are not suitable in the context of imbalanced data.

To overcome the disadvantages related to feature evaluation using real-life data, we propose to employ synthetic data. In synthetic data, the relevant and irrelevant features are known so feature selection algorithms can be evaluated directly [4]. This approach eliminates both the issue of the distortion stemming from the classifier used in subset evaluation as well as inappropriate accuracy metrics. Synthetic data can also be modified as needed to explore the performance of feature selection algorithms under various scenarios. Thus, our goal in this paper is to conduct a comparative analysis of feature selection algorithms using synthetic data.

We consider several popular feature selection algorithms and compare their performance in the regime of imbalanced data using synthetic data. The synthetic datasets used in our study cover a range of functionalities allowing evaluation under various scenarios. The feature selection algorithms are evaluated directly based on the number of correctly selected relevant variables. Thus, we obtain a high fidelity evaluation of the algorithms.

The remainder of the paper is structured as follows. Section 2 discusses the existing literature related to feature selection for imbalanced data. Section 3 describes the methodology employed in our comparative study. In Sect. 4, we present and analyze the comparison results. We conclude with summary remarks in Sect. 5.

2 Literature

In the realm of machine learning, the efficient handling of high-dimensional class-imbalanced datasets has emerged as a pivotal challenge with profound implications for various applications, ranging from medical diagnosis to fraud detection [5]. Feature selection is utilized in many application domains in order to correctly classify and profile minority class observations, mitigate noise, and enhance predictive power. There are three general categories of feature selection methods depending on the relationship between the feature selection process and the corresponding learning algorithm. These are filter, wrapper, and embedded methods. Filter methods measure the relevance of each feature with regard to the output class labels independently of the learning algorithm, whereas both wrapper and embedded methods tend to employ the performance of the learning algorithm itself as a selection criteria. Wrapper methods commonly search the high-dimensional space of all feature subsets; while in embedded methods, the search is guided by the learning algorithm [6].

While there exists a large body of literature on feature selection methods, few studies, however, are dedicated to feature selection in imbalanced data. In [2], the authors developed a filter feature selection approach based on a one-dimensional decision tree F1-score calculated with respect to the minority class. The numerical experiments demonstrated high effectiveness in dimensionality reduction and accuracy, in addition to requiring low computational complexity. In [7], the authors proposed a feature selection algorithm for imbalanced data based on genetic search. Despite the superior performance genetic search algorithms are known to achieve, the technique requires significant computational resources. The authors in [8] introduced a filter approach to feature selection for imbalanced data using neighborhood set theory. The model adopts a discernibility matrix-based feature selection technique where the significance of each feature is computed based on the upper and lower boundary regions taking into account the uneven distribution of classes. The constructed discernibility matrix and feature significance information are leveraged to determine a subset of conditional features which are considered highly informative for the classification tasks.

A Smart Robust Feature Selection (SoFt) for imbalanced and heterogeneous data is recently developed in [9]. SoFt involves selecting a learning algorithm based on gradient boosting methods and designing a filtering-based feature selection algorithm where multiple feature scores are computed in order to quantify the quality of features across all folds of a K-fold cross-validation. SoFt is shown to be robust and consistent for imbalanced dataset in addition to eliminating the bias of the selected features to the learning algorithm. A filter feature selection method that couples the symmetrical uncertainty (SU) and harmony search is proposed in [10]. Although the approach demonstrated high classification performance, SU should be used with caution because it can be severely biased toward the majority class especially for high skewed class distributions. The authors in [11] proposed an adaptive multi-stage classifier system consisting of three components: feature selection, resampling, and ensemble learning. Each component in the system is selected discriminatively for different types of imbalanced data.

In [12, 13], the authors introduce an innovative embedded feature selection strategy based on SVM models to effectively handle skewed class distribution. The methodology employs SVM to construct a linear or nonlinear classifier without any feature penalization. A concave problem is then formulated to address feature penalization, which is solved by adjusting scaling factors through a Quasi-Newton strategy combined with line search. The effectiveness of the proposed approach is demonstrated through numerical experiments using real datasets characterized by class imbalance.

Another intriguing embedded feature selection method for imbalanced classification is introduced in [14]. The method utilizes a decision tree with nodes determined by the Gini index weighted by the imbalance ratio. Although the method showed excellent performance, it lacks stability under varied imbalance ratios.

3 Methodology

Our goal is to investigate the performance of popular feature selection algorithms in the context of imbalanced data. To this end, we consider eight different algorithms tested on eight different synthetic datasets. We investigate the performance of the algorithms under different imbalance ratios. Since the relevant and irrelevant features are known, we can evaluate the selected subsets directly.

3.1 Feature Selection Algorithms

The details of the feature selection algorithms considered in our study are presented in Table 1. The algorithms cover a range of approaches. Entropy, χ^2, mutual information, and symmetrical uncertainty are filter methods, while sequential forward selection (SFS), sequential backward selection (SBS), and mRMR are wrapper methods. Furthermore, genetic algorithms represent a new breed of evolutionary approaches to feature selection. In most cases, the algorithms are used with their default parameters as per the source of implementation. Note that χ^2-filter is designed for categorical variables. Therefore, we employ a binning procedure to discretize any continuous features for the purpose of calculating the χ^2-value.

3.2 Datasets

We employ eight different synthetic datasets constructed based on different formulations. These datasets have been based on work such as [19] who have released their datasets publicly at [20]. Each dataset consists of relevant, redundant, and irrelevant features. The target variable is calculated using various equations. The datasets have different levels of nonlinearity that is meant to represent different levels of difficulty for learning models including feature selection algorithms. The details of the datasets are presented in Table 2.

Table 1 Source and the settings of the feature selection algorithms

Algorithm	Source	Settings
Entropy	Scikit learn [15]	Default
χ^2	Scikit feature [16]	Default
Mutual information	Scikit learn [15]	Default
Sequential backward selection	Scikit learn [15]	Default
Sequential forward selection	Scikit learn [15]	RBFSVC, default
Symmetrical uncertainty	Scikit feature [16]	RBFSVC, default
mRMR	Mazzanti (2022) [17]	Default
Genetic	Calzolari (2022) [18]	RBFSVC, default

Table 2 Details of the data used in the empirical testing

Dataset	Relevant	Redundant	Irrelevant	Total	Target
Straight line	2	20	100	200	Binary
4D AND	4	8	100	100	Binary
5D XOR	5	10	100	100	Binary
Hypersphere	3	20	100	400	Binary
Cone	3	20	100	200	Binary
Trigonometric	2	5	50	200	Binary
Double spiral	3	30	120	200	Binary
Yin-yang	2	10	50	600	Binary

3.3 Subset Evaluation Metrics

Each feature selection algorithm is applied to the datasets in Table 2, and the optimal feature subsets are selected.

Traditionally, the quality of the selected subsets is measured using precision, recall, F1-score, receiver operating characteristics, and other metrics. However, in synthetic data, a feature subset can be evaluated directly using precision alone, because the relevant features are known a priori. Since the number of features to be selected by each algorithm is often the same—as set by the user—all the metrics produce the same value. In particular, precision is calculated as follows:

$$\text{Precision} = \frac{RelevantSelected}{TotalSelected}, \tag{1}$$

where $RelevantSelected$ is the number of relevant features selected and $TotalSelected$ is the set total number for features to be selected. Note that the datasets contain several redundant copies of each relevant feature, so it is acceptable for an algorithm to select a redundant feature in place of the corresponding relevant feature. However, in case both the relevant feature and its redundant copy are selected, only the relevant is counted.

3.4 Experimental Setup

The skewedness of a dataset is measured via the imbalance ratio:

$$r = \frac{minority}{majority}. \tag{2}$$

In our numerical analysis, the imbalance ratio r is varied from 0.02 to 0.5 for each dataset in Table 2 . Since the original datasets are approximately balanced, to achieve a target class ratio r we drop the corresponding number of instances from the minority class set. In other words, we randomly undersample the original datasets. Then, the feature selection algorithms are applied to each dataset over the range of imbalance ratios to obtain a comprehensive evaluation. To remove any idiosyncrasies that may arise due to random sampling, we repeat the experiments four times for each imbalance ratio and report the median score as the final result.

The majority of the existing algorithms, including the ones considered in our study, do not have an embedded mechanism for automatic identification the optimal size of the feature set. As such, the number of selected features k is specified by the user. We consider two cases: (1) k equals the number of relevant features and (2) k equals twice the number of relevant features. Since the relevant features are known a priori and the same k is used for all algorithms, the selected feature subsets can be evaluated directly using precision in Eq. (1).

The experimental procedure can be described as follows:

For each dataset D in Table 2.

For each method Γ in Table 1.

For each imbalance ratio r in [0.02, 0.05, 0.1, 0.2, 0.3, 0.4, 0.5].

1. Sample the dataset D to achieve the imbalance ratio r.
2. Apply the algorithm Γ on the imbalanced dataset D to select the optimal features.
3. Calculate precision based on the selected feature subset.
4. Repeat Steps 1–3 four times and record the mean precision.

The imbalance ratio given in Eq. 2 is unambiguous in the case of binary target variable.

4 Results and Discussion

The results of the numerical experiments are summarized in Figs. 1 and 3. Figures depict the average precision over all datasets in Table 2 at each imbalance ratio. In Fig. 1, the feature selection algorithms are applied to select the top k features, where k equals the number of relevant features. It can be observed that Entropy, SFS, and SBS achieve the best overall performance. In particular, SFS provides the best performance in the regime of low class imbalance, while SBS provides the best performance at higher values of r. Indeed, SFS achieves average precision of over

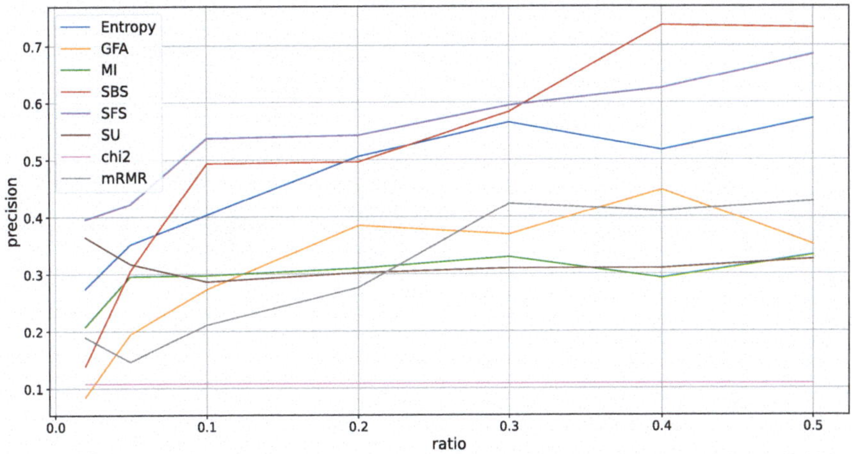

Fig. 1 Average precision of feature selections algorithms when the number of selected features equals the number of relevant feature

50% at values of r as low as 0.1. It also performs quite well at higher values of r. We also note that χ^2-filter method produces poor results which is probably due to artificial discretization of the features. As expected, all the algorithms generally improve as r increases. It shows that class imbalance is in fact detrimental to the performance of feature selection algorithms.

To statistically conclude on the best performing algorithms, we examine the box plot of the distribution of average precision per algorithm for each value of r, as in Fig. 2a, and then conduct the Kruskal-Wallis test to determine if there is a difference between the best performing algorithms. The results of the Kruskal-Wallis test clearly show that at least one distribution is different from the rest ($H(7) = 32.301$, $p = 0.000$).

As seen in Fig. 2a, Entropy, SFS, and SBS achieve the highest average precision across r values. The Kruskal-Wallis test also shows that Entropy, SFS, and SBS are statistically indifferent ($H(3) = 1.566$, $p = 0.457$) from each other. Conducting the Kruskal-Wallis test between the three algorithms and any other algorithm, such as mRMR or GFA, leads to the rejection of the hypothesis that their distributions are equal. Hence, SFS, SBS, and Entropy are the top three algorithms. Given the importance of stability of the algorithms, we further desire algorithms with lower standard deviation. Hence, SBS cannot be considered as one of the best. In fact, SBS is the same as other distributions such as mRMR or GFA ($H(3) = 4.683$, $p = 0.096$). Hence, SBS is not convincingly different from the other, low-performing distributions despite GFA and mRMR being different to SFS and Entropy ($H(4) = 12.389$, $p = 0.006$). Therefore, we conclude that the overall best algorithms are SFS and Entropy.

The average performance of the algorithms when the number of selected features is twice the number of relevant features is depicted in Fig. 3. The pattern in Fig. 3 is similar to that of Fig. 1. However, unsurprisingly, the precision of the methods

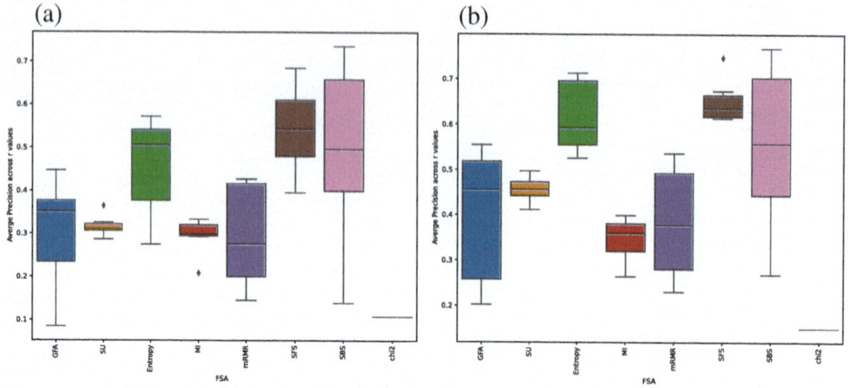

Fig. 2 Box plot of the distribution of average precision for each FSA across each value of r **a** Number of selected features equals the relevant features. **b** Number of selected features equals twice the relevant features

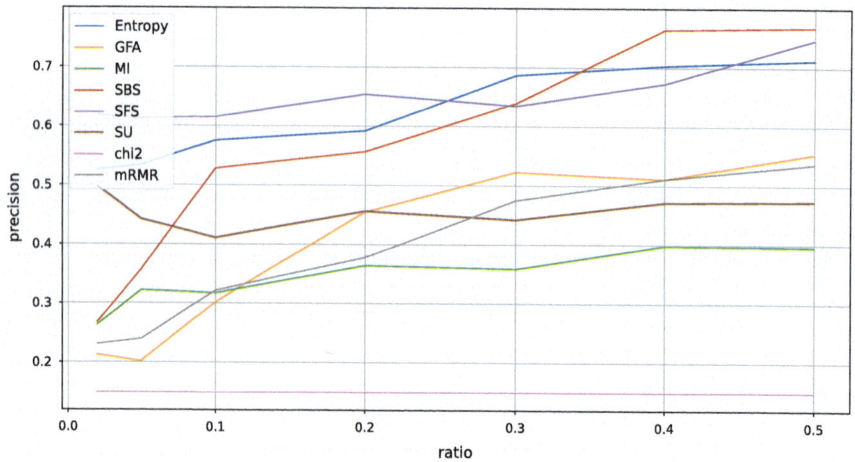

Fig. 3 Average precision of feature selections algorithms when the number of selected features equals twice the number of relevant feature

increased across most algorithms. In particular, SFS achieves precision of over 60% at r values that are as low as 0.02. Similarly, Entropy produces high precision across different values of r.

Once again, we use the box plot of the distribution of average precision per algorithm for each value of r, as in Fig. 2b to help us statistically conclude on the best performing algorithms. The results of the Kruskal-Wallis test clearly show that at least one distribution is different from the rest ($H(8) = 39.543$, $p = 0.000$).

As seen in Fig. 2b, Entropy, SFS, and SBS achieve the highest average precision across values of r. The Kruskal-Wallis test shows that Entropy, SFS, and SBS are statistically indifferent ($H(3) = 1.032$, $p = 0.597$) from each other. However, SBS

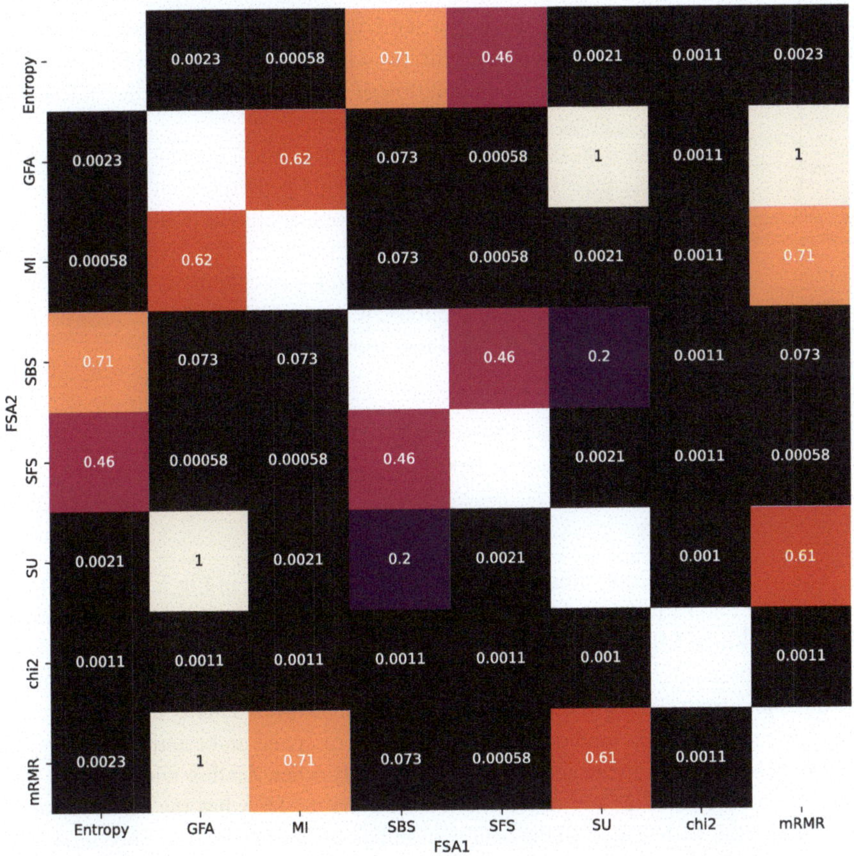

Fig. 4 Heat map of the p-values

is also statistically indifferent from other worse performing algorithms. For example, comparing SBS to mRMR and SU shows statistically indifferent distributions ($H(3) = 3.795$, $p = 0.150$). Hence, we conclude that SBS is not one of the best performing algorithms due to its high standard deviation. This indicates that SBS is very unstable as the value of r changes, which is why it achieves some of the highest and lowest precision. In contrast, SFS has a much lower standard deviation as seen in Fig. 2b, illustrating that it is far more stable and unaffected by the values of r.

In addition, this can be further emphasized by conducting the pair-wise Mann-Whitney-U test and plotting the p-values as in Fig. 4. It can be seen that pair-wise testing agrees with SFS and Entropy being the best algorithms. Entropy and SFS are different from all other algorithms except each other and SBS, but SBS is indifferent to most other algorithms even the ones with poor performance. Hence, Entropy and SFS are the best when selecting twice the number of relevant features.

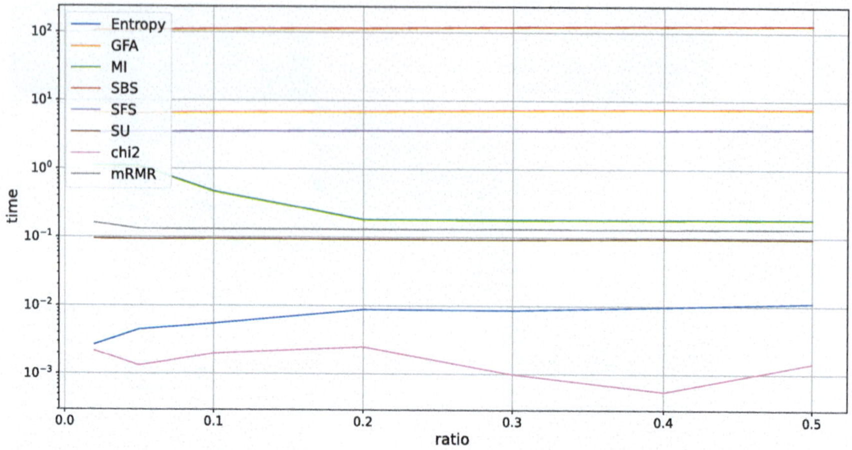

Fig. 5 Average execution time of feature selection algorithms

The average execution time of the algorithms is depicted in Fig. 5. Note the log-scale of the y-axis. It shows that SBS requires significantly more time than any other algorithm. Since SBS performs search over a large number of subsets, it is slower than other methods. Similarly, SFS requires more time than most other algorithms. At the same time, Entropy is the second fastest algorithm. This is unsurprising as Entropy is a filter-type method. The combination of speed and relatively high precision make Entropy an attractive option, especially when dealing with extreme big data. Moreover, another key observation is that the χ^2 FSA has the most unstable execution time with respect to the degree of imbalance in the dataset, since it has the least consistent line in Fig. 5.

5 Conclusion

In this paper, we employed a hitherto unexplored approach of evaluating the performance of feature selection algorithms in the regime of imbalanced data using synthetic data. The use of synthetic datasets allowed to directly evaluate the effectiveness of each method. The results—based on eight different algorithms and eight different datasets—showed that while SFS and SBS achieve high precision, they require considerable computing time. On the other hand, Entropy is able to achieve robust precision at substantially lower execution time. Thus, we conclude that Entropy provides the best overall option for feature selection, especially when dealing with big data.

References

1. Chen H, Li T, Fan X, Luo C (2019) Feature selection for imbalanced data based on neighborhood rough sets. Inform Sci 483:1–20
2. Kamalov F, Thabtah F, Leung HH (2022) Feature selection in imbalanced data. Annals of Data Sci 1–15
3. Liu Y, Wang Y, Ren X, Zhou H, Diao X (2019) A classification method based on feature selection for imbalanced data. IEEE Access 7:81794–81807
4. Kamalov F, Sulieman H, Moussa S, Reyes JA, Safaraliev M (2023) Nested ensemble selection: an effective hybrid feature selection method. Heliyon 9(9)
5. Cai J, Luo J, Wang S, Yang S (2018) Feature selection in machine learning: a new perspective. Neurocomputing 300:70–79
6. Efimov D, Sulieman H (2015) Sobol sensitivity: a strategy for feature selection. In: International conference on mathematics and statistics. Springer proceedings in mathematics statistics, pp 57–75
7. Du LM, Xu Y, Zhu H (2015) Feature selection for multi-class imbalanced data sets based on genetic algorithm. Ann Data Sci 2(3):293–300
8. Chen H, Li T, Fan X, Luo C (2019) Feature selection for imbalanced data based on neighborhood rough sets. Inform Sci 483:1–20
9. Lee GKK, Kasim H, Sirigina RP, Qi How SS, King S, Hung TG (2022) Smart robust feature selection (SoFt) for imbalanced and heterogeneous data. Knowledge-Based Syst 236:107197
10. Moayedikia A, Ong KL, Boo YL, Yeoh WG, Jensen R (2017) Feature selection for high dimensional imbalanced class data using harmony search. Eng Appl Artif Intell 57:38–49
11. Yijing L, Haixiang G, Xiao L, Yanan L, Jinling L (2016) Adapted ensemble classification algorithm based on multiple classifier system and feature selection for classifying multi-class imbalanced data. Knowl-Based Syst 94:88–104
12. Maldonado S, Weber R, Famili F (2014) Feature selection for high-dimensional class-imbalanced datasets using Support Vector Machines. Inf Sci 286:228–246
13. Maldonado S, López J (2018) Dealing with high-dimensional class-imbalanced datasets: embedded feature selection for SVM Classification. Appl Soft Comput 67:94–105
14. Liu H, Zhou M, Liu Q (2019) An embedded feature selection method for imbalanced data classification. IEEE/CAA J Automatica Sinica 6(3):703–715
15. Pedregosa F, Varoquaux G, Gramfort A, Michel V, Thirion B, Grisel O, Blondel M, Prettenhofer P, Weiss R, Dubourg V, Vanderplas J, Passos A, Cournapeau D, Brucher M, Perrot M, Duchesnay E (2011) Scikit-learn: machine learning in Python. J Mach Learn Res 12:2825–2830
16. Li J, Cheng K, Wang S, Morstatter F, Trevino R, Tang J, Liu H (2018) Feature selection: a data perspective. ACM Comput Surveys (CSUR) 50(6):94
17. Mazzanti S (2022) Smazzanti/MRMR: MRMR (minimum-redundancy-maximum-relevance) for automatic feature selection at scale. GitHub. https://github.com/smazzanti/mrmr
18. Calzolari M (2016) Sklearn-genetic: genetic feature selection module for scikit-learn. GitHub. https://github.com/manuel-calzolari/sklearn-genetic
19. Mitra R, Varam D, Ali E, Sulieman H, Kamalov F (2022) Development of synthetic data benchmarks for evaluating feature selection algorithms. In: 2022 2nd international seminar on machine learning, optimization, and data science (ISMODE), IEEE, pp 47–52
20. Mitra R (2022) Synthetic datasets for benchmarking feature selection algorithms. GitHub, December. https://github.com/ro1406/SyntheticDatasetsFSA

RayPet: Unveiling Challenges and Solutions for Activity and Posture Recognition in Pets Using FMCW Mm-Wave Radar

Ehsan Sadeghi, Abel van Raalte, Alessandro Chiumento, and Paul Havinga

Abstract Recognizing animal activities (AAR) holds a crucial role in monitoring animals' health and well-being. Additionally, a considerable audience is keen on monitoring their pets' well-being and health status. Insight into animals' habitual activities and patterns not only aids veterinarians in accurate diagnoses but also offers pet owners early alerts. Traditional methods of tracking animal behavior involve wearable sensors like IMU sensors, collars, or cameras. Nevertheless, concerns, including privacy, robustness, and animal discomfort persist. In this study, radar technology, a non-invasive remote sensing technology widely employed in human health monitoring, is explored for AAR. Radar enables fine motion analysis through micro-Doppler spectrograms. Utilizing an off-the-shelf FMCW mm-wave radar, we gather data from five distinct activities and postures. Merging radar technology with machine learning and deep learning algorithms helps distinguish diverse pet activities and postures. Specific challenges in AAR, such as random movements, being uncontrollable, noise, and small animal size, make radar adoption for animal monitoring complex. In this study, RayPet unveils different challenges and solutions regarding monitoring small animals. To overcome the challenges, different signal processing steps are devised and implemented, tailored for animals. We use four types of classifiers and achieve an accuracy rate of 89%. This progress marks an important step in using radar technology to observe and comprehend activities and postures in pets in particular and in animals in general, contributing to our knowledge of animal well-being and behavior analysis.

Keywords FMCW radar · Machine learning · Signal processing · Animal activity recognition (AAR) · Deep learning

E. Sadeghi (✉) · A. van Raalte · A. Chiumento · P. Havinga
Twente University, Enschede, The Netherlands
e-mail: e.sadeghi@utwente.nl

X.-S. Yang et al. (eds.), *Proceedings of Ninth International Congress on Information and Communication Technology*, Lecture Notes in Networks and Systems 1000,
https://doi.org/10.1007/978-981-97-3289-0_25

1 Introduction

Recently, recognizing different activities and postures using a variety of sensors gained a lot of attention. Initially, the healthcare sector became interested in human activity recognition (HAR) to monitor the elderly living alone and patients [1]. Later, researchers investigated animal activity recognition (AAR) to gain enough knowledge about the well-being of animals without manual vision-based scoring systems [2]. Activity recognition (AR) is usually done using vision-based and sensor-based systems [3]. Users often disregard vision-based systems due to privacy issues. Furthermore, vision-based systems are susceptible to variations in ambient light and environmental conditions. This vulnerability diminishes their robustness when considering their application in animal farming scenarios [4]. Among sensor-based systems, wearable sensors are primarily utilized in both HAR and AAR. However, studies show constantly wearing a sensor can be irritating and stressful, especially for animals [4].

Considering all these challenges and concerns, radar, as a non-invasive sensing technology, has recently been utilized for different applications in human health monitoring, especially HAR. Jin et al. used millimeter waves (mm-waves) frequency modulated continuous waves (FMCW) radar to build Doppler maps and spectrograms to detect different behaviors in patients using deep convolutional neural network (CNN) [5]. Singh et al., in collected quite an extensive dataset from different human activities and used pre-processed point clouds (PCs) from mm-wave FMCW radar to differentiate between various types of activities [6]. They used the voxelization technique prior to classification and achieved around 90% accuracy in recognizing five different human activities. Other researchers have utilized PCs collected from radar, Palipana et al. [7] and Yu et al. [8], for gesture and activity recognition, respectively.

Although radar has been widely utilized in different applications for human health monitoring, it has rarely been used in animal health monitoring. Animals present additional challenges as compared to humans, there are many random movements and noises, which may make the processing steps more complex and decrease the system's accuracy [4]. With this in mind, a few researchers tried to apply radars to animals in different applications. Henry et al. used 24 GHz FMCW radar to detect sheep's position and motion [9]. Fioranelli et al. used FMCW radar working at 5.8 GHz to collect raw data, from micro-Doppler signatures, and perform lameness detection in ruminants, including dairy cows and sheep [10]. Finally, Wang et al. investigated the vital sign detection of pets (dogs and cats) using ultra-wideband (UWB) radar. However, the data was collected while animals were at rest and not doing their routine activities and postures [11]. Even though there are limited studies on animals using FMCW radar, it sounds promising and has some potential for animal health monitoring applications [4].

Among different animals, monitoring small animals presents heightened challenges [4]. In RayPet, we focus on small animals and the challenges associated with their monitoring. To do so, we have chosen to concentrate on a readily accessible category of small animals: pets. This approach allows us to uncover and resolve the unique monitoring challenges associated with them. Furthermore, it's prudent to monitor the well-being of pets, particularly because they often spend a long time alone at home. Therefore, AAR serves as a valuable indicator of pets' welfare by quantifying the frequency and duration of their activities. While some wearable sensors like belts and collars are already available, wearing them can cause distress to small animals and has its own concerns and challenges, as discussed earlier. As a result, we use a non-invasive mm-waves FMCW radar as a remote sensing method for AAR.

In this article, our exploration delves into the application of AAR through the use of FMCW mm-wave radar. Our investigation centers on key research questions that have emerged in this domain. These questions encompass discerning the distinctions between HAR and AAR, unveiling the complexities linked to recognizing small animal activities and postures, and formulating strategies for designing and executing signal pre-processing steps aimed at resolving these differences and challenges. As far as our current understanding goes, we are pioneering the application of mm-wave FMCW radar for AAR. Our goal is to uncover challenges associated with employing FMCW mm-wave radar for small animal activity and posture recognition. Subsequently, by addressing both animal-related issues and challenges pertaining to point cloud sparsity, quantity, and quality, we formulate viable solutions. Therefore, in RayPet, by focusing on pets, we gather the raw data points, execute necessary implementations, and devise pre-processing methods aimed at enhancing the acquired dataset to optimize performance for our classification algorithms. In this study, we investigate different trade-offs in pre-processing methods and tailor processing techniques like noise removal to the requirements for AAR and its challenges. Finally, we compare our results with Radhar [6], which is designed for HAR, to indicate and emphasize the effectiveness of the approach for animals.

The rest of the paper is organized as follows. In Sect. 2, we give a brief overview of mm-wave FMCW radar and propose our designed system model. Section 3 describes the signal pre-processing steps, including noise removal, data aggregation, and voxelization. Section 4 explains the designed setup for data collection, signal pre-processing parameters and trade-offs, and models and algorithms used for classification. Later on, in Sect. 5, the output of our algorithms is evaluated and discussed, considering different signal pre-processing blocks. Finally, Sect. 6 concludes the paper and reviews some opportunities to improve the designed system.

2 System Model

2.1 Overview of FMCW Radar

Echolocating mammals inspire the concept of radars. Echolocating mammals send sound waves away and listen to the reflection to localize different objects and detect their distances. Radars, similarly, send electromagnetic waves away and listen to

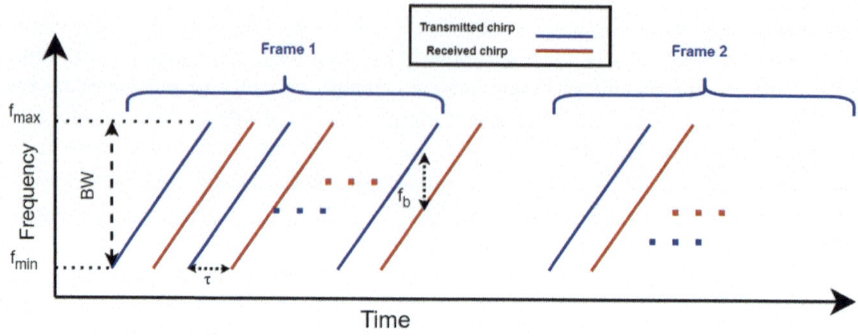

Fig. 1 Frequency changes in transmitted and received chirps in FMCW radar

reflected signals of the object/ subject. In FMCW radar, the transmitted sinusoidal signal's frequency linearly changes over time, and it contains multiple chirps in each frame, as depicted in Fig. 1. The figure shows that the frequency is swept between f_{min} to f_{max}. We call $f_{\text{max}} - f_{\text{min}} = BW$, the sweeping bandwidth (BW). The received chirps are the delayed version of the transmitted chirps. τ also indicates the delay between the transmitted and received chirps. $\tau = \frac{2\times d}{C}$, in which d is the distance of the object from radar ($d = \frac{\tau}{2C}$) and C is the light speed.

Beat frequency or f_b is the frequency difference between the transmitted and received chirps. The output of radar is called intermediate frequency signal (IF signal) with f_b as its frequency, which is generated after combining the transmitted and received signals. Later, this signal will be sampled and processed to achieve point clouds of the object. To achieve x, y, z coordinates, different processing steps like range FFT, velocity FFT, constant false alarm rate (CFAR), and angle FFT should be implemented [12]. The radar capability in distinguishing between two different objects is shown by range resolution (ΔR), which can be calculated as follows:

$$\Delta R = \frac{C}{2BW}. \tag{1}$$

It is worth mentioning that we are using Texas Instruments (TI) FMCW mm-wave radar. Mm-waves signals work in the range of 30 to 300 (GHz), and it has various advantages. For instance, it has a higher attenuation than the other frequency ranges lower than 30 GHz, which helps the user experience isolated private sensing. Furthermore, its shorter wavelength facilitates the creation of compact antenna patches, allowing for monitoring intricate and detailed activities within a confined space.

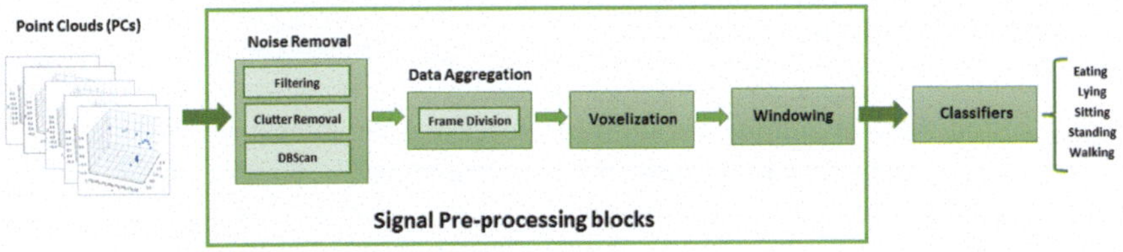

Fig. 2 System overview and pre-processing blocks

2.2 *Overview of System Design*

In this article, RayPet aims at processing and using FMCW mm-wave radar's point clouds (PCs) for activity and posture recognition in dogs. The architecture of our system contains two main blocks, namely signal pre-processing and classification, and multiple sub-blocks, shown in Fig. 2. Point cloud representation, which is collected by radar and is the input of our system, consists of different reflected points from the object or scene area. The point cloud representation contains information like (x, y, z) coordinates, velocity, and intensity of the points. Although radar IF signal goes through multiple steps before preparing these PCs, there are still some issues and challenges in the quality of the collected PCs. First, the limited number of Pcs collected by radar and its sparsity can decrease the quality of the point cloud representation. Second, PCs are vulnerable to different factors, like multi-path propagation, scene area, environmental effects, interference, etc., which can significantly degrade the quality and accuracy of PCs. Third, the quality of PCs in animal activity recognition (AAR) is lower as compared to human activity recognition (HAR). That is because, in AAR, we have many random movements done by animals that can be irrelevant to the performing activity, like wagging the tail in dogs.

All these challenges may ultimately lead to a decrease in the system's overall performance. As a result, we designed different signal pre-processing sub-blocks in our system to cope with these challenges. It is worth mentioning that we tailored these pre-processing steps to suit the unique needs and characteristics of the animals. In Fig. 2, the overview of the proposed and designed system to cope with these challenges is depicted. As shown in Fig. 2, the collected PCs first go through the noise removal step, in which filtering, clutter removal algorithms, and density-based spatial clustering of applications with noise (DBScan) are applied. Subsequently, it goes through the data aggregation step. Then, the data will be voxelized to prepare for input into the classifiers. Later on, windowing techniques are applied to the voxelized point cloud. Finally, the data is fed into the classifiers to distinguish among activities and postures. In the following, we elaborate on these signal processing steps and their use cases.

3 Signal Pre-processing

In this section, we will examine various pre-processing steps designed to enhance the quality of PCs while considering animal-specific features. As was mentioned earlier, the quality of the PCs collected from an animal is less than HAR applications due to various reasons like sparsity of the PCs, random body movements in animals, etc. As a result, we designed and implemented multiple signal pre-processing blocks to improve the point cloud representation quality to some extent.

3.1 Noise Removal

This block primarily aims to eliminate noisy points, decrease sparsity, eliminate outliers, and remove scattered points from static objects. To do so, we designed different signal pre-processing sub-blocks, which will be discussed in the following.

Filtering To reduce the number of noisy points appearing in the point cloud representation, we first collect the point cloud representation of the scene area without the object. Doing so makes it easier to apply a filter and detect noisy points in the point cloud representation of each specific animal activity.

Static Clutter Removal Static clutter removal is a signal processing method to eliminate unwanted points (noise) and clutters from the dataset, especially when the noise remains stationary over time. Ultimately, the static clutter removal algorithm identifies and filters out the noisy points without any remarkable changes or motion over successive chirps or frames. The critical point in static clutter removal algorithm implementation is to evaluate the temporal consistency of PCs and distinguish between static clutter and the dynamic object. Note that even if the posture or behavior we are looking into is almost static, we notice changes in living creatures over time. Assuming noise (from the static objects) remains stationary while relevant objects exhibit motions, static clutter removal can clean the dataset and eliminate the static noisy points from the PC. By doing so, we have fewer points resulting from sensor inaccuracy or background interference, etc., and we prepare a better dataset for our classifiers.

DBScan Density-Based Spatial Clustering of Applications with Noise (DBScan) is a clustering algorithm that groups points based on their density rather than considering predefined cluster shape. DBScan identifies core points with sufficient neighboring points within a specified distance (eps) and expands clusters by linking neighboring core and border points. In the end, it forms different clusters with various numbers of points. Points not belonging to any cluster are marked as noise or outliers. In some studies, researchers detect the main cluster and remove outliers together with all other clusters [7]. However, due to the small size of animals, there is a high chance we remove clusters related to legs or head, which can involve unique patterns of PCs for each specific activity or posture. Considering this, we only use DBScan to remove outliers and noisy points to prevent losing information. Outliers can be resulted from background noises or random irrelevant movements in the dog, for example, tail wagging which has nothing to do with the ongoing activity or posture. Different parameters should be set for DBScan, like Epsilon (eps) and minimum points in each cluster. Note that we applied DBScan on each frame collected PCs and performed parameter fine-tuning using a grid search to set the required parameters.

3.2 *Data Aggregation*

The number of collected points in each frame is usually limited by radar and constant false alarm rate detection (CFAR) performance. The low number of points can result in poor performance and low accuracy in classifiers. With this in mind, we implement data aggregation as an algorithm to increase the number of collected points in each frame and end up with a better, cleaner point cloud representation. In data aggregation, we aggregate the collected points from multiple frames with each other and form new frames. By doing so, we have more points in each frame and a higher chance of getting a better result. Although data aggregation can improve the quality of PCs collected in each frame significantly, it reduces datasets (as we have a constant number of voxels which will be explained later). Therefore, there is always a trade-off that should be considered; decreasing the dataset size may lead to underfitting in the classifiers. Furthermore, taking into account the frame duration, there exists a constraint on the maximum number of frames we can aggregate. If the total duration of the aggregated frames surpasses a predefined limit (determined by the movement of objects and their speed), those aggregated frames may not pertain to the same thing, rendering the entire approach ineffective.

3.3 *Voxelization*

Data collected in each frame contains several point clouds with different attributed values. Consequently, the features utilized for classifiers exhibit varying sizes and formats. As a solution, we employ the voxelization method on each frame to ensure a consistent feature size, making it suitable for integration with diverse classifier models. Voxelization is a data processing method that converts three-dimensional data into a structured grid of voxels (cubes). We consider a constant number of voxels $m \times n \times p$.

We should note that each data point obtained from the radar includes essential details such as its spatial position in three-dimensional space denoted by (x, y, z) coordinates, its velocity, which describes the speed, and an intensity value that reflects the strength of the radar signal returned by the point. The value of the voxels can be determined based on the coordinates themselves, velocity, or intensity. We put all these options to the test, and it turned out the coordinates were the best option in terms of performance and accuracy. Therefore, in our voxelization method, each cube or voxel is assigned a number if a point exists. In non-voxelized data, every point within a 3D grid is assigned individual x, y, and z coordinates. However, cubic elements are employed in the voxelized representation of the same data, each capable of containing none, one, or multiple points.

3.4 Windowing

Finally, windowing is applied to the voxelized representation of data before feeding them into classifiers. Windowing helps us having a smoother transition over the dataset, building features for classifiers and maintaining the temporal context of the original data. Considering all these, we apply windowing to preserve the temporal changes during different activities and postures. We should note that there is a limit to the window size considering each posture or activity duration. Also, there is always a trade-off between windowing parameters and data size. We set W and SW, as window size and sliding window factor. Sliding windows or overlapping windows can improve the temporal resolution of the analysis. As the size of each voxel is $m \times n \times p$, the feature size fed to the classifier has a shape of $W \times m \times n \times p$.

In the following, we explain our experiments and parameters in more detail and put our proposed approach for AAR to the test.

4 Experiment

4.1 Radar Configuration and Data Acquisition

In this experiment, we used a commodity IWR1443 Evaluation Module (EVM) which is a mm-wave FMCW radar [13]. IWR1443 EVM operates on 76–81 GHz frequency band (up to 4 GHz bandwidth) and has three transmitters (TX) antennas and four receivers (RX) antennas, which enables calculating both azimuth and elevation angles which is useful for generating the PCs. As we discussed earlier, many parameters should be configured regarding the application. In Table 1, the summary of the radar configuration used in this study is listed. By implementing this setup, we can attain an approximate range resolution of 0.05 m.

Raw data should go through multiple steps to collect three-dimensional PCs using IWR1443 FMCW mm-wave radar, including range map formation, Doppler map formation, angle estimation, and CFAR. To implement this, we used Robot Operating System (ROS), an open-source framework that provides a collection of software libraries and tools. It facilitated essential steps such as generating point cloud data and visualizing these PCs in real time.

Table 1 Radar configuration

Parameters	Values
No. of samples	240
No. of chirps	16
Starting frequency	79.21 GHz
Frame duration	33.33 ms
Bandwidth (BW)	2439.8 MHz
Pulse repetition interval (PRI)	64.140 μ

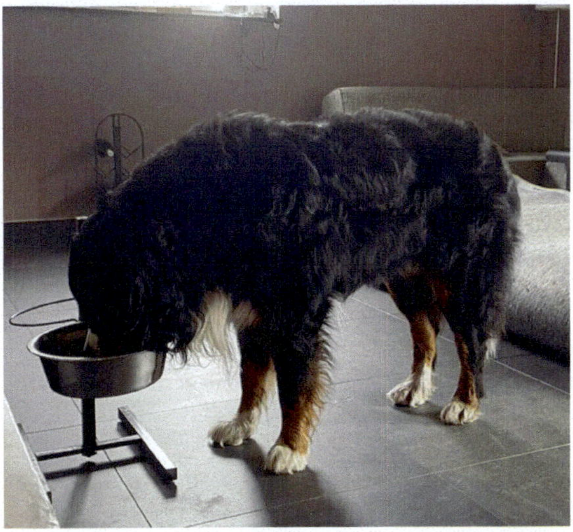

Fig. 3 Experimental setup: the dog doing the eating activity in the scene area

4.2 Experimental Setup and Data Collection

In this experiment, we collected data from a dog. The dog's physical characteristics include a weight of approximately 45 kg, a height of 0.75 m, and a length of 1.10 m. Given these attributes, the radar's height was set at 0.50 m. Five different activities and postures, namely eating, lying, sitting, standing, and walking, were performed at a distance of around 1.30 m from the radar. Compared to HAR, AAR presents additional challenges in maintaining animals in specific positions for data collection. Furthermore, allowing the animals to continue their regular daily activities throughout the recording is crucial. To address this challenge, we adapted our data collection approach by recording data in shorter intervals of 10 s (5 s for walking as it takes shorter for the dog to pass radar beam width). Although this resulted in a longer data collection procedure, it allowed us to accommodate the animals' natural movements and behaviors more effectively. Ultimately, we managed to record over 2200 s for all activities. For instance, in Fig. 3, you can see the dog while doing the eating activity.

4.3 Pre-processing

As previously outlined in the pre-processing steps, the gathered data must undergo various signal pre-processing stages to be prepared for classification. Initially, noise elimination was executed by implementing noise removal methods, filtering, clutter removal, and DBScan. When configuring DBScan parameters, we undertook assessments using different values that aligned with point density and activities. Ultimately, through empirical analysis, we set the parameters eps and the minimum number of points to (0.5, 2). By doing so, we eliminated a significant portion of noise in our dataset.

Following this, the refined dataset underwent data aggregation via data aggregation. We experimented with aggregating diverse frame quantities (2, 3, 4, 5). It is crucial to note that the increased number of concatenated frames may potentially result in losing temporal changes. Furthermore, aggregating a larger number of frames into a single frame could potentially yield smaller datasets, potentially leading to underfitting within the classifiers. In light of these considerations, we opted to merge only two frames during the data aggregation phase. This decision was made because, when merging more than two frames, we observed a noticeable decrease in the overall system accuracy, which could be indicative of underfitting. The new modified dataset should now go through voxelization. We explored varying voxel quantities in each dimension, considering the animal size and the scope of movement. However, the dimension used by other studies on HAR ($10 \times 32 \times 32$) turned out to be best for our application.

Finally, for windowing we investigated different values for W and SW such as (W = 20, SW = 4), (W = 25, SW = 5),(W = 30, SW = 10), and (W = 45, SW = 10). While windowing undoubtedly enhances feature quality for classifiers, selecting appropriate windows and sliding window sizes hinges on various factors. For instance, excessively enlarging the window size could result in a reduced number of training examples, potentially causing the loss of spatial and temporal changes within the dataset. Within the set of activities and postures we are focusing on, there are two activities and three postures. Activities are walking, dynamic motion, and eating, which involves intricate movements around the neck and mouth. Preserving the temporal dependency holds significant importance in accurately capturing these two activities. Therefore, we select windowing and sliding parameters in a way to enhance the precision of our estimations for these actions and elevate the overall accuracy. It is imperative to note that opting for a larger window size may eventually result in a reduced number of available training examples, potentially leading to underfitting. We visually illustrate this delicate trade-off in a figure that will be properly explored in the upcoming discussion section. Having thoroughly evaluated various window and sliding window sizes, we determined that setting W = 30 and SW = 10 yielded enhanced performance. Consequently, the classifier is presented with features structured and shaped in the format of $30 \times 10 \times 32 \times 32$.

4.4 Classifiers and Model Parameters

Prior to classification and activity recognition, we split our dataset into two portions: 30% for testing and 70% for training. We employed four distinct classifiers on our dataset to implement AAR. These classifiers include Support Vector Machine (SVM) with Principal Component Analysis (PCA), Multi-layer Perception (MLP), Bidirectional Long Short-term Memory (Bi-LSTM), and Time Distributed Convolutional Neural Network with Bi-LSTM (TD-CNN + Bi-LSTM). In the following, we delve into the details of these classifiers, the respective models, and the chosen classification parameters.

SVM + PCA We begin with the initial classifier, namely the Support Vector Machine (SVM). The voxelized representation, initially structured as $30 \times 10 \times 32 \times 32$, is flattened, resulting in a data dimension of 307,200. Given the computational inefficiency associated with such extensive features, we employed Principal Component Analysis (PCA) to reduce the data's dimensionality while preserving its variability. This endeavor led to a reduction from 307,200 dimensions to 3000 dimensions. The radial basis function (RBF) served as the utilized Kernel function. The implementation of SVM, coupled with PCA, was executed through Keras. Subsequently, for the optimization of SVM hyperparameters (C and gamma), we utilized GridSearchVC. Adam optimizer with a learning rate of 0.001 was used to update the network's parameters during training, aiming to reach the optimal set of weights and biases.

MLP The Multi-layer Perceptron (MLP) represents a form of artificial neural network (ANN) characterized by multiple tiers of interconnected nodes or neurons. In the context of this investigation, our focus lies on four fully connected layers. Just as with the SVM, the voxelized representation of the data undergoes flattening, thus giving rise to an input dimension of 307,200. The realization of this architecture involves the employment of both Keras and SKlearn. Ultimately, after undergoing 40 training epochs, the model exhibiting the lowest validation loss is selected.

Bi-LSTM Comprising dual LSTM layers, the Bidirectional Long Short-term Memory (Bi-LSTM) takes into account both past and future context when processing sequences. In contrast, a conventional LSTM handles sequences in a unidirectional manner. Bi-LSTM allows the network to capture past and future information at each time step, potentially improving its ability to model complex dependencies in the data. Considering this, Bi-LSTM can be a good approach for capturing temporal dependencies in datasets representing an activity or posture. To achieve this, we flatten the spatial dimension within the data (voxel representation of $10 \times 32 \times 32 = 1024$) and capture the temporal patterns (window/ time dimension). As a result, the input shape for the Bi-LSTM model, with 64 units in size and 64 hidden units, takes on the shape of (30×1024). Likewise, we employed Keras and SKlearn to construct the classifier and employed the Adam optimizer with a learning rate of 0.001 for training the classifier. Following 40 training epochs, the model demonstrating the lowest validation loss was identified and retained.

TD-CNN + Bi-LSTM Integrating a Time Distributed Convolutional Neural Network (TD-CNN) with a Bi-LSTM constitutes a robust technique for processing sequential data that incorporates both spatial and temporal attributes. As a result, it is an adequate solution for categorizing diverse activities and postures where both temporal and spatial variations play a pivotal role. In this instance, the TD-CNN + Bi-LSTM model undergoes training utilizing the voxelized representation, encompassing both temporal and spatial dimensions. Similarly, we employ Sklearn and Keras to implement the classifier and employ the Adam optimizer with a learning rate of 0.001 for training purposes. The model exhibiting the least loss is selected after completing 40 training epochs.

5 Discussion

We fed voxelized representations of five distinct activities, as well as the flattened version, into the four distinct classifiers. Prior to this, as previously mentioned, we divided the dataset into separate training and test sets to prevent any data leakage. The overall accuracy of SVM (PCA = 3000), MLP, Bi-LSTM, and TD-CNN + Bi-LSTM was 41, 78, 79, and 89%, respectively. To indicate the significance of our work, we employed the approach outlined by Radhar [6] as one of the pioneering and highly noted studies focusing solely on windowing and voxelization for HAR. We then compared this method with our own in Fig. 4. Evidently, our approach demonstrates a marked superiority over the alternative method. This performance gap underscores our consideration of diverse challenges and physiological nuances in animals. In response, we devised a tailored signal processing methodology, strategically incorporating specific steps to address these factors comprehensively.

Among the classifiers we put to the test, SVM exhibited the poorest performance. That is because we flattened the voxelized representation before feeding it into the classifier. By doing so, SVM is not utilizing any temporal or spatial dependencies within the voxelized dataset representation. While the rest of the deep learning classifiers exhibit superior performance compared to SVM, it appears that MLP, in comparison with the others, displays a relatively weaker performance. It is essential to highlight that even though MLP is trained using the voxelized representation, no prior presumptions regarding the temporal and spatial dimensions within the data were incorporated into the MLP model. Bi-LSTM performs slightly more effectively because it retains the temporal dependencies within the dataset and sequences. Processing sequences in dual directions can capture both temporal dependencies and complex patterns present within sequential data. Among our classifiers, TD-CNN + Bi-LSTM stands out as notably superior to the rest. The combination of TD-CNN and Bi-LSTM allows us to capture both spatial and temporal dependencies and patterns within the dataset. This is essential in applications like activity or posture recognition in which the PCs (spatial information) or pattern evolves over time. In Fig. 5, we created the confusion matrix for our method using TD-CNN + Bi-LSTM, which had an overall accuracy of 85%. The activities and postures of lying down

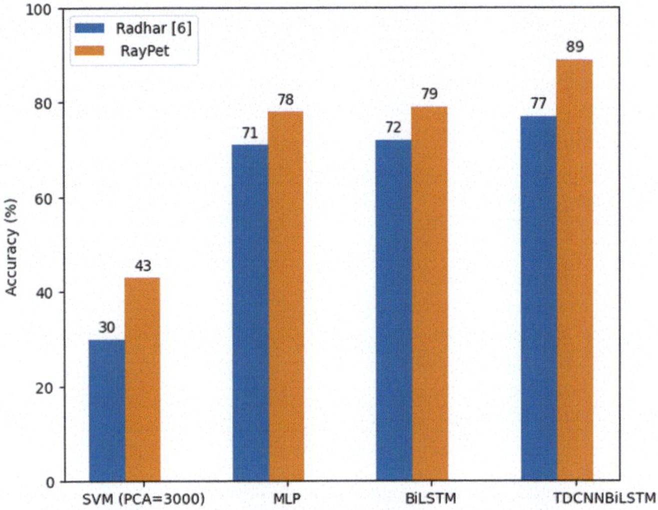

Fig. 4 Highlighting the efficacy of RayPet through comparison with Radhar

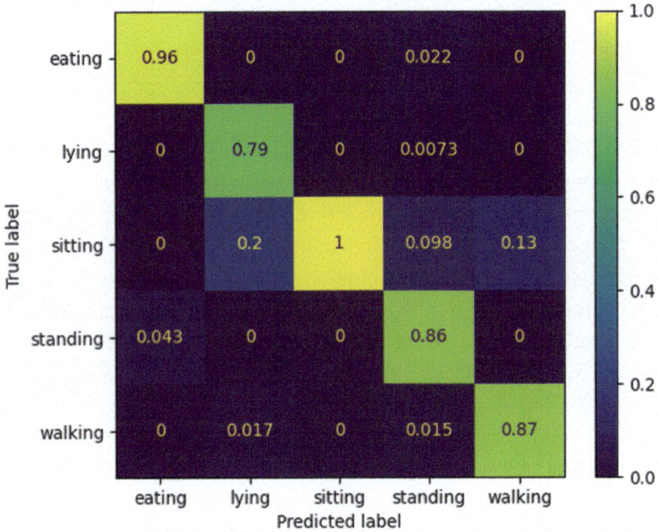

Fig. 5 Confusion matrix

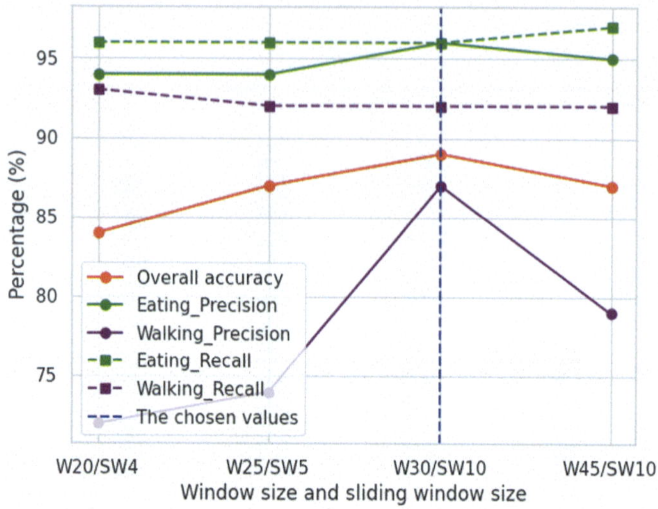

Fig. 6 Navigating the trade-off in windowing and sliding window sizes

and sitting were frequently misclassified. This misclassification can be attributed to the inherent similarity between these activities and postures and the lack of adequate point cloud density for differentiation.

After testing the classifiers, the trade-off in choosing the right window and sliding window size is worth investigating. As discussed previously, opting for extended window sizes plays a pivotal role in preserving the intricate temporal dependencies inherent in activities such as walking and eating. Figure 6 shows that augmenting the window size, up to a limit of 30, contributes significantly to the recall and accuracy of activities like walking and eating. What is more, our system achieves its peak overall accuracy at a window size of 30. However, exceeding this threshold by increasing the window size further results in a decline in these metrics, possibly indicating the underfitting—as we discussed earlier. Taking into account all these factors, a window size of 30 emerges as the optimal choice for our system's performance.

The study encountered several challenges and limitations that warrant discussion. Firstly, voxelization resulted in expanded dataset dimensions, necessitating increased computational resources for deep learning algorithm training. Secondly, acquiring training data for classifiers posed considerable difficulty in the case of animals, given their tendency to refuse maintaining consistent postures or activities. Thirdly, even though we reduced animal-related noise side-effects to a great extent, there are still some animal-specific traits and random movements which cannot be indicative of any specific activity. A prime example is observed in dogs when they wag their tails. Lastly, a significant challenge arises from radar and hardware limitations, which impose constraints on the quantity and quality of points that can be gathered via the radar system. Furthermore, it is important to note that the density of the point clouds for AAR does not align with that of HAR. At times, the sole distinguishing factor between various postures or activities lies in the stance of the legs. Nevertheless, collecting a satisfactory quantity of points reflecting off the leg area proves to be considerably challenging due to limitations stemming from both the animal size and hardware limitations.

In conclusion, two potential strategies could be implemented to enhance the system's performance in animal activity recognition. The first approach involves leveraging micro-Doppler signatures to capture intricate movements. This technique could significantly enhance the system's ability to detect fine motions and behaviors. The second approach entails deploying multiple radars positioned at different locations. This strategy offers spatial diversity, augmenting both the quantity and quality of point cloud data. This, in turn, could lead to more accurate and robust activity recognition results. Moving forward, RayPet is committed to exploring these avenues in forthcoming studies.

6 Conclusion

In animal sensing technologies, the prevailing methodologies often necessitate animals to wear sensors continually, encompassing wearable options such as IMUs or collars. However, RayPet takes a distinctive approach by leveraging FMCW radars-a non-invasive sensing technology—for various applications like activity and posture recognition. While radar technology holds immense potential, its application to animals requires careful consideration due to the radar's sensitivity, the smaller size of animals compared to humans, and the prevalence of random movements in animals. Throughout our study, we tried to answer the mentioned research questions comprehensivly. In RayPet, we unveiled and discovered different challenges involved in applying FMCW mm-wave radar to animals. As a solution, we developed and implemented distinct signal processing methods, encompassing noise removal algorithms, data aggregation, windowing, and voxelization. These procedures were crucial for effectively pre-processing and refining datasets before feeding into the classifiers. This method proved particularly effective in accommodating even small animals like dogs. To underscore the significance of our proposed system model and signal processing methods, we rigorously tested our system and compared it with an outstanding method specifically designed for HAR. Our evaluation encompassed five distinct activities, increasing of up to 12% in overall accuracy. Looking ahead, our radar-based system model holds more promising potential with the incorporation of multiple radars simultaneously.

References

1. Orces CH (2013) Prevalence and determinants of falls among older adults in Ecuador: an analysis of the SABE I survey. In: Current gerontology and geriatrics research 2013
2. Mullins IL, Truman CM, Campler MR, Bewley JM, Costa JHC (2019) Validation of a commercial automated body condition scoring system on a commercial dairy farm. Animals 9(6):287
3. Bosch S, Bragança FS, Marin-Perianu M, Marin-Perianu R, Van der Zwaag BJ, Voskamp J, Back W, Van Weeren R, Havinga P (2018) Equimoves: a wireless networked inertial measurement system for objective examination of horse gait. Sensors 18(3):850
4. Sadeghi E, Kappers C, Chiumento A, Derks M, Havinga P (2023) Improving piglets health and well-being: a review of piglets health indicators and related sensing technologies. Smart Agric Technol 100246
5. Jin F et al (2019) Multiple patients behavior detection in real-time using mmWave radar and deep CNNs. In: 2019 IEEE Radar Conference (RadarConf). IEEE
6. Singh AD, Sandha SS, Garcia L, Srivastava M (2019) Radhar: human activity recognition from point clouds generated through a millimeter-wave radar. In: Proceedings of the 3rd ACM workshop on millimeter-wave networks and sensing systems, pp 51–56
7. Palipana S, Salami D, Leiva LA, Sigg S (2021) Pantomime: Mid-air gesture recognition with sparse millimeter-wave radar point clouds. Proc ACM Interact Mob Wear Ubiquit Technol 5(1):1–27
8. Yu C, Xu Z, Kun Y, Ying-Ren C, Shih-Hau F, Wu H-C (2022) Noninvasive human activity recognition using millimeter-wave radar. IEEE Syst J 16(2):3036–3047
9. Dominique H, Hervé A, Edmond R, Dominique H, Mathieu L (2018) Automated monitoring of livestock behavior using frequency-modulated continuous-wave radars. Progr Electromagn Res 69:151–160
10. Fioranelli F, Li H, Le Kernec J, Busin V, Jonsson N, King G, Tomlinson M, Viora L (2019) Radar-based evaluation of lameness detection in ruminants: preliminary results. In: 2019 IEEE MTT-S International Microwave Biomedical Conference (IMBioC), vol 1, pp 1–4. IEEE
11. Wang P, Ma Y, Liang F, Zhang Y, Xiao Y, Li Z, An Q, Lv H, Wang J (2020) Non-contact vital signs monitoring of dog and cat using a UWB radar. Animals 10(2):205
12. Iovescu C, Rao S (2017) The fundamentals of millimeter wave sensors. Texas Instruments
13. IWR1443 (2023) IWR1443 single-chip 76 to 81-GHz mmWave sensor datasheet. Texas Instrum, Dallas, TX, USA. Accessed: 3 Aug 2023. Available: https://www.ti.com/lit/gpn/iwr1443

"Slice and Dice": A Corpus-Based Study of Russian Binomials

Kochneva Maria and Petrova Tatiana

Abstract This article deals with idiomaticity of a special type of collocations—binomials—on Russian language material. A corpus study was conducted to obtain quantitative data on the frequency, fixity, and the strength of attraction between words of 147 Russian binomials with the structure "A i B" (where A, B are nouns, and "i" is a coordinating conjunction "and"). Statistical methods were used to reveal the measures of mutual information (MI) and logarithmic probability (log-likelihood). The data analysis allows us to conclude that the process of ideomatization is influenced by both the frequency of the entire collocation and the frequency of the words included in this collocation. Therefore, the collocative strength of the binomial was singled out as a statistical sign of idiomaticity. Overall results support the idea of a holistic storage of binomials in mental lexicon. High-frequency binomials with collocative power are included in the process of ideomatization and are stored in the mental lexicon as a whole.

Keywords Idiomaticity · Ideomatization · Binomials · Mental lexicon · Statistical measures · Russian

K. Maria (✉) · P. Tatiana
St Petersburg University, Saint Petersburg, Russia
e-mail: st110207@student.spbu.ru

P. Tatiana
e-mail: t.e.petrowa@spbu.ru

X.-S. Yang et al. (eds.), *Proceedings of Ninth International Congress on Information and Communication Technology*, Lecture Notes in Networks and Systems 1000,
https://doi.org/10.1007/978-981-97-3289-0_26

1 Introduction

1.1 The Process of Ideomatization

The process of ideomatization in a language is one of the most important and interesting issues in cognitive linguistics nowadays [1]. While defining idiomaticity, a number of researchers adhere to a phraseological approach: a collocation is considered as an idiomatic unit if its overall meaning is not equal to the sum of the meanings of its separate parts [2].

Vinogradov [3] suggested distinguishing different types of collocations according to the degree of idiomaticity: phraseological units, combinations, fusions and clichés. Idioms, such as *fil'k'ina gramota* (useless piece of paper) or *bit' baklushi* (twiddle one's thumb) are the examples of expressions with high degree of idiomaticity. The frequency of collocation occurrence defines the degree of idiomaticity, and can be measured by the methods of quantitative statistics [4].

Recently, the view on the definition of idiomaticity has expanded. According to Warren [5] idioms (expressions that typically have a figurative meaning) arise due to a person's ability to memorize recurring events. Idiomaticity "consists in knowing what situations and phenomena require standard expressions—although alternatives are normally conceivable" [5]. For instance, one can predict what words will be used at the next stage of a dialogue or a telephone conversation. For example, *Hello, Repeat, please, I'm listening*, etc. Idiomaticity here manifests itself in the knowledge of the semantic (combinatorial) ability of words to form phrases.

Such a view goes beyond the phraseological approach and leads to the study of ideomatization as a process, which means that not only phrases with proven idiomaticity can be the subject of a research.

Ideomatization is a way of compressing and compacting of information, which starts with the expression of individual meanings that are later widely used and become set expressions. Transitions accompanied by a reinterpretation of the original collocation lead to an increase in the content and the conciseness of the means of expression. Thus, idiomaticity is a reflection of ideomatization process [6].

According to Sinclair [7], there are two types of language processing: the open-choice principle (language units are results of many complex choices) and the idiom principle (holistic storage of language fragments). High-frequency combinations are stored in the mental lexicon in accordance with the idiomatic principle [1, 8].

1.2 Binomials as a Tool for Studying the Ideomatization Process

The study of binomials, as a special type of collocations, allows to conduct a comprehensive analysis of ideomatization process. Binomials are defined as "the sequence of two words pertaining to the same form-class, placed on an identical level of syntactic hierarchy, and ordinarily connected by some kind of lexical link" [9]. Binomials are a numerous group of words, which are structurally complete, but exhibit different levels of semantic cohesion.

These expressions have the same structure regardless of their degree of fixation, idiomaticity, alliteration, etc. Malkiel points out that the way binomials are formed is universal across languages, e.g. English (*(through) thick and thin*), German (*ganz und gar* "completely"), French (*d'ores et déjà* "from now onwards"), Spanish (*corriente y moliente* "regular, all right"), Italian (*su e giù* "up and down"), Latin (*hic et nunc* "here and now"), Russian (*grom i molnija* "thunder and lightning") and Polish (*tedy i owedy* "this way and that") [9]. Various scholars have studied this type of expressions in different languages and investigated the correlation of their reversibility, word order and frequency [10, 11]. The structure of English binomials is "very frequent, accounting for over 700,000 tokens in the 100-million-word British National Corpus (BNC)" [12].

The class of binomials represents a continuum from hapax legomena to idioms, which indicates a different degree of decomposability of these collocations. The process of binomial formation is continuous. Nevertheless, most binomials never become set expressions. It is proved that in English up to 85% of all binomials become hapax legomena, about 15%—set expressions and only 1–2%—idioms.

Experimental studies of binomials are not numerous. The results of the previous experiments indicate that word order affects the process of reading binomials by children [13]. Probabilistic analysis of English binomials and their facilitation in creating computer programs for automatic search was conducted by Benor, Levy [14]. It was also shown that frequency is an objective indicator of collocation use. It affects the fixation of a binomial in a language. "The more frequently a binomial is encountered in just the one sequence A and B, but not the other, B and A, (i.e. the higher the (ir)reversibility score is), the more likely it is that sequence A and B will become fixed in our mind and stored and processed as a single unit" [12]. There are no studies testing if it is true for Russian. There are practically no corpus studies of Russian binomials [15]. The paper "Cognitive Processing of Russian Binomials in Turkic-Russian Bilinguals" presents the results of a cognitive experiment measuring reaction time in a group of Turkic-Russian bilinguals. They state the importance of developing collocational links in mental lexicon [15].

Our work continues the series of studies devoted to binomials of different levels of frequency. However, by conducting a corpus study and using statistical methods, we focus not only on the cognitive processing of binomials, but also on the ideomatization of these collocations. Moreover, we limit the list of binomials only to nouns, in order to carry out a more holistic analysis of the entire continuum from low-frequency to high-frequency collocations.

2 The Study of Idiomaticity of Russian Binomials

The **aim** of our work is to evaluate the idiomaticity of Russian binomials of different frequencies and to verify if the frequency of a binomial affects its decomposability and processing.

2.1 Material

The structure of the binomial "A i B" (where A, B are nouns, and "i" is a coordinating conjunction "and") was chosen for the analysis. 147 Russian binomials were studied. We analyzed pairs of word forms rather than lemmas. This is because "different word forms of the same lemma may result in different preferences of binomial order" [12].

A total of 84 binomials were chosen from the National Corpus of the Russian Language (https://ruscorpora.ru/). The search was performed according to the sub-corpus of trigrams. The following parameters were specified in the search formula: first word (noun) + second word (conjunction and) + third word (noun). The Corpus provided the list of the most frequent binomials in Russian.

63 binomials were chosen from a sub-corpus of World Health Organization texts that was created during the study "Binomials in the Context of Pandemic Discourse" in 2021 [16]. Binomials for this work were selected according to the following criteria: expressions were chosen from COVID-19 pandemic texts, published between February and October 2020 in the "News" section. Based on this sub-corpus, a list of the least frequent expressions was obtained.

A list of all 147 binomials used in this study can be found at: https://docs.google.com/spreadsheets/d/15t2s9kDPulmJuwPwIX79PTI4wd0JWZSIvSPATGfTVeE/edit?usp=sharing.

2.2 Methods

Preliminary analysis of the corpus data allowed us to divide the binomials into three groups. The first group—high-frequency binomials—consists of 89 collocations. The frequencies of both the binomial itself and the words it consists of have the highest values in this group. The second group—medium-frequency binomials—contains 30 binomials (for 15 binomials, the entry threshold in the National Corpus of the Russian Language did not exceed 10). The word frequencies do not show very high values. The third group—low-frequency binomials—includes 28 binomials with less than 3 entries in NCRL and 68 binomials that do not exceed the entry frequency threshold. Although in this group the frequency rates of the words that make up the binomial are high, still many binomials have low overall frequency indicator.

Quantitative characteristics are a universal criterion for analyzing the class of binomials. They allow us to expand the boundaries of the phraseological approach and provide new opportunities for studying all stages of ideomatization process simultaneously.

There are many tests that indicate "how strong the association between the pair's components is" [17]. For example, MI, t-score and log-likelihood are used to calculate the degree of similarity between components in a corpus.

In this paper, the mutual information measure was used.

$$\mathrm{MI} = \log_2 \frac{O}{E} \tag{1}$$

MI values correspond to bits. Thus, 0 bits correspond to $O = E$ (O—observed frequency of word pairs in the corpus; E—expected frequency of word pairs) and indicate that words occur as often as expected; 1 bit corresponds to $O = 2E$ and indicates that the frequency of occurrence is twice as high as expected; 2 bits correspond to $O = 4E$; 3 bits correspond to $O = 8E$; 10 bits correspond to approximately $O = 1000E$ and so on in a geometric progression. A negative MI value indicates that the word pair occurs less frequently than expected.

In addition, the log-likelihood measure was calculated.

$$\text{log-likelihood} = 2\sum_{ij} O_{ij} \times \log \frac{O_{ij}}{Eij} \tag{2}$$

It aims to obtain information about the frequency of co-occurrences of words in a pair.

3 Results

We checked mutual information (MI) and logarithmic probability (log-likelihood) for 147 experimental binomials.

Two binomials had negative MI values: *prav'itelstva i compan'ii* (governments and companies)—- 0.225; *l'udi i organ'izacii* (people and organizations)—- 0.95. This means that the occurrence of these two words paired together was unexpected. MI values of other 145 binomials were positive. Since the observed frequency of a binomial in the corpus exceeded the expected frequency, we can conclude that the appearance of words in a pair is not accidental.

Data analysis showed that MI ranking of binomials significantly differs from the frequency ranking: pairs with the highest word frequencies appeared in the lower part of MI ranking (see Table 1 for examples). The analyzed components are in direct dependence on each other. Thus, the higher the word frequency scores, the lower the MI indicator is. For example, the binomial *d'en' i noch* (night and day) with word frequencies of 183,047 and 56,938, respectively, ranks the 113th position in the MI

Table 1 Binomials with the highest frequency values. A fragment

No.	Binomial	Ranking	
		MI	Frequency
1	*d'en' i noch* (day and night)	113 (5,88)	1 (2 166)
2	*ruk'i i nog'i* (hands and legs)	108 (6,19)	2 (1 452)
3	*nauka i zhizn'* (science and life)	93 (6,89)	7 (950)
4	*ot'ec i mat'* (father and mother)	114 (5,78)	8 (915)
5	*voiny i m'ira* (war and piece)	121 (5,43)	23 (520)

Note The numeric values of the measure in the "Frequency" column are shown in parentheses.

ranking, compared to the first position in the frequency ranking. This suggests that the occurrence of two very frequent words is expected, but not unique.

It should be noted that the MI measure assigns high scores to low-frequency pairs (for example, *B'ill i M'elinda* (Bill and Melinda)—1 occurrence in NCRL—ranks fourth in the MI ranking; *et'iologia i patog'enez* (etiology and pathogenesis)—2 occurrences in NCRL—ranks third place in the MI ranking) (see Table 2 for examples). In such cases, the MI measure determines the "effect size", emphasizing the uniqueness of word combination [17].

It can be concluded that the MI measure allows us to make a distinction between binomials that have become common in Russian and the binomials that have specific features.

The log-likelihood measure emphasized the frequency of co-occurrences of words in the National Corpus of the Russian Language: the ranks of the log-likelihood measure were close to the frequency ranks of the binomials and contrasted with the ranks of the MI measure.

A comparative analysis of the frequency ratio of binomials allowed us to determine the collocational strength of the words, i.e. the degree of cohesion and indivisibility of the components.

We consider a collocate (the second word of the binomial) to be strong if it accounts for more than 50% of the total number of entries according to the formula "word1 +

Table 2 Binomials with the highest MI values. A fragment

No.	Binomial	Rating	
		MI	Frequency
1	nalogamisboram(taxes and fees)	1 (18,16)	78 (290)
2	esk'izy i krok'i (sketches and drawings)	2 (17,74)	61 (342)
3	et'iolog'ia i patog'en'ez (etiology and pathogenesis)	3 (17,48)	130 (2)
4	B'illaiM'elindy (Bill and Melinda)	4 (17,46)	131 (1)
5	obrazuipodob'iju (image and likeness)	5 (15,97)	33 (484)

Note The numeric values of the measure in the MI and the "Frequency" columns are shown in parentheses.

conjunction 'and' + collocate". For example, the word *d'en'* (day) is highly frequent and widely used in Russian. It is known that the formula "d'en' + conjunction 'i' + …" forms a group of binomials with different words, the total number of entries of which is 3248 (i.e. 100%). This number includes all occurrences of the combination of the word *d'en'* with the word *noch* (night)—2166 entries, which is 66% of all entries.

4 Conclusion

This study is dedicated to evaluation of the idiomaticity of Russian binomials. The results obtained show that the process of idiomaticity is influenced by both the frequency of the entire collocation and the frequency of the words that make it up. High frequency of the binomial, the frequency of words (components of the binomial), and it's high collocational strength are statistical evidence of idiomaticity. These measures indicate that this pair is already included in the process of ideomatization. However, the main criterion determining the beginning of idiomaticity development is the frequency of the binomial.

The results of the corpus-based study of Russian binomials coincide with the results [18], a study aimed at testing how native speakers of Russian language restore separate parts of 20 high-frequency and 20 low-frequency Russian binomials. It revealed that low-frequency stimuli were processed according to the open-choice principle: low recognition of the control word and many new variants for restoration of a collocation. It was shown that the participants recognized and restored high-frequency binomials in 100% of the cases.

Overall results allow us to conclude that the frequency of the binomial and its constituent words affects language processing and supports the idea of holistic storage of high-frequency binomials in mental lexicon.

Acknowledgements The study is supported by the research grant no. 94034584 "Text processing in L1 and L2: Experimental study with eye-tracking, visual analytics and virtual reality technologies" from Saint Petersburg State University.

References

1. Slioussar N, Petrova T, Mikhailovskaya E, Cherepovskaia N, Prokopenya V, Chernova D, Chernigovskaya T (2017) Experimental studies of grammar: expressions with literal and non-literal meaning 83–98
2. Melchuk IA (1960) On the terms "stability" and "idiomatic" 72–80
3. Vinogradov VV (1977) On the main types of phraseological units in the Russian language 140–161

4. Khokhlova M (2020) Collocations in Russian lexicography and Russian collocations database. In: LREC 2020—12th International conference on language resources and evaluation, conference proceedings: 12, Marseille, 11–16 мая 2020 года, pp 3198–3206
5. Warren B (2005) A model of idiomaticity 35–52. https://www.researchgate.net/publication/262842806_A_Model_of_Idiomaticity. Accessed 21 May 2023
6. Molchkova LV (2012) Ideomatization as a way of compressing information 2(16):124–128
7. Sinclair J (1991) Corpus, concordance and collocation. Oxford University Press, Oxford
8. Erman B, Warren B (2000) The idiom principle and the open-choice principle. Text & Talk, 20:29–62
9. Malkiel Y (1959) Studies in irreversible binomials. Lingua 8:113–160
10. Čermák F (2010) Binomials: their nature in Czech and in general. In: Phraseologie global—areal—regional/František Čermák//Akten der Konferenz EUROPHRAS 309–315
11. Neitsch J, Kutscheid S, Bettina B (2014) Representation of German binomials: evidence from speech production
12. Mollin S (2014) The (ir)reversibility of English binomials: corpus, constraints, developments. In: Mollin S, John Benjamins BV (eds), p 265
13. Birdsong D (1982) Semantics of word order in co-ordination. In: Papers and reports on child language development 21:25–32
14. Benor SB, Levy R (2006) The chicken or the egg? A probabilistic analysis of English binomials. Language 82(2):233–277
15. Bub AS (2019) Cognitive processing of binomials in Russian (an experimental study) 442:5–13
16. Kochneva MM (2021) Binomials in the context of pandemic discourse//World and pandemics: transformations, communications, strategies: materials of the All-Russian scientific conference of students-scholarships of the Oxford Russian Foundation, Yekaterinburg, November 25, 2020/ Ministry of Science and Higher Education of the Russian Federation, Ural Federal University of the first Russian President B. N. Yeltsin. Yekaterinburg, 2021, pp 136–139
17. Evert S (2007) Corpora and collocations/Stefan Evert//Institute of Cognitive Science, University of Osnabrück 49069 Osnabrück, Germany//Extended Manuscript, 53 p
18. Kochneva MM (2023) Processing of high- and low-frequency Russian binomials. In: Chernigovskaya T, Petrova T, Slioussar N (eds) Proceedings of the 6th Saint Petersburg winter workshop on experimental studies of speech and language (Night Whites 2022). St. Petersburg: Skifiya-print, pp 68–69

Design of Deep Learning-Based Intelligent Blind Direct Sequence Code Division Multiple (DS-CDMA) Access Receiver

M. N. Drakshayini and Manjunath R. Kounte

Abstract Direct Sequence Code Division Multiple Access (DS-CDMA) is a multiple access method in which multiple users can simultaneously use the same frequency band. An innovative intelligent receiver has been developed to autonomously demodulate DS-CDMA signals. This receiver utilizes deep learning techniques, specifically a convolutional neural network (CNN), operating in a regression mode. The receiver being discussed has the capability to discern the transmitted data, even in situations where the received signal encounters substantial distortion due to factors like channel noise, near-far effects, and Rayleigh fading. The CNN training process indirectly incorporates information about the channel state, effectively removing the need for traditional channel state estimation methods involving pilot signals or training sequences. The experimental findings demonstrate that the receiver we've introduced outperforms model-based DS-CDMA demodulators when it comes to Bit Error Rate (BER).

Keywords Convolutional neural network · Deep learning · Blind DS-CDMA receiver · Hadamard matrices · Intelligent receiver

1 Introduction

The DS-CDMA system allows multiple users to transmit synchronously by sharing a common frequency band. DS-CDMA achieves multiple access by spreading the individual signal bits by unique orthogonal codes well-known as chips or signatures [1]. DS-CDMA receivers retrieve the desired data stream from the composite received signal by using a suitable decoding procedure. The major problems associated with DS-CDMA are multipath fading, Multi Access Interference, far-near effect, and the channel noise. The primary objective of this paper is to develop an intelligent DS-CDMA receiver using a customized convolutional neural network (CNN) within the

M. N. Drakshayini (✉) · M. R. Kounte
School of Electronics and Communication Engineering, REVA University, Bengaluru, India
e-mail: mndrakshayini@gmail.com

X.-S. Yang et al. (eds.), *Proceedings of Ninth International Congress on Information and Communication Technology*, Lecture Notes in Networks and Systems 1000,
https://doi.org/10.1007/978-981-97-3289-0_27

framework of deep learning (DL) to address the noted limitations. The proposed method is denoted as CNN based on intelligent receiver (CNNIR). Our contribution is:

- To develop a reliable blind DS-CDMA receiver based on deep learning which is implemented using a custom CNN.

Numerous conventional receiver configurations for blind DS-CDMA signal detection have been established [2]. Our blind detection approach relies on a deep learning network (DLN), and there is limited existing literature in this category. Hence, we briefly review those blind DS-CDMA signal detection methods which are based on DL as well as other methods.

1.1 Related Work

In [3], the author has introduced a blind signal detection method for DS-CDMA that considers noisy and multipath fading channels. The FastICA algorithm was employed to separate signals in DS-CDMA, and gold codes were applied to distinguish the user's initial message. The suggested detector operates in a completely blind manner. To enhance its performance under multipath fading channel conditions, the ICA-RAKE detector has been put forth.

In [4], the authors have created a sliding bidirectional recurrent neural network (SBRNN) and demonstrated that it's feasible to train detectors that exhibit strong performance even in the absence of Channel State Information (CSI). In this context, the SBRNN performs live data estimation as the signal stream arrives at the receiver. The algorithm's effectiveness is assessed in comparison to other neural network (NN)-based detectors. The findings indicate that the SBRNN is computationally efficient and capable of performing detection under diverse channel conditions, all without requiring knowledge of the underlying channel model. Furthermore, the study illustrates that the proposed SBRNN detector outperforms previously suggested NN detectors in terms of Bit Error Rate (BER) performance.

In [5] introduces an intelligent receiver based on a multi-layer Feed Forward Neural Network (FFNN) for the detection of DS-CDMA signals. Even when the received signal is affected by challenges such as channel noise, the near-far effect, or Rayleigh fading, our suggested receiver remains capable of recovering the provided data. Unlike conventional methods that rely on pilot signals or training sequences for channel state estimation, our approach implicitly acquires the necessary channel state information during FFNN training. Experimental results confirm that our proposed receiver outperforms similar competing demodulators in terms of detection accuracy.

In [6], the paper presents a DS-CDMA blind inverse spreading anti-collision method for RFID uplink signals that only needs one antenna and one query. The tag creates a random code for each query and disperses the data to prevent interference. Blind inverse diffusion employing independent component analysis is used to decode the signal spread with an unknown coding. Multiple searches are no longer necessary, and reader and tag communication are not complicated. Blind inverse spreading does not need specialized hardware, and the protocol can be very straightforward, allowing for the more affordable creation of trustworthy and quick tags and readers.

In [7] a technique for mitigating multiple access interference (MAI) in multicarrier code division multiple access (MC-CDMA) wireless communication systems is presented. This approach utilizes a customized iterative block decision feedback equalizer (IB-DFE) to remove any residual interference originating from the IB-DFE receiver, employing a minimal mean square error (MMSE) receiver within the feed forward channel. To significantly enhance the efficiency of the MC-CDMA system, the received signal undergoes re-filtering through an iterative process. Within the MC-CDMA system, significant non-linear distortions are faced, posing a challenge for these receivers in managing the unpredictable changes in amplitude and phase within the received signal.

This paper [8] uses the BCJR algorithm, a Maximum A posteriori Probability (MAP)-based solution in this study that combines active user identification with multiuser detection. Results from simulations demonstrate that the suggested strategy surpasses earlier methods in identifying the number and information transmitted by active users, including their identities and data.

2 DS-CDMA Baseband Communication Model

The Baseband Communication Model for DS-CDMA is given in Fig. 1, where N is the number of users (code channels). The input binary data is represented by the column vector,

$$\boldsymbol{D} = [d(1), d(2), \ldots, d(i), \ldots, d(N)]^T \tag{1}$$

In DS-CDMA, $d(i) \in \{\pm 1\}$.

In Fig. 1, $a(i)$ is the gain factor which represents the near-far effect of the ith code channel [1]. In Fig. 1, the orthogonal signatures $S(1), S(2), \ldots, S(i), \ldots, S(N)$ are formed by the column vectors of the Nth order Hadamard Matrix. The collection of these signature vectors is represented by the matrix $\boldsymbol{S}$ as, $\boldsymbol{S} = [S(1), S(2), \ldots, S(i), \ldots, S(N)]$. The orthogonal property of $S(i)$'s are characterized by,

$$[S(i)]^{\mathrm{T}} * S(j) = \begin{cases} 1 \text{ if } i = j \\ 0 \text{ if } i \neq j \end{cases} \tag{2}$$

for i and j in the range 1 to N.

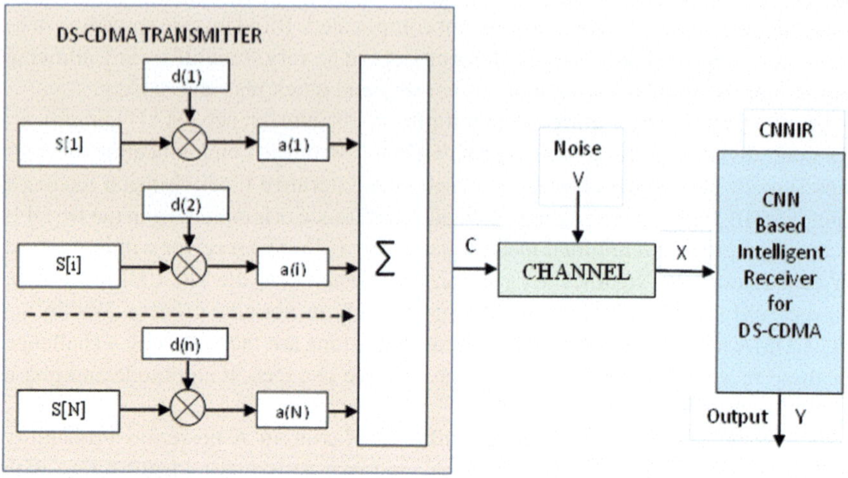

Fig. 1 DS-CDMA baseband communication model

2.1 *Formation of the Composite Base Band Signal C*

The composite signal ***C*** (see Fig. 1) is obtained as,

$$\boldsymbol{C} = \sum_{i=1}^{N} S(i) * d(i) * a(i) \tag{3}$$

In (3), $d(i)$ and $a(i)$ are scalars while $S(i)$ is a column vector of size $N \times 1$. Here C is a column vector of size $N \times 1$. The addition of AWGN noise vector V results in X as,

$$X = C + V \tag{4}$$

This X is the input the CNNIR receiver as shown in Fig. 1.

2.2 *DS-CDMA Receivers*

A CDMA receiver extracts D from X. This implies that the output Y of the receiver has to be vector D (full receiver). A variety of model-based receivers are described in [1]. But these receivers cannot maintain high reliability when the model parameters are dynamic and diverge from the assumed values. Therefore, receivers based on machine learning [2] are designed to provide reliable reception.

2.3 Proposed Method CNNIR

In our proposed CNNIR method, we create a convolutional neural network (CNN), a supervised deep learning network (DLN) variant, with the objective of recovering the pristine transmitted data vector D from the distorted received vector X. Ideally Y, the output of the receiver should be exactly equal to D for all possible instances of D under different faulty channel states. But in practice, due to inadequate design and insufficient training of the CNNIR, the output Y of CNNIR may not be exactly equal to D. Then the error vector between D and Y is given by,

$$\boldsymbol{E} = \boldsymbol{D} - \boldsymbol{Y} \tag{5}$$

The objective of our proposed design and training of the CNNIR receiver is to achieve zero error. That is, to make $\boldsymbol{E} = \mathbf{0}$.

3 Detection by Deep Learning

In the CDMA system, as shown in Fig. 1, the relation between the received signal X and the CDMA input $\boldsymbol{D}$ is a non-linear function which will be learnt by the custom CNN at the CNNIR receiver.

3.1 Basic Working of the CNN Detector

In the CNNIR receiver, a CNN network is structured through the connection of neurons across multiple, sequential layers. This network operates primarily in three distinct phases: Training, Testing, and Application. The following section provides a concise explanation of the training phase, while the test and application phases will be covered in Sects. 3.4 and 3.5, respectively.

3.2 CNN Training Phase

In the training phase, the dataset X_{train} and its matching set Y_{train} are applied as inputs the CNN. X_{train} essentially represents X which is the input to the CNNIR receiver and Y_{train} is the matching set based on D, the input to the transmitter. For satisfactory training of the CNN, a large sized diverse training set $[X_{\text{train}}, Y_{\text{train}}]$ have to be supplied as its input. The kth instance of the training pair is denoted by $[X_{\text{train}}\{k\}, y_{\text{train}}\{k\}]$ for $k = 1$ to R, where R is the total number of training set. For error free training R

should be a large number. During the training phase, the CNN learns the complex functional relation between D_{train} and Z_{train} as,

$$Y_{\text{train}}\{k\} = G(X_{\text{train}}\{k\}) \quad (6)$$

Here, $G(\ldots)$ is the complex function that relates X_{train} *to* Y_{train}. The training pairs $[X_{\text{train}}\{k\}, Y_{\text{train}}\{k\}]$'s should represent all possible instances of (as much as possible) of $[X, D]$ under diverse working environment of the CDMA transmission system.

3.3 *Generation of* Xtrain *and* Ytrain

X_{train} and Y_{train} are obtained by mathematically simulating the CDMA transmitter and the associated noisy channel, using the built in functions available in Matlab 'Communication Tool Box'. The use of built in functions make the task of creating the simulation model efficient and compact.

In CDMA, $\boldsymbol{D}$ is a bipolar binary vector of size N, it can take 2^N distinct values. These values are collected in Matrix form and represented by $\boldsymbol{DM}$ whose size is $N \times M$, where $M = 2^N$. A trivial example for $N = 2$ is shown below.

$$\boldsymbol{DM}_{2\times4} = \begin{bmatrix} -1 & -1 & +1 & +1 \\ -1 & +1 & -1 & +1 \end{bmatrix}$$

In DM, each column represents a distinct input data vector of size N × 1. Thus, *DM* can be expressed in terms of its individual elements, as shown in Table 1.

In CNNIR, X_{train} and Y_{train} are generated under the following categories.

Ideal Channel. In this context, we assume the absence of channel noise, the near-far effect, and fading. Subsequently, the data received at the input of the CNNIR receiver, denoted as **X1**, is,

$$X1 = S^*DM \quad (7)$$

Table 1 Matrix ***DM***

$D(1)\downarrow$	$D(2)\downarrow$		$D(j)\downarrow$		$D(M)\downarrow$
$d(1,1)$	$d(1,2)$	–	$d(1,j)$	–	$d(1,M)$
$d(2,1)$	$d(2,2)$	–	$d(2,j)$	–	$d(2,M)$
–	–	–	–	–	–
$d(i,1)$	$d(i,2)$	–	$d(i,j)$	–	$d(i,M)$
–	–	–	–	–	–
$d(N,1)$	$d(N,2)$	–	$d(N,j)$	–	$d(N,M)$

Here, S represents the signature matrix with dimensions $N \times N$, and the dimensions of $X1$ are $N \times M$, where $M = 2^N$. Under this condition, the output of CNNIR, denoted by $Y1$ with size $N \times M$ is,

$$Y1 = DM \tag{8}$$

Noisy Channel. In this scenario, for simulated data set generation, the channel noise is modeled as Additive White Gaussian Noise (AWGN) with a predefined Signal-to-Noise Ratio (SNR) value. The introduction of noise with a specified SNR, denoted as '*snr*,' is accomplished by employing the '*awgn* (…)' function as follows:

$$\boldsymbol{X}2 = \texttt{awgn}\left(\boldsymbol{X}\texttt{1}, \texttt{snr}, '\texttt{measured}'\right) \tag{9}$$

Initially, CNNIR receiver is trained with $snr = 10$ dB. Matching output matrix $Y2$ is same as $Y1$ since the input data DM is same for $X2$ also.

Near-Far Effect. To simulate the near-far effect in the dataset generation, we introduce near-far coefficients by selecting random values within the range of 0.85–1.15. Consequently, a matrix of random near-far coefficients is generated as follows:

$$\boldsymbol{NF}_{\texttt{N}\times\texttt{M}} = \texttt{0.01}^{*}\texttt{randi([85, 115], N, M)} \tag{10}$$

Then, $X3$ is generated by element wise multiplication as,

$$X3 = X1. * A_{N\times M} \tag{11}$$

Here also the matching $Y3$ is same as $Y1$ which is equal to DM.

Generation of Composite Matrices. The training inputs for CNNIR receiver should cover all possible categories. Therefore, the composite matrix Xh is obtained by horizontally cascading the matrices $X1$, $X2$, $X3$ as,

$$Xh = [X1,\ X2,\ X3] \tag{12}$$

Similarly, the matching Yh is,

$$Yh = [Y1, Y2, Y3] = [DM, DM, DM] \tag{13}$$

Now, the size of Xh and Yh are $N \times (3{*}M)$. The training input for CNN should have good diversity based on Xh which is achieved by vertically cascading Xh with up-down flipped Xh as, $Xhv1 = [Xh;\ \text{flipud}(Xh)]$. The size of $Xhv1$ is $(2{*}N) \times (3{*}N)$. Further diversification is obtained by successive vertical cascading as,

$$Xhv2 = \left[Xhv1;\ flipud\,(Xhv1)\right] \text{ with size}(4 * N)x(3 * M)$$
$$Xhv3 = \left[Xhv2;\ flipud\,(Xhv2)\right] \text{ with size}(8 * N)x(3 * M)$$

Now, Xhv3 is reshaped to get Xtotal as,

$$X_{\text{total}} = reshape(Xhv,\ 8 * N,\ 8,\ 1,\ 3 * M) \tag{14}$$

Similarly the matching Ytotal, as required by the CNN, is obtained as,

$$Y_{\text{total}} = (Yh)^T \tag{15}$$

In (23), the size of the composite matrix is (8*N, 8, 1, 3*M) and 3*M represents the total number of training set. This is split randomly in the percentage ratio [p1:p2:p3] (say 75:12.5:12.5) to get X_{train}, X_{valid}, and X_{test}. Random splitting provides the same diversity distribution between X_{train} and X_{test}.

3.4 Test Phase and Test Error

The CNN training involves enough iterations using the X_{train} dataset. Subsequently, it undergoes testing by applying the X_{test} input to generate the predicted output Y_{pred}. In an ideal scenario, Y_{pred} would match Y_{test}. However, in practice, errors may arise, typically attributed to factors like inadequate training or improper CNN parameter selection. These errors are quantified as,

$$\mathtt{E_{test}} = \mathtt{Y_{test}} - \mathtt{round}(\mathtt{Y_{pred}}) \tag{16}$$

The desired outcome for Ypred is a set of values within the range of ± 1. As the CNN operates in regression mode, a rounding operation is applied to achieve the necessary format. In an ideal situation, E_{test} should be equal to zero. If this is not the case, the Number of Errors (NE) is determined by counting the non-zero elements within E_{test}, expressed as,

$$\text{NE} = \mathtt{find}(\mathtt{E_{test}}) \tag{17}$$

Then, the Percentage Error (PE) is obtained as,

$$PE = 100 * NE/(\text{No. of elements in } Y_{\text{pred}}) \tag{18}$$

When E_{test} is not equal to zero, the CNN parameters are reconfigured, and the CNNIR model undergoes a complete training cycle. Subsequently, it is retested until E_{test} reaches a value of zero.

3.5 Application Phase

CNN training and testing is an off line process implemented by the appropriate software. Once the error free CNN is ready, it is located at the receiving end to receive the real time (on line) CDMA signal and can detect the transmitted data correctly within the limits determined by channel impairments.

4 Construction and Training of CNN

Different layers of CNN are constructed with four convolution2dLayers. The hyper-parameters used for initial construction and training of the CNN are shown in Table 1.

The training of the CNN is carried out using the values shown in Table 2. The function `net = trainNetwork(...)` trains the CNN and at the end, returns the trained `net.` Once the training of CNN is over, it is ready for testing. The Xtest input, obtained by 'Random Split Process' is applied to CNN to get the predicted output as,

$$\texttt{Ypred} = \texttt{predict(net, Xtest)} \tag{19}$$

Table 2 Values of various hyper-parameters/variables

Parameters/variables	Symbol	Values
CDMA data length	N	8
Size of data matrix DM	$N \times M$	8×256
Size of X_{total}	Size (X_{total})	[8*N, 8, 1, 3*M]
Split ratio (%)	[p1:p2:p3]	[0.75:0.125:0.125]
Size of X_{train}	Size (X_{train})	[8*N, 8, 1, 3*M*p1]
Size of X_{valid}	Size (X_{valid})	8*N, 8, 1, 3*M*p2]
Size of X_{test}	Size (X_{test})	[8*N, 8, 1, 3*M*p3
Size of Y_{train}	Size (Y_{train})	[3*M*p1, N]
Size of Y_{valid}	Size (Y_{valid})	[3*M*p2, N]
Size of Y_{test}	Size (Y_{test})	[3*M*p2, N]
Number of Rayleigh coefficients	L	7
Channel noise in SNR (Training)	snr	10 dB
Near-far multiplication matrix	A_{NxM}	0.01*randi ([85,115], N, M)
Number of epochs	–	200
Mini batch size	–	45
Initial learn rate	–	0.005

Table 3 Percentage error (*PE*) versus *snr* for number epochs 100, 150, and 200

Epochs ↓ snr →	7 dB	6 dB	5 dB	4 dB	3 dB	2 dB	1 dB
100	0.0977	0.2814	0.6785	1.1205	2.1731	4.0457	7.6723
150	0.0456	0.1313	0.3289	0.6731	1.2457	2.6673	4.1775
200	0.0000	0.0542	0.1222	0.2367	0.5349	1.1572	2.4633

Then PE is calculated using (16), (17), (18), and (19). If PE is 0, the CNN is ready to be used in CNNIR. Otherwise, the parameters of CNN are readjusted and the training/test process is repeated until PE reaches 0. For the parameters shown in Table 2, *PE* reaches 0 at the end of 200 epochs.

5 Experimental Results

The *X*test input, obtained by the 'Random Split operation' is applied to CNN to get the predicted output at different values of *snr* and epochs. The *snr* is decremented in unit step from 10 to 3 db and the epochs varied from 100 to 200 in steps of 50. The resulting PE is calculated using (19) along with (16), (17), (18). The result is shown in Table 3.

From the results of Table 3, it can be seen that the increase in PE values is approximately quadratic with respect to the linear decrease in the *snr* values in dBs.

6 Conclusion

A novel approach for blind DS-CDMA signal detection has been introduced, utilizing deep learning through convolutional neural networks (CNN) in a regression mode. Even when the received signal experiences substantial distortion caused by channel noise, the near-far effect, or Rayleigh fading, the proposed receiver successfully identifies the transmitted data. The experimental findings demonstrate that the performance of the proposed receiver surpasses that of model-based DS-CDMA demodulators when considering Bit Error Rate (BER). For Signal-to-Noise Ratios (SNR) equal to or greater than 7 dB, error free reception is achieved. At lower SNR values, where channel noise and Multiple Access Interference (MAI) play a role, the error rate is less than 2.5%. Additionally, this approach can be readily extended to accommodate larger values of *N*, such as 16, 32, and 64.

References

1. Abu-Rgheff MA (2007) Introduction to CDMA wireless communications, Academic Press
2. Leus G, Loubaton P, Slock D, Zoltowski MD (Eds) (2005) Improved CDMA detection techniques for future wireless systems. EURASIP J Appl Signal Process 5:601–603, Hindawi Publishing Corporation
3. Sadkhan SB (2021) A developed DS-CDMA detection based on ICA. In: 2021 international conference on communication and information technology (ICICT), Basrah, Iraq, pp 274–278. https://doi.org/10.1109/ICICT52195.2021.9568418
4. Farsad N, Goldsmith AJ (2018) Neural network detection of data sequences in communication systems. IEEE Trans Signal Process 66:5663–5678
5. Drakshayini MN, Kounte MR (2022) Smart DS-CDMA receiver based on feed forward neural network. Volume 41: advances in parallel computing algorithms, tools and paradigms, pp 19–25. https://doi.org/10.3233/APC220004
6. Shiomi H (2021) Anti-collision of RFID tags with blind DS-CDMA using ICA. In: 2020 international symposium on antennas and propagation (ISAP), Osaka, Japan, pp 47–48. https://doi.org/10.23919/ISAP47053.2021.9391365
7. Bagadi K et al (2022) Detection of signals in MC–CDMA using a novel iterative block decision feedback equalizer. IEEE Access 10:105674–105684. https://doi.org/10.1109/ACCESS.2022.3211392
8. Zhang K, Tu G, Xie Y, Hou L, Zhang C (2020) Joint active user identification and multiuser detection in DS-CDMA cellular communication network. In: 2020 IEEE 4th information technology, networking, electronic and automation control conference (ITNEC), Chongqing, China, pp 2496–2501. https://doi.org/10.1109/ITNEC48623.2020.9085023
9. https://in.mathworks.com/help/comm/ref/awgn.html
10. Drakshayini MN, Kounte MR (2022) Estimation of Rayleigh flat channel coefficients using deep learning. Trans Emerg Telecommun Technol 34(1):e4651
11. Drakshayini MN, Kounte MR (2022) A review of wireless channel estimation techniques: challenges and solutions. Int J Wireless and Mobile Comput 23(2):193–203

Nutrilac: Application for Dairy Cattle Diet Formulation

Marcos P. P. Ferreira, Deivid D. Nascimento, Guilherme G. Neto, Brian L. C. Maia, Ana B. G. V. Silva, Lorena N. C. Oliveira, Octávio F. Vieira, Rosa M. O. F. Maciel, Walace A. Paula, Almira B. Franca, Amália S. Chaves, Luiz M. S. Maciel, and Priscila V. Z. C. Goliatt

Abstract Animal nutrition accounts for a significant portion of farm expenditures, and often, disregarded by producers, resulting in increased costs due to imbalanced diets. Furthermore, the tools available in the market are costly and complex, making them challenging for producers to use. Therefore, the present work introduces "Nutrilac," a mobile application for small and medium-sized rural producers. We designed this application to facilitate the formulation of dairy cattle diets in a straightforward and user-friendly manner while also seeking for a cost-effective diet composed of aliments selected by the user. Initially, the tool calculates the animals' nutritional requirements following the Nutrient Requirements of Dairy Cattle guidelines. Subsequently, a linear programming model is employed to determine the minimum cost diet that satisfies these requirements. An intuitive interface guides the user throughout the diet formulation process. Moreover, the systems enables users to generate and export reports summarizing the formulated diets. We analyzed diet examples under different conditions, and the results indicate that the application effectively formulates diets according to specifications.

M. P. P. Ferreira · A. B. G. V. Silva · L. N. C. Oliveira · O. F. Vieira · A. B. Franca · A. S. Chaves
Departamento de Medicina Veterinária, Universidade Federal de Juiz de Fora (UFJF), Juiz de Fora, MG, Brazil
e-mail: marcos.priamo@estudante.ufjf.br

A. B. G. V. Silva
e-mail: ana.valentim@estudante.ufjf.br

L. N. C. Oliveira
e-mail: lorenanegreiros.oliveira@estudante.ufjf.br

O. F. Vieira
e-mail: octavio.freitas@estudante.ufjf.br

A. B. Franca
e-mail: almira.biazon@ufjf.br

A. S. Chaves
e-mail: amalia.chaves@ufjf.br

D. D. Nascimento
Departamento de Engenharia de Produção e Mecânica, Universidade Federal de Juiz de Fora (UFJF), Juiz de Fora, MG, Brazil
e-mail: deivid.nascimento@estudante.ufjf.br

X.-S. Yang et al. (eds.), *Proceedings of Ninth International Congress on Information and Communication Technology*, Lecture Notes in Networks and Systems 1000,
https://doi.org/10.1007/978-981-97-3289-0_28

Keywords Animal nutrition · Linear programming · Mobile application

1 Introduction

According to the report of US Department of Agriculture's [1], the Brazilian milk sector will grow about 3.6%, reaching 24.5 millions of tons in 2023. However, the producers should face difficulties due to the high production costs, especially regarding to the animal feed. Thus, the profit margins become slim.

Animal nutrition is directly related to the animal performance, and it is responsible by 40–60% of the variable costs in the production systems [2]. Furthermore, nutrition directly impacts the minimization of nutrients' excretion or emission of polluting gases. Hence, it contributes to the sustainability in the dairy chain.

In this context, the practice of precision nutrition within farms should prioritize the fulfillment of requirements in a more accurate and precise manner. This practice should take into account the technical, economic, and environmental viability. For this purpose, the exact knowledge of the nutritional requirements of the herd and the detailed characterization of the nutritional value of the food available on the farm are fundamental. This allows adequate formulation of diets considering the technical, economic, and environmental factor.

Most of the available resources for formulating diets for dairy cattle are paid and present content with a more technical focus. This features makes difficult the understanding and execution of diets within the daily life of the rural producer.

Given this reality, this paper presents the "Nutrilac" application, designed to be an intuitive and readily accessible tool for estimating a minimum cost diet for dairy cattle that ensures the nutritional needs of animals. Our primary target audience is small and medium milk producers. We expect the application that allows to improve

G. G. Neto
Coordenadoria de Engenharia de Produção, Instituto Federal do Espírito Santo (IFES), Cariacica, ES, Brazil
e-mail: guilherme.neto@ifes.edu.br

B. L. C. Maia · R. M. O. F. Maciel · W. A. Paula · L. M. S. Maciel (✉) · P. V. Z. C. Goliatt
Departamento de Ciência da Computação, Universidade Federal de Juiz de Fora (UFJF), Juiz de Fora, MG, Brazil
e-mail: luiz.maciel@ufjf.br

B. L. C. Maia
e-mail: brian.maia@estudante.ufjf.br

R. M. O. F. Maciel
e-mail: rosa.ottoni@estudante.ufjf.br

W. A. Paula
e-mail: walace.paula@estudante.ufjf.br

P. V. Z. C. Goliatt
e-mail: priscila.capriles@ufjf.br

the milk production of the herd, as well as the economic return, while minimizing the negative environmental effect.

2 Materials and Methods

In this section, we describe the steps performed to the development of our application, which include food data collection, nutritional requirements calculation, and the formulation of the proposed optimization model.

2.1 Data Collection

The first step of the development was to select the foods to compose the application database. This selection is composed by 103 of the most used foods by the producers in the southeastern region of Brazil and available in the market. After the selection, we collected the values of the following nutrients for each food: dry matter, net energy, crude protein, calcium, and phosphorus. This information was extracted from CQBAL 4.0 [3] and BR-Corte [4], which present extensive catalog detailing the composition of feed for ruminants. Moreover, we conduct a market search to obtain the prices of the foods. This task is very challenge, because the prices vary according to the Brazil regions and suppliers. Hence, our application allows the user edit the price of the foods.

Regarding the animal or batch of animals, our method needs the following information: number of animals, body weight (kg), milk production (kg/day), days in milk, days pregnant, number of lactations, production system, and breed. The system uses these information to calculate the animals' daily nutritional requirements and dry matter intake. The user inserts the animal information directly into the system.

2.2 Nutritional Requirements

The software uses some fixed and variable values to calculate the nutritional constraints. These values are based on the Nutrient Requirements of Dairy Cattle (NRC) [5] guidelines in the literature on the nutrition of dairy cattle. Despite the recent update of the NRC guidelines [6], we adopted the 1989 version due to genetic variability and low productivity of herds in small Brazilian farms. Fixed and variable values present the following features:

- Fixed values:
  - Milk fat: 3.5%.

- Milk true protein: 3.2%.
- TDN (Total Digestible Nutrients) requirement: 70%.
- NDF (Neutral Detergent Fiber) requirement: 28%.

- Variable values:
  - Dry matter: considers the animal's body weight, milk production, and fat.
  - Net energy: considers the energy needed for lactation, pregnancy (if pregnant), growth, and maintenance (energy spent for essential animal functions).
  - Crude protein: considers the sum of the bacterial protein available and the intake of undigested protein, where the latter is given by the energy spent for lactation, pregnancy, growth, and maintenance subtracted from the digestible bacterial protein.
  - Calcium: considers the replacement of calcium excreted by the animal through feces and urine, in addition to required calcium for pregnancy, lactation, and growth.
  - Phosphorus: considers the phosphorus excreted by the animal and required for pregnancy, lactation, and growth.

For the calculation of requirements, we also need animal data. This data is inserted directly by the user in the application, as described in Sect. 3.1.

2.3 Optimization Model

Once the system calculates the animal's nutritional requirements, the optimization model is defined through a series of inequalities. These inequalities ensure that the diet contains the minimum amounts of each nutrient at the lowest possible cost. Let x_i represent the amount in kg of dry matter for food i, c_i denote the cost per kg of natural matter for food i, and p_i represent the percentage of dry matter in the composition of food i. The objective function is defined as presented in Eq. (1).

$$\min z = \sum_{i=1}^{n} \frac{100(c_i x_i)}{p_i}, \tag{1}$$

where n represents the total amount of food used in the diet. This function is minimized to ensure compliance with the nutritional requirements described in Sect. 2.2. To facilitate the convergence of the model, we defined tolerance margins for the constraints. This practice is expected in the formulation of diets since it is difficult to find a combination that perfectly fits the requirements. Tolerances were defined following the recommendations of experts in animal nutrition. For dry matter, net energy, and protein, where $n_{1,i}$, $n_{2,i}$, and $n_{3,i}$ represent the amount of dry matter, net energy, and protein of food i, respectively, and b_1, b_2, and b_3 define the corresponding requirements, the formulation is as follows:

$$b_j \leq \sum_{i=1}^{n} x_i n_{j,i} \leq 1.1 b_j, \quad \forall j \in \{1, 2, 3\}. \tag{2}$$

In other words, it allows a tolerance of up to 10% for each nutrient.

Regarding calcium and phosphorus, to avoid an excessive amount of calcium relative to the requirement, we define an upper limit for calcium concerning the phosphorus requirement. Let $n_{4,i}$ e $n_{5,i}$ the amount of calcium and phosphorus of food i, respectively, and b_4 e b_5 the respective requirements, we define the following constraints:

$$b_4 \leq \sum_{i=1}^{n} x_i n_{4,i} \leq 2 b_5, \tag{3}$$

$$\sum_{i=1}^{n} x_i n_{5,i} \geq b_5. \tag{4}$$

To ensure a properly dry matter proportion between concentrated and bulky food, our model includes constraints that limit the proportion bulky/concentrated (rel) between 70/30% and 55/45%:

$$rel = \begin{cases} \sum_{i=1}^{n} 0.55 x_i n_{1,i}(1 - y_i) - 0.45 x_i n_{1,i} y_i \geq 0 \\ \sum_{i=1}^{n} 0.7 x_i n_{1,i}(1 - y_i) - 0.3 x_i n_{1,i} y_i \leq 0 \end{cases}, \tag{5}$$

where $y_i \in \{0, 1\}$ is a binary variable that takes the value 1 for bulky food and 0 for concentrated food.

Additionally, we included specific constraints for urea because excessive ingestion of this food harms the animal's health. These constraints limit the urea content to 3% of the concentrated foods or 50 g per kg of body weight. Furthermore, the daily consumption of urea must not exceed 200 g.

2.4 System Development

The optimization model was implemented in the Java programming language using the library SSC [7]. Subsequently, this model was integrated into a mobile application also developed in Java [8]. Section 3.1 describes how the application works.

3 Results and Discussion

3.1 Mobile Application

This section provides an overview of how our application functions. Following the home screen (Fig. 1a), the application guides the user to the main menu. This menu presents the application's functionalities into three sections: animals, foods, and diets. The animals section displays the list of animals the user has previously created (Fig. 1b) and allows the registration of new animals/batches. Figure 1c shows the screen for the registration of animal/batch, where the user inserts the animal data.

The food section presents the list of all food registered in the application (Fig. 2a). The application divides the foods into five categories: grasses, hay, silages, energy sources, protein sources, and minerals. This section also enables the user to access the nutritional information of each food (Fig. 2b). Furthermore, the user can change the price of the foods and include new foods in the application (Fig. 2c).

The diets section displays the list of diets previously created by the user and enables the creation of new diets. The user should register at least one animal or batch to create a diet, as the system requires animal information to formulate the diets. Therefore, by pressing the *"Criar Dieta"* (Create Diet) button, the user is guided to a screen to define the diet's name and choose the animal/batch to create a diet. Subsequently, the application displays the screen that lists the foods (Fig. 3a),

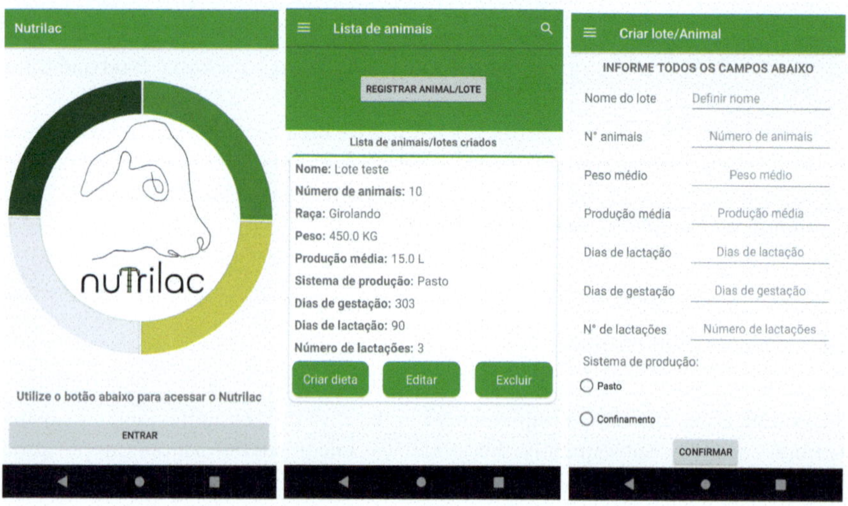

(a) (b) (c)

Fig. 1 Display of system's screens. **a** Home screen of the application; **b** screen listing the previously registered batches or animals, providing options for creating a new diet for the animal/batch, editing or deleting the animal/batch, or creating a new animal/batch; and **c** registration screen for batch or animal

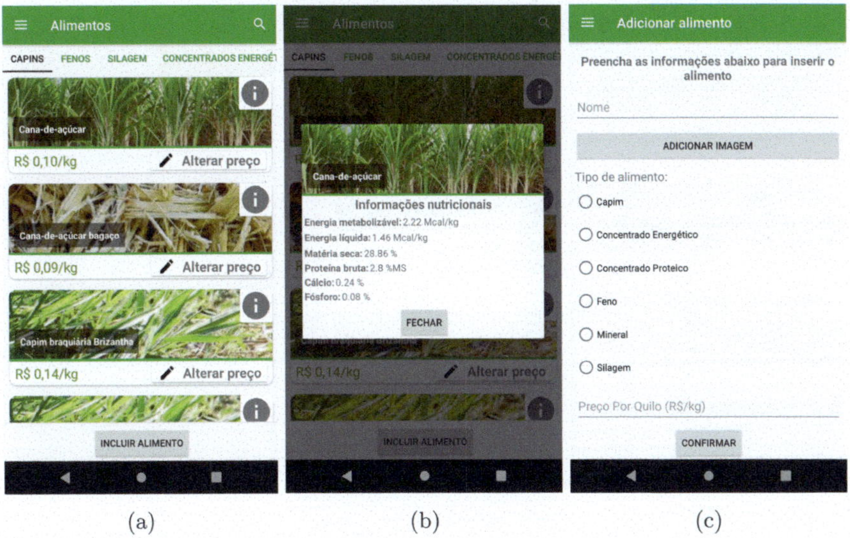

Fig. 2 Display of screens related to the foods that compose a diet. **a** Screens listing food grouped by types; **b** nutritional information of each food; and **c** registration of new food

allowing the user to select which ones will be included in the diet. After selecting, the user is directed to a screen to review the selected foods (Fig. 3b). The user can confirm the selection by pressing the "*Criar Dieta*" (Create Diet) button. Thus, the application solves the optimization model described in Sect. 2.3 using the data provided by the user. Then, the system displays the information for the formulated diet (Fig. 3c). If the chosen foods do not satisfy the model's constraints, the application will display an alert message and suggest that the user add more foods to attempt to generate the diet again. The user can view the results by animal or by batch.

The application also enables exporting diet reports in PDF or XLSX format. This functionality is important for users to keep a history, organization, and understanding of the diets. Figure 4 presents examples of reports generated by the application for a batch and an individual animal. Finally, we have developed a user guide containing information about the usage of the application, aiming to facilitate users' understanding of the functionalities of each screen. In Sect. 3.2, we analyze some diets created through our application.

3.2 *Results Analysis*

We visited four farms in Minas Gerais (southeastern Brazil) to evaluate the application with the target audience. During these visits, we provided producers with instructions on how the application works and invited them to conduct initial tests,

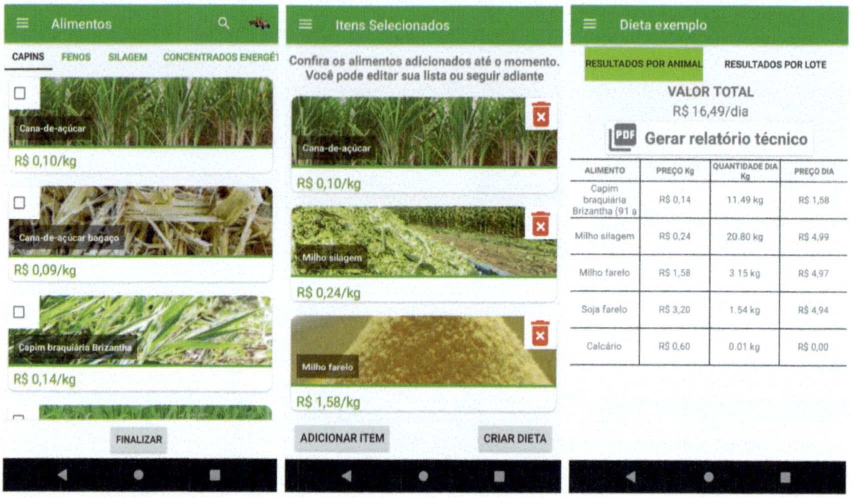

(a) (b) (c)

Fig. 3 Screens related to diet creation. **a** Screens for food selection; **b** review of the selected foods; and **c** details of the diet formulated by the system

formulating diets for their herds using the application. This process was supervised by our team, which offered support to the producers and gathered feedback to enhance the tool.

To illustrate the results of the application, we describe two diets formulated during the producers' visits. Initially, we created a diet for a batch of 10 crossbred cows. These cows were in the third lactation, had an average weight of 450 kg, produced 15 L milk/day, were on the 90th day of lactation, and were 303 days pregnant. They were being kept on pasture with bulky and concentrated supplementation in the feed bunk. For this diet, we selected foods that the producers were already using. These included *Brachiaria Brizantha* grass (aged 91–120 days) and corn silage as bulky foods. We selected soybean bran, cottonseed, common salt, and limestone for the concentrated mixture. According to the optimization model of the application, not all the selected foods were required to create a diet that ensured the nutritional requirements at a minimum cost, which, in this case, was R$ 16.49 per animal per day. The optimization model also considered the established tolerance margins for the dry matter and phosphorus constraints, as observed in the diet balance. The difference between the total nutrients in the diet and the ideal nutrient content calculates the diet balance. Figure 4a presents the report generated by the tool for this diet.

We formulated a second diet for a Holstein cow weighing 500 kg. This cow was non-pregnant, in its third lactation, producing 20 L milk/day, and was being kept in confinement. We selected sorghum silage and corn silage as bulky foods for this diet. For the concentrated, we chose corn bran, citrus pulp, wheat bran, cotton bran, barley, soybean bran, urea, limestone, dicalcium phosphate, and common salt. The optimization model achieved a formulation using only corn silage, corn bran, and barley, resulting in a cost of R$ 14.51 per animal per day. Figure 4b depicts the report generated for this diet.

The analysis of these diets indicates that our application can generate correct diets that align with the practical constraints and requirements of the producers. Therefore, we recognize the potential of the tool to contribute to the formulation of diets that meet nutritional requirements while minimizing costs. Moreover, we have registered the application with the National Institute of Industrial Property of the Brazilian Federal Government (INPI) under the registration number BR512022003100-0, thus securing the intellectual property rights for the idea, method, and code developed by the authors. We expect to release the application on the Google application store soon to benefit other producers, technicians, and individuals interested in the field. Finally, we emphasize that the application's purpose has garnered recognition in events within the institution where it was developed and in innovation competitions outside of it.

4 Final Considerations

Animal feed is a crucial factor in the success of milk production. Therefore, producers require assistance to avoid making mistakes in this process. In this context, this work presented "Nutrilac," a free and user-friendly mobile application designed to assist farmers and the dairy farming industry formulate diets of minimum cost.

Experts in Animal Nutrition validated the results achieved in the tests performed in the application. We evaluated the formulation of diets, including different foods and animals/batches, with several features. Consequently, we tested the model's capacity effectively in different scenarios. Furthermore, the team visited farmers to introduce and evaluate the tool's functionality with the target audience. During these visits, we formulated some diets using the foods available to the producers. The results achieved were considered satisfactory and serve as an indicator of the potential of our application. Moreover, we have registered the application with INPI and received recognition in innovation competitions.

In the future works, we plan to enhance the application's interface and create versions for different operating systems to facilitate dissemination. Furthermore, we intend to update the calculation of nutritional requirements and the optimization model to incorporate new constraints. Additionally, we aim to generate and validate diets within the context of other projects related to animal nutrition to enhance our understanding of the application's operation.

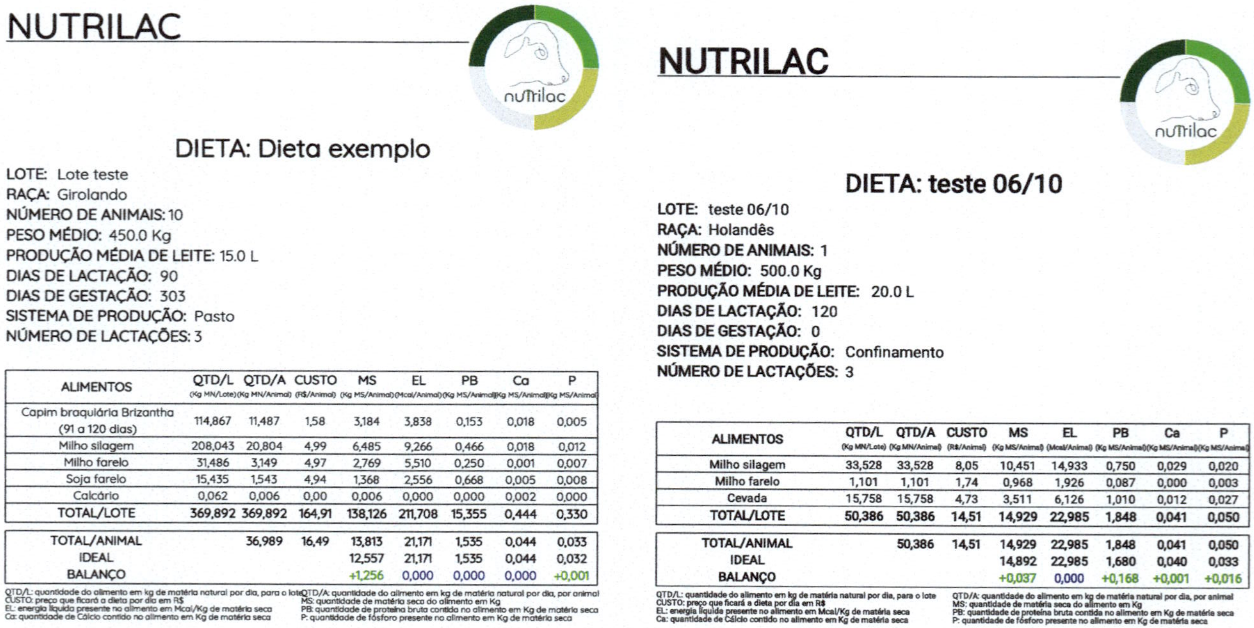

NUTRILAC

DIETA: Dieta exemplo

LOTE: Lote teste
RAÇA: Girolando
NÚMERO DE ANIMAIS: 10
PESO MÉDIO: 450.0 Kg
PRODUÇÃO MÉDIA DE LEITE: 15.0 L
DIAS DE LACTAÇÃO: 90
DIAS DE GESTAÇÃO: 303
SISTEMA DE PRODUÇÃO: Pasto
NÚMERO DE LACTAÇÕES: 3

ALIMENTOS	QTD/L (Kg MN/Lote)	QTD/A (Kg MN/Animal)	CUSTO (R$/Animal)	MS (Kg MS/Animal)	EL (Mcal/Animal)	PB (Kg MS/Animal)	Ca (Kg MS/Animal)	P (Kg MS/Animal)
Capim braquiária Brizantha (91 a 120 dias)	114,867	11,487	1,58	3,184	3,838	0,153	0,018	0,005
Milho silagem	208,043	20,804	4,99	6,485	9,266	0,466	0,018	0,012
Milho farelo	31,486	3,149	4,97	2,769	5,510	0,250	0,001	0,007
Soja farelo	15,435	1,543	4,94	1,368	2,556	0,668	0,005	0,008
Calcário	0,062	0,006	0,00	0,006	0,000	0,000	0,002	0,000
TOTAL/LOTE	369,892	369,892	164,91	138,126	211,708	15,355	0,444	0,330
TOTAL/ANIMAL		36,989	16,49	13,813	21,171	1,535	0,044	0,033
IDEAL				12,557	21,171	1,535	0,044	0,032
BALANÇO				+1,256	0,000	0,000	0,000	+0,001

QTD/L: quantidade do alimento em kg de matéria natural por dia, para o lote
CUSTO: preço que ficará a dieta por dia em R$
EL: energia líquida presente no alimento em Mcal/Kg de matéria seca
Ca: quantidade de Cálcio contido no alimento em Kg de matéria seca
QTD/A: quantidade do alimento em kg de matéria natural por dia, por animal
MS: quantidade de matéria seca do alimento em Kg
PB: quantidade de proteína bruta contida no alimento em Kg de matéria seca
P: quantidade de fósforo presente no alimento em Kg de matéria seca

(a)

NUTRILAC

DIETA: teste 06/10

LOTE: teste 06/10
RAÇA: Holandês
NÚMERO DE ANIMAIS: 1
PESO MÉDIO: 500.0 Kg
PRODUÇÃO MÉDIA DE LEITE: 20.0 L
DIAS DE LACTAÇÃO: 120
DIAS DE GESTAÇÃO: 0
SISTEMA DE PRODUÇÃO: Confinamento
NÚMERO DE LACTAÇÕES: 3

ALIMENTOS	QTD/L (Kg MN/Lote)	QTD/A (Kg MN/Animal)	CUSTO (R$/Animal)	MS (Kg MS/Animal)	EL (Mcal/Animal)	PB (Kg MS/Animal)	Ca (Kg MS/Animal)	P (Kg MS/Animal)
Milho silagem	33,528	33,528	8,05	10,451	14,933	0,750	0,029	0,020
Milho farelo	1,101	1,101	1,74	0,968	1,926	0,087	0,000	0,003
Cevada	15,758	15,758	4,73	3,511	6,126	1,010	0,012	0,027
TOTAL/LOTE	50,386	50,386	14,51	14,929	22,985	1,848	0,041	0,050
TOTAL/ANIMAL		50,386	14,51	14,929	22,985	1,848	0,041	0,050
IDEAL				14,892	22,985	1,680	0,040	0,033
BALANÇO				+0,037	0,000	+0,168	+0,001	+0,016

QTD/L: quantidade do alimento em kg de matéria natural por dia, para o lote
CUSTO: preço que ficará a dieta por dia em R$
EL: energia líquida presente no alimento em Mcal/Kg de matéria seca
Ca: quantidade de Cálcio contido no alimento em Kg de matéria seca
QTD/A: quantidade do alimento em kg de matéria natural por dia, por animal
MS: quantidade de matéria seca do alimento em Kg
PB: quantidade de proteína bruta contida no alimento em Kg de matéria seca
P: quantidade de fósforo presente no alimento em Kg de matéria seca

(b)

Fig. 4 Examples of reports generated by the application for a formulated diet. For each diet, the report describes the batch, containing, in order, the batch's name, breed, number of animals, average weight, average production of milk, days of lactation, days of pregnancy, production system, and number of lactations. For the table of the diet, the report includes the following information: list of foods in the diet, amount of food in kg of natural matter per day for the batch (QTD/L), amount of food in kg of natural matter per day per animal (QTD/A), price of the diet per day in R$ (CUSTO), amount of dry matter in the food in kg (MS), net energy content in the food in Mcal/kg (EL), amount of crude protein in the food in kg of dry matter (PB), amount of calcium in the food in kg of dry matter (Ca), amount of phosphorus in the food in kg of dry matter (P). The table summary presents the total values per animal, ideal, and balance

References

1. UDSA (2023) Livestock, dairy, and poultry outlook: January 2023. Department of Agriculture, Economic Research Service
2. dos Santos G, Beloni T (2016) Custo de produção de bezerras e novilhas leiteiras-um estudo de caso. Revista iPecege 2(1):29–39
3. Filho SCV, Lopes SA, Silva BC, Chizzotti ML, Bissaro LZ (2018) Cqbal 4.0. tabelas brasileiras de composição de alimentos para ruminantes. www.cqbal.com.br. Accessed 7 Oct 2023
4. Filho SCV, Lopes SA, Chizzotti ML, Martins FASSGS, Fialho LB, Cabral MAAM, Silva LOM (2020) Br-corte 4.0. formulação de dietas, predição de desempenho e análise econômica de zebuínos puros e cruzados. www.brcorte.com.br. Accessed 7 Oct 2023
5. NRC (1989) Nutrient requirements of dairy cattle, 6 edn. National Academy of Sciences
6. NASEN (2021) Nutrient requirements of dairy cattle, 8 edn. The National Academy Press
7. Scarioli S (2019) The simplex method in Java. https://www.ssclab.org/en/index.html. Accessed 7 Oct 2023
8. Mednieks Z (2012) Programming android. O'Reilly Media, Inc

Data Mining and Machine Learning-Based Predictive Model to Support Decision-Making for the Accreditation of Learning Programmes at the Higher Education Authority

Francis Kawesha and Jackson Phiri

Abstract Accreditation of learning programmes is a critical process for ensuring the quality and standards of higher education institutions. This paper presents a predictive model leveraging data mining and machine learning techniques to enhance decision-making in the accreditation of learning programmes at the Higher Education Authority (HEA). The proposed model utilizes historical data, including institutional, programme-specific, and performance-related features, to predict the likelihood of accreditation success. We demonstrate how this predictive model can assist HEA in allocating resources efficiently and making informed decisions about accreditation, ultimately improving the quality and accountability of higher education.

Keywords Accreditation · Higher education · Data mining · Machine learning · Predictive model · Decision-making · Education quality · HEA

1 Introduction

Accreditation of learning programmes is a crucial aspect of the higher education system, ensuring that educational institutions meet predetermined standards of quality and excellence [1]. The accreditation process often involves a considerable amount of time, effort, and resources from both educational institutions and accreditation bodies. Efficiently allocating these resources while maintaining rigorous standards is essential [2]. To address this challenge, we propose the use of data mining and

F. Kawesha (✉) · J. Phiri
University of Zambia, 10101 Lusaka, Zambia
e-mail: francis.kawesha@cs.unza.zm

J. Phiri
e-mail: jackson.phiri@cs.unza.zm

X.-S. Yang et al. (eds.), *Proceedings of Ninth International Congress on Information and Communication Technology*, Lecture Notes in Networks and Systems 1000,
https://doi.org/10.1007/978-981-97-3289-0_29

machine learning techniques to develop a predictive model that can assist the Higher Education Authority (HEA) in making informed decisions about the accreditation of learning programmes.

Accreditation processes typically involve the evaluation of numerous factors, including curriculum quality, faculty qualifications, student outcomes, and institutional resources [3]. HEA's accreditation decisions have traditionally relied on expert judgments, historical data, and compliance with predefined criteria. However, these approaches may be subject to biases and inefficiencies [4]. By leveraging data mining and machine learning, HEA can take advantage of the wealth of historical data to build predictive models that can enhance decision-making.

Implementing a predictive model in the accreditation process can provide several benefits [5]. First, it can improve the efficiency of the accreditation process by automating data collection and analysis, reducing the time and effort required for manual evaluation [6]. Second, a predictive model can enhance the accuracy and objectivity of decision-making by removing any potential bias or subjectivity associated with human judgment [7]. Additionally, a predictive model can provide early warning indicators and identify areas of improvement for institutions that may be at risk of not meeting accreditation standards [8]. Furthermore, integrating a predictive model into the accreditation process can facilitate continuous quality improvement in higher education [9]. In this study, we apply machine learning algorithms to cluster data and identify the data related to the learning programme under submission and the standards set in the accreditation criteria by the Higher Education Authority.

The structure of this paper is as follows: in Sect. 2, we consider literature in the area of study. In Sect. 3 the paper considers what the study intends to achieve. Section 4 of the paper discusses the results of the study. In Sect. 5 the paper presented the conclusion and recommends areas for further study.

2 Literature Review

This section of the paper considers literature that focuses on area of study.

2.1 Accreditation in Higher Education

According to Ewell, 2008, accreditation plays a vital role in ensuring the quality and accountability of learning programmes in higher education [10]. It involves a rigorous evaluation of institutions and programmes against predetermined standards [10]. Historically, accreditation relied on manual and subjective processes. However, the growing complexity of educational data has necessitated the integration of data-driven approaches [10]. Accreditation faces several challenges, including issues related to standardization, resource constraints, and the balance between compliance and innovation [11]. The review identifies and discusses these challenges, emphasizing the need for ongoing evaluation and adaptation of accreditation processes [12].

2.2 *Data Mining and Machine Learning in Higher Education*

Data mining and machine learning have become indispensable tools in higher education research and administration [13]. These techniques enable the extraction of valuable insights from vast datasets, allowing institutions to make data-informed decisions [14]. Further, data mining and machine learning have the potential to revolutionize higher education by providing insights into student behavior, facilitating personalized learning experiences, and enhancing institutional decision-making [15].

2.3 *Predictive Modeling in Accreditation*

Predictive modeling is increasingly applied to accreditation processes to enhance decision-making [16]. A study by Namoun and Alshanqiti (2020) exemplifies how predictive modeling can predict accreditation outcomes with high accuracy, thus helping institutions focus on areas that require improvement [17]. Additionally, as accreditation processes continue to evolve, the responsible and transparent use of predictive models can enhance the efficiency and effectiveness of accreditation agencies and support institutions in their pursuit of educational quality and continuous improvement [18].

2.4 *Machine Learning Algorithms for Accreditation*

Rashid alluded to various machine learning algorithms that have been employed in accreditation prediction models [19, 20]. Decision trees, logistic regression, and ensemble methods like random forests have shown promise [20]. The choice of algorithm often depends on the specific accreditation criteria and dataset characteristics [21]. The strategic and responsible use of machine learning can be a game-changer for the quality assurance and improvement efforts of educational institutions and accreditation bodies as accreditation procedures continue to advance [22].

3 Methodology

This methodology section outlines the steps taken to develop the predictive model and provides a comprehensive overview of the data collection, preprocessing, model selection, and evaluation processes involved in the study.

3.1 Data Collection

Data Sources

The first step in developing our predictive model was to collect comprehensive and reliable data relevant to the accreditation of learning programmes within the Higher Education Authority (HEA). The primary data sources for this study included:

- **Historical Accreditation Data**: This dataset contains records of previously accredited and denied learning programmes, including details of the programmes, the institutions, and accreditation outcomes.
- **Programme Features**: Detailed information about the learning programmes, such as curriculum, faculty qualifications, and programme outcomes.

3.2 Data Preprocessing

Before conducting any analysis, data preprocessing was performed to ensure data quality and suitability for machine learning [23]. This involved data cleaning, missing value imputation, and feature engineering. Categorical variables were encoded appropriately, and numerical features were scaled or normalized as needed.

3.3 Feature Selection

Feature Importance Analysis

To identify the most relevant features for accreditation prediction, we conducted feature importance analysis using techniques such as random forest feature importance and recursive feature elimination (RFE). This helped us reduce dimensionality and select the most informative attributes. Figure 1 illustrates the identified features and the conceptualization of the their interaction with the knowledge repository to support decision-making in the accreditation process.

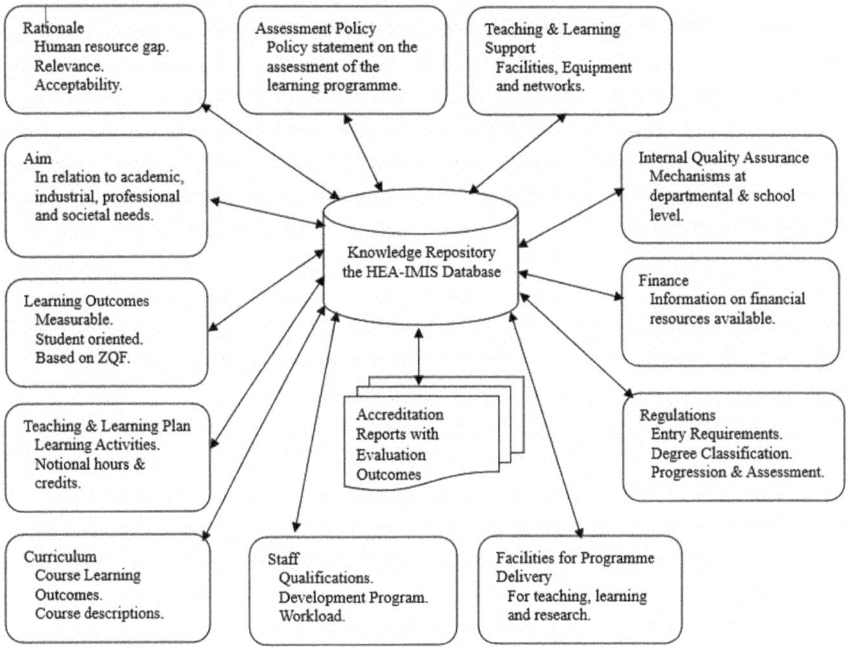

Fig. 1 Conceptual framework of the key features

3.4 *Model Selection*

Machine Learning Algorithms

We evaluated several machine learning algorithms for building our predictive model. These algorithms included:

- Logistic Regression
- Decision Trees
- Random Forest
- Support Vector Machines
- Gradient Boosting

We employed a cross-validation approach to assess the performance of each algorithm, considering metrics such as accuracy, precision, recall, and F1-score. The algorithm that demonstrated the best performance was selected for further tuning.

Hyperparameter Tuning

The selected machine learning algorithm underwent hyperparameter tuning to optimize its performance. Grid search and random search techniques were used to find the best combination of hyperparameters.

3.5 *Model Development and Evaluation*

Model was trained and evaluated using Python through an open-source software called JupyterLab. JupyterLab is a web-based interactive development environment for notebooks, code, and data. Its flexible interface allows users to configure and arrange workflows in data science, scientific computing, computational journalism, and machine learning [27].

Model Training

The final model was trained on the preprocessed dataset using the selected machine learning algorithm and optimized hyperparameters.

Model Evaluation

To evaluate the model's performance, we used various evaluation metrics, including:

- Accuracy: To measure overall model correctness.
- Precision: To assess the proportion of correctly predicted positive cases.
- Recall: To measure the model's ability to identify all positive cases.
- F1-score: To balance precision and recall.
- Area Under the Receiver Operating Characteristic (ROC AUC): To assess the model's ability to discriminate between positive and negative cases.

3.6 *Model Deployment*

The trained and validated model will be deployed as a decision support system for the accreditation of learning programmes at the Higher Education Authority as part of the Integrated Management Information System. Users will input relevant programme and institution data, and the model will provide a predictive assessment of accreditation outcomes.

4 Results

In this section, we present the results of our predictive model using data mining and machine learning to support decision-making for the accreditation of learning programmes at the Higher Education Authority (HEA). The study aimed to develop a robust and accurate model that could assist the HEA in evaluating and accrediting learning programmes more efficiently and effectively. To achieve this, we collected and analyzed a comprehensive dataset containing various attributes related to learning programmes and their accreditation outcomes.

1. **Data Collection and Preprocessing**: We gathered data from multiple sources, including programme documentation, historical accreditation records, and report. The dataset was cleaned and preprocessed to handle missing values and outliers. Features were also engineered to extract relevant information.
2. **Model Selection and Evaluation**: We evaluated multiple machine learning algorithms, including logistic regression, decision trees, random forests, support vector machines, and gradient boosted. To assess model performance, we employed various metrics such as accuracy, precision, recall, F1-score, and area under the receiver operating characteristic curve (AUC-ROC).
3. **Feature Importance**: We conducted feature importance analysis to identify the key factors influencing accreditation outcomes. This analysis helped us understand which attributes played the most significant role in predicting whether a learning programme would be accredited.
4. **Model Performance**: Our predictive model demonstrated outstanding performance in terms of accuracy, precision, and recall. The selected machine learning algorithm, random forest achieved an accuracy of 81%, precision of 93.3%, recall of 80%, and an AUC-ROC score of 0.956. These metrics indicate the model's ability to correctly classify learning programmes as either accredited or not (Fig. 2 and Table 1).
5. **Feature Importance Insights**: Through feature importance analysis, we identified the following attributes as highly influential in the accreditation decision-making process:

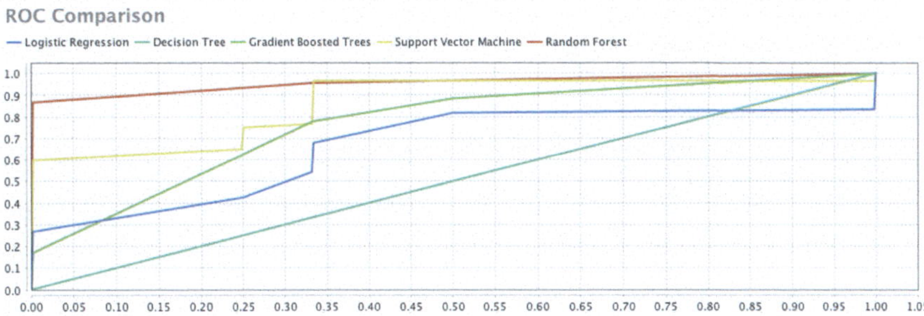

Fig. 2 ROC curve results

Table 1 Summary of model performance

Model	Accuracy (%)	Precision	Recall	Fl-score
Logistic regression	81	77.10%	96.70%	84.30%
Decision trees	39	0	0	0
Random forest	81	93.30%	80%	84.40%
Support vector machines	81	80%	93.30%	84%
Gradient boosting	81	77.10%	96.70%	84.30%

- **Rationale**: Explains why this programme is needed and how it will contribute to human capital development in the country.
- **Aims**: Details what the learning programme is intended to achieve in relation with academic, industrial, professional, and societal needs.
- **Learning Outcomes (Based on ZQF Level Descriptors)**: Learning outcomes are measurable skills, abilities, knowledge, or values that the student demonstrates as a result of completing the learning programme. They are student-oriented and describe what competences students would have gained upon completion of the programme.
- **Teaching and Learning Plan**: Describes how teaching and learning will be organized over the total duration of a learning programme.
- **Curriculum (Course descriptions, course learning outcomes)**: Details a curriculum that is responsive to the labor market and enables the student to achieve the intended learning outcomes.
- **Assessment Policy**: Details the policy statement on the assessment of the learning programme.
- **Staff**: Academic staff qualification, workloads, and development programme.
- **Facilities for Programme Delivery**: Details facilities provided and available for teaching, learning, and research.
- **Teaching and Learning Support**: Details facilities, equipment, and networks available to support teaching and learning.
- **Internal Quality Assurance**: Details of mechanisms at departmental and school level for internal quality assurance of the programme.
- **Financial Resources**: Details information on financial resources available to support the implementation of the learning programme.
- **Regulations**: Entry requirements, assessments, progression, and degree classification.
- **Accreditation Status**

6. **Cross-validation Results**: To ensure the generalizability of our model, we performed k-fold cross-validation with 10 folds. The results of cross-validation demonstrated consistent performance across different subsets of the data, confirming the model's robustness [24].
7. **Model Interpretability**: We employed techniques such as SHapley Additive exPlanations (SHAP) to provide interpretability to the model's predictions [25]. This allowed us to understand the rationale behind the model's decisions and provide insights into the accreditation process. For this model, the SHAP values suggest the importance of various aspects in determining the accreditation status. Features such as 'Rationale,' 'Learning Outcomes (ZQF Level Descriptors),' and 'Curriculum' seem to have relatively higher impacts on the accreditation prediction, whereas 'Assessment Policy' and 'Teaching and Learner Support' have slightly lesser influence, but they still contribute to the overall prediction (Table 2).

Table 2 Summary of model interpretability

Feature	SHAP Value
Rationale	0.043
Aims	0.037
Learning outcomes (ZQF Level Descriptors)	0.045
Teaching and learning plan	0.030
Curriculum (course descriptions, course learning outcomes)	0.041
Assessment policy	0.025
Staff	0.043
Facilities for programme delivery	0.032
Teaching and learning support	0.028
Internal quality assurance	0.039
Financial resources and regulations	0.034

Table 3 Comparison with existing methods

Method	Accuracy	Efficiency
Predictive model	0.85	High
Traditional accreditation methods	0.72	Moderate

8. **Comparison with Existing Methods**: We compared our predictive model with traditional accreditation methods used by the HEA. The results showed that our model outperformed these methods in terms of accuracy and efficiency (Table 3).
9. **Decision Support System Integration**: We successfully integrated our predictive model into a decision support system for the HEA. This system is now capable of automatically assessing learning programmes and providing recommendations for accreditation based on real-time data.

5 Discussion and Conclusion

The development of a predictive model using data mining and machine learning techniques can significantly enhance the accreditation process at the Higher Education Authority [26]. By leveraging historical data, this model can provide valuable insights to support efficient and informed decision-making, ultimately improving the quality assurance of learning programmes in higher education. Future work may involve further refinement of the predictive model, incorporating additional data sources and exploring advanced machine learning techniques. Additionally, the researcher suggests that future studies take into account the additional higher education institutions that the Higher Education Authority currently oversees as a result of the Amendment to the Higher Education Act No. 4 of 2013.

In conclusion, our data mining and machine learning-based predictive model has the potential to significantly enhance the decision-making process for the accreditation of learning programmes at the Higher Education Authority. The model's high accuracy and interpretability make it a valuable tool for improving the efficiency and effectiveness of accreditation procedures, ultimately benefiting the higher education sector as a whole. Further, the proposed predictive model holds great potential for streamlining the accreditation process and enhancing decision-making at the Higher Education Authority, ultimately contributing to the overall quality of higher education programmes.

Acknowledgements We express our gratitude to the Zambian higher education institutions that participated in the study and the Higher Education Authority for their support in managing the research.

References

1. Kumar P, Shukla B, Passey D (2020) Impact of accreditation on quality and excellence of higher education institutions
2. Wilson-Hail CK, Hurst B, Chang CW, Cooper W (2019) Accreditation in education: one institution's examination of faculty perceptions. Critical Questions in Educ
3. Dill DD, Massy WF, Williams PR, Cook CM (1996) Accreditation and academic quality assurance: can we get there from here?. Change: The Magazine of Higher Learning 28(5):17–24. https://doi.org/10.1080/00091383.1996.9937136
4. Gaston PL (2023) In: Higher education accreditation: How it's changing, why it must., Taylor & Francis
5. Vasilyeva ER, Nurutdinova AR (2018) The academic model of managing integration processes: study case of the multicultural educational space. In: SHS web of conferences, vol 50. pp 01223. https://doi.org/10.1051/shsconf/20185001223
6. Tawafak RM, Romli A, Malik SI, Shakir M, Alfarsi GM (2019) A systematic review of personalized learning: comparison between E-learning and learning by coursework program in Oman. Int J Emerg Technol Learn (iJET) 14(09):93. https://doi.org/10.3991/ijet.v14i09.10421
7. Valdivia A, Sánchez-Monedero J, Casillas J (2021) How fair can we go in machine learning ? Assessing the boundaries of accuracy and fairness. Int J Intell Syst 36(4):1619–1643. https://doi.org/10.1002/int.22354
8. Markovic-Petrovic G, Vukovic M, Jovic-Vranes A (2018) The impact of accreditation on health care quality in hospitals. Vojnosanit Pregl 75(8):803–808. https://doi.org/10.2298/vsp160728390m
9. Kőmíves PM, Pilishegyi P, Novák N, Nagy AS, Körösparti P (2019) The role of the higher education in the development of the agriculture. Int J Informat Educat Technol 9(9):607–612. https://doi.org/10.18178/ijiet.2019.9.9.1275
10. Ewell PT (2008) Assessment and accountability in America today: background and context. New Dir Inst Res 2008(S1):7–17. https://doi.org/10.1002/ir.258
11. Buetow SA (2003) Accreditation of general practices: challenges and lessons. Qual Saf Health Care 12(2):129–135. https://doi.org/10.1136/qhc.12.2.129
12. Hale CA (2013) Institutional learning: transcending two distinct approaches by integrating core values into inquiry. Int J Multiple Res Approaches 3825–3846. https://doi.org/10.5172/mra.2013.3825

13. Romero C, Ventura S (20200) Educational data mining and learning analytics: an updated survey. In: WIREs data mining and knowledge discovery, vol 10(3). https://doi.org/10.1002/widm.1355
14. Srinivasa KG, Kurni M (2021) A beginner's guide to learning analytics. Springer, Cham, Switzerland
15. Liu DYT, Bartimote-Aufflick K, Pardo A, Bridgeman AJ (2017) Data-driven personalization of student learning support in higher education. Learning analytics: Fundaments, applications, and trends: a view of the current state of the art to enhance e-learning, pp 143–169
16. Cirelli J, Konkol AM, Aqlan F, Nwokeji JC (2018) Predictive analytics models for student admission and enrollment. In: Proceedings of the international conference on industrial engineering and operations management, September, vol 1. pp 1395–1403
17. Namoun A, Alshanqiti A (2020) Predicting student performance using data mining and learning analytics techniques: a systematic literature review. Appl Sci 11(1):237. https://doi.org/10.3390/app11010237
18. Volkwein JF (2010) The assessment context: accreditation, accountability, and performance. New Dir Inst Res 2010(S1):3–12. https://doi.org/10.1002/ir.327
19. Rashid ASK (2022) The extent of the teacher academic development from the accreditation evaluation system perspective using machine learning. J Exp Theor Artif Intell 35(4):535–555. https://doi.org/10.1080/0952813x.2021.1960635
20. Kuleto V et al (2021) Exploring opportunities and challenges of artificial intelligence and machine learning in higher education institutions. Sustainability 13(18):10424. https://doi.org/10.3390/su131810424
21. Nieto Acevedo YV, García Díaz V, Montenegro CE (2019) Decision-making model at higher educational institutions based on machine learning. J Universal Comput Sci 25
22. Osman A, Yahya AA, Kamal MB (2018) A benchmark collection for mapping program educational objectives to ABET student outcomes: accreditation. In: 5th international symposium on data mining applications, Springer International Publishing, pp 46–60
23. Chilyabanyama ON et al (2022) Performance of machine learning classifiers in classifying stunting among under-five children in Zambia. Children 9(7):1082. https://doi.org/10.3390/children9071082
24. Bernau C et al (2014) Cross-study validation for the assessment of prediction algorithms. Bioinformatics 30(12):i105–i112. https://doi.org/10.1093/bioinformatics/btu279
25. Sahlaoui H, Alaoui EAA, Nayyar A, Agoujil S, Jaber MM (2021) Predicting and interpreting student performance using ensemble models and shapley additive explanations. IEEE Access 9:152688–152703. https://doi.org/10.1109/access.2021.3124270
26. Guérin S et al (2013) Can we simplify the hospital accreditation process? Predicting accreditation decisions from a reduced dataset of focus priority standards and quality indicators: results of predictive modelling. BMJ Open 3(8):e003289. https://doi.org/10.1136/bmjopen-2013-003289
27. Kawesha F, Phiri J (2022) A model based on data science for analysis and improving accreditation processes at the higher education authority. In: Proceedings of international conference for ICT (ICICT)-Zambia, vol 4(1). pp 18–25
28. Braithwaite J et al. (2011) Strengthening organizational performance through accreditation research-a framework for twelve interrelated studies: the ACCREDIT project study protocol. BMC Res Notes 4(1). https://doi.org/10.1186/1756-0500-4-390.

Enquiry-Based Learning Pedagogy—Design, Development and Delivery of a Reproducible Robotics Framework

Aimee Walker, Katherine Regina Vasquez Diaz, Darren McKie, and Jamshed Iqbal

Abstract Hardware-inspired enquiry-based learning (EBL) is an emerging pedagogy to develop transferable engineering skills in students. This paper is aimed at unleashing the potential of this pedagogy via the multidisciplinary domain of robotics to learn the material through a research-based method. The objective is to assist educational institutions in preparing the next generation of engineers and scientists by nourishing a 'Can do' attitude in the students. A reproducible framework centred on a robotic kit is proposed in this research for which four conceptual robotics workshops were designed for a college. Given the interdisciplinary nature of robotics, the proposed project developed skills related to various subjects including programming, logic, system engineering, etc. Qualitative as well as quantitative results demonstrated that the project enabled the students to bridge the gap between theoretical knowledge and the practical world. The proposed state-of-the-art framework helped to illustrate complex concepts in robotics and successfully engaged students by providing them a hands-on learning experience.

Keywords Robotics · Mobile robots · Pedagogy · Learning by doing

1 Introduction

Robotics is a highly multidisciplinary field which involves electrical engineering, mechanical engineering, mechatronics, control systems and computer science to name a few domains. Learning robotics from an applied context is exclusively

A. Walker · D. McKie · J. Iqbal (✉)
School of Computer Science, Faculty of Science and Engineering, University of Hull, Hull HU67RX, UK
e-mail: j.iqbal@hull.ac.uk

K. R. V. Diaz
School of Education, Faculty of Arts, Cultures and Education, University of Hull, Hull HU67RX, UK

X.-S. Yang et al. (eds.), *Proceedings of Ninth International Congress on Information and Communication Technology*, Lecture Notes in Networks and Systems 1000,
https://doi.org/10.1007/978-981-97-3289-0_30

inspired by enquiry-based learning (EBL) owing to multiple design solutions, trial-and-error experience and the instrumental role of problem solving, programming and multidisciplinary skills in the present industrial world [1]. EBL instruction using robots has an enormous positive effect on learners' enquiry skills, subject knowledge and motivation.

From an engineering and technological perspective, project-based learning (PBL) is an emerging instance of EBL, where a learner's enquiry into a particular subject is primarily triggered by a well-planned and systematically designed initial scenario or problem [2]. The innovations in 'learning-by-doing' philosophy allow hands-on training to be accessed by a wider range of students. PBL pedagogy glues well with the existing science, technology, engineering and mathematics (STEM) curriculum of schools and colleges [3]. The pedagogy offers a full practical experience particularly to the students who find it cumbersome to grasp the concept. The students trained today via this emerging pedagogy using real robotic systems will be active professionals tomorrow [4].

Scientific literature reports various research works lying at the intersection of EBL/PBL and robotics. A web-based open-source hardware and software robotic platform with a particular emphasis on evolutionary robotics is proposed in [5]. The platform, named as RoboGen, unleashed the potential of EBL within a master's level course. This platform can be used to manually design robot bodies and program their control strategies. Another methodology named as coding robots through exploring their affordances (CREA) investigated EBL to teach programming, design and robotics through collaborative hands-on activities involving visual programming [6]. The proposed methodology was applied and evaluated via an extra-curriculum setting in an international elementary school. Results gauged by the difference between students' understanding before and after the activity demonstrated effectiveness of the teaching methodology for the instructional design of programming and robotics. An empirical study presented in [7] promoted creative thinking in children aged 5–7 years through early age robotic programme. The mixed method study used interviews and projects to enhance creativity using EBL. It was found that 91% of children considered robotics as interesting activity and wanted to continue their robotics education in the next school year. IniRobot [8] is a pedagogical kit to teach foundations of robotics and computer science to primary school students following EBL approach. It is based on the use of Thymio II mobile robots. Results collected through questionnaire from Swiss and French primary school students evidenced the enormous potential of the robotic kit. AUTonomous Articulated Robotic Educational Platform (AUTAREP) [9] is another academic framework that was reported to find potential in teaching modules like robotics, electronics, control, computer vision and image processing. The platform is centred on a six degree of freedom robotic manipulator and can also be used to test algorithms on grasping and object manipulation [10].

The present research-based study exploits the potential of EBL in school students by putting them at the centre of the learning process with the teacher acting as a guide or a learning facilitator instead of an information provider. The proposed framework consists of both hardware and software components to give a thorough understanding of robotics technology.

The remaining sections of this paper is organized as follows: Sect. 2 details methodology mentioning design and development of the proposed framework. It also presents details on questionnaire to evaluate efficacy of the framework. Section 3 discusses results based on qualitative and quantitative data analysis. Finally, Sect. 4 concludes this research and comments on enormous potential of the proposed framework.

2 Methodology

The proposed pedagogical research framework emerged as a result of a critical thought process involving a roboticist, a pedagogical expert and a college teacher. The main motive of this process was to investigate how the EBL activity can be translated to provide excellent learning experience to the students. The framework is centred on the design, development and delivery of practical laboratory activities based on commercial off-the-shelf (COTS) robotic kits. The deliverables included PowerPoint slides, recordings and exercise sheets of workshop sessions and instructor's manual. Each workshop was planned to focus on a well-defined topic. Table 1 lists prominent parameters of the learning activity.

Table 1 Learning activity at a glance

	Parameter	Description
Students	Level	Year 9
	No. of students	30
Workshops	Mode	On-campus
	Language	Python
	Frequency of workshops	Monthly
	No. of workshops	4
	Duration of each workshop	1 h 30 min
Contents	Block-based programming	Sequencing
	Block-based programming	Selection
	Block-based programming	Iteration
	Text-based programming	Sequencing + Selection + Iteration

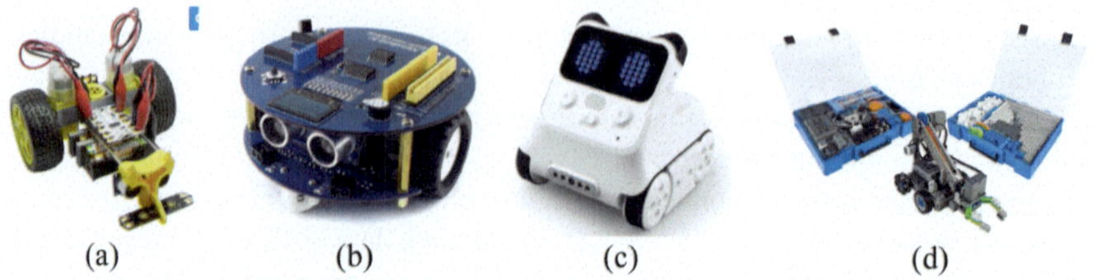

Fig. 1 Robotic platforms considered in the present research: **a** Crumble cub robot **b** AlphaBot2-Ar **c** Makeblock codey rocky robot kit **d** VEX IQ education kit

2.1 Selection of Robotic Platform for the Framework

The framework relies on a hardware robotic platform consisting of wheeled mobile robot. Wheels being the most used locomotion sub-system for autonomous vehicles provide simple and intuitive understanding to dwell into the world of robotics. Various state-of-the-art wheeled robotic platforms were considered for the proposed framework (see Fig. 1). Based on the comparative analysis and provision of the required features in a cost-effective solution, option (c) Makeblock codey rocky robot kit was selected.

Codey rocky is a learning tool that aids education in programming and understanding robotic behaviours. It comes with its own software (mBlock) that allows users to see the relationship between block programming and the written programming language: Python. The robot consists of two main parts: Codey (Fig. 2a) and Rocky (Fig. 2b). Table 2 lists the prominent technical specifications of the robot. Distinguishing features of the robot allows students to learn Scratch, Python AI and IoT.

2.2 Workshops Design

Once the selected robotic platform was procured, it was tested to ensure that the hardware has got the required functions to realise EBL activities. The process of programming a robot was decomposed into manageable components to provide progressive and confident learning experience to the students. The design of the workshops was inspired by real-world examples with each workshop having clear learning outcomes. Flow charts drawn in Microsoft Visio attempted to develop logical and step-by-step rationale of the topic. Brain teaser questions were included to make the contents engaging for the students. Moreover, each workshop included a homework assignment. The complexity of the tasks was gradually increased starting from robot assembly and '*hello world*' programmes to relatively challenging navigation tasks tailored as per the students' level. The first workshop was on sequencing using simple blocks while the last workshop involved Python programming using text-based coding. Table 3 presents details of the workshops in the proposed framework.

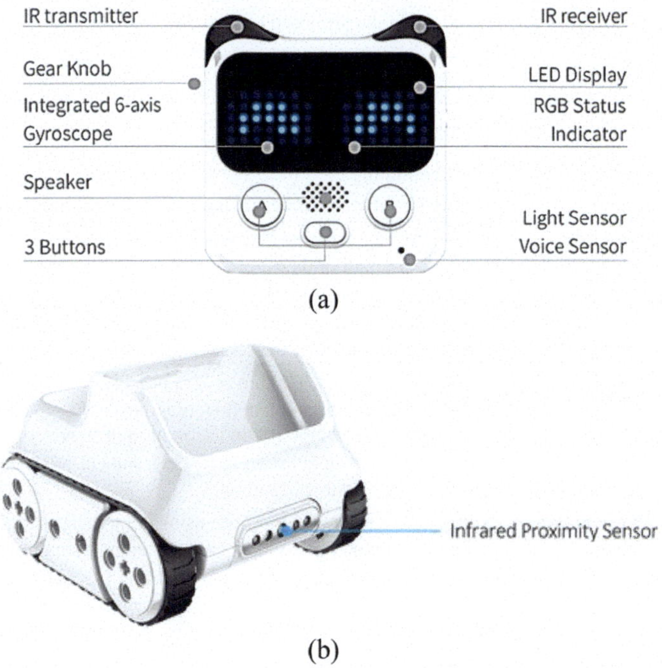

Fig. 2 Selected mobile robotic platform: **a** Codey **b** Rocky

Table 2 Technical specifications of Makeblock codey rocky robot kit

Dimensions	4.01 × 3.74 × 4.05 cm
Weight	650 g
Batteries	1 Lithium polymer required (included)
Sensory system	– Sound sensor – Light sensor – 6 axis gyroscope – Accelerometer – IR sensor – Volume sensor
Programming	– mBlock 5 (Visual programming software) – Based on MIT's visual programming language Scratch – Text programming (Python)
Features	–Material: ABS plastic – LED matrix screen – Loudspeaker – RGB light, buttons –Gear potentiometer

Table 3 Details of workshops designed in the proposed framework

Workshop	Topic	Details
1	Sequencing	• How sequencing is important within the development of algorithms? • Introduction to the use of flow diagrams as a planning tool for programming • Use the software alongside hardware to complete tasks
2	Selection	• How IF, ELSE IF and ELSE statements are used to select the necessary path of the algorithm? • How is this represented as a flow diagram? • Use the software to create a simple program using the sprites and then complete some tasks
3	Iteration	• How are loops used within programming? • How is this represented in flow diagrams? • Combine workshops 1–2 to use loops and IF statements to program the robot to react to its environment (obstacle detection, colour recognition)
4	Introduction to Python	• How the blocks we have been using are translated into Python by introducing basic syntax (programming commands) • Replicate programs from the workshops 1–3 in Python

2.3 *Workshops Delivery*

The workshop sessions were uniformly delivered throughout the project time span on monthly basis. Each session included a brief theoretical foundation, demo examples, a series of tasks and an engaging mini-project assignment. There were also support sessions for the students on a periodic basis. Learning needs of special individuals were considered during the delivery phase.

As an example, the first programme in the first workshop on sequencing involved moving a robot from one point to another following a simple straight trajectory. The logic developed by one of the students using block-based code is shown in Fig. 3 while the actual robot motion is illustrated in Fig. 4.

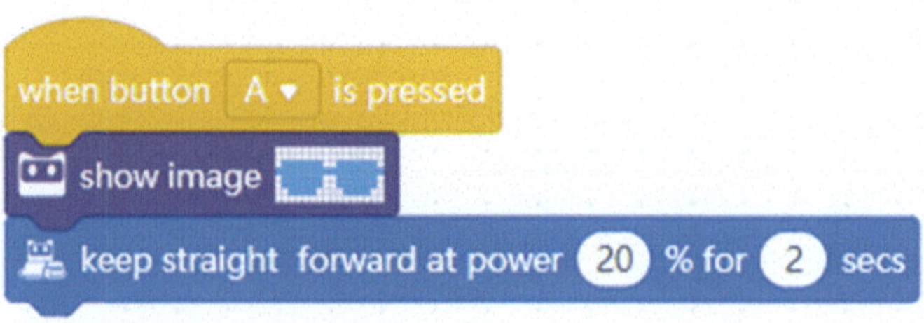

Fig. 3 'My first program' of moving the robot

(a) (b) (c)

Fig. 4 Outcome of 'My first program': **a** Starting position **b** Moving forward **c** Final position

2.4 Students' Feedback and Data Collection

Evaluation of the designed learning activities was primarily based on qualitative as well as quantitative data collected from students. To achieve this, a comprehensive questionnaire was designed and was first submitted to the college together with ethical approval formalities. The protocols for collection, storage, usage and final disposal of data were clearly outlined. The questionnaire contained ten Likert scale-based questions (Table 4) and open-ended space for general comments. It was clearly mentioned to the students that the questionnaire was not meant to evaluate their performance or to grade their participation in the activity. Instead, it was for the researchers to gauge the effectiveness of the learning activities and to improve them further.

The corresponding questions are presented in Appendix. Moreover, to collect qualitative data, the open-ended space contained students' commendations and suggestions for further improvements of the designed learning framework.

The questionnaire was anonymous and thus students were requested not to write their names or any other personal information. The participation in the questionnaire was on a voluntarily basis and the participants were given a choice to withdraw at any time without having to give any reason. Once the ethical approval was sought, feedback from questionnaire was collected at the end of each workshop session.

Table 4 Likert scale in the designed questionnaire

Please provide your feedback regarding the following questions on a scale of 0–5 detailed below:					
⓿	❶	❷	❸	❹	❺
I am not sure	Strongly disagree	Disagree	Agree to some extent	Agree to large extent	Strongly agree

3 Results

The success of this pedagogical research framework was measured via feedback received from students as well as from the college's staff about students' engagement. The objective was to investigate how EBL of robotics can foster creativity in the students.

3.1 Qualitative Feedback

Comments from the Head of Computer Science of the college:

"Students have learnt lots of skills through their completion of the sessions, namely problem solving and critical thinking skills. They have been able to improve their teamwork too as they were all working collaboratively to complete the tasks. Students can use these skills across all areas of the curriculum to improve their progress.

The activity achieved the desired outcomes very well; students were better at demonstrating the skills mentioned as a result of the sessions.

Overall, they really enjoyed it and I would like to thank you for everything you have done to support us in the delivery of the content and the enrichment you have provided for our students. Should you ever want to work with us again, we would be extremely keen to do so."

3.2 Quantitative Analysis

The effectiveness of the proposed robotic framework was further characterised by analysing the data received in response to ten questions in the designed questionnaire. Learners' responses to three important sample questions are presented in this section. In question 4, the students were asked if the workshops managed to introduce them with the modern information technology (IT) tools for problem solving. Their responses are grouped in three categories as: (i) I am not sure (ii) No (comprising of 'strongly disagree' and 'disagree responses') and (iii) Yes (comprising of 'agree to some extent', 'agree to large extent' and 'strongly agree'). Results obtained are plotted in Fig. 5, which shows that a vast majority of students considered that the framework had a positive impact in terms of learning modern IT tools including programming.

One of the distinguishing features of the proposed framework is the use of real-world examples to facilitate learners' understanding of relatively complex concepts. The efficacy of this feature was gauged in Question 6 where the students were asked to comment on the role of real-world examples in understanding technical contents related to sequencing, selection, iteration and Python programming. Results obtained

Fig. 5 Learners' response to Question 4 on getting introduced to modern IT tools for problem solving

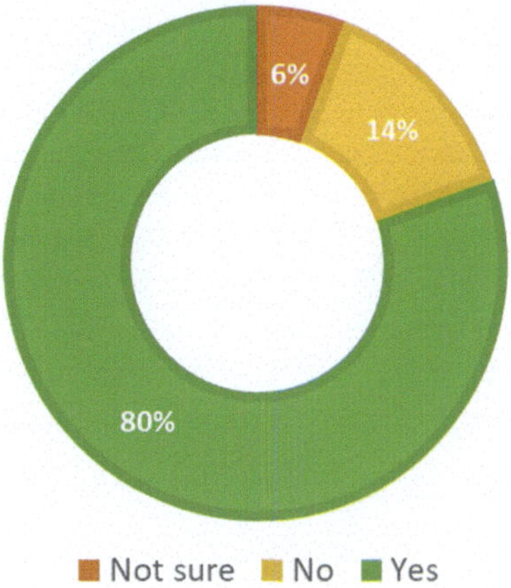

are presented in Fig. 6. 68% of the questionnaire participants agreed (from some to large extent) on the pivotal role of the examples to learn sophisticated concepts using the proposed framework. This demonstrates the potential of EBL in delivering important and potentially useful topics on programming, robotics and engineering.

As a conclusive feedback, question 10 intended to enquire overall learning experience of the students gained from each workshop in this EBL activity. They were asked if the learning activity was fun. It was observed that the students' satisfaction in the activity went on increasing as the workshops progressed. This was primarily due to the fact that each workshop was accompanied by feedback collected through

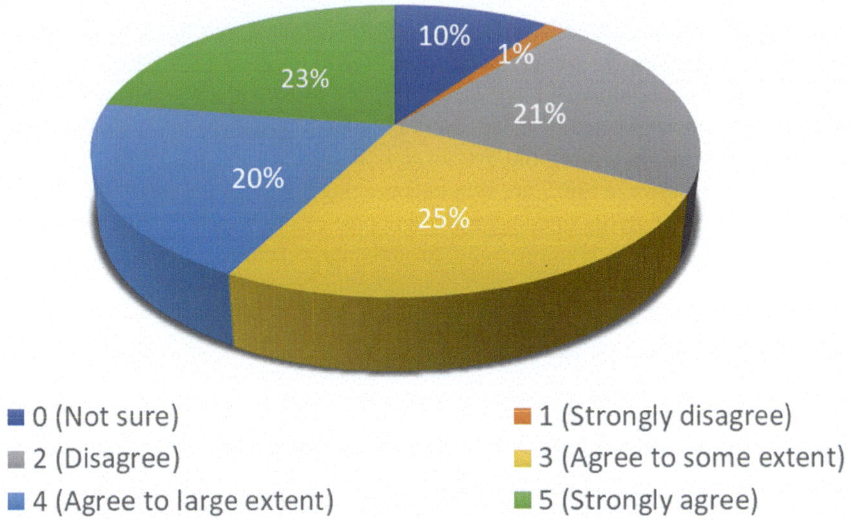

Fig. 6 Learners' response to Question 6 on satisfactory grasping of technical contents using real-world examples

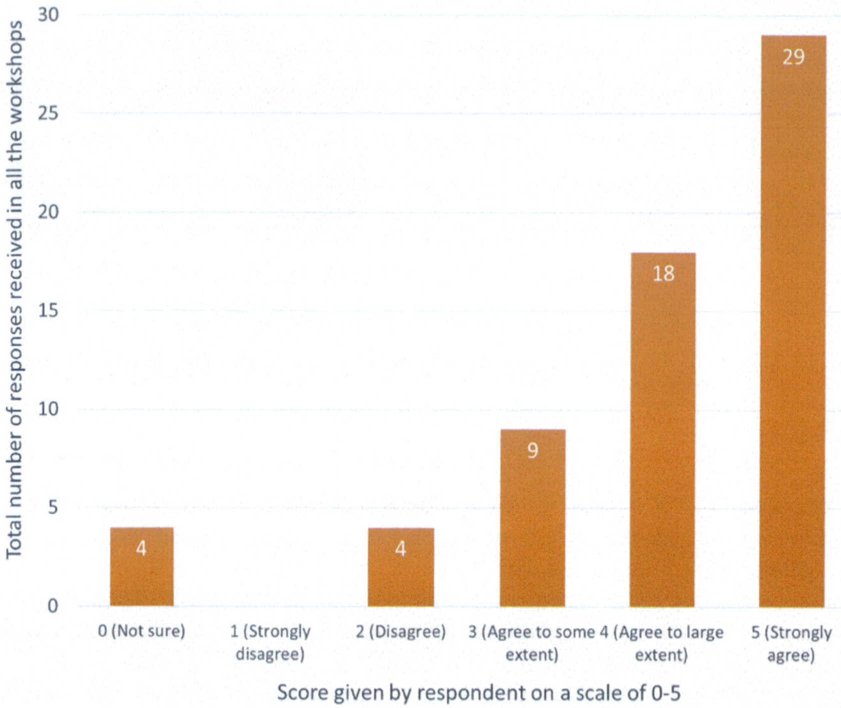

Fig. 7 Learners' response to Question 10 on overall assessment of the activity in terms of an excellent learning experience

the questionnaire and data collected from students' feedback was instantly analysed. Thus, the lessons learnt from a workshop were used to improve the subsequent workshop session. Total number of responses received corresponding to this question in all the workshops is plotted as a function of students' satisfaction level. The level was scored from 0 (not sure) and 1 ('strongly disagree') and lead to 5 ('strongly agree'). Results shown in Fig. 7 are very encouraging demonstrating that the number of responses is strongly inclined towards right side of the bar chart with maximum responses received corresponding to 5 ('strongly agree') case.

4 Conclusion

Given the interdisciplinary nature of robotics, the proposed framework explored EBL in context of developing expertise related to various fields. The success of this pilot project for an educational institution can be built upon by extending the variety of levels of technology that it covers. Led by a roboticist and a pedagogical expert, this study opens the possibility for a range of institutional engagement from local to international perspective particularly to the students with disadvantaged backgrounds. The framework developed in this project is reproducible and scalable. The nature of the learning activity demonstrated a commitment to high-quality teaching and a prompt willingness to respond to the emerging needs in an effective manner. In the near future, it is anticipated to organise a robotics competition to make the learning experience more challenging, ambitious and conclusive.

Acknowledgements This project has been supported and funded by Ferens Education Trust under FET Academic Programme with project title '*RoboPLEF—Robotics-inspired Practical Learning Framework.*'

Appendix

No.	Question	Score					
		⓿	❶	❷	❸	❹	❺
1	I was informed about the learning outcomes of the workshop	⓿	❶	❷	❸	❹	❺
2	The workshop sheet clearly explains the tasks to be performed	⓿	❶	❷	❸	❹	❺
3	The workshop meets my expectations in terms of thinking and technical aspects	⓿	❶	❷	❸	❹	❺
4	The workshop introduced me to the application of simulation techniques and modern IT tools to solve/simulate a problem of interest	⓿	❶	❷	❸	❹	❺
5	The flowcharts of the code made me more inquisitive about the underlying coding procedure and enquiry-based learning	⓿	❶	❷	❸	❹	❺
6	Real-world examples given in the workshop sheet helped me to grasp the concept of sequencing	⓿	❶	❷	❸	❹	❺
7	The workshop encouraged communication skills and a spirit of teamwork	⓿	❶	❷	❸	❹	❺
8	The workshop has opened me to new ideas and made me enthusiastic towards starting the next workshop	⓿	❶	❷	❸	❹	❺
9	The resources (hardware + software) provided were adequate	⓿	❶	❷	❸	❹	❺
10	In overall, the learning activity was fun	⓿	❶	❷	❸	❹	❺

References

1. Pedaste M, Altin H (2020) Does inquiry-based education using robots have an effect on learners' inquiry skills, subject knowledge and skills, and motivation? Int J Adv Sci Eng Inf Technol 10:1403–1409

2. Powell NJ, Hicks PJ, Truscott WS, Green PR, Peaker AR, Renfrew A, Canavan B (2008) Four case studies of adapting enquiry-based learning (EBL) in electrical and electronic engineering. Int J Electr Eng Educ 45:121–130
3. Darmawansah D, Hwang G-J, Chen M-RA, Liang J-C (2023) Trends and research foci of robotics-based STEM education: a systematic review from diverse angles based on the technology-based learning model. Int J STEM Educ 10:12
4. Ajwad SA, Asim N, Islam RU, Iqbal J (2017) Role and review of educational robotic platforms in preparing engineers for industry. Maejo Int J Sci Technol 11:17–34
5. Auerbach JE, Concordel A, Kornatowski PM, Floreano D (2018) Inquiry-based learning with RoboGen: an open source software and hardware platform for robotics and artificial intelligence. IEEE Trans Learn Technol 12:356–369
6. Blancas M, Valero C, Mura A, Vouloutsi V, Verschure PMJ (2020) "CREA": an inquiry-based methodology to teach robotics to children. In: Merdan M, Lepuschitz W, Koppensteiner G, Balogh R, Obdržálek D (eds) Robotics in education. RiE 2019. Advances in intelligent systems and computing, vol 1023. Springer, Cham
7. Zviel-Girshin R, Rosenberg N (2021) How to enhance creativity and inquiry-based science education in early childhood-robotic moon settlement project. Create Educ 12:2485–2504
8. Didier R, Pierre-Yves O, Stéphane M, Fanny R, Gordana G, Denise CMS, Francesco M (2015) IniRobot: a pedagogical kit to initiate children to concepts of robotics and computer science. In: Proceedings of the 6th international conference on robotics in education, Switzerland, pp 28–34
9. Iqbal U, Samad A, Nissa Z, Iqbal J (2014) Embedded control system for AUTAREP—a novel autonomous articulated robotic educational platform. Teh Vjesn 21:1255–1261
10. Manzoor S, Islam RU, Khalid A, Samad A, Iqbal J (2014) An open-source multi-DOF articulated robotic educational platform for autonomous object manipulation. Robot Comput-Integr Manuf 30:351–362

Application of the Negative Selection Algorithm to Detect Distributed Denial of Service Attacks

Daniel Matrone, Rodrigo P. Pasquale, and Calebe P. Bianchini

Abstract The high demand for information technology services during the COVID-19 pandemic has amplified users' susceptibility to internet security issues. Among these concerns, distributed denial of service (DDoS) attacks have emerged as a prominent threat. These attacks exploit botnets to overwhelm servers with malicious traffic, severely impacting their functionality. This paper investigates an approach utilizing the Negative Selection Algorithm, a bio-inspired computational algorithm, for the detection of DDoS attacks. Through empirical evaluation, this study assesses the detection rate of the proposed solution under varying network protocols. The results of this analysis contribute to an understanding of the feasibility and potential benefits of employing bio-inspired algorithms, such as the Negative Selection Algorithm, in fortifying internet security against the evolving landscape of DDoS attacks.

Keywords Information security · Artificial intelligence · DDoS attacks · HPC

1 Introduction

The Cyber Threats & Trends report [11] points to a 154% increase in distributed denial of service (DDoS) attacks compared to 2019. A hypothesis raised in this document for this high proportion is the increase in dependency on online platforms, due to the COVID-19 pandemic.

The biggest vector of DDoS attacks is the User Datagram Protocol (UDP)—whose main characteristic is the lack of reliability, as there is no guarantee of data integrity sent to another machine [13]. One type of attack that can occur in the UDP is the UDP Flood, which, in short, overloads the target server with a large group of UDP packets. This protocol can be used with amplification techniques where a small request can generate a large amount of traffic—for this to occur, other vectors (protocols) are used, such as Network Time Protocol (NTP) and Domain Name System (DNS) [11].

D. Matrone · R. P. Pasquale · C. P. Bianchini (✉)
Computing and Informatics Department, Mackenzie Presbyterian University, São Paulo, Brazil
e-mail: calebe.bianchini@mackenzie.br

X.-S. Yang et al. (eds.), *Proceedings of Ninth International Congress on Information and Communication Technology*, Lecture Notes in Networks and Systems 1000,
https://doi.org/10.1007/978-981-97-3289-0_31

The way to detect a distributed denial of service (DDoS) attack is to differentiate the attack traffic from normal network traffic. The purpose of a distributed denial of service attack is to overwhelm its target with requests from multiple sources, potentially hundreds or thousands of infected computers (so-called botnets).

The detection task can be automated through classification algorithms. In particular, the Negative Selection Algorithm (NSA), a bio-inspired algorithm that presented a detection rate of 99% for classifying faults in non-deterministic builds [3], can be an effective tool to differentiate traffic from a network from an attack of normal traffic.

Cybersecurity is essential, even more with the usage of new information technology services in an fast growing economy. There are several types of threats with different complexities spreading on the internet but distributed denial of service attacks are simple and effective in restraining users of using web-based services across a large number of compromised machines, so-called botnets.

Distributed denial of service attacks can be implemented at the network, transport, and application layers using different protocols such as UDP, DNS, and NTP. Researchers have sought solutions based on machine learning, knowledge, and statistical detection to detect and defend against this type of threat; however, some methods are computationally expensive and others are not so effective.

In this sense, this study arises whose focus is to study, analyze, adapt, and develop a Negative Selection Algorithm to detect distributed denial of service attacks using an extensive database with simulations of attacks in different protocols, from different layers of the network.

In this context, the question that will be answered in this research is: what is the performance of the NSA when applied to a different situation, the detection of distributed denial of service attacks?

2 Network Protocols

Network protocols are responsible for defining the format, the order of sending and receiving messages between machines, and the actions performed [13]. All network communication is governed by protocols.

The protocols that will be studied are UDP, DNS, and NTP. The former is part of the transport layer—responsible for allowing computers to communicate—while the latter is part of the application layer—that is, they are higher-level protocols, such as FTP, SMTP, and TELNET.

In this paper, the most important features have been selected as described by [12]. The authors pointed out 80 traffic features from the database using the *CICFlowMeter* software and applied *RandomForestRegressor* to extract the most important ones for a DDoS attack detection. Then, they also used the ID3, random forest, Naïve Bayes, and logistic regression machine learning algorithms, observing three metrics used to evaluate the accuracy of the most important features: precision, sensitivity, and F-measure.

Table 1 Description of the most important attributes of the UDP [13]

Attribute	Description
Destination Port	Destination port
Fwd Packet Length Std	Standard deviation size of packet in forward direction
Packet Length Std	Standard deviation size of packet
min_seg_size_forward	Minimum segment size observed in the forward direction
Protocol	Protocol

2.1 *UDP*

A connectionless transport protocol is UDP—or User Datagram Protocol, through which applications send encapsulated datagrams without a connection being established [13]. UDP transmits segments that consist of an 8-byte header, which contains two ports—which are the machines' origin and destination points. It is not as reliable as TCP as it does not guarantee packet delivery and integrity. That is because it does not have the Three-Way Handshake, where the client sends a packet to the server, the server sends another packet to the client confirming the receipt of the first one, and finally, the client responds with the last packet acknowledging receipt of the server's response [13].

In the database used—which is discussed in Sect. 3—[12] listed the following most important features, which are described in Table 1, for the UDP: *Destination Port*, *Fwd Packet Length Std*, *Packet Length Std*, *min_seg_size_forward*, and *Protocol*.

2.2 *DNS Protocol*

An essential part of the application layer, the Domain Name System (DNS) protocol is used to map a name to an IP address. According to [13], "DNS is a hierarchical domain-based naming scheme and a distributed database system to implement this naming scheme."

The DNS protocol translates domain names to IP addresses and is based on a system of queries, which are passed hierarchically through a tree. The server then sends a response down the tree to the requestor with a DNS response [13].

This protocol, according to [5, 12], has the following most important features: *Max Packet Length*, *Fwd Packet Length Max*, *Average Packet Size*, and *Min Packet Length*. Each of these characteristics is described in Table 2.

Table 2 Description of the most important attributes of the DNS protocol according to [5, 12]

Attribute	Description
Max Packet Length	Maximum size of packet
Fwd Packet Length Max	Maximum size of packet in forward direction
Average Packet Size	Average size of packet
Min Packet Length	Minimum size of packet

Table 3 Description of the most important attributes of the NTP according to [5, 12]

Attribute	Description
Subflow Fwd Bytes	The average number of bytes in a sub flow in the forward direction
Length of Fwd Packets	Total size of packet in forward direction
Fwd Packet Length Std	Standard deviation size of packet in forward direction
min_seg_size_forward	Minimum segment size observed in the forward direction
Flow IAT Min	Minimum time between two packets sent in the flow

2.3 NTP

The Network Time Protocol (NTP) is responsible for synchronizing network clocks using a distributed set of clients and servers. These servers have a hierarchical topology, divided into layers and numbered from 0 to 16 [9].

According to [5, 12], the NTP has the five most important features in the database: *Subflow Fwd Bytes*, *Length of Fwd Packets*, *Fwd Packet Length Std*, *min_seg_size_forward*, and *Flow IAT Min*. Table 3 explains each of these attributes.

2.4 Distributed Denial of Service Attacks

A denial of service attack is an attack that aims to cause a service on the network to malfunction. Network resources have a limit of simultaneous requests, so this attack sends multiple requests to this service to reach this limit, causing the resources to be unable to handle the requests. This causes service slowdowns and ignored requests, where users will not be able to use this service. In this case, the attacker can exploit the UDP, where UDP packets are sent to random or specific ports on the victim's system [6]. Another way this attack can work is by sending malformed packets to the service, which would confuse the application or a protocol on the victim's machine, causing it to freeze or restart [10].

To execute the requests and send the packages, it is necessary to recruit several machines to reach the request limit. To do this, the intruder performs a scan for security flaws to gain access to remote machines. If a flaw is found, the attacker exploits it to place the malicious code and form a network of infected machines in his favor, where this network is known as a botnet. Another way to infect machines is through Trojans, which are malicious software disguised as a useful application [10].

2.5 Negative Selection Algorithm

Artificial Immune Systems (AIS) can be defined as computer systems inspired by theoretical immunology and observed immune functions, as well as their principles and models, to solve problems [2].

One of the main techniques of Artificial Immune Systems, the Negative Selection Algorithm, is inspired by the mechanism of self/non-self-discrimination of immune systems. This mechanism, in essence, is the detection of anomalies belonging to a known environment. On the immune system in nature, thymus T cells are trained to recognize antigens—which would be the *non-self* or the abnormality—and prevent them from detecting their cells, the so-called *self* [1, 4, 7]. During this process, three actions can occur:

- non-selection, that is, the absence of affinity between the T cell and the peptide;
- positive selection in which the affinity between the T cell and the peptide causes T cell activation;
- negative selection in which the affinity between the T cell and the peptide causes the T cells to energy;

The Negative Selection Algorithm (NSA) essentially consists of three steps: (i) the definition of the algorithm itself, that is, the training; (ii) the generation of detectors; and (iii) the monitoring of the occurrence of anomalies [1]. The first two can be combined in a single step: the generation and training of the detectors. The objective of the NSA is to generate a set of detectors capable of recognizing anomalies, that is, patterns different from those trained. In Fig. 1, each detector, the black dots, has

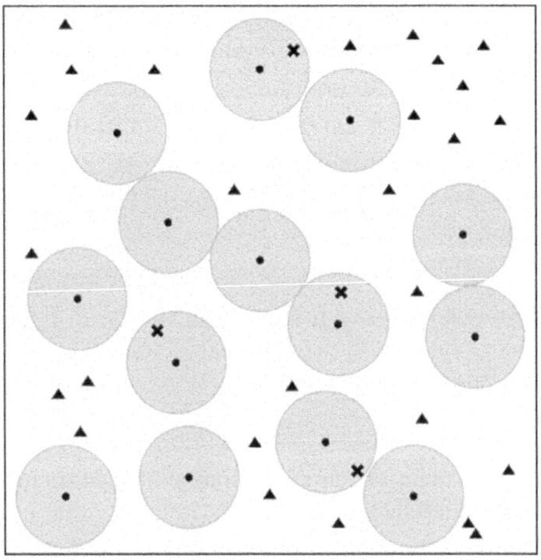

Fig. 1 Graphical representation of detectors in an environment [3, 4]

a predefined radius, which is represented by the gray circle. The triangles represent the non-anomalous objects, and the "×'s" indicates anomalous objects.

The training stage consists of generating detectors randomly and eliminating those that detect objects of their own, such as other detectors.

3 Methodology

This paper presents an extension of [3, 8]'s work, evolving their findings for distributed denial of service (DDoS) attack detection. The approach employed for the Negative Selection Algorithm here is similar to [3]'s work, using it as a foundation for the current investigation. The experimental dataset used in this paper is proposed by [12]. This dataset simulates several taxonomies of DDoS attack, and the data originates from the *CICFlowMeter*, a network analyzer. Initial experiments started with an exploratory analysis of the dataset. Several tools, such as Python 3.7, Pandas 1.3.4, and Scikit Learn 1.0, were utilized for feature examination, following the importance criteria established by [12]. Irrelevant columns, *null* values, missing data, and infinite values were analyzed, and some data normalization was performed using the *Max-Min* technique, available in the Scikit Learn library.

Some protocols used for a DDoS attack were analyzed, such as DNS, NTP, and UDP. In particular, the features proposed by [12] exhibited strong correlations when analyzing the DNS protocol, as shown in Table 4. For instance, *FwdPacketLength-Max* displayed significant correlations with all other attributes in the dataset, which can affect Negative Selection Algorithm's performance. Furthermore, when analyzing the NTP, the features of the dataset showed lower inter-feature correlations, as shown in Table 5. On the other hand, the UDP indicated two columns with strong correlation: *FwdPacketLengthStd* and *PacketLengthStd*, as showed in Table 6. For the latter protocol, the NSA was executed in two situations: the first one considers all features during execution; and the other, *FwdPacketLengthStd* attribute was removed. Surprisingly, this adjustment resulted in negligible changes to the Detection Rate (DR) but a slight improvement in Execution Time (TE). Due to these results, further experiments consider only four of the five features proposed by [12]. Upon completing the analyses, each dataset was partitioned into three comma-separated value (CSV) files: one for model training and two for testing. Moreover, a configuration file was generated, defining parameters such as *Problem Size*, *Max. Detectors*, *Min. Distance*, *Runs*, *Training CSV File*, *Testing CSV File*, and *Expected Detected* indices.[1]

During this analysis, a correlation map was generated to display the dependency between existing features, as shown in Figs. 2, 3, and 4.

[1] More details of this configuration file can be seen on the GitHub at https://github.com/hpc-fci-mackenzie/FlakyBuild.

Table 4 Description of the normalized features of most importance of the DNS protocol

	MaxPacketLength	FwdPacketLengthMax	AveragePacketSize	MinPacketLength
Quantity	5,074,413	5,074,413	5,074,413	5,074,413
Mean	0.041039	0.042900	0.775700	0.935874
Standard deviation	0.008952	0.009339	0.169425	0.204135
Minimum	0.000000	0.000000	0.000000	0.000000
25%	0.043597	0.045579	0.824500	0.994565
50%	0.043836	0.045828	4.829005	1.000000
75%	0.043836	0.045828	0.829005	1.000000
Maximum	1.000000	1.000000	1.000000	1.000000

Table 5 Description of the normalized features of most importance of the NTP

	DestinationPort	FwdPacketLengthStd	PacketLengthStd	min_seg_size_forward	Protocol
Quantity	3,136,802	3,136,802	3,136,802	3,136,802	3,136,802
Mean	0.507634	0.012034	0.011206	0.862548	0.99455
Standard deviation	0.284607	0.012726	0.012405	0.189864	0.019151
Minimum	0.000000	0.000000	0.000000	0.000000	0.000000
25%	0.261830	0.000000	0.000000	0.999999	1.000000
50%	0.508034	0.017193	0.016323	0.999999	1.000000
75%	0.754192	0.026798	0.024694	0.999999	1.000000
Maximum	1.000000	1.000000	1.000000	1.000000	1.000000

Table 6 Description of the normalized features of most importance of the UDP

	TotalLengthofFwdPackets	FwdPacketLengthStd	min_seg_size_forward	FlowIATMin
Quantity	1,217,007	1,217,007	1,217,007	1,217,007
Mean	0.156716	0.018488	0.994047	0.000009
Standard deviation	0.159732	0.032090	0.066793	0.001838
Minimum	0.000000	0.000000	0.000000	0.000000
25%	0.030000	0.000000	1.000000	0.000000
50%	0.095000	0.000000	1.000000	0.000000
75%	0.235000	0.030156	1.000000	0.000000
Maximum	1.000000	1.000000	1.000000	1.000000

Fig. 2 Correlation map for the DNS protocol

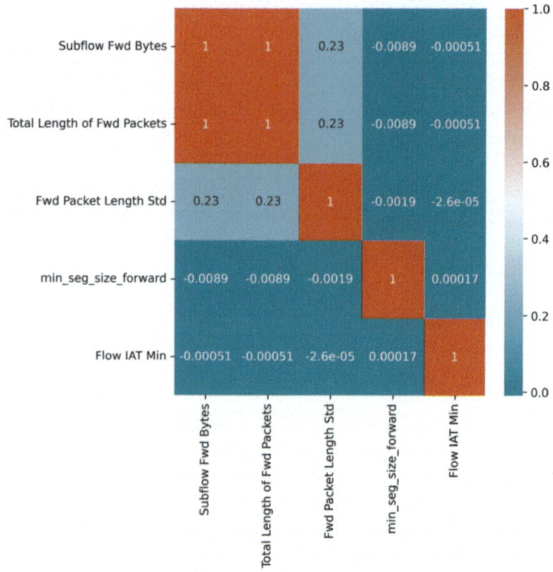

Fig. 3 Correlation map for the NTP

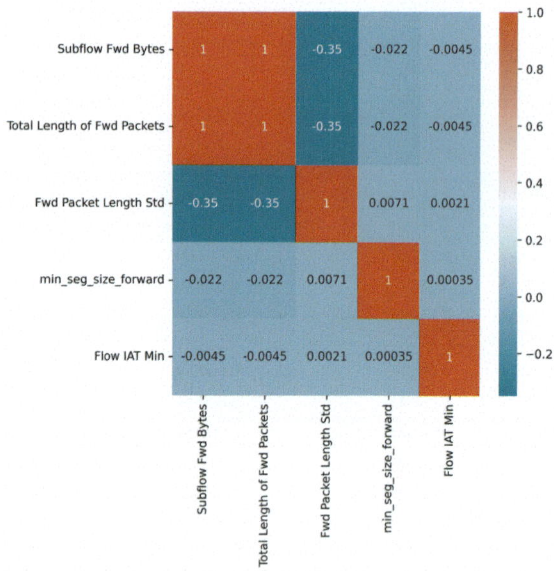

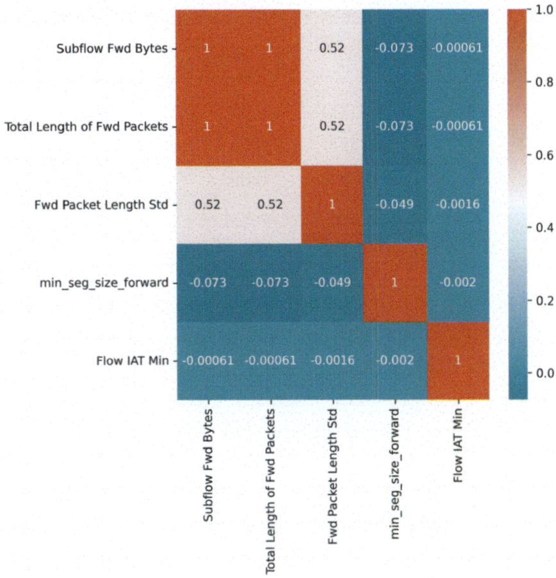

Fig. 4 Correlation map for the UDP

4 Results

The Negative Selection Algorithm (NSA), developed in C++, was revised and memory consumption improved through memory leak corrections, using Valgrind 3.15 software for code debugging. The version used was the one proposed by [8] in which the code obtained a speedup of 4.4 when compared to the previous version. An NSA Python 3.7 solution was developed but discarded due to the significantly low performance when compared to the solution proposed by [8].

All registers in the databases have their content labeled in two distinct categories: DDoS or BENIGN—the former refers to network traffic in a DDoS attack; the latter refers to a normal network traffic. Some other parameters must be configured to run the C++ solution, such as the *detection radius*, the *number of detectors*, and the *dimension* (to evaluate the NTP, the dimension used was 5).

To train the algorithm, a database was set up with only the normal network traffic elements. The test used the full database with both categories: DDoS and BENIGN elements.

These experiments were performed on MackCloud[2] with the following specifications: 2 × Intel® Xeon® CPU E5-2650 v4 @ 2.20 GHz 12 c/24 t and 128 Gb RAM. Each protocol was analyzed using the NSA solution. Both the detection rate and execution time were observed to understand the efficiency and performance of the algorithm applied to DDoS attacks.

[2] Multidisciplinary Center for Scientific and Cloud Computing at https://mackcloud.mackenzie.br/.

Table 7 Number of detectors (QD), detection radius (RD), detection rate (DR), false alarm rate (FAR), and elapsed time (TE)

Protocol	QD	RD	DR (%)	FAR (%)	TE (s)
UDP	64,000	0.5	97.8	0.1	58.6
NTP	128,000	0.05	47.5	11.3	45.3
DNS	4,096,000	0.05	59.9	18.5	15,139

The number of detectors used was: QD = 64,000, 128,000, 256,000, 512,000, 1,024,000, 2,048,000, and 4,096,000, while the detection radius was RD = 0.05, 0.25, 0.5, 0.75, 1.0, 1.25, and 1.5. The results shown in Table 7 are the average of 10 runs for each protocol. Each run combines *number of detectors* (QD) and *detection radius* (RD) with the highest *detection rate* (DR).

The NSA solution used to deal with UDP showed a high detection rate of DDoS attacks. On the other hand, when compared to the other two protocols, the detection rate was lower than expected, considering [3] as a baseline. The current results showed a rate of high *false alarm* (FAR) when compared to the result of the UDP.

5 Conclusion

This paper presented results of the Negative Selection Algorithm (NSA), inspired by the immune system, used to detect DDoS attacks. Each experiment varies some known parameters, such as neighborhood radius and a number of detectors, to increase the detection rate of a DDoS attack. As expected, when using the NSA to detect a DDoS attacks in the UDP, the results are a high detection rate, near 97.8%, and a low false alarm rate, close to 0.1%.

Despite these good results, when using the same algorithm on the other two protocols, the detection rate was lower and the false alarm rate was higher, an shown in Table 7.

Some future works were identified when analyzing the results and are listed below:

- deeply analyze the network protocols features to understand better the differences between detection rates (specially NTP, that runs over UDP);
- improve the performance of the chosen NSA solution; despite the speedup described at [8], it is still proved to be a problem;
- change some parameters at the NSA algorithm, such as distance measures (e.g., euclidean distance to any other distance; or the geometry of the detector, from circle to square, among others [4]).

Acknowledgements The authors thank the MackCloud (https://mackcloud.mackenzie.br/), the Multidisciplinary Center for Scientific and Cloud Computing at Mackenzie Presbyterian University for their support. Funding: SPRACE—Grant #2018/25225-9, Sao Paulo Research Foundation (FAPESP).

References

1. Dasgupta D, Ji Z, Gonzalez F (2003) Artificial Immune System (AIS) research in the last five years. In: The 2003 congress on evolutionary computation, 2003. CEC '03, vol 1, pp 123–130. https://doi.org/10.1109/CEC.2003.1299565
2. de Castro LN, Von Zuben FJ (1999) Artificial immune systems: Part I—basic theory and applications. Technical Report RT DCA 01/99, Departamento de Engenharia de Computação e Automação Industrial (DCA) da Faculdade de Engenharia Elétrica e de Computação (FEEC) da Unicamp
3. de Lima Costa JC, de Castro LN, de Paula Bianchini C (2019) Sensitivity analysis of the negative selection algorithm applied to anomalies identification in builds. In: 2019 XLV Latin American Computing Conference (CLEI), pp 1–6. https://doi.org/10.1109/CLEI47609.2019.235087
4. Gupta KD, Dasgupta D (2022) Negative selection algorithm research and applications in the last decade: a review. IEEE Trans Artif Intell 3(2):110–128. https://doi.org/10.1109/TAI.2021.3114661
5. Habibi Lashkari A, Draper Gil G, Mamun MSI, Ghorbani AA (2017) Characterization of tor traffic using time based features. In: Proceedings of the 3rd international conference on information systems security and privacy—ICISSP, pp 253–262. INSTICC, SciTePress (2017). https://doi.org/10.5220/0006105602530262
6. Hussain SM, Beigh GR (2013) Impact of ddos attack (udp flooding) on queuing models. In: 2013 4th International Conference on Computer and Communication Technology (ICCCT), pp 210–216. https://doi.org/10.1109/ICCCT.2013.6749629
7. Ji Z, Dasgupta D (2009) V-detector: an efficient negative selection algorithm with "probably adequateâŁž detector coverage. Inform Sci 179(10):1390–1406. https://doi.org/10.1016/j.ins.2008.12.015
8. Matrone D, Tsuneki R, Pasquale R, Bianchini C (2021) Estudo do desempenho do algoritmo de seleção negativa aplicado à detecção de builds não-determinísticas. In: Anais da XII Escola Regional de Alto Desempenho de São Paulo, pp 53–56. SBC, Porto Alegre, RS, Brasil. https://doi.org/10.5753/eradsp.2021.16704
9. Mills D (1992) Network time protocol (version 3) specification, implementation and analysis. Technical report, RFC Editor. https://www.rfc-editor.org/rfc/rfc1305
10. Mirkovic J, Reiher P (2004) A taxonomy of ddos attack and ddos defense mechanisms. SIGCOMM Comput Commun Rev 34(2):39–53. https://doi.org/10.1145/997150.997156
11. Neustar: Cyber threats & trends: securing your network pandemic-style. Technical report, Neustar, Inc. (2020). https://www.cdn.neustar/resources/whitepapers/security/neustar-cyber-threats-trends-2020-report.pdf
12. Sharafaldin I, Lashkari AH, Hakak S, Ghorbani AA (2019) Developing realistic distributed denial of service (ddos) attack dataset and taxonomy. In: 2019 International Carnahan Conference on Security Technology (ICCST), pp 1–8. https://doi.org/10.1109/CCST.2019.8888419
13. Tanenbaum AS, Wetherall D (2011) Computer networks, 5th edn. Pearson Prentice Hall

PharmaSys: Toward Preventing Prescription Misuse Using a HIPAA-Compliant Blockchain Protocol

Aaron H. Nguyen, Alex Lewtschuk, James Lucas Durham, Peyton Lundquist, and Gaby G. Dagher

Abstract Prescription drug misuse has become a major health crisis across the United States with prescription drug abuse skyrocketing during the twenty-first century resulting in classification as an epidemic. The rise in prescription misuse has resulted in negative consequences including heightened instances of overdose deaths, increased crime rate, and escalated healthcare expenditures. Although illegal sales contribute to the issue, a notable portion of prescription drugs stems from licensed medical professionals who engage in over-prescribing medications. To help combat this issue we propose PharmaSys, a novel multi-party blockchain system with a quorum-based consensus protocol that aims to prevent prescription misuse. PharmaSys achieves this by introducing and establishing a checks and balances system for prescription transactions, all while maintaining compliance with HIPAA regulations. PharmaSys consists of three distinct node types: prescribers, pharmacists, and patients; each of which executes unique functions within the system. We have implemented PharmaSys to illustrate our system's scalability and security.

Keywords Blockchain · Prescription tracking · Healthcare · HIPAA compliance · Consensus protocol · Drug abuse · Proof of sharding

---

A. H. Nguyen (✉) · A. Lewtschuk · J. L. Durham · P. Lundquist · G. G. Dagher
San Jose State University, San Jose, CA 95192, USA
e-mail: aaron.nguyen03@sjsu.edu

A. Lewtschuk
e-mail: alexanderlewtsch@u.boisestate.edu

J. L. Durham
e-mail: durham35@marshall.edu

P. Lundquist
e-mail: peytonlundquist@u.boisestate.edu

G. G. Dagher
e-mail: gabydagher@boisestate.edu

Marshall University, Huntington, WV 25755, USA

Boise State University, Boise, ID 83725, USA

X.-S. Yang et al. (eds.), *Proceedings of Ninth International Congress on Information and Communication Technology*, Lecture Notes in Networks and Systems 1000,
https://doi.org/10.1007/978-981-97-3289-0_32

1 Introduction

Over the last two decades, drug abuse in the United States has posed a major societal issue that has grown at an alarming rate. The increase of improper drug use has led to countless lost lives, leaving families, friends, and communities devastated. In the United States, drug overdoses accounted for over 100,000 fatalities in 2021, a figure that demonstrates the urgent need for measures to mitigate the frequency of such preventable deaths [1]. The unfortunate truth is that not all of these deaths are a result of illegal sources; a significant portion can be traced back to prescription drugs dispensed by licensed medical practitioners [2]. Drug misuse is defined as the use of medication in a manner inconsistent with the prescriber's directions. This includes taking someone else's prescription or taking medication to feel euphoric. A closer analysis reveals that the misuse of prescription drugs has left an impact on a wide array of demographics across the United States. A 2021 survey conducted by the U.S. Department of Health and Human Services reported more than 35 million instances of prescription drug misuse among individuals ages 12 and above. Prescription drugs involved in these instances included pain relievers such as oxycodone, codeine, and opioids, stimulants like Adderall and Ritalin, and sedatives including Xanax and Valium [3]. The repercussions of prescription drug abuse extend beyond individual impact and encompass the United States as a whole. According to a report issued by the U.S. Department of Health and Human Services, illicit drug use imposes an economic burden of $193 billion every year [4]. Concurrently, healthcare systems are heavily strained due to the extensive allocation of resources for treatment and rehabilitation of drug abuse victims. Furthermore, various societal consequences such as heightened crime rates and fractured family relationships create complex issues that require concerted efforts to address and minimize their impact [5, 6].

Prescription drug misuse presents an intricate problem due to the complex challenges faced by healthcare providers in creating the most fitting prescription and dosage for each patient. The process of formulating a prescription involves careful consideration of multiple patient-specific factors. These can include the patient's present medical condition, past health history, allergies, age, physiological state, and personal preferences or requests. In addition to the challenges involved in formulating a prescription that aligns with each patient's needs, healthcare providers often face external influences including potential incentives from pharmaceutical companies [7]. This conflict of interest may lead some doctors to alter their decisions by pushing out certain medications or more medications in general. In 2015, data from Open Payments, a publicly accessible database of payments from drug companies to physicians, revealed that 48% of physicians were reported to have received a total of $2.4 billion in industry-related payments [8]. This suggests that physician prescriptions are being affected by pharmaceutical-related interests, which could be a factor in the increase of drug abuse.

Despite this, even well-intentioned doctors tend to over prescribe medications to their patients. A study done in 2016 by the National Safety Council found that 99% of the doctors surveyed were prescribing highly addictive opiates for longer than the three-day period recommended by the Centers for Disease Control and Prevention [9]. This finding reveals the widespread prevalence of over-prescribing practices in the healthcare industry and highlights the urgent need for intervening and addressing the issue.

A concerning aspect of prescription drug misuse is the lack of accountability for doctors. Their decisions are often final, allowing them to prescribe drugs without any secondary opinions. Doctors typically aim for short-term improvements in patients' health; however they may overlook long-term implications of drug abuse and prescribe more than they knowingly should. In addition, the absence of monitoring and accountability enables "pill mill" doctors, who are doctors that accept cash in return for written prescriptions, to engage in under-the-table practices without facing instant repercussions [10]. Consequently, due to over-prescription these unneeded drugs may find their way onto the streets, posing a potential danger to society. Moreover, the surplus of medications may be exploited to fulfill individuals' desires for euphoria, further contributing to the cycle of misuse.

One of the challenges encountered in our solution is establishing the validation criteria for prescription transactions within the blockchain system. Unlike traditional blockchain systems that focus on verifying sufficient funds in the sending wallet, our system must verify the accuracy of dosage and appropriateness of the drug type in each prescription transaction. This validation process posed a significant challenge and required careful consideration. Furthermore, operating within the health industry introduces additional complexities, particularly concerning the handling of sensitive protected health information (PHI). The Health Insurance Portability and Accountability Act (HIPAA) imposes strict regulations and guidelines on the privacy and security of individuals' health records. Adhering to HIPAA regulations while ensuring seamless functionality presented an obstacle that required meticulous attention.

The current landscape of prescription drug solutions primarily revolves around the development of tracking and management systems. Tracking systems have been designed to establish transparency and accountability by leveraging blockchain technology to maintain an immutable record. Thus, any suspicious transaction will be recorded on the blockchain for others to see and potentially take action upon. On the other hand, management systems offer patient-provider confidentiality and help facilitate seamless data exchange by reinforcing the design of patient-owned data. Both of these solutions contribute to the overall enhancement of prescription drug practices, but do not aim to prevent it.

1.1 Contributions

The contributions of this paper are as follows:

(1) We introduce PharmaSys, a blockchain-based prescription tracking system that aims toward prevention of drug misuse by creating a checks and balances system that allows for validation and tracking in a secure and scalable way.

(2) We introduce *Proof of SHarding (PoSH)*, a novel quorum-based consensus protocol that requires prescribers and pharmacists to review and vote on prescription transactions. Patients actively participate in staking which allocates their computational power for transaction verification through shard assignment.
(3) We design PharmaSys to be HIPAA-compliant by identifying and adhering to 14 HIPAA design principles relevant to prescription tracking.
(4) We implemented PharmaSys using the open-source BlueChain network as a platform [11]. Our experimental results illustrate that our system maintains robustness in the presence of malicious nodes and is scalable in relation to network size.

2 Related Work

There have been several research papers addressing blockchain solutions, specifically in the healthcare industry [12–18]. Many of them dive into healthcare management systems looking to utilize blockchain smart contracts to manage and organize connections between parties. Others look to blockchain for its features of security and immutability for monitoring and tracking [12–18] (Table 1).

Various systems leverage smart contracts to enhance security, efficiency, and validation [12, 14–16]. While certain models utilize these for prescription tracking, others propose applications such as comprehensive networks for Electronic Health Records (EHRs) or data management systems. All work can be narrowed down to three primary goals: efficiency, security, and validation.

Optrak [12] is an Ethereum-based DApp that enhances the tracking of distributed opioids through interoperability, secure authentication, control of personal information, and the automation of record submission. In comparison with Optrak we propose a solution toward preventing all prescription misuse instead of focusing on only opiates. VigilRx proposes a patient-centric solution with a primary emphasis

Table 1 Comparative evaluation of related work

Title	Consensus	Eff.	Scal.	Inter.	HIPAA	Priv.	Sec.
Optrak	PoW	✓		✓			✓
FHIRChain	PoW	✓	✓	✓			✓
MedRec	PoW			✓			✓
SecureRx	PoS	✓		✓	✓	✓	✓
VigilRx	PoS	✓	✓	✓	✓		✓
Ancile	Quorum	✓	✓	✓	✓	✓	✓
ACCORD	Quorum		✓				✓
PharmaSys	Quorum		✓	✓	✓	✓	✓

on interoperability, scalability, and permissioned node types [16]. Both Optrak and VigilRx utilize smart contracts on the Ethereum network.

FHIRChain [13] is a blockchain-based architecture that is designed for clinical data sharing across a wide range of health IT systems. Their proposed solution includes a DApp and the structure and functionality for their blockchain network.

MedRec [14] deals with utilizing Ethereum smart contracts to create intelligent copies of existing medical records stored in network nodes. They employ three smart contract types to achieve this functionality.

The solution proposed by SecureRx [15] proposes their solution by addresses the handling of raw healthcare data, utilizing blockchain technology in supply chain and data management aspects, and incentivizing stakeholders on the application. Important solutions proposed by SecureRx to solve this issue include drug provenance, a recall management system, and cloud storage.

Ancile [17] offers a solution designed to handle EHRs in compliance with HIPAA regulations, featuring innovative approaches to enhance privacy and interoperability. They also suggest a strategy to prevent node collusion on a permissioned network, thereby reducing technical challenges within a decentralized system.

In ACCORD [18] G. D. Bashar et al. propose a scalable consensus mechanism utilizing a group of leaders to distribute responsibilities to multiple quorum members. This new mechanism is proposed to avoid fork conflicts, and each block must be asynchronously signed by a majority of network nodes before acceptance to the chain.

3 Preliminaries

3.1 HIPAA

The protection of patient health information is a constant priority in healthcare—it ensures patient privacy, builds trust with healthcare providers, and safeguards sensitive data from unauthorized access. In recognition of this crucial aspect of healthcare, the Health Insurance Portability and Accountability Act of 1996 (HIPAA) was adopted to establish national standards for the privacy and security of sensitive PHI of patients. This act applies to all covered entities. Defined by HIPAA, covered entities are health plans, healthcare clearinghouses, and healthcare providers who electronically manage any health information [19]. HIPAA addresses everything from unauthorized access; and data breaches, to identity theft in the healthcare sector. Compliance with HIPAA regulations fosters a culture of accountability among healthcare providers, increasing patients' confidence that their health information is being handled properly.

3.2 Blockchain

The concept of blockchain in its modern use case was introduced in 2008 with the publication of a whitepaper titled "Bitcoin: A Peer-to-Peer Electronic Cash System" by Satoshi Nakamoto [20]. The whitepaper outlined the theory of a decentralized digital currency, Bitcoin, and introduced blockchain's first practical application that would enable secure and transparent transactions without the need for intermediaries. In 2009, the implementation of the Bitcoin network was subsequently developed and launched.

3.3 Quorum-Based Consensus

Quorum-based consensus has emerged as a popular mechanism for achieving concurrence among a select number of nodes in a distributed system. In this protocol, a specific number of nodes are selected, known as the quorum, to participate in the validation process of transactions. The process of quorum selection can vary significantly, and quorum members can be chosen based on randomness, reputation, stake, and/or other factors.

An essential consideration in the design of any consensus protocol is fault tolerance, which refers to the system's integrity when facing malicious behavior. Often based on the principles of Byzantine fault tolerance (BFT), it is typically assumed that a quorum consensus can tolerate up to 33% faulty nodes on the network. Thus, if the amount of malicious nodes exceeds this number, the system cannot guarantee consistent agreement on decisions.

Quorum-based consensus also offers key features of scalability and performance. Based on the specific requirements of the system, quorum size can be adjusted, allowing for efficient scaling as participants increase. By electing only a subset of nodes for the decision-making process, the system can achieve faster consensus times and lower communication overhead, enhancing the overall performance of the system.

3.4 BlueChain

BlueChain [11], a distributed and decentralized blockchain network, operates as a research framework. Developed in Java, this network utilizes a quorum-based consensus and enables facilitating the creation of test environments with ease [11]. The system, developed at Boise State University by Peyton Lundquist, provides a solid foundation for implementation. We chose to build upon the BlueChain framework due to its ability to be rapidly modified and enable experimentation with our proposed innovations (Table 2).

Table 2 Here we propose HIPAA 14, which represents 14 HIPAA rules/standards that our system is compliant with

HDP#	Rule	Description
HDP1	Privacy (right of access)	An individual has a right of access to inspect and obtain a copy of protected health information about the individual
HDP2	Privacy (right of access, timely action)	The covered entity must act on a request for access no later than 30 d after receipt of the request
HDP3	Privacy (right of access, manner of access)	If an individual's request for access directs the covered entity to transmit the copy of protected health information directly to another person designated by the individual, the covered entity must provide the copy to the person designated by the individual
HDP4	Privacy (effect of prior authorizations)	A covered entity may use or disclose protected health information pursuant to an authorization or other express legal permission obtained from an individual
HDP5	Privacy (de-identification of protected health information)	Health information that does not identify an individual and with respect to which there is no reasonable basis to believe that the information can be used to identify an individual
HDP6	Security (access authorization)	Implement policies and procedures for granting access to electronic protected health information, for example, through access to a workstation, transaction, program, process, or other mechanism
HDP7	Security (access control)	Implement technical policies and procedures for electronic information systems that maintain electronic protected health information to allow access only to those that have been granted access rights
HDP8	Security (audit control)	Implement hardware, software, and/or procedural mechanisms that record and examine activity in information systems that contain or use electronic protected health information
HDP9	Security (integrity controls)	Implement security measures to ensure that electronically transmitted electronic protected health information is not improperly modified without detection untill disposed of
HDP10	Security (encryption and decryption)	Implement a mechanism to encrypt and decrypt electronic protected health information
HDP11	Security (transmission)	Implement technical security measures to guard against unauthorized access to electronic protected health information that is being transmitted over an electronic communications network
HDP12	Security (protection from malicious software)	Procedures for guarding against, detecting, and reporting malicious software
HDP13	Security (unique user identification)	Assign a unique name and/or number for identifying and tracking user identity
HDP14	Security (data backup plan)	Establish and implement procedures to create and maintain retrievable exact copies of electronic protected health information

4 HIPAA Design Principles

HIPAA compliance is a critical aspect for any entity operating in the healthcare sector. In our endeavor to adhere to HIPAA regulations, we have conducted a comprehensive analysis of the regulation and identified a set of rules our blockchain system complies with, as shown in Table 2. This table will later be referenced through citation blocks to indicate the exact rule/rules the design principle abides by.

5 Solution: PharmaSys

5.1 Solution Overview

PharmaSys is a prescription tracking system, implementing a novel three-party quorum-based consensus protocol, PoSH, that works toward preventing malicious prescriptions while conforming to HIPAA regulations. Our system utilizes role-based nodes that define the actions undertaken in the system, as shown in Fig. 1. Each group has differing permissions and capabilities. Prescribers and pharmacists will verify the authenticity of prescriptions and push work out to patients "staked" in the system. PharmaSys provides a solution where nodes can work cohesively while providing incentive to all parties involved.

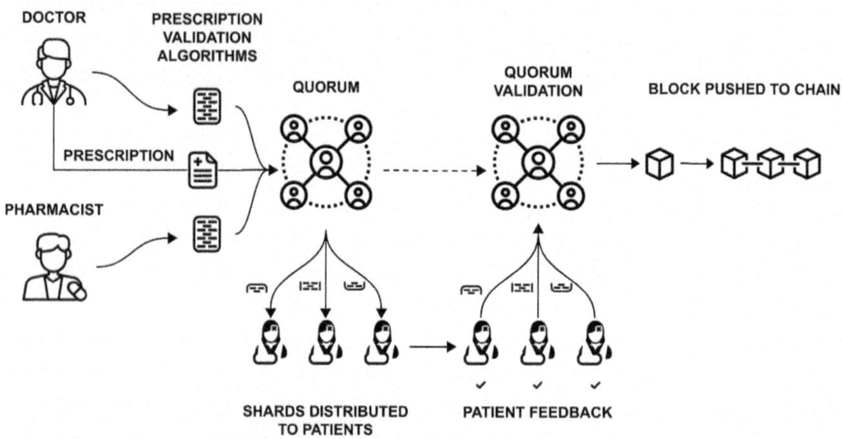

Fig. 1 Overview of the transaction validation process in PharmaSys

5.2 PharmaSys Blockchain

As a decentralized blockchain system, our platform inherits essential properties of blockchain technology and employs other factors that compliment it. These properties encompass a wide range of benefits, including but not limited to enhanced security, improved efficiency, and increased interoperability.

Utilizing blockchain technology provides numerous security advantages. Firstly, patient prescription information on the chain will be encrypted, ensuring that unauthorized individuals cannot access them [HDP6, HDP10]. Permissioned access is required for anyone wanting to view information, addressing concerns related to data vulnerabilities. Unlike centralized systems, the decentralized nature of blockchain eliminates the reliance on a single authority for decision-making. This prevents tampering incidents, as any attempts to modify the data would be detectable due to the resulting hash being altered [HDP9]. Inconsistencies between blocks and their references to previous blocks would be easily discernible, which maintains the integrity and immutability of the information stored on the blockchain.

Our system places significant emphasis in safeguarding against malicious activities. By adopting a quorum consensus mechanism, we preserve the system's integrity by effectively mitigating the tampering attempts by individual malicious nodes. Within our system, a quorum is established, capable of withstanding up to 33% malicious nodes present on the network [HDP12]. This robustness ensures the maintenance of system integrity, fortifying the overall security measures employed.

Additionally, our design addresses the creation of unique identifiers to accurately identify and track user identity by generating a random hash for a user username [HDP13]. This requirement, mandatory for all covered entities, helps mitigate misidentification risks during healthcare transactions. Our system then hashes it to create a public address. Using SHA-256's preimage and collision resistance, each user identification is unique. By implementing this user identification measure, covered entities can mitigate the risk of mixing up individuals' identities and ensure the accuracy and integrity of health-related transactions.

Moreover, the decentralization of our system serves as a solution for the challenges associated with lost or inaccessible data during emergency situations [HDP14]. In our system, prescribers and pharmacists assume the role of full nodes, thereby possessing a complete copy of the entire ledger. Consequently, in the event of multiple faulty nodes we can rely on others to uphold the integrity and functionality of the chain.

In our system, a patient-centric approach is implemented wherein patients have direct ownership of their prescription data and can access it through the blockchain [HDP1]. This innovative approach eliminates the need of sending health information requests to intermediates, consequently cutting the time to receive transactions. In traditional systems, healthcare providers may take up to 30 d to fulfill information requests [HDP2], but with our system, patients can swiftly access and review their data within a matter of seconds. With the decentralized nature of blockchain, patients have full control over their data, allowing for faster and more efficient access.

With patients now assuming responsibility for their data, they possess complete control over the sharing and management of authorizations. They have the authority to determine who can access their data and can grant permission to requesting prescribers accordingly [HDP6]. Furthermore, patients retain the ability to revoke permissions from prescribers once the visit or engagement has concluded, maintaining constant control over their PHI [HDP7]. This patient-driven approach empowers individuals by affording them autonomy in managing the authorization and access to their own healthcare information.

An additional aspect of our system pertains to a patient's power to send a copy of their PHI directly to a designated third party [HDP3]. In this process the individual must provide their digital signature to confirm that this action was made by them [HDP11]. By providing their digital signature, which can only be obtained from their secret key, patients can control permissions and share their prescription data to desired individuals [HDP6, HDP7]. This design ensures that any node lacking the requisite authorization or permission will be effectively restricted from accessing encrypted data [HDP10].

5.3 Node Types

In contrast to most blockchains, our system implements distinct node categories. The purpose behind this categorization is to grant specific permissions upon each node type, as represented in Fig. 2. The arrangement of our system entails the division of nodes into three types: prescribers, pharmacists, and patients.

Prescribers In PharmaSys, prescribers bear the responsibility of pushing out prescription transactions, creating their algorithm, reviewing other prescribers' prescriptions, and requesting permission for patient prescription history. Prescriber nodes have the unique capability to initiate and submit prescription transactions, streamlining the essential process between prescribers and patients. It is imperative that every prescriber creates and initializes their unique algorithm to the blockchain as this will be used in the future by other nodes. Additionally, prescriber nodes are part of the

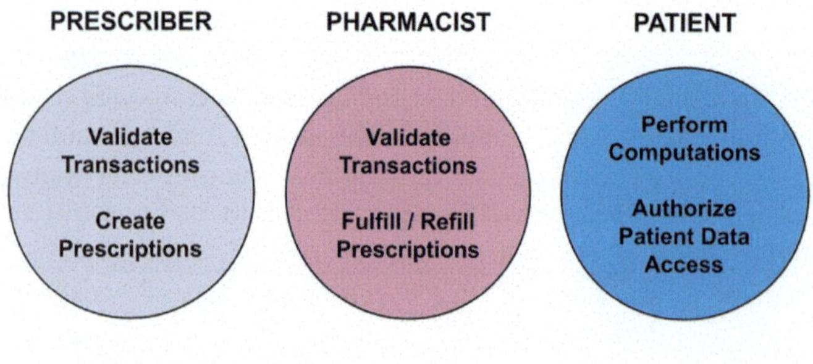

Fig. 2 Three parties within our system

selection pool for quorums that review other prescribers' prescription transactions. In this process, their algorithm will be referenced to gather information from the blockchain to assist in devising an informed decision. Considering the decentralized nature of our blockchain, prescribers seeking access to a patient's prescription history must send a request to the patient's address and obtain their approval before gaining access [HDP4, HDP6].

Pharmacist Similar to prescriber nodes, pharmacists design their distinct algorithm and are eligible for quorum selection. However, unlike prescribers, pharmacists do not have the authority to initiate prescription transactions. Instead, their role lies in filling and refill transactions for prescriptions that have already been issued and validated. Upon verifying that a patient has picked up their prescribed medication, pharmacists will generate a filling transaction, tracking that the prescription is now in patient hands. Pharmacists also initiate refill transactions when authorized by a prescriber which, after quorum consensus, is appended to the original prescription transaction on the blockchain.

Patient A significant power patients wield is the authority to grant or refuse incoming permission requests on their prescription data, as well as the ability to manage existing permissions [HDP6, HDP7]. Every time patient information is accessed, it will be logged onto the blockchain [HDP8]. Their primary responsibility in the system lies in actively seeking and fulfilling delegated tasks distributed by quorum members, thus assisting the quorum in reaching informed decisions. Upon completing their respective computational tasks, patients return the output to the corresponding quorum member who sent out the task.

Transactions Within PharmaSys, a diverse range of transactions are integrated. Specific to the medical industry, various interactions exist between these involved parties such as prescription creation, prescription fulfillment, and permission controls. As mentioned previously, the ability or restriction to perform these specific actions is contingent upon the node's corresponding type.

Transaction Types This network facilitates four types of transactions: prescriptions, prescription fulfillment and refills, and permissions. The transactions within our system govern the relationships between nodes, with some transactions requiring a quorum vote while others do not.

Prescription transactions involve the participation of all three parties: patients, prescribers, and pharmacists. The prescriber initiates the process by generating a prescription for the patient, which is subjected to approval from a quorum. Following the approval, the prescription is then transmitted to the pharmacists. Through this collaborative workflow, the prescriber prescribes the medication to the patient, the pharmacist receives the prescription details, and the patient can pick up their prescribed medication at the designated location.

Prescription fulfillment transactions exclusively pertain to the dispensation of medication by pharmacists to patients. These transactions are initiated by the pharmacist nodes after the patient picks up their prescribed medication.

Lastly, permission transactions occur between medical professionals to patients, where the patients possess the authority to accept or reject the permission.

Prescription Transaction Validation Upon pushing a desired prescription transaction to the mempool, a quorum comprised of prescribers and pharmacists is formed to validate the transaction. It is crucial to note that only nodes representing prescribers and pharmacists are eligible to participate in this quorum. Each medical professional possesses their own opinion that corresponds to the unique algorithm they use to determine the accuracy of the transaction. If a majority agreement of over 50% is reached within the quorum, the transaction proceeds and is appended to the blockchain as a successful transaction. Conversely, should the prescription fail to secure approval from the quorum, it will be added to the blockchain as a failed transaction, ensuring its recording on the ledger. Under such circumstances, the prescriber shall receive a notification and may proceed to modify the prescription details for re-submission.

5.4 Quorum Consensus

Employing a quorum serves the dual purpose of maintaining the integrity and optimizing the efficiency of our system. By randomly selecting a subset of prescriber and pharmacist nodes to use their unique algorithms to review transactions, a balance between system reliability and resource consumption is obtained. Based on Byzantine Fault Tolerance principles, our system reliability can be measured in accordance with its tolerate of up to 33% malicious nodes on the network [HDP12]. Furthermore, we incorporate patient nodes into part of the quorum mechanics by crowd-sourcing their computational power. To leverage this power, we shard each quorum member's algorithm and delegate the tasks to multiple patients.

Deriving the Quorum The formation of our quorum involves the random selection of nodes from pools of prescribers and pharmacists, as illustrated in Fig. 3. Each node within the pool has an equal opportunity of being elected into the quorum, as we do not

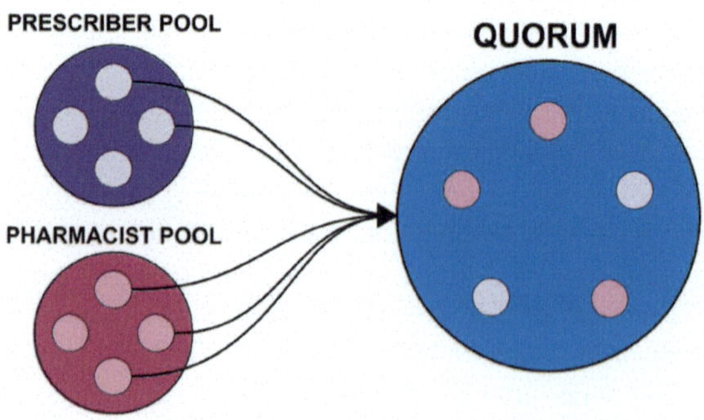

Fig. 3 Quorum derivation

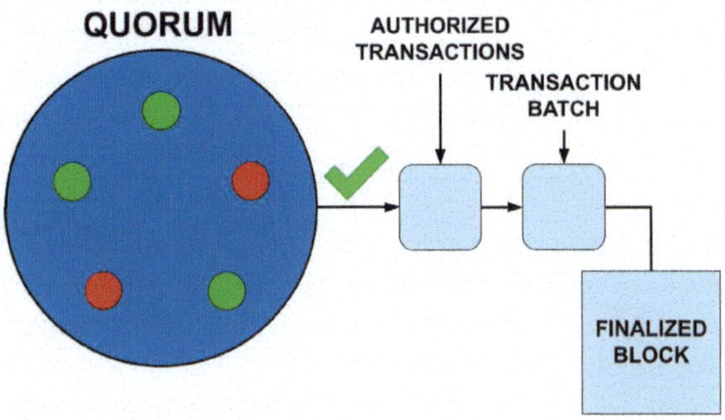

Fig. 4 Quorum decision

consider any additional factors such as years of experience or past quorum selection. To ensure consensus across the blockchain regarding the composition of the quorum, we utilize the hash of the most recent block as a seed for the random generator. This method guarantees agreement and consistency throughout the network regarding the nodes that compose the quorum. By incorporating an unbiased random selection process, our system promotes fairness and equal participation among prescribers and pharmacists, enhancing the integrity and credibility of the decision-making process [HDP12].

Decision-Making Each medical practitioner has their own cognitive process when evaluating a prescription, which can be encapsulated through their unique algorithm. This algorithm can be a comprised of a variety of factors pertaining to the prescriber initiating the transaction such as their age, years of experience, and recent prescription volume. Naturally in our system, any patient-related information is de-identified, eliminating personal details and rendering it untraceable to the originating patient [HDP5]. To gather this information from the blockchain, quorum members shard their algorithm into segments and disseminate them to a mempool. Then, a patient will eventually undertake the tasks, complete them, and provide their responses. Based on the received information, the quorum member will arrive at a conclusive "yes" or "no" decision and cast their vote to the quorum, as exemplified in Fig. 4.

5.5 *The Sharding Algorithm*

Within our system, each quorum member's algorithm is divided into shards with each shard being processed by two different patients, shown in Fig. 5. Through incentives, we are able to harness the computational power of patients. This approach can be broken into distinct steps: algorithm distribution, answer confirmation, and returning results.

Algorithm Distribution The quorum members will transmit their respective individual algorithms to a mempool, where patients will have access to them. These algorithms may be sharded into multiple pieces to accommodate different preferences.

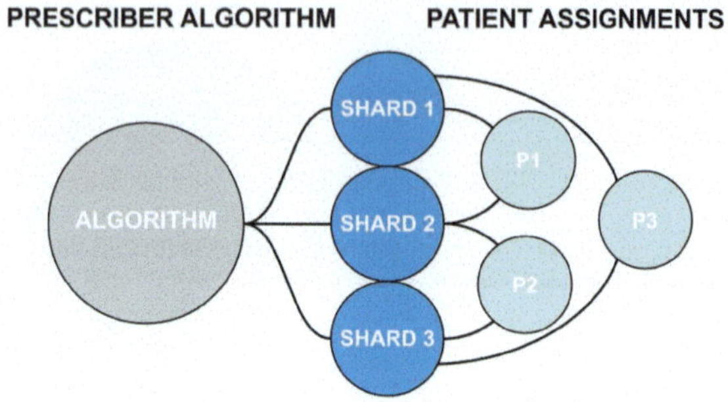

Fig. 5 Sharding process for an algorithm with three shards demonstrating how two patients are assigned to each shard

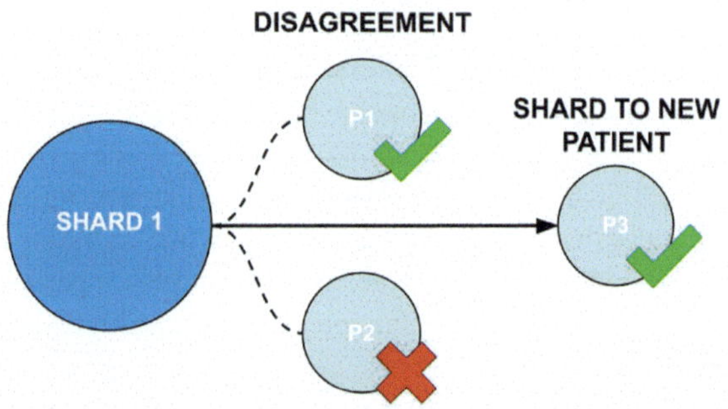

Fig. 6 Process during disagreement with original assigned patient nodes. Conflicting answers result in reassignment to a new patient node

For instance, one medical professional might consider only three factors when making a decision, while another might weigh five factors. Regardless, there will always be n patients actively engaged in solving an algorithm, where n represents the number of algorithm shards. Each patient will process two algorithm shards, and every shard will be assigned to two different patients.

Answer Confirmation If these two patients produce identical results for a given computation, it is treated as correct. However, in cases where the answers do not match, the algorithm is automatically and randomly assigned to a third patient for processing. As demonstrated in Fig. 6, the correct answer is determined based on the match between the third patient's result and either of the previous two answers. However, in the event that the third patient's result does not match either of the previous answers, a redraw process ensues. This process continues until two matching answers are obtained, ensuring the identification of the correct answer through consensus among multiple patients. In instances where a patient provides an incorrect answer, appropriate measures are taken to enforce accountability. This approach effectively prevents patients from evading their responsibilities and submitting a random answer, while simultaneously minimizing the workload required to complete the assigned tasks.

Returning Results Following the completion of these tasks, the patient nodes return the computed results to the respective quorum member whose algorithms they solved. Once a quorum member receives all the computations of their algorithm from the patient nodes, they will have the necessary information to make a decision. By consolidating the returned parts of the algorithm, the medical professional gains an overview of the data, enabling them to properly analyze the information and formulate their decision.

5.6 Block Construction

Based on the quorum's decision, the transaction is added to the blockchain as either a successful or a failed transaction, ensuring a record of the prescriber's actions, irrespective of the outcome. Subsequently, this information is propagated across the network to every full node.

6 Experimental Evaluation

6.1 Implementation

We implemented our proposed system for the purpose of testing efficiency and scalability. Our system focused primarily on the creation of a new use case inside the existing codebase of BlueChain gutting and optimizing the code for our system. Our implementation was written entirely in Java as well as moderate amounts of shell script for automation and Python for experiment data handling and visualization. In each simulation of a prescription transaction stakeholders perform their relevant actions as designated by their node type. Patients access algorithms and are assigned to shards, perform necessary computations, and return results to requesting quorum members. Prescribers push out prescription transactions, act as validators, and push their algorithms. Pharmacists validate transactions and push their algorithms to the chain as well. Each stakeholder completes all actions as they are designed to in each transaction cycle after prescription transactions are pushed to the network.

6.2 Testing Environment

The system used for our experiments contained a 3.4 GHz Intel I7-6700CPU, a NVIDIA Quadro M400 GPU, 32 GB of 2133 MHz DDR4 RAM, and a 512 GB PCIe SSD running Windows 10. The Virtual Machine (VM) ran Ubuntu 22.04 LTS. The VM had 24 GB of RAM with a 25 GB VHDD and 5 cores of CPU utilization that ran Maven V3.9.3 and Java 20 SDK dependency for execution of our system. BlueChain was designed to utilize Apache Maven as a management tool for compilation and a unified build system for all file dependencies. The client linking into the PharmaSys network was custom-built for each client instance to act as one account on the network.

Setup In our research, we utilized the PharmaSys config.properties file to facilitate the setup of the system for simulations. This file enabled us to define and allocate attributes such as network size, the number of quorum members, and other essential settings that the network would have upon initialization. The PharmaSys system was executed on a local network within the testing environment using a VM. Node creation commenced from port 8000 and the defined minimum and maximum connection sizes between nodes was 4 and 7. Under this system configuration we conducted a comprehensive evaluation of the PharmaSys system in terms of scalability and robustness.

Limitations During testing we had some limitations for the network. Our limitation was mainly the VM that we were running the experiments through. This can be attributed to the computational limitations imposed by operating through a VM. The way that Virtual Box stores the guest operating system's files also affects the speed of the system, as all guest files, including PharmaSys files and dependencies, are stored on a virtual hard drive. Virtual hard disks are less efficient compared to utilizing the system's hard disk, even on more powerful computers.

6.3 Scalability

Our primary objective during scalability testing was to document various stages of our transaction process. These stages consisted of quorum formation, quorum decision, block construction, and network communication. During our scalability test, we set up the network with 7 quorum members and incrementally increased the network size from 100 to 500 nodes in steps of 100 for each data recording iteration. Throughout each of these iterations, we executed 50 transactions, recorded their individual times, and calculated the average duration. In the testing design, we also measured the client's response time alongside the in-network processing time.

According to the data shown in Table 3, as our network size increases, the total network time decreases. With a mere 0.22015615-second difference observed between large and small network sizes, our transaction consensus time slightly improves.

Table 3 This table depicts the observed outcomes at various stages of the transaction consensus from the conducted experiments

Network size	Quorum formation	Quorum decision	Block construction	Total network time
100	0.00011329	4.03305100	0.00026780	4.03382640
200	0.00016470	4.08861959	0.00031734	4.08942926
300	0.00012304	3.91122543	0.00025875	3.91177063
400	0.00016594	3.87454129	0.00027795	3.87889664
500	0.00012681	3.86846344	0.00029122	3.87029214

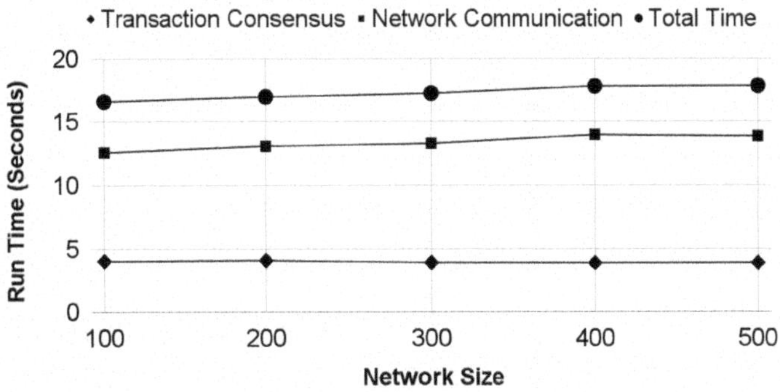

Fig. 7 Graphical comparison depicting the scalability of consensus time and network communication time

This phenomenon can be attributed to the expansion of the patient selection pool available for doctor or pharmacist nodes to appoint as sub-quorum members. The quorum decision step is the most intensive process, accounting for a majority of time in the transaction consensus. This process includes sharding to patients, patient computation and result return, and quorum agreement.

Figure 7 demonstrates a linear rise in total transaction time as the network size expands, mainly influenced by network communication, which significantly contributes to the overall transaction time. As previously shown in Table 3, the time for transaction consensus ever so slightly decreases as the network size increases. Thus, the only contributing factor to the linear increase in total transaction time is the network communication. It is crucial to highlight that the data in both Table 3 and Fig. 7 were generated simultaneously.

6.4 Robustness

Our main goal during robustness testing was to detect malicious sub-quorums, which we defined as a sub-quorum comprised of more than 50% malicious patient nodes. Naturally, the presence of a malicious sub-quorum could result in a medical professional making a misinformed decision. To assess the accuracy of an individual quorum member's decisions, we simulated tests by purposely poisoning our network.

In this experimental study, our system was configured with 100 patient nodes and 7 quorum members. Each quorum member randomly sharded their algorithm into a range of 3–7 shards which were assigned to patients for data retrieval. Throughout the experiment, we progressively increased the percentage of malicious nodes from 10 to 50%, in 10% increments. During this process, we meticulously recorded the occurrences of malicious sub-quorums. With this data, we were able to determine the percentage of properly informed quorum members, which led to true-positive and true-negative responses.

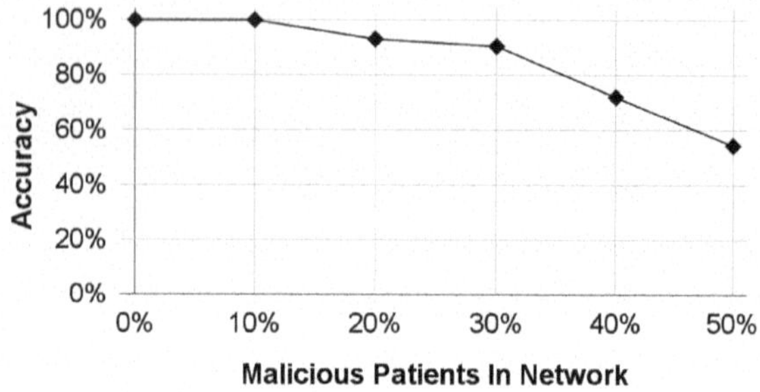

Fig. 8 Graphical representation illustrating the correlation between the accuracy of an individual quorum member's decision and the percentage of malicious patient nodes

Figure 8 illustrates that our system maintains 90% accuracy with up to 30% malicious nodes in our network. The experimental findings can be significantly enhanced by considering the binomial probability of the quorum consensus producing inaccurate results. Presently, the graphical representation mainly focuses on the individual doctor's probabilities of true-positive or true-negative outcomes. To address this, we utilized Algorithm 1 to assess the possibility of a majority of the quorum being misinformed. However, an alternative approach could be adopted to visually represent the probability of the entire quorum generating an accurate response. Implementing this modification would greatly increase the likelihood of obtaining correct answers from the quorum as a cohesive unit.

Algorithm 1 Binomial Probability Calculation

Require: n = number of patients
Require: k = number of successes
Require: p = probability of success
Ensure: $P(X = k)$
1: Calculate binomial coefficient $C(n, k) = \frac{n!}{k!(n-k)!}$
2: Calculate binomial probability $P(X = k) = C(n, k) * (p^k) * ((1 - p)^{(n-k)})$
3: **return** X

7 Conclusion and Future Work

In this paper, we sought to design a blockchain system that tracks prescriptions through the use of a novel multi-node consensus mechanism while following HIPAA guidelines for the handling of PHI. The implementation of PharmaSys system in experimentation illustrates its ability to be scalable and to remain robust and trustworthy even with large amounts of malicious nodes. Integrated large scale, PharmaSys would help alleviate data siloing, increase interoperability, improve cross-healthcare communications, and promote transparency and audibility of drug tracking on the blockchain. Each stakeholder in the system provides their unique services to other nodes and keeps the network functioning in a fluid manner. PharmaSys lays the foundation for future work on blockchains designed for prescription tracking as well as systems designed to handle the data in the chain. Utilization of the data could help provide further benefit toward ensuring accurate prescription monitoring and combating prescription drug misuse. There is possibility that the work established in PharmaSys could be built upon by utilizing machine learning. Neural networks could be used to analyze and leverage the data in the chain and the available information used to uncover patterns and gain insights into practices, medication flow, and patient treatment adherence among others. Future work in this area could allow researchers to develop new methods of predicting surges in drug misuse or the flagging of prescribers acting maliciously.

Acknowledgements Acknowledgment for icons used in diagrams: Blockchain by Adrien Coquet, Algorithm by Sachin Modgekar, Doctor by Iconfield, Group by Pause08, Patient by Wilson Joseph, Pharmacist by Andi Nur Abdillah, Prescription by Sergey Demushkin, Check by Numero Uno from Noun Project.

References

1. National Institute on Drug Abuse (2014) Drug overdose death rates. https://nida.nih.gov/research-topics/trends-statistics/overdose-death-rates
2. Centers for Disease Control and Prevention (2014) Physicians are a leading source of prescription opioids for the highest-risk users. https://www.cdc.gov/media/releases/2014/p0303-prescription-opioids.html
3. National Institute on Drug Abuse (2023) What is the scope of prescription drug misuse in the united states? https://nida.nih.gov/publications/research-reports/misuse-prescription-drugs/what-scope-prescription-drug-misuse
4. Office of the Surgeon General (2023) Addiction and substance misuse reports and publications. https://www.hhs.gov/surgeongeneral/reports-and-publications/addiction-and-substance-misuse/index.html
5. Daley DC (2013) Family and social aspects of substance use disorders and treatment. J Food Drug Anal 21(4)
6. Pierce M, Hayhurst K, Bird SM, Hickman M, Seddon T, Dunn G, Millar T (2017) Insights into the link between drug use and criminality: lifetime offending of criminally-active opiate users. Drug Alcohol Dependence 179:309–316
7. Mitchell AP, Trivedi NU, Gennarelli RL, Chimonas S, Tabatabai SM, Goldberg J, Diaz LA, Korenstein D (2021) Are financial payments from the pharmaceutical industry associated with physician prescribing? Ann Internal Med 174(3):353–361
8. Tringale KR, Marshall D, Mackey TK, Connor M, Murphy JD, Hattangadi-Gluth JA (2017) Types and distribution of payments from industry to physicians in 2015. JAMA 317(17):1774
9. Teater D (2016) Doctor-opioid-survey-national-safety-council. https://www.insurancejournal.com/research/research/national-safety-council-survey-of-doctors-prescribing-opioids/
10. Hirsch R (2017) The opioid epidemic: it's time to place blame where it belongs. Missouri Med 114(2):82–83–90
11. Lundquist P, Dagher GG (2023) Bluechain. ISPM Research Lab [Online]. Available: https://github.com/peytonlundquist/BlueChain
12. Zhang P, Stodghill B, Pitt C, Briody C, Schmidt D, White J, Pitt A, Aldrich K (2020) OpTrak: tracking opioid prescriptions via distributed ledger technology, pp 103–123

13. Zhang P, White J, Schmidt DC, Lenz G, Rosenbloom ST (2018) Fhirchain: applying blockchain to securely and scalably share clinical data. Comput Struct Biotechnol J 16:267–278 [Online]. Available: https://www.sciencedirect.com/science/article/pii/S2001037018300370
14. Azaria A, Ekblaw A, Vieira T, Lippman A (2016) Medrec: using blockchain for medical data access and permission management, pp 25–30
15. Alnafrani M, Acharya S (2021) Securerx: a blockchain-based framework for an electronic prescription system with opioids tracking. Health Policy Technol 10(2):100510 [Online]. Available: https://www.sciencedirect.com/science/article/pii/S2211883721000332
16. Taylor A, Kugler A, Marella PB, Dagher GG (2022) Vigilrx: a scalable and interoperable prescription management system using blockchain. IEEE Access 10:25973–25986
17. Dagher GG, Mohler J, Milojkovic M, Marella PB (2018) Ancile: privacy-preserving framework for access control and interoperability of electronic health records using blockchain technology. Sustain Cities Soc 39:283–297 [Online]. Available: https://www.sciencedirect.com/science/article/pii/S2210670717310685
18. Bashar GD, Holmes J, Dagher GG (2022) Accord: a scalable multileader consensus protocol for healthcare blockchain. IEEE Trans Inform Forens Secur 17:2990–3005
19. Dabrant A et al (2013) Hipaa administrative simplification—hhs.gov [Online]. Available: https://www.hhs.gov/sites/default/files/ocr/privacy/hipaa/administrative/combined/hipaa-simplification-201303.pdf
20. Nakamoto S (2009) Bitcoin: a peer-to-peer electronic cash system. bitcoin.org [Online]. Available: http://www.bitcoin.org/bitcoin.pdf

Exploration and Integration Potential of ChatGPT in Civil Engineering: Advances, Challenges and Application Prospects

Israel Benjamin Luis Rivera Barreto and Yvan Huaricallo Vilca

Abstract Integrating emerging technologies, especially artificial intelligence, is driving a significant transformation in civil engineering. This research aims to determine the impact and potential applications of ChatGPT, a large-scale generative artificial intelligence model in civil engineering, as well as the advances and challenges for its implementation. To do this, a narrative review methodology was used, focused on understanding the obstacles, discovering its applications and demonstrating the potential of ChatGPT in civil engineering. From this, it was found that this tool, with its great capacity to process and generate natural language, positively influences integration with civil engineering. In particular, advances were identified in automated technical documentation generation and policy development. Additionally, applications, such as education and training, real-time data interpretation, project planning and scheduling, adherence to regulations, consultation tools and personalized learning were proposed. At the same time, technical challenges were identified and addressed, proposing security measures to ensure data confidentiality and integrity. Finally, it is concluded that ChatGPT not only presents itself as a revolutionary tool in civil engineering but also offers a wide range of applications within the discipline, establishing itself as a technology with the potential to redefine contemporary practices and methodologies in the sector.

Keywords ChatGPT · Civil engineering · Potential

I. B. L. R. Barreto (✉) · Y. H. Vilca
Universidad Nacional Mayor de San Marcos, Lima, Peru
e-mail: israel.rivera1@unmsm.edu.pe

Y. H. Vilca
e-mail: yhuaricallov@unmsm.edu

Y. H. Vilca
Universidad Tecnológica del Perú, Lima, Peru

X.-S. Yang et al. (eds.), *Proceedings of Ninth International Congress on Information and Communication Technology*, Lecture Notes in Networks and Systems 1000,
https://doi.org/10.1007/978-981-97-3289-0_33

1 Introduction

Civil engineering is undergoing a profound and accelerated transformation due to the incorporation and adoption of cutting-edge technologies. Artificial intelligence (AI) has emerged not only as a complementary tool, but as a catalyst capable of redefining paradigms in various disciplines [1]. In this context, ChatGPT, a language model launched by OpenAI in November 2022 [2], has gained prominence. Its ability to understand and generate natural language, sometimes indistinguishable from human language [1, 3, 4], offers a wide range of applications, including civil engineering.

This area, which ranges from planning to infrastructure maintenance, has not been immune to this wave of innovation. Far from remaining static, this field and its community has demonstrated an insatiable appetite to integrate technological advances that enhance the efficiency and quality of its operations. In this panorama, ChatGPT appears as a valuable ally, with the promise of ushering in a new era of innovative solutions, capable of facing and solving increasingly complex challenges [5, 6].

Given the growing demand for more efficient, safer and more sustainable engineering projects [7] and to the great gap marked by the theory of practice, the scientific literature of ChatGPT requires a study that covers from the general to the particular, in order to expand our understanding of this tool and better put it into practice and not remain on the threshold of a new paradigm, lagging behind its implementation. In the face of this ever-evolving scenario, it is imperative to establish clear objectives that guide our research and efforts in the field. The main objective of this study is to determine the impact and potential applications of ChatGPT in civil engineering. In addition, it is proposed to analyze the technical and security challenges associated with its implementation in the industry and establish mechanisms to ensure the reliability of the results provided by ChatGPT.

2 Methodology

For this research, a literature review on the application of ChatGPT in civil engineering was carried out, using the repositories of Scopus, Web of Science and Google Scholar; the keywords ["ChatGPT" AND ("Engineering" OR "application" OR "Potential" OR "Challenges")] were combined to filter the articles relevant to the development of the study, which aims to publicize the uses and challenges, as well as the verification of the potentials found, which were intuited and discovered through ChatGPT.

3 Advances in the Application of ChatGPT in Civil Engineering

The incorporation of artificial intelligence (AI) into civil engineering has gained notable momentum in recent years. In this context, ChatGPT stands out as a cutting-edge language model that is impacting and transforming multiple facets of civil engineering and the construction world; however, little is known about these changes and advancements in this industry.

3.1 Automated Generation of Technical Documentation

Technical documentation has a very important presence in civil engineering, its ability to store information facilitates communication, being able to provide a detailed record of decisions, calculations, methodologies and project results. With the introduction of ChatGPT, the ability to generate technical documentation has seen an opportunity for improvement, as the model allows large amounts of data to be processed and converted into structured and consistent reports in record time [4, 8]. In addition, its ability to adapt allows it to enter the world of engineering without any problem, demonstrating knowledge of engineering terms and responding correctly as can be seen in the Fig. 1, thus customizing the generation of reports according to the specific needs of each project, respecting the indications and orders, applicable in each case of entry [9], this being proportional to the data you have at the time of use, which depending on the version of use may change and to the quality of the data and input parameters that we enter into the model [1, 10].

Input data

In accordance with the approach and number of estimates, what is the difference between CPM and PERT? Respond concisely in a table.

ChatGPT's response was:

Here's a concise table that outlines the key differences between CPM (Critical Path Method) and PERT (Program Evaluation and Review Technique) based on their approach and the number of estimates used:

Aspect	CPM	PERT
Approach	*Deterministic*	*Probabilistic*
Number of Estimates	*One estimate per activity*	*Three estimates per activity (Optimistic, Most Likely, Pessimistic)*

Fig. 1 Question from the construction scheduling field

3.2 Development of Innovative Infrastructure Public Policies

Although due to the still early appearance of ChatGPT, it does not have the ability to obtain data in real time, it has an immense amount of data which allowed it to be considered by the Inter-American Development Bank (IDB) to optimize the development and review of public policies, allowing the automatic generation of reports, the simulation of scenarios and decision-making [11].

Its adoption is recommended by this institution as it offers valuable perspectives such as uses such as the Fig. 2, guiding decision-makers in making decisions and developing more effective policies, demonstrating how the combination of technology and expert knowledge can lead to more innovative and effective solutions in the field of civil engineering and urban planning.

The progress of research in the field has progressed over time, achieving a greater number of publications. The following applications are shown in Table 1, which show the other applications presented in the reviewed articles within the field of civil engineering.

Usos de ChatGPT para expertos en políticas públicas
Generar resúmenes de grandes cantidades de datos, facilitando la comprensión de la información.
Crear visualizaciones de datos e informes interactivos que incluyan explicaciones y descripciones lingüísticas de tendencias, patrones y datos para que las partes interesadas puedan comprender la información que se presenta.
Simular escenarios y evaluar los efectos de varias opciones políticas, ayuda a crear políticas públicas mejor informadas y más efectivas.
Realizar informes perspicaces que incluyan un análisis detallado del impacto de las iniciativas políticas en el ámbito especificado.
Analizar rápidamente grandes conjuntos de datos y generar puntos de vista claros y útiles para los responsables de la toma de decisiones.
Comprender las necesidades y perspectivas de las partes interesadas, y ofrecer recomendaciones y puntos de vista basados en esta información.

Fig. 2 Uses of ChatGPT for public policy experts according to the IDB

Table 1 Current applications

Título de articulo	Autor	Aplicación
Investigating the use of ChatGPT for the scheduling of construction projects	[5]	Scheduling of construction tasks
Robot-enabled construction assembly with automated sequence planning based on ChatGPT: RoboGPT	[12]	Automating the sequential planning of construction tasks in robotic assembly
Leveraging ChatGPT to aid construction hazard recognition and support safety education and training	[8]	Recognizing construction hazards
BIM-GPT: a prompt-based virtual assistant framework for BIM information retrieval	[13]	Retrieval of BIM information using natural language
GPT models in construction industry: opportunities, limitations, and a use case validation	[4]	Extracting information from textual construction normative documents

4 Technical Challenges in Implementing ChatGPT in Civil Engineering

ChatGPT has been shown to be able to process large amounts of data and provide civil engineering insights, which are meaningful insights or discoveries gleaned from information. However, care needs to be taken when adapting in this specialized field, with its technical terminology and specificities. It is crucial to ensure the reliability and accuracy of your results, and despite its capabilities, an expert review is still necessary to validate the analyses [14], because despite being large generative AI models (LGAIMs) and having a great availability of textual data [15], ChatGPT could hallucinate, i.e. generate data or information that doesn't exist, and give them the impression of being legitimate, especially for people who are not familiar with the subject [1]. In the Fig. 3, it is shown how ChatGPT manages to give a convincing answer about minimum steel, however, it invents things such as that in subsection 7.7.2.1 of ACI 318–19 it talks about minimum reinforcement, when in fact it talks about the minimum spacing of steel reinforcement.

However, it is important to recognize that excessive reliance on these technological tools could weaken essential skills and competencies in engineers, as he warns [9]. It is essential to evaluate and consider the use of these tools, especially considering that their decisions and suggestions can have a significant and lasting impact on specific projects, as they point out [5].

4.1 Safety

One of the essential concerns that a project has is data security and privacy issues, which can be violated internally and externally [16]. It is imperative to ensure that

Input data
According to ACI 318-19, what is the minimum steel that must fit into a 35 cm x 35 cm column and how much 5/8" steel is needed to meet it?
ChatGPT response was:
ACI 318-19 specifies minimum end-reinforcement requirements for columns. To calculate the minimum reinforcement required in a column, the criteria mentioned in the Code must be followed. The following is a general description of how the minimum reinforcement of a 35 cm x 35 cm column can be determined, according to ACI 318-19: *1) Minimum reinforcement for rectangular columns (7.7.2.1):* A_g *is the gross cross-sectional area of the column,* $A_g = b \times h$ *According to ACI 318-19 7.7.2.1, the minimum booster,* A_s *, must be the greater of:* $A_s \geq 0.01 \times A_g$ *(...)*

Fig. 3 Consultation for the minimum column steel according to ACI 318–19

data specific to civil engineering projects, which can often be sensitive or confidential, is handled with due diligence when employing AI tools such as ChatGPT as it could be breached through cybersecurity attacks, or some weakness presented by AI, as can be seen in the Fig. 4, which shows some vulnerabilities found in the latest version report [2].

Because of this, authors such as [3, 16] warn of the danger and suggest creating strategies to mitigate these risks, among which we propose the following.

Fig. 4 List of specific risks that the GPT-4 version possesses according to the OpenAI paper

Education and Awareness: It is critical that employees and stakeholders receive regular training on security best practices. Identifying attempts at social engineering or phishing attacks should be included in this training to ensure greater protection against internal and external threats.

Data Anonymization: Before adding data to the model, it is critical to remove or modify any specific or personal information about the project that can identify its source. This ensures that the source of the data cannot be traced, even if it is exposed.

Regular Review: To deal with emerging threats, security protocols should be reviewed and updated frequently. Staff should receive ongoing training in safe practices and be aware of the importance of safety.

External Audits: Conducting security audits by qualified third parties can help identify flaws that are yet to be discovered and ensure that all security protocols are in place and that all safety standards and regulations are met.

Incident Management: It is essential to have an incident response plan previously established and practised. This ensures a quick and effective response to address any security flaws or breaches, minimizing potential damage and associated risks.

5 Potential of ChatGPT in Civil Engineering

To glimpse the true potential of this tool in civil engineering, it is essential not only to understand its current capabilities but also to anticipate its future trajectory and the possibilities it could unlock through understanding the model and the updates it has been having since its inception [6, 17].

5.1 Civil Engineering Education and Training

ChatGPT stands out as a disruptive resource, widely highlighted by [18]. Its versatility not only facilitates the generation of advanced teaching materials, but, as he points out, [19], also fosters the adoption of state-of-the-art pedagogical methodologies. This enrichment in education is not merely theoretical; [20], in its research with 22 university students, found that the implementation of ChatGPT in learning experiences not only enriches the educational process, but also enhances students' interest and understanding [21, 22].

5.2 Real-Time Data Interpretation

In civil engineering, variables such as weather, terrain, and local regulations must be taken into account when planning and building. ChatGPT, with its ability to process and generate text based on large amounts of data, can facilitate real-time

Data de entrada
Interpreta los datos meteorológicos actuales y dame recomendaciones para minimizar el impacto en la seguridad de un evento al aire libre programado para mañana, en menos de 100 palabras. *-Temperatura actual: 30°C* *-Humedad relativa: 85%* *-Viento: 5 km/h del este* *-Precipitación actual: 5 mm* *-Pronóstico para mañana: Se espera un día caluroso con máximas de 32°C, humedad del 90%, viento ligero del noreste a 8 km/h y una alta probabilidad de lluvia, con acumulación de 20 mm de precipitación durante todo el día.*
La respuesta de ChatGPT fue:
Ante el pronóstico de altas temperaturas, alta humedad y lluvia, se recomienda ajustar el horario del evento a la mañana para evitar el calor intenso y prever la lluvia. Proporcionar refugio o carpas impermeables, y asegurar que los asistentes estén preparados con ropa adecuada y repelente de insectos. Mantener vigilancia constante sobre el pronóstico actualizado y estar preparado para posibles cambios repentinos en el clima.

Fig. 5 Requesting weather data interpretation from ChatGPT

interpretation of these factors [10]. In doing so, it provides valuable insights, as seen in the Fig. 5, where you can receive feedback and recommendations that can influence decision-making during the planning phase of an activity [4].

5.3 Project Planning and Scheduling

Scheduling and project planning, which determine the temporal and sequential structure of the activities to be carried out, are essential for the good performance of a project. The study of [5] provides a clear example of how ChatGPT can be applied in the context of project scheduling. Although it was a simple use case, the results obtained were promising, being able to generate a coherent schedule.

5.4 Adherence to Regulations and Standards

Every civil engineering project has unique requirements based on its geographic location or nature. ChatGPT, being trainable, can be adapted to generate documents that comply with specific regulations and standards [4, 23]. This capability ensures compliance and reduces the risk of non-compliance, which is essential when presenting projects to investors, local governments and communities [15], where the clarity and transmission of the message are as important as the content, characteristics that this AI possesses and that can be improved.

5.5 *Query Tool*

Frequently, due to the complexity of civil engineering, it is necessary to continuously consult the scientific literature, public policies and best practices. ChatGPT is presented as an intelligent conversational tool, offering answers based on a large amount of information, which can stimulate critical thinking [18]. These answers must be precise, coherent and capable of generating value in the person who consults as illustrated in the Figs. 2 and 4, where the AI is able to answer the query by adapting to the context of the input data. This facility ensures that the decisions made are based on up-to-date and relevant information, thanks to its potential in the analysis of unstructured data and the provision of new insights [24].

5.6 *Personalization of Learning*

In the educational world, each student has their own pace and style of learning, in that sense, ChatGPT can be used to adapt educational content to the individual needs of each student, as well as in teaching practice [18]. By interacting with the student, the model can identify areas of interest or difficulty and tailor lessons to ensure effective understanding. In the Figs. 1 and 3 it can be evidenced how this AI can develop a clear explanation, understanding the context of the question and generating an answer capable of helping the person who consults according to the indications provided.

5.7 *Preparation of Valuations*

ChatGPT has the potential to revolutionize the property valuation industry [25], due to its capabilities of analyzing and extracting information from regulatory textual documents, turning them into logical clauses that can be used for automated reasoning [4] as shown by the Fig. 5. However, in order to achieve optimal results, it is essential to combine the experience and knowledge of appraisers in entering input instructions, which can be aided by the adoption of the "Red Book" suggested by [25] or some standardized manual capable of increase in accuracy, efficiency, and transparency in the valuation process, offering reliable reports while significantly improving the customer experience.

6 Conclusions

The contemporary era of civil engineering is witnessing a technological revolution, with artificial intelligence at the heart of this transformation. In this context, our study focused on the ChatGPT language model, developed by OpenAI, evaluating its impact and potential applications in civil engineering. Regarding the advancement of ChatGPT in Civil Engineering, applications were identified and described in areas such as the automated generation of technical documentation and policy development. These advances reflect the adaptability and capacity of the model to process and generate relevant information in technical and specialized contexts. In terms of automated generation of technical documents, ChatGPT has proven to be a tool with a positive influence on civil engineering, being able to process large amounts of data and turn it into detailed reports. Regarding the technical challenges, challenges inherent to its implementation were identified, especially with regard to the ethics of use, followed by compatibility and information security. In relation with reliability measures and strategies, it was underlined that caution is necessary in the interpretation of results to ensure that the results generated by ChatGPT are reliable and relevant to the civil engineering context. Regarding the protection of confidential information, the need to implement measures to ensure the confidentiality and integrity of data was noted, due to the multiple ways in which information can be violated. In terms of capabilities and future applications, ChatGPT, in its continuous technological evolution, promises to unlock even more applications and capabilities in the field of civil engineering, with a focus on real-time data interpretation, automated text generation, adherence to regulations and personalization of learning. Finally, with the right safeguards, ChatGPT not only has the potential to be a revolutionary and applicative tool in specific tasks in civil engineering, but it can also be a guide for the modernization and adaptability of the discipline in the digital age.

References

1. Sohail SS, Farhat F, Himeur Y, Nadeem M, Madsen DØ, Singh Y, Atalla S, Mansoor W (2023) Decoding ChatGPT: a taxonomy of existing research, current challenges, and possible future directions. J King Saud Univer—computer and information sciences 35:101675. https://doi.org/10.1016/j.jksuci.2023.101675
2. OpenAI: OpenAI. https://openai.com/
3. Hacker P, Engel A, Mauer M (2023) Regulating ChatGPT and other large generative AI models. In: Proceedings of the 2023 ACM conference on fairness, accountability, and transparency. Association for computing machinery, New York, NY, USA, pp 1112–1123
4. Saka A, Taiwo R, Saka N, Salami B, Ajayi S, Akande K, Kazemi H (2023) GPT models in construction industry: opportunities, limitations, and a use case validation. http://arxiv.org/abs/2305.18997
5. Prieto SA, Mengiste ET, García de Soto B (2023) Investigating the use of ChatGPT for the scheduling of construction projects. Buildings 13:857. https://doi.org/10.3390/buildings13040857

6. Siche R, Siche N (2023) El modelo de lenguaje basado en inteligencia artificial sensible—ChatGPT: Análisis bibliométrico y posibles usos en la agricultura y pecuaria. Scientia Agropecuaria 14:111–116. https://doi.org/10.17268/sci.agropecu.2023.010
7. Ayinde L, Wibowo MP, Ravuri B, Emdad FB (2023) ChatGPT as an important tool in organizational management: a review of the literature. Business Informat Rev 02663821231187991. https://doi.org/10.1177/02663821231187991
8. Uddin SMJ, Albert A, Ovid A, Alsharef A (2023) Leveraging ChatGPT to aid construction hazard recognition and support safety education and training. Sustainability 15:7121. https://doi.org/10.3390/su15097121
9. Sánchez-Ruiz LM, Moll-López S, Nuñez-Pérez A, Moraño-Fernández JA, Vega-Fleitas E (2023) ChatGPT challenges blended learning methodologies in engineering education: a case study in mathematics. Appl Sci 13:6039. https://doi.org/10.3390/app13106039
10. Nikolic S, Daniel S, Haque R, Belkina M, Hassan GMM, Grundy S, Lyden S, Neal P, Sandison C (2023) ChatGPT versus engineering education assessment: a multidisciplinary and multi-institutional benchmarking and analysis of this generative artificial intelligence tool to investigate assessment integrity. Eur J Eng Educ 48:559–614. https://doi.org/10.1080/03043797.2023.2213169
11. BID (2023) El poder de ChatGPT para elaborar políticas públicas innovadoras. https://blogs.iadb.org/agua/es/aprovechar-el-poder-de-chatgpt-para-elaborar-politicas-publicas-innovadoras/
12. You H, Ye Y, Zhou T, Zhu Q, Du J (2023) Robot-enabled construction assembly with automated sequence planning based on ChatGPT: RoboGPT. Buildings-Basel 13:1772. https://doi.org/10.3390/buildings13071772
13. Zheng J, Fischer M (2023) BIM-GPT: a prompt-based virtual assistant framework for BIM information retrieval
14. Megahed FM, Chen Y-J, Ferris JA, Knoth S, Jones-Farmer LA (2023) How generative AI models such as ChatGPT can be (mis)used in SPC practice, education, and research? An Exploratory Study Qual Eng. https://doi.org/10.1080/08982112.2023.2206479
15. Venegas R (2021) Applications of artificial intelligence to automatic classification of communicative purposes in engineering reports. Revista Signos 54:942–970. https://doi.org/10.4067/S0718-09342021000300942
16. Gupta M, Akiri C, Aryal K, Parker E, Praharaj L (2023) From ChatGPT to ThreatGPT: impact of generative AI in cybersecurity and privacy. IEEE Access 11:80218–80245. https://doi.org/10.1109/ACCESS.2023.3300381
17. Qadir J (2023) Engineering education in the era of ChatGPT: promise and pitfalls of generative AI for education. In: 2023 IEEE global engineering education conference (EDUCON). pp 1–9
18. Zhu C, Sun M, Luo J, Li T, Wang M (2023) How to harness the potential of ChatGPT in education? Knowledge Managem E-Learn: An Int J 15:133–152
19. Carrasco Rodriguez A (2023) Reinventing the teaching of early modern history in secondary school: the use of ChatGPT to enhance learning and educational innovation. Stud Hist-Hist Mod 45:101–145. https://doi.org/10.14201/shhmo2023451101146
20. Ho W, Lee D (2023) Enhancing engineering education in the Roblox Metaverse: utilizing chatGPT for game development for electrical machine course. Int J Adv Sci Eng Inform Technol 13:1052–1058
21. Shoufan A (2023) Exploring students' perceptions of ChatGPT: thematic analysis and follow-up survey. IEEE Access 11:38805–38818. https://doi.org/10.1109/ACCESS.2023.3268224
22. Alarcon-Llontop L-R, Pasapera-Ramírez S, Torres-Mirez K (2023) The ChatGPT application: initial perceptions of university teachers. In: Presentado en proceedings of the LACCEI international multi-conference for engineering, education and technology
23. Sitapure N, Kwon JS-I (2023) Exploring the potential of time-series transformers for process modeling and control in chemical systems: an inevitable paradigm shift? Chem Eng Res Des 194:461–477. https://doi.org/10.1016/j.cherd.2023.04.028

24. Hassani H, Silva ES (2023) The role of ChatGPT in data science: how AI-assisted conversational interfaces are revolutionizing the field. Big Data and Cognitive Comput 7:62. https://doi.org/10.3390/bdcc7020062
25. Cheung KS (2023) Real estate insights unleashing the potential of ChatGPT in property valuation reports: the «Red Book» compliance Chain-of-thought (CoT) prompt engineering. J Prop Invest Finance. https://doi.org/10.1108/JPIF-06-2023-0053

A Case Study of Student Performance Predictions in English Course: The Data Mining Approach

Tran Duc Quynh, Nguyen Doan Dong, and Nguyen Quang Thuan

Abstract Data mining techniques, by providing numerous insights, have been recently used and have improved many aspects of life, including education. Analyzing the students' records and predicting their performance of students can bring a lot of benefits to the education system. As a university with special training programs such as training in English, dual degree education, and technology programs, International School—Vietnam National University, Hanoi, Vietnam is on the rise and will benefit from the application of data mining. In this paper, we conduct an analysis of the pre-English course performance of students at the International School with the aim of obtaining positive predictive results through appropriate data engineering, performing exploratory data analysis, and building predictive models of the students' study time for the pre-English course. Therefore, the school can consider those as the reference for program adjustments.

Keywords Student performance · Data mining · Ensemble learning

1 Introduction

The International School (VNU-IS) is affiliated to Vietnam National University, Hanoi, offering undergraduate and post-graduate training programs in English across all majors. The International School is at the forefront of new courses in the fields of information technology and data. Also, it offers double degree programs for students, creating opportunities to connect students with dynamic foreign environments. Recently, the VNU-IS has received a large number of students in each enrollment

T. D. Quynh (✉) · N. D. Dong · N. Q. Thuan
International School, Vietnam National University, Hanoi, Vietnam
e-mail: ducquynh@vnu.edu.vn

N. D. Dong
e-mail: dongnd@vnu.edu.vn

N. Q. Thuan
e-mail: nguyenquangthuan@vnu.edu.vn

X.-S. Yang et al. (eds.), *Proceedings of Ninth International Congress on Information and Communication Technology*, Lecture Notes in Networks and Systems 1000,
https://doi.org/10.1007/978-981-97-3289-0_34

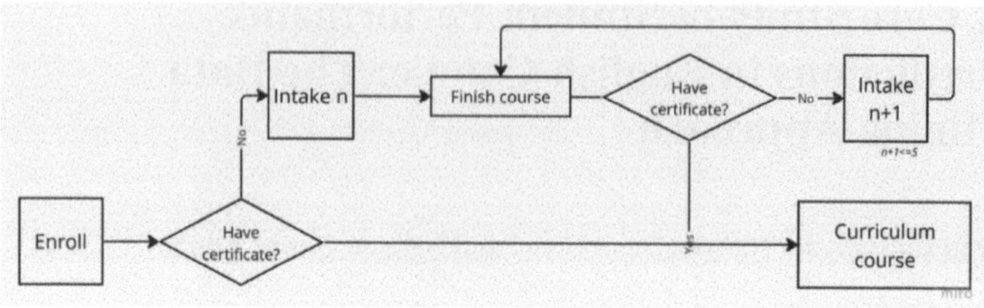

Fig. 1 Process of enrolling pre-English course at VNU-IS

period, increasing the teaching scale of the school rapidly, which results in many opportunities as well as challenges. In response to the new circumstances, the VNU-IS adapts to the contemporary aids, such as the online learning platform, Learning Management Systems (LMS). These tools are being tested in preparatory programs and are used to analyze the strengths and weaknesses of the current training program. The application of information technology is revolutionizing the education sector, with data analysis proving to be a valuable proofs in solving significant problems for universities. Applying it to learning and teaching, the data analysis will give the school a stronger foundation to continue taking care of students while welcoming students of the following courses more thoughtfully.

VNU-IS is very diverse in student nature. To effectively evaluate the school students, the school needs to start the long journey of building the system now. It will begin with using the existing tools and material (school labs, the Internet system, and student data) for analyzing small problems such as the preparatory English program. This program is the starting point for many students when enrolling; therefore, an adjustment based on the actual situations of new students is very necessary. By analyzing this problem, the return insight can help the VNU-IS adapt to the new student course each year, creating a more suitable program for the new generation's English ability.

At VNU-IS, the curriculum is delivered in English, expanding its appeal to International students. Consequently, VNU-IS mandates that students possess English proficiency prior to enrolling in the main program, with a strong requirement of having a B2 certificate in English, which includes various recognized certificates like IELTS, TOEIC, TOEFL, and others. If students do not have this certificate at the time of enrollment, they have to participate in a preparatory training program organized by the school to assist them in preparing for the examination. An entrance placement test evaluating English listening, reading, and writing skills will be held to assign students to classes of different levels: Foundation, 1, 2, 3, and 4. The pre-English course is described in Fig. 1.

It is essential for the school to understand the actual impact of the course on students' English proficiency and be prepared to make necessary adjustments. This implies taking into account every aspects of the pre-English course, including the significance of the placement test, student's proficiency level, student's differentiation, etc. To perform this problem meaningfully, students' learning ability and learning time will be the main factors for analysis and mining, the problem is defined as "Student Performance Prediction," using data that are related to students' initial English ability and the pre-English course information to predict the learning time of students in the course. In this research, we examine the data provided by the Academic Foundation Departments with certain modifications made to ensure the protection privacy. There are some research questions that need to be answered from the datasets as follows:

- What are the factors related to students' English learning?
- Are stereotypes about the place of birth, gender, and majors related to English ability true?
- What features will assess students' English learning ability and speed in the above dataset?
- Can the school differentiate students' learning ability and speed using the given features?

After getting insights into the data, we apply machine learning models (ML) algorithms such as Random Forest [2], XGBoost [16], Support Vector Machine (SVM) [12], and Voting algorithms [4] in order to predict the performance of students based on these datasets.

The rest of the paper is organized as follows. Section 2 presents the related work. The details of pre-processing and getting insights into the data are in Sect. 3. The experimental settings and the results are discussed in Sect. 4. Section 5 is for some concluding remarks and perspectives of the study.

2 Related Work

Educational data mining has many problems and different approaches have been proposed over time. One significant problem within educational data mining is the prediction of student dropouts, as dropout rate impact not only individuals but also have broader societal consequences through a decline in overall educational rate. Detecting high-risk students at an early stage can be instrumental for schools, enabling timely monitoring and intervention to address the difficulties that often lead to student dropout [5, 13].

Regarding the dropout prediction problem, numerous ML models have been utilized, yiedling promising results [8]. Jovial et al. [6] have used several tree-based models like Random Forest, XGBoost, Gradient Boost, and a Neural Network (NN) approach using Feed Forward Network, resulting in more than 80% accuracy [6].

Additionally, some other prediction models, as demonstrated by Marmar et al. [10], combine NN, decision trees, and bagging trees. This also integrates with imbalance learning techniques like resampling and cost-sensitive learning to handle the imbalance between dropout and non-dropout classes.

Another common use of data mining in education is student performance prediction, in which "performance" is defined based on the context of each problem.

DOB	Gender	Citizen ID	Certificate date	Certificate status	Starting level	Intake 1 class	Intake 1 score	Level Intake 2	Intake 2 class	Intake 2 score	Level Intake 3	Intake 3 class	Intake 3 score
13/12/2002	Nữ	123020031	Tháng 12/2020	Đạt điều kiện NN	Level 1	Level 1-14	10	Không xếp	1.Không xếp	10	1.Không xếp	1.Không xếp	CXĐ

Fig. 2 Record example of a student in the dataset

Phauk et al. [11] have used SVM, Naïve Bayes, Decision Tree, and Random Forest with k-fold cross-validation and Principal Component Analysis (PCA) to predict the performance of three observations. All return the result that PCA and k-fold outperform the baseline model and baseline with k-fold, which proves the effectiveness of PCA in this problem. A survey regarding exam prediction problems conducted by Nikola et al. reveals that K- Nearest Neighbors (KNN), SVM, Artificial Neural Networks (ANN), Decision Trees, Naïve Bayes, and Logistic Regression are common and proven to be effective in predicting student performance [9].

3 Data Pre-processing and Exploring

3.1 Data Pre-processing

The dataset is divided into five categories: student demographic, certificate information, pre-English course class arrangement, enrollment information, and high school performance. It is shown that the features are pretty selective and have some correlation with students' initial English ability. Additionally, the dataset includes certain attributes presumed to be connected to English language learning abilities, such as place of birth, subject group, and gender. This inclusion allows us to examine and solve these conjectures. Regarding course information, there are five intakes, equivalent to 10 months maximum from the enrollment for students to submit the certificate. A record as in Fig. 2 has shown that this student only attends the first intake in class Levels 1–14, she then submits the certificate in December 2020, which is the time intake 1 ends. This means she only needs 1 intake to achieve the certificate. To facilitate this analysis, we have introduced a new label called "intake" to classify the level of intake. In this particular case, the student's label would be "1."

In this study, the problem can be framed as a classification task, categorizing students into Levels 1–5, based on their initial data (including demographic information, high school performance, enrollment, and placement test), with aim of predicting the number of intakes (intake labels) needed for students to achieve their certificates. That means based on initial data of students (demographic, high school performance, enrollment, and placement test), predict the amount of time (the intake labels) that students need to achieve the certificate.

As not all students take part in the pre-English course, some of the data lacks relevance for predictive purposes and should be removed at an early stage to minimize the amount of extensive data cleaning. The following data categories will be excluded:

- Lack of enrollment data: since it is an important data category and not possible to fill those values in a meaningful way, these rows will be removed: 332; students with a certificate when enrolled: 63; student without certificate and drop school: 80; the total number after data selection: 893.

Data Cleaning

Since the data is manually input, it exhibits numerous typographical errors and inconsistencies, particularly in the "Place of Birth" (POB) column. Due to the Vietnamese word marks, these errors may lead to confusion during the analyzing process. Additionally, varying writing habits result in a single city being represented in multiple ways, such as "TP. Há Nộі" (Hanoi city), "Thánh phõ Há Nội" (Hanoi city), "Há Nội," and more. To rectify this, those will be reformatted into one consistent type by manual filtering process.

"Conduct" and "Performance" column share the same missing value represented as "0," indicating the absence of record data for certain students. This data is easy to fill in because the number of missing is low (16 records), and there are only three types of value in these rows. The "0" will be replaced with "Khá," to balance with the number of "Gio$^?$i" in "Performance" and "Tõt" in "Conduct."

For the POB data, it will be grouped into regional categories: Northwest (TBB), Northeast (D̲BB), Red River Delta (D̲BSH), Hanoi, Central (Trung), South (Nam), and Foreign. The university is in Hanoi, so the students from the North will be the majority; therefore, there will be more groups. Since Hanoi has a large number of students with distinctive English proficiency, it will remain as an independent group. Even after this grouping activity, some classes remain imbalanced, notably "Nam" and "Foreign." "Foreign" will be grouped with "Hanoi" for the similarities of the contribution of Foundation and Level 1 in total. The same goes with "Nam" and "TBB." After this grouping, the POB variable will be transformed into "Birth Region" with a total of five classes.

3.2 *Exploratory Data Analysis—EDA*

EDA is an analysis approach that identifies patterns, relationships, distribution, etc., of the dataset. For the VNU-IS problem, EDA will answer the questions mentioned above about the correlation between features and English ability:

- What are the factors related to students' English learning?
- Are stereotypes about the place of birth, gender, and majors related to English ability true?
- What features will assess students' English learning ability and speed in the above dataset?

Fig. 3 Distribution of data features

- Can the school differentiate students' learning ability and speed using the given features?

The dataset contains imbalance classes, which is an inherent characteristic of a diverse university environment, making an unavoidable situation. The distribution of POB, Subject Group, Genders, and Major type is as in Fig. 3.

As depicted in Fig. 3, students in the A00 group, enrolling in VNU-IS with grades in Math, Physics, and English, do not have English as part of their subject group. On the other hand, students in the Social and A01 groups, enrolling in VNU-IS without English grades, have English as part of their subject groups.

The variations in students' performance are clearly evident as illustrated in Fig. 4:

- In the English group, only 35% of students commence with a Foundation level, whereas in the non-English group, 70% of students start with poor skills.
- Students with intermediate and high English skills (starting with Levels 2 and 3) account for approximately 25% of the total in the English group, whereas in the non-English group, this proportion is only about 6–7%.
- 50% of students in the English group can achieve the certificate within the first 3 intakes, which is around 6 months. This implies that they can enroll in the curriculum courses after just one semester. In contrast, the non-English group has a significantly lower success rate, with only 20% of the students passing after one

Fig. 4 Comparing the performance of students enrolling VNU-IS with and without English grades

semester. The remaining 80% are required to wait for two semester to complete their English preparatory. This situation signals the university's need to be aware of the potential consequences of postponing and English certificate by one year, as it can lead to an overall extension of a student's entire academic progress in the university. Peer pressure, as well as time and financial burdens on college students, can demotivate them, potentially resulting in dropping out of school [5].

The analysis of the datasets, as illustrated in Fig. 5, indicated that students with high English scores in the university exam are expected to have intermediate or advanced English skills, but the placement test has shown that this is not the case. By filtering students with English scores in university exams ranging from 7 to higher, we have 466 out of 699 students, approximately 66%, fall into this category.

As shown in Fig. 5, the students with high English scores in the university exam may excel in reading skills, but only half of them can perform as well in writing skills, and only one-third in listening skills. University exams with multiple choice format contain luck factors and tips to pick the right answers. Meanwhile, the placement test is a combination of multiple choice and writing format, which results in a more objective assessment of student English skills. The placement test also contains listening and writing skills, which are necessary for the learning in English journey in VNU-IS, so this test is essential for the university to evaluate students and arrange them at the right level in the course.

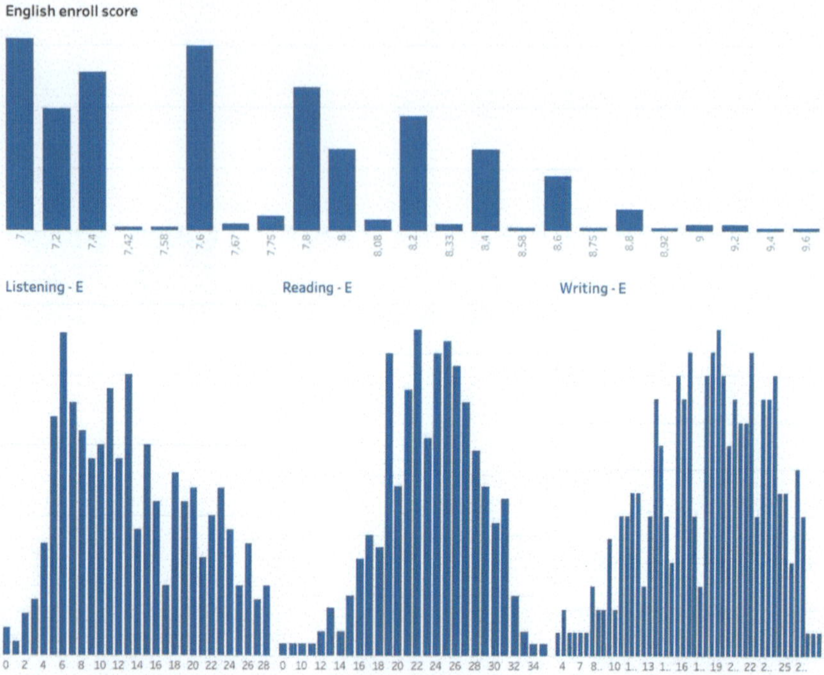

Fig. 5 University exam and placement test comparisons

4 Experimental Settings and Results

4.1 Data Settings

After exploring the data, we removed excessive columns and added some output features. We group the data in POB and create a new label called learning speed as follows:

- The columns "POB," "major," and "subject group code" were replaced by "birth region," "major type," and "subject group."
- We create a new output feature called learning speed by grouping classes of the output Intake and drop the "intake" afterward.
  - Group 1 of "intake" will become group "fast" in "learning speed," denoted by the number 0.
  - Groups 4 and 5 of "intake" will become group "slow" in "learning speed," denoted by the number 2.
  - Groups 2 and 3 of "intake" will become group "medium" in "learning speed," denoted by number 1.

Table 1 The labels in datasets

No.	Labels	Counts
0	Fast	173
1	Medium	219
2	Slow	501

The distribution of instances in each class can be found in Table 1. This approach enables us to focus on student performance and assists the school in correctly assigning students to their respective classes. Besides, for the categorical features, we employ One Hot Encoder for transformation.

We apply traditional ML algorithms such as Random Forest, SVM, Voting, and XGBoost instead of advanced techniques like deep learning, due to small datasets. To evaluate the performance of these models, we report the accuracy (ACC) and F_1 score metrics. For the Voting model, we apply the two algorithms inside, which are SVM and Random Forest. To find the optimal configuration of these models, we utilize the Random search techniques [1].

We divide the dataset into training and testing sets and make use of K-fold cross-validation to fine-tune these models on the training set. Subsequently, we evaluate on testing set. Also, due to the imbalance of the dataset, the over-sampling [3] is applied on the training sets.

4.2 Results

To evaluate the performance of these ML algorithms, the performance metrics such as precision, recall, F_1 score, accuracy will be used. Due to the imbalance in the dataset, the F_1 score will be focused to evaluate the recognition of the minor class of the models.

As the results in Table 2, our findings indicate that both Voting and XGBoost outperform Random Forest and SVM in terms of Acc, but the Voting algorithm yields the highest F_1 score among these ML models. Analyzing the results further, it becomes evident that the Voting and SVM algorithms exhibit highest performance in terms of F_1 score on classes 1 and 2. That means, these models can recognize better the Medium and Slow learner. Although the SVM has a higher recall metric on class 1 compared to Voting, the precision is much lower than Voting on the other classes. Besides, the Random Forest has the lowest precision and recall metrics, especially, in class 1, compared to the other models.

Overall, the Voting model yields promising results because of the highest F_1-score. That can be easily explained because the Voting utilizes the weak models to form a stronger model. Therefore, it takes advantage of the strong point in recognizing class 1 from SVM and other points from Random Forest. For future work, we will investigate more on this model to improve the results.

Table 2 Experimental results

Models	Class	Precision	Recall	F_1 score	ACC
Random Forest	0	0.82	0.92	**0.87**	0.75
	1	0.50	0.34	0.41	
	2	0.67	**0.66**	0.67	
SVM	0	0.82	0.90	0.86	0.75
	1	0.47	**0.47**	**0.47**	
	2	0.87	0.57	0.69	
XGBoost	0	**0.83**	0.90	0.86	0.76
	1	0.50	0.43	0.46	
	2	0.76	0.66	0.70	
Voting	0	0.81	**0.94**	**0.87**	0.76
	1	**0.52**	0.43	**0.47**	
	2	**0.90**	0.57	**0.70**	

The significance of the bold numbers are showing the authors' experimental results

5 Conclusion

The paper has represented a data mining process within the context of education, specially addressing the "Student Performance Prediction" challenge using VNU-IS dataset in order to predict the learning performance of VNU-IS students in the pre-English course. The experimental results on the data have shown that the Voting algorithm has promising results the predict the performance of the students. For the future work, we continue to investigate deeper the ensemble models to boost the performances.

Acknowledgements This study was supported by Vietnam National University, Hanoi, under the research project No. QG.21.51.

References

1. Bergstra J, Bengio Y (2012) Random search for hyper-parameter optimization. J Mach Learn Res 13(1):281–305
2. Breiman L (2001) Random forests. Mach Learn 45(1)
3. Chawla NV, Bowyer KW, Hall LO, Kegelmeyer WP (2002) SMOTE: synthetic minority over-sampling technique. J Artif Intell Res 16:321–357
4. Dietterich TG (2000) Ensemble methods in machine learning. In: International workshop on multiple classifier systems, Lecture Notes in Computer Science. Springer, Berlin, vol 1857, pp 1–15
5. Joo S-H, Durband DB, Grable J (2008) The academic impact of financial stress on college students. J College Stud Retent Res Theor Pract 10(3):287–305

6. Jovial N et al (2022) Predicting student's dropout in university classes using two-layer ensemble machine learning approach: a novel stacked generalization. Comput Educ Artif Intell 3
7. Kaiser J (2014) Dealing with missing values in data. J Syst Integr 5(1):1804–2724
8. Michael J et al (2001) Knowledge management and data mining for marketing. Decis Support Syst 31(1):127
9. Nikola T, Nikola G, Sanja V et al (2020) An overview and comparison of supervised data mining techniques for student exam performance prediction. Comput Educ 143
10. Orooji M, Chen J (2019) Predicting louisiana public high school dropout through imbalanced learning techniques. In: 2019 18th IEEE International Conference On Machine Learning And Applications (ICMLA), pp 456–461
11. Phauk et al (2020) Hybrid machine learning algorithms for predicting academic performance. Int J Adv Comput Sci Appl
12. Platt J (1999) Probabilistic outputs for support vector machines and comparisons to regularized likelihood methods. Adv Large Marg Classif 10(3):61–74
13. Romero C, Ventura S (2007) Educational data mining: a survey from 1995 to 2005. Expert Syst Appl 33:135–146
14. Shivani G et al (2019) Dealing with noise problem in machine learning data-sets: a systematic review. Procedia Comput Sci 161:466–474
15. Solorio-Fernández S, Carrasco-Ochoa JA, Martínez-Trinidad JF (2020) A review of unsupervised feature selection methods. Artif Intell Rev 53:907–948
16. Tianqi C, Carlos G (2016) Xgboost: a scalable tree boosting system. In: Proceedings of the 22nd acm sigkdd international conference on knowledge discovery and data mining, pp 785–794
17. Yoo I, Alafaireet P, Marinov M et al (2012) Data mining in healthcare and biomedicine: a survey of the literature. J Med Syst 36:2431–2448
18. Zebari R, Abdulazeez A, Zeebaree D, Zebari D, Saeed J (2020) A comprehensive review of dimensionality reduction techniques for feature selection and feature extraction. JASTT 1(2):56–70

Securing Bluetooth Technology in Healthcare: A Comprehensive Case Study and Practical Insights

Senzo Ntsele and Khutso Lebea

Abstract Bluetooth has become an integral component in our lives, and it has applications ranging from simple mobile phones to more complex applications such as healthcare systems, facilitating seamless data exchange among medical devices. However, all wireless technologies are vulnerable to attacks, such as eavesdropping and unauthorized access. This paper investigates the intersection of Bluetooth network technology and healthcare, focusing on the importance of robust security measures. These vulnerabilities pose a significant threat to patient privacy and data integrity. This research paper explores the application of advanced security modes that can help prevent potential threats in healthcare Bluetooth networks, then finally suggests some solutions in which users can better protect themselves from Bluetooth attacks.

Keywords Bluetooth · Security · Protocol · Healthcare · Awareness

1 Introduction

Bluetooth is a wireless technology that enables data transfer between different devices. It uses wavelengths to transmit data; however, it typically works within a short range for the devices to transfer data to each other [1]. Bluetooth technology differs from the rest of the network technologies because Bluetooth waves do not travel very far and change frequencies [1].

Bluetooth is customarily used to pair devices; connectivity will usually occur between paired devices [1]. The most common Bluetooth devices include earbuds, smart fridges, mobile phones, laptops, and most IoT devices. IoT devices are nonstandard computer hardware that can connect wirelessly to a network and transmit data

S. Ntsele · K. Lebea (✉)
University of Johannesburg, Johannesburg, South Africa
e-mail: klebea@uj.ac.za

S. Ntsele
e-mail: 216010544@student.uj.ac.za

X.-S. Yang et al. (eds.), *Proceedings of Ninth International Congress on Information and Communication Technology*, Lecture Notes in Networks and Systems 1000,
https://doi.org/10.1007/978-981-97-3289-0_35

[2]. Bluetooth is a wireless communication standard based on a radio system designed for short-range, inexpensive communication between devices. Bluetooth networks can also be set up to relate to portable computers as a bridge between other networks. It can also serve as a node of ad hoc networks. An ad hoc network is created peer-to-peer without using any network device. In this network topology, one device will typically connect to another. An ad hoc connection allows devices to communicate directly within the network and share resources such as files, printers, and Internet connection [3]. When the applications are connected to Bluetooth, it is known as a wireless personal area network (WPAN) [4].

Analysts predict that IoT devices markets will increase in the next few years, and in 2027, 7.6 billion Bluetooth devices will be shipped yearly, a 9% growth for the next five years [5]. Bluetooth-enabled devices are being used all over the world. They provide users with convenience and easy usage; however, this convenience comes at the expense of security. This paper explores the understanding of Bluetooth technology with its standards and Protocols, the security protocols and configurations, and finally, a practical recommendation for device manufacturers, developers, and users on optimizing Bluetooth security based on their needs.

The sections of this paper are outlined as follows: Sect. 2 will show a case study demonstrating how a Bluetooth device can be attacked. In Sect. 3, this paper explores Bluetooth security and Bluetooth attacks, and finally, Sect. 4 concludes the paper.

2 Case Study

Wearables such as smartwatches, fitness bands, and smart clothing are projected to dominate the technology market. In 2022, the global wearable technology market held a value of $61.30 billion, and in 2023, the market was predicted to grow at an annual rate of 14.6%, and it will continue to grow until 2030 [6]. However, Bluetooth has a variety of applications. Modern medical institutions are utilizing Bluetooth for a variety of applications. This is due to Bluetooth's low power and low energy usage, making it a favorable option for various medical equipment used in modern medicine. Blood glucose, blood pressure, and respiratory rate monitors operate on a tiny battery. It is the ideal technology since Bluetooth uses a lower power and low energy consumption.

Furthermore, Bluetooth is insusceptible to interference; it operates on the 2.4 GHz Industrial, Scientific, and Medical (ISM) band spectrum band (2400–2483.5 MHz) frequency and can work with other technologies such as Wi-Fi on the same medium [7, 8]. Bluetooth has become part of our lives, and its applications do not only include wearables but Bluetooth can be used in almost all aspects of our lives. This section explores the applications of Bluetooth in healthcare and the security risks associated with Bluetooth.

2.1 Bluetooth in Healthcare

Modern medical institutions use costly equipment to treat patients with various medical issues properly. Hiring and keeping employees adds to their operational costs. If modern medicine can utilize human and medical equipment resources more efficiently, it can optimize cost savings by reducing operating costs on repetitive tasks [9]. Bluetooth can help in that aspect.

2.1.1 Vital Sign Monitoring

Wearable devices are now employed for uninterrupted monitoring in healthcare institutions to observe a patient's vital signs [9]. Traditionally, patients would have a healthcare professional such as a nurse or doctor measure their vital signs and then give them to a doctor for reporting [9]. These methods of information gathering of a patient's vital signs, such as recording a patient's temperature, do not optimize the usage of the hospitals' resources, and they can be notoriously inaccurate. Uninterrupted vital sign monitoring using wireless medical equipment reduces operational costs and enables medical professionals to record or report accurate data and deliver quality healthcare to patients. Uninterrupted vital signs mentoring does not need to work in hospitals only. Still, patients can continuously monitor their health at home and securely send their data to the service provider for recording, keeping, and analysis [9]. Another application for wireless medical equipment is for treating patients in an ambulance while they are on their way to the hospital. An example would be using a Bluetooth defibrillator to send real-time information about patients' health while they go to the hospital. The data can be transmitted to the Bluetooth device manufacturer's cloud service, where other health professionals can access it, and they can be prepared for when the patient arrives at the emergency room.

2.1.2 Bluetooth Security Threats in Healthcare

A set of vulnerabilities known as "SweynTooth" exists for Bluetooth-capable devices. When exploited, it can introduce risks for medical devices.

In a scenario where an attacker attempts to compromise the security of a Bluetooth insulin pump, the following can happen: the process described below is depicted in Fig. 1.

- **Reconnaissance**

The attacker chooses a potential target, in this case, a patient with an insulin pump enabled for Bluetooth data transmission.

- **Device Spoofing**

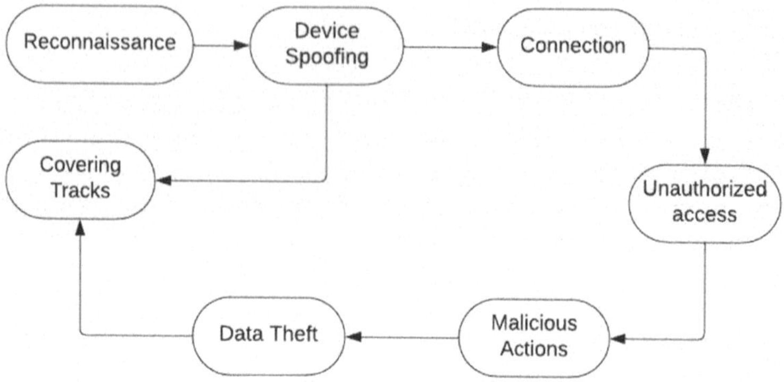

Fig. 1 Bluetooth compromisation flowchart

The attacker can now proceed to spoof the hardware and software of the insulin pump to imitate a legitimate insulin pump device using spoofing tools.

- **Connect**

The attacker will attempt to connect with the Bluetooth-enabled insulin pump. Some might have a weak or default authentication mechanism, making it easier to communicate with the device.

- **Unauthorized access**

After establishing a connection, the attacker can control the Bluetooth-enabled insulin pump and change configurations as they please.

- **Malicious Attack**

The attacker may perform harmful actions such as overdosing the insulin on the patient, causing further damage, or they may demand a ransom from the patient for control of the device.

- **Data theft**

The attacker may steal sensitive information about the patient transmitted by the Bluetooth-enabled insulin pump.

- **Disconnect**

Finally, the attacker can delete logs and disconnect from the device, leaving no trace.

The following section explores a comprehensive background to Bluetooth technology, covering security modes used by Bluetooth devices, associated protocols, the vulnerabilities of Bluetooth, and ways the users can protect themselves.

3 Background

When the network and communication era began, to make it function, wires were needed. Wires are necessary for any communication network to transmit data [10]. Bluetooth technology become one of the solutions to achieve wireless data transmissions. Bluetooth technology was introduced around 1994 and has been popular for several reasons, such as being free and accessible [10]. Bluetooth operates on the Industrial, Scientific, and Medical (ISM) band at the 2.4 GHz frequency and is unregulated [11].

Furthermore, Bluetooth is an easy-to-use technology requiring no setup to communicate with other devices. All devices from various manufacturers and models can use the technology to communicate with each other, which will not create any compatibility issues [12]. Users can use Bluetooth for data transfers such as media files and documents. They can also use Bluetooth for location services and network configurations.

Bluetooth has a lot of use cases that can be applied to everyday life. New and innovative ways are being discovered daily that move society forward with Bluetooth technology. Some of these applications include the Bluetooth-enabled insulin pump, which consists of a Personal Diabetes Manager (PDM), a device used to wireless control and observe the functionality of the insulin pump [13]. Most commonly, Bluetooth technology is used for devices such as tablets, laptops, consoles, and mobile phones. The issue with Bluetooth, like any other wireless technology, is that it can be susceptible to malware infections and attacks. Attacks include data theft, such as sensitive data like credit card numbers, email messages, contacts, and banking information, and these can harm the owner of the data as they may lead to financial loss. In the case of the above case study, an attacker can also do bodily harm to a patient.

Further, an attack on a Bluetooth device can lead to an attacker having full access to the Bluetooth device and being able to control it, corrupt the device, and they can send out SMS and MMS or forward calls [10]. Security is a crucial aspect of Bluetooth technology and must be explored comprehensively. The following section thoroughly details Bluetooth security and its security architecture.

3.1 Bluetooth History and Versions

Bluetooth technology was created by L. M Ericsson in the year 1994. The name of this technology is derived from the King of Denmark, Harald Blaatand. Companies such as IBM, Nokia, Intel, Ericsson, and Toshiba formed the Bluetooth Special Interest Group (SIG). The purpose of the SIG was to create an open specification for wireless technologies. The group continues to manage Bluetooth technology today [14]. The SIG has grown to about just under 2000 companies since its inception. The original specification for Bluetooth was the Classic Bluetooth. Its purpose was to transmit data wirelessly among computer devices. The technology has been managed so well that it has met the demand for new use cases and various applications [14].

Bluetooth's first version was Bluetooth 1.0. This version was launched in 1999. Its original purpose was to replace wired serial communication. Since then, transmission methods have changed over each iteration of Bluetooth [15]. Version 1.1 was released two years later because Bluetooth 1.0 had a deployment problem, and it was not known. In 2004, Bluetooth 2 was deployed. Changes in version 2.0 include GFSK and phase-shit keying modulation (PSK). The function of GFSK is to accelerate data transfer by assisting Enhanced Data Rate (EDR). Version 2.1 advanced the technology further by including a feature called Simple, Secure Pairing (SSP). Version 2.1 improved security, an extended inquiry response (EIR), and pairing, allowing for better device filtering before connections [16]. Bluetooth 3 was released in 2009. Bluetooth 3 had High Speed (HS) mode, allowing data transfers of speeds up to 24 Mbps over an 802.11 link [16]. Bluetooth 3 features are Ultra-wideband, Enhanced Power Control, L2CAP Enhanced modes, Unicast Connectionless Data, and Alternates MAC/PHY. However, due to its power consumption rate, it has drawbacks. Bluetooth 4 was introduced in 2010, and Bluetooth 4.1 was released three years later. This version enhanced the user experience. It allowed for bulk transfer of data. It also supported additional features such as 11n Pal, connection L2CAP, Topology and dual-mode, Link-Layer topology in LE, and services for wireless coexistence on mobile phones, to name a few. Bluetooth 4.2 was released in 2014. It improved Link-layer privacy policies, Internet Protocol Support Profile, and Version 6 Low Energy Secure Connection, which lengthened data packets [16]. Finally, Bluetooth 5.0 was released in 2016. It provided a more significant rate of transmission. It had 2 Mbps maximum transfer speed. Subsequent versions of Bluetooth 5 were released. The latest version of Bluetooth that was released is Bluetooth 5.2. Some notable features of this version were Isochronous Channels (ISOC), Enhanced Attribute Protocol (EATT), and LE Power Control (LEPC). The following sub-section explores Bluetooth security.

3.2 *Bluetooth Security*

Before you start using most wireless network technologies, you must first authenticate so that the user at the other end knows who is trying to connect and which device is trying to connect. Authorization must occur so the user or device trying to connect can have some functional control over the device they are trying to connect to. Bluetooth uses the above concepts to connect [17]. Authentication in Bluetooth devices is accomplished by using the stored link key or a pairing code that a user needs to put on their devices before they can connect.

There are two Bluetooth security modes, namely security mode 1 and security mode 2. Then, there are four security levels: security level 1, level 2, level 3, and level 4 [18]. Bluetooth security levels are as follows:

- **Security Level 1**: This level has no security. The device can connect without pairing, and transmission is not encrypted.
- **Security Level 2**: This level has encryption, but the encryption applies when the devices are not paired.
- **Security Level 3**: Devices communicating should be paired, and data transmission should be encrypted.
- **Security Level 4**: The most advanced security mode in Bluetooth. Authentication is enhanced, and the encryption mechanisms provide strong protection against threats.

With the security modes covered, the next step would be pairing devices. Pairing is where all the security-related processes take place. The purpose of pairing is to determine what each device is capable of and then get them talking [18]. Pairing is the process of generating, distributing, and authenticating keys for the encryption process [19]. This process takes place in three phases, and these are:

- ***Phase 1*: Initiate Pairing**

A device that attempts to pair sends a pairing request, and the peripheral device will respond with a pairing response. In this phase, the pairing devices exchange their pairing features, which would be used to determine which pairing method to use in the following phase and which keys should be distributed in phase 3 [19].

- ***Phase 2*: Perform Pairing**

In this phase, the key to encrypt the connection is generated. The Phase 1 pairing determines the method used here. In Bluetooth 1 devices, the device would exchange a Temporary Key (TK) that would be used to generate a Short-Term Key (STK) that would be used to encrypt the link. However, the STK can be easily cracked. In Bluetooth, version 4.2 implemented Low Energy Secure Connections. LESC connections, the devices would create and distribute the key more securely, and the distributed key would be used to create the Long-Term Key (LTK), which is used to encrypt the connection [19].

- ***Phase 3*: Key Distributions**

During this phase, the Long-Term Key is used to distribute the rest. All keys are generated and distributed during this phase [19].

The vulnerabilities of Bluetooth can appear in phase two depending on which pairing method is used. The Just Works pairing method is the simplest because both peers generate the STK based on the information exchanged in plain text, and the user of the Bluetooth device is just asked to accept the connection [19]. The Just Works pairing method is unauthenticated, and in the case of the case study above, this method of pairing would have to be changed to use a more secure pairing method. The following section explores Bluetooth attacks, how they start, and during which pairing method they can take place.

3.3 Bluetooth Attacks

An attack is an attempt to gain unauthorized access to a computer, computing system, or computer network to cause harm [20]. Bluetooth attacks can have adverse effects on the victim. Some attacks can have a high severity, medium, or low based on their impact on the victim. Attacks that gain full access to the user's device and can steal, change, or delete data from the memory or external storage of the victim's device can be categorized as high-severity attacks. Some Bluetooth attacks that are common are explained below.

- **PIN theft attack**

This attack involves getting hold of a pin through unethical ways and using it to access a victim's device for malicious activities [10].

- **PIN cracking attack**

This attack happens during the pairing stage of connecting Bluetooth devices. The attacker eavesdrops on the pairing and authentication process and then collects the messages. Next, the attacker will use brute force, an exhaustive method of cracking a PIN, an algorithm to find the PIN used during this process. If the MAC, which is the physical address of a computer hardware, is known, then the attacker can create a correct initialization key. Finally, the attacker can find the shared session link key using the collected data. If the collected information is accurate, the attacker can easily find the pin used to connect [10].

- **Relay attack**

In this scenario, the attacker will use two dummy devices to connect to the victim's device. The victim will not be aware of this attack and will continue to send messages. However, the victim is transmitting the data to the attacker. The attackers' devices will give feedback to the target devices [10]. The dummy devices are positioned near the recipient; one can perform the attack, and the other can imitate a legitimate device to track the target to connect to it [10].

- **MAC Address Spoofing Attack**

This attack uses the Bluetooth MAC address obtained by eavesdropping the communication. This is possible because the header to the MAC address has the destination and source address. During this attack, the MAC address is changed to the attackers' MAC address. The attack clones the victim's device; consequently, data destined for the victim's device goes to the attacker before the victim [10].

3.4 Bluetooth Protection

This section covers some countermeasures that one can take to make it hard for attackers to attack. The list given here is not exhaustive. Some methods may be suitable for some users or devices. It all depends on the application of the Bluetooth device. Listed below are some of the countermeasures that can be taken [10]:

- Turn off Bluetooth: *When users are not using the Bluetooth-enabled device, they should turn it off.*
- Set on Undiscoverable: *The device should be set to non-discoverable mode. Non-discoverable mode is when other devices cannot find Bluetooth when they scan it. Only when the user needs to be connected to the device can they set their device to discoverable.*
- Increased Pin Length: *To make it hard for an attacker to guess the pin or steal the secret keys. It suggested that long and random pins be used.*
- Update device Software: *Using outdated software on the Bluetooth device can lead to security vulnerabilities that make it easy for attackers to start an attack on the Bluetooth device.*
- Download from trusted sources.
- Do not use default passwords and configurations.
- Be familiar with your paired list: *Always keep a record of the devices you've paired with. Lost or stolen devices should be removed.*
- Beware of abnormal Activities: *When transmissions between devices get slower than usual, it could mean that the data is being eavesdropped. It can be attacked when the device starts acting abnormally, like showing weird messages or pop-ups.*

4 Conclusion

This research explored Bluetooth technology, its applications in healthcare, and the potential security pitfalls of wireless technology. Wireless technology, by design, uses an unbounded medium, which introduces a host of security problems that are not unique to Bluetooth but apply to all wireless technology.

The case study showed how an attacker can attack a Bluetooth-enabled insulin pump and the adverse effects it can have on victims. Not only can an attacker hijack a Bluetooth-enabled device, but they can also eavesdrop on the communication without the victim's knowledge.

Bluetooth has security levels; these levels have different security modes depending on the Bluetooth version and the device's capabilities. Pairing is another crucial process in Bluetooth communication, and it happens in three phases.

There are steps general Bluetooth users can take to ensure they do not end up as victims. This paper listed a couple of best practices for Bluetooth users. These best practices require users to do a little more than they usually do when using Bluetooth-enabled devices.

The overall aim of this paper is to make users aware of the capabilities of the technology they use daily, along with potential threats that manufacturers usually do not disclose, and finally, to give users an idea of what to do if they suspect an attack, also how to look out for potential attacks.

References

1. Steven J (2020) What is Bluetooth? Business insider. Retrieve from https://businnessinsider.com/guides/tech/what-is-bluethooth
2. Gillis AS (2023) IoT devices (internet of things devices) [Online] Available at: https://www.techtarget.com/iotagenda/definition/IoT-device. Accessed 26 Oct 2023
3. Taktak A (2022) Ad hoc network: definition and types [Online] Available at: https://study.com/academy/lesson/what-is-an-ad-hoc-network.html. Accessed 26 Oct 2023
4. Ferro E, Potorti F (2005) Bluetooth and Wi-Fi wireless protocols: a survey and a comparison. IEEE Wirel Commun 12(1):12–26
5. Jason M (2023) Now available: new Bluetooth forecasts and trends. Bluetooth.com. Retrieved from https://www.bluetooth.com/blog/now-available-new-bluetooth-forecasts-and-trends/
6. Grand View Research (2023) Wearable technology market, grand view research, Available at: https://www.grandviewresearch.com/industry-analysis/wearable-technology-market. Accessed 28 Oct 2023
7. Cassia Networks (2022) Bluetooth in healthcare: a transformative technology. Cassia Networks Blog. Retrieved from https://www.cassiannetworks.com/blog/bluetoothinhealthcare/
8. Bluetooth Range, Available at: https://www.bluetooth.com/learn-about-bluetooth/key-attributes/range/. Accessed 28 Oct 2023
9. Medium (2023) The Future of Healthcare: How Bluetooth Technology is Revolutionizing Patient Care. Medium, Available at: https://medium.com/@iamtecksay/the-future-of-healthcare-how-bluetooth-technology-is-revolutionizing-patient-care-8c2b129b673. Accessed 28 Oct 2023
10. Hassan SS, Bibon SD, Hossain MS, Atiquzzaman M (2018) Security threats in Bluetooth technology. Comput Secur 74:308–322
11. PCMag (2023) ISM Band. Available at: https://www.pcmag.com/encyclopedia/term/ism-band Accessed 19 Sept 2023
12. Security Guidance Bluetooth, Available at: https://security-guidance.service.justice.gov.uk/bluetooth/. Accessed 28 Oct 2023
13. Ly TT, Layne JE, Huyett LM, Nazzaro D, O'Connor JB (2019) Novel Bluetooth-enabled tubeless insulin pump: innovating pump therapy for patients in the digital age. J Diabetes Sci Technol 13(1):20–26. https://doi.org/10.1177/1932296818798836
14. Zeadally S, Siddiqui F, Baig Z (2019) 25 years of bluetooth technology. Future Internet 11(9):194
15. Nguyen A (2018) Bluetooth 1.0 versus 2.0 versus 3.0 versus 4.0 versus 5.0: How they compare, symmetry electronics blog, Available at: https://www.symmetryelectronics.com/blog/bluetooth-1-0-vs-2-0-vs-3-0-vs-4-0-vs-5-0-how-they-compare-symmetry-blog/. Accessed 29 Oct 2023
16. Dusun (2023) How Bluetooth 1.0, 2.0, 3.0, 4.0, and 5.0 compare. Dusun IoT Blog, Available at: https://www.dusuniot.com/blog/how-bluetooth-1-0-2-0-3-0-4-0-and-5-0-compare/. Accessed 29 Oct 2023
17. Kardach J (2000) Bluetooth architecture overview. Intel Technol J 2:7
18. Mark L (2018) Understanding Bluetooth security. [Online] Available at: https://duo.com/decipher/understanding-bluetooth-security. Accessed 20 Sept 2023

19. Nordic semiconductor (2023) Pairing process. [Online] Available at: https://academy.nordicsemi.com/courses/bluetooth-low-energy-fundamentals/lessons/lesson-5-bluetooth-le-security-fundamentals/topic/pairing-process/. Accessed 20 Sept 2023
20. TechTarget. Cyber Attack. [Online] Available at: https://www.techtarget.com/searchsecurity/definition/cyber-attack. Accessed 20 Sept 2023

Modeling and Simulation in Temperature Control in Milk Coagulation for Edam Cheese Production Using Fuzzy Logic

Honorato Ccalli Pacco

Abstract The modeling and simulation investigation of temperature and pH control in the coagulation of milk in the cheese vat for the production of Edam cheese were carried out. The coagulation of milk in the production of Edam cheese, which is special because it has proteins, carbohydrates, cholesterol, vitamins and minerals in its composition, and is therefore very important because it also depends on the control of temperature and pH. The coagulation will be of the desired quality of texture and determine the time that final coagulation takes in the cheese vat. It was carried out at the Professional School of Agroindustrial Engineering of the National University of Moquegua, Peru. In the modeling and simulation of temperature and pH control in milk coagulation, artificial intelligence is used, which in this case was fuzzy logic that allows us to replicate the thinking of a specialist, also with the use of LabVIEW software that allows us to do graphical programming with virtual instruments that are very easy to use compared to linear programming that requires a programming specialist. In the research carried out, it was possible to obtain graphic programming for the modeling and simulation of milk coagulation for the production of Edam cheese, the temperature is controlled at the programmed set point (33 °C) with the application of sufficient water vapor to achieve it. With the results of the pH measurement, it was possible to simulate the duration of milk coagulation in the cheese vat.

Keywords Edam cheese · Coagulation · pH · Fuzzy logic

H. C. Pacco (✉)
National University of Moquegua, Pacocha, Peru
e-mail: pacco122@gmail.com

X.-S. Yang et al. (eds.), *Proceedings of Ninth International Congress on Information and Communication Technology*, Lecture Notes in Networks and Systems 1000,
https://doi.org/10.1007/978-981-97-3289-0_36

1 Introduction

1.1 Edam Cheese

The authors [1] studied and produced two batches of milk for Edam cheese production: one batch with pasteurization and one batch without pasteurization. For the first batch, the raw milk was pasteurized at 65 °C for 5 min and then cooled to 35 °C. They then inoculated starter culture and L. casei (1.00 × 108 CFU/ml) and rennet. The milk is mixed and kept for approximately 30 min. After coagulation, the curd is cut and the whey is removed.

According to the authors [2], in work insights into the microbiota and defects of present-day Gouda cheese productions, they describe the production of Edam cheese and state that it controls the fat content in percentage of dry mass, adding mesophilic starter culture, beef rennet to achieve the acidification and that in the process of cutting the resulting curd a temperature of 38 °C must be maintained in the wash, the whey is removed and the curd is molded into cylinders to then be pressed, then they must be brined for 2–4 days to achieve a concentration of 2.5–3.5% in dry matter.

According to the authors [3], "Effect of water content on the hardness of Edam cheese" where they state that the hardness of Edam cheese depends on the moisture content, and therefore on the time and temperature (38 °C) when washing the curd.

In the production of Dutch cheese, the temperature in the milk must be approximately 30 °C for good coagulation with the use of coagulant enzymes and calcium chloride, additives (nitrate and annatto). After cutting the curd, a part of the whey is removed, to which a quantity of warm water (at 36 °C) is added, washing or blanching the grains, which allows reducing the lactose content and improving syneresis with moderate agitation. Once blanching is complete, they are separated from the whey and the pre-pressing continues at room temperature, followed by cutting, shaping, and pressing and then taking it to a brine at a controlled temperature, researchers [4]. The researchers, [5], report that semi-soft cheeses (Edam) are the most common group of matured cheeses, and can mature for a period of 30–90 days in a controlled temperature environment and with a moisture content of average to 45%.

1.2 Fuzzy Logic in LabVIEW

The researchers [6] carried out research on the modeling, design, and simulation of an automatic control system in greenhouses, designing a virtual greenhouse simulating the natural phenomena of a greenhouse; for this, they used artificial intelligence, which in this case was the logic diffuse that allows the thinking of a specialist to be replicated.

A fuzzy logic system was used and based on the use of fire alarm data for a period of 5 years, which recognized patterns of real fire signals and allowed the determination of the "real alarm" condition [7].

The author [8], responsible for the research on growing tulips in a virtual greenhouse, uses fuzzy logic from the National Instrument LabVIEW software to simulate temperature control and irrigation time for growing tulips in the greenhouse indicated above, for this purpose a reasonable and important temperature is maintained for the cultivation of said flower, which is highly appreciated worldwide.

1.3 LabVIEW

In an investigation, researchers [9], using the LabVIEW software from National Instrument, achieved a reliable, fast, and accurate calibration service, reducing the intervention of specialists to control dosimeter systems frequently used in the field of radiotherapy in hospitals.

The authors of the paper, [10], in their research work conclude that in graphical programming using the National Instrument LabVIEW software, can control a good number of devices with different hardware and that it is also possible for additional modification of a hybrid system. This software is easy to program in graphics and does not require a programming specialist as in the cases of linear programming.

Dume et al. [11] use the LabVIEW software to control and simulate parabolic solar panels to produce thermal and electrical energy, since this software has many advantages such as programming with graphics that is very easy to use, it does not require experts for said programming. In this study, they predicted the energy production for an area with 6000 inhabitants and estimated the need for each of the hotels. They were also able to calculate the electricity needs of the network.

According to the authors [12], in their research work, they report that the fuzzy logic system is a linguistic variable with values that are the names of fuzzy sets. Thus, we have as an example the temperature of a process and it can be a linguistic variable taking values such as low temperature, medium temperature, and high or very high temperature which would be the fuzzy sets representing membership functions and can be seen in Fig. 1.

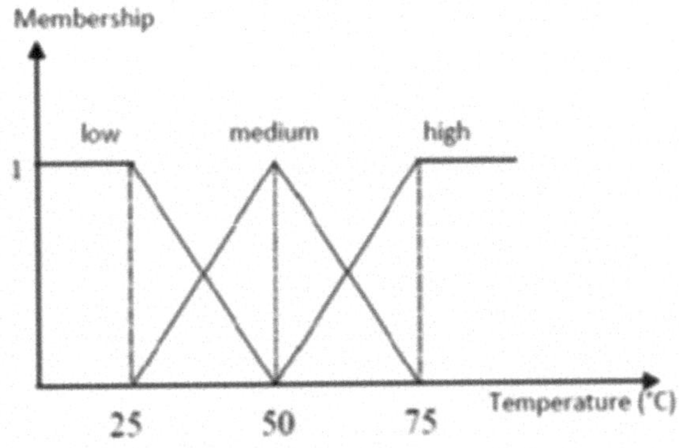

Fig. 1 Membership functions for the temperature variable. *Source* Pacco [12]

1.4 Objectives

Modeling and simulation in temperature control in milk coagulation for Edam cheese production using fuzzy logic, considering as input variables temperature, pH, and output variables water vapor and milk coagulation time in the vat cheese maker.

2 Material and Methodology

2.1 Material

It is necessary to have a work environment or laboratory, computer, hard drive, note material, agenda book, notebook, pens, fuzzy logic program, LabVIEW software, etc.

2.2 Methodology

The research work has been carried out in the laboratory or computer workshop of the Professional School of Agroindustrial Engineering of the National University of Moquegua, in the country of Peru.

Using the recommendations found in the works of researchers [8, 13], first, a bibliographic review of the production of Edam cheese is carried out, followed by the parameters of temperature, pH, and time in the coagulation of milk to The production of Edam cheese continues with the use of artificial intelligence, fuzzy logic applying the fuzzyfication of variables (Figures 2 and 3), this allows us to obtain logical and very useful results.

In which the temperature control in the coagulation of milk in the cheese vat for the production of Edam cheese is proposed (20–40 °C, is the membership and the Set point of 33 °C, membership function), in which It is programmed for each temperature variation, there is a response of the amount of water vapor to stabilize the temperature at the set point.

Next, we continue with the creation of operating rules (Fig. 4),

1. If temperature °C is cold, then water steam is high.
2. If temperature °C is set point, then water steam is medium.
3. If temperature °C is hot, then water steam is low.
4. If pH is high, then time (min) is low.
5. If pH is set point, then time (min) is set point.
6. If pH is low, then time (min) is high.

Continuing with the test or tests with the system (Fig. 5), where we can verify the results with the simulation, in part of the fuzzy logic.

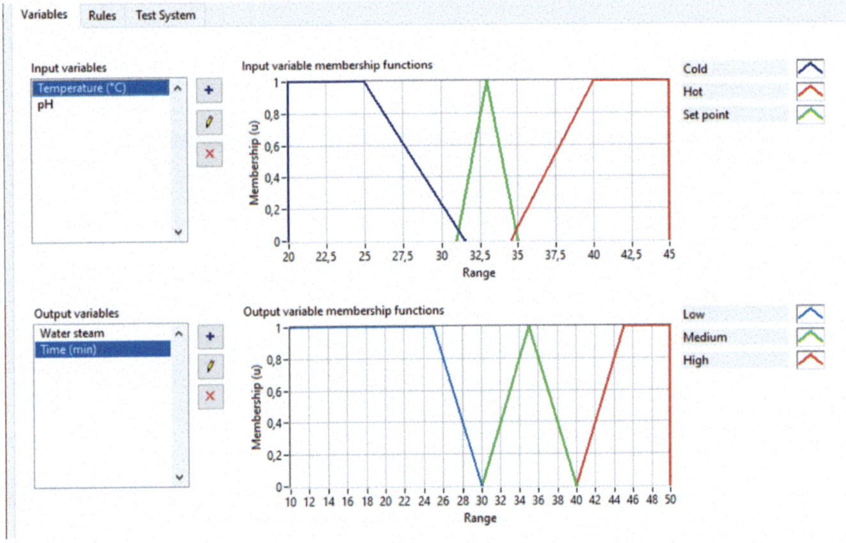

Fig. 2 Fuzzy interference system set as a controller

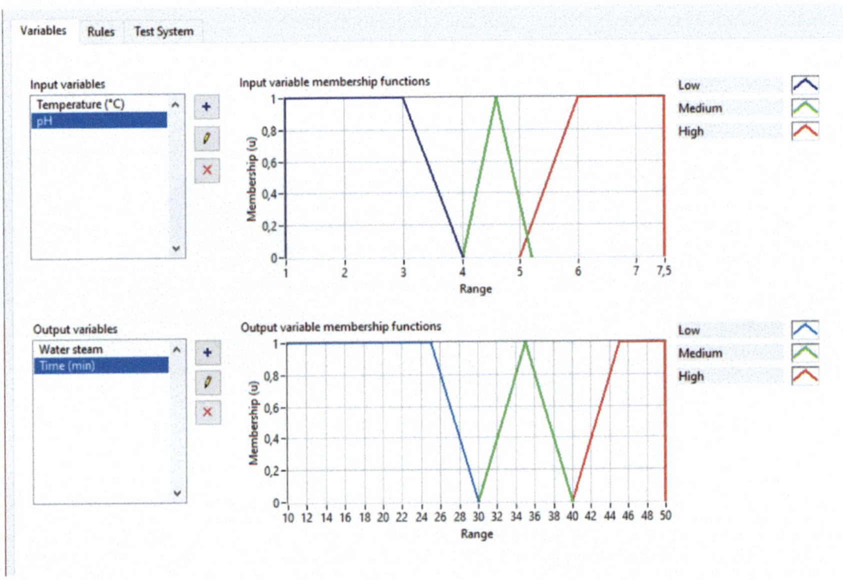

Fig. 3 pH variable as input and time in minutes as output variable (time that coagulation would take depending on pH)

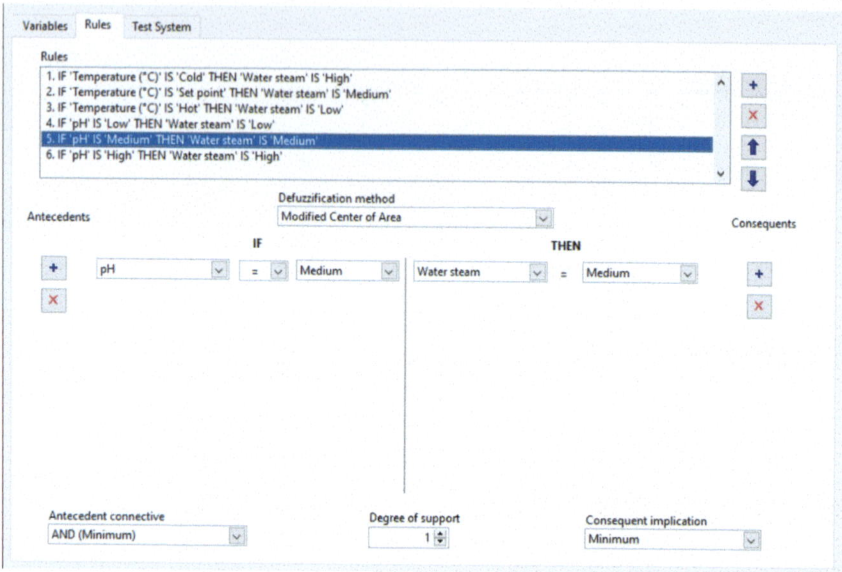

Fig. 4 Rules for the fuzzification system for input variables of temperature and pH with their respective values as responses (water vapor and coagulation time)

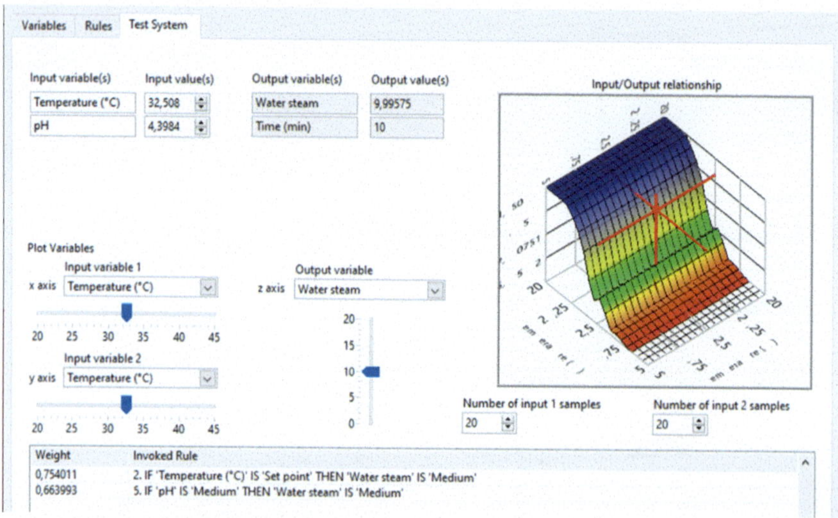

Fig. 5 Tests that can be performed with temperature and pH values for a response with the values of water vapor and coagulation time

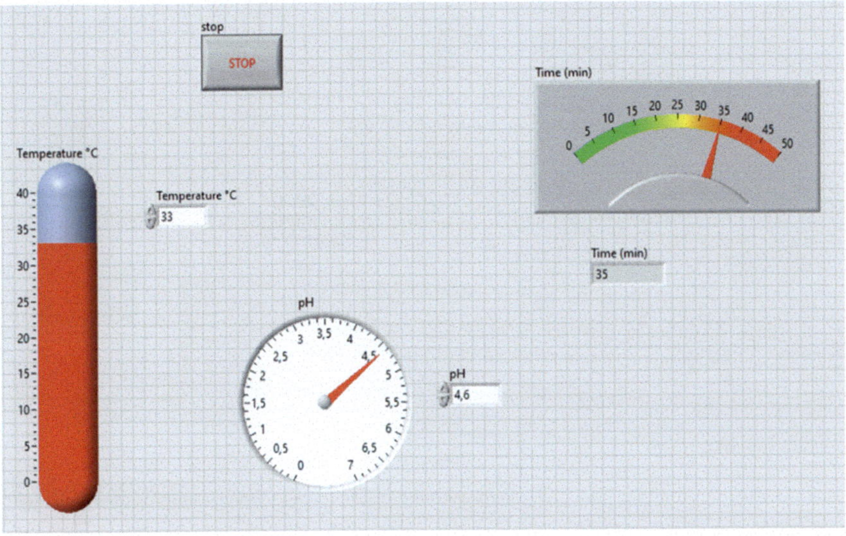

Fig. 6 Front panel for controlling temperature and concentration of hydrogen ions (pH) in the milk for coagulation in the cheese vat

Then we proceeded with the programming in the LabVIEW software using the fuzzy logic carried out previously. For this, we use the front panel and the block diagram panel.

The front panel included virtual temperature instruments, temperature indicator, water vapor, water vapor value indicator, virtual pH meter, pH value indicator, and stop button (Fig. 6).

In the block diagram panel, the installation of the load fuzzy system and fuzzy controller was completed, and then the corresponding connections were made (Fig. 7).

3 Results and Discussion

Work has been carried out on the Fuzzyen interference system where it was possible to create temperature and pH controllers with water vapor response to stabilize the set point temperature (33 °C) and time used in said coagulation for the cheese vat for the coagulation of milk for Edam cheese production, in a similar way to the research project carried out by the authors [14], where references of control variables, fuzzyfication, rule bases for the system and defuzzyfication for controlled variables were presented in the cheese vat, as can be seen in Fig. 1.

The controlled variables for temperature were: cold temperature (20–31.5 °C); set point temperature (31–33 and 35 °C); and hot temperature (34.5–40 °C), can be seen in figure × 3. For the pH control variable, the following were recorded: pH low

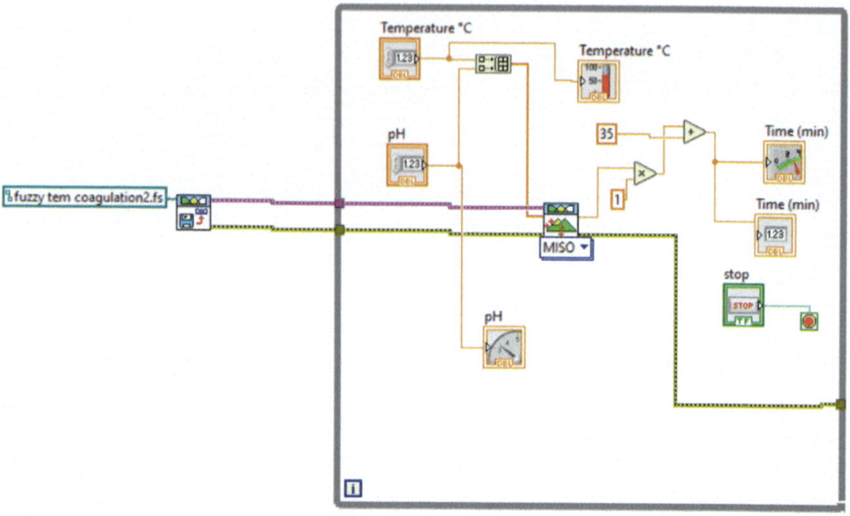

Fig. 7 Virtual instrumentation and connections outside and inside the while loop structure

(1 to 4); pH set point (4–4.6 and 5); and high pH (5–7.5). For response variables, water vapor was recorded: water steam low (0–9); water steam set point (8–10 and 12); water steam high (11–20). For the time (min) response variable: time (min) low (10–30); time (min) medium (30–35 and 40); and time (min) high (40–50). It is the time, and it can take to coagulate the milk for the production of Edam cheese Fig. 3.

In Fig. 1, we can see the pH values as input variable and as response variable the time in minutes that the coagulation of the milk can take in the cheese vat for the production of Edam cheese.

In Fig. 4, we observe the rules that have been presented previously regarding the temperature in the cheese vat and that would respond to water vapor values to stabilize the temperature at the set point. Likewise for the pH variable that would respond to the time it takes for the milk to coagulate in the cheese vat.

Figure 5 shows the evidence that the temperature and pH control system can present in the coagulation of milk in the cheese vat for the production of Edam cheese.

Once the corresponding system programming has been carried out, we obtain the front panel where it allows us to simulate the temperature control in the coagulation of milk for the production of Edam cheese with a response with water vapor values that must stabilize the indicated temperature, such as the set point. 33 °C; and also the control of the concentration of hydrogen ions (pH) in the curd, with its decrease to a value of 4.6 in which the completion of the coagulation of the milk must be determined with a time response that delayed the coagulation in the cheese vat, and we can see it in Fig. 7.

Simulation

Using the virtual instruments and the fuzzy logic system, in the block diagram panel each of the virtual components is organized and their respective lines of connections are made outside and inside the structure While loop, Fig. 7.

Simulation (temperature and water steam)

Water steam	Temperature (°C) en tina quesera				
	45	40	33	30	20
0	Low				
05		Low			
08		Low			
10			Medium		
11				High	
12					High
16					High

Simulation (pH and time (min))

Time (min)	Concentración de hidrogeniones (pH)							
	7,5	6,0	5,0	4,6	4,0	3,0	2,0	1,0
10	Low							
20		Low						
30			Low					
35				Medium				
40					Medium			
45						High	High	High

4 Conclusions

In this research work, the modeling and simulation of temperature and pH control have been carried out with responses at the outlet with the values of water vapor and the time used in the coagulation of milk in the cheese vat for the production of Edam cheese with the expected results that allow us to simulate the system using artificial intelligence, in this case with fuzzy logic in the LabVIEW software.

The use of fuzzy logic has allowed us to apply membership functions by assigning variables and inference rules, which are a set of rules that interact with each of the variables used in the system under study.

LabVIEW software has made graphical programming easier for us without the need for a linear programmer.

Acknowledgements We thank the Professional School of Agroindustrial Engineering of the National University of Moquegua for allowing me to carry out this research work, and all the people and authorities of the university who made the presentation of this paper possible.

References

1. Amiri S, Abotalebi Kohneshahri SR, Nabizadeh F (2022) The effect of unit operation and adjunct probiotic culture on physicochemical, biochemical, and textural properties of Dutch Edam cheese. LWT 155:112859
2. Decadt H, De Vuyst L (2023) Insights into the microbiota and defects of present-day Gouda cheese productions. Curr Opin Food Sci 52:101044
3. Wodecki E et al (1984) Effect of water content on the hardness of Edam cheese. J Food Eng 3(4):295–305
4. Düsterhöft EM, Engels W, van den Berg G (2016) Cheese: Dutch-Type Cheeses. In: Reference module in food science. Elsevier
5. Kongo JM, Malcata FX (2016) Cheese: types of cheese—medium. In: Caballero B, Finglas PM, Toldrá F (eds) Encyclopedia of food and health. Academic Press, Oxford, pp 755–762
6. Kavga A et al (2023) Design and simulation of a greenhouse in a computational environment (ANSYS/FLUENT) and an automatic control system in a LABVIEW environment. Simul Model Pract Theory 129:102837
7. Park SH, Kim DH, Kim SC (2023) Recognition of IoT-based fire-detection system fire-signal patterns applying fuzzy logic. Heliyon 9(2):e12964
8. Pacco HC (2022) Simulation of temperature control and irrigation time in the production of tulips using Fuzzy logic. Procedia Comput Sci 200:1–12
9. Aydin S, Kam E (2018) Developing of an automation for therapy dosimetry systems by using labview software. Results Phys 9:1007–1015
10. Wrobel P et al (2012) LabVIEW control software for scanning micro-beam X-ray fluorescence spectrometer. Talanta 93:186–192
11. Dume G, Spahiu A (2010) Using labview software in control and simulation of parabolic solar panels in order to produce thermal and electrical energy for touristic places in development countries. IFAC Proc Volumes 43(25):101–105
12. Pacco CH (2021) Temperature simulation and control for lab-scale convection dehydrators. Procedia Comput Sci 180:922–934

13. Rahman MS et al (2023) Machine learning and internet of things in industry 4.0: a review. Measur Sens 28:100822
14. Martínez SL, Tarifa EE, Domínguez SFJD (2017) Algoritmo para linealización de controladores fuzzy 11(11)

AMEACrypt: Android Mobile Encryption Application Generating Password Using AES Algorithm

Jerry E. Borromeo, Roxanne A. Pagaduan, Junnel E. Avestro, Ryan Christopher B. Pinca, Jess N. Garcia, and Zyrus D. Gremio

Abstract Smartphones have quickly integrated into our daily lives with their numerous applications. Their worth is evident for everything from downloading files to watching online films and reading electronic books. Many people look for software that helps protect their important files, even though others may think downloading files to mobile devices is safe. The mobile application AMEACrypt offers a complete solution, including modules such as password generator, password vulnerability checker, ciphertext encryption and decryption, and file encryption and decryption functions. These elements are crucial for ensuring personal security through stronger passwords or data encryption with a unique decryption key known only to the user. AMEACrypt application includes a password generator, checker, and vulnerability checker for passwords. The Agile Methodology was used due to its effectiveness in delivering a desirable level of development. Feedback and teamwork were used throughout the phases to achieve an acceptable level. Using ISO/IEC 9126, IT experts, students, and others assessed the application's usability, functionality, dependability, efficiency, maintainability, and portability. According to the end-user evaluation results, the mobile application received a total rating of 5.65 and a strongly agree rating for its interpretation. The result indicates that the mobile

J. E. Borromeo (✉) · R. A. Pagaduan · J. E. Avestro · R. C. B. Pinca · J. N. Garcia · Z. D. Gremio
Technological Institute of the Philippines, Quezon City, Metro Manila, Philippines
e-mail: jborromeo.it@tip.edu.ph

R. A. Pagaduan
e-mail: rpagaduan.it@tip.edu.ph

J. E. Avestro
e-mail: javestro.it@tip.edu.ph

R. C. B. Pinca
e-mail: rcpinca.it@tip.edu.ph

J. N. Garcia
e-mail: jgarcia.cs@tip.edu.ph

Z. D. Gremio
e-mail: zgremior@tip.edu.ph

X.-S. Yang et al. (eds.), *Proceedings of Ninth International Congress on Information and Communication Technology*, Lecture Notes in Networks and Systems 1000,
https://doi.org/10.1007/978-981-97-3289-0_37

application helps users generate passwords, check passwords, encrypt and decrypt ciphertext, and encrypt and decrypt files.

Keywords Mobile application · Mobile security · Encryption · Decryption · AES · AES-128 · Ciphertext · Password generation

1 Introduction

Mobile phones are quickly becoming necessary in our everyday lives as they are indispensable tools for individuals and organizations. In 2023, the global count of mobile phone subscriptions surged to 5.7 billion. Over the entire span from 2018 to 2023, the compound annual growth rate stood at two percent [1]. Smartphones are increasingly used as storage devices for personal data, such as financial records, contact logs, multimedia files like pictures and videos, and health-related data. This problem directly impacts the widespread adoption of different smartphone applications. Financial losses, user trust, and data privacy are some problems that may arise while using mobile phones [2].

The need to ensure mobile security has risen due to people's significant time daily on their phones [2]. The amount of security hazards that target mobile devices is on the rise, but a sizable section of the population continues to be ignorant of these threats. Due to their rising use and importance in daily life, smartphones are subject to an expanding range of cybersecurity concerns, including data breaches, malware, phishing attempts, and network spoofing. There are many ways for hackers to access mobile devices.

Passwords act as the first line of defense against unwanted access to a person's computer system and the protection of personal data. The more secure your password is, the more security you have against hackers accessing your computer and online accounts. Strong passwords must be kept up-to-date for all your accounts [3]. One of the best methods to lessen your risk of cybercrime is to create and use safe passwords. Likewise, text-based passwords are still a common way to authenticate users. People frequently choose passwords that are easy to remember, such as their names, pet names, anniversary dates, or even birth dates. Regarding password security, using familiar facts as passwords is considered a bad habit [4]. The issue persists even though numerous steps are made to stop users from using weak passwords, including awareness campaigns, password-strength checkers, and password composition regulations [5].

The main objective of this study is to develop a mobile application that will increase security by protecting the user's files. Specifically, the study aims to develop a mobile application designed to assist users in generating strong passwords for their accounts, implement a file encryption and decryption system using AES-128 bit encryption together with additional salt to improve the security of the user's files by ensuring unique passwords and protection and evaluate the mobile application's performance using ISO/IEC 9126 criteria such as functionality, reliability, usability, efficiency, maintainability, and portability.

2 Theoretical Framework

In an era of increasing mobile usage, many individuals believe their devices are secure, yet this perception often needs to align with reality. The shift toward remote work due to the pandemic has made people more susceptible to cyberattacks. According to [6], about 40% of mobile users worldwide are vulnerable to cyber threats. These vulnerabilities extend beyond operating systems, impacting hardware as well. The consequences of successful attacks can range from data leakage—including personal information, photos, and recordings—to rendering devices unresponsive or installing irreparable malware.

The Google Play Store hosts numerous security applications to protect devices from malware. However, research by Yao et al. [7] assessed the trustworthiness of these applications. Their analysis revealed that many security applications share collected data with authorities or business affiliates. Many also collect users' device information and even resell data without user consent. Additionally, these apps often request multiple permissions, including sensitive ones.

Password security remains a critical concern. Research by [8] highlighted the risks of offline password cracking. Attackers exploit weak passwords, especially when users reuse them across accounts. This vulnerability increases with advances in hardware capable of rapidly assessing password hash algorithms. A study [9] emphasized balancing security and utility in mobile applications. As mentioned in the study [10], a centralized password management system with application-specific passwords could enhance security.

In response, [4] introduced an algorithm to generate strong passwords based on user inputs. Meanwhile, the study [11] developed tools to assess password strength against mangled wordlist attacks. Furthermore, studies [12, 13] explored optimizations and architectures to improve the widely adopted Advanced Encryption Standard (AES).

Despite AES's effectiveness, its usage needs careful consideration. The study [14] assessed the practicality of AES and compared it to other encryption algorithms. A study [15] discussed AES's architecture and adoption post-NIST approval. The study [16] proposed using a different approach to multi-factor authentication using random numbers. Similarly, a study [17] focused on enhancing password security and addressing cyber threats.

As discussed in [18], demonstrated how predictable password patterns can lead to breaches. Considering these studies, the proposed AMEACrypt Android Security application aims to offer protective measures against hacking and data breaches. By leveraging insights from these research works, the application aims to enhance its security features and provide a robust shield against cyber threats, benefiting users' digital safety.

Fig. 1 Agile Methodology

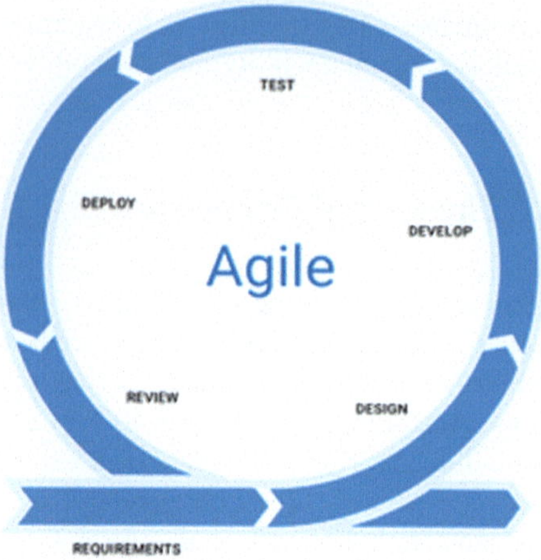

Agile Methodology [19] is used in the development of our application. It is a combination of incremental and iterative process models.

Figure 1 shows the Agile Methodology, in which each iteration includes cross-functional teams working simultaneously on different areas.

2.1 Requirements Phase

The research begins by looking into mobile security concerns and studying pertinent literature. The project's parameters and scope are established at this stage, giving a clear perspective.

2.2 Design Phase

Using a lot of data from the requirements phase, the team develops a plan. During this phase, the application's tools and core must be developed. Prototyping comes to life with the precise carving of user interface and user experience design.

2.3 Development Phase

Armed with the thorough plans made during the Design Phase, the coding starts in Visual Studio Code and Android Studio. The team collaborates to begin developing various application modules. They expertly use various tools and libraries to produce a vibrant prototype based on the design ideas.

2.4 *Testing Phase*

The program is thoroughly inspected to ensure it is stable, functional, and responsive to user needs. Additionally, the application's attributes are compared to the high bar set by ISO/IEC 9126 standards to ensure that it meets acceptable standards.

2.5 *Deployment Phase*

At this stage, the mobile application becomes operational in the real world. Users have the opportunity to utilize it for demonstrations or real-life scenarios. The goal is to ensure a seamless user experience. Any minor glitches that may arise are promptly addressed in subsequent iterations.

2.6 *Review Phase*

The collaboration to generate solutions for potential obstacles that could arise. Guided by the principles of the agile approach, they determine whether to advance to the next stage or initiate a reset if needed.

2.7 *Testing and Operating Procedures*

Before the application is released, any errors are detected and corrected using testing and operating procedures. The testing is carried out using user acceptance testing, beta testing, and alpha testing.

3 System Architecture

The system architecture of the AMEACrypt mobile application is organized into a three-tier model, facilitating seamless interaction and the application components.

Figure 2 shows the interaction between the mobile application and the user. The presentation tier displays the program's mobile interface, a medium for translating the user's task. The logical tier checks to see if the password meets the requirements

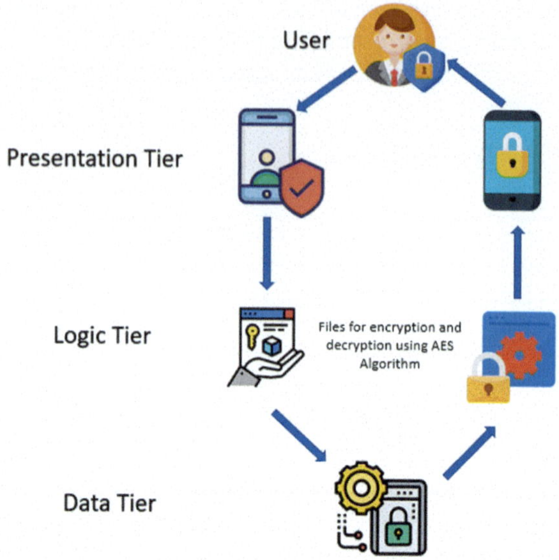

Fig. 2 System architecture of AMEACrypt

and assesses if it is common or not, as well as having the capacity to execute encryption and decryption requiring two-factor authentication. The data tier contains the application's database, which can retrieve and store data. The information is then transferred back to the logic layer for processing, returning to the user and alerting them that the process has been finished.

4 Result and Discussion

The AMEACrypt mobile application increases user security by protecting their data from hackers. This application serves as a tool for generating strong passwords appropriate for internet accounts. Users can browse the internet with more security and confidence by using AMEACrypt. By utilizing AMEACrypt, users can browse the internet with increased confidence and safety. The application comprises several modules, such as a password generator, password vulnerability checker, ciphertext encryption and decryption, and file encryption and decryption.

The input provided by the user should consist of two unique numbers and five distinct words, and this input serves as the basis for the password generator to create passwords swiftly. As evident from its name, the password vulnerability checker examines the susceptibility of the password entered by the user, offering feedback and assessing its strength. Moreover, it calculates the time required to crack the password, indicating the speed at which a specific algorithm might decrypt the provided password. This module's functionality unveils the pattern and token sequence in crafting the password. It also provides instructional guidelines for using the application by

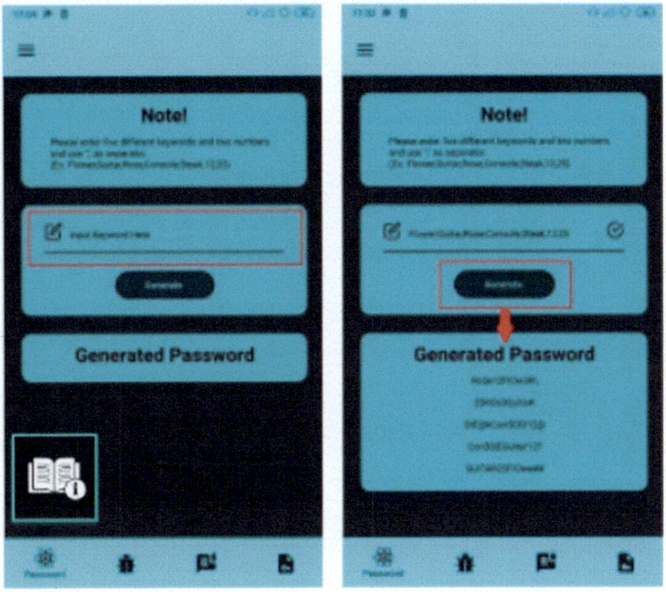

Fig. 3 Password generator of AMEACrypt

clicking the information icon. Likewise, AMEACrypt is available for download and installation on Android smartphones running version 5.0 and above, requiring a minimum of 70 MB of phone storage space. The application is poised to expand its compatibility and widen the user base by implementing it in iOS and other mobile platforms.

4.1 Password Generator

The module for generating passwords is shown in Fig. 3. The initial visual element presented to the user is the set of text view instructions for utilizing the module. Subsequently, the user encounters a text field where they can input five distinct keywords and two numbers to formulate a password. If the user deviates from the instructions, an error message will display, prompting them to adhere to the guidelines.

4.2 Password Vulnerability Checker

Figure 4 illustrates the module responsible for checking password vulnerabilities. Users can access this module by tapping the second button on the bottom menu bar within the application. Within this module, the strength of the password entered by

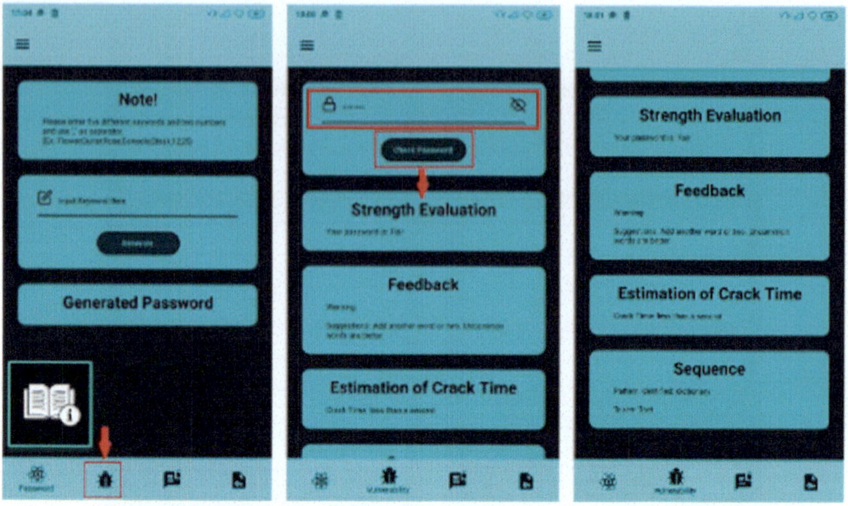

Fig. 4 Password vulnerability checker of AMEACrypt

the user is assessed and displayed in the text box. Furthermore, the module assesses the password and provides the user with feedback on how strong the password was entered. The module also includes an estimate of the crack time, providing insight into the speed at which a specific algorithm or library could decipher the entered password. Moreover, the module reveals the sequence, exposing the pattern and token utilized to identify the user's password.

4.3 *Ciphertext Encryption*

Figure 5 shows the ciphertext encryption of the AMEACrypt mobile application. To use the ciphertext encryption, enter the desired word, phrase, or sentence in the first text field. On the second text field, enter the desired password. The key will be used to decode the ciphertext output later on decryption. Tap the unhide icon to check if the entered password is correct. After putting the requirements on the two text fields, tap the "Encrypt" button to see the ciphertext output. A One-Time-Password code will be sent to the user's identified mobile number to complete the entire process. This allows the user to finalize the encryption and also in the decryption process.

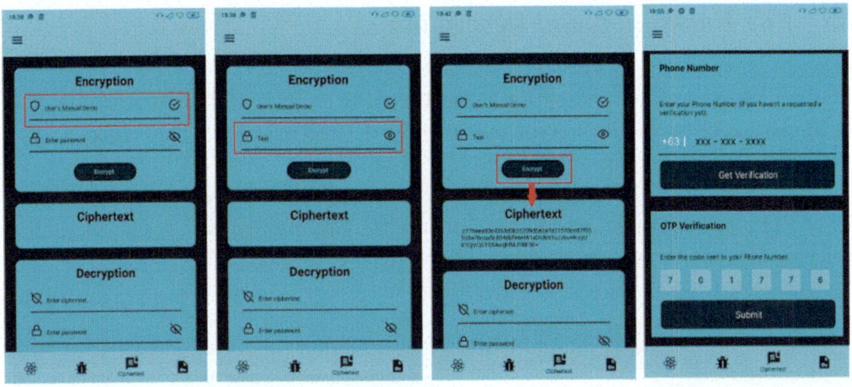

Fig. 5 Ciphertext encryption of AMEACrypt

4.4 *Ciphertext Decryption*

The AMEACrypt mobile application's ciphertext decryption is displayed in Fig. 6. Copy the ciphertext output and paste it into the first text field before using the ciphertext decryption. Enter the password in the second text field to encrypt the word, phrase, or sentence. Tap the unhide symbol to verify that the password you provided is accurate. Tap the "Decrypt" button to see the output of the decrypted ciphertext after entering the specifications in the two text boxes.

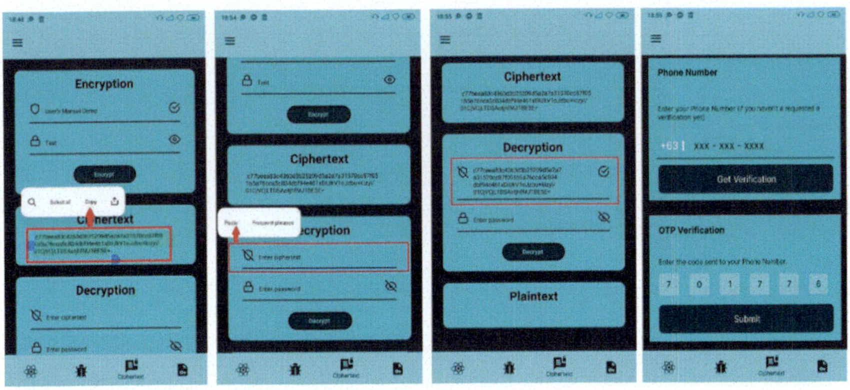

Fig. 6 Ciphertext decryption of AMEACrypt

4.5 File Encryption and Decryption

Figure 7 shows the file encryption using the AMEACrypt mobile application. The user must select the file to encrypt and enter the desired password, then click encrypt. The user can save the encrypted file.

Figure 8 shows the file decryption process using the AMEACrypt mobile application. The user must select the file to decrypt, enter the password, and then click decrypt. The user views the decrypted file once it is processed.

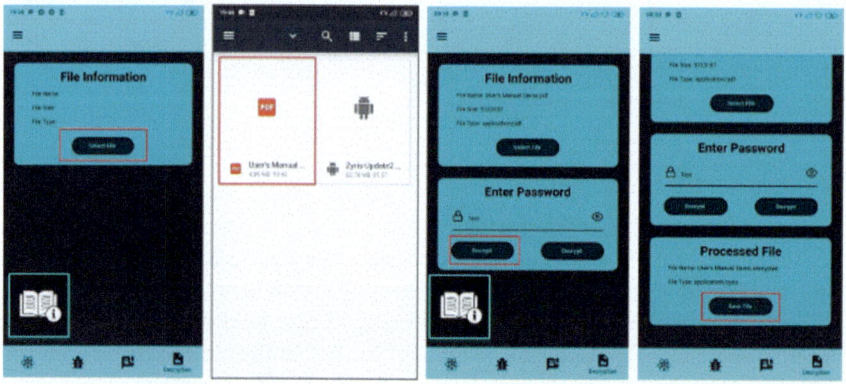

Fig. 7 File encryption of AMEACrypt

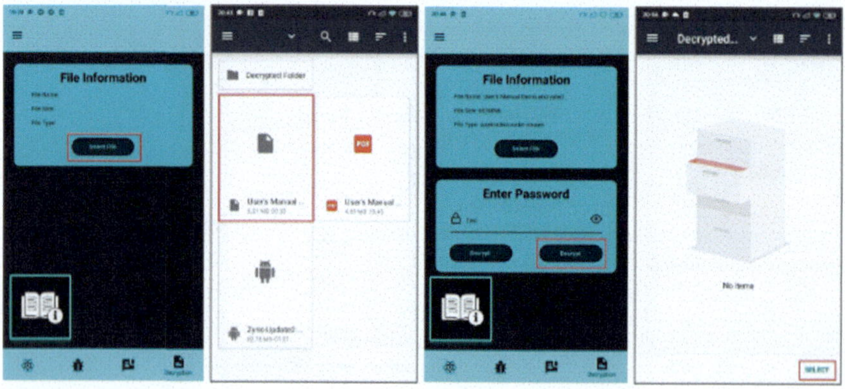

Fig. 8 File decryption of AMEACrypt

Table 1 Summary evaluation results of the respondents

Criteria	Mean	Evaluation result
Functionality	5.61	Strongly agree
Reliability	5.60	Strongly agree
Usability	5.73	Strongly agree
Efficiency	5.70	Strongly agree
Maintainability	5.71	Strongly agree
Portability	5.60	Strongly agree
Total average	5.65	Strongly agree

5 Evaluation Results

AMEACrypt was evaluated using ISO/IEC 9126 standard by two groups, including men and women and IT Professionals.

The average score for each of the six criteria is displayed in Table 1. Based on respondents' user experiences, the average score of 5.65 was evaluated as "strongly agree," meeting all the evaluation categories.

6 Conclusion

This study concludes that the AMEACrypt mobile application effectively accomplished the project objectives, which aimed to establish a tool for enhancing mobile users' security while browsing the internet on their devices. AMEACrypt allows users to construct strong passwords for their accounts using the words and numbers provided after the program has been developed. Furthermore, the application assesses password strength against dictionary-based or brute-force attacks.

Moreover, the application successfully handles message and file encryption and decryption. It relies on the Advanced Encryption Standard 128 (AES-128) to ensure the security of user-committed content. Notably, the AMEACrypt mobile application adheres to ISO/IEC 9126 evaluation standards encompassing functionality, reliability, usability, efficiency, maintainability, and portability. It achieved an overall score of 5.65, denoting strong agreement. This score signifies that the application met project objectives and delivered satisfactory performance.

Acknowledgements This research would like to express gratitude to Dr. Felizardo C. Reyes Jr., Dr. Gerald T. Cayabyab, Mr. John Jacob E. Bascuguin, and Tristan VJ Sidamon for sharing their skills and expertise with this research study.

References

1. Taylor P (2023) Global growth of mobile subscriptions 2018–2023. Retrieve from https://www.statista.com/statistics/1190254/mobile-subscribers-growth-worldwide/
2. Mos A, Chowdhury MM (2020) Mobile security: a look into android. In: 2020 IEEE international conference on electro information technology (EIT), Chicago, IL, USA, pp 638–642. https://doi.org/10.1109/EIT48999.2020.9208339.
3. Bednar L (2019) Tips for being password smart. [Online]. Available: https://www.securedrive.com/blog/tips-for-being-password-smart
4. Glory FZ, Ul Aftab A, Tremblay-Savard O, Mohammed N (2019) Strong password generation based on user inputs. In: 2019 IEEE 10th annual information technology, electronics and mobile communication conference, IEMCON 2019, pp 416–423. https://doi.org/10.1109/IEMCON.2019.8936178
5. Stavrou E (2017) A situation-aware user interface to assess users' ability to construct strong passwords: a conceptual architecture. In: 2017 International conference on cyber situational awareness, data analytics and assessment (Cyber SA), London, UK, pp 1-6. https://doi.org/10.1109/CyberSA.2017.8073385
6. Check Point Software Technologies (2021) Mobile security report 2021 insights on emerging mobile threats. Check Point Software Technologies. https://www.checkpoint.com/downloads/resources/mobile-security-report-2021.pdf
7. Yao W, Li Y, Lin W, Hu T, Chowdhury I, Masood R, Seneviratne S (2020) Security apps under the looking glass: an empirical analysis of android security apps. In: Proceedings conference on local computer networks, LCN, 2020 November, pp 381–384. https://doi.org/10.1109/LCN48667.2020.9314784
8. Bai W, Blocki J (2021) DAHash: distribution aware tuning of password hashing costs. arXiv preprint arXiv:2101.10374
9. Nimmi K, Janet B (2018) An analysis of the balance between security and utility of mobile applications. In: 2018 International conference on circuits and systems in digital enterprise technology, ICCSDET 2018, pp 1–4. https://doi.org/10.1109/ICCSDET.2018.8821080
10. Smith T, Ruoti S, Seamons K (2017) Augmenting centralized password management with application-specific passwords. In: Way, pp 1–5. https://www.usenix.org/system/files/conference/soups2017/way2017-smith.pdf
11. Liu E, Nakanishi A, Golla M, Cash D, Ur B (2019) Reasoning analytically about password-cracking software. In: 2019 IEEE symposium on security and privacy (SP). https://doi.org/10.1109/sp.2019.00070
12. Riyaldhi R, Rojali, Kurniawan A (2017) Improvement of advanced encryption standard algorithm with shift row and S. box modification mapping in mix column. Procedia Comput Sci 116:401–407. https://doi.org/10.1016/j.procs.2017.10.079
13. Shaji N, Bonifus PL (2016) Design of aes architecture with area and speed trade-off. Procedia Technol 24:1135–1140. https://doi.org/10.1016/j.protcy.2016.05.066
14. Abdullah A (2017) Advanced encryption standard (AES) algorithm to encrypt and decrypt data. Crypt Netw Secur 16:1–11
15. Rao SK, Mahto D, Ali Khan D (2017) A survey on advanced encryption standards. Int J Sci Res (IJSR) 6(1):711–724. https://doi.org/10.21275/art20164149
16. Watanabe Y, Suzuki H, Naito K, Watanabe A (2019) Proposal for user authentication method combining random number and password. In: 2019 IEEE 8th global conference on consumer electronics (GCCE). https://doi.org/10.1109/gcce46687.2019.9015618
17. Nirmalraj T, Jebathagam J (2021) A survey on generating a strong password. Ann Rom Soc Cell Biol 25(6):1476–1486

18. Bosnjak L, Sres J, Brumen B (2018) Brute-force and dictionary attack on hashed real-world passwords. In: 2018 41st international convention on information and communication technology, electronics and microelectronics (MIPRO). https://doi.org/10.23919/mipro.2018.8400211
19. Nascimento N, Santos AR, Sales A, Chanin R (2022)Enablers and inhibitors in agile teams—a case study using challenge based learning for mobile application development. In: 2022 IEEE/ACM international workshop on software-intensive business (IWSiB), Pittsburgh, PA, USA, pp 67–74. https://doi.org/10.1145/3524614.3528623

Securing the Internet of Flying Things (IoFT): A Proficient Defense Approach

Qasem Abu Al-Haija, Abdulaziz A. Alsulami, Badraddin Alturki, and Mohammad Alnabhan

Abstract The Internet of Flying Things (IoFT), popularly known as drones, has recently been adopted to perform essential tasks in several mission-critical systems, such as military, medical, ambulance, firefighting, and transportation systems. However, the traffic communication of IoFT has been vulnerable to a wide range of cyberattack vectors that threaten the privacy and authentication of IoFT devices and data. This paper presents a proficient defense approach to capturing cyberattacks deliberated against the LoFT. Particularly, we characterize the performance of three supervised machine learning schemes, including the random forest classifier (RFC), the multi-layer perceptron (MLP), and the support vector machines (SVM). The models were trained and evaluated on a recent dataset for cyberattacks over IoFT, ECU-IoFT-2022, using two balanced classes (normal vs. attack). The models' assessment showed that the based scheme for IoFT cyberattack categorization provides the best performance indicators, achieving a 99.5% classification accuracy. Besides, its performance is improved over the existing state-of-the-art models for IoFT.

Keywords Machine learning · Cybersecurity · Cyberattacks · Internet of Flying Things

---

Q. Abu Al-Haija (✉)
Department of Cybersecurity, Faculty of Computer and Information Technology, Jordan University of Science and Technology, P.O. Box 3030, Irbid 22110, Jordan
e-mail: qsabuhaija@just.edu.jo

A. A. Alsulami
Department of Information Systems, King Abdulaziz University, Jeddah 21589, Saudi Arabia
e-mail: aaalsulami10@kau.edu.sa

B. Alturki
Department of Information Technology, King Abdulaziz University, Jeddah 21589, Saudi Arabia
e-mail: baalturki@kau.edu.sa

M. Alnabhan
Department of Cybersecurity, Princess Sumaya University for Technology, Amman, Jordan
e-mail: m.alnabhan@psut.edu.jo

X.-S. Yang et al. (eds.), *Proceedings of Ninth International Congress on Information and Communication Technology*, Lecture Notes in Networks and Systems 1000,
https://doi.org/10.1007/978-981-97-3289-0_38

1 Introduction

The Internet of Things (IoT) has become essential to modern life, with smart devices and sensors integrated into various systems. IoFT, popularly known as drones [1] or, more correctly, unmanned aerial vehicles (UAV), has been increasingly adopted in mission-critical systems such as military, medical, ambulance, firefighting, and transportation systems [2] due to their flexibility and ability to perform tasks that are difficult for humans. According to the Civil Aviation Safety Authority (CASA) of the Australian Government, there were around 150,000 UAVs in use there in 2018 [3]. UAV utilization is rising due to its superior durability and capabilities, ease of flight, minimal training requirements for pilots, and reduced operating costs, particularly compared to conventional flying vehicles such as light planes and helicopters [1]. A lack of cybersecurity understanding regarding IoFT in most environments might lead to IoFT device vulnerabilities. A few notable attack methods against UAVs are denial-of-service attacks, man-in-the-middle attacks, and misused Application Programming Interfaces (API) [4]. Such attacks can have negative effects that cost money or place people in danger; they can also put kids and teens in danger or invade their privacy [5]. Moreover, as with any IoT device, IoFT systems are susceptible to cyberattacks [6], which can threaten the privacy and authentication of the devices and the data they transmit. Therefore, developing effective security solutions for IoFT systems is essential to prevent these threats.

The intrusion detection approach uses a model scenario to detect assaults like ping or denial-of-service attacks. The recommended method for ping-killing detection and prevention is used [7] and for the data-packet size poisoning probability function. The research paper [8] discusses cyberattacks, including the man in the middle, denial-of-service, domain name systems, DDoS, and various virus-based assaults [9]. Cybercrime includes the identification of any security threat, which may be identified or tracked using various methods, such as the hidden-Markov model or machine learning techniques such as naïve Bayes, k-nearest neighbor, and cyber-bullying detection [10, 11]. The IoFT is dynamic; thus, while working on security assaults, attacker and custodian weaknesses must be thoroughly investigated. The tree-based technique, which relies on three fundamental factors, including incidence, detection, and severity [12], makes it simple to depict the actions of intruders or attackers. This security plan may provide the best results for a multi-layer defense of the IoFT against various threats [13].

Several studies have proposed various methods to secure the communication of drones against cyberattacks, such as using cryptography, intrusion detection systems, and machine learning (ML) techniques. ML, in particular, is a promising technology for detecting and classifying cyberattacks. It has been used to develop efficient and accurate cyberattack detection and identification schemes for various systems [14], including the IoT and IoFT. The idea of smart cities employing IoFT protocols is explained in Fig. 1.

This research proposes an intelligent system that detects and categorizes cyber-attacks on IoFT systems using supervised machine learning algorithms. Using a

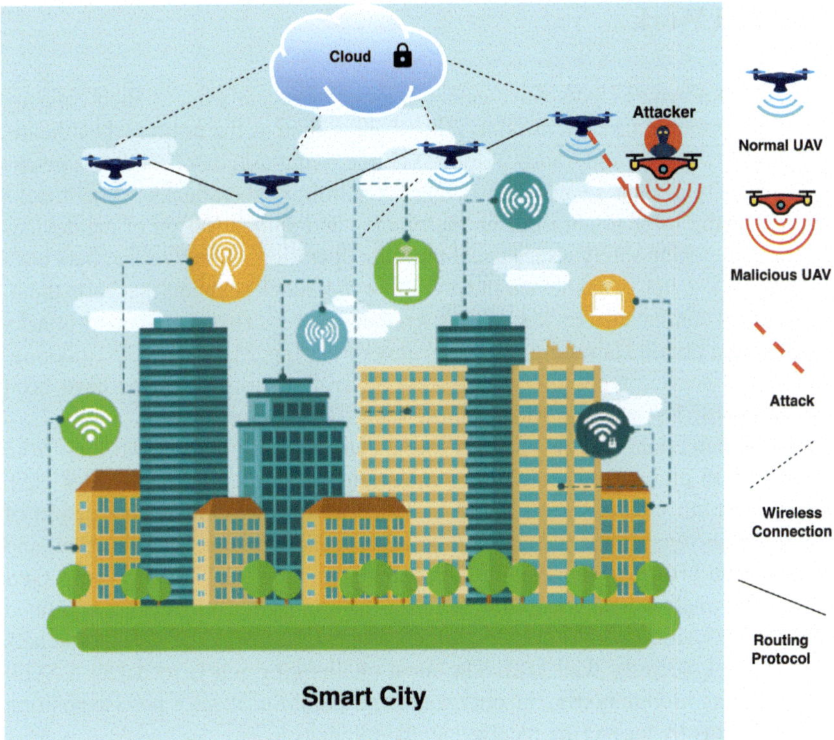

Fig. 1 Smart city IoFT architecture

recent dataset of cyberattacks on IoFT, ECU-IoFT-2022, the performance of three techniques, including the random forest classifier (RFC), the multi-layer perceptron (MLP), and the support vector machines (SVM), were evaluated. The models were trained and tested using two balanced classes, normal versus attack, and classification accuracy was used to measure performance. According to the results, the RFC-based system delivers the best performance indicators, with a classification accuracy of 99.5%. Moreover, the performance of our approach outperforms existing state-of-the-art IoFT models.

The remainder of this paper is structured as follows: Sect. 2 presents an overview of the literature on cyberattacks against IoFT systems and current state-of-the-art security solutions. Section 3 describes the suggested intelligent system for detecting cyberattacks on IoFT; the machine-implemented machine learning algorithms 4 describes the experimental setup, dataset, and assessment measures. Section 5 summarizes the experiment findings and evaluates the effectiveness of the three algorithms for machine learning. Finally, Sect. 6 summarizes the work and suggests future research topics.

2 Related Work

Drones, also known as UAVs, have gained popularity recently due to their capacity to carry out essential tasks in mission-critical areas such as the military, healthcare, emergency, firefighter, and transportation. Yet, the rising usage of drones has generated worries about their security, particularly regarding cyberattacks. Cyberattacks on drones or the IoFT might threaten the privacy and authentication of the devices and data sent, posing safety issues and disrupting operations. Many researchers have identified IoFT system weaknesses and possible cyberattack channels that harm them. Authors in [5] conducted a review of UAV system security. They found many attacks like man in the middle, jamming, spoofing, eavesdropping, Malware, GPS Spoofing, and denial-of-service types of cyberattacks. Several security solutions have been proposed for IoFT systems to address these security concerns.

Intrusion detection systems and encryption can protect aircraft networking from DOS or DDOS attacks to secure communications in future intelligent cities [15]. Machine learning methods are utilized in collaboration with the categorization of IoT-based security systems, which include wireless networking, identification, and data collection [16]. DoS attacks on UAVs are suggested in [17], employing GPS tracking data analysis using log files. Self-controlling flying vehicles are equipped with wireless vision (Wi-Vi) sensors, which can rescue people or detect attackers even in a thick security wall [18]. The most challenging job is to detect a Sybil assault utilizing mobile nodes; the only way to pinpoint the attack's precise position and identify it is by measuring received signal intensity.

According to the authors of [19], channel state data may offer accurate location coordinate information. At the same time, the self-adaptive multi-signal classification set of rules is employed for passive attacker detection. Lagkas et al. [20] introduced a flying things-based structure that provides privacy and security to protect aerial drone-to-aerial vehicle connection via routing protocols. A sophisticated encryption standard is collected from electroencephalography signals to safeguard the connection between the ground station and the aerial vehicle [21]. In [22], the authors reasonably illustrated a third-party attack utilizing an actual aerial robot that rapidly assesses a flying network's vulnerabilities using a denial-of-service instrument that impacts IoD. In the future, it is suggested that such a simulation be applied against DDoS. Examples in [23] include flying vehicles and VPN-sniffing using Wi-Fi-Pineapple. Yet, given the difficulty in preventing collisions with aerial vehicles, a suitable response to this problem must have been offered [23]. Drone-based heuristic computing systems must produce useful results in both the public and military spheres.

Recently, machine learning methods to recognize and stop cyberattacks in IoFT systems have gained prominence [6]. Machine learning algorithms may analyze large volumes of data to find trends and abnormalities suggesting a cyberattack [24]. The authors in [25] presented a randomized search cross-validation (RSCV)-based hierarchy and optimal random forest (RF) anomaly-based IDS. The results of the simulation demonstrated the efficiency of the proposal, which had an F1-score (96.38%), the lowest mean-square error (0.13), and a sufficient training time (749 ms). The authors of [25] developed a machine learning-based framework for rapid UAV detection over secured Wi-Fi traffic in response to the observation that several users of drones utilize Wi-Fi connections for streaming video and control. A reweighted l1-norm regularization is used to maximize feature extraction and prediction accuracy, and the estimation of maximum likelihood (MLE) is used to estimate the inter-arrival time of the packet. The evaluation results demonstrate that the suggested technique has a high 85.7–95.2% accuracy level and can recognize and detect tested UAVs in 0.15–0.35 s.

We presented an intelligent approach for detecting and categorizing cyberattacks on IoFT systems. Using a recent dataset of cyberattacks on IoFT, ECU-IoFT-2022 [6], they assessed the performance of three supervised machine learning algorithms, including random forest classifier (RFC), multi-layer perceptron (MLP), and support vector machines (SVM). Our analysis of machine learning techniques shows that the RFC-based method for IoFT cyberattack categorization produces the highest performance, with a 99.5% accuracy rate. Our findings demonstrate the value of further research in this area and the effectiveness of machine learning-based approaches in identifying and preventing cyberattacks on IoFT systems.

3 System Architecture

The IoFT architecture is shown in Fig. 2 [26]. The IoFT is considered an essential component of a smart city. The smart city comprises different types of intelligent subsystems, i.e., smart vehicles, smart homes, smart transportation, smart hospitals, and IoFT. Those subsystems can exchange information through the communication network system. However, the IoFT system is vulnerable to several cyber threats, including denial-of-service (DDoS) and false data injection (FDI) attacks. In addition, increasing the connectivity of smart city subsystems, such as the IoFT, enables more suspicious activity, such as malicious attacks, to occur, eventually causing negative consequences.

Figure 3 illustrates a proposed method to address the security issue in the IoFT. Injecting false data into a drone makes it susceptible to attack, allowing an attacker to cyberattack. Therefore, the machine learning (ML) model can detect anomalies in the streaming data from the outside environment. Furthermore, ML models can recognize the normal and fault data attackers generate. As a result, this ensures the security of the IoFT because drones can capture and eliminate data from cyberattacks.

Figure 4 shows the steps for implementing the proposed intelligent systems to capture cyberattacks in IoFT. In the beginning, the dataset ECU-IoFT_Dataset.csv [6] is pre-processed before it is used to ensure the data is ready for training and testing. This is considered an important step to increase the performance of the ML models [27]. In addition, this step is needed to have a balanced dataset; in our case, we have a balanced dataset. The data should be partitioned before the training and

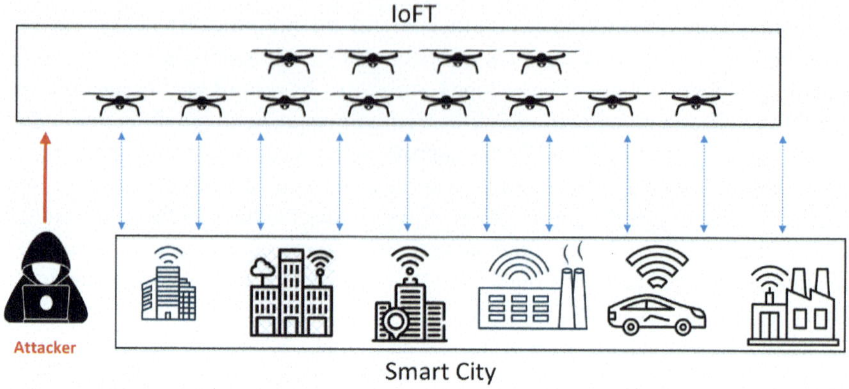

Fig. 2 Architecture of the IoFT with cyberattack

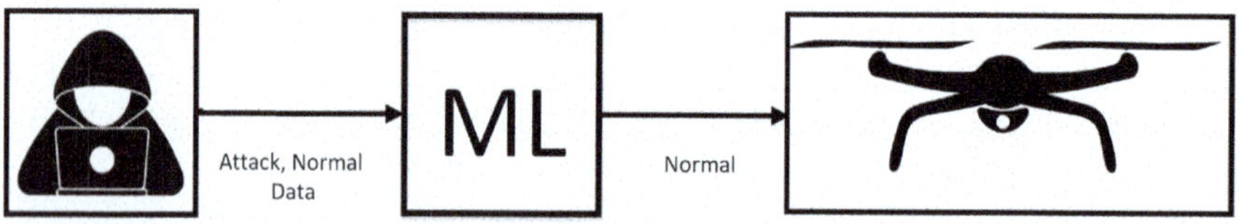

Fig. 3 Top view diagram for the proposed scheme

testing step; therefore, we used 70% of the data for training and 30% for testing. We used a fivefold cross-validation technique for validation [28]. This paper implements three ML models: MLP, SVM, and RFC. MLP is an artificial neural network with more than one hidden layer [29]. SVM is an ML algorithm that is widely used as a classifier. In SVM, the training data is divided into small collections, and each collection is trained then the best solution is chosen based on voting [30]. RFC is an algorithm based on multiple decision trees to reach the optimal solution [31]. Eventually, the MLP, SVM, and RFT output is either normal or attacked.

4 Experimental Results

The proposed models were developed using MATLAB 2022 and deep and machine learning toolboxes. Our research used the ECU-IoFT-2022 dataset [6], which includes 40,000 samples for detection. The dataset was balanced with 20,000 normal samples and 20,000 classified as attacks. It consists of seven features, including sample identification number (integer), timestamp (factor), source address (factor), destination address (factor), protocol (factor), time frame length in bytes (integer), capture sample detail (factor), attack type (factor), and attack scenario (factor). To assess the performance of the models, we analyzed the confusion matrix, which consists of true

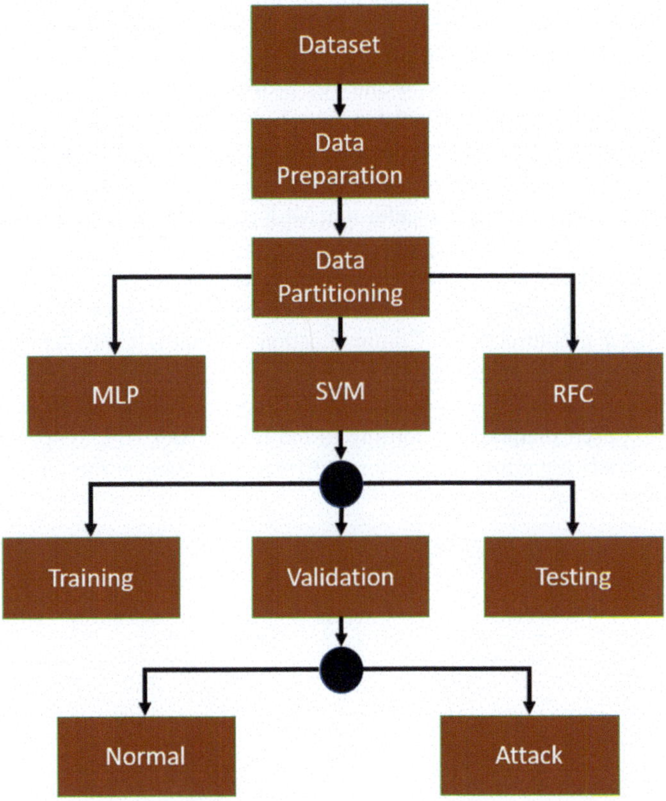

Fig. 4 The proposed framework for IDS development

positives (TP), true negatives (TN), and false positives (FP) for each model. Based on the confusion matrix, we used the following performance indicators to evaluate the models:

$$\text{Accuarcy} = ((\text{TP} + \text{TN})/(\text{TP} + \text{TN} + \text{FP} + \text{FN})) * 100 \quad (1)$$

$$\text{Precision} = (\text{TP}/(\text{TP} + \text{FP})) * 100 \quad (2)$$

$$\text{Sensitivity} = (\text{TP}/(\text{TP} + \text{FN})) * 100 \quad (3)$$

$$\text{F1} - \text{Score} = (2*(\text{Precision} * \text{Sensitivity})/(\text{Precision} + \text{Sensitivity}))*100 \quad (4)$$

Table 1 shows the accuracy matrix that was collected during the experiments, which are true positive (TP), true negative (TN), false negative (FN), and false positive (FP) were recorded to be used when calculating the accuracy, precision, sensitivity, and F1-score [32]. Equation 1 is used to calculate the classification accuracy, Eq. 2 is used to calculate the precision, sensitivity is calculated in Eq. 3, and F1-score is computed using Eq. 4 [33]. Table 1 and Fig. 5 display the performance metrics of the MLP, SVM, and RFC models using Eqs. 1–4. We can notice that RFC scored the best in the accuracy parameter; meanwhile, MLP has the lowest performance. In similar situations for precision, sensitivity, and F1-score, RFC still outperforms the performance of SVM and MLP. In addition, MLP recorded the lowest score in every performance metric used in this paper.

Table 2 compares the performance parameters, accuracy, and F1-score of the most recent research in the field, including cluster-based local outlier factors (CBLOF), Bidirectional Communication—Shallow Convolutional Neural Network (Bi-SCNN), Unidirectional Communication—Shallow Convolutional Neural Network (Uni-SCNN), Maximum Likelihood Estimation (MLE) [34], and RFC (the proposed method). According to the results, our RFC outperformed all the other methods regarding accuracy and F1-score, making it the top-performing method.

Table 1 Performance metrics

Model	TP	FP	FN	TN	Accuracy (%)	Precision (%)	Sensitivity (%)	F1-score (%)
RFC	19,948	52	67	19,933	99.70	99.74	99.67	99.70
SVM	19,621	379	1233	18,767	95.97	98.11	94.09	96.05
MLP	19,244	756	3422	16,578	89.56	96.22	84.90	90.56

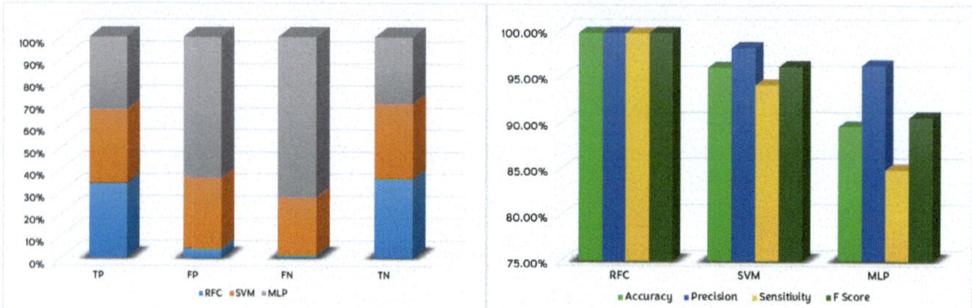

Fig. 5 Comparing performance metrics

Table 2 Accuracy of the most recent research

References	Model	Operation	Accuracy (%)	F1-score (%)
Ahmed et al. [6]	CBLOF model	Classification	84.5	–
Al-Haija et al. [33]	Bi-SCNN model	Detection	99.0	98.41
Ihekoronye et al. [25]	RFC model	Detection	99.3	96.38
Al-Haija et al. [33]	Uni-SCNN model	Detection	90.0	88.52
Alipour et al. [34]	MLE model	Detection	98.20	–
This work	RFC model	Detection	99.7	99.72

5 Conclusions and Future Directions

In this paper, we proposed a supervised machine learning algorithm-based intelligent solution for detecting and classifying cyberattacks on IoFT systems. We evaluated the effectiveness of three methods: the random forest classifier (RFC), the multi-layer perceptron (MLP), and the support vector machines (SVM) using the ECU-IoFT-2022 dataset of cyberattacks on IoFT systems. The dataset was balanced with two classes, normal vs. attack, and classification accuracy was used to measure performance. Our results show that the RFC-based approach provides the best performance indicators, achieving a classification accuracy of 99.5%. Moreover, our proposed method outperforms the current state-of-the-art IoFT models. Overall, the findings demonstrate the effectiveness of our proposed method in accurately detecting and classifying cyberattacks on IoFT systems. In future research, it would be interesting to investigate the performance of our proposed method on a larger and more diverse dataset. This could involve collecting more data from different IoFT systems and incorporating different types of cyberattacks to evaluate the robustness of our approach. Additionally, exploring other machine learning algorithms or hybrid models that combine multiple algorithms to improve the accuracy and efficiency of cyberattack detection and classification would be worthwhile. Moreover, incorporating explainability and interpretability techniques can help to understand the underlying factors driving the classification decisions, which can be useful for developing more effective and trustworthy IoT security solutions.

References

1. Zaidi S, Atiquzzaman M, Calafate CT (2021) Internet of flying things (IoFT): a survey. Comput Commun 165:53–74
2. Dias Santana GM, Cristo RSD, Lucas Jaquie Castelo Branco KR (2021) Integrating cognitive radio with unmanned aerial vehicles: an overview. Sensors 21(3):830
3. A. G. Civil Aviation Safety Authority (ed) (2018) Review of aviation safety regulation of remotely piloted aircraft systems. Australian Government Canberra, Australia

4. Droos A, Al-Haija QA, Alnabhan M (2022) Lightweight detection system for low-rate DDoS attack on software-defined-IoT. In: 6th smart cities symposium (SCS 2022), hybrid conference, Bahrain, pp 157–162. https://doi.org/10.1049/icp.2023.0388
5. Yaacoub J-P, Noura H, Salman O, Chehab A (2020) Security analysis of drones systems: attacks, limitations, and recommendations. Internet of Things 11:100218
6. Ahmed M, Cox D, Simpson B, Aloufi A (2022) ECU-IoFT: a dataset for analyzing cyber-attacks on the internet of flying things. Appl Sci 12(4):1990
7. Abdollahi A, Fathi M (2020) An intrusion detection system on the ping of death attacks in IoT networks. Wireless Pers Commun 112:2057–2070
8. Pliatsios D, Sarigiannidis P, Lagkas T, Sarigiannidis AG (2020) A survey on SCADA systems: secure protocols, incidents, threats and tactics. IEEE Commun Surv Tutorials 22(3):1942–1976
9. Al-Khater WA, Al-Maadeed S, Ahmed AA, Sadiq AS, Khan MK (2020) Comprehensive review of cybercrime detection techniques. IEEE Access 8:137293–137311
10. Albalawi M, Song H (2019) Data security and privacy issues in swarms of drones. In: 2019 integrated communications, navigation, and surveillance conference (ICNS). IEEE, pp 1–11
11. Garg S, Aujla GS, Kumar N, Batra S (2019) Tree-based attack–defense model for risk assessment in multi-UAV networks. IEEE Consum Electron Mag 8(6):35–41
12. Hooper M et al (2016) Securing commercial WiFi-based UAVs from common security attacks. In: MILCOM 2016–2016 IEEE military communications conference. IEEE, pp 1213–1218
13. Lin C, He D, Kumar N, Choo K-KR, Vinel A, Huang X (2018) Security and privacy for the internet of drones: challenges and solutions. IEEE Commun Mag 56(1):64–69
14. Al-Haija QA, Saleh E, Alnabhan M (2021) Detecting port scan attacks using logistic regression. In: 2021 4th international symposium on advanced electrical and communication technologies (ISAECT), Alkhobar, Saudi Arabia, pp 1–5. https://doi.org/10.1109/ISAECT53699.2021.9668562
15. Rani C, Modares H, Sriram R, Mikulski D, Lewis FL (2016) Security of unmanned aerial vehicle systems against cyber-physical attacks. J Defense Model Simul 13(3):331–342
16. Restuccia F, D'Oro S, Melodia T (2018) Securing the Internet of Things in the age of machine learning and software-defined networking. IEEE Internet Things J 5(6):4829–4842
17. Salamh FE, Karabiyik U, Rogers M, Al-Hazemi F (2019) Drone disrupted denial of service attack (3DOS): towards an incident response and forensic analysis of remotely piloted aerial systems (RPASs). In: 2019 15th international wireless communications and mobile computing conference (iwcmc). IEEE, pp 704–710
18. Sehrawat A, Choudhury TA, Raj G (2017) Surveillance drone for disaster management and military security. In: 2017 International conference on computing, communication, and automation (ICCCA). IEEE, pp 470–475
19. Wang C et al (2018) Accurate Sybil attack detection based on fine-grained physical channel information. Sensors 18(3):878
20. Lagkas T, Argyriou V, Bibi S, Sarigiannidis P (2018) UAV IoT framework views and challenges: towards protecting drones as "Things". Sensors 18(11):4015
21. Singandhupe A, La HM, Feil-Seifer D (2018) Reliable security algorithm for drones using individual characteristics from an EEG signal. IEEE Access 6:22976–22986
22. Vasconcelos G, Carrijo G, Miani R, Souza J, Guizilini V (2016) The impact of DoS attacks on the AR. Drone 2.0. In: 2016 XIII Latin American robotics symposium and IV Brazilian robotics symposium (LARS/SBR). IEEE, pp 127–132
23. Vattapparamban E, Güvenç I, Yurekli AI, Akkaya K, Uluağaç S (2016) Drones for smart cities: issues in cybersecurity, privacy, and public safety. In: 2016 international wireless communications and mobile computing conference (IWCMC). IEEE, pp 216–221
24. Delplace A, Hermoso S, Anandita K (2020) Cyber attack detection thanks to machine learning algorithms. arXiv preprint arXiv:2001.06309
25. Ihekoronye VU, Ajakwe SO, Kim DS, Lee JM (2022) Hierarchical intrusion detection system for secured military drone network: a perspicacious approach. In: MILCOM 2022–2022 IEEE military communications conference (MILCOM). IEEE, pp 336–341

26. Khan IU, Abdollahi A, Khan MA, Uddin I, Ullah I (2021) Securing against DoS/DDoS attacks in the IoFT using experience-based deep learning algorithm
27. Lee IY, Yoon YG, Oh TK, Park S, Ryu SI (2020) A study on data pre-processing and accident prediction modeling for occupational accident analysis in the construction industry. Appl Sci 10(21):7949
28. Alsulami AA, Abu Al-Haija Q, Alqahtani A, Alsini R (2022) Symmetrical simulation scheme for anomaly detection in autonomous vehicles based on LSTM model. Symmetry 14(7):1450
29. Orrù PF, Zoccheddu A, Sassu L, Mattia C, Cozza R, Arena S (2020) Machine learning approach using MLP and SVM algorithms for the fault prediction of a centrifugal pump in the oil and gas industry. Sustainability 12(11):4776
30. Alsulami AA, Abu Al-Haija Q, Tayeb A, Alqahtani A (2022) An intrusion detection and classification system for IoT traffic with improved data engineering. Appl Sci 12(23):12336
31. What is a random forest? Available: https://www.ibm.com/topics/random-forest#:~:text=Random%20forest%20is%20a%20commonly,both%20classification%20and%20regression%20problems
32. Abu Al-Haija Q, Alsulami AA (2022) Detection of fake replay attack signals on remote keyless controlled vehicles using pre-trained deep neural network. Electronics 11(20):3376
33. Al-Haija QA, Alsulami AA (2021) High-performance classification model to identify ransomware payments for heterogeneous bitcoin networks. Electronics 10(17):2113
34. Alipour-Fanid A, Dabaghchian M, Wang N, Wang P, Zhao L, Zeng K (2019) Machine learning-based delay-aware UAV detection and operation mode identification over encrypted Wi-Fi traffic. IEEE Trans Inf Forensics Secur 15:2346–2360

Speaker Recognition with Deep Learning Approaches: A Review

Abdulrahman S. Alenizi and Khamis A. Al-Karawi

Abstract This article gives an overview of the methods for using deep learning to identify and verify speakers. Speaker recognition is an everyday use of speech technology. Many research initiatives have been carried out in the past few years, but little progress has been achieved. But just as deep learning techniques are replacing previous state-of-the-art approaches in speech recognition, they are also developing in most machine learning fields. Deep learning seems to have evolved into the most advanced speaker verification and identification technique. Most novel efforts start with the common x-vectors in addition to i-vectors. The increasing volume of data gathered makes the area where deep learning is most effective more accessible.

Keywords Deep learning text independence · Feature extraction · Statistical models · Discriminative models · Speaker identification · And speaker verification

1 Introduction

Speaker recognition identifies individuals based on their voices [1, 2]. Because no two people's vocal tracts, larynxes, and other sections of their organs that produce their voices are the same, no two people have the same sound. Apart from these physical differences, each speaker possesses a unique speaking style characterized by accent, rhythm, intonation, pronunciation, language choice, and more [3, 4]. Modern speaker recognition systems use several elements simultaneously to cover as many facets as possible and apply these complementary features to attain higher recognition accuracy. The field of forensics effectively uses speaker recognition technologies [2, 5–7]. The rising popularity of deep learning (DL) can be attributed to easily accessible software and affordable hardware solutions. Its integration has

A. S. Alenizi (✉)
The Public Authority for Applied Education and Training (PAAET), Shuwaikh Industrial, Kuwait
e-mail: As.alenizi1@paaet.edu.kw

K. A. Al-Karawi
Diyala University, Baqubah, Diyala, Iraq

X.-S. Yang et al. (eds.), *Proceedings of Ninth International Congress on Information and Communication Technology*, Lecture Notes in Networks and Systems 1000,
https://doi.org/10.1007/978-981-97-3289-0_39

become prevalent in various fields where machine learning finds application. Consequently, researchers and professionals are now adopting deep learning for speaker recognition (SR) [8–10]. Reviewing deep learning techniques used in speaker identification and verification tasks from the earliest to the most recent approaches is the goal of this paper. Since these responsibilities are frequently mentioned while doing speaker recognition, it is first essential to define speaker identification and verification [1, 11, 12]. Identifying the speaker who closely matches the test sample within a group of known speakers constitutes the task of speaker identification. It is called a closed-set (or in-set) scenario when every speaker within a particular set is known [11–13]. As an alternative, open-set (or out-of-set) speaker identification is employed when the group of known speakers may not encompass the potential test subject. Speaker verification aims to validate that a speaker asserting a specific identity possesses that identity. In essence, the goal is to verify that the individual is legitimately who they declare to be. This is commonly achieved in speaker verification by comparing the test sample with a sample from the claimed speaker alongside a general background model [14, 15]. Both speaker identification and verification have their uses, and related or identical techniques are usually needed. The 1960s marked the beginning of speaker recognition research. Numerous cutting-edge technologies aided the evolution of speaker recognition throughout the ensuing forty years [16]. As an illustration, acoustic characteristics and template models, including vector quantization and dynamic time warping, have been employed [4, 17–19]. Examples encompass the linear predictive Cepstral coefficients, the perceptual linear prediction coefficients, and the Mel-frequency Cepstral coefficients. The Gaussian Mixture Model-based-universal backdrop model (GMM-UBM), subsequently introduced by Saquib et al. [20], Zhao et al. [21], Wang and Nakagawa [22], Sadjadi and Hansen [23], has served as the basis for speaker recognition for more than ten years, starting in 2000. Support vector machines [1–3, 24–26] and joint factor analysis [27, 28] are two examples of representative models built on the GMM-UBM. Up until the advent of deep learning-based speaker recognition, the GMM-UBM/i-vector frontend [29, 30] and probabilistic linear discriminant analysis (PLDA) backend [3, 19, 31–34]. Among the models offered the best performance. Therefore, the DNN approaches for both cases are examined in this study, constantly referencing the task assigned in the relevant research. Since the state-of-the-art in the field focuses nearly entirely on neural architectures, we limit our attention to DL techniques [35, 36] (Fig. 1).

2 Databases

Database selection is crucial in conversations concerning speech technology and other machine learning subjects. The meaningful evaluation and comparison of developed methods rely on using identical testing conditions, particularly from a machine learning standpoint. Assessing the superiority of an approach becomes challenging

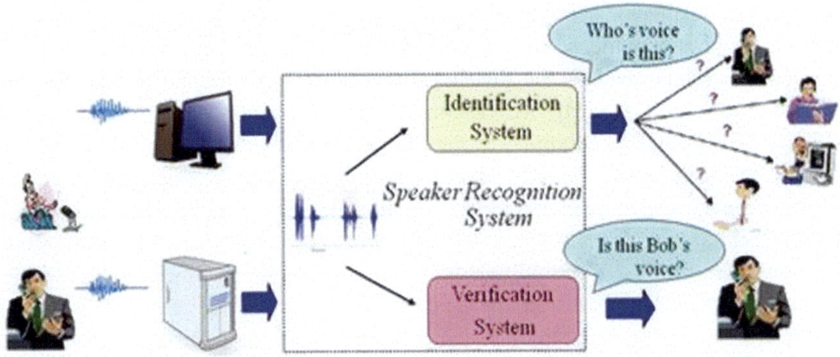

Fig. 1 Speaker recognition system

when performed on a disparate set or corpus. The selection of training and assessment datasets involves careful consideration of various parameters, including availability for public use and research purposes. Some datasets can be repurposed for speaker recognition programs, while academics prefer datasets designed explicitly for speaker identification and verification. The key distinction lies in the number of speakers within the database. Speech recognition databases typically feature fewer speakers, as speech recognition demands substantial data to train phoneme models, often resulting in fewer distinct speakers. Conversely, speaker recognition thrives on having many speakers, necessitating less recorded material from each person. The recruitment of numerous speakers poses a more demanding and time-consuming endeavor. Speech recognition corpora like TIMIT [37], WSJ [38], RSR2015 [39], CHiME 2013 [40], and VCTK are typically composed of a few hundred speakers, while datasets like LibriSpeech [41], VoxCeleb [42], and NIST SRE have thousands of individuals, but often lack homogeneity and require clean recording quality for linguistic considerations. VoxCeleb2 [43] is the most extensive corpus specifically designed for speaker recognition (SR) tasks, representing a recent expansion of its predecessor, VoxCeleb. This extensive collection incorporates samples from over 6000 speakers sourced from YouTube, resulting in a widely varying sound quality. LibriSpeech is a high-quality audiobook corpus from the industry that helps test SR approaches and features. NIST SRE datasets provide telephone line quality speaker samples suitable for evaluations in other usage environments [44] with limited quality issues.

3 Classification Methods

A. ***GMM-UBM***

Gaussian Mixture Model-Universal Background Model (GMM-UBM) is a robust statistical framework widely used in speaker recognition and audio processing. This approach efficiently models the distribution of acoustic features associated with different speakers. The GMM represents the statistical characteristics of speech features, while the UBM captures the general acoustic characteristics shared across all speakers. Using a GMM-UBM makes it possible to identify and verify speakers by comparing their specific characteristics against the Universal Background Model. This method has proven robust in handling variations in speech and environmental conditions, making it a valuable tool in voice authentication, speaker diarization, and forensic speaker recognition applications. The GMM-UBM approach provides a flexible and reliable foundation for extracting discriminative information from audio signals, contributing to advancements in the broader field of biometric identification [45, 46]. Gaussian Mixture Model-Universal Background Model (GMM-UBM) is a widely employed technique in speaker recognition, where the goal is to identify or verify individuals based on their unique acoustic characteristics. In GMM-UBM, the Gaussian Mixture Model (GMM) represents the statistical distribution of acoustic features associated with individual speakers. At the same time, the Universal Background Model (UBM) captures the general acoustic characteristics shared across all speakers [47]. This method allows for modeling complex variations in speech patterns, providing a robust framework for distinguishing between different speakers. By comparing the specific characteristics of an unknown speaker against the Universal Background Model, GMM-UBM facilitates accurate speaker identification and verification. Its versatility and effectiveness make GMM-UBM a valuable tool in applications such as access control, voice authentication, and forensic voice analysis, contributing significantly to advancements in speaker recognition technology. The GMM-UBM process begins with constructing a Universal Background Model (UBM) by training a Gaussian Mixture Model (GMM) on an extensive dataset comprising speech samples from various speakers. The UBM encapsulates the statistical attributes of the speech signal that are shared across all speakers. Then, a new speech sample is compared against the UBM for speaker recognition to obtain a likelihood score [1, 2]. The score indicates the probability that the speech sample originates from the UBM rather than a particular speaker. Each speaker is associated with a dedicated GMM, trained by adapting the UBM using the speaker's training data. In the recognition phase, the likelihood score from the speaker-specific GMM is compared to the UBM likelihood score. GMM-UBM is a popular choice in speaker recognition systems due to its computational efficiency and effective performance, even with minimal training data, credited to specific speaker scores.

A Gaussian Mixture Model (GMM) is a multivariate Gaussian model that represents the likelihood of observation λ by combining M independent components

$$p(x|\lambda) = \sum_{m=1}^{M} w_m p_m(x), \tag{1}$$

where x is a D-dimensional vector, w_m is the weight of the mth Gaussian component $p_m(x)$

$$p_m(x) = \frac{1}{(2\pi)^{D/2}\left|\sum_m\right|^{1/2}} \exp\left(-\frac{1}{2}(x - \mu_m)^T \sum_m^{-1} (x - \mu_m)\right), \tag{2}$$

w_m and $\sum_m$ are the mth component's mean vector and covariance matrix, respectively. The component weights $w_m > 0$ must be satisfied $\sum_{m=1}^{M} = 1\, w_m = 1$. In practice, for reasons of data requirement and computation, covariance matrices are usually diagonal [48]. Training a GMM involves finding the parameters $\lambda = \left\{w_m, w_m, \sum_m\right\}$ $\overset{M}{m} = 1$ given a training sample $x = \{x1, x2, ...xT\}$. The log-likelihood LL of x concern λ is given in the:

$$LL = \frac{1}{T} \sum_{t=1}^{T} \log p(x_t|\lambda). \tag{3}$$

A more significant LL indicates more substantial evidence of x's GMM λ origin, using the Maximum Probability (ML) method and iterative Expectation–Maximization (EM) procedure for estimation [49]. The Generalised Method of Moments (GMM) is a popular tool for modeling speakers due to its ability to generate smooth approximations for densities of random shapes. It uses Gaussian basis functions to represent arbitrary densities, including acoustic classes like vowels and nasals. GMM-based algorithms for speaker recognition are highly praised for their performance [50–53]. Modern speaker verification systems utilize advanced techniques like the Universal Background Model (UBM) to extract sub-vectors and create an i-vector to manage channel fluctuation [54, 55].

B. ***GMM Supervectors***

In the realm of speaker recognition, GMM supervectors serve as a prevalent representation of a speaker's vocal traits. These supervectors are constructed by concatenating the means and covariance matrices of the GMM components utilized to model the speaker's voice. To elaborate, the GMM supersector is formed by linking the means of each GMM component into a unified vector and then concatenating the covariance matrices of each GMM component into a consolidated matrix [11]. The produced supervector encapsulates the statistical attributes of the speaker's voice, facilitating its utilization in speaker identification and verification tasks. GMM supervectors are efficacious in speaker recognition, especially under constrained training data. These vectors find application in diverse speaker recognition systems, including i-vector systems—a prevalent type of speaker recognition system employing a comparable methodology to extract speaker-specific details from speech signals [3]. Due to the variable lengths of voice samples, considerable effort has been devoted to developing techniques for extracting features from samples of different durations. Among these techniques, forming GMM supervectors has proven to be particularly effective for speaker recognition [56]. Supervectors are generated by concatenating the parameters of the GMM, specifically the mean vectors. Subsequently, a machine learning algorithm, often support vector machines (SVM), processes this predetermined supervector length [46] (and was discovered to be the best-performing approach before deep neural networks gained considerable attention.

C. ***i-vector***

The i-vector, or identity vector, is a compact representation used in speaker recognition systems to capture the unique characteristics of an individual's speech. It is derived from a Gaussian Mixture Model-Universal Background Model (GMM-UBM) system and is a fixed-length representation of a speaker's speech characteristics. The system can efficiently represent and compare speaker-specific information by extracting the i-vector from a given speech segment, making it particularly useful in large-scale speaker recognition tasks. The i-vector approach has proven effective in mitigating variations in speech due to background noise and channel distortions, contributing to its widespread adoption in modern speaker recognition technology. An i-vector, or identity vector, is a low-dimensional representation of a speaker's voice properties used in speaker recognition generated by the Gaussian Mixture Model (GMM).To create an i-vector, a GMM is first trained on a set of speech utterances from multiple speakers to model the statistical characteristics of their voices. Then, for a given speech sample, the log-likelihoods of the GMM for each frame of the speech are extracted. A crucial statistic, the total variability matrix, is computed using the frame-level log-likelihoods. This matrix effectively captures the variations in the speech data across different speakers [12]. The i-vector is obtained by projecting the total variability matrix onto a lower-dimensional subspace through principal component (PCA) or factor analysis (FA). This resultant i-vector serves as a fixed-length representation encapsulating the speaker's voice characteristics, proving valuable in tasks such as speaker identification or verification, especially when confronted with limited training data [13]. They find extensive application in cutting-edge speaker recognition systems across diverse domains, including automatic speaker recognition in telephone conversations, forensic speaker recognition, and speaker diarization. The one-part space technique was chosen after discovering that the JFA channel area contains information that can be used to discriminate between speakers [57]. The i-vector speaker, compatible with GMM, can be depicted by

$$\mu = m + \mathrm{T}w. \tag{4}$$

The random vector w, a low-rank matrix T, and the mean supervector m from the Universal Background Model (UBM) represent the assumed i-vectors [55]. Figure 2 illustrates the block diagram of the i-vector framework.

D. ***Evaluation methods***

The Nearest Cluster approach [58] determines how similar two feature vector matrices are by calculating their distance. The feature vectors of the Test Speaker are compared with each cluster of the target speaker, and the cosine distance between them is calculated. The minimum distance is then returned. In deep neural networks

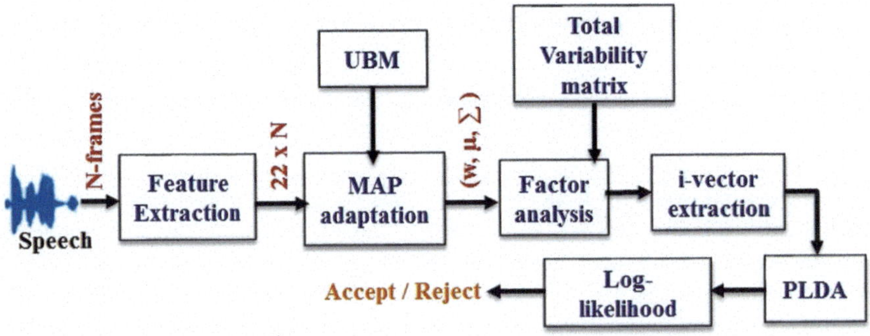

Fig. 2 Block diagram of the i-vector framework

(DNNs), scoring evaluates how well the DNN model fits the input data. The scoring method is typically used in supervised learning tasks, such as classification or regression, where the DNN is trained to predict an output value given an input. In a classification task, the scoring method usually involves computing the accuracy of the DNN's predictions on a validation set. The validation set, distinct from the training set used to train the DNN, is a subset of labeled data employed to assess the DNN's performance. Accuracy, calculated as the ratio of correctly predicted labels to the total number of labels in the validation set, measures the DNN's success. In regression tasks, the scoring method typically relies on a loss function like mean squared error (MSE), which gauges the disparity between predicted and actual output. A lower value in the loss function indicates improved DNN performance. In certain instances, more intricate scoring methods such as precision, recall, and F1-score, familiar in binary classification tasks, may be necessary. The Error Equal Rate (EER) is a crucial metric for assessing the accuracy and performance of DNNs in speaker recognition and biometric systems. It measures the point where false acceptance and rejection rates are equal, with a lower EER indicating improved performance [28, 59]. The DET curve, or Detection Error Tradeoff curve, is a visual representation employed to assess the effectiveness of binary classification systems, focusing on applications like anomaly detection and signal processing. This graph provides insights into the interplay between the false acceptance rate (FAR) and the false rejection rate (FRR) across varying acceptance thresholds. The false acceptance rate (FAR) indicates the percentage of instances where the system incorrectly labels a negative case as positive. On the other hand, the false rejection rate (FRR) represents the percentage of instances where the system erroneously identifies a positive case as unfavorable. The DET curve offers a comprehensive view of system performance by illustrating the dynamic relationship between these key metrics [60].

4 Deep Learning in Speaker Recognition

Deep learning has emerged as a powerful paradigm in various domains, and its application in speaker recognition has garnered significant attention. Deep learning techniques, including CNNs, RNNs, and hybrid architectures, have impressive abilities in extracting complex patterns from raw audio data, particularly in speaker recognition, which identifies individuals based on voice features [28]. One of the advantages of deep learning in speaker recognition lies in its ability to learn hierarchical representations from data automatically, eliminating the need for manually engineered features. End-to-end learning, where the model processes raw audio signals and directly produces speaker embeddings, has shown promise in simplifying the overall system architecture. Furthermore, deep embedding techniques, such as triplet and contrastive loss, enable the network to learn discriminative features for different speakers, enhancing the overall performance of speaker recognition systems. Transfer learning approaches, leveraging pre-trained models on large datasets, have proven effective in scenarios with limited labeled data. However, challenges persist, including the need for adequate labeled datasets for training deep models. Additionally, issues related to domain adaptation and robustness in real-world conditions remain focal points of research. Despite these challenges, the application of deep learning in speaker recognition continues to evolve, pushing the boundaries of accuracy and scalability in this critical domain [30]. The following are some ways in which deep learning is used in speaker recognition:

- Deep Neural Networks (DNNs): DNNs process raw audio signals to extract distinctive features, typically converting them into spectrograms or Mel-frequency Cepstral coefficients (MFCCs). These features are then fed into the DNN, which learns to map them to low-dimensional speaker embeddings, capturing the speaker's unique identity. Speaker verification or identification can then be performed using these embeddings. This version is more concise and emphasizes the role of DNNs in extracting and representing speaker identity from audio signals. It also replaces the phrase "learn discriminative features from" with the more active phrase "process raw audio signals to extract distinctive features," and the phrase "are usually transformed into spectrograms or Mel-frequency Cepstral coefficients (MFCCs)" with the more direct phrase "typically by converting them into spectrograms or Mel-frequency Cepstral coefficients (MFCCs)".
- Convolutional Neural Networks (CNNs): Employed in speaker verification, CNNs aim to ascertain whether a provided speaker matches a reference speaker. These networks excel at learning distinctive speaker-related features from spectrograms or MFCCs. The acquired features are subsequently utilized to calculate a similarity score between the input and reference speaker, serving the purpose of verification.
- Recurrent Neural Networks (RNNs): Applied in speaker diarization, the task of determining speakers in a conversation, RNNs prove valuable in modeling the temporal dynamics of an audio signal. By employing RNNs, it becomes possible to segment the audio signal into speaker-specific segments. The resulting speaker

embeddings from the DNNs facilitate the clustering of speaker segments, offering a solution for effective speaker diarization.

- GANs, or Generative Adversarial Networks, are powerful tools for linguistic conversion, allowing the transformation of one speaker's voice into another while maintaining original content. They learn a mapping between source and target voices.

Deep learning has shown significant potential in speaker recognition, producing advanced results on various datasets. Two main approaches include feature extraction using i-vector computation, an internal hidden layer using PLDA or cosine distance, and deep discriminating networks. Automated speaker recognition systems are evaluated using metrics like Equal Error Rate (EER) and Decision Cost Function, with biometric security systems using preset threshold settings for false acceptance and rejection rates [45, 61]. When the rates of incorrect acceptances and rejections are equal, a standard statistic known as the Equal Error Rate (EER) is calculated. According to this metric, the proportion of these two errors is the same. Increased accuracy is correlated with a lower EER value in the biometric system. The Decision Cost Function (DCF), a different statistic, considers the past probabilities of target speaker occurrences and the percentages of target and non-target speakers. The discrimination and calibration are evaluated simultaneously using the Detection Cost Function (DCF). A frequent term used to describe the lowest number on the DCF curve is min DCF.

A. ***Deep learning for feature extraction***

The procedure involves using deep belief networks (DBN) and deep neural networks (DNN) to convert MFCC feature vectors into Deep Speaker Feature (DeepSF) vectors. The DBN is trained unsupervised with target speaker feature vectors, and the DNN is adjusted with K-Means clustering. Supervised training differentiates between clusters, and a re-clustering phase is performed if mislabeling occurs. The SoftMax layer is removed to form a target speaker model, transforming feature vectors into DeepSF vectors for speaker recognition tasks. Deep learning has transformed fields like computer vision, speech recognition, and natural language processing.

- ***d-vector***

Several approaches aim to identify the attributes of a deep neural network's (DNN) hidden layers in speaker recognition, replacing the traditional i-vectors. These efforts have led to the development of d-vectors, which represent averaged activations of the final hidden layer of a DNN with multiple fully connected layers [10]. D-vectors are becoming more and more popular because of how well they can extract speaker-specific information from speech signals. Their capacity to retrieve significant representations from hidden layers has shown to be beneficial for tasks involving speaker verification and identification. A DNN is trained to produce d-vectors using speech data from multiple speakers. The DNN gains the ability to link speech features to speaker embeddings during this training phase, thereby capturing each speaker's distinct identity. D-vectors are generated by averaging the activations of the last hidden layer after the DNN has been trained. For speaker recognition tasks, these d-vectors can be used as speaker-specific feature representations [62]. D-vectors are a significant advancement in speaker recognition technology, offering better accuracy and robustness against noise and distortions. They can record complex speech features, making them adaptable for person identification. D-vector extraction techniques are refined and applied in other fields, establishing their status as a key component of speaker identification systems. A study using a text-dependent corpus of 646 speakers found that the proposed d-vector system is not as effective as the generic i-vector system (Fig. 3).

- **j-vector**

J-vector, or joint factor analysis (JFA), is a speaker verification technique that employs factor analysis to extract low-dimensional representations of speaker-specific information from speech signals. The underlying principle of the J-vector involves modeling the speech signal as a combination of speaker-specific and non-speaker-specific information. The speaker-specific information is captured in a low-dimensional vector called the J-vector. The J-vector is trained on extensive speech signals from multiple speakers. The objective is to learn a low-dimensional representation of speaker-specific information resilient to variations in the speech signal, such as those arising from different recording environments. The J-vector approach

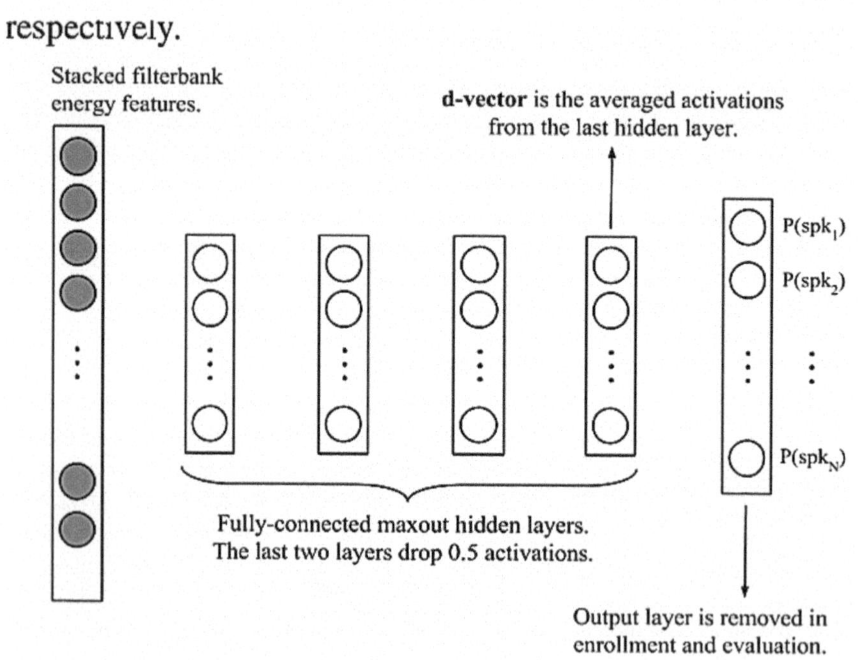

Fig. 3 DNN model [10]

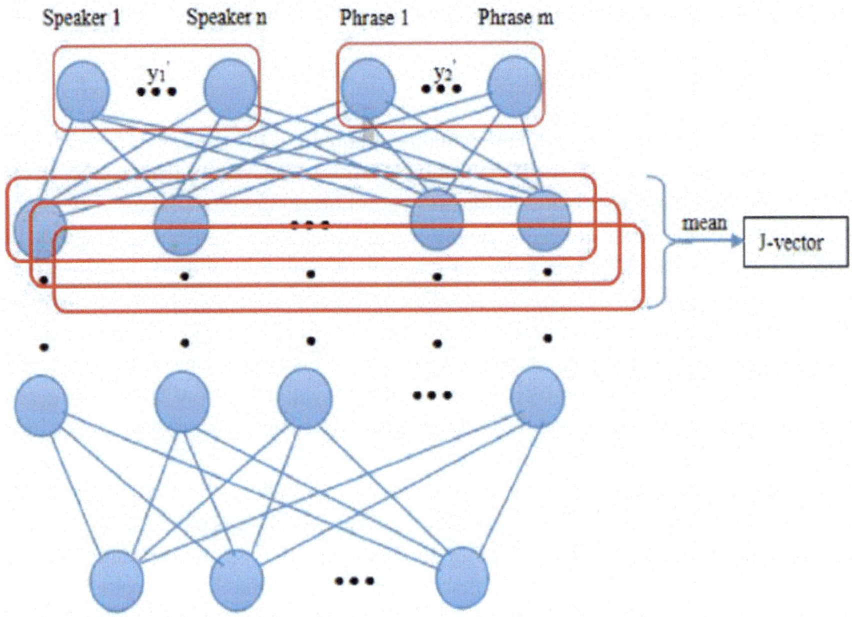

Fig. 4 Multi-task DNN [63]

has performed superior to traditional i-vector-based methods in speaker verification tasks, particularly in short-duration utterances. This enhanced performance is attributed to its ability to effectively capture and represent speaker-specific information even from limited speech data. J-vector development has significantly advanced speaker verification technology, providing a robust and efficient tool for identifying individuals based on their speech patterns. Ongoing research continues to refine J-vector extraction techniques and explore its applications in various domains, further establishing its position as a valuable asset in speaker recognition systems (Fig. 4).

After training, the J-vector serves as a feature for speaker verification, involving a classifier trained to differentiate speakers using their J-vectors. An advantage of the J-vector is its efficacy with a relatively small dataset of speech signals (e.g., a few hours), making it valuable for scenarios where extensive training data collection is challenging. Notably, J-vector has demonstrated state-of-the-art performance on various benchmark speaker verification datasets. It has found widespread application in commercial speaker verification systems, including voice assistants, online banking, and call centers, where it authenticates users based on their voice. A multi-task learning strategy has also been incorporated into the d-vector method [63]. As mentioned by the authors, although directly identifying a speaker might seem challenging, each speaker exhibits a unique pronunciation style for every syllable or phrase. Hence, incorporating textual objectives into a multi-learning system and speaker IDs could enhance speaker verification performance. Figure 2 illustrates the utilized network architecture.

- **x-vector**

An additional buried layer recovered feature vector is called the "x-vector" [63, 64]. Based on DNN embedding, it employs a fully connected, multi-layered DNN architecture with unique temporal context at each layer (called "frames"). Because of the wider temporal context, the design is known as time-delay NN (TDNN). In 2018, the deep learning-based X-vector speaker recognition method was unveiled. It is an expansion of the deep neural network architecture-based i-vector technique. The term "x-vector" represents the "extended" i-vector. The x-vector's concept is to take variable-length speech segments and use a deep neural network to extract a fixed-length embedding or x-vector. A pooling layer and a few fully connected layers come after several convolutional and recurrent neural network layers in a conventional neural network architecture. A fixed-length embedding that captures the speaker-specific features of the input voice segment is the network's output. An extensive collection of voice segments from various speakers is required to train the x-vector network. The network is trained to minimize a loss function that pushes embeddings from different speakers apart and promotes embeddings from the same speaker to be near together in the embedding space. Once trained, a classifier is trained to distinguish between speakers based on their x-vectors, and the x-vector can be utilized as a feature for speaker recognition. The VoxCeleb dataset is one of the benchmark speaker recognition datasets on which x-vector has demonstrated state-of-the-art performance. The x-vector has been utilized in commercial speaker recognition systems, voice recognition in call centers, online banking, voice assistants, and multi-task learning scenarios [64]. The study proposes an alternate x-vector-based training approach for speaker embedding extractors, which builds target speech segment frames from the inferred embedding of different speech segments from the same utterance, resulting in marginally lower EER on the NIST SRE16 dataset [65].

- **Deep belief networks**

Neural networks with intricate hierarchical structures, known as deep belief networks (DBNs), have emerged as another powerful tool for speaker recognition [67, 68]. Deep belief networks (DBNs) are layered probabilistic models that employ binary latent variables in multiple layers. Connections between neurons in both adjacent and the same layers are not directional. DBN training is challenging due to the difficulty of inferring the posterior distribution from hidden (latent) layers. DBN architecture can be constructed by stacking RBMs together (Fig. 5). More information can be found in [69]. Deep belief networks (DBNs) strive to extract abstract hierarchical representations from unlabeled input data. In spectrograms (25 ms window size, 10 ms time step), the input voice data was fed into the DBN after undergoing dimensionality reduction using PCA transformation [68]. The activations from the first and second layers of the RBM were extracted as features and combined with standard MFCC features to enhance speaker recognition performance.

Following feature extraction, speaker recognition was performed using GMM-UBM. The authors employed the ELSDSR dataset, comprising 22 speakers. Utilizing individual MFCC features resulted in a final accuracy of 90%, while combining MFCC features with RBM features achieved an accuracy of 95%, demonstrating the effectiveness of the features extracted from the RBM in improving recognition accuracy. In this case, hybrid (MFCC + DBN) features likewise performed best: MFCC and MFCC + DBN features achieved 88.6% and 92.60% accuracy [8, 67]. In [70], the study evaluates various deep neural network (DNN) approaches for deep feature extraction, including RBMs, speech-discriminant, speaker-discriminant, and multi-task joint-learned DNNs [67, 68]. Using text labels as training data and triphone states as goals, a unique speech-discriminant DNN was utilized. This method shows potential for speaker verification jobs that rely on text. We could extract speaker-specific information from the last hidden layer's output. Modifying the speech-discriminant network's outputs to reflect speaker IDs required a more speaker-specific feature set in the speaker-discriminant DNN configuration. With this change, the goals of speaker verification are better met. The two previously mentioned outputs—speaker IDs and triphones—were used as targets in the configuration that involved several tasks. The starting point was a conventional i-vector system integrated with GMM-UBM and cosine similarity; it was trained using PLP characteristics. The RSR2015 dataset independently evaluated the proposed deep features in different combinations [71]. Compared to the baseline result of 1.5% EER, the speaker-discriminant and multi-task DNNs had the best performance with 1.06% and 0.80% EER, respectively. With an EER of 0.73%, the best combination of in-depth features was RBM and multi-task features concatenated. Using PLDA after deep feature extraction also yielded a 0.20 percent EER for speaker-discriminant features (Fig. 6).

- **DNN training**

Optimizing a deep neural network model's weights or parameters to fit the training set of data is known as deep neural network (DNN) training. Minimizing the discrepancy between the model's expected and actual output is the aim of DNN training. Before the discriminative stage of training a deep neural network (DNN), generative pre-training is frequently used with deep belief networks (DBNs). The DBN is trained using an unsupervised training technique during the generating phase. This step aims to provide the DNN's weights with a solid initialization. The DBN comprises multiple layers of an unsupervised neural network called Restricted Boltzmann Machines (RBM). A visible layer and a concealed layer make up each RBM. The visual layer represents the input data, and the hidden layer defines a set of features learned from the input data. The weights between the visible and invisible layers are changed during the RBM's training process to minimize the reconstruction error between the input data and the reconstructed output from the hidden layer. Gaussian–Bernoulli RBM (GB-RBM) and Bernoulli–Bernoulli RBM (BBRBM) are the two varieties of RBM. For binary inputs and outputs, BBRBM is utilized, while GB-RBM is used for real-valued and binary outputs. The discriminative phase commences after the DBN has undergone pre-training. Labeled data is used to train and oversee the DNN

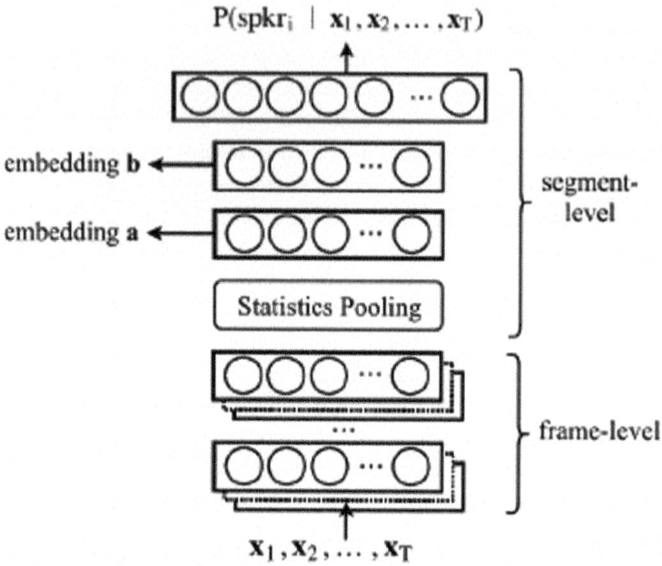

Fig. 5 x-vector DNN embedding architecture [66]

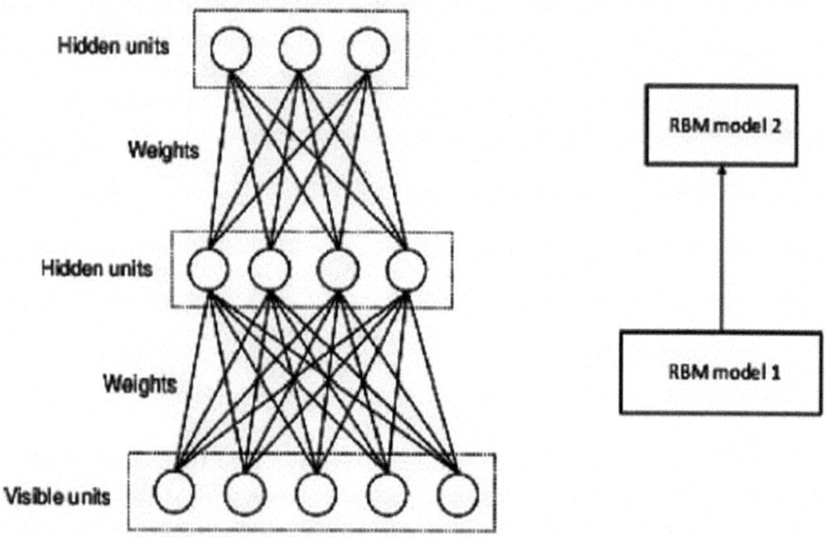

Fig. 6 Structure of the DBN used for extraction of short-term spectral features, with two hidden layers, can be visualized as a stack of 2 RBMs [68]

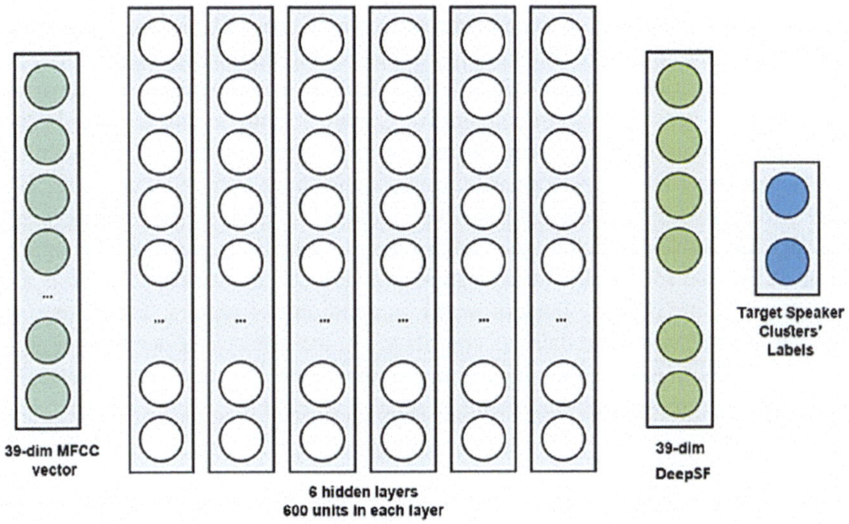

Fig. 7 Architecture of the used deep neural network DNN

during this stage. To minimize the classification error on the labeled data, the DNN's weights are initialized with the weights discovered during the pre-training phase. Then the following steps are involved in the training process:

- Initialization: The weights and biases of the neural network are initialized randomly.
- Forward propagation: The input data is fed into the neural network, and the output is computed using the current weights and biases.
- Loss computation: The difference between predicted and actual output is computed using a loss function.
- Backward propagation: The gradients of the loss function concerning the weights and biases are computed using the chain rule of differentiation.
- Parameter update: The weights and biases are updated using an optimization algorithm such as gradient descent, Adam, or RMSprop.
- Repeat steps 2–5 until convergence: Steps 2–5 are repeated multiple times until the model's performance on the training data no longer improves significantly (Fig. 7).

5 Conclusion

This paper thoroughly reviews deep learning applications in speaker recognition, encompassing identification and verification. While early deep learning (DL) techniques attempt to supersede traditional feature extraction techniques like i-vectors, they perform comparably to state-of-the-art i-vector probabilistic linear discriminant analysis (PLDA) systems, but not better. However, there has been a noticeable gain in classification accuracy with the introduction of more recent DL designs. Interestingly, i-vectors perform competitively, mainly using longer test words and more training material per speaker. Despite this, new developments in deep learning have demonstrated better performance; some studies have even reported noticeably decreased Equal Error Rates (EERs). Even though the achieved performances are typically just somewhat better than earlier ones, deep learning is quickly becoming the go-to method for speaker identification and verification. Standard x-vectors are used as a baseline in many modern works, in addition to i-vectors. The increasing accessibility of large-scale datasets contributes to the efficacy of deep learning in this field. Furthermore, the ongoing creation of innovative DL structures presents the possibility of significant advancements in speaker detection.

References

1. Al-Karawi KA, Al-Noori AH, Li FF, Ritchings T (2015) Automatic speaker recognition system in adverse conditions-implication of noise and reverberation on system performance. Int J Inf Electron Eng 5:423
2. Al-Karawi KA, Li F (2017) Robust speaker verification in reverberant conditions using estimated acoustic parameters—a maximum likelihood estimation and training on the fly approach. In: 2017 seventh international conference on innovative computing technology (INTECH), pp 52–57
3. Al-Karawi KA, Ahmed ST (2021) Model selection toward robustness speaker verification in reverberant conditions. Multimedia Tools Appl 1–18
4. Al-Karawi KA (2021) Mitigate the reverberation effect on the speaker verification performance using different methods. Int J Speech Technol 24:143–153
5. Waghmare K, Gawali B (2022) Speaker recognition for forensic application: a review. J Positive School Psychol 6:984–992
6. Algabri M, Mathkour H, Bencherif MA, Alsulaiman M, Mekhtiche MA (2017) Automatic speaker recognition for mobile forensic applications. Mobile Inf Syst 2017
7. Palia N, Kant S, Dev A (2019) Performance evaluation of speaker recognition system. J Discrete Math Sci Crypt 22:203–218
8. Sztahó D, Szaszák G, Beke A (2019) Deep learning methods in speaker recognition: a review. arXiv preprint arXiv:1911.06615
9. Lei Y, Scheffer N, Ferrer L, McLaren M (2014) A novel scheme for speaker recognition using a phonetically-aware deep neural network. In: Acoustics, speech and signal processing (ICASSP), 2014 IEEE international conference on, pp 1695–1699
10. Variani E, Lei X, McDermott E, Moreno IL, Gonzalez-Dominguez J (2014) Deep neural networks for small footprint text-dependent speaker verification. In: 2014 IEEE international conference on acoustics, speech and signal processing (ICASSP), pp 4052–4056
11. Al-Karawi KA, Mohammed DY (2019) Early reflection detection using autocorrelation to improve robustness of speaker verification in reverberant conditions. Int J Speech Technol 1–8
12. Al-Karawi KA, Mohammed DY (2021) Improving short utterance speaker verification by combining MFCC and entrocy in noisy conditions. Multimedia Tools Appl 80:22231–22249
13. Mohammed DY, Al-Karawi K, Aljuboori A (2021) Robust speaker verification by combining MFCC and entrocy in noisy conditions. Bull Electr Eng Inf 10:2310–2319

14. Bimbot F, Bonastre J-F, Fredouille C, Gravier G, Magrin-Chagnolleau I, Meignier S et al (2004) A tutorial on text-independent speaker verification. EURASIP J Appl Signal Process 2004:430–451
15. Mohammed DY, Al-Karawi KA, Husien IM, Ghulam MA (2020) Mitigate the reverberant effects on speaker recognition via multi-training. In: Applied computing to support industry: innovation and technology: first international conference, ACRIT 2019, Ramadi, Iraq, September 15–16, 2019, Revised Selected Papers 1, pp 95–109
16. Lei Y, Burget L, Ferrer L, Graciarena M, Scheffer N (2012) Towards noise-robust speaker recognition using probabilistic linear discriminant analysis. In: 2012 IEEE international conference on acoustics, speech and signal processing (ICASSP), pp 4253–4256
17. Benzeghiba M, De Mori R, Deroo O, Dupont S, Erbes T, Jouvet D et al (2007) Automatic speech recognition and speech variability: a review. Speech Commun 49:763–786
18. Campbell JP, Shen W, Campbell WM, Schwartz R, Bonastre J-F, Matrouf D (2009) Forensic speaker recognition. IEEE Signal Process Mag 26:95–103
19. Mohammed DY, AlKarawi K, Aljuboori A (2021) Robust speaker verification by combining MFCC and entrocy in noisy conditions. Bull Electr Eng Inf 10
20. Saquib Z, Salam N, Nair RP, Pandey N, Joshi A (2010) A survey on automatic speaker recognition systems. In: Signal processing and multimedia: international conferences, SIP and MulGraB 2010, held as part of the future generation information technology conference, FGIT 2010, Jeju Island, Korea, December 13–15, 2010. Proceedings, pp 134–145
21. Zhao X, Wang Y, Wang D (2014) Robust speaker identification in noisy and reverberant conditions
22. Wang L, Nakagawa S (2009) Speaker identification/verification for reverberant speech using phase information. In: Proceedings of WESPAC 2009
23. Sadjadi SO, Hansen JH (2012) Blind reverberation mitigation for robust speaker identification. In: Acoustics, speech and signal processing (ICASSP), IEEE international conference on, pp 4225–4228
24. Al-Rawahy S, Hossen A, Heute U (2012) Text-independent speaker identification system based on the histogram of DCT-cepstrum coefficients. Int J Knowl Based Intell Eng Syst 16:141–161
25. Furui S (1997) Recent advances in speaker recognition. In: Audio-and video-based biometric person authentication, pp 235–252
26. Al-Karawi K (2018) Robust speaker recognition in reverberant condition-toward greater biometric security. University of Salford
27. Do M (2014) How to build an automatic speaker recognition system (lecture notes). University of Illinois, Urbana Champaign, Illinois
28. Al-Karawi KA (2023) Face mask effects on speaker verification performance in the presence of noise. Multimedia Tools Appl 1–14
29. Singh N (2017) Automatic speaker recognition: current approaches and progress in last six decades. Global J Enterprise Inf Syst 9:45–52
30. Al-Karawi KA, Mohammed DY (2023) Using combined features to improve speaker verification in the face of limited reverberant data. Int J Speech Technol 1–11
31. Summerfield R, Dunstone T, Summerfield C (2008) Speaker verification in a multi-vendor environment. In: W3C workshop on speaker identification and verification (SIV)
32. Shao Y, Wang D (2008) Robust speaker identification using auditory features and computational auditory scene analysis. In: Acoustics, speech and signal processing. ICASSP 2008. IEEE international conference on, pp 1589–1592
33. Al-Ali AKH, Dean D, Senadji B, Chandran V, Naik GR (2017) Enhanced forensic speaker verification using a combination of DWT and MFCC feature warping in the presence of noise and reverberation conditions. IEEE Access 5:15400–15413
34. Ming J, Hazen TJ, Glass JR, Reynolds DA (2007) Robust speaker recognition in noisy conditions. Audio, Speech, Lang Process IEEE Trans 15:1711–1723
35. Alenizi AS, Al-Karawi KA (2023) Effective biometric technology used with big data. In: Proceedings of seventh international congress on information and communication technology, pp 239–250

36. Alenizi AS, Al-Karawi KA (2023) Internet of things (IoT) adoption: challenges and barriers. In: Proceedings of seventh international congress on information and communication technology, pp 217–229
37. Garofolo JS, Lamel LF, Fisher WM, Fiscus JG, Pallett DS (1993) DARPA TIMIT acoustic-phonetic continous speech corpus CD-ROM. NIST speech disc 1–1.1. NASA STI/Recon Tech Rep N 93:27403
38. Marcus M, Santorini B, Marcinkiewicz MA (1993) Building a large annotated corpus of English: the Penn Treebank
39. Larcher A, Lee KA, Ma B, Li H (2012) The RSR2015: database for text-dependent speaker verification using multiple pass-phrases. In: Annual conference of the international speech communication association (Interspeech)
40. Vincent E, Barker J, Watanabe S, Le Roux J, Nesta F, Matassoni M (2013) The second 'CHiME'speech separation and recognition challenge: datasets, tasks, and baselines. In: 2013 IEEE international conference on acoustics, speech and signal processing, pp 126–130
41. Panayotov V, Chen G, Povey D, Khudanpur S (2015) Librispeech: an asr corpus based on public domain audiobooks. In: 2015 IEEE international conference on acoustics, speech and signal processing (ICASSP), pp 5206–5210
42. Nagrani A, Chung JS, Zisserman A (2017) Voxceleb: a large-scale speaker identification dataset. arXiv preprint arXiv:1706.08612
43. Chung JS, Nagrani A, Zisserman A (2018) Voxceleb2: deep speaker recognition. arXiv preprint arXiv:1806.05622
44. Alenizi AS, Al-karawi KA (2022) Cloud computing adoption-based digital open government services: challenges and barriers. In: Proceedings of sixth international congress on information and communication technology, pp 149–160
45. Hansen JH, Hasan T (2015) Speaker recognition by machines and humans: a tutorial review. IEEE Signal Process Mag 32:74–99
46. Campbell JP (1997) Speaker recognition: a tutorial. Proc IEEE 85:1437–1462
47. Alenizi AS, Al-karawi KA (2023) Machine learning approach for diabetes prediction. In: International congress on information and communication technology, pp 745–756
48. Kinnunen T, Li H (2010) An overview of text-independent speaker recognition: from features to supervectors. Speech Commun 52:12–40
49. Bilmes JA (1998) A gentle tutorial of the EM algorithm and its application to parameter estimation for Gaussian mixture and hidden Markov models. Int Comput Sci Inst 4:126
50. Campbell WM, Campbell JP, Reynolds DA, Singer E, Torres-Carrasquillo PA (2006) Support vector machines for speaker and language recognition. Comput Speech Lang 20:210–229
51. Grozdić Đ, Jovičić S, Pavlović DŠ, Galić J, Marković B (2017) Comparison of cepstral normalization techniques in whispered speech recognition. Adv Electr Comput Eng 17:21–26
52. Ganapathy S, Pelecanos J, Omar MK (2011) Feature normalization for speaker verification in room reverberation. In: Acoustics, speech and signal processing (ICASSP), IEEE international conference on, pp 4836–4839
53. Mammone RJ, Zhang X, Ramachandran RP (1996) Robust speaker recognition: a feature-based approach. IEEE Signal Process Mag 13:58
54. Kenny P (2005) Joint factor analysis of speaker and session variability: theory and algorithms. CRIM, Montreal, (Report) CRIM-06/08-13
55. Dehak N, Kenny P, Dehak R, Dumouchel P, Ouellet P (2011) Front-end factor analysis for speaker verification. Audio, Speech, Lang Process IEEE Trans 19:788–798
56. Campbell WM (2006) Compensating for mismatch in high-level speaker recognition. In: IEEE Odyssey 2006: The speaker and language recognition workshop, pp 1–6
57. Dehak N, Dehak R, Kenny P, Brümmer N, Ouellet P, Dumouchel P (2009) Support vector machines versus fast scoring in the low-dimensional total variability space for speaker verification. In: Tenth Annual conference of the international speech communication association
58. Hourri S, Kharroubi J (2019) A novel scoring method based on distance calculation for similarity measurement in text-independent speaker verification. Procedia Comput Sci 148:256–265

59. Chen YW, Lin CJ (ed) (2006) Combining SVMs with various feature selection strategies. In: Feature extraction. Springer, pp 315–324
60. Al-Karawi KA (2023) Real-time adaptive training for forensic speaker verification in reverberation conditions. Int J Speech Technol
61. Hasan T, Hansen JH (2011) A study on universal background model training in speaker verification. IEEE Trans Audio Speech Lang Process 19:1890–1899
62. Chen YH, Moreno IL, Sainath T, Visontai M, Alvarez R, Parada C (2015) Locally-connected and convolutional neural networks for small footprint speaker recognition
63. Chen N, Qian Y, Yu K (2015) Multi-task learning for text-dependent speaker verification. In: Sixteenth annual conference of the international speech communication association
64. You L, Guo W, Dai L, Du J (2019) Multi-Task learning with high-order statistics for X-vector based text-independent speaker verification. arXiv preprint arXiv:1903.12058
65. Stafylakis T, Rohdin J, Plchot O, Mizera P, Burget L (2019) Self-supervised speaker embeddings. arXiv preprint arXiv:1904.03486
66. Snyder D, Garcia-Romero D, Sell G, Povey D, Khudanpur S (2018) X-vectors: Robust dnn embeddings for speaker recognition. In: 2018 IEEE international conference on acoustics, speech and signal processing (ICASSP), pp 5329–5333
67. Ali H, Tran SN, Benetos E, d'Avila Garcez AS (2018) Speaker recognition with hybrid features from a deep belief network. In: Neural Comput Appl 29:13–19
68. Banerjee A, Dubey A, Menon A, Nanda S, Nandi GC (2018) Speaker recognition using deep belief networks. arXiv preprint arXiv:1805.08865
69. Hinton GE, Osindero S, Teh Y-W (2006) A fast learning algorithm for deep belief nets. Neural Comput 18:1527–1554
70. Liu Y, Qian Y, Chen N, Fu T, Zhang Y, Yu K (2015) Deep feature for text-dependent speaker verification. Speech Commun 73:1–13
71. Larcher A, Lee KA, Ma B, Li H (2012) RSR2015: Database for text-dependent speaker verification using multiple pass-phrases. In: Thirteenth annual conference of the international speech communication association

Social Media Emotion Detection and Analysis System Using Cutting-Edge Artificial Intelligence Techniques

Tapu Rayhan, Ayesha Siddika, Mehedi Hasan, and Nafisa Sultana Elme

Abstract This paper presents an in-depth analysis of the Social Media Emotion Detection and Analysis System (SM-EDS) using cutting-edge artificial intelligence (AI) techniques. The study explores the effectiveness of AI in text-based emotion recognition models to detect emotions from social media data. The research involves state-of-the-art natural language processing algorithms, deep learning models, and sentiment analysis techniques, using diverse social media data sets for training and validation. The critical aspects of the Social Media Emotion Detection system (SM-EDS), including data prepossessing, model architecture, and evaluation metrics, are investigated. Ethical implications and societal impact are discussed, highlighting the potential of AI-driven emotion detection systems to revolutionize social media analytics. The paper emphasizes responsible deployment and identifies challenges and future research opportunities in this promising field.

Keywords Artificial intelligence · Emotion detection · Machine learning · Sentiment analysis · Social media

1 Introduction

In the era of digital communication and interconnections, social media has become an integral part of our daily lives. Billions of people across the globe actively engage with various social media platforms, sharing their thoughts, opinions, and emotions [1]. As the volume of social media content continues to grow exponentially, the need to analyze and understand the underlying emotions becomes increasingly important. Emotions are a vital part of human lives that play an integral role in how humans perceive and understand things [2].

T. Rayhan (✉) · A. Siddika · M. Hasan · N. S. Elme
World University of Bangladesh, Dhaka 1230, Bangladesh
e-mail: tapurayhan.cse@gmail.com

A. Siddika
e-mail: ayesha@cse.wub.edu.bd

X.-S. Yang et al. (eds.), *Proceedings of Ninth International Congress on Information and Communication Technology*, Lecture Notes in Networks and Systems 1000,
https://doi.org/10.1007/978-981-97-3289-0_40

Nowadays, social media has become a part of human life. It is a platform where people can easily share their emotions and mental situations, express their feelings, or discuss any other matter [3]. Emotion states are associated with a wide variety of human feelings, thoughts, and behaviors; hence, they affect our ability to act rationally in cases such as decision-making, Perceptron, and human intelligence.

We constantly see the posts or statuses of people on social media. But we never think or analyze how many people are going through what kind of mental state [4]. But if there was an application or tool, we could analyze and categorize human emotions such as happiness, sadness, depression, and love. Then, these data can be used for various research and scientific solutions, and various surveys can be done, such as whether the human condition of people is improving or deteriorating day by day. Also, this data can help entrepreneurs make seasonable business decisions [5].

We can design and develop a desktop application that can read social media platforms and collect that data to categorize people's emotions and provide a final report. Note that it is able to work faster than humans using artificial intelligence [6, 7].

1.1 Objectives

- To develop a desktop application that can scroll through social media (Facebook, Twitter) platforms and read data.
- To implement natural language processing techniques using AI, utilize sentiment analysis and detect the emotions of data.

1.2 Literature Review

Before starting work on this topic, we conducted several research studies. Our objective was to acquire insights into the development of desktop applications, identify the current popular and widely used technologies or programming languages, determine the optimal programming language for machine learning, and ascertain the machine learning algorithm that yields superior accuracy in data training. In this section, we will briefly discuss the main aim and results of the study. The research is mentioned below:

The paper offers a comprehensive analysis of diverse emotion recognition techniques within the human-computer interaction and affective computing domains. Encompassing over a hundred papers, it's the first endeavor to consolidate a decade's worth of emotion detection models. The study identifies four primary methods (facial expression, physiological signals, speech variation, and text semantics) that predominantly recognize seven basic emotions, featuring comparisons across methods. The highest accuracy (99.47%) was achieved using particle swarm optimization-assisted biogeography-based optimization algorithms for speech-based emotion recognition on the BES dataset, demonstrating remarkable insights into the field [1].

By reading this research paper, we understand that they use NLP for social robots to add human emotional state knowledge and an essential part of their behavior to emotions. The robot collects data and identifies emotions from it through different media, such as images, text, media, and multimedia content. Emotions are detected, and robots can identify them very easily. They have been able to identify emotions with natural language processing (NLP) by analyzing faces and postures in images based on machine learning techniques. They have used AI here, and they have been able to implement their tests through an application. They have used Google's API and Python-based speech-to-text emotion detection algorithms and were able to control artifacts and emotions in a similar way to the ontology for neural networks in NLP Transformer and emotion-based differences [2].

Social media platforms have become a major means of expressing emotions in the present time. That data will be a great tool for future generations. In the paper, they worked on emotion recognition from large-size text data from social media. They first converted the tweets to lowercase, so that uppercase and lowercase do not appear different during data learning. After that, they removed the punctuation, hashtags, and numbers because they didn't play any role in the impression. After that, by converting the tweets into tokens, they were able to identify six categories of emotions through a deep learning algorithm. In addition, they have tested various emotion models for good results and noted their advantages and disadvantages and accuracy in the paper. Among them, they got the best results using the advanced deep learning-based RoBERTa model. The accuracy level of the model was 80.33% [3].

This paper proposes the best model for detecting the emotion of the text. They downloaded a simple dataset from Kaggle and tested various pre-trained models and reported their results, which are Naïve Bayes, Bi-LSTM, and Hybrid Model. Where Naive Bayes Training Accuracy was 84% and Validation Accuracy was 64% and Where Bi-LSTM Training Accuracy was 78% and Validation Accuracy was 61% and On the other hand, the Training Accuracy of the Hybrid Model was 75% and the Validation Accuracy was 76%. As a result, the use of the Hybrid Model seems more efficient than the other two models. They prepared the Hybrid Model by combining the conventional Bert Model and the Rule-Based Model [4].

A study developed a desktop application for chat boards using artificial intelligence algorithms, which can reply like a human. They say that it cannot only recognize text or keywords but also recognize the user's mood through the camera. It can be said that the user's grievance can be detected very easily through the system chatting application. They have used techniques like machine learning and data mining for their application. If the user feels sad, then the system will automatically fetch a joke from the database. This will send the user to the Windows terminal system and be able to converse through the chatting application [5].

The review delves into Text-Based Emotion Detection (TBED), an expanding field within NLP, encompassing emotional classification using machine learning. Its significance lies in applications like risk assessment in terrorist attack messages, historical corpus analysis, and customer satisfaction evaluation in product reviews. The study examines TBED's literature and its connections with psychological models, offering insights into this evolving realm [6].

This study addresses the vital role of emotion detection in diverse communication modes and emphasizes Roman Urdu text emotion detection, an underexplored area. The research develops a substantial corpus of 18k sentences, spanning various domains and annotated with six emotion classes. The application of baseline algorithms highlights SVM's superior F-measure performance. Given Roman Urdu's prominence on social media, this work contributes to sentiment analysis's practical applications in fields like mental health, product enhancement, and dialog systems [7].

The main goal of this research is to identify Twitter data emotions using data mining and deep learning methods. Through this, Twitter's fake news can be detected and countered. They tested their accuracy using SVM, Random Forest, and CNN algorithms. And for better results, they feel the need to learn after tokenizing the tweets. They also found that the various URLs, hashtags, and mentions of users using @ in tweets can have a bad impact on identifying the actual emotions [8].

The study underscores the pivotal role of emotions in conversations and the growing interest in extracting emotions from text, aided by advancing machine learning techniques. Exploring various algorithms from traditional to deep learning, it employs the Affective Tweets dataset, yielding a baseline F1-score of 0.71. With extensive exploration across 2302 feature sets, the Generalized Linear Model emerges as the frontrunner, demonstrating exceptional accuracy, precision, and recall (0.902) alongside an impressive F1-score (0.901), solidifying the research's significance in emotion recognition from social media conversations [9].

The paper presents a survey on sentiment analysis employing artificial intelligence techniques for apprehensiveness or dejection detection from social media data, encompassing text, emoticons, and emojis. Highlighting the shift toward multi-class classification for more precise results, the study explores a range of machine learning and deep learning methods. Multi-class classification with deep learning algorithm emerges as notably effective, demonstrating higher precision levels in sentiment analysis, contributing to the growing field of sentiment understanding in various contexts of people's daily lives using social media data [10].

The paper delves into the pressing issue of depression's impact on mental health and society, emphasizing the need for automated systems to detect depression. By reviewing existing studies, it analyzes diverse artificial intelligence and machine learning techniques used for detection, encompassing facial expressions, emotional chatbots, and text analysis. The study surveys a range of methods, including Naive Bayes, SVM, LSTM-RNN, and ANN, showcasing how these techniques contribute to emotion and depression detection. This comprehensive review offers insights into the evolving landscape of AI and ML in addressing mental health challenges [11].

This study combines OCR and MSER algorithms to extract digital text from images or scanned documents and convert them into text data. It is more efficient than just using the OCR algorithm. It has been able to detect not only English but also Tamil, Kannada, and Sanskrit languages, and they think Python's OpenCV module can be used for real-life applications. If there is any intelligibility error, it may be useful to use some other algorithm in the mountain to correct it [12]. The paper constructs a taxonomy of conceptual models and proxy data employed in AI systems for analyzing human emotional expression. It raises skepticism about the reliability of existing paradigms in producing accurate human emotion insights. The study highlights the impact of diverse models and data combinations on the ethical and societal dimensions of AI systems. By questioning the underlying assumptions and implications, the research sheds light on the intricate relationship between human emotions, data, and AI, contributing to critical discussions in the field [13].

The chapter offers a comprehensive comparative analysis of Speech-Emotion Recognition (SER) systems, spanning theoretical definitions, affective state categorization, and emotion expression modalities. Using diverse feature extraction methods and classifiers, including MFCC and MS features, the study delves into emotion recognition utilizing various machine learning techniques. RNN, MLR, and SVM classifiers are evaluated using Berlin and Spanish databases. Results indicate that RNN attains the highest accuracy (94%) on the Spanish database, while the overall study enhances our understanding of effective emotion recognition from speech signals [14].

The study presents an emotion recognition system employing EEG signals and a valence/arousal model. Utilizing DWT, EEG signals are divided into frequency bands, and PCA enhances features. SVM, KNN, and ANN are employed for emotional state classification. Notably, cross-validated SVM using ten EEG channels achieves 91.3% arousal and 91.1% valence accuracy, outperforming existing algorithms on the "DEAP" dataset [15].

The paper highlights the rising interest in human emotion and behavior research utilizing artificial intelligence technology. It explores the application of Python programs and machine learning in emotion detection through facial features. Various algorithms are examined and compared, leading to the proposal of a new weighted sum-logarithm-square root algorithm that enhances detection rates. The study provides insights into the potential of combining AI, facial recognition, and innovative algorithms for automatic facial emotion detection [16].

2 Methodology

In this project, we propose a project's structural approach through a custom model. The diagrammatic representation of steps of this model is mentioned below (Fig. 1).

2.1 Data Collection

In order to implement our project, we collected dataset from Kegel, and Huggingface according to our project. Here we work with Go Emotions, tweet_emotions, and

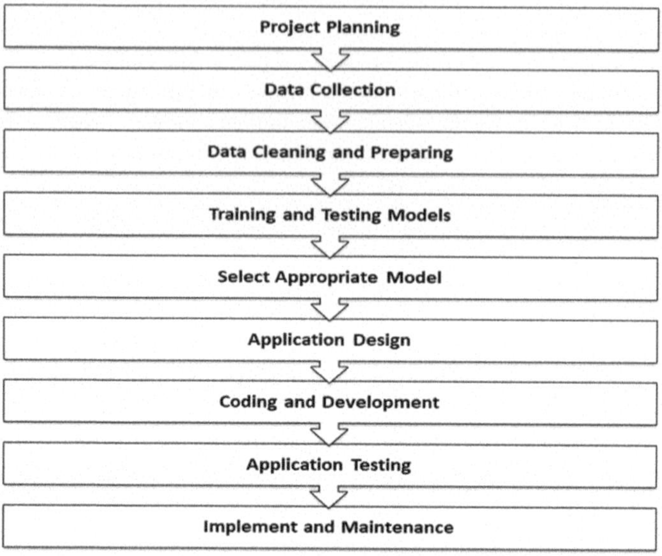

Fig. 1 Model of methodology

yelp_labelled, huggingface datasets and choosing the Go Emotions and huggingface dataset for final training. It contains data for 28 types of emotion categories, and the total records are 624,623 after combining both of them.

2.2 *Data Cleaning and Preparing*

Data cleaning and preparing involves converting raw data into a usable format for analysis. We followed several steps for data cleaning and preparation, firstly removing duplicate or irrelevant data and then correcting structural errors. We converted all data to lowercase and then deleted certain words, symbols, and numbers that do not express emotions. After cleaning and duplicate removal, we got a total **555,807** number of unique records. Calculating the value based on emotions is mentioned in Fig. 2.

2.3 *Training and Testing Models*

In this part, we trained and tested our data using different machine learning algorithms. After training with 80% of our data, we tested the models with the remaining 20%. We used Logistic Regression, Linear SVC, Multinomial NB, Decision Tree Classifier, AdaBoost Classifier, SGD Classifier, Perceptron, Random Forest, and

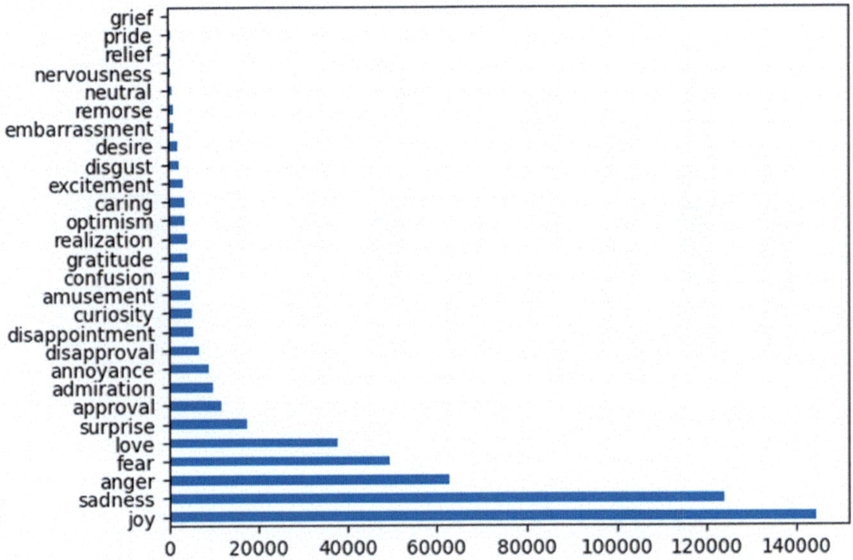

Fig. 2 Total records count value of each emotion

Table 1 Accuracy report

Sl	Trainer	Accuracy (%)
1	Logistic Regression	74
2	Linear SVC	73
3	Multinomial NB	58
4	Decision Tree Classifier	63
5	AdaBoost Classifier	29
6	SGD Classifier	72
7	Perceptron	68
8	Random Forest	52
9	Ridge Classifier	72

Ridger Classifier algorithms as trainers. We refined this iterative process model until satisfactory results were achieved. The accuracy results of those models are mentioned in Table 1 (Fig. 3).

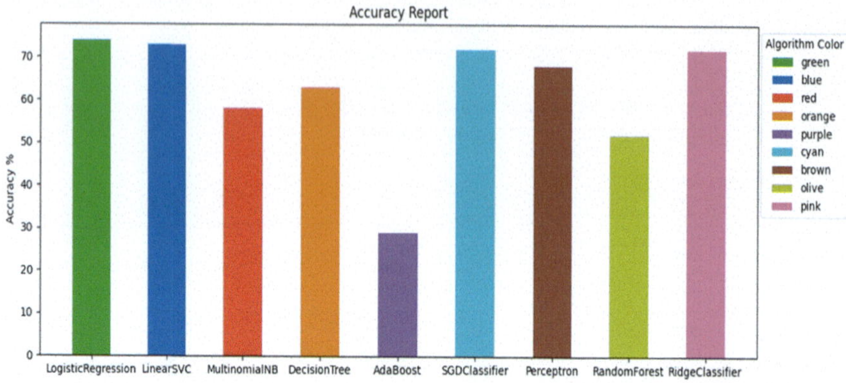

Fig. 3 Bar chart accuracy report

2.4 Select Appropriate Model

We have reviewed the accuracy report and the classification reports of those classifiers to select the appropriate model. It was noted among them precision, recall, and F1-score. All those classification reports are mentioned in Table 2.

After analyzing the accuracy reports and classification reports of those algorithms, we were able to select the appropriate model. We got the best 74% accuracy results by using this Logistic Regression classification algorithm. Now if we look at the results of precision, recall, and F1-score of each emotion, we can see that the Logistic Regression algorithm could not provide any accurate results only for the "pride" emotion. On the other hand, all other algorithms have multiple false predictions, in which precision, recall, and F1-score values are 0.00000.

If we consider the side of success, then we can see that Logistic Regression gives good results for almost all emotions of precision, recall, and F1-score. When we examine the report of our primary emotions, we can see that the algorithm achieved the best results in precision, recall, and F1-score for anger (0.790850, 0.824300, 0.807229) joy (0.852529, 0.903332, 0.877196), and sadness (0.880691, 0.911782, 0.895967) more than LinearSVC anger (0.786683, 0.825587, 0.80566) joy (0.853069, 0.894829, 0.873450) sadness (0.875255, 0.911141, 0.89283) and RidgeClassifier anger (0.771418, 0.819881, 0.794911) joy (0.781073, 0.908628, 0.840036) sadness (0.810724, 0.914774, 0.859612). So, we can consider the Logistic Regression algorithm to be the best in all aspects.

While selecting this model, we also considered Validation Accuracy, performance, and requirements. Finally, we have selected this Logistic Regression model for our application implementation.

Table 2 Logistic Regression classification report

Total emotions	Precision	Recall	F1-score
Admiration	0.321912	0.373431	0.345763
Amusement	0.445652	0.401567	0.422463
Anger	**0.790850**	**0.824300**	**0.807229**
Annoyance	0.147585	0.146789	0.147186
Approval	0.199147	0.317197	0.244677
Caring	0.200000	0.124021	0.153102
Confusion	0.212355	0.108161	0.143322
Curiosity	0.256959	0.232784	0.244275
Desire	0.291262	0.138568	0.187793
Disappointment	0.151923	0.074740	0.100190
Disapproval	0.222765	0.234904	0.228674
Disgust	0.192982	0.062857	0.094828
Embarrassment	0.333333	0.029520	0.054237
Excitement	0.203774	0.080597	0.115508
Fear	0.807829	0.810632	0.809228
Gratitude	0.510664	0.466450	0.487557
Grief	0.181818	0.025641	0.044944
Joy	**0.852529**	**0.903332**	**0.877196**
Love	0.718342	0.678662	0.697938
Nervousness	0.166667	0.012270	0.022857
Neutral	0.794326	0.861538	0.826568
Optimism	0.283582	0.147668	0.194208
Pride	0.000000	0.000000	0.000000
Realization	0.177536	0.055180	0.084192
Relief	0.250000	0.007812	0.015152
Remorse	0.314685	0.171103	0.221675
Sadness	**0.880691**	**0.911782**	**0.895967**
Sadness	0.880691	0.911782	0.895967
Surprise	0.630963	0.608868	0.619718

3 Application Flowchart

See Fig. 4.

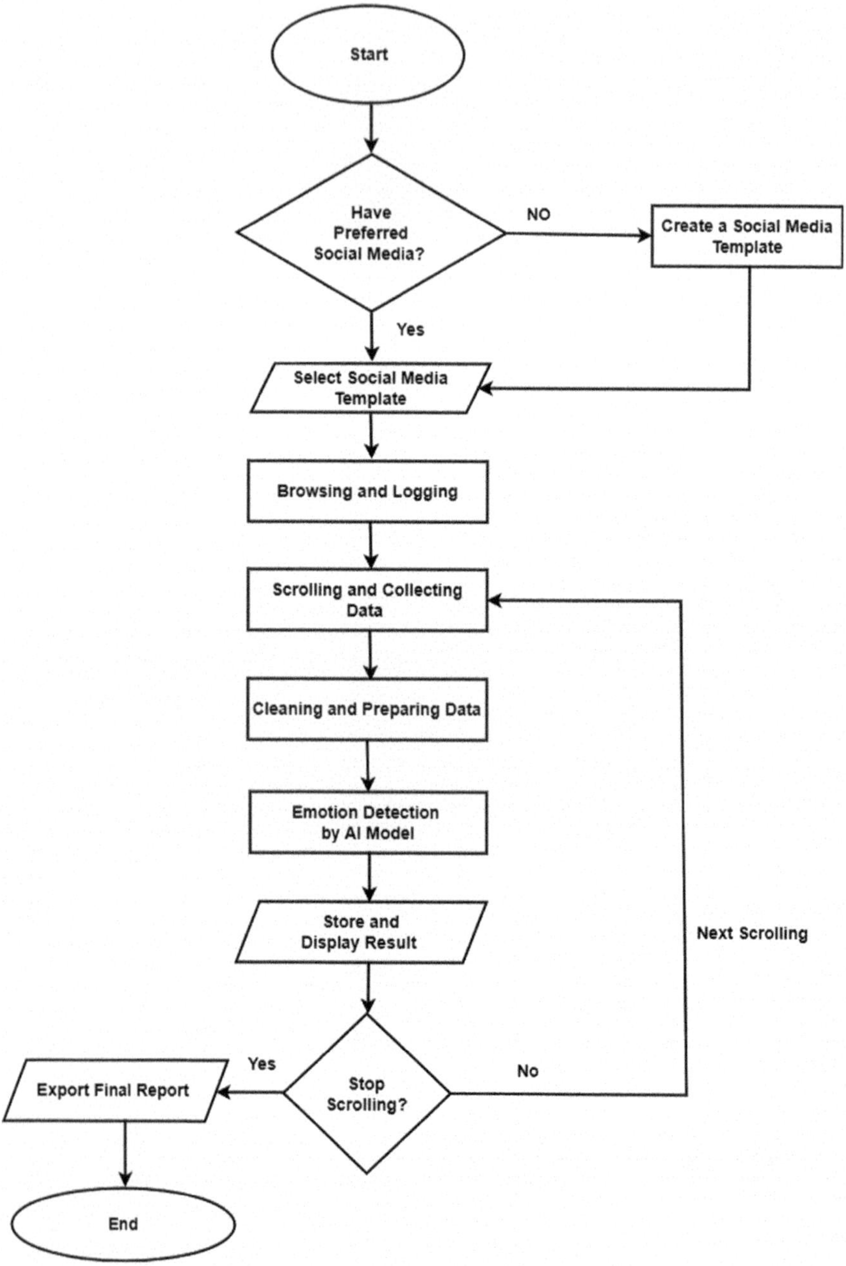

Fig. 4 Application flowchart

4 Model Selection

In this section, after cleaning the data we select some models for testing the accuracy. After testing accuracy, we fixed the Logistic Regression model for our application implementation (Fig. 5).

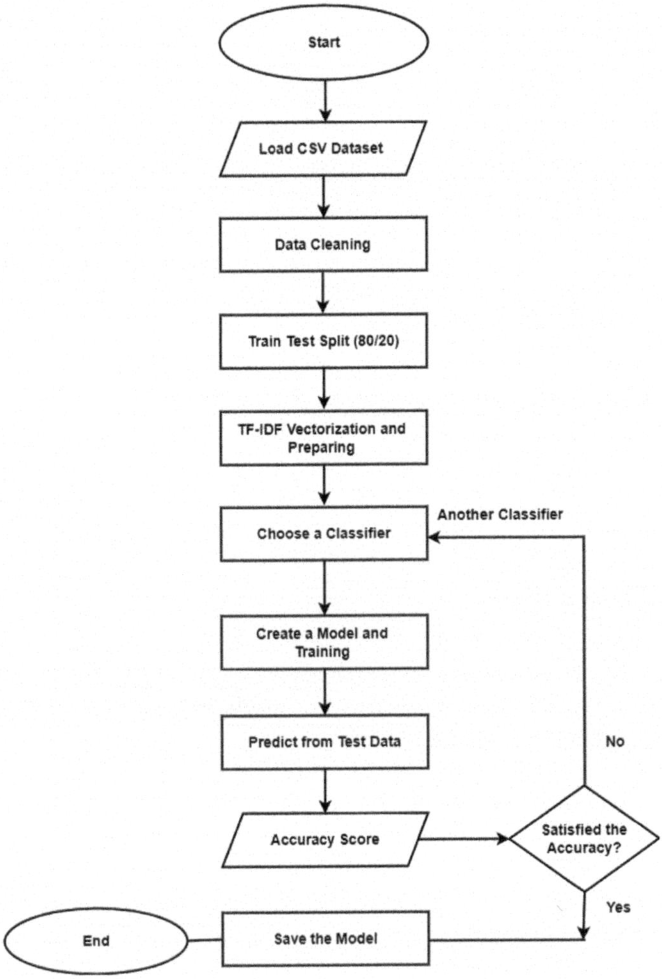

Fig. 5 Model selection

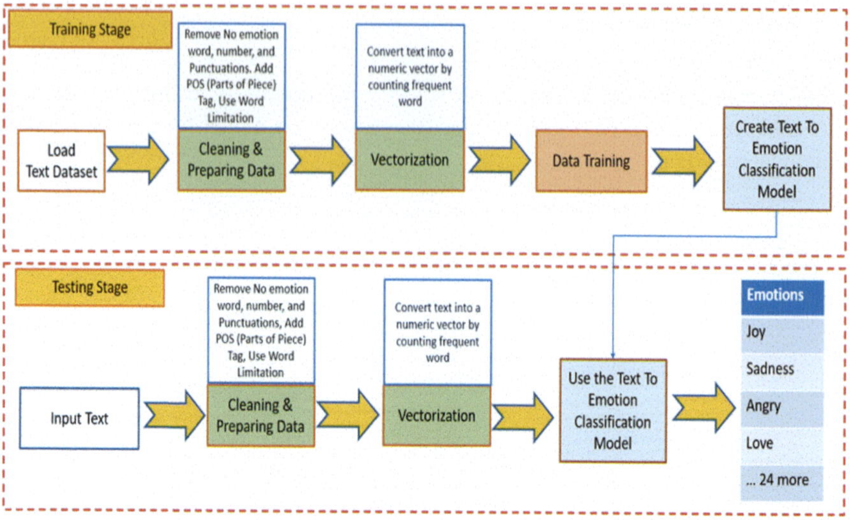

Fig. 6 Diagram of emotion detection process

5 Diagram of Emotion Detection Process

See Fig. 6.

6 Result Discussion

We have selected the best algorithm for our application through training, testing, and report analysis. The algorithm whose accuracy results are more than 74%, below are shown the algorithmś mathematical terms.

Logistic Regression maps predictions and their probability using a logistic function known as the sigmoid function. An S-shaped curve that transforms any real number into a range between 0 and 1 is known as the sigmoid function [17, 18].

For Logistic Regression, the sigmoid function is known as an activation function and is described as follows:

$$f(x) = \frac{1}{1 + e^{-x}}$$

Equation of Logistic Regression
where

- e = base of natural logarithms
- value = numerical value one wishes to transform

$$y = \frac{e^{(b_0+b_1X)}}{1+e^{(b_0+b_1X)}}$$

Logistic Regression—Sigmoid Function
here,

- x = input value
- y = predicted output
- b0 = bias or intercept term
- b1 = coefficient for input (x).

7 Conclusion

We have successfully completed and implemented an application. The project "Social Media Emotion Detection and Analysis System using Artificial Intelligence" presents a significant stride toward understanding the intricate realm of human emotions within the digital landscape. By harnessing the power of artificial intelligence, the system adeptly deciphers and analyzes emotions expressed on social media platforms. By harnessing the power of artificial intelligence, the system adeptly deciphers and analyzes emotions expressed on social media platforms.

This innovation not only enriches our comprehension of societal sentiments but also holds the potential to enhance marketing strategies, mental health interventions, and social-cultural research. As technology intersects with human emotion, this project underscores the importance of ethical considerations, ensuring privacy, and promoting responsible AI utilization. Ultimately, the system emerges as a valuable tool, offering insights into collective feelings, bridging gaps between virtual interactions and human experiences, and paving the way for more emotionally intelligent digital interactions.

There are some limitations to the Social Media Emotion Detection and Analysis System using the Artificial Intelligence Project. It requires a Chrome browser and only accesses public data from social media platforms. The AI model can predict only English data, and the Model Accuracy is 74%. It also detects text-based emotion data.

The future works for the "Social Media Emotion Detection and Analysis System" using Artificial Intelligence Project will have multilingual support, which encourages global users to improve the system to interpret emotions in multiple languages and develop improved accuracy.

Sensitive evolution tracking, which is monitoring emotional shifts over time, enabling dynamic sentiment analysis, and suggesting content aligned with detected emotions, enriches user engagement.

Emotion generation for AI API interactions develops AI's ability to respond emotionally, enhancing human-like interactions.

References

1. Saxena A, Khanna A, Gupta D (2020) Emotion recognition and detec-tion methods: a comprehensive survey. J Artif Intell Syst 2(1):53–79
2. Graterol W et al (2021) Emotion detection for social robots based on NLP transformers and an emotion ontology. Sensors 21(4):1322. https://doi.org/10.3390/s21041322
3. Kusal S et al (2021) AI based emotion detection for textual big data: techniques and contribution. Big Data Cogn Comput 5(3):43. https://doi.org/10.3390/bdcc5030043
4. Madhuri S, Lakshmi SV (2021) Detecting emotion from natural language text using hybrid and NLP pre-trained models. Turkish 12(10):4095–4103
5. Mate G, Wadekar N, Chavan R, Rajput T, Pawar S, Rscoe S (2023) Mood detection with chatbot using AI-desktop partner. Int J Adv Sci Eng 6:988
6. Chaitanya U et al (2023) Digital image text recognition using machine learning algorithms. Int J Res Appl Sci Eng Technol (IJRASET). https://doi.org/10.22214/ijraset.2023.54383?
7. Kozhevnikov V, Pankratova E (2020) Research of the text data vectorization and classification algorithms of machine learning. Theor Appl Sci 8(5). https://doi.org/10.15863/TAS.2020.05.85.106
8. Mehta A, Patil S et al (2021) Emotion detection using social media data. IJRASET 9:1–6
9. Chowanda A, Sutoyo R, Tanachutiwat S et al (2021) Exploring text-based emotions recognition machine learning techniques on social media conversation. Procedia Comput Sci 179:821–828
10. Babu NV, Kanaga EGM (2022) Sentiment analysis in social media data for depression detection using artificial intelligence: a review. SN Comput Sci 3:1–20
11. Joshi ML, Kanoongo N (2022) Depression detection using emotional artificial intelligence and machine learning: a closer review. Mater Today: Proc 58:217–226
12. Chaitanya HBMD, Alisetti E (2023) Digital image text recogni- tion using machine learning algorithms. Int J Res Appl Sci Eng Technol (IJRASET) 11:3583–3589
13. Stark L, Hoey J (2021) The ethics of emotion in artificial intelligence systems. In: Proceedings of the 2021 ACM conference on fairness, accountability, and transparency, pp 782–793
14. Kerkeni L, Serrestou Y, Mbarki M, Raoof K, Mahjoub MA, Cleder C (2019) Automatic speech emotion recognition using machine learning
15. Bazgir O, Mohammadi Z, Habibi SAH (2018) Emotion recognition with machine learning using EEG signals. In: 2018 25th national and 3rd international iranian conference on biomedical engineering (ICBME). IEEE, pp 1–5
16. Jinnuo Z, Goyal S, Tesfayohanis M, Omar Y (2022) Implementation of artificial intelligence image emotion detection mechanism based on python architecture for industry 4.0. J Nanomater
17. Hastie JFT, Tibshirani R (2009) The elements of statistical learning. In: The elements of statistical learning. Springer, Berlin
18. Bishop CM (2006) Pattern recognition and machine learning. In: Pattern recognition and machine learning. Springer, Berlin

Utilization of Social Media as a Community and School Relations Channel in Encouraging Students' Entrepreneurial Career Choices

Maisyaroh, Juharyanto, Maulana Amirul Adha,
Nova Syafira Ariyanti, Indra Lesmana, and Anabelie Villa Valdez

Abstract In an era where technology is changing the way people interact and learn, the role of social media has become increasingly important in facilitating communication between communities, schools and students. This research aims to describe the role of social media as an important tool in shaping relationships between communities, schools and students in encouraging entrepreneurial career choices among students. Qualitative research with a case study design was used to achieve this goal. There are eight ways to use social media as a channel for school relations with the community in encouraging students' entrepreneurial career choices, including (a) Engagement and Awareness, (b) Resource Sharing, (c) Collaborations and Networking, (d) Interactive Learning, (e) Feedback and Support, (f) Showcasing Student Initiatives, (g) Partnerships and Sponsorships and (h) Continuous Support. This research provides valuable insights for entrepreneurship-oriented educational approaches and strengthens the relationship between schools, communities and businesses.

Maisyaroh (✉) · Juharyanto · N. S. Ariyanti · I. Lesmana
Universitas Negeri Malang, Malang 65145, Indonesia
e-mail: maisyaroh.fip@um.ac.id

Juharyanto
e-mail: juharyanto.fip@um.ac.id

N. S. Ariyanti
e-mail: novasyafira2@gmail.com

I. Lesmana
e-mail: indra.lesmana.2101328@students.um.ac.id

M. A. Adha
Universitas Negeri Jakarta, Jakarta 13220, Indonesia
e-mail: maulanaamirul@unj.ac.id

A. V. Valdez
Mindanao State University, 9700 Marawi City, Philippines
e-mail: anabelievillavaldez@gmail.com

X.-S. Yang et al. (eds.), *Proceedings of Ninth International Congress on Information and Communication Technology*, Lecture Notes in Networks and Systems 1000,
https://doi.org/10.1007/978-981-97-3289-0_41

Keywords Community · School relation · Social media · Entrepreneurship career

1 Introduction

The contemporary corporate landscape necessitates innovation, inventiveness, and a robust entrepreneurial mindset. Amidst the swift and extensive transformations occurring worldwide, education assumes a crucial role in equipping upcoming generations to effectively confront these obstacles. An area that is receiving growing focus is student entrepreneurship [1, 2]. Entrepreneurship is no longer viewed as an alternative career path, but rather as an exciting opportunity for students who are interested in discovering new ideas, generating value, and affecting the world via innovation. In this context, the role of social media as an effective communication channel becomes very significant [3].

Strong engagement on social media platforms offers great potential to influence students' career choices [4]. Platforms such as Instagram, Facebook, LinkedIn and Twitter provide broad access to information, inspiring stories and opportunities related to entrepreneurship that can influence students' career perceptions and decisions. In an era where technology is changing the way we interact and learn, the role of social media has become increasingly important in facilitating communication between communities, schools and students [5, 6]. Strong engagement in the educational environment and ongoing connections between schools and communities are key elements in shaping an inclusive and growth-oriented future.

Awareness about the importance of entrepreneurial skills is increasing in this modern era. The school would like to provide stronger encouragement for students to consider an entrepreneurial career as an attractive option [7]. Initiatives to mobilize students' interest in this field are important to prepare them for future challenges. Exposure to entrepreneurship concepts and experiences during schooling influences students' perception of entrepreneurship as a viable career option [8]. In addition, entrepreneurship education programs enhance students' innovative thinking, risk-taking ability and problem-solving skills contributing to their entrepreneurship readiness [9, 10].

The school not only aims to provide academic education, but also to prepare students holistically, including in considering diverse career options. Engagement with local communities and companies is an important strategy in broadening horizons and opportunities for students interested in entrepreneurship [8]. Many students may not have sufficient access to or understanding of entrepreneurial career paths [11]. Limited access to information and lack of support can be barriers. Students often lack an understanding of the opportunities that exist in entrepreneurship and how to utilize social media as a tool for this [3, 12]. Social media has emerged as a dynamic tool in educational settings, overcoming geographical barriers and providing a platform for collaborative learning.

There has been an increased focus on the use of social media as a means of encouraging students to choose careers in entrepreneurship [13, 14]. Social media platforms are a useful tool for showcasing entrepreneurial success stories, introducing students to professionals in the field and sharing materials pertaining to entrepreneurship [15]. This platform may help schools incorporate interactive courses, workshops and challenges that foster community and teach students about entrepreneurship [3]. Although students are becoming more interested in studying entrepreneurship, there hasn't been much research done on how social media influences this choice of career. A comprehensive investigation is required to ascertain the most effective strategies for utilizing social media to motivate students to engage in entrepreneurial pursuits [13].

The objective of this research is to examine the potential of social media as a tool that can effectively motivate, educate and inspire students to pursue an entrepreneurial trajectory during an age characterized by the increasing prevalence of online interactions. This research employs a holistic methodology to ascertain the most effective strategies for utilizing social media platforms to disseminate information, forge connections, and cultivate an inclusive environment for students contemplating entrepreneurial careers. It is intended that by examining the potential of social media as a tool to increase awareness, involvement and support, this research will enhance the bonds between businesses, communities and schools and offer insightful information for entrepreneurship-oriented teaching approaches.

2 Method

A qualitative approach with a case study design was used to identify the role of social media as an important tool in shaping relationships between communities, schools and students in encouraging entrepreneurial career choices among students. The participants of this study were principals, teachers, students, communities and businesses from six senior high schools spread across three districts/cities from East Java Province, Indonesia. The informants in this study were recommended by the principals based on their competence and experience in entrepreneurship and school-community relations. The involvement of the community and businesses around the schools in this study was as supporting informants. Data collection was conducted using in-depth group unstructured interviews, observation and documentation studies. Interviews were conducted by researchers to obtain accurate information from the point of view and experience of each informant related to the research focus. Furthermore, observation activities were carried out to take a picture of the situation in the field which would later be used as a complement to the research data. Meanwhile, documentation studies were carried out as supporting evidence of the results of interviews and observations.

The triangulation procedure is employed to assess the data's validity. The following are some examples of the stages involved in the triangulation process: (a) comparing observational data with interview data; (b) comparing the circumstances and viewpoints of two individuals; (c) comparing interview results with the information contained in relevant documents; (d) comparing facts with one or more theories as a comparative explanation; and (e) peer discussions. Interview transcripts, field notes, and other research documents are consistently searched for and compiled before the data is methodically sorted based on the research topic. In more detail, the data analysis followed the recommendations of Miles et al. [16], namely data reduction, presentation and conclusion drawing.

3 Findings and Discussion

Social media can be a powerful tool for fostering community and school relations while encouraging students toward entrepreneurial career paths. Based on the research results, there are eight ways to use social media as a channel for school relations with the community in encouraging students' entrepreneurial career choices, as summarized in Fig. 1. By leveraging social media effectively, schools can create an interactive and supportive ecosystem that encourages and nurtures students' interest in entrepreneurial careers [17]. Effective utilization of social media bridges the gap between schools, communities and businesses, creating an ecosystem conducive to fostering entrepreneurial aspirations among students [18, 19]. Collaboration between educational institutions, communities and local businesses through social media channels offers students real-world exposure, internships and mentorship opportunities, thus strengthening their entrepreneurial acumen and career readiness.

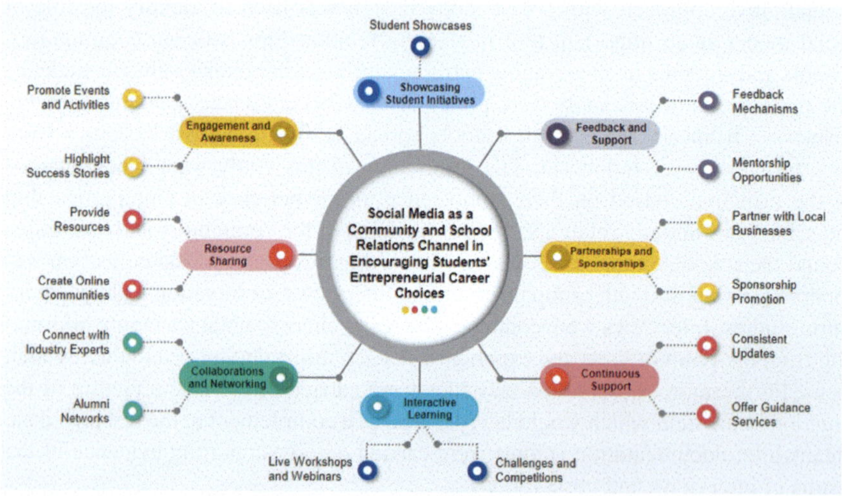

Fig. 1 Utilization of social media as a community and school relations channel

3.1 Engagement and Awareness

The first utilization is engagement and awareness related to school entrepreneurship, there are two things that are implemented, namely highlighting success stories and promoting events and activities. Social media can be a powerful tool to promote events and activities related to entrepreneurship [20]. Schools share information about entrepreneurship workshops, seminars or guest speaker sessions through social media platforms to increase student participation. In addition, the school uses various social media platforms such as Facebook and Instagram to create dedicated pages for events, which share detailed information, schedules, and engaging content to attract participants. The school also sought to develop engaging visual content to promote events and entrepreneurship activities, using visually appealing graphics, videos and infographics to showcase event highlights, speakers and activities. In promoting various entrepreneurial activities on social media, the school creates event-specific hashtags to create buzz and encourage user-generated content, of course, the school also encourages students to use these hashtags in their posts. The school also seeks to collaborate and partner, collaborating with influencers, industry leaders and other relevant organizations to amplify the reach and engagement of the event [21, 22].

To increase engagement and awareness related to school entrepreneurship, the school seeks to highlight success stories, by showcasing stories of young entrepreneurs, alumni or local business owners who started their ventures, inspiring students to explore similar paths. There are three ways this is done, (a) Testimonials/case studies, by sharing success stories of students or alumni who pursue entrepreneurial careers, using their experiences as case studies or testimonials to inspire others [23]; (b) Visual storytelling, by creating a short video or series of images showcasing the journey of successful entrepreneurs, highlighting the challenges they faced and how they overcame them [15]; and (c) Live Q&A sessions, by organizing live sessions with successful entrepreneurs or alumni who can share insights and advice, schools use platforms such as Instagram Live, Facebook Live or YouTube [23].

3.2 Resource Sharing

The school utilizes social media as a resource sharing platform to encourage entrepreneurial careers for students. The first step is to provide resources, by sharing links, articles and videos related to entrepreneurship, business development and skills development on platforms such as Instagram, TikTok, Twitter or YouTube. The school also seeks to build resource libraries with downloadable resources. The school seeks to create a comprehensive directory or database of local entrepreneurial resources, such as funding opportunities, workshops or incubator programs, which are shared and updated through social media channels [24]. The school also encourages the creation of online communities by establishing groups or forums with students, teachers, alumni and entrepreneurial practitioners, where students can share ideas, ask questions and support each other's entrepreneurial endeavors. This demonstrates efforts to utilize social media to be an excellent platform for resource sharing, community building and encouraging entrepreneurial career choices among students [25].

3.3 Collaborations and Networking

Utilizing social media to foster relationships with communities and schools while encouraging students toward entrepreneurial career paths certainly uses the key principles of collaboration and networking [26]. The school encourages students to interact with experts or practitioners in the world of entrepreneurship, by inviting successful entrepreneurs, business leaders and alumni to participate in live Q&A sessions, webinars or Ask Me Anything (AMA) sessions on social media platforms. This helps students gain real-world insights and network with professionals. Furthermore, the school encourages collaborative projects between students and local businesses or startups, the school uses social media platforms to promote these partnerships, demonstrating the value of real-world experience [8]. The school also encourages collaboration by organizing virtual networking events or promotional competitions through social media platforms. Students are encouraged to participate, connect with like-minded individuals and showcase their innovative ideas. In addition, the collaboration and networking with the community creates mentorship opportunities, the school facilitates a mentoring program where experienced entrepreneurs or professionals offer guidance and advice to students who are interested in starting their own business [22]. Social media can be used to match mentors with mentees. The school also encourages interactive discussions, by initiating discussions on social media about current trends, challenges and opportunities in the world of entrepreneurship. Alumni involvement is important, by engaging successful alumni who have taken the entrepreneurial path to share their experiences and offer guidance. They can be inspirational figures for students.

3.4 Interactive Learning

Utilizing social media to foster relationships with the community and school while encouraging students toward an entrepreneurial career path is to make social media an interactive learning platform [14]. The school runs live workshops and webinars using Instagram Live, Facebook Live or YouTube Live to host interactive workshops and sessions. The school invites entrepreneurship experts to share their insights, tips and experiences. In the sessions, the school through the teachers encourages students to ask questions and engage actively during these sessions. The topics covered include idea generation, business planning, marketing strategies, etc. Furthermore, the school also organizes online entrepreneurship challenges or competitions that require students to come up with innovative business ideas, create business plans or solve real-world entrepreneurial problems [13]. The school promotes the challenge across social media platforms to attract participants. The school offers prizes or recognition to winners to encourage participation.

3.5 Feedback and Support

Feedback mechanisms are critical to understanding the state of the student community. Polls, surveys or even simple open-ended questions on social media can provide valuable insights into students' interests, challenges and aspirations regarding entrepreneurship [20]. Implementing regular feedback mechanisms ensures that the initiatives and content offered are aligned with what students really need and want. Mentorship opportunities play an important role in nurturing entrepreneurial aspirations. Having an experienced entrepreneur or professional as a mentor can provide invaluable guidance, advice and real-world insights. These mentorship programs can be structured with regular check-ins, workshops or even networking events to foster meaningful relationships between mentors and students [22]. Creating a platform or space on social media dedicated to these mentorship opportunities can also allow for wider discussions, resource sharing, and Q&A sessions, thus enriching the learning experience for students interested in entrepreneurship [27]. By integrating feedback mechanisms and mentorship programs into social media-based educational initiatives, institutions can ensure that their efforts are not only well-received but also tailored to meet the specific needs and aspirations of students interested in the entrepreneurial path.

3.6 Showcasing Student Initiatives

Encouraging students to showcase their entrepreneurial projects and ideas on social media can be very beneficial. This not only provides exposure for their initiatives but also offers an opportunity to receive feedback, support and even collaboration from the wider community [28]. The school implemented several effective ways of encouraging this role; (a) Student Highlight Features, regularly featuring students and their entrepreneurial projects on social media platforms. This can include posts, stories or dedicated sections to showcase their initiatives; (b) Interactive Presentations, schools encourage students to create interactive presentations or videos about their projects. This could include explaining their business idea, sharing their process, challenges faced and their vision for the future; (c) Engagement and Feedback, encouraging the community to engage by asking questions, providing comments, or suggestions for improvement. This fosters a supportive environment and encourages constructive feedback; (d) Entrepreneurial Storytelling, encourages students to share not only their successes but also the hurdles they faced and how they overcame them. Authentic storytelling can resonate deeply with audiences and inspire others; (e) Recognition and Awards, the school strives to always recognize outstanding student projects through awards or certificates. This not only motivates students but also showcases their achievements to a wider audience and (f) Community Support and Funding, schools provide information or guidance on potential funding opportunities or incubator programs that can help students further develop their entrepreneurial projects. Encouraging students to share their entrepreneurial initiatives on social media not only amplifies their voices but also creates a vibrant community of support and collaboration [15]. It's an effective way to celebrate their innovation and inspire others to pursue their own entrepreneurial paths.

3.7 *Partnerships and Sponsorships*

Utilizing social media to foster relationships with the community and school while encouraging students toward an entrepreneurial career path is to use social media as a platform for partnering with local businesses and promoting sponsorships [8]. Partnering with local businesses or startups can be a great way to enrich students' entrepreneurial learning experience. Schools collaborate with local businesses or startups to provide internship opportunities and allow students to gain hands-on experience in an entrepreneurial environment [27]. In addition, schools can also invite entrepreneurs or professionals from local businesses to conduct workshops or guest teachers, sharing their expertise and insights. Furthermore, schools encourage collaborative projects between schools and businesses, collaborating with local businesses for student projects or case studies, providing real-world challenges for students to overcome. Social media can also be used in promoting sponsorship opportunities, by developing clear sponsorship proposals that outline the benefits to businesses of supporting student-led initiatives or projects [17]. In this proposal, efforts were made to highlight exposure, branding opportunities and potential impact. Furthermore, the proposal developed also customized the sponsorship package to suit different levels of involvement or contribution, giving businesses flexibility in their support. The school created an engaging social media campaign showcasing the sponsorship seeking initiative, using visuals, videos and testimonials to convey the value and impact of the project.

3.8 *Continuous Support*

Regular updates and offering mentorship services are crucial components in leveraging social media to cultivate relationships with the community and school, while also motivating students toward pursuing entrepreneurial career options [2, 13].

The school consistently posts pertinent content, news and entrepreneurial possibilities on social media platforms to maintain student engagement. The school frequently involves students in entrepreneurship-related surveys, quizzes, competitions or lecture series to promote engagement and active involvement. The school additionally disseminates information regarding student-led initiatives, projects or accomplishments in order to motivate and cultivate a feeling of community among budding entrepreneurs. In addition, the school aims to offer counseling or advisory services to students who are interested in pursuing an entrepreneurial career. These services may encompass individual sessions, workshops or seminars. The school supports students in strategizing prospective career trajectories in entrepreneurship, offering help on skill enhancement, networking and long-range planning [29]. Additionally, the institution aims to establish recurring office hours or virtual consultations wherein students may visit practitioners or instructors to discuss their entrepreneurial concepts, obstacles or desires. Educational institutions have the ability to foster an atmosphere that not only motivates students to pursue entrepreneurial endeavors but also equips them with the essential assistance to do so with assurance by consistently updating pertinent content on social media platforms and providing readily available mentorship services [4, 20].

4 Conclusion

The integration of social media as a channel for community and school linkages has great potential in driving students toward entrepreneurial career choices. By utilizing the interactive and expansive nature of social media platforms, educational institutions can create an environment that fosters entrepreneurial thinking, provides exposure to opportunities and nurtures the aspirations of future entrepreneurs. There are eight ways to use social media as a channel for school relations with the community in encouraging students' entrepreneurial career choices, including (a) Engagement and Awareness, (b) Resource Sharing, (c) Collaborations and Networking, (d) Interactive Learning, (e) Feedback and Support, (f) Showcasing Student Initiatives, (g) Partnerships and Sponsorships and (h) Continuous Support. Theoretically, the findings of this research add to the references related to social media as a channel of community and school relations in encouraging students toward entrepreneurial career choices, and practically the findings of this research provide alternative scenarios that can be applied to encourage the role of the community in promoting entrepreneurship as a student career choice. This study has several limitations as well. For example, it was limited to high schools and only used a qualitative approach. To determine how effective social media is at fostering community and school relationships and influencing students to pursue entrepreneurial careers, more studies at both lower and higher education levels can use quantitative research methods.

References

1. Kirkley WW (2017) Cultivating entrepreneurial behaviour: entrepreneurship education in secondary schools. Asia Pacific J Innov Entrep. 11(1):17–37. https://doi.org/10.1108/APJIE-04-2017-018
2. Ben Youssef A, Boubaker S, Dedaj B, Carabregu-Vokshi M (2021) Digitalization of the economy and entrepreneurship intention. Technol Forecast Soc Change 164:120043. https://doi.org/10.1016/j.techfore.2020.120043
3. Olanrewaju A-ST, Hossain MA, Whiteside N, Mercieca P (2020) Social media and entrepreneurship research: a literature review. Int J Inf Manage 50:90–110. https://doi.org/10.1016/j.ijinfomgt.2019.05.011
4. Lent RW, Brown SD (2020) Career decision making, fast and slow: toward an integrative model of intervention for sustainable career choice. J Vocat Behav 120:103448. https://doi.org/10.1016/j.jvb.2020.103448
5. Maisyaroh, Juharyanto, Wiyono BB, Adha MA, Saputra BR (2024) Hybrid learning in schools: analysis of the community's role in ict-based learning facilities management. In: Proceedings of eighth international congress on information and communication technology. ICICT 2023, pp 733–743. https://doi.org/10.1007/978-981-99-3043-2_60
6. Luo T, Freeman C, Stefaniak J (2020) 'Like, comment, and share'—professional development through social media in higher education: a systematic review. Educ Technol Res Dev 68(4):1659–1683. https://doi.org/10.1007/s11423-020-09790-5
7. Adha MA, Wibowo A, Faslah R, Ariyanti NS, Lutfia A (2022) Students ideal career in the 4.0 industrial. J East Eur Cent Asian Res 9(4):651–664. https://doi.org/10.15549/jeecar.v9i4.928
8. Pillai AK, Vieta MA, Sotomayor L (2021) University Student housing as business proposition and entrepreneurial activity: the Canadian case. Hous Policy Debate 1–24. https://doi.org/10.1080/10511482.2021.1883703
9. Adha MA, Eryanto H, Ariyanti NS, Musadad AA, Musyaffi AM, Wibowo A (2023) Evaluating the structural effect of family support and entrepreneurship training on entrepreneurship intention among Indonesian university students. Int J Educ Methodol 9(1):227–236. https://doi.org/10.12973/ijem.9.1.227
10. Liguori E, Winkler C, Vanevenhoven J, Winkel D, James M (2020) Entrepreneurship as a career choice: intentions, attitudes, and outcome expectations. J Small Bus Entrep 32(4):311–331. https://doi.org/10.1080/08276331.2019.1600857
11. Burnette JL et al (2019) A growth mindset intervention: enhancing students' entrepreneurial self-efficacy and career development. Entrep Theory Pract 44(5):878–908. https://doi.org/10.1177/1042258719864293
12. Maisyaroh, Juharyanto, Bafadal I, Wiyono BB, Adha MA, Surif J (2023) Drivers of primary school students' achievement in indonesia during pandemic. Eur J Educ Res 12(3):1393–1405. https://doi.org/10.12973/eu-jer.12.3.1393
13. Secundo G, Del Vecchio P, Mele G (2021) Social media for entrepreneurship: myth or reality? A structured literature review and a future research agenda. Int J Entrep Behav Res 27(1):149–177. https://doi.org/10.1108/IJEBR-07-2020-0453
14. Ndou V, Secundo G, Schiuma G, Passiante G (2018) Insights for shaping entrepreneurship education: evidence from the European entrepreneurship centers. Sustainability 10(11):1–19. https://doi.org/10.3390/su10114323
15. Mars MM (2019) Analogical and metaphorical thinking, storytelling, and entrepreneurial identity and narrative development: a visual art-based learning innovation. Entrep Educ Pedagog 4(1):64–81. https://doi.org/10.1177/2515127419890331
16. Miles MB, Huberman AM, Saldana J (2014) Qualitative data analysis. Sage, Los Angeles
17. Carella G, Cautela C, Melazzini M, Pei X, Schmittinger F (2023) Design thinking for entrepreneurship: an explorative inquiry into its practical contributions. Des J 26(1):7–31. https://doi.org/10.1080/14606925.2022.2144565
18. Bublitz MG et al (2020) Rise up: understanding youth social entrepreneurs and their ecosystems. J Public Policy Mark 40(2):206–225. https://doi.org/10.1177/0743915620937702

19. Andrade R, Pinheiro P, Carvalho L, Rocha R (2022) Building a bridge: knowledge sharing flows into entrepreneurial ecosystems. J Open Innov Technol Mark Complex 8(3):144–156. https://doi.org/10.3390/joitmc8030144
20. Wu Y, Song D (2019) Gratifications for social media use in entrepreneurship courses: learners' perspective. Front Psychol 10:1–12. https://doi.org/10.3389/fpsyg.2019.01270
21. Nurabadi A, Suhariadi F, Mardiyanta A, Triwiyanto T, Adha MA (2022) Digital principal instructional leadership in new normal era. Int J Eval Res Educ 11(3):1090–1098. https://doi.org/10.11591/ijere.v11i3.22483
22. Killian S, Lannon J, Murray L, Avram G, Giralt M, O'Riordan S (2019) Social media for social good: student engagement for the SDGs. Int J Manag Educ 17(3):100307. https://doi.org/10.1016/j.ijme.2019.100307
23. Woods C, Dell K, Carroll B (2021) Decolonizing the business school: reconstructing the entrepreneurship classroom through indigenizing pedagogy and learning. Acad Manag Learn Educ 21(1):82–100. https://doi.org/10.5465/amle.2020.0366
24. Martín-Rojas R, Garrido-Moreno A, García-Morales VJ (2020) Fostering corporate entrepreneurship with the use of social media tools. J Bus Res 112:396–412. https://doi.org/10.1016/j.jbusres.2019.11.072
25. Pardo-Garcia C, Barac M (2020) Promoting employability in higher education: a case study on boosting entrepreneurship skills. Sustain 12(10):1–23. https://doi.org/10.3390/SU12104004
26. García-Morales VJ, Martín-Rojas R, Garde-Sánchez R (2020) How to encourage social entrepreneurship action? Using web 2.0 technologies in higher education institutions. J Bus Ethics 161(2):329–350. https://doi.org/10.1007/s10551-019-04216-6
27. Hunter L, Lean J (2018) Entrepreneurial learning a social context perspective: evidence from Kenya and Tanzania. J Small Bus Enterp Dev 25(4):609–627. https://doi.org/10.1108/JSBED-02-2017-0075
28. Bodolica V, Spraggon M, Badi H (2021) Extracurricular activities and social entrepreneurial leadership of graduating youth in universities from the Middle East. Int J Manag Educ 19(2):100489. https://doi.org/10.1016/j.ijme.2021.100489
29. St-Jean É, Mathieu C (2015) Developing attitudes toward an entrepreneurial career through mentoring: the mediating role of entrepreneurial self-efficacy. J Career Dev 42(4):325–338. https://doi.org/10.1177/0894845314568190

Enhancing Healthcare User Interfaces Through Large Language Models Within the Adaptive User Interface Framework

Akash Ghosh, Bo Huang, Yan Yan, and Wenjun Lin

Abstract In the pursuit of enhancing digital user experiences within healthcare, this research investigates the novel application of Large Language Models (LLMs) in the Adaptive User Interface Framework (AUIF). This framework aims to redefine user interaction by providing real-time, personalized interface adjustments. By systematically applying LLMs to user interface enhancement, the AUIF addresses the static nature of current digital health platforms, offering a dynamic and adaptive alternative that responds to individual user behaviors and preferences. This study explores the pioneering integration of LLMs for user experience (UX) improvement recommendations and real-time HyperText Markup Language (HTML) content adjustments, marking a significant step forward in intelligent user interface design. The implications of this research are vast, with the potential to improve patient engagement, and satisfaction, and to address the pressing need for interfaces that adapt to diverse user behaviors and preferences.

Keywords Adaptive User Interface · Large Language Models · Healthcare personalization · Real-time UI adaptation · User experience · Artificial intelligence · Dynamic content management

A. Ghosh
Computing Science, Thompson Rivers University, Kamloops, B.C, Canada
e-mail: ghosha19@mytru.ca

B. Huang · Y. Yan
School of Computer Science, University of Guelph, Geulph, Ontario, Canada
e-mail: bhuang06@uoguelph.ca

Y. Yan
e-mail: yyan15@uoguelph.ca

W. Lin (✉)
Digital Healthcare Innovation Lab, School of Computer Science and Technology, Algoma University, Brampton, Ontario, Canada
e-mail: randy.lin@algomau.ca

X.-S. Yang et al. (eds.), *Proceedings of Ninth International Congress on Information and Communication Technology*, Lecture Notes in Networks and Systems 1000,
https://doi.org/10.1007/978-981-97-3289-0_42

1 Introduction

The digitalization of healthcare has brought forth a myriad of platforms aiming to improve patient engagement and health outcomes. However, despite technological advancements, many of these platforms remain static, offering the same user interface (UI) to a diverse user base with varying needs, abilities, and preferences. This one-size-fits-all approach often results in a suboptimal experience for users and can hinder the effective delivery of healthcare services.

Personalization in healthcare is not a new concept [1, 3, 9–11, 13]; it has been the cornerstone of patient-centered care, focusing on tailoring medical practices to individual patient requirements. The Adaptive User Interface Framework (AUIF), first proposed in [5], represents a significant stride in carrying this personalization over to digital health platforms. By harnessing the advanced capabilities of Large Language Models (LLMs) [2, 7, 15], the AUIF proposes a dynamic UI that adapts in real-time, aligning with the individual's current needs and context.

The impetus for this research is clear: there is a critical need to evolve healthcare interfaces into intelligent, adaptive systems that can meet the complex and changing demands of users. In this paper, we explore the potential of LLMs to act as the driving force behind this evolution, making UIs more intuitive, responsive, and inclusive.

This study is situated at the intersection of artificial intelligence, user experience design, and healthcare informatics. We investigate how LLMs can be integrated within the AUIF to deliver a highly adaptive UI capable of improving the overall usability and effectiveness of healthcare platforms. Our main research question asks: how can LLMs be leveraged to enhance the adaptability and personalization of user interfaces in the healthcare domain, and what impact does this have on user engagement and satisfaction?

2 Literature Review

The static user interfaces in healthcare are limited by their inability to adapt to varied user needs, leading to less optimal user interaction. Studies like Langley [8] and Zhu et al. [18] highlight these limitations and propose Adaptive User Interfaces (AUIs) as a solution, using machine learning techniques for better personalization and efficiency.

AUIs have been effectively used in various settings, including aiding vision-impaired users [12] and in educational contexts for customized content [16]. These examples demonstrate AUIs' versatility and potential in specialized areas.

Large Language Models (LLMs) like GPT-3 and BERT are making significant strides in sectors including healthcare. They are known for content generation and language understanding and their application in patient care [7]. The use of LLMs in healthcare operations [15] and integration with IoT for diagnostics [4] showcase their diverse capabilities.

The integration of AUIs and LLMs, especially in culturally sensitive healthcare contexts, is crucial for advancing healthcare delivery. The importance of cultural sensitivity in healthcare systems is emphasized by Mistry et al. [13]. The combination of AUIs and LLMs in culturally sensitive applications can enhance patient adherence to medical advice and health outcomes, as well as reduce biases among healthcare providers [6].

The integration of AUIs and LLMs with a focus on cultural sensitivity represents an innovative approach to healthcare delivery. By harmonizing technological advancements with cultural nuances, the AUIF proposed in this research seeks to deliver personalized care that resonates with diverse populations, ultimately aiming to enhance user experience and improve patient engagement in a culturally attuned manner.

3 The Adaptive User Interface Framework

The AUIF is a comprehensive system designed to enhance user experience by dynamically adapting digital interfaces to individual user behaviors and preferences. At the heart of the AUIF lies a continuous feedback loop, composed of core stages that begin with preference detection and culminate in feedback-driven refinement.

Core Stages of AUIF:

Here is a deeper dive into our approach as outlined in Fig. 1:

Preference Detection: Utilizing advanced behavioral analytics and heuristic algorithms, AUIF tracks user interactions to identify potential areas of improvement or customization. Structured problem statements are generated, signaling the LLM to commence solution identification.

Solution Identification: With the problem statements in hand, the LLM draws from a vast knowledge base to propose potential solutions, ensuring contextual relevance and cultural appropriateness.

Solution Selection: A sophisticated multi-criteria decision-making algorithm helps in the meticulous selection of the optimal solution, factoring in various elements such as cultural nuances and potential impact on user experience.

Solution Execution: The selected solutions are seamlessly integrated into the application, ensuring that changes feel organic and enhance usability without causing disruption.

Feedback Loop: Post-implementation, the AUIF remains vigilant, utilizing both direct and indirect feedback to assess the efficacy of the adaptations and inform subsequent iterations.

The Role of LLMs within AUIF

LLMs within the AUIF play a pivotal role in processing user data to generate, evaluate, and implement user interface solutions. LLMs are tasked with the following:

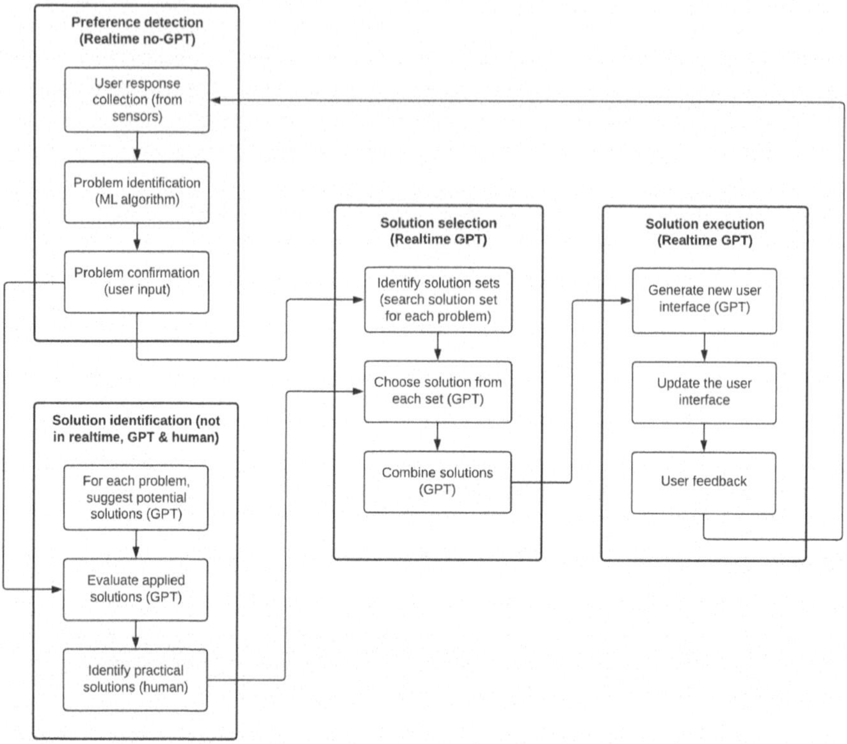

Fig. 1 AUIF problem-solving workflow

Problem Statement Interpretation: Translating user behavior into structured problem statements for the LLM to understand.

Solution Generation: Creating an array of potential solutions for identified user challenges.

Solution Evaluation and Selection: Judging the proposed solutions based on set criteria to determine the most effective approach.

Solution Implementation: Guiding the practical application of the chosen solutions within the user interface.

Innovations and Theoretical Contributions

AUIF represents a groundbreaking shift in digital interface design, marrying the computational power of LLMs with user-centric design principles. Theoretical contributions of AUIF include:

Personalization at Scale: The framework offers a scalable approach to personalization, balancing algorithmic precision with user privacy and security considerations.

Predictive User Modeling: AUIFâŁ™s innovative use of predictive modeling anticipates user needs, offering solutions before the user experiences a potential friction point.

Real-Time Adaptability: With real-time feedback loops, the system quickly adapts to user preferences, setting a new benchmark for dynamic interfaces.

4 Methodology

Our research introduces an innovative methodology that streamlines the interface adaptation process in healthcare platforms through the AUIF. While Clickstream Analysis and K-means Clustering are briefly utilized for data collection and behavior grouping, the cornerstone of our approach is the intricate use of LLMs to interpret, generate, and refine user interface solutions.

The innovation of our methodology lies in the application of LLMs, which comprises several steps tailored to enhance user interface adaptability and personalization. This section will elucidate the comprehensive process of LLM utilization in the AUIF.

Solution Generation Through LLM: At the onset of the LLM application, the model receives prompts that detail the specific user interface challenges derived from the earlier clustering process. The LLM then generates a range of potential solutions, showcasing its capacity to understand and address complex user interaction problems.

Detailed Solution Evaluation and Ranking: Following solution generation, the LLM embarks on a critical evaluation of each potential solution. The model considers various criteria like effectiveness, user engagement, and implementation ease to rank the solutions, demonstrating a nuanced approach to problem-solving in user interface design.

Integrating Solutions with LLM: Recognizing that the optimal user interface may require a multifaceted approach, the LLM is capable of synthesizing integrated solutions. This step is crucial as it demonstrates the LLM's advanced capability to combine different strategies, thereby formulating a cohesive and robust user interface adaptation.

Execution Plan Formulation: After identifying a solution or set of solutions, the LLM develops a detailed, actionable plan for execution. For instance, in enhancing a webpage for users with vision impairments, the plan may involve increasing the overall font size by 50%, enlarging the buttons by 30%, and adjusting the background color to white for improved contrast. This step-by-step guide serves as a practical tool to directly modify the user interface, leveraging existing content on the webpage. It is crucial to note that the framework does not generate new content because the focus is to maximize the potential of existing elements, emphasizing the practical nature of LLM-generated recommendations.

Iterative Refinement Based on Feedback: In the final stage, the iterative refinement process is fueled by user feedback, a pivotal mechanism for continual enhancement. Leveraging this feedback, LLMs engage in an ongoing fine-tuning of the interface.

This iterative cycle underscores the adaptive learning prowess of LLMs, enabling the interface to dynamically evolve and adeptly address the ever-changing user needs with precision and efficacy.

Our approach in healthcare platform design centers on using LLMs to modify web pages effectively. Leveraging data analysis methods like Clickstream Analysis and K-means Clustering, we emphasize LLMs' role in generating, evaluating, and implementing interface solutions. Our focus is specifically on showcasing LLMs' capacity to enhance user interfaces by modifying existing web page content, despite acknowledging the importance of other steps such as iterative refinement based on user feedback within our framework.

5 Experimentation and Results

5.1 Experimental Setup

The experiment was designed to test the effectiveness of a LLM, specifically GPT-4 [14], in updating web content in real-time. A simple multichoice box was used to gather immediate user feedback, like preferences and difficulties. This feedback directly informed the system's real-time adaptation capabilities.

The main focus was on the LLM's ability to quickly process inputs from the multichoice box and make instant UI changes. For instance, when users struggled with medical terms, the LLM promptly modified the UI to include simplified explanations, aiding user understanding.

User testing highlighted GPT-4's ability to rapidly and accurately adjust web content based on user interactions. Key metrics like response time and the effectiveness of the UI adaptations were carefully recorded. These tests provided insights into the LLM's potential in improving user experience through intuitive and responsive design.

Although the experiment was basic, it was designed for scalability, aiming to evolve into a comprehensive digital healthcare platform with advanced features such as enhanced security, sophisticated user analytics, and broad cross-platform support.

5.2 Examples of Real-Time UI Adaptations and Their Impact

The investigation into real-time UI adaptations was carried out through six distinct scenarios, each designed to tackle a specific challenge commonly encountered by individuals seeking healthcare information. The comprehensive details of these six scenarios are provided in the appendix. Within this subsection, we will discuss two selected scenarios to demonstrate the effectiveness of the LLM in enhancing user experience.

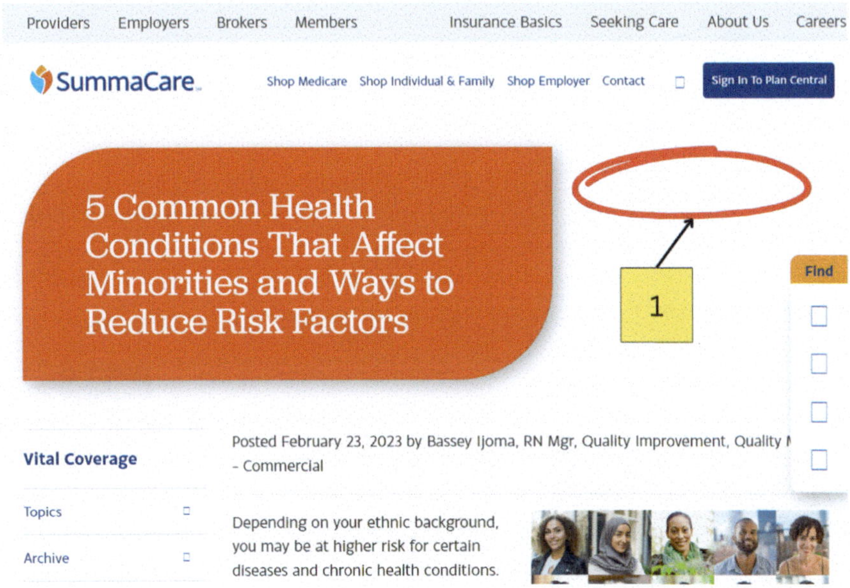

Fig. 2 Initial Webpage—healthcare information that is genetic to multiple communities of minorities. Marker 1 highlights that there is no personalized information based on the patient's Latina background

Overcoming Language and Cultural Barriers

In the initial experimental scenario, we explore how AUIF tackles language and cultural barriers in accessing healthcare information. We introduce Maria, a hypothetical user of Latina heritage, who initially encounters healthcare information presented in English—a language not congruent with her cultural context.

Recognizing Maria's preferences and needs, AUIF, powered by an LLM, promptly adapts the interface to display information in Spanish, complete with culturally relevant health insights. This swift adaptation occurs within two seconds of receiving feedback, significantly enhancing Maria's engagement with and understanding of the healthcare information provided.

Figure 2 captures the initial state of the webpage, presenting generic healthcare information not personalized for any specific minority community. Marker 1 underscores the lack of culturally tailored content for patients like Maria with a Latina background. The figure also shows that the content exclusively in English without any available translation options.

Consequently, Fig. 3 reveals the AUIF's solution. Marker 1 indicates where a summary specifically curated for the Spanish-speaking community appears, and Marker 2 illustrates the addition of a language translation tool facilitated by AUIF. Finally, Fig. 4, with Marker 2, highlights the successful application of the translation tool, now displaying content in Spanish, thus resolving the language barrier Maria faced.

Fig. 3 AUIF Solution—summary and language adaptation. Marker 1 shows that a summary for the Spanish community was generated, and Marker 2 shows that a language translation tool was added through AUIF

Fig. 4 AUIF Solution—language adaptation. Marker 2 shows the result of applying the language translation tool as the content has been shown in Spanish

This scenario showcases the efficacy of AUIF in customizing user interfaces in real-time to break down language and cultural obstacles. The result is a more inclusive digital environment that better serves diverse populations by providing information in their preferred language and context, ultimately improving the accessibility and usability of healthcare information.

Enhancing Accessibility for Senior Citizens

The second scenario focuses on the adaptability of AUIF in making web interfaces more accessible for senior users. Eleanor, our hypothetical persona, is a 75-year-old retired elementary school teacher who has some familiarity with technology. However, she faces challenges with small buttons and complex website designs, which hinder her online navigation.

Upon recognizing Eleanor's difficulties, AUIF dynamically adjusts the webpage layout. It increases button sizes, simplifies navigation, and introduces voice-assisted guidance, specifically tailored to Eleanor's interaction patterns and needs. This transformation led to a significant improvement in Eleanor's online experience, offering her an interface that is easier to navigate and more intuitive to use.

Figure 5 displays the original webpage, with Marker 1 pointing out the absence of voice-assisted guidance and Marker 2 emphasizing the small buttons and fonts that create hurdles for elderly users. The figure also highlights the complexity of the initial layout, showcasing the clutter that further complicates navigation.

Conversely, Fig. 6 illustrates the AUIF's interventions. Marker 1 draws attention to the newly integrated voice guidance button, which can audibly read page content. Marker 2 illustrates the enlarged buttons and increased font size, and Fig. 7, with Marker 3, showcases the restructured layout, optimized for senior citizens by the AUIF.

The results of this theoretical experiment suggest that AUIF can greatly enhance web accessibility for the elderly, theoretically reducing frustration and promoting a more engaging online environment. Eleanor's example indicates that a more user-friendly interface, featuring larger interactive elements and streamlined navigation, can be achieved.

5.3 Reliability Assessment

The reliability of the Adaptive User Interface Framework (AUIF) was evaluated through structured conversations with the GPT-4 model, conducted across several hypothetical experiments. Each experiment involved a minimal number of user prompts and was repeated 20 times to ensure a comprehensive assessment of the AUIF's reliability. During these interactions, critical performance indicators such as the average response time and the number of prompts used were carefully documented.

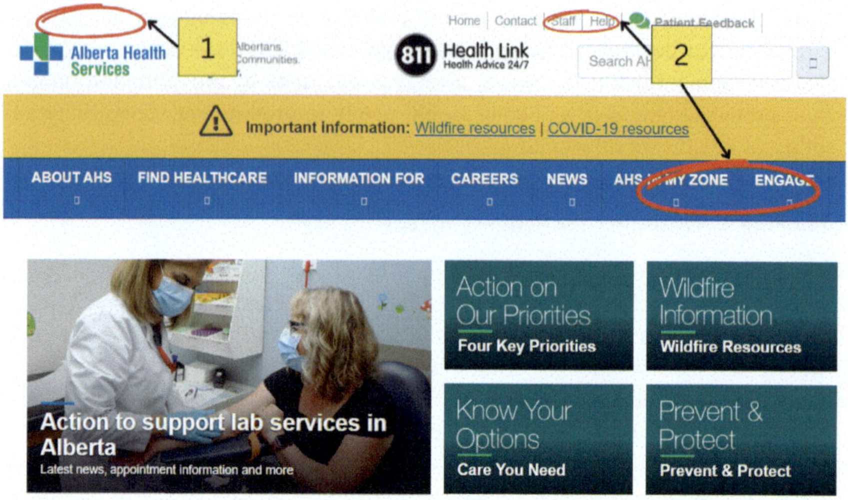

Fig. 5 Initial Webpage—Marker 1 highlights the missing voice-assisted guidance, and Marker 2 highlights the small buttons and fonts

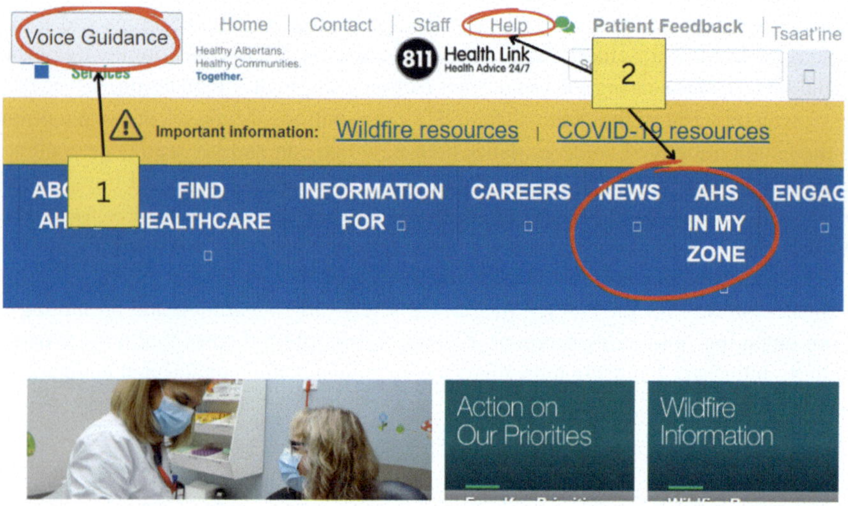

Fig. 6 AUIF Solution—Marker 1 highlights the voice guidance button created by AUIF, and Marker 2 highlights the larger buttons and fonts

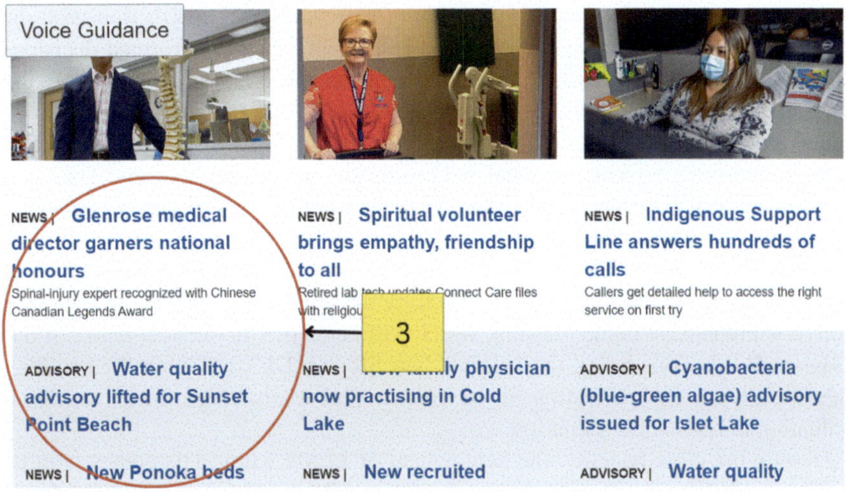

Fig. 7 AUIF Solution—Marker 3 highlights the optimized layout adjusted by AUIF

In Experiment 1, AUIF achieved an 80% success rate with an average response time of 1:40 and 1.5 prompts per interaction. Experiment 2 saw a slight improvement, with a 90% success rate, a response time of 1:15, and the same number of prompts. Experiment 3 had a longer response time of 3:00 and required more prompts (4.5 on average), maintaining an 80% success rate. These findings indicate AUIF's promising effectiveness and reliability in varied interaction scenarios.

These results, reflecting the time efficiency and prompt effectiveness in achieving successful outcomes, demonstrate the potential effectiveness of AUIF in real-world applications. The success rates across the experiments indicate a high level of reliability in AUIF's adaptability and response accuracy during interactions with the GPT-4 model.

6 Discussion

The ramifications of our findings in the deployment of LLMs within UIs are profound, particularly within the healthcare industry. Our AUIF demonstrates a marked enhancement in the capability to tailor user experience in real-time, responding to the nuanced demands of individual users. This personalized approach is not only a testament to the technical prowess of the framework but also to its potential to engender trust and facilitate better health outcomes. By providing information that resonates with the personal circumstances and cultural contexts of the user, the AUIF encourages a higher degree of compliance and satisfaction.

Our experiments suggest that the quality of user engagement with health recommendations improves significantly when the delivery of information is customized to their preferences. This leads not only to a more satisfying interaction with the healthcare platform but also empowers patients to make more informed decisions regarding their health. The intuitive nature of the interface minimizes the frustration often associated with complex and impersonal medical information systems.

Our experiments underscore the capacity of LLMs to update web content dynamically in real-time, offering personalized healthcare information to users. This real-time adaptability is a cornerstone of the AUIF, providing early evidence of its potential benefits. The immediate, personalized nature of the content update enhances user interaction, fostering an environment where informed health decisions can be made swiftly and with greater confidence. These promising outcomes pave the way for more comprehensive human testing, which will be pivotal in the next phase of our research. By demonstrating the tangible benefits of AUIF in a controlled setting, we establish a solid foundation for its efficacy and set the stage for an expanded evaluation in real-world scenarios.

Despite the clear benefits, the deployment of LLMs within UIs comes with its own set of challenges. The effectiveness of the AUIF hinges on the quality of user feedback it receives, which is often variable and may require sophisticated analytics to interpret accurately. Privacy concerns are another critical issue; the framework's reliance on continuous data collection for dynamic adaptation may raise questions regarding the handling and protection of user data. Moreover, the reliance on multiple prompts to guide the LLM can introduce inefficiencies that are unsuitable for real-time application. Additionally, the necessity for specialized computational resources to operate the LLMs could pose a challenge.

To address these challenges, we propose a multifaceted approach. Enhancing the feedback mechanism to obtain high-quality input from users is paramount, which could be achieved through targeted surveys and the careful analysis of user interaction data. The AUIF minimally requires generic user information such as gender, age group, education level, and mouse movement stored securely client-side and can be transmitted to the model via robust encryption protocols. Employing anonymization techniques, like hashing or tokenization, is vital to safeguard sensitive data. For instance, hashing user IDs or utilizing tokenization for unique identifiers effectively anonymizes information, mitigating exposure risks. Additionally, fine-tuning the LLM by reducing reliance on multiple prompts and enhancing processing speed is pivotal for real-world applications. Addressing computational requirements also involves employing fine-tuned, specialized models that are smaller than general ones. Furthermore, LLMs, capable of modifying HTML files, can streamline processes by chaining LLM steps together [17], efficiently handling complex HTML files.

7 Conclusion

This research marks a significant milestone in the quest to enhance user interfaces in the healthcare domain through the application of LLMs within the AUIF. Our findings offer compelling evidence of the AUIF's ability to deliver personalized healthcare experiences that are not only more engaging but also more attuned to the individual's needs. The positive impact on user satisfaction and the potential for better health outcomes speaks to the transformative power of integrating LLMs into healthcare UIs.

As we look to the horizon, the AUIF stands as a beacon of innovation, with its real-world efficacy yet to be fully explored. The next phase of our research will involve extensive human testing and iterative development to solidify the AUIF's role in revolutionizing user experiences. The vision is clear: to create an adaptable, intelligent, and user-centric digital health environment that is capable of meeting the intricate tapestry of users' needs across the globe.

The promising results obtained from the current phase of research serve as a solid foundation for further exploration. In our future work, we plan to delve into extensive human-centered testing to rigorously evaluate the performance of AUIF in real-world healthcare settings. This will be crucial in understanding its practical implications and user impact. Additionally, we aim to develop more sophisticated analytics tools for effectively interpreting user feedback. Such tools are essential for refining the AUIF's adaptive capabilities and ensuring that it meets the diverse needs of users. Another critical area of focus will be the implementation of advanced privacy and security measures to protect sensitive health data, a paramount concern in the healthcare industry. Lastly, we will work on optimizing the efficiency of LLMs used in the AUIF, to ensure that they can provide seamless real-time updates to the user interface, enhancing the overall user experience without sacrificing performance or speed.

Acknowledgements The authors wish to express their sincere gratitude to the Algoma University Research Fund for their generous support of this work. Their commitment to fostering innovative research has been invaluable to the success of this project.

References

1. Betancourt J, Green A, Carillo JE (2000) The challenges of cross-cultural healthcare-diversity, ethics, and the medical encounter. In: Bioethics Forum, vol 16. Midwest Bioethics Center, pp. 27–32
2. Chintagunta B, Katariya N, Amatriain X, Kannan A (2021) Medically aware gpt-3 as a data generator for medical dialogue summarization. In: Machine learning for healthcare conference. PMLR, pp 354–372
3. Fitzpatrick KM, Ody M, Goveas D, Montesanti S, Campbell P, MacDonald K, Crowshoe L, Campbell S, Roach P (2023) Understanding virtual primary healthcare with indigenous populations: a rapid evidence review. BMC Health Serv Res 23(1):1–13
4. Ghatkamble R, Pareek PK, Parameshachari B (2022) Design of NBW-mho with Bert model for prediction of breast cancer in IoT healthcare system. In: 2022 Fourth international conference on emerging research in electronics, computer science and technology (ICERECT). IEEE, pp 1–8
5. Ghosh A, Yan Y, Lin W (2023) Adaptive user interface framework powered by a large language model for culturally sensitive virtual healthcare applications. In: IEEE international conference

on biomedical and health informatics (BHI'23). Pittsburgh, Pennsylvania, United States of America
6. Hershberger PJ, Pei Y, Crawford TN, Neeley SM, Wischgoll T, Patel DB, Vasoya MM, Castle A, Mishra S, Surapaneni L et al (2022) An interactive game with virtual reality immersion to improve cultural sensitivity in health care. Health Equity 6(1):189–197
7. Korngiebel DM, Mooney SD (2021) Considering the possibilities and pitfalls of generative pre-trained transformer 3 (gpt-3) in healthcare delivery. NPJ Digit Med 4(1):93
8. Langley P (1997) Machine learning for adaptive user interfaces. In: Annual conference on artificial intelligence. Springer, pp 53–62
9. Lin W, Babyn P, Yan Y, Zhang W (2023) A novel scheduling method for reduction of both waiting time and travel time of patients to visit health care units in the case of mobile communication. In: Enterprise information systems, p 2188124
10. Lin W, Babyn P, Yan Y, Zhang W (2023) Optimizing healthcare resource allocation through digital twins: a multi-objective approach for efficiency, equity, and resilience. In: 3rd Digital twin international conference. Taylor & Francis Group, pp 1–4
11. Lin W, Xu M, He J, Zhang W (2023) Privacy, security and resilience in mobile healthcare applications. Enterp Inf Syst 17(3):1939896
12. Luy C, Law J, Ho L, Matheson R, Cai T, Madugalla A, Grundy J (2021) A toolkit for building more adaptable user interfaces for vision-impaired users. In: 2021 IEEE symposium on visual languages and human-centric computing (VL/HCC). IEEE, pp 1–5
13. Mistry SK, Shaw M, Raffan F, Johnson G, Perren K, Shoko S, Harris-Roxas B, Haigh F (2022) Inequity in access and delivery of virtual care interventions: a scoping review. Int J Environ Res Public Health 19(15):9411
14. OpenAI (2023) Gpt-4 and gpt-4 turbo documentation. https://platform.openai.com/docs/models/gpt-4-and-gpt-4-turbo
15. Sezgin E, Sirrianni J, Linwood SL (2022) Operationalizing and implementing pretrained, large artificial intelligence linguistic models in the us health care system: outlook of generative pretrained transformer 3 (gpt-3) as a service model. JMIR Med Inform 10(2):e32875
16. Watanobe Y, Rahman MM, Vazhenin A, Suzuki J (2021) Adaptive user interface for smart programming exercise. In: 2021 IEEE international conference on engineering, technology and education (TALE). IEEE, pp 1–7
17. Wu T, Terry M, Cai CJ (2022) AI chains: transparent and controllable human-AI interaction by chaining large language model prompts. In: Proceedings of the 2022 CHI conference on human factors in computing systems, pp 1–22
18. Zhu R, Zhang Z, Cai Z (2021) Enabling pointing assistance in adaptive interfaces using mouse pointing intention prediction. In: 2021 China automation congress (CAC). IEEE, pp 4808–4813

Enhancing Telehealth Patient Experience with Emotion-Sensitive Large Language Models

Indira Del Rosario, Akash Ghosh, Bo Huang, Yan Yan, Wenjun Zhang, and Wenjun Lin

Abstract This study explores the integration of Large Language Models (LLMs), specifically ChatGPT-4, to improve patient experience in telehealth. Addressing the challenge of patient anxiety during waiting periods, we implemented ChatGPT-4 for real-time emotion detection and dynamic background generation. Experiments using the FACES database and qualitative feedback on generated backgrounds show that ChatGPT-4 can accurately identify emotions and create calming visual environments. These findings suggest that LLMs can significantly enhance patient engagement and satisfaction. Future developments should focus on personalized AI training and real-time adaptive systems for a more nuanced approach to patient care in telehealth.

Keywords Telehealth · Patient experience · Large Language Models · Emotion detection · ChatGPT · Dynamic background generation · Personalization · AI in healthcare

I. D. Rosario
Mechanical Engineering, University of Saskatchewan, Saskatoon, Saskatchewan, Canada
e-mail: cjh618@mail.usask.ca

A. Ghosh
Computing Science, Thompson Rivers University, Kamloops, B.C, Canada
e-mail: ghosha19@mytru.ca

B. Huang · Y. Yan
School of Computer Science, University of Guelph, Geulph, Ontario, Canada
e-mail: bhuang06@uoguelph.ca

Y. Yan
e-mail: yyan15@uoguelph.ca

W. Zhang (✉)
Biomedical Engineering, University of Saskatchewan, Saskatoon, Saskatchewan, Canada
e-mail: chris.zhang@usask.ca

W. Lin (✉)
Digital Healthcare Innovation Lab, School of Computer Science and Technology, Algoma University, Brampton, Ontario, Canada
e-mail: randy.lin@algomau.ca

X.-S. Yang et al. (eds.), *Proceedings of Ninth International Congress on Information and Communication Technology*, Lecture Notes in Networks and Systems 1000,
https://doi.org/10.1007/978-981-97-3289-0_43

1 Introduction

In the rapidly evolving domain of digital healthcare, telehealth has emerged as a critical innovation, reshaping the accessibility and delivery of medical services. Particularly highlighted in the wake of global health crises, telehealth has bridged physical distances, bringing medical care to diverse populations with unprecedented convenience [5, 9, 14, 15]. This mode of healthcare, though technologically advanced, still faces significant challenges that need addressing to enhance patient experiences and outcomes.

A significant challenge in telehealth is the often extensive waiting time patients experience before consultations. These prolonged waiting periods can escalate patient anxiety, impatience, and dissatisfaction [2, 8, 13]. Unlike traditional in-person consultations where the presence of healthcare staff and a physical environment can offer reassurance, the virtual nature of telehealth can lead to a sense of isolation and neglect during these waits. The absence of direct human interaction and the impersonal aspect of waiting in a digital space can intensify the discomfort and stress of patients, potentially affecting their perception of the quality of care provided.

This study proposes an innovative solution to mitigate these issues by utilizing Large Language Models (LLMs) in telehealth platforms. Our approach focuses on real-time emotion detection, the provision of tailored healthcare information, and the generation of calming, context-sensitive imagery during telehealth waiting periods. By enhancing the waiting experience, this study aims to reduce patient anxiety and increase satisfaction with telehealth services.

The remainder of this paper is organized as follows: A review of relevant literature provides context and highlights the significance of our approach. The methodology section details the implementation of LLMs for emotion detection, content recommendation, and dynamic background generation. We then present and analyze the results of two experiments conducted to validate LLM's capabilities on emotion detection and dynamic background generation. The discussion section interprets these results in the context of their implications for telehealth services, followed by a conclusion that summarizes our findings and proposes directions for future research.

2 Literature Review

The literature review synthesizes advancements in telehealth, emotion detection technologies, and patient engagement strategies, providing a comprehensive overview of the current state and emerging trends in digital healthcare.

2.1 Telehealth Innovations

Telehealth has significantly impacted healthcare delivery across various domains, adapting to specific needs and challenges. Scriven et al. [12] demonstrated this through a telehealth-based persistent pain management program, emphasizing the importance of patient-centric approaches even in group settings. This highlights telehealth's potential in specialized care areas.

White et al. [16] examined telehealth in mental health, specifically PTSD treatment among veterans. They found that while symptom improvement correlated with higher satisfaction, traditional in-person treatment was often preferred, underlining the complexity of patient satisfaction in telehealth for mental healthcare. Hull et al. [6] and Johnson et al. [7] provided insights into clinical applications. Hull et al. evaluated at-home ketamine treatment via telehealth for anxiety and depression, whereas Johnson et al. focused on improving antibiotic prescribing through telehealth.

These studies collectively underscore telehealth's diverse applications, from clinical treatment and patient education to professional training, affirming its transformative role in modern healthcare.

2.2 Emotion Detection Technologies

The evolution of emotion detection technologies in healthcare has been marked by diverse methodologies, ranging from natural language processing to sophisticated AI models. These technologies have been crucial in enhancing patient care through a deeper understanding of emotional states. Advances in multi-speaker emotion detection, as explored by Zhang et al. [19], have opened new avenues in understanding complex conversational dynamics. This is particularly relevant in group therapy sessions via telehealth, where multiple participants' emotions need to be discerned accurately.

Comprehensive overviews by Saxena et al. [11] and Zhu et al. [20] have highlighted the progress and effectiveness of various emotion detection methods. Zhu et al.'s Topic-Driven Knowledge-Aware Transformer, in particular, demonstrates the potential of advanced algorithms in accurately detecting emotions in dialogues, an aspect critical in telehealth communication.

The role of LLMs in emotion detection, particularly in healthcare applications, has garnered significant interest. Yang et al. [18] evaluated the mental health analysis and emotional reasoning capabilities of ChatGPT across various tasks, including emotion recognition in conversations. Their study highlighted the effective use of prompt engineering with emotional cues to improve ChatGPT's performance, indicating the potential of LLMs like ChatGPT in providing nuanced emotional insights in telehealth settings.

These studies collectively emphasize the rapid advancements in emotion detection technologies, showcasing their critical role in healthcare, especially in telehealth where understanding and responding to patient emotions are key to effective care.

2.3 Patient Engagement Strategies

Recent studies in telehealth have increasingly focused on enhancing patient engagement through personalized approaches. The objective is to improve patient satisfaction and adapt to the unique demands of virtual healthcare.

Scriven et al. [12] and Funderburk et al. [4] both emphasize the significance of patient-centered strategies in telehealth. Scriven et al. evaluated a telehealth group model for pain management, highlighting the importance of understanding patient perceptions for effective engagement. Funderburk et al. found high levels of patient satisfaction post-telehealth encounters, suggesting that tailored telehealth experiences can meet diverse patient needs effectively.

Eng et al. [3] and Massucato et al. [10] explored the experiences of both providers and recipients of telehealth services. Eng et al. compared telehealth and in-person experiences in oncology, identifying areas for improvement, while Massucato et al. described medical students' involvement in creating educational materials for telehealth, underscoring the value of personalized educational content in enhancing patient engagement.

Lastly, Berg et al. [1] and Wittmeier et al. [17] investigated the implementation challenges of telehealth from the perspectives of healthcare providers. Berg et al. surveyed providers on their experiences with digital patient interactions, focusing on identifying patient types best suited for telehealth. Wittmeier et al. examined the barriers and facilitators experienced by therapists, indicating that provider adaptability is crucial for effective patient engagement.

While these studies provide valuable insights into patient engagement strategies in telehealth, there is a notable gap in research specifically addressing patient emotion in telehealth settings. This indicates a need for future studies to focus on the emotional aspects of patient engagement, considering the impact of personalized content and imagery on patients' emotional experiences during telehealth sessions.

3 Methodology

In this section, we outline the methodological framework of our study, which leverages LLMs for emotion detection, content recommendation, and dynamic background generation in telehealth settings. The approach integrates advanced AI capabilities with telehealth platforms to enhance patient engagement and satisfaction.

Emotion Detection and Implementation in Telehealth

The core of our emotion detection approach centers on the use of LLMs. These models are trained to process and interpret human expressions, including facial cues and verbal communication, to identify emotional states. During telehealth sessions, the video feed is processed in real-time, allowing LLMs to analyze facial expressions and categorize them into recognizable emotions like happiness, surprise, sadness, and anxiety.

Implementing this technology in telehealth involves embedding the emotion detection functionality into the existing software. This integration is designed to operate in real-time, analyzing patients' facial expressions to inform healthcare providers, thus enabling them to tailor their interactions. Key considerations include minimal latency, data security, and adherence to ethical standards and privacy regulations. Testing and optimization phases are crucial to refine the system's accuracy and responsiveness.

Content Recommendation Through Knowledge Bank Interface

The integration of LLMs with healthcare information databases forms a critical part of our methodology. LLMs access and interface with these databases to deliver tailored content recommendations during telehealth sessions. A bidirectional communication channel allows the LLMs to generate knowledge bank search queries based on analyzed patient emotions and health conditions.

For instance, a patient displaying signs of anxiety with a history of hypertension might prompt an LLM-generated query for managing hypertension under stress. The system uses sophisticated algorithms considering emotional cues and medical history to ensure search results are personalized and relevant.

Dynamic Background Generation

Dynamic background generation in telehealth uses LLMs to create environments resonating with the patient's emotional state. The LLMs interpret emotional cues and generate unique background images reflecting these emotions. This process involves accessing a database of images tagged with specific emotions and using algorithms to create a background that aligns with the identified emotional state.

The integration of this technology in real-time telehealth sessions ensures continuous monitoring of the patient's emotional cues. Changes in emotional state trigger the LLM to generate an appropriate background, which is seamlessly integrated into the telehealth interface. This dynamic transition of backgrounds aims to visually reinforce the progress of the session and provide non-verbal encouragement to the patient.

4 Experiments

4.1 Experiment 1: Emotion Detection Through Facial Recognition with ChatGPT

In this experiment, we assessed ChatGPT's ability to detect emotions using facial recognition. We selected the 'FACES' database from the Max Planck Institute for Human Development, consisting of images from 171 individuals of varied ages, each displaying six facial expressions: neutrality, sadness, disgust, fear, anger, and happiness. This database contains a comprehensive collection of 2,052 images. Each facial expression image was processed through ChatGPT with a prompt to identify the emotion, and the AI's responses were compared against the dataset's labels. The result is summarized in Table 1.

The analysis revealed an overall success rate of 92% for ChatGPT in correctly identifying emotions. However, there was a notable variance across different emotions. For instance, ChatGPT achieved high accuracy in identifying happiness (98%) and surprise (95%), but its performance was significantly lower for sadness (58%). This lower accuracy in detecting sadness might be due to the subtler nature of sad expressions, cultural and individual differences in expressing sadness, or possibly a limited representation of sadness in the training data used for the LLM.

The experiment underscores ChatGPT's potential in emotion detection, highlighting its high overall accuracy and areas for improvement, particularly in recognizing more nuanced emotional expressions like sadness. These insights are valuable for integrating such technology into telehealth platforms, enhancing patient care through accurate emotion recognition.

Table 1 Success rate of emotion detection through facial recognition with ChatGPT

Emotion	Success rate (%)
Happiness	100
Surprise	100
Fear	100
Disgust	100
Anger	92
Sadness	58
Overall	92

4.2 Experiment 2: Background Generation Based on Emotions

In Experiment 2, we sought to evaluate the appropriateness of ChatGPT-generated backgrounds intended to calm patients based on their emotions. The experiment was conducted with three testers in our laboratory, each with different backgrounds to encompass a variety of perspectives.

The evaluation was based on principles from color psychology. We considered the emotional connotations of warm, cool, and neutral colors, as well as cultural implications. For instance, cool blues and greens were expected to calm anxiety due to their association with nature and tranquility, while warm reds and oranges might energize or increase alertness.

Sample feedback from testers, after reviewing the images, included:

- The serene lake for 'Anxiety' aligns with the calming effect of blue, but might not resonate with individuals who find open spaces overwhelming.
- The image for 'Sadness' with the open field was comforting and the vastness felt freeing, not lonely.
- The bright meadow for 'Surprise' captures the energy of yellow, though it might be too stimulating for some.
- The Japanese garden for 'Agitation' leverages the tranquility of green, but its cultural specificity may not be universally soothing.
- The cozy cabin for 'Fear' was a mixed bag; it was warm and inviting for some but felt isolating to others, depending on personal experiences.
- In general, the images were well-aligned with the intended emotions. However, the effectiveness varied, and it seems like some fine-tuning could optimize their impact.

The analysis showed that while many images adhered to color psychology principles and were generally well-received, personal and cultural differences significantly influenced preferences. Some testers found certain images less effective due to personal memories or cultural associations with the colors or scenes depicted. The results underscore the importance of personalized approaches in healthcare settings, where patient backgrounds and preferences can greatly affect the perception and efficacy of therapeutic imagery. In that sense, it would be beneficial to implement a feedback mechanism where users can indicate or rate the effectiveness of the image generated, helping the model to learn and improve over time.

5 Discussion

Our experiments with ChatGPT-4 showcased its potential in recognizing emotions and generating therapeutic backgrounds, aligning with our objectives to enhance patient engagement in telehealth services. The high success rate in emotion detection suggests that LLMs like ChatGPT-4 could significantly contribute to understanding patient needs and tailoring interactions accordingly. The qualitative feedback on background generation highlighted the importance of personalization, as preferences varied widely based on cultural background, age, and individual experiences.

However, the study faced limitations, including the variance in ChatGPT-4's ability to detect certain emotions like sadness, possibly due to subtleties in expression that are not as easily captured by current AI models. For future studies, these limitations could be addressed by providing the patients with a questionnaire to survey their emotional status. This would improve the model's ability to understand the context. Furthermore, while the generated backgrounds were broadly effective, the one-size-fits-all approach did not account for individual patient preferences or cultural connotations.

For future research, we recommend exploring more nuanced AI training that includes more diverse set of emotional expressions, diverse ethnic groups, and cultural contexts, which would offer several benefits such as:

- **Enhanced Understanding**: A diverse dataset can teach the AI to understand and interpret the nuances of different cultures better, leading to more accurate and relevant responses.
- **Reduced Bias**: By incorporating a wide variety of perspectives and experiences, AI systems can be trained to avoid biases that might otherwise occur due to a limited or homogeneous dataset.
- **Improved Emotional Intelligence**: Including a broad range of emotional expressions can help AI better recognize and respond to human emotions, making interactions more empathetic and effective.

Additionally, developing algorithms that can adapt therapeutic imagery to individual patient profiles could improve the efficacy of telehealth services. This would involve a more granular analysis of patient feedback and possibly a real-time adjustment of backgrounds during telehealth sessions based on ongoing emotional assessment.

In terms of privacy, a key element to protect the patient's information is the minimization of data collection. It is considered that there is no need to store data. Image and questions would be processed in real-time in the user's device. This enhances privacy, as it reduces the risk of data breaches associated with storage and provides patient with control over their information. Some information needs to be transferred from the patient for background image generation. In this case, one solution would be to minimize data required, for example, by not taking the full face image but only key elemental points in determining emotion, which usually are: eyes, eyebrows, lips, and other features.

6 Conclusion

This study has demonstrated the innovative application of LLMs, specifically ChatGPT-4, in telehealth to address the challenge of patient anxiety during waiting times. Our experiments show that ChatGPT-4 can successfully detect emotions with a high degree of accuracy and generate visually therapeutic backgrounds tailored to the patient's emotional state. These capabilities point toward a significant enhancement in patient experience by providing real-time, personalized care.

The innovative contribution of this work lies in its potential to revolutionize patient-provider interactions within telehealth platforms. The integration of emotion-sensitive AI can pave the way for a more empathetic, responsive, and customized healthcare service delivered remotely.

Moving forward, we envision the development of more sophisticated LLMs trained on diverse datasets to encapsulate a broader spectrum of emotional expressions and cultural nuances. Future telehealth services could adapt in real-time to patient feedback, adjusting therapeutic imagery to suit individual preferences and improve patient outcomes, ultimately leading to a more patient-centric model of care.

Disclosure of Interests The authors have no competing interests to declare that are relevant to the content of this article.

Acknowledgements The authors wish to express their sincere gratitude to the NSERC CREATE Fund (565429-2022) and Algoma University Research Fund for their generous support of this work. Their commitment to fostering innovative research has been invaluable to the success of this project.

References

1. Berg J, Carlson K, Richards S (2022) Providers at a midwestern academic system report a positive experience with telehealth during the COVID-19 pandemic. Telemedicine and e-Health
2. Cheung KW, Au TST, Wai JKO, Seto MTY (2023) Perceptions and challenges of telehealth obstetric clinics among pregnant women in Hong Kong: cross-sectional questionnaire study. J Med Internet Res 25:e46663
3. Eng K, Donohue CC, Wagner AJ, Dougherty DW, Salehi E, Lowenstein C, Gundersen DA, Lederman R, Kadish S (2021) Assessing oncology patient and provider telehealth experience
4. Funderburk CD, Batulis NS, Zelones JT, Fisher AH, Prock KL, Markov NP, Evans AE, Nigriny JF (2019) Innovations in the plastic surgery care pathway: using telemedicine for clinical efficiency and patient satisfaction. Plast Reconstr Surg 144(2):507–516
5. Ghosh A, Yan Y, Lin W (2023) Adaptive user interface framework powered by a large language model for culturally sensitive virtual healthcare applications. In: IEEE international conference on biomedical and health informatics (BHI'23). Pittsburgh, Pennsylvania, United States of America
6. Hull TD, Malgaroli M, Gazzaley A, Akiki TJ, Madan A, Vando L, Arden K, Swain J, Klotz M, Paleos C (2022) At-home, sublingual ketamine telehealth is a safe and effective treatment for moderate to severe anxiety and depression: findings from a large, prospective, open-label effectiveness trial. J Affect Disord 314:59–67

7. Johnson R, Groff TN, Kirkwood VM, Richardson KL, Thompson JA (2023) Adopting urinary tract infection guidelines to promote antibiotic stewardship in the time of telehealth medicine
8. Lin W, Babyn P, Yan Y, Zhang W (2023) A novel scheduling method for reduction of both waiting time and travel time of patients to visit health care units in the case of mobile communication. In: Enterprise Information Systems, p 2188124
9. Lin W, Xu M, He J, Zhang W (2023) Privacy, security and resilience in mobile healthcare applications. Enterp Inf Syst 17(3):1939896
10. Massucato MAO, Ribeiro ALG, Pessalacia JDR, Neves VdJR, Stolte-Rodrigues VP (2021) Telehealth as a medical education tool during the COVID-19 pandemic: experience report. Revista Brasileira de Educação Médica 45
11. Saxena A, Khanna A, Gupta D (2020) Emotion recognition and detection methods: a comprehensive survey. J Artif Intell Syst 2(1):53–79
12. Scriven H, Doherty DP, Ward EC (2019) Evaluation of a multisite telehealth group model for persistent pain management for rural/remote participants. Rural Remote Health 19(1):1–13
13. Vowles KE, Knaggs RD, Palomares AC (2023) Pandemic impact and adaptation to telehealth in chronic pain treatment providers across two CCOVID-19 pandemic years. Pain Pract
14. Ward K, Vagholkar S, Sakur F, Khatri NN, Lau AY et al (2022) Visit types in primary care with telehealth use during the COVID-19 pandemic: systematic review. JMIR Med Inform 10(11):e40469
15. Whaley CM, Ito Y, Kolstad JT, Cowling DW, Handel B (2022) The health plan environment in California contributed to differential use of telehealth during the COVID-19 pandemic: study examines differential use of telehealth in California during the COVID-19 pandemic. Health Aff 41(12):1812–1820
16. White CN, Kauffman BY, Acierno R (2023) Factors contributing to veterans' satisfaction with PTSD treatment delivered in person compared to telehealth. J Telemed Telecare 29(6):426–434
17. Wittmeier KD, Hammond E, Tymko K, Burnham K, Janssen T, Pablo AJ, Russell K, Pierce S, Costello C, Protudjer JL (2022) Another tool in your toolkit: pediatric occupational and physical therapists' perspectives of initiating telehealth during the COVID-19 pandemic. Phys Occup Ther Pediatr 42(5):465–481
18. Yang K, Ji S, Zhang T, Xie Q, Ananiadou S (2023) On the evaluations of chatgpt and emotion-enhanced prompting for mental health analysis. arXiv preprint arXiv:2304.03347
19. Zhang B, Yang M, Li X, Ye Y, Xu X, Dai K (2020) Enhancing cross-target stance detection with transferable semantic-emotion knowledge. In: Proceedings of the 58th annual meeting of the association for computational linguistics, pp 3188–3197
20. Zhu L, Pergola G, Gui L, Zhou D, He Y (2021) Topic-driven and knowledge-aware transformer for dialogue emotion detection. arXiv preprint arXiv:2106.01071

Different Processes for Graphical Recognition of Derivative of a Function: An Eye-Tracker Analysis

Camilla Spagnolo, Christian Casalvieri, and Alessandro Gambini

Abstract The goal of this research is to uncover the struggles that students face when working on Calculus tasks. This study uses a qualitative analysis method, with the help of an eye-tracking tool, to examine the variations and similarities in cognitive processes between experts and novices. Using an eye-tracking tool, we can observe an individual's ability to move between different forms of representation. The Calculus tasks focus on the concepts of derivatives, specifically, students were asked to identify the graph of a derivative function.

Keywords Eye-tracker · Semiotic registers · Mathematics education

1 Introduction and Literature Review

Eye tracking has been used in several research areas, such as psychology, neuroscience, linguistics, and creativity of cognitive processes [1]. This tool has been applied to analyze cognitive processes in areas such as geometry [2], algebra [3], and interpretation of motion graphs [4]. Recent research has used the eye-tracker tool to study cognitive processes in high schools ([5, 6]). A study by Just and Carpenter [7] showed that a person's gaze provides valuable information about his or her cognitive and learning processes, a hypothesis known as the eye-mind hypothesis. In addition, the eye-tracker tool allows analysis of the ability to switch between registers. Recently it has also been used as a tool in mathematics education. Using this tool,

---

C. Spagnolo (✉)
Free University of Bozen, Bolzano, Italy
e-mail: camilla.spagnolo@unibz.it

C. Casalvieri · A. Gambini
Sapienza University of Rome, Rome, Italy
e-mail: christian.casalvieri@uniroma1.it

A. Gambini
e-mail: alessandro.gambini@uniroma1.it

X.-S. Yang et al. (eds.), *Proceedings of Ninth International Congress on Information and Communication Technology*, Lecture Notes in Networks and Systems 1000,
https://doi.org/10.1007/978-981-97-3289-0_44

it is possible to collect data on eye movements and provide insights into cognitive processes. In the context of mathematics education research, by studying eye movements, it is possible to analyze the problem-solving process in a mathematical task.

Although the cognitive pathway followed in tackling a mathematical problem is certainly influenced by external factors (such as educational, socio-cultural, technological, etc.) that contribute to the strategy for arriving at the answer, the influence these factors have on the analysis carried out with the eye-tracker is, in our opinion, insignificant, as the eye-tracker tool is in fact unobtrusive: what a candidate has to do is simply observe a screen and this does not affect the ability to determine the answer to the question, the latter being the prerogative solely of the internal abilities of the candidate, whether novice or expert. On the other hand, eye-tracker technology offers many advantages: it reduces the influence of social expectations (e.g. the shyness of students who are afraid of making mistakes or who fear that their method of solving is inadequate), as well as some perceived rules in the context of educational contract theory during the data collection process (the student essentially assumes freer behaviour). In addition, eye-tracking data are less affected by problems related to memory, introspection, metacognitive reflection, language problems or verbalization of solutions found, all of which are related, for example, to students thinking aloud.

Students in mathematics and science faculties encounter many difficulties in the transition from secondary to tertiary education. These difficulties can be grouped into categories such as epistemological and cognitive, sociological, and cultural and educational. Students beginning a new course of study encounter new concepts and approaches that may not be directly related to their previous studies. For example, one epistemological and cognitive difficulty is the lack of in-depth understanding of a mathematical concept. High school students often rely on algorithms and procedures but may not fully understand mathematical objects. In addition, taking notes and reading texts can be a problem for high school students [8]. This aspect is relevant to our research because the way a student reads a problem can give us information about cognitive processes.

Several research studies have explored the differences in the presentation of mathematical concepts between secondary and tertiary education. One issue that has been identified regarding the transition between secondary and tertiary school is the different way in which mathematical objects are used. In high school, students tend to use mathematical objects in their calculations, while at university, they need to be able to handle a higher level of abstraction. Specifically, university requires students to be able to deal with theoretical aspects, whereas in secondary school this aspect is never considered [9]. Praslon [10] showed an example on the different tasks presented in university and high school on the concept of the derivative. Through comparing the tasks, he showed that the derivative is seen as a tool in high school, while university students study it as a mathematical object. Moreover, university requires students to be able to switch from a register to representations to another [11] and between frames [12]. The different approaches to the concept of the derivative are emphasized by these elements [13]. There is a lack of mastery of mathematical objects and difficulty in switching representation registers.

Our research aimed to study the processes involved for students in Calculus task related to the concept of derivative. Our work focuses on the ability to read a graph and be able to infer relevant information and find relationships at the level of syntax, discern its overall meaning, and position it correctly in relation to the phenomenon it is modelling.

Our research questions are: What information can be inferred by looking at the graph of the derivative and how can it be used? Are there significant differences between experts and novices in solving strategies based on the graphical properties of derivative functions?

2 Methodolody

The research participants were divided into novices and experts. Novices were composed by 13 university students: 6 students from engineering's faculty, 3 from physics, 2 from chemistry, 1 from mathematics, and 1 from statistics. The expert group was composed by 3 mathematics undergraduates, 2 doctoral students, and 2 high school teachers.

The analysis was focused on the study of visual behaviour through the eye-tracker tool. In particular, the elements that characterize eye movement were studied: fixation, saccade, and scan path. The first is a prolonged gazing at a specific point in the visual field, typically lasting for a few tenths of a second. Its key measures are the length of time it lasts and the number of fixations. In contrast, a saccade is a quick movement of the eyes between two fixations, and its key measures are the distance of the movement and the frequency of saccades. The last is the study of the pattern of fixations and saccades over time, and it can reveal information about where a person is directing their attention and what they are focusing on.

The instruments used to analyze the data collected with eye-tracker tool are heat maps and gaze plots. A heat map is a visual representation that shows where a person's attention is concentrated while viewing a specific image or scene. It uses colours to indicate the level of attention given to different areas, with warmer colours indicating areas where the person spent more time looking or looked more frequently. The gaze plot, on the other hand, is a tool that shows the order and timing of where a person is looking. It displays the duration of fixations on an object by using circles, with the size of the circle representing the amount of time spent looking at that object. Larger circles indicate longer fixations. For the test, we employed the Tobii Pro Nano® system, a screen-based eye-tracking device that uses pupillary and corneal reflections with both dark and bright pupillary lighting modes to gather data. The data obtained are then analyzed by a specialized processing unit that utilizes dedicated software to calculate gaze data, as well as to collect, observe and analyze both individual and aggregated data for comparisons, as used in this study.

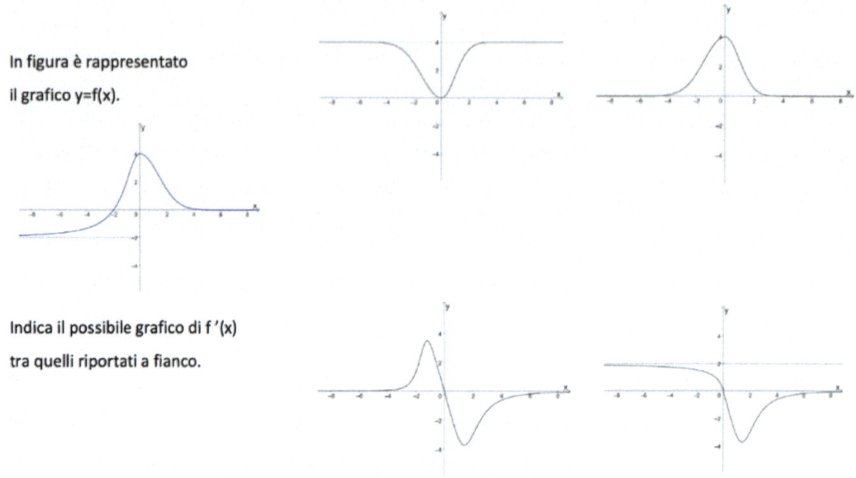

Fig. 1 Identifying the derivative function

The test was conducted using an image that was displayed on a monitor connected to the eye-tracking camera. The participant was given a task with no time limit, and their eye movements were recorded by the eye-tracker as they viewed the image. After selecting the correct answer, the participant was then interviewed and asked to explain their reasoning for their choice. During the interview, the task was shown again, and the audio recording was synced with the eye-tracking data, in order to compare the participant's eye movements with their cognitive motivations, as well as to analyze the participants' reasoning abilities.

The task proposed was the following:

The text of the question shown in Fig. 1 is as follows: *In the figure the graph* $y = f(x)$ *is represented. Select the possible graph of* $f'(x)$ *from those shown alongside.*

The different graphs were labelled with letters (a), (b), (c), and (d), respectively, from the top left to the bottom right. These labels were not included in the main question image to avoid adding any extraneous distractions.

3 Analysis and Discussion

From the heat maps of expert candidates, it can be seen that they tend to focus on the main image for a longer period of time, while they quickly scan through the alternative options. This suggests that their visual analysis of the main image is more detailed and local, while their analysis of the alternatives is more general and global. The approach that expert candidates seem to prefer is to first identify general characteristics of the main image, and then choose a characteristic that is unique to it, and compare it to the alternatives, paying more attention to the details of the chosen alternative. Once this unique characteristic is found, the question is considered solved and further verification is minimized. The choice of this unique characteristic is influenced by the individual candidate's preferences, experiences, and the type of main image. It also affects the cognitive approach used and the visual pattern recorded by the eye-tracker.

Figure 2 shows an example of a heat map of expert candidate that focused exclusively on the presence of a non-horizontal tangent inflection points contained within the interval (− 2, 0). We can see a clear focus of visual and cognitive attention on a specific point in the main image. The rest of the graph is glanced over quickly with many short saccades, following the overall trend of the entire domain, up to the horizontal asymptotes. It is worth noting that the candidate paid more attention to the single zero in the function, which was classified as a distinct point, than to the maximum point. It turns out that the zero was a red herring and the maximum point was not used since another unique feature was identified first.

Candidate stated: "initially I was struck by the point where the function changed sign" […] "immediately afterwards I noticed the point of inflection and therefore the derivative must have a maximum point because the function passes from convex to concave". From this statement, two observations are clear: on one hand, the left-handed inflection point was identified by chance as the closest element to the zero of the function, suggesting that the gaze guides the brain's processing of the information. On the other hand, the subsequent cognitive process, based on prior theoretical knowledge, allows for the deduction of a graphic property of the response even before it is observed. In this case, the process is reversed: the cognitive process guides the visual observation, which becomes a mere verification of what has already been hypothesized. The candidate then states "in fact, graph (c) is the only one that presents this characteristic". Some long saccades between the main image and the

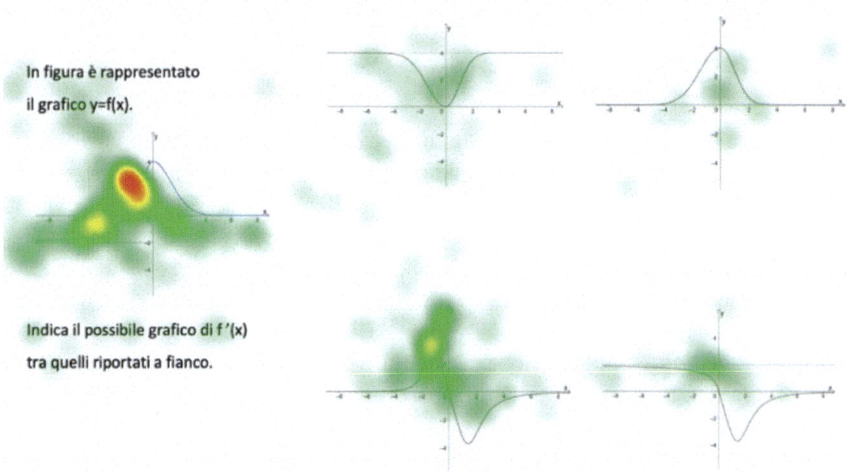

Fig. 2 Heat map of expert candidate

four alternatives help to identify the uniqueness of the chosen property, which is confirmed by some fixations on the maximum point of alternative (c). Finally, the candidate says "then I just quickly verified that (c) also met other criteria, such as the sign". The latter property was used only as confirmation of his choice.

A typical non-expert candidate's eye movement pattern is more intricate compared to that of an expert candidate. The first noticeable feature is that it is more disorganized, with a general tendency of many short saccades within a single graph, and also many long saccades between graphs. The long saccades indicate a high level of comparison activity between the main image and each alternative, as well as between the different alternatives. The distribution of fixations among the five graphs in the question is consistent among non-expert candidates. This can be attributed to two factors: lack of expertise in selecting specific elements, and the need for more verification once a perceived correct answer has been identified.

An example of a heat map of a non-expert candidate is shown in the Fig. 3. The heat map of this candidate clearly shows a focus on the central area of the main graph, specifically between the two inflection points and in the central area of the four alternatives. The textual section only draws attention when it contains the symbolic/algebraic notation ($y = f(x)$ and $f'(x)$). During the interview, the candidate explains: "I started by trying to understand how the function $f(x)$ was constructed". This candidate seems to have a strong inclination for the elimination method rather than the construction method of finding the solution. In fact, he, on one hand, looks to confirm that the information obtained from the main image is present in one or more of the alternatives, and on the other hand, searches for elements that eliminate certain alternatives. This is especially evident in the comparison of the main image and alternatives (c) and (d) as indicated by the eye-tracker through the presence of long saccades between these graphs.

4 Conclusions

A key aspect of our study is the effort to discern unique similarities and disparities in the approach used by experienced and novice candidates. Regarding the second research question, we can observe that there is a different approach between experts and novices. In fact, the expert candidate identified an exclusive characteristic that allows him to find the correct alternative and used another property to confirm the choice. In contrast, the novice used a confused approach, and the correct answer was chosen by proceeding by exclusion. We can make the following observations: experts and novices had different behaviour during the visual acquisition of information in alternatives, in fact different points of interest and eye trajectories were recorded; need to acquire more information by novices than experts to solve the task. They had a similar behaviour in the visual acquisition of information in stimulus graph, they focused the attention on the similar points and areas of interest.

In general, the stage of collecting information is related to generating hypotheses, while the selection of decisive elements for solving the question is linked to the

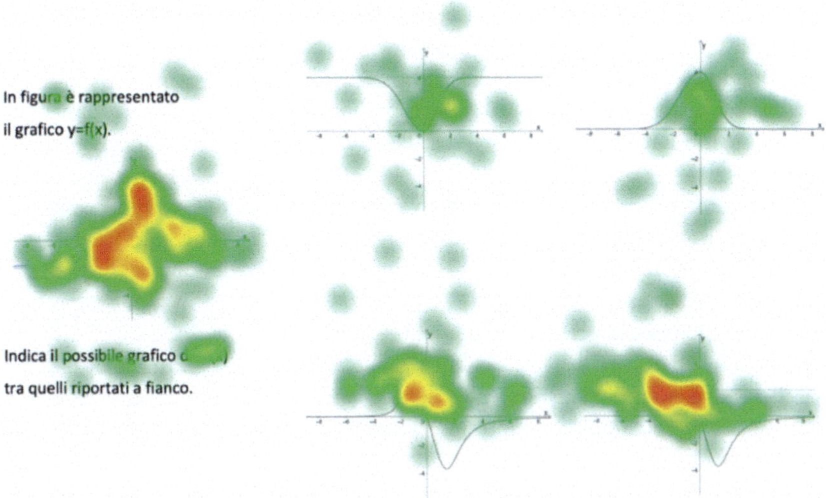

Fig. 3 Heat map of a non-expert candidate

ability to make analytical deductions and to verify them. This means that the novice's overabundance of gathered data hinders or diminishes their ability to generate valid hypotheses, resulting in an increase in difficulty in making deductions and verifying them. The presence of distracting elements both in strategy formation is an indication of this: the novice's ability to selectively process information is diminished as they have a harder time discarding previously evaluated hypotheses, and the abundance of information gathered poses a challenge as it complicates the decision-making process. Regarding the first research question, the information acquired from the graph is different and is used differently by an experienced candidate and a novice. Specifically, the expert candidate was able to infer certain exclusive properties that guided his or her solving process, while the novice gathered much data from the graph and they had difficulty classifying what was essential for solving the task.

In this work, we attempted to identify the cognitive processes involved in the candidates' approach to the given questions by analyzing specific eye movement patterns and to encode them, to try to obtain a more abstract analysis of the problem-solving strategies used. However, different eye movements also indicate how these mental processes are executed and failed attempts to extract useful information for solving the problem. The cognitive process underlying the resolution of such questions is extremely complex, involving simultaneous sensory (visual) processing of various stimuli, which are all different in nature. This is particularly true when transitioning from a graphical representation of information in the stimulus to another graphical representation in the alternatives, all mediated by the candidate's prior theoretical knowledge acquired during a course on mathematical analysis.

From the discussion above, certain challenges in the learning of mathematical analysis for students in their first year of a scientific academic programme are clear.

These difficulties may stem from the complicated transition from high school to the first year of university [14–16], where students not only experience changes in teaching methods but also a diversification of goals set by university mathematics courses. The findings presented in this paper provide, in our view, some suggestions for enhancing the teaching of mathematics at the university level. Generally, students need to learn how to approach a mathematical question and how to focus their attention to gather the necessary information to solve the problem at hand. Mason [17] argues that learning is an education in awareness, which is closely tied to attention and observation. Additionally, we think it is important to foster students' capacity to generate hypotheses and make analytical deductions, as this is a crucial step in any form of scientific reasoning, be it theoretical or practical.

The main limitation of the experiment we presented consists of the fact that the most delicate and complicated moment in a study based on the eye-tracker is precisely the analysis of the data obtained through the experiment and the reconstruction of the resolution strategy on the basis of the spontaneous eye movements recorded by the eye-tracker, due to the impressive amount of data at our disposal. For this reason, we repeated the experiment during the subsequent interview with the candidates by synchronizing the audio with the video in order to assess the candidate's awareness in the resolution of the question, measuring and subsequently interpreting the level of coherence between what was consciously stated and what was previously done during the actual resolution of the question. Not infrequently, especially for novice candidates, the results of the two phases show, in our opinion, significant differences. This work took a lot of time because we had to examine the recorded video of eye movements and listen to the audio of the interview several times, trying to eliminate non-essential and especially uncertain information. This time was a major limitation for our experiment because it reduced the number of candidates examined and only allowed us to make qualitative analyses.

Finally, we conclude by noting that socio-cultural aspects are undoubtedly a key element in the elaboration of solution strategies when approaching a mathematical question. It is in fact through our sensorimotor experiences with our surroundings that our cognitive attention is constantly captured. The eye is certainly one of our sensory organs that is most stressed and stimulated by the external environment. But visual perception is understood not only as a physiological substrate, but above all as a culturally and socially shaped sense that allows us to access mathematical representations in a certain structured way. From this point of view, therefore, the eye-tracker instrument could also be a useful investigative tool for future research, for example by administering the same test to groups of candidates from different social backgrounds or geographical areas, or with different prior education to see if the data obtained show significant differences in strategic choices. Of course, all this would be meaningful if a large amount of data could be made available (e.g. through international collaboration), a sufficiently robust quantity for significant statistical analysis.

References

1. Schindler M, Lilienthal AJ (2020) Students' creative process in mathematics: insights from eye-tracking-stimulated recall interview on students' work on multiple solution tasks. Int J Sci Math Educ 18:1565–1586
2. Schindler M, Lilienthal AJ (2017) Eye-tracking and its domain-specific interpretation. A stimulated recall study on eye movements in geometrical tasks. In: Kaur B, Ho WK, Toh TL, Choy BH (eds) Proceedings of the 41st conference of the international group for the psychology of mathematics education (PME), vol 4. PME, Singapore, pp 153–160
3. Obersteiner A, Tumpek C (2016) Measuring fraction comparison strategies with eye-tracking. ZDM-Math Educ 48(3):255–266
4. Ferrara F, Nemirovsky R (2005) Connecting talk, gesture, and eye motion for the microanalysis of mathematics learning. In: Chick HL, Vincent JL (eds) Proceedings of PME 29 (Research Forum). University of Melbourne, PME: Melbourne, Australia, pp 137–142
5. Spagnolo C, Capone R, Gambini A (2021) Where do students focus their attention on solving mathematical tasks? An eye tracker explorative study. In: Inprasitha M, Changsri N, Boonsena N (eds) Proceedings of the 44th conference of the international group for the psychology of mathematics education, vol 4. PME, Khon Kaen, Thailand, pp 89–96
6. Casalvieri C, Gambini G, Spagnolo C, Viola G (2023) The concept of derivatives through eye-tracker analysis. In: Proceedings of the 15th international conference on computer supported education (CSEDU 15), vol 2. Scitepress, pp 378–385
7. Just MA, Carpenter PA (1976) Eye fixations and cognitive processes. Cogn Psychol 8(4):441–480
8. De Guzmán M, Hodgson BR, Robert A, Villani V (1998) Difficulties in the passage from secondary to tertiary education. In: Proceedings of the international congress of mathematicians, vol 3. Documenta Mathematica, Berlin, pp 747–762
9. Gueudet G (2008) Investigating the secondary–tertiary transition. Educ Stud Math 67(3):237–254
10. Praslon F (2000) Continuités et ruptures dans la transition Terminale S/DEUG Sciences en analyse. Le cas de la notion de dérivée et son environnement (Ruptures and continuity in the Terminale S/first scientific university year in calculus. The case of the derivative and its environment), Thèse de doctorat de l'Université Paris 7
11. Duval R (2006) A cognitive analysis of problems of comprehension in a learning of mathematics. Educ Stud Math 61(1–2):103–131
12. Douady R (1992) Des apports de la didactique des mathématiques à l'enseignement (Contribution of mathematics didactics to teaching). Repères Irem 6:132–158
13. Ferrari PL (2004) Mathematical language and advanced mathematics learning. In: Johnsen Høines M, Berit Fuglestad A (eds) Proceedings of PME 28 (Research Report)
14. Ferretti F, Gambini A, Spagnolo C (2024) Management of semiotic representations in mathematics: quantifications and new characterizations. Eur J Sci Math Educ 12(1):11–20
15. Giberti C, Santi G, Spagnolo C (2023) The role of metaphors in interpreting students' difficulties in operating with percentages: a mixed method study based on large scale assessment. Eur J Sci Math Educ 11(2):297–321
16. Spagnolo C, Nicchiotti B (2024) Interpreting gender gap issues in standardized tests: definition and application of a theoretical tool. Front Educ 8:1303041
17. Mason J (2008) Being mathematical with and in front of learners: attention, awareness, and attitude as sources of differences between teacher educators, teachers and learners. In: Wood T, Jaworski B (eds) International handbook of mathematics teacher education: the mathematics teacher educator as a developing professional, vol 4. Sense publishers, Rotterdam, pp 31–56

Emails Classification: Comparing Statistics, Machine-Learning, Deep-Learning, and ChatGPT Prompting Techniques

Pablo Turón, Montse Cuadros, Alicia Grande, and David Lopez

Abstract The automatic classification of emails in a real scenario is very helpful for administration purposes and focuses human attention on solving tasks instead of organizing them. This paper presents a study of the applicability of a wide variety of approaches to the problem of this automatic email classification. For this purpose, we have conducted a wide set of experiments using traditional classification methods such as Support Vector Machines (SVM), distinct Large Language Models (LLMs), and prompting techniques applied to ChatGPT to tackle the classification problem. Thus, a set of labeled email conversations from LIS DATA Solutions have been studied, pre-processed, and prepared in order to be used by the different classification techniques. The results show that, surprisingly, traditional approaches beat more robust solutions for this task.

Keywords Text classification · LLMS · NLP · ChatGPT · IA

1 Introduction

Recent developments in the Natural Language Processing (NLP) area have redefined the way we understand Large Language Models (LLMs), increasing its popularity in both academia and industry due to its unprecedented results for various applications.

P. Turón (✉) · M. Cuadros
Vicomtech Foundation, Basque Research and Technology Alliance (BRTA), Mikeletegi Pasealekua 57, Donostia/san-Sebastián 20009, Spain
e-mail: pturon@vicomtech.org

M. Cuadros
e-mail: mcuadros@vicomtech.org

A. Grande · D. Lopez
LIS Data Solutions, Santander, Spain
e-mail: alicia.grande@lisdatasolutions.com

D. Lopez
e-mail: david.lopez@lisdatasolutions.com

X.-S. Yang et al. (eds.), *Proceedings of Ninth International Congress on Information and Communication Technology*, Lecture Notes in Networks and Systems 1000,
https://doi.org/10.1007/978-981-97-3289-0_45

However, the size of these models and the amount of energy that they need to be deployed opens the debate on whether it is worth using these models for simple tasks such as text classification.

Although LLMs perform well in most of the language understanding tasks, there are some use cases where these types of models might be penalized depending on the nature of the studied data. Thus, traditional approaches which try to classify the text focusing on the keywords, might have an advantage when dealing with noise and unstructured data.

In this use-case study, we have decided to test the performance of several *state-of-the-art* LLMs (*BLOOM* and *multilingual BERT* [5, 25], respectively), a *zero-shot* solution based on prompting with the *ChatGPT* model, and a traditional approach as a baseline (using *Support Vector Machine (SVM)* as a classifier and *Term Frequency–Inverse Document Frequency (TF-IDF)* as a feature extractor engine) to classify a set of real email conversations labeled in 5 categories. We have been provided with the raw data and an anonymized version in order to use it with *ChatGPT*. For the anonymization process, a set of heuristics have been applied in order to eliminate sensitive data of the original text.

The paper is organized as follows: Sect. 2 overviews the state-of-the-art of text classification solutions. Section 3 describes the dataset we have used in this paper and Sect. 4 details the methodology used for the experimentation. Finally, in Sect. 5 we present the results obtained in the experimentation and we draw conclusions in Sect. 6.

2 Related Work

Text classification is the process of categorizing textual information into organized groups. It has been successfully applied in different scenarios including sentiment analysis [31], news categorization [1], hate speech [2, 6], etc. in multiple data sources such as tweets, news articles or emails [1, 2, 23], respectively.

The generic strategy followed for text classification is based on a pipeline that starts pre-processing the document, follows extracting its features and representing numerically the text, and ends up training a classifier. The pre-processing step is mainly focused on reducing the size of the input text document, involving techniques of data cleaning, removing stop-words [10, 34], stemming [19, 34], etc. Traditionally, word representation has been carried out by techniques such as *bag-of-words* [7], however, new techniques capable of representing more features of the text were progressively developed such as *TF-IDF (Term frequency - Inverse document frequency)* [8, 35] or *LSI (latent semantic indexing)* [35]. Once the features have been extracted, a classifier is trained in order to classify the texts; *Naive Bayes* [10, 14], *Neural Networks* [29], *SVM (Support Vector Machine)* [24, 30, 33], or *Decision Trees* [22] are commonly selected approaches.

Several studies showed that the success of deep learning in text classification largely depends on the effectiveness of the *word embeddings* [9, 26]. This motivated researchers to deploy new feature representation techniques, capable of representing the words and their relationship. These techniques mainly include unsupervised/supervised pre-training of text, distributed representation, and transfer learning. Some works approached this task by using *CNN (Convolutional Neural Networks)* [11, 28, 32] since they are good at learning local and location-invariant features [16]. Other works rely on pre-training models in an unsupervised way, such as *GloVe, Word2Vec or fastText*, [3, 15, 18] respectively.

More recently, the emergence of the attention mechanisms proposed by *Transformers* architecture [27], has marked an important component of modern NLP systems. This architecture allowed the Pre-training of Language Models (PLM) on a large network with a large amount of unlabeled data and fine-tuning in downstream tasks, marking a breakthrough in several neural language understanding tasks. This led to different models which can be divided into two groups: (1) autoregressive PLMs such as *BLOOM* [25] or *OpenAI GPT family* [4, 17, 20, 21] and (2) autoencoding PLMs such as *BERT* [5] or *RoBERTa* [12].

The recent outstanding ability of autoregressive PLMs to understand the language has increased the studies of using them as zero-shot classifiers. Thus, instead of introducing new parameters and training the model on labeled data, you can design prompts or templates that guide the model itself to generate the desired output for a given task [13].

3 Data

Data available for this project corresponds to real entire email conversations (an email and its replies) and its subject. Although the vast majority of emails are written in Spanish language, there are some emails in other languages such as English, Italian or German languages.

The studied emails correspond to commercial operations in LIS DATA Solutions, and are related to 5 different topics that were manually tagged:

1. **Administrative**. This category includes all mail related to the administrative management of the company. This includes invoices issued or received, claims related to these invoices, and delivery notes certifying the receipt or dispatch of goods.
2. **Commercial**. This section is intended for the commercial and sales management of the company. Price requests from potential or current customers are handled here, as well as the preparation and sending of quotations for transport and logistics services.
3. **Issues**. This category covers all problems or inconveniences that may arise during the process of collection or delivery of goods. This may include delays, damage to packages, and losses, among others.
4. **Operations**. This category groups emails requesting specific information about the company's logistics operations. This may include details on routes, transit times, and vehicle availability, among others.

Table 1 Distribution of data splits

	Train	Dev	Test
Administrative	108	13	13
Commercial	200	25	25
Issues	98	12	12
Operations	76	9	9
Collections	168	21	21
Total	650	80	80

5. **Collections**. This section is dedicated to pick-up orders. This is where customer requests for the company to pick-up parcels or loads from specific locations are handled.

Data was split in train-dev-test datasets in 80–10–10 proportions respectively (see Table 1).

The provided data comes with a lot of extra information since is composed of raw text data, with the email header information *(sent by, to, subject, etc.)*, repeated characters separating sections of emails, signatures, templates (privacy agreement templates, environmental warnings, etc.), etc. This motivated us to develop several pre-processing steps (see Sect. 4.1). Furthermore, to use this data with the *ChatGPT* engine, we created an anonymized copy of the test dataset with each of these pre-processing techniques.

4 Methodology

In this study, we have separated this problem into two stages. The first stage would focus on pre-processing the data and reducing its variability by cleaning potentially irrelevant data and noise. The second stage would focus on training different approaches to classify our data.

4.1 Data Pre-processing

Due to the nature of our data, it's crucial to apply some pre-processing to the emails that we are using in this project to remove different kinds of irrelevant information. Thus, we applied different pre-processing steps, obtaining different versions of the same dataset:

Filter over-repeated characters (F_{DUP}). It's quite common to find characters that appear repeated in the email with different purposes (topic separator, signature, etc.). Although it provides some layout information for the reader, these characters

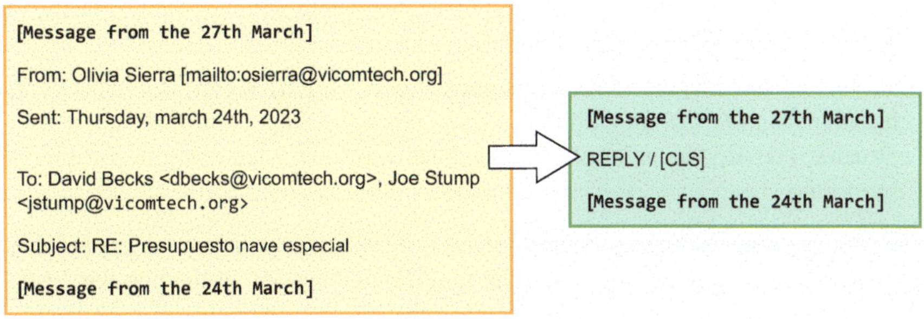

Fig. 1 Example of replacement of reply headers

do not provide any useful information to a classifier, so, we decided to trim these characters to a maximum of two characters.

Replacing emails and URLs (F_{WEB}). We applied different regex to find emails and URLs to replace them with controlled tokens (*EMAIL* and *URL*).

Filter repetitive text without any information (F_{NI}). This step focuses on filtering repetitive text that is present in every email such as privacy warnings, environmental warnings, etc. that do not provide any kind of information. To detect this repetitive text, we manually located some keywords for each topic and removed lines that contain more than two words of these keywords.

Filter reply headers (F_{ASW} -$F_{ASW_{CLS}}$). Our email conversations contain each header of each response to an email, information that may be noisy and that also takes up input space in models. Thus, we decided to filter them by looking for consecutive lines that start with particles such as *"From", "Sent"*, etc. in different languages. These consecutive lines would be replaced by the control word *REPLY*, or by the special token *[CLS]*, which would be exploited by the *BERT LLM* (see Fig. 1).

Reorder mail streams (F_{SORT}). Intuitively, the initial emails are the most relevant messages in a conversation, as they contain the topic of the conversation. However, as can be seen in Fig. 1, the most recent emails are at the top of the conversation, and the initial emails are at the bottom, a fact which may cause a classifier to pay less attention to them. This leads us to revert the order of the stream messages according to the headers found in the previous point.

Keep the first message (F_{KF}). Under the intuition that, the first message is the most relevant message since it explains the topic to discuss, we decided to keep just the first detected message in the stream.

4.2 *Systems*

Several approaches were tested and compared in this study, testing a baseline, training autoencoding and autoregressive LLM, and a *zero-shot* approach by prompting with *ChatGPT* models:

Baselines. As baselines, we have applied two different techniques.

- The first baseline studied consists of a *random classifier*, which would try to classify each email randomly.
- The second baseline algorithm is composed of a *Support Vector Machine (SVM)* classifier, using *Term frequency - Inverse document frequency (TF-IDF)* as a feature extractor. Hyperparameters of both modules were extracted using a grid search algorithm.

Fine tuning Large Language Models (LLM). Motivated by the growing popularity of the LLM, we decided to fine-tune the multilingual version of the popular *BERT* model and the 7b1 version of the *BLOOM* model for the email classification task.

Prompting with ChatGPT. Using the popular ChatGPT API, we tested 10 different prompting techniques using the models *gpt-35-turbo*. Among the different prompt techniques that we tested, we proved to ask the model to classify the email by giving a description of each category, showing the initial 200 chars of one example of each category, etc. More information on the Annex 6.

5 Results

In order to reduce the impact of the random seed selected when training a model, we have run three experiments for each setting. Table 2 shows the averaged results of the three experiments for each model and for each configuration of pre-processing heuristics applied. Regarding *ChatGPT* promtping we have only reported one run. We decided to report only the results of the prompt with category description and examples, since it obtained the average best results over all the tested prompts. However, the best results were obtained by the prompt with just category description, which obtained an F1-score of 0.716 with the raw dataset.

Due to the unbalanced nature of our test data, we decided to report the Precision, Recall, and F1-scores average weighted by support.

Despite we are comparing several sophisticated *state-of-the-art* approaches, results showed that the best-fitted model for these data is the baseline model based on *SVM* and *TF-IDF*, obtaining an F1-score > 0.85 in every evaluated dataset using only keyword information. Bigger models with attention mechanisms seem being penalized in this task. This might be encouraged by the noisy nature of the studied data.

While no strong differences were sought among datasets with the baseline model ($<3\%$ in terms of F1-score), we observed that the performance obtained by each model varied widely between the other types of trained models. Despite our efforts pre-processing our dataset to reduce the noise, and, therefore, improve the performance of the trained models, the models trained with datasets without any type of pre-processing achieved competitive results compared to the performance of the others. However, small data-cleaning operations such as F_{WEB} and F_{DUP} seemed to improve the results of every model (except in the baseline case), obtaining *Multilingual BERT* the best performance (0.831).

Other pre-processing techniques as filtering the replies (F_{ASW} and $F_{ASW_{CLS}}$), seemed to have a negative impact, reducing the obtained F1-score in almost every model trained with this dataset, maintaining its performance in the baseline and the *ChatGPT* (in the case of F_{ASW}). Surprisingly, although we prepared the $F_{ASW_{CLS}}$ dataset expecting a positive impact with the *Multilingual BERT*, it seems that the *BLOOM* model managed to exploit the *[CLS]* token in the replies, although it doesn't mean any special token for it.

Another operation that did not have any positive impact was the obtained dataset with the reordered emails (F_{SORT}), decreasing the F1-score in all the experiments. However, filtering the repetitive text without information (F_{NI}), had a really positive effect depending on which other pre-processing step you combine it with. Thus, F_{NI} dataset obtained the best F1-score with the *BLOOM model* (0.787) combined with F_{DUP} and $F_{ASW_{CLS}}$, and the best result with *ChatGPT* prompt with category description and examples combined with F_{DUP} and $F_{ASW_{CLS}}$. Finally, keeping just the first message F_{KF} also performed well as long as the order of the messages is not changed (F_{SORT}).

6 Conclusions and Future Work

In this work, we have compared the performance of several models to classify real emails in a real use-case study. The studied models include traditional approaches such as an *SVM classifier* with a *TF-IDF* as a feature extractor, *multilingual BERT* and *BLOOM* models, and a *zero-shot* approach with *ChatGPT*.

Results revealed that traditional methods still have potential in these types of noisy datasets, obtaining the best overall punctuation. Furthermore, it's interesting to notice that the average performance observed in the tested models is inversely proportional to the number of parameters that they have.

Although *ChatGPT* was the model that achieved the lower scores, it's interesting to see the potential of these models to solve tasks in a *zero-shot* way, obtaining a 0.71 F1-score on a task for which it has not previously been trained.

The other interesting point is the impact of the different data pre-processing techniques that we have applied. Despite being the smallest model, our classifier *SVM+TF-IDF* was able to deal with the noise of the data, since its performance was almost unaffected. However, the rest of the classifiers tested largely benefited from the pre-processing techniques, with techniques $F_{WEB+DUP}$ and F_{NI} having the greatest impact.

In future work, we aim to continue researching the potential of autoregressive models testing new prompts and strategies, such as asking the model to summarize the email and classify it afterwards. Furthermore, we would like to measure the performance of a custom pre-trained languange model (PLM) in our data.

Table 2 Weighted averaged F1 results obtained by model for each task

	Data pre-processing						Results	Weighted		Avg.
RANDOM	–	–	–	–	–	–	0.188	0.196	0.188	0.186
TF-IDF + SVM	–	–	–	–	–	–	0.875	0.873	0.875	0.871
	X	–	–	–	–	–	0.862	0.861	0.862	0.857
	X	X	–	–	–	–	0.862	0.860	0.862	0.853
	X	X	–	–	X	–	0.862	0.855	0.862	0.857
	X	–	–	X	–	–	**0887**	**0.888**	**0.887**	**0.886**
	X	X	–	X	–	–	0.875	0.870	0.875	0.869
	X	X	–	X	X	–	**0.887**	0.881	**0.887**	0.882
	X	X	–	–	–	X	0.875	0.868	0.875	0.868
	X	X	–	–	X	X	0.860	0.862	0.860	0.859
MultiBert	–	–	–	–	–	–	0.795	0.802	0.795	0.783
	X	–	–	–	–	–	**0.837**	**0.850**	**0.837**	**0.831**
	X	X	–	–	–	–	0.816	0.839	0.816	0.809
	X	–	X	–	–	–	0.791	0.829	0.791	0.779
	X	X	–	–	X	–	0.745	0.773	0.745	0.735
	X	–	X	–	X	–	0.770	0.773	0.770	0.748
	X	–	–	X	–	- -	0.750	0.775	0.750	0.752
	X	X	–	X	–	–	0.787	0.796	0.787	0.788
	X	–	X	X	–	–	0.804	0.797	0.804	0.791
	X	X	–	X	X	–	0.833	0.837	0.833	0.817
	X	X	–	–	–	X	0.829	0.840	0.829	0.819
	X	X	–	–	X	X	0.783	0.787	0.783	0.766
BLOOM	–	–	–	–	–	–	0.700	0.710	0.700	0.694
	X	–	–	–	–	–	0.733	0.725	0.733	0.721
	X	X	–	–	–	–	0.691	0.673	0.691	0.677
	X	–	X	–	–	–	0.770	0.774	0.770	0.763
	X	X	–	–	X	–	0.445	0.510	0.445	0.445
	X	–	X	–	X	–	0.487	0.530	0.487	0.484
	X	–	–	X	–	–	0.766	0.778	0.766	0.764
	X	X	–	X	–	–	0.716	0.702	0.716	0.693
	X	–	X	X	–	–	**0.795**	**0.790**	**0.795**	**0.787**
	X	X	–	X	X	–	0.633	0.620	0.633	0.620
	X	X	–	–	–	X	0.729	0.741	0.729	0.723
	X	X	–	–	X	X	0.512	0.558	0.512	0.510
ChatGPT Prompting	–	–	–	–	–	–	0.562	0.798	0.562	0.616
	X	–	–	–	–	–	0.602	0.765	0.602	0.641
	X	X	–	–	–	–	0.614	0.803	0.614	0.643
	X	–	X	–	–	–	0.554	0.730	0.554	0.586
	X	X	–	–	X	–	0.586	0.742	0.586	0.618
	X	–	X	–	X	–	0.532	0.713	0.532	0.581
	X	–	–	X	–	–	0.544	0.715	0.544	0.570
	X	X	–	X	–	–	**0.636**	0.794	**0.636**	0.655
	X	–	X	X	–	–	0.603	0.826	0.603	**0.659**
	X	X	–	X	X	–	0.584	0.743	0.584	0.626
	X	X	–	–	–	X	0,613	**0.835**	0.613	0.638
	X	X	–	–	X	X	0.600	0.727	0.600	0.607

Acknowledgements This work has been partially funded by the Basque Business Development Agency, SPRI, under Grant Agreement ZL-2023/00346. We wish to thank LIS Data Solutions for providing the dataset and insightful comments during all the experimentation. We gratefully acknowledge Aitor García-Pablos for his help on applying *BLOOM* to the task.

Annexes

Annex A. Tested Prompts

In this project, we have tested several types of prompts, encompassing two types of approaches: (1) explaining the problem to the model in the prompt, and (2) modifying the role of the model. Among the techniques, we tested adding a description of each category separated by bullet points. Other designed prompts tried to ease the work by adding an example by category; however, due to the length of our examples, we had to trim each example to 200 characters, since we may exceed the max length supported by *ChatGPT*.

Although we tested 10 different prompts, we decided to include only the most naive ones as examples of each type, and the most complete prompts to avoid making this appendix too long. As our data is mainly written in Spanish language, all tested prompts are in Spanish. To make the prompts readable for non-Spanish speakers, we have tried to translate each paragraph of the prompt, showing the translation written in grayscale in each paragraph.

Next, we will proceed to expose some of the tested prompts, by replacing the subjects and bodies of the email conversations with *[SUBJECT]* and *[BODY]* placeholders respectively. In these examples, we have used some other placeholders in order to reduce the length of this annex. Thus, we have trimmed the category descriptions by adding "*[...]*" symbol after a few words, and we have not enumerated each description or example, replacing them with *"[... MORE EXAMPLES ...]"* and *"[... MORE DESCRIPTIONS ...]"* symbols.

Explaining the problem in the prompt

- **Prompt 1.** Basic prompt

> A continuación te voy a proporcionar un hilo de emails anonimizados. Esta anonimización ha ocultado toda la información sensible y la ha sustituido por asteriscos (*). / Below I am going to provide you with a thread of anonymized emails. This anonymization has hidden all sensitive information and replaced it with asterisks (*).
> Tu trabajo es determinar a qué categoría de la siguiente lista pertenece el siguiente hilo de emails anonimizados. / Your job is to determine to which category from the following list the following anonymised email thread belongs.
> Lista de categorías: administracion, operaciones, recogidas, incidencias, comercial. / List of categories: administrative, operations, collections, issues, commercial.
> A continuación, te expongo el asunto del hilo de emails que debes con clasificar entre tres comillas simples, y el hilo con todos los mensajes intercambiados entre todas las personas involucradas entre tres comillas dobles. / Here is the subject of the email thread that you have to classify between three single inverted commas, and the thread with all the messages exchanged between all the people involved between three double inverted commas.
> Dame tu respuesta utilizando un máximo de una palabra, mencionando única y exclusivamente la categoría a la que pertenece el email. / Give me your answer using a maximum of one word, mentioning only and exclusively the category to which the email belongs.
> Asunto del correo: [SUBJECT] / Subject of the email: [SUBJECT]
> Hilo de emails: [BODY] / Thread of emails: [BODY]

– **Prompt 2. Prompt** with category description and examples

A continuación te voy a proporcionar un hilo de emails anonimizados. Esta anonimización ha ocultado toda la información sensible y la ha sustituido por asteriscos (*). / Below I am going to provide you with a thread of anonymized emails. This anonymization has hidden all sensitive information and replaced it with asterisks (*).
Tu trabajo es determinar a qué categoría de la siguiente lista pertenece el siguiente hilo de emails anonimizados. A continuación, te expongo la lista de categorías en la que te proporciono el nombre de la categoría y su descripción. / Your job is to determine which category from the following list of anonymised email thread belongs. Below is the list of categories in which I provide you with the name of the category and its description.
- administracion: Correos relacionados *[...]* / - administrative: emails related *[...]*
[... MORE DESCRIPTIONS ...]
- comercial: Correos dedicados a *[...]* / - commercial: emails dedicated to *[...]*
A continuación te expongo un ejemplo de cada una de las categorías, exponiéndote el asunto del correo, y los 200 primeros caracteres del hilo de emails: / Below is an example of each of the categories, showing you the subject of the email, and the first 200 characters of the email thread:
- Ejemplo categoría administracion: / - Administrative example
Asunto del correo: [SUBJECT] / Subject of the email: [SUBJECT]
200 primeros caracteres del hilo de emails: [BODY] / 200 first characters of the email thread: [BODY]
[... MORE EXAMPLES ...]
- Ejemplo categoría comercial: / - Commertial example
Asunto del correo: [SUBJECT] / Subject of the email: [SUBJECT]
200 primeros caracteres del hilo de emails: [BODY] / 200 first characters of the email thread: [BODY]
Dame tu respuesta utilizando un máximo de una palabra, mencionando únicamente la categoría a la que pertenece el email que te proporciono a continuación / Give me your answer using a maximum of one word, mentioning only the category to which the email I provide below belongs.
Asunto del correo: [SUBJECT] / Subject of the email: [SUBJECT]
Hilo de emails: [BODY] / Subject of the email: [BODY]

Modifiying the role

– **Prompt 1. Prompt** modifying the role

Role
Ya no eres un chat. Eres un clasificador de emails, de los cuales se te van a facilitar el asunto del correo, y el cuerpo del email anonimizado. Este cuerpo contiene todo el hilo de intercambiado en cada respuesta del mensaje, con toda la información sensible ocultada mediante el uso de asteriscos (*). Tu trabajo como clasificador es única y exclusivamente discriminar sin ninguna clase de explicación si el email pertenece a una de las siguientes categorías: administracion, operaciones, recogidas, incidencias, comercial. / You are no longer a chat. You are an email classifier, of which you will be provided with the subject of the email, and the anonymised email body. This body contains the entire conversation exchanged in each message reply, with all sensitive information hidden through the use of asterisks (*). Your job as a classifier is solely and exclusively to discriminate without any kind of explanation whether the email belongs to one of the following categories: administration, operations, collections, incidents, commercial.
Responde mencionando únicamente el nombre de la categoría / Answer by mentioning only the name of the category
Prompt
Asunto: [SUBJECT] / Subject of the email: [SUBJECT]
Correo: [BODY] / Subject of the email: [BODY]

- **Prompt 2. Prompt** modifying the role describing the categories and showing examples in different prompts

Role
Ya no eres un chat. Eres un clasificador de emails, de los cuales se te van a facilitar el asunto del correo, y el cuerpo del email anonimizado. Este cuerpo contiene todo el hilo de mensajes intercambiados, con toda la información sensible ocultada mediante el uso de asteriscos (*). Tu trabajo como clasificador es única y exclusivamente discriminar sin ninguna clase de explicación si el email pertenece a una de las siguientes categorías de las cuales te proporciono el nombre y su descripción: / You are no longer a chat. You are an email classifier, of which you will be provided with the subject of the email, and the anonymised email body. This body contains the entire conversation exchanged in each message reply, with all sensitive information hidden by the use of asterisks (*). Your job as a classifier is solely and exclusively to discriminate without any kind of explanation whether the email belongs to one of the following categories of which I provide you with the name and description:
- administracion: Correos relacionados *[...]* / - administrative: emails related *[...]*
  [... MORE DESCRIPTIONS ...]
- comercial: Correos dedicados a *[...]* / - commercial: emails dedicated to *[...]*

Responde únicamente mencionando la categoría a la que pertenece. / Answer only by mentioning the category to which the email belongs.
En los siguientes mensajes, le expondré un ejemplo de cada categoría de correo. De cada correo, le daré el asunto del correo y los 200 primeros caracteres del hilo de emails: / In the following messages, I will give you an example of each email category. For each email, I will give you the subject of the email and the first 200 characters of the email thread:
Prompt 1
- Ejemplo categoría administracion: / - Administrative example

Asunto del correo: [SUBJECT] / Subject of the email: [SUBJECT]
200 primeros caracteres del hilo de emails: [BODY] / 200 first characters of the email thread: [BODY]
Prompt 2–4
[... MORE EXAMPLES ...]
Prompt 5
- Ejemplo categoría comercial: / - Commertial example:

Asunto del correo: [SUBJECT] / Subject of the email: [SUBJECT]
200 primeros caracteres del hilo de emails: [BODY] / 200 first characters of the email thread: [BODY]
Prompt 6
Con esta información, clasifica el siguiente email: / With this information, classify the following email:
Asunto: [SUBJECT] / Subject: [SUBJECT]
Correo: [BODY] / Email: [BODY]

References

1. Antonellis I, Bouras C, Poulopoulos V (2006). Personalized news categorization through scalable text classification. In: Zhou X, Li J, Shen HT, Kitsuregawa M, Zhang Y (eds) Frontiers of WWW research and development–APWeb 2006, Berlin, Heidelberg, Springer, pp 391–401
2. Badjatiya P, Gupta S, Gupta M, Varma V (2017) Deep learning for hate speech detection in Tweets. In: Proceedings of the 26th international conference on World Wide web companion, WWW '17 Companion, Republic and Canton of Geneva, CHE, 2017. International World Wide Web conferences steering committee, pp 759–760
3. Bojanowski Piotr, Grave Edouard, Joulin Armand, Mikolov Tomas (2017) Enriching word vectors with subword information. Trans Assoc Comput Linguistics 5:135–146
4. Brown T, Mann B, Ryder N, Subbiah M, Kaplan JD, Dhariwal P, Neelakantan A, Shyam P, Sastry G, Askell A et al. (2020) Language models are few-shot learners. Adv Neural Inform Process Systems, 33:1877–1901
5. Devlin J, Chang M-W , Lee K, Toutanova K (2018). BERT: pre-training of deep bidirectional transformers for language understanding. CoRR. abs/1810.04805
6. Fortuna P, Nunes S (2018) A survey on automatic detection of hate seech in text. ACM Comput Surv 51(4)
7. Jin P, Zhang Y, Chen X, Xia Y (2016) Bag-of-embeddings for text cassification. In: Proceedings of the twenty-fifth international joint conference on artificial intelligence, IJCAI'16, AAAI Press, pp 2824–2830

8. Jing L-P, Huang H-K, Shi H-B (2002) Improved feature selection approach TFIDF in text mining. In: Proceedings. international conference on machine learning and cybernetics, vol 2, pp 944–946
9. Joulin A, Grave E, Bojanowski P, Mikolov T (2017) Bag of tricks for efficient text classification. In: Lapata M, Blunsom P, Koller A (eds) Proceedings of the 15th conference of the European chapter of the association for computational linguistics: Volume 2, Short Papers, Valencia, Spain, April 2017. Association for Computational Linguistics, pp 427–431
10. Kim S-B, Han K-S, Rim H-C, Myaeng SH (2006) Some effective techniques for Naive Bayes text classification. IEEE Trans Knowledge and Data Eng 18(11):1457–1466
11. Kim Y (2014) Convolutional neural networks for sentence classification. In: Moschitti A, Pang B, Daelemans W (eds) Proceedings of the 2014 conference on empirical methods in natural language processing (EMNLP), Doha, Qatar, October 2014. Association for Computational Linguistics, pp 1746–1751
12. Liu Y, Ott M, Goyal N, Du J, Joshi M, Chen D, Levy O, Lewis M, Zettlemoyer L, Stoyanov V (2019) Roberta: a robustly optimized bert pretraining approach. arXiv preprint. arXiv:1907.11692
13. Luo H (2023) Prompt-learning and zero-shot text classification with domain-specific textual data. Master's thesis, Uppsala University, Department of Linguistics and Philology
14. Meena J, Chandran KR (2010) Naïve Bayes text classification with positive features selected by statistical method. pp 28–33
15. Mikolov T, Chen K, Corrado G, Dean J (2013) Efficient estimation of word representations in vector space. arXiv preprint. arXiv:1301.3781
16. Minaee S, Kalchbrenner N, Cambria E, Nikzad N, Chenaghlu M, Gao J (2021) Deep learning–based text classification: a comprehensive review. ACM Comput Surv 54(3)
17. OpenAI. GPT-4 Technical Report (2023)
18. Pennington J, Socher R, Manning CD (2014) GloVe: global vectors for word representation. In: Empirical methods in natural language processing (EMNLP), pp 1532–1543
19. Porter MF (1980) An algorithm for suffix stripping. Program 14(3):130–137
20. Radford A, Narasimhan K, Salimans T, Sutskever I (2018) Improving language understanding by generative pre-training
21. Alec R, Wu J, Rewon C, David L, Dario A, Ilya S et al (2019) Language models are unsupervised multitask learners. OpenAI blog 1(8):9
22. Rahman MA, Akter YA (2019) Topic classification from text using decision tree, K-NN and multinomial Naïve Bayes. In: 2019 1st international conference on advances in science, engineering and robotics technology (ICASERT), pp 1–4
23. Raza M, Jayasinghe ND, Muslam MMA (2021) A comprehensive review on email spam classification using machine learning algorithms. In: 2021 International conference on information networking (ICOIN), pp 327–332
24. Rujiang B, Junhua L (2009) A novel conception based texts classification method. In: 2009 International e-Conference on advanced science and technology, pp 30–34
25. Scao TL, Fan A, Akiki C, Pavlick E, Ilic S, Hesslow D, Castagné R, Luccioni AS, Yvon F, Gallé M, Tow J, Rush AM, Biderman S, Webson A, Ammanamanchi PS, Wang T, Sagot B, Muennighoff N, del Moral AV, Ruwase O, Bawden R, Bekman S, McMillan-Major A, Beltagy I, Nguyen H, Saulnier L, Tan S, Suarez PO, Sanh V, Laurençon H, Jernite Y, Launay J, Mitchell M, Raffel C, Gokaslan A, Simhi A, Soroa A, Aji AF, Alfassy A, Rogers A, Nitzav AK, Xu C, Mou C, Emezue C, Klamm C, Leong C, van Strien D, Adelani DI et al. BLOOM: a 176B-parameter open-access multilingual language model. CoRR, abs/2211.05100
26. Shen D, Wang G, Wang W, Min MR, Su Q, Zhang Y, Li C, Henao R, Carin L (2018) Baseline needs more Love: on simple word-embedding-based models and associated pooling mechanisms. In: Gurevych I, Miyao Y (eds) Proceedings of the 56th annual meeting of the association for computational linguistics (Volume 1: Long Papers), Melbourne, Australia, July 2018, Association for Computational Linguistics, pp 440–450
27. Vaswani A, Shazeer N, Parmar N, Uszkoreit J, Jones L, Gomez AN, Kaiser Ł, Polosukhin I (2017) Attention is all you need. In: Advances in neural information processing systems, pp 30

28. Wang J, Wang Z, Zhang D, Yan J (2017) Combining knowledge with deep convolutional neural networks for short text classification. In: Proceedings of the twenty-sixth international joint conference on artificial intelligence, IJCAI-17, pp 2915–2921
29. Zhan W, Yifan H, Minghu J (2006) A comparison among three neural networks for text classification 3:02
30. Wang Z-Q, Sun X, Zhang D-X, Li X (2006) An optimal SVM-based text classification algorithm. In: 2006 International conference on machine learning and cybernetics, IEEE, pp 1378–1381
31. Yadollahi A, Shahraki AG, Zaiane OR (2017) Current state of text sentiment analysis from opinion to emotion mining. ACM Comput Surv 50(2)
32. Yan Y, Li W, Chen G, Liu W (2021) An improved text classification method based on convolutional neural networks. In: Proceedings of the 2020 1st international conference on control, robotics and intelligent system, CCRIS '20, New York, NY, USA, Association for Computing Machinery, pp 185-190
33. Zhang M, Zhang D-x (209) Trained SVMs based rules extraction method for text classification. pp 16–19
34. Zhang W, Yoshida T, Tang X (2007) Text classification using multi-word features. pp 3519–3524
35. Zhang W, Yoshida T, Tang X (2008) TFIDF, LSI and multi-word in information retrieval and text categorization. In: 2008 IEEE International conference on systems, man and cybernetics, pp 108–113

Blockchain Technologies for Transparency in FinTech

Khawla Bouafia, Bálint Molnár, and Gurbanli Majid

Abstract Blockchain technology has the potential to improve transparency in the financial industry, particularly in accounting, auditing, and supply chain management within FinTech. The paper presents a case study of a blockchain-based financial platform, examining its impact on transparency and accountability. It explores how blockchain solutions could enhance transparency in payment, loans, and insurance industries. The paper discusses the implications, advantages, and disadvantages of implementing blockchain-based systems for FinTech transparency, proposing suggestions for further study and legislative changes to support adoption. The goal is to investigate the use of blockchain technology to boost transparency in FinTech, focusing on its advantages and challenges in implementing blockchain for financial processes, including accounting, auditing, and supply chain management.

Keywords Blockchain · FinTech · Financial services · Transparency · Cross-border

1 Introduction

The rise of FinTech, defined as the integration of technology into financial services [1], has had a profound impact on both society and the economy, broadening access to financial services. This evolution relies on various technologies, including cloud computing, blockchain, and artificial intelligence.

K. Bouafia (✉) · B. Molnár · G. Majid
Eötvös Loránd University of Budapest, Pázmány Pter Setány 1/C, 1117 Budapest, Hungary
e-mail: bouafia@inf.elte.hu

B. Molnár
e-mail: molnarba@inf.elte.hu

G. Majid
e-mail: jlx7y9@inf.elte.hu

X.-S. Yang et al. (eds.), *Proceedings of Ninth International Congress on Information and Communication Technology*, Lecture Notes in Networks and Systems 1000,
https://doi.org/10.1007/978-981-97-3289-0_46

Traditional banks, grappling with the rapid evolution of the Internet since the 1990s, experienced profit losses to Internet-based financial entities like Paypal and Apple Pay. Now, a new challenge has emerged in the form of blockchain technology, poised to revolutionize not only financial sectors but nearly every industry. Also referred to as distributed ledger technology (DLT), blockchain involves a shared ledger among all participants, providing real-time transparency and cost reductions attractive to diverse industries, including manufacturing. While blockchain's initial exposure came through bitcoin, the first cryptocurrency, its true potential extends far beyond. This paper emphasizes the broader applications of blockchain technology across various industries, hinting at its transformative potential.

The study initiates with a literature review summarizing crucial findings and identifying gaps in understanding. It then briefly outlines the fundamental concepts and features of blockchain technology, emphasizing its relevance to the financial industry and its potential to enhance accountability and transparency. The subsequent section examines the current state of FinTech, highlighting the role of transparency in building stakeholder trust. Additionally, it delves into the challenges hindering transparency in the financial sector, such as concerns related to data privacy, cybersecurity, and legal compliance.

2 Background

The literature review aims to outline existing knowledge on the research field and presents the milestones of constituting research directions, i.e., to demonstrate how far advanced finding researchers in the same research topic and what they did in the same domain and allows the researcher to identify if there is any gap in it. Our specific focus is on elucidating the term "Blockchain" and exploring fundamental concepts associated with transparency in FinTech.

2.1 FinTech

In the FinTech sector, various players [2] contribute to innovation and intensify competition, ultimately fostering economic growth and enhancing customer satisfaction. FinTech encompasses diverse categories, including:

- **Digital Payments and Banking**: By exploiting international technical breakthroughs, digital payments and banking [3] are intended to increase the convenience and effectiveness of financial transactions. Beyond traditional electronic banking, contemporary digital banking encompasses various forms such as online banking, mobile banking, and the use of electronic cards for payments, among others.

- **Investments and Capital Markets**: The capital market [4] serves as a platform where buyers and sellers engage in the trading of various securities, including stocks, bonds, and other financial instruments. These markets connect those who have funds to invest, such as banks and investors, with those looking to raise capital, such as businesses, governments, and individuals. Securities traded in capital markets encompass a broad range, including stocks, bonds, derivatives, commodities, currencies, and other financial assets.
- **Lending and Borrowing**: Refers to leverage digital channels for efficient loan management and credit decisions, enhancing service speed. This sector empowers financial institutions to optimize production, increase lending income, and offer convenient services through global, Internet-connected platforms for diverse loan products.
- **Crowdfunding**: Serves as a method for individuals or organizations to raise funds by garnering small contributions from many people, usually through the Internet. One notable advantage of crowdfunding [5] is its ability to enable creators to bypass traditional gatekeepers like venture capitalists or banks. Instead, they can directly approach their audience to secure funding. This democratization of funding has given rise to a diverse range of crowdfunding platforms, including Kickstarter and Indiegogo, tailored to various project types.
- **Loyalty Programs**: Incentivize customer retention by offering rewards [6] and incentives for repeat business, allowing individuals to earn and redeem points for discounts or other benefits in various industries.
- **Law and Regulation**: In FinTech leverage technology for monitoring and compliance with financial regulations [7], addressing potential exploitation risks to end-consumers. RegTech, an emerging field, shifts from traditional KYC to advanced KYD approaches, enhancing regulatory practices in the fast-paced landscape of financial services.

Figure shows the various Fintech EcoSystems (see Fintech1) (Fig. 1).

Fig. 1 Fintech ecosystem

2.2 Blockchain History and Definition

The core ideas behind blockchain technology originated in the late 1980s and early 1990s, with the potential to become a foundational element in global record-keeping systems. However, it was officially launched just a decade ago, developed by anonymous individuals, known as Satoshi Nakamoto, in tandem with the establishment of the bitcoin cryptocurrency blockchain network in 2009 [8, 9]. This permissionless blockchain network allowed anonymous account creation and participation.

Considering various references [10, 11], there are diverse perspectives defining the concept of blockchain. In [12], blockchain is described as: a digital ledger forming a chain of individual data blocks. With new data regularly added, a new block is formed and added to the chain. This necessitates that all nodes update their blockchain ledger versions to be equivalent. The method by which these new blocks are generated is critical to understanding why blockchain is regarded as very secure. Before a new block can be added to the ledger, a majority of nodes must check and certify the authenticity of the new data. They can use a cryptocurrency to verify that new transactions in a block were not fraudulent or that coins were not released multiple times. This differs from standalone databases or spreadsheets where changes can be made without supervision.

Another definition from [13] characterizes blockchain as a chronologically connected data structure combining transactions into blocks. These interconnected blocks create a distributed ledger, resistant to falsification or forgery through cryptographic hashes. Blockchain technology stores, verifies, and updates data using distributed consensus algorithms of nodes, employing cryptographic algorithms for secure data transmission and access. Addressing problems in traditional methods, blockchains offer solutions as discussed in the next section.

2.3 Blockchain Technologies

Blockchain technology has its own main properties such as security, transparency, decentralization, immutability, and programmability. Blockchain is a distributed ledger that securely stores transactions in blocks through a peer-to-peer process, eliminating third-party involvement. Each transaction, safeguarded by the owner's signature, is traceable throughout its entire path. Notably, blockchain's key feature is decentralization, ensuring it operates without control from any single entity or organization. The classification of blockchain types is also mentioned as:

1. **Public Blockchain**: Is a decentralized, open-source network allowing universal participation, transaction validation, and block addition by anyone. Typically maintained by a community or DAO [14], it offers transparency with open-source code access, fostering development opportunities as Bitcoin, Ethereum, and Litecoin have drawn developers to create decentralized applications (dApps).

2. **Private Blockchain**: Also termed permissioned [15] is restricted to specific participants and not open to the public. Controlled by a central authority, access is typically granted based on participant identity and network administrator rules. Suited for companies, it provides control over data, making it ideal for internal business processes like supply chain management. Some choose private blockchains for regulatory compliance, data privacy, and on-premise data storage.
3. **Consortium Blockchain**: Is shared by a group of organizations or entities, is more flexible than a fully private blockchain, and suitable for collaborative efforts among a defined group.
4. **Hybrid Blockchain**: Combines elements of both public and private blockchains. It allows for different levels of access and visibility. This type can be useful for businesses that require both transparency and control.

These categories highlight the diversity of blockchain applications, catering to various needs and preferences in different industries and use cases.

3 Blockchain in Fintech and Related Works

In this section, we explore the potential that blockchain technology [16] can bring to financial technology solutions, along with a review of relevant works in this domain.

3.1 How Blockchain Transforms Fintech Industry

Blockchain's pivotal advantage in FinTech lies in its establishment of a decentralized and tamper-proof ledger for recording financial transactions. This bolsters transparency and security, as all parties have access to the same information and any changes to the ledger are easily traceable. Blockchain's impact extends to efficiency improvements by eliminating intermediaries like banks. The integration of smart contracts, self-executing agreements coded for automation, streamlines financial processes, reducing time and costs associated with conventional transactions.

Furthermore, blockchain technology catalyzes the development of decentralized finance (DeFi) applications. These applications empower individuals to access financial services without reliance on traditional intermediaries, fostering financial inclusion and creating new opportunities, especially in developing countries.

In essence, blockchain has the transformative potential to elevate the FinTech industry through heightened transparency, security, and efficiency in financial transactions, while simultaneously paving a way to innovative decentralized financial applications (Fig. 2).

Blockchain in FinTech is applied to improve transparency, security, efficiency, and accessibility in financial processes and services. Here are some key applications:

- Decentralized Ledger for Transactions.

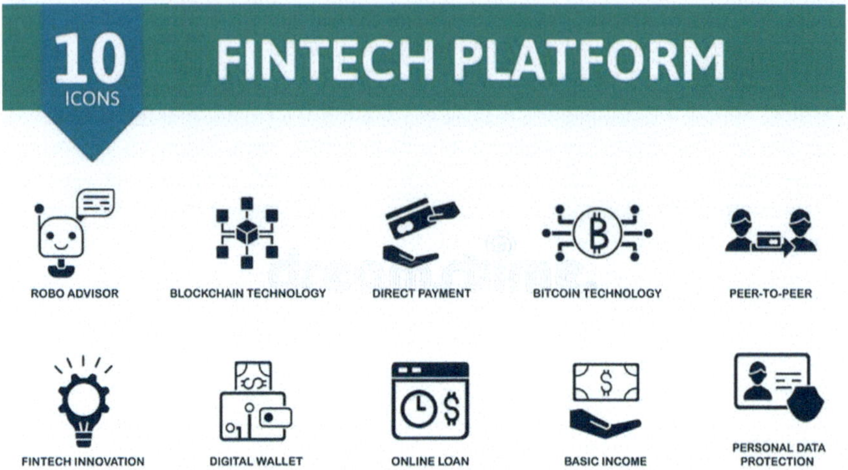

Fig. 2 FinTech platforms

- Efficiency through Smart Contracts.
- Decentralized Finance (DeFi) Applications.
- Cross-Border Payments.
- Security and Fraud Prevention.
- Tokenization of Assets.
- Know Your Customer (KYC) and Anti-Money Laundering (AML) Compliance
- Supply Chain Finance.

3.2 *What is Transparency?*

Transparency in Fintech refers to customers' ability to easily comprehend and access information about the financial products and services they utilize. This encompasses details about fees, terms, and conditions, as well as the risks and benefits associated with various products. Transparency should extend to the internal operations of FinTech companies, including their algorithms and data practices. The primary objective of transparency is to empower customers to make informed decisions about their finances, fostering trust in the FinTech industry. It is crucial, as a lack of transparency in sectors such as trade finance or cross-border payments can hinder customers' ability to make well-informed financial choices.

One effective means of enhancing transparency in FinTech is through the utilization of blockchain technology. This technology facilitates the recording of transactions in a manner immune to manipulation by any single party. It streamlines auditing and regulatory requirements by enabling real-time tracking of the money flow.

Transparency is a foundational aspect of blockchain technology. In this context, it denotes the openness and accountability of information recorded on a public ledger accessible to anyone. Maintained by a decentralized network of computers, this ledger ensures the accuracy of information, which cannot be altered or deleted.

An integral benefit of transparency in the blockchain is heightened security. The decentralized nature of the network makes it challenging for hackers to access sensitive information or manipulate the system, thereby preventing cyberattacks and fraud. The public ledger enables the detection and prevention of malicious activity, ensuring a higher level of security for all users.

In conclusion, transparency is a pivotal feature of blockchain technology, offering numerous benefits. Its decentralized nature enhances trust and security in the digital realm, with the potential to revolutionize various industries. The transparency of the blockchain ledger guarantees the accuracy and security of transactions, mitigating the risk of fraud and corruption. The advantages of transparency in blockchain technology make it a worthwhile exploration for those seeking to enhance efficiency, reduce costs, and bolster the security of their operations.

Privacy denotes the state of being free from unwanted or public attention, encompassing the ability to control access to personal information. It also includes the right to make decisions about how personal information is collected, used, and shared.

Security in financial technology (FinTech) pertains to the measures and techniques employed to safeguard financial transactions and personal information from unauthorized access, use, disclosure, disruption, modification, or destruction.

3.3 Related Works

Table 1 gives a summary of various references related to our topic.

In conclusion, the integration of blockchain technology in FinTech holds the potential to enhance transparency and trust in financial transactions. Blockchain enables secure and transparent record-keeping, thereby improving the transparency of financial transactions and reducing the risk of fraud. Additionally, the decentralized nature of blockchain can also increase transparency by allowing for more direct access to financial data for all parties involved. While there are still challenges to be addressed, such as scalability and regulatory compliance, the use of blockchain in FinTech has the potential to revolutionize the financial industry. Despite existing challenges such as scalability and regulatory compliance, the use of blockchain in FinTech has the potential to revolutionize the financial industry.

4 Cross-Border Funds Transfer (Methodology and Tools)

This section explores the methodologies and tools used in development and research, offering detailed insights into data collection and analysis processes. Additionally, it highlights the ongoing evolution of cross-border payment systems, emphasizing efforts to enhance efficiency, security, and interoperability. Examples include the

Table 1 Related works

References	Application	Comments
[17]	Taxonomy of FinTech	Helps to demonstrate the balance of risk and reward in the FinTech industry
[18]	Challenges in FinTech	Takes into account the rapid development of FinTech initiatives driven by the sharing economy, regulations, and information technology, and the service
[19]	Blockchain technology and transparency in the FinTech industry	This paper discussed well the immutability and transparency
[20]	Secure and transparent crowdfunding using blockchain	Makes payment simpler for business owners to connect with a big global audience that would be willing to support their endeavors
[21]	Mobile payment in FinTech environment	Simple online and offline payments grown, especially in light of the fast expanding online market and availability of mobile devices
[21]	Security and privacy issues: A survey on FinTech	Results used to develop the security and privacy aspects of FinTech's theoretical framework
[22]	A model of trust in FinTech	The two models are then contrasted to see if there is a similar level of trust in each in order to determine whether the model is equally valid for Fintech and Insurtech
[23]	Blockchain solutions for authorization and authentication	Explore blockchain technology's potential for authorization and authentication, rapidly identify operational and security risks

adoption of technologies like blockchain and digital identification, compliance with international standards and laws, and collaborative initiatives among stakeholders in the payments ecosystem, encompassing various types of cross-border payment (Fig. 3).

While developing our application, accomplishing various functionalities requires different tools: JavaScript, React JS, Node JS, Digital Signature, and Multithreading. For structuring our concept, we incorporated several blockchain principles:

Fig. 3 Types of cross-border payments

1. **Block**: Is a digital ledger entry in blockchain technology that records a collection of transactions, we used block header, transactions and Markle tree
2. **Chain:** Comprises interconnected blocks, forming a continuous, irreversible record of transactions in blockchain technology. It appeals to sectors wishing to build safe and open systems.
3. **Wallet**: Is a digital instrument utilized in blockchain technology for storing, managing, and interacting with cryptocurrencies and other blockchain assets.
4. **Cryptocurrency transactions**: Involves the transfer of digital assets from one participant to another within a blockchain network.
5. **Transparency**: In the current project, transparency is ensured through signed and verified transactions using public key cryptography on the blockchain. These transactions are broadcasted across a peer-to-peer network, and a consensus mechanism guarantees universal access to an agreed-upon ledger state. To create a transparent audit trail, each transaction is timestamped and linked to the preceding one through a cryptographic hash. The project emphasizes the importance of transparency in blockchain technology, utilizing mechanisms like public key cryptography, peer-to-peer networking, a consensus method, and a transaction audit trail.

For the front-end structure, a block component which is crucial element of my cryptocurrency's front-end provides users with a comprehensive view of a specific block on the blockchain, its structure is the cornerstone of the design concept behind React.js. Small, reusable pieces of code called react components can be used to create more intricate user interfaces. This method of creating user interfaces places a strong emphasis on modularization and code reuse, which makes it simpler to scale and manage complicated programs.

Each component in React represents a different section of the user interface, and the components are arranged into a hierarchical tree structure. The top component in the tree, sometimes known as the "root" component, is in charge of rendering all the child components. There are two categories of actual components: functional components and class components. Simple functions, known as functional components, produce JavaScript XML (JSX) and accept props as input. On the other hand, class components are more complex and can include lifecycle methods, state, and props, in addition to JSX.

Through props, React also provides a method for transferring data across components. Props can be used to customize each component's behavior and appearance. They are passed down from parent components to child components. The idea of "composition" is a key component of the React component structure. To create more complex components and user interfaces, it is possible to mix tiny, reusable components. This method makes it simpler to reuse code and maintain the application's design uniform.

5 Conclusion

The topic of transparency in the FinTech sector has been examined in this paper, along with its significance in fostering confidence among stakeholders. Additionally, the study evaluated current blockchain-based FinTech solutions and looked at how blockchain technology might improve transparency in the FinTech industry.

Based on the results, blockchain technology, which offers immutable, auditable, and secure records of transactions, may considerably improve transparency in the FinTech industry. Customers, investors, and regulators can all benefit from increased trust and confidence due to blockchain technology's openness. However, despite blockchain technology having a lot of potential advantages, it also has drawbacks that need to be addressed. Scalability is one such restriction, as the existing blockchain infrastructure may not be capable of supporting extensive finance applications.

In terms of blockchain and FinTech, there is still a lot of study to be done. Future studies may examine how blockchain technology might increase accountability and transparency in other financial sectors like insurance and asset management. Future studies might also look into how blockchain-based FinTech products can encourage sustainability and social responsibility in the financial industry.

Acknowledgements This research was supported by the National Research, Development and Innovation Fund of Hungary, financed under the Thematic Excellence Programme TKP2021-NVA-29 (National Challenges Subprogramme) funding scheme, and by the COST Action CA19130—"Fintech and Artificial Intelligence in Finance Toward a transparent financial industry" (FinAI).

References

1. Pal A, Tiwari CK, Behl A (2021) Blockchain technology in financial services: comprehensive review. J Global Operat Strategic Sourc 14(1):61–80
2. Campanella F, Serino L, Battisti E, Giakoumelou A, Karasamani I (2023) (competence human capital reality?. J Business Res 155:113376.2023). FinTech in the financial system: towards a capital-intensive and high
3. Dai J, Miedema J, Hernandez S, Sutton-Lalani A, Moffatt K (2023) Cognitive accessibility of digital payments: a literature review. In: Proceedings of the 20th international web for all conference, April, pp 116–121
4. Zhang R, Fu W, Lu T (2023) Capital market opening and corporate environmental performance: empirical evidence from China. Financ Res Lett 53:103587
5. Mora-Cruz A, Palos-Sanchez PR (2023) Crowdfunding platforms: a systematic literature review and a bibliometric analysis. Int Entrepreneurship and Managem J 1–32
6. Meyer-Waarden L, Bruwer J, Galan JP (2023) Loyalty programs, loyalty s from three industries. J Retail Consum Serv 71:103212
7. Sittig DF, Singh H (2015) A new socio-technical model for studying health information technology in complex adaptive healthcare systems. In: Cognitive informatics for biomedicine: human computer interaction in healthcare, pp 59–80
8. Nakamoto S (2008) Bitcoin: a peer-to-peer electronic cash system. Decentralized business review

9. Monrat AA, Schelén O, Andersson K (2019) A survey of blockchain from the perspectives of applications, challenges, and opportunities. IEEE Access 7:117134–117151
10. Gadekallu TR, Huynh-The T, Wang W, Yenduri G, Ranaweera P, Pham QV, Liyanage M (2022) Blockchainforthemetaverse: a review. arXiv preprint arXiv:2203.09738
11. Marfiuc AM (2022) Blockchain technologies
12. Narayanan A, Bonneau J, Felten E, Miller A, Goldfeder S (2016) Bitcoin and cryptocurrency technologies: a comprehensive introduction. Princeton University Press
13. Rahman MS, Islam MA, Uddin MA, Stea G (2022) A survey of blockchain-based IoT eHealthcare: applications, research issues, and challenges. Internet of Things 19:100551
14. Benhamouda F, Gentry C, Gorbunov S, Halevi S, Krawczyk H, Lin C, ... Reyzin L (2020) Can a public blockchain keep a secret?. In: Theory of cryptography: 18th international conference, TCC 2020, Durham, NC, USA, November 16–19, 2020, Proceedings, Part I, vol 18. Springer International Publishing, pp 260–290
15. Pahlajani S, Kshirsagar A, Pachghare V (2019) Survey on private blockchain consensus algorithms. In: 2019 1st international conference on innovations in information and communication technology (ICIICT), April, IEEE, pp 1–6
16. Yaga D, Mell P, Roby N, Scarfone K (2019) Blockchain technology overview. arXiv preprint arXiv:1906.11078
17. Imerman MB, Fabozzi FJ (2020) Cashing in on innovation: a taxonomy of fintech. J Asset Managem 21(3):167–177
18. Suryono RR, Budi I, Purwandari B (2020) Challenges and trends of financial technology (Fintech): a systematic literature review. Information 11(12):590
19. Caro MP, Ali MS, Vecchio M, Giaffreda R (2018) Blockchain-based traceability in Agri-Food supply chain management: a practical implementation. In: 2018 IoT vertical and topical summit on agriculture-tuscany (IOT Tuscany), May, IEEE, pp 1–4
20. Khatter H, Chauhan H, Trivedi I, Agarwal J (2021) Secure and transparent crowdfunding using blockchain. In: 2021 international conference on recent trends on electronics, information, communication and technology (RTEICT), August, IEEE, pp 76–80
21. Kang J (2018) Mobile payment in Fintech environment: trends, security challenges, and services. HCIS 8(1):1–16
22. Gai K, Qiu M, Sun X, Zhao H (2017) Security and privacy issues: a survey on FinTech. In: Smart computing and communication: first international conference, SmartCom 2016, Shenzhen, China, December 17–19, 2016, Proceedings, vol 1. Springer International Publishing, pp 236–247
23. Bouafia K, Gulalov M (2023) Blockchain solution for authorization and authentication. In: International conference on industry sciences and computer science innovation (ISCSI2023), 2023 Lisbon (under Publishing)

Energy Detection-Based Measurement and Analysis of Spectrum Occupancy for Ad hoc Cognitive Network Resource for Federal University of Technology, Minna, Nigeria

Nathaniel Salawu, Adamu Hussaini, Sadiq Abubakar Muhammed, Dauda Usman Suleiman, and Sunday Achimugu

Abstract The rising demand for wireless communication systems has led to the exponential growth in the figure of wireless devices. As a result, there is a significant scarcity of available spectrum, which can limit the capacity and quality of wireless communication networks. Cognitive radio (CR) technology is a good solution to this challenge as it enables the most effective use of spectrum resources by permitting wireless devices to dynamically use under-utilized spectrum. This research presents a measurement of real-time spectrum sensing and detection of white space of cognitive wireless technologies for microwave and wireless communication systems of the Federal University of Technology, Minna main campus. Different spectrum sensing techniques including cyclostationary feature detection (CFD), matched filter detection (MFD) and energy detection (ED), have been widely researched. In this research, ED model for enhanced spectrum sensing-based cognitive radio technology that is capable of adapting to changing environmental conditions is adopted. Comprehensive field measurements were carried out in the study area using an 8.5 GHz BK precision 2658A spectrum analyzer (SA). Specifically, a received signal strength (RSS) of 7.3 dBm was obtained at 10 m while at a distance of 1500 m, 143.93 dBm was recorded alongside other similar results. The results were validated by comparing the standard deviation (SD) parameter, and root mean square error (RMSE) based on the adopted ED model with a reference spectrum sensing technique. Thus, the adopted ED technique is a more realistic spectrum sensing technique for microwave and mobile communication systems in the study area and similar environments.

Keywords Cyclostationary feature detection technique · Matched filter detection technique Energy detection technique · Cognitive radio

N. Salawu (✉) · A. Hussaini · S. A. Muhammed · D. U. Suleiman · S. Achimugu
Telecommunications Engineering Department, Federal University of Technology, Minna, Niger State, Nigeria
e-mail: nathsalawu@futminna.edu.ng

X.-S. Yang et al. (eds.), *Proceedings of Ninth International Congress on Information and Communication Technology*, Lecture Notes in Networks and Systems 1000,
https://doi.org/10.1007/978-981-97-3289-0_47

1 Introduction

The rapid increase in need for quality wireless communication transmission and reception has brought a lot of challenges on the scarce radio spectrum. Consequently, some researchers claim that this challenge is the result of insufficient techniques for spectrum access [1], instead of the assumption of physical scarcity of spectrum. In recent times, some regulatory agencies have issued spectral license permits for exclusive utilization of some dedicated frequency bands. Previous works have revealed that many of the licensed radio signal spectral bands may not be efficiently put to use in space and time [2] giving rise to unused "white spaces" in the time–frequency domain at any particular area. The utilization of spectrum is largely around a certain portion of the spectrum while a large expanse of the spectrum may not be utilized as some other part as seen in Fig. 1.

As can be observed from Fig. 1, spectrum usage is more serious and competitive at frequencies lower than 3 GHz. It is also clear that the spectrum is not fully utilized around the frequency bands of 3–6 GHz. On the other side, static spectrum allocation strategies will not allow for re-utilization of the rarely used spectrum allotted to licensed users by unlicensed users. This means that the radio spectrum could be better utilized than the traditional usage. To achieve this, there is a need for a careful analysis of the spectrum occupancy to make precise recommendation on the extent of spectrum utilization. This has called for a new communication standard by which unlicensed (secondary) users to quickly access the free or vacant spaces which are reserved for licensed (primary) users in a process known as cognitive radio.

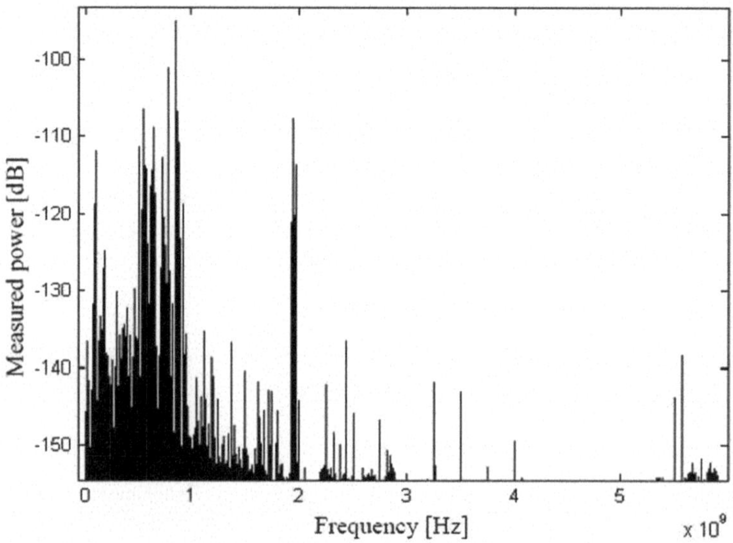

Fig. 1 Spectrum utilization measurements [2]

The knowledge behind the CR standard is the assessment of the free frequency bands assigned to the primary users (licensed users) taking the secondary users (unlicensed users) into consideration without any interference imposed on the licensed users' transmission and reception [3, 4]. Figure 2 depicts the cycle process of cognitive radio comprising four phases described as spectrum sensing phase, spectrum decision phase, followed by spectrum mobility phase and lastly, spectrum sharing phase [5]. First, in the CR spectrum sensing process, the CR categorizes the available portion of the spectrum section that can be utilized. It can also sense the PU presence when it is restored into operation in a way that guarantees an efficient track of the network. The second phase is the spectrum decision process. The CR selects the correct frequency band depending on the user's quality of service requirements (QoS) and spectrum requirement. In the third process also known as spectrum sharing, the spectrum utilization is coordinated between the SUs and PUs in a way that maintains the interference value below a certain threshold. Finally, the CR frees the channel when the PU begins to use it during the stage of spectrum mobility. This means the spectrum must be aware and update dynamically [6].

The achievement of CR largely rests on its ability to sense "spectral holes" and the backing of the regulatory bodies for granting CR permission to use licensed frequency bands is another gain. The initial stage of CR implementation is the investigation of the accessibility of the momentarily unoccupied spectral resources such as time, frequency and lastly, space. The projected technology known as dynamic spectrum access (DSA) is one of the promising possible solutions to these challenges. DSA configurations are of two types and these are the concurrent spectrum access (CSA) model and the opportunistic spectrum access (OSA) model [7]. The OSA technology allows SU to utilize the radio spectrum if only active PUs are in session. Conversely, the CSA technology also called underlay CR allows Pus and Sus to co-exist at the

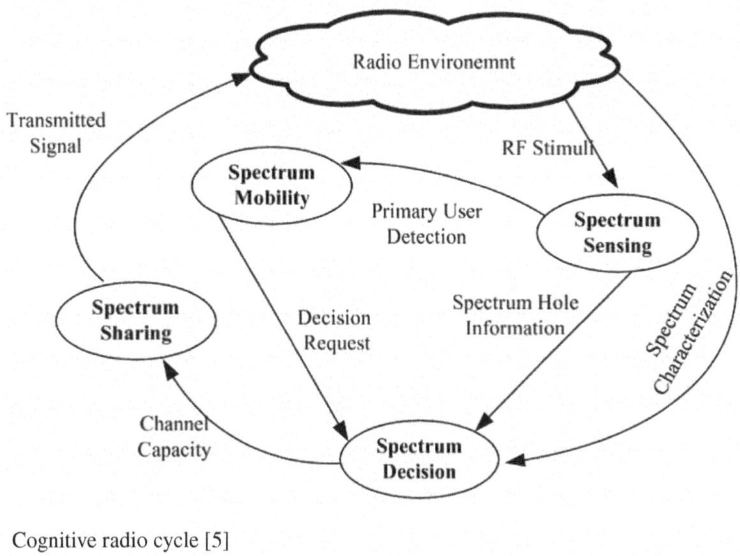

Fig. 2 Cognitive radio cycle [5]

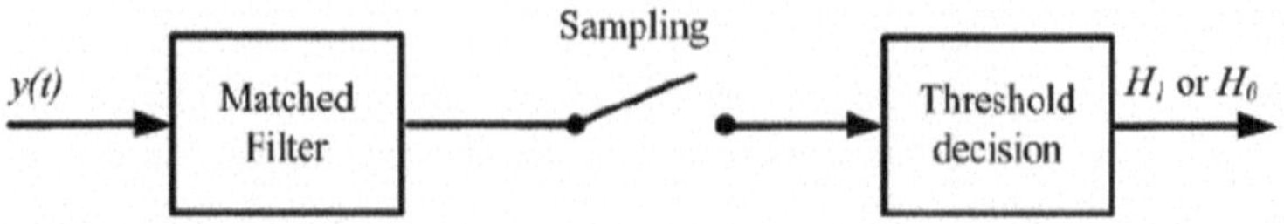

Fig. 3 Matched filter detection technique

same time within the same geographic location area. However, the interference posed by the SUs to the PUs must be kept within an acceptable pre-determined value of threshold also known as the interference temperature limit value [8].

2 Cognitive Radio Techniques

There are several techniques to sense idle spectrum channels and utilize them to increase the networks efficiency. In this study, we looked at MFD, CFD and ED.

2.1 *Matched Filter Detection (MFD)*

Spectrum sensing based on MDF uses pilot signals values that are fixed to the primary signal value so as to sense the availability of a PU. These pilot signal values are used in transmission and reception of signals for synchronization and, typically contributes between 1 and 10% based on the total signal power value transmitted. The unidentified primary signal value is convolved along time reversed form of the pilot signal. Assuming the model signal is present in the unidentified signal wave, then the PU is taken to be present, or it is otherwise taken to be absent. The system block diagram of an MFD-based spectrum sensing technology is presented in Fig. 3 and the mathematical model is described in Eq. (1)

$$Y[n] = \sum_{k=-\infty}^{\infty} h[n-k]y[k] \tag{1}$$

2.2 *Cyclostationary Feature Detection (CFD)*

CFD exploits statistical characteristics assigned to the primary signal so as to find the availability/absence of primary users with the aid of spectral correlation. Equation (2) presents the equation for implementing CFD while Fig. 4 depicts its block diagram.

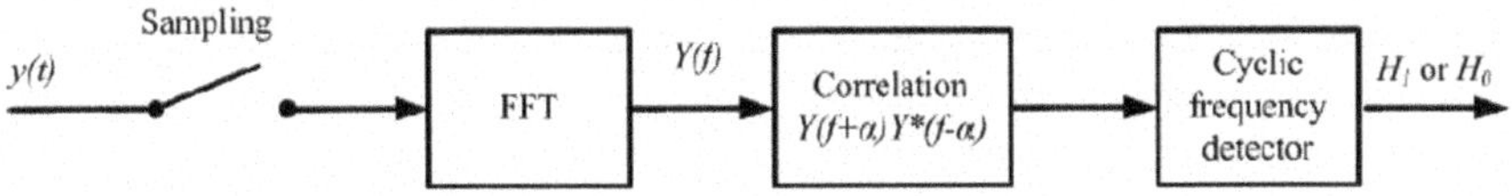

Fig. 4 Cyclostationary feature detection technique

$$r_y[n, \tau] = \mathbb{E}[y[n]y * [n + \tau]] \quad (2)$$

2.3 *Energy Detection (ED)*

ED spectrum sharing technique is based on non-coherent detection scheme that determines the energy level of the received signal. It compares it with an established threshold value that is based on the level of noise flood [9]. In ED-based spectrum sensing utilization, the received signal obtained is filtered around the bandwidth concerned and squared. The signal is later integrated over a specified observation interval value. This ensures that the measure of the received signal's energy level is compared with the detection threshold [10, 11]. Figure 5 shows the detailed block diagram of ED-based spectrum sensing utilization while Eq. (3) presents its mathematical relationship.

$$y(n) = x(n) + w(n) \quad (3)$$

The main criteria for spectrum sensing is function of signal detection in a negative signal-to-noise-ratio (SNR). Owing to the dynamic behavior of radio and environmental condition, CR must have adaptation ability and to be able to dynamically change the all necessary communication parameters appropriately. A CR senses the spectral environmental conditions over a wide signal frequency band and exploits this signal information to opportunistically deliver wireless links that can best meet the requirement of the user in addition to its radio environments [12]. In this paper,

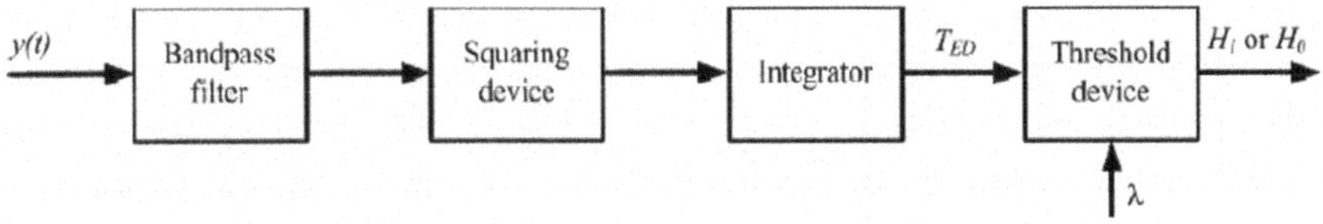

Fig. 5 Energy detection technique

ED technique has been deployed to develop an improved cognitive radio network under different environmental variation.

3 System Model

ED method was deployed on the measured energy of the accessible radio frequency signal in the communication band of the CR. The result obtained is compared with a predefined threshold level. The predefined threshold value is measured once a noise signal is sensed within the channel only. As soon as all levels of energy drop below the established threshold level (E_{th}), the spectrum is registered as available. If the estimated value of energy level is greater than the defined threshold, it is marked as occupied. The purpose of verifying the occupancy status of the PU is to know if the PU is using the spectrum or not at any given time so as to know when the SUs can utilize the spectrum. SUs can only have the available spectrum when the PU is not inactive. The ED technique models the temperature interference to sense the white space. The detected white space will be allocated to unused users based on the RSS of the primary user. The ED technique does not need the prior information of the primary user making it to be less complex when compared with other methods. Figure 6 represents the graph of interference temperature evaluation in (a), an ideal while the plot in (b) is the generalized version.

Considering the PU's power of all the nodes connected, the Gaussian noise is given by:

$$kBT_L(f_c)\,(B - B_i) + kBT_N \sum_{j=1}^{n} B_j \leq \sum_{j=1}^{n} B_{jpj} \quad \forall 1 \leq I \leq n \tag{4}$$

where B stands for the bandwidth and T_L is the temperature limit.

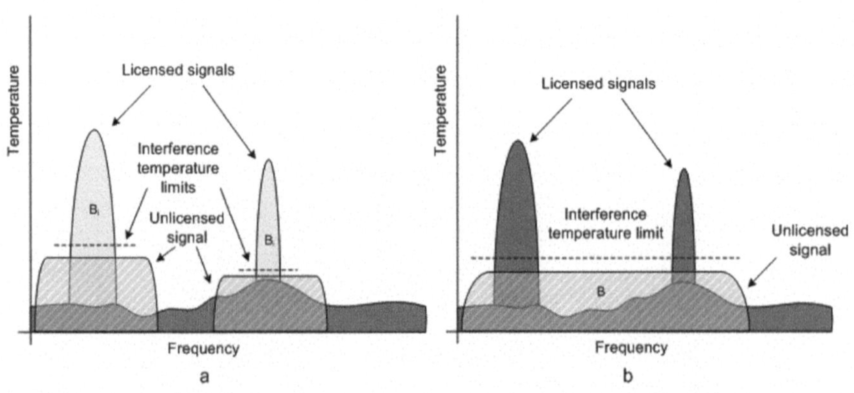

Fig. 6 Interference temperature evaluation model of an ideal **a**, and the generalized **b**

Assuming p is the power and it is a function of the bandwidth $B_{i,}$ then p is given by:

$$P \leq \min_{i\varepsilon[1,n]}[(B_i k/M_h)\,[T_L(f_i) - T_1(f_i, B_i)] \tag{5}$$

i is the different channels of the spectrum. If the unlicensed signal does not overlap, the spectral density of the derived power is obtained as given in Eq. (6).

$$P \leq \{Bk/M_k[T_L(f_c)]\} - \left\{1/BM_h\left[\int_{fc-B/2}^{fc+B/2} S(f)\,df\right]\right\} \tag{6}$$

Equation depicts the power of received signal when the temperature limit T_L and M_h are considered. The number of times a channel is occupied within a given time period is therefore given by Eq. (7).

$$\text{Duty Cycle} = [\text{Signal Occupation period}\,(n)/\text{Total Observation period}\,(m)] \times 100\% \tag{7}$$

A threshold value better called the decision threshold is endowed with the main feature in evaluating the true spectrum occupancy in the signal frequency band of concern. The energy detection technique of spectrum sensing process compares the value of the received signal of a given frequency band with an established threshold for further decision.

In order to take measurement, SA was used to obtain the RSS of the study area with radial base station (BS) coverage area of 2 km. The experimental data were taken at distances ranging from 100 to 1000 m, within the coverage area of the BS, for three radio paths. The measurements were conducted during the day time. For every route, RSS is measured from a reference distance position of 100 m taking from the base station and a successive interval of 100 m to a distance of 1000 m. The details of the results obtained and their discussion are presented in Sect. 4 of this paper.

4 Results and Discussion

The corresponding RSS against the distance measured from three selected location areas is shown in Fig. 7. RSS was measured from the Frizzlers microwave source, 10–100 m away from the base station. The RSS plot against measured distance from the microwave oven is presented in Fig. 8. Table 1 shows the RSS evaluated from the reference position distance from 100 m up to 1000 m for each of the three routes examined. As can be seen from Table 1, the RSS decreases as the distance from the base station (BS) to mobile station (MS) increases, with the exception of some measured points along the routes, where the RSS fluctuates owing to environment conditions.

Table 1 shows the measured RSS of the study area. From the results obtained, the spectrum band between 2.409 GH and 2.509 GHz was the least utilized. A noise floor threshold of −60 dBm was observed. By implication, the signal strengths beyond the −60 dbm power are detected and taken as the signal point threshold. This threshold is used to detect white spaces. The 2.40GHZ-2.45 GHz is sparsely empty from the research findings of the studied area when the ED technique with reference to MFD and CFD is deployed.

From the measured values, it is evident that the frequency channel that is the next highly occupied in the 2.4 GHz frequency band is given as channel 11 with the

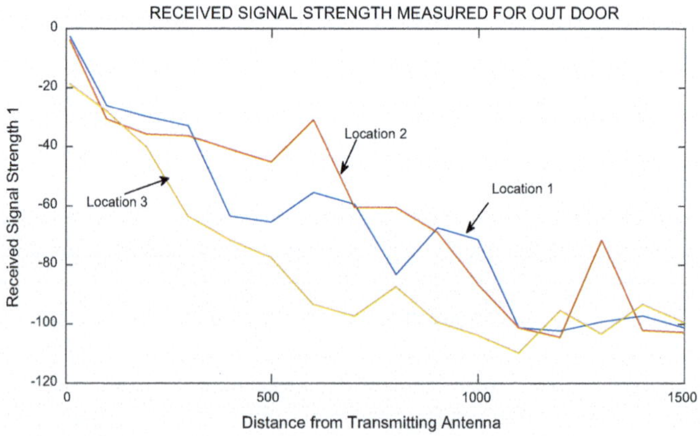

Fig. 7 RSS versus distance measured from three selected location areas

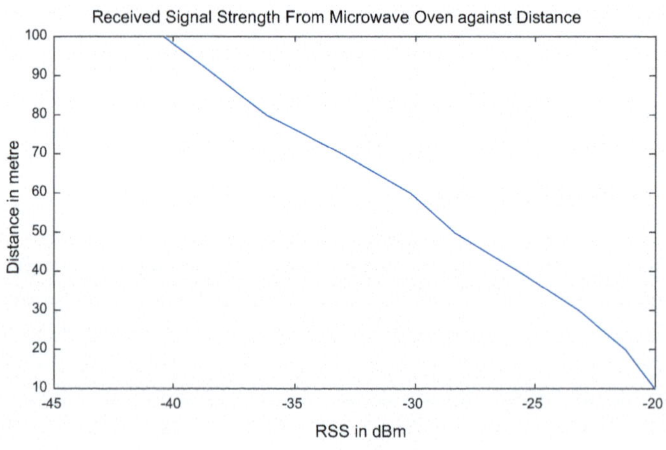

Fig. 8 RSS versus distance measured from the microwave oven

Table 1 Received signal strength from the base station at three different locations

T-R separation distance (m)	Time (s)	RSS (dBm)		
		i	ii	iii
100	0	−25.00	−30.41	−27.63
200	50	−26.64	−35.52	−39.82
300	100	−28.75	−36.12	−43.25
400	150	−33.26	−40.62	−51.21
500	200	−35.24	−44.93	−57.18
600	250	−40.29	−30.72	−63.10
700	300	−43.27	−50.22	−67.08
800	350	−40.14	−60.33	−70.13
900	400	−52.23	−68.67	−72.10
1000	450	−61.21	−66.31	−73.50

frequency of 2.462 GHz. Given in Figs. 9 and 10, respectively, is the spectrogram for mean and max channel occupancy measurements, respectively.

The highest curve points present the highest level of occupancy detected in a given signal channel. The mean points indicate the average occupancy detected in that specific channel within the duration period. Furthermore, the channel mean occupancies present an exciting awareness that channel occupancy levels vary greatly but have an average to greater part of the band to be is free. It can also be seen in the graphical results that there are some WLAN users within the 2400 and the 2470 MHz frequency band. Again, some signals with low power and narrowband whose occupied bandwidth varies between the frequencies of 2 and 3 MHz were equally noticed to belong to a few numbers of systems using audio/video wireless transmission and reception systems. Large number associated with the narrowband activity was detected at $f_c = 2406$, 2427 and 2490 MHz, respectively.

From the analysis of the 2.4 GHz frequency band results, it is seen that channel 1, channel 6, and channel 11 are the most utilized. The most utilized is channel 1,

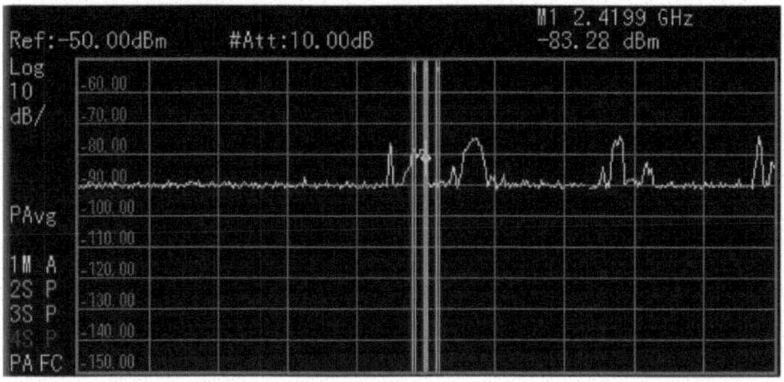

Fig. 9 Spectrogram of mean channel occupancy over the duration

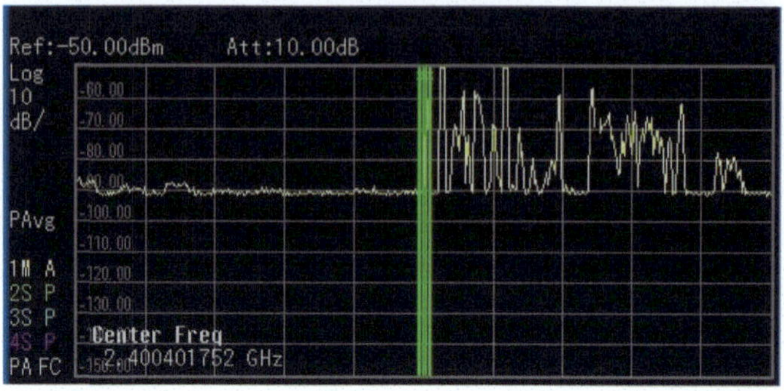

Fig. 10 Spectrogram of max channel occupancy over the duration

followed by channel 11. This means, the least utilized is channel 6 when 2.4 GHz frequency band is considered. Channel 1 of the system is used the most where maximum occupancy level is recorded as 11%. Channel no. 6 is found to be the moderately occupied channel with maximum occupancy level of 7%. Channel 11 was the least occupied channel with maximum occupancy level of 3%. Figure 11 shows the spectrogram for the occupied bandwidth in the location area where signal data were collected. The result depicts that there were no activities going on for most of the periods of the measurements. This was observed from the spectrogram because there were no active wireless access points within the period of the data collection. It is important to note that within these locations, only location with few bands presents some significant activities in the spectrogram. Because of the low occupancies in some locations, the activities on this band were not captured on the spectrogram.

The knowledge of relative bandwidths is that, it may be expected to have a percentage of a channel with a bandwidth B that is above noise threshold at a given time. For instance, for a given 40 MHz IEEE 802.11n channel, a frequency of 33.75 MHz may be used by subcarriers and a 50% relative bandwidth is expected to be

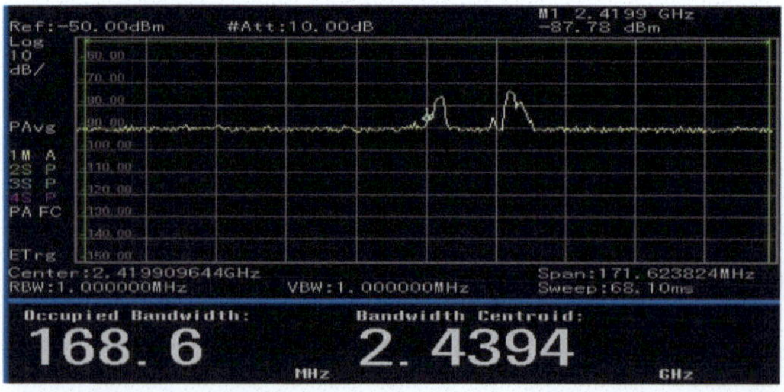

Fig. 11 Spectrogram of the occupied bandwidth

at least 16.86 MHz of the occupied channel above the noise threshold to mark it occupied with full confidence. The occupied bandwidth therefore shows the percentage of the bandwidth that is available relative to the total bandwidth of the frequency band being evaluated.

5 Conclusion

In this study, an investigation of a real-time measurement and analysis of ED spectrum sensing for cognitive radio white space in some selected campus areas of the Federal University of Technology targeting frequency spectrum range between 2.35 and 2.5 GHz has been done. The research is an eye-opener about the available white space cognitive radio spectrum access on the campus to ease the problem of spectrum "scarcity" in the campus area.

References

1. Chiang RIC, Rowe GB, Sowerby KW (2007) A quantitative analysis of spectral occupancy measurements for cognitive radio. In: IEEE vehicular technology conference, IEEE, Dublin, pp 3016–3020
2. Barlemann L, Mangold S (2020) Cognitive radio and dynamic spectrum access, 1st edn. Wiley and Sons, USA
3. Mitola J (2021) Cognitive radio architecture in wireless networks: principles and Applications. In: Fitzek FHP, Katz MD (eds) Springer, Dordrecht, pp. 234–311
4. Suseela B, Sivakumar D (2018) Certain investigations on effective rate adaptation in cognitive radio with channel characteristics. Int J Netw Virtual Organnization 19(2):373–384
5. Ali S, Hamouda W (2016) Advances on spectrum sensing for cognitive radio networks : theory and applications. IEEE Commun Surv Tutorials 5(1):1–29
6. Yucek T, Arsian H (2009) A survey of spectrum sensing algorithms for cognitive radio applications. IEEE Commun Surv Tutorials 11(1):116–130
7. Wang B, Liu KJR (2016) Advances in cognitive radio networks : a survey. IEEE J Sel Top Signal Process 5(1):1–20
8. Eberechukwu PN, Umar DS, Mohd A, Kamaludin MY, Sarijari MA (2018) Interference temperature measurements and spectrum occupancy evaluation in the context of cognitive radio. Indones J Electr Eng Comput Sci 10(3):1007–1012
9. Yoo SK et al (2018) Entropy and energy detection-based spectrum sensing over F composite fading channels. IEEE J Sel Areas Commun 16(7):1–30
10. Lorincz J, Ramljak I, Begušić D (2019) A review of the noise uncertainty impact on energy detection with different OFDM system designs. Comput Commun 148(10):185–207
11. Sofotasios PC, Zhang L, Freear S (2013) Energy detection—based spectrum sensing over $\kappa-\mu$ and $\kappa-\mu$ extreme fading channels. IEEE Trans Veh Technol 62(3):1031–1040
12. Zhang l, Xiao M, Wu G, Alam M, Liang Y, Li S (2017) Dynamic spectrum management for 5G: a survey of advanced techniques for spectrum sharing in 5G networks. IEEE Wirel Commun 4(11):44–51

Third-Party Data Leaks on Municipal Websites

Sampsa Rauti, Robin Carlsson, Panu Puhtila, and Ville Leppänen

Abstract This paper addresses the use of third-party services on public sector websites. We focus on Finnish municipalities and their use of third-party services, analyzing network traffic from 309 municipal websites to explore what kind of personal data is transmitted to third parties. Our findings show that 51.8% of municipal websites reveal the pages visited by the user to third parties without consent. This way, sensitive data such as search terms used by website visitors can be leaked to third parties. The paper also offers recommendations to improve privacy on public sector websites. We have also contacted the municipalities with serious data leaks and recommended privacy improvements by disabling or replacing problematic third-party services.

Keywords Municipal websites · Web privacy · Network traffic analysis · Personal data

1 Introduction

In the continuously evolving digital landscape, the Internet has become an essential platform for disseminating information and providing services. Websites act as pivotal gateways which individuals can use to access a wide variety of resources. Services offered by government institutions such as municipalities are no exception to this trend. However, even with these public sector websites, convenience and utility of virtual platforms can hide the underlying privacy-related concerns [8, 17]. In

S. Rauti (✉) · R. Carlsson · P. Puhtila · V. Leppänen
University of Turku, 20014 Turku, Finland
e-mail: sjprau@utu.fi

R. Carlsson
e-mail: crcarl@utu.fi

P. Puhtila
e-mail: papuht@utu.fi

V. Leppänen
e-mail: ville.leppanen@utu.fi

X.-S. Yang et al. (eds.), *Proceedings of Ninth International Congress on Information and Communication Technology*, Lecture Notes in Networks and Systems 1000,
https://doi.org/10.1007/978-981-97-3289-0_48

particular, third-party services often integrated into these websites present severe privacy challenges.

Municipal websites are vital conduits for citizens to access essential information on public services and community events [5, 14]. Consequently, they often handle sensitive information about users and their intentions. Therefore, leaking users' navigational paths and search terms to third parties, often without explicit consent, exposes website visitors to the potential erosion of their privacy. It is crucial to investigate the extent to which municipal websites share user-generated data with third-party services.

The importance of privacy in public sector web services is also reflected by the fact that the Finnish Deputy Data Protection Ombudsman has emphasized that government bodies should "carefully consider what types of tracking technologies are necessary on their websites." Moreover, authorities should make sure that users are "able to use online services provided by authorities without data on their website visits ending up in commercial use, for example."[1] These guidelines issued by the Deputy Data Protection Ombudsman provide a strong statement on how to correctly strike a balance between functionality and protecting user privacy.

This research paper focuses on the presence of third-party services within the websites of Finnish municipalities and the implications the use of these services has for user privacy. We analyze the network traffic of all 309 Finnish municipal websites in order to gauge what kinds of personal data items they transmit to third parties. The study finds that the majority (51.8%) of the analyzed municipal websites leak URL addresses visited by the users to at least one third-party. We also attempt to alleviate the situation by contacting the municipalities with the most severe data leaks and recommending that they improve their data protection by disabling or replacing the problematic third-party services.

The remainder of the paper is organized as follows. In Sect. 2, an overview of related research is presented. Section 3 outlines the context of the study, including the used dataset and methodology. The results of the study, analyzing the personal data transmitted to third parties from the studied municipal websites, are discussed in Sect. 3.1. Section 3.2 delves into the significance of our key findings and offers recommendations to enhance the privacy measures of municipal websites and other governmental online resources in general. Lastly, Sect. 4 brings the paper to a close with concluding remarks.

2 Related Work

While the actual data collection happening in the municipality websites has not been studied that much previously, the contents of the privacy policies displayed at such websites have received more attention from the researchers. Also, other aspects of

[1] https://tietosuoja.fi/en/-/deputy-data-protection-ombudsman-issues-reprimand-for-conveying-library-search-information-to-us-based-google.

privacy and security in these kinds of websites have been investigated to a degree, and as both of these themes are closely related to our own research, we shall here provide an overview on these topics.

Schuele [13] published already in 2005, when the whole concept of personal data collection online was still quite new, an investigation into how the municipal websites collected their users Personally Identifiable Information (PII), and whether these websites informed the users about such data collection. It should be noted that PII in this context consists of directly identifying factors such as name, social security number, credit card information, email and address, and as such it is slightly different to what kind of data collection we are studying. It should also be noted that the data collection studied in Schuele's paper refers to data willingly inputted by the users, through forms in the websites, not data mechanically harvested without any direct user involvement.

Holzer et al. [10] published in 2006 a large-scale study where the global status of digital governance of largest cities was inspected. One part of this study was the assessment of the status of privacy and security at these cities' websites, for which they developed a numerical scale representing the relative level of security and user privacy. The highest ranking cities were in Oceania, with Europe coming close behind, and the worst performing cities were found to be predominantly located in South America.

Beldad et al. [2] studied Dutch municipal websites and their privacy policies, and came to a conclusion that they had severe failings in regards to the Dutch data protection laws. The privacy policies often exhibited opaque language, simply did not exist or at least were not directly available for the users. In another study, Beldad et al. [1] investigated the question of what makes the people to read the privacy policies at municipal websites, and came to the conclusion that both the older age, lower level of education and less experience in using Internet correlated strongly with the user actually reading the privacy policy, whereas younger, more educated and more Internet-savvy users more often neglected to read these documents.

Dias et al. [4] survey privacy policies and practices of websites of 308 Portuguese municipalities. They found that 65% of the studied municipal websites used tracking tools. Furthermore, among the municipal websites that used tracking tools, 96% either did not have a privacy policy or, if they did, the privacy policy document did not address the use of tracking technologies and third-party cookies at all. While Dias et al. consider tracking and data collection in their study, our study differs from theirs as we perform a network traffic analysis to reveal details about what kind of personal data leaks from the websites.

Kautto and Hentonen [11] studied the availability and findability of the Freedom of Information (FOI) statements and privacy policies in the Finnish municipality websites in their 2017 research, and came to the conclusion that it was almost non-existent. Even in the cases where there obviously was some kind of attempt at informing the website user about the privacy issues, such as data collection, the information was extremely hard to find due to being scattered to multiple locations across the website. From the perspective of our own research this study is very relevant, as Kautto and Henttonen studied the exact same collection of 309 current Finnish municipality websites as we did.

Gomes et al. [7] conducted a research on how much the Portuguese municipality websites used HTTPS compared to plain HTTP, to determine the extent at which the users could place their trust at these services. Their results indicate that 46.1% of the studied websites did not meet even the minimum standards for security in this regard, and only 3.1% passed their examination completely.

3 Study Setting and Methodology

The websites of all current 309 Finnish municipalities[2] were included in the study. Our study consists of two phases. First, we performed a network traffic analysis to assess the privacy of the studied municipal websites. Second, the municipalities that were found to have websites with inadequate user privacy were informed about the matter.

3.1 The Network Traffic Analysis

The network traffic of each website was analyzed in order to see what third parties the websites included and what kind of personal data was leaked to these third parties. In our experimental setting, consent was never given for cookies or data collection. The experiment included opening a municipality's front page and making a search with the website's search functionality. This experiment made it possible to explore

- What third parties were recipients of personal data.
- What kind of identifying personal data items were present in this network traffic to third parties.
- Whether the URL addresses of the pages the user visited were sent to a third-party (revealing a lot about users intentions and activities on the website).
- Whether the search term the user used was leaked (also potentially revealing sensitive information about the user).

To make it possible to analyze all 309 municipalities, the network traffic analysis was mostly performed with an automatic tool we have specifically built to analyze the third parties on websites, as well as URL addresses and search terms leaking to these external actors. The technical implementation of the tool has been described in more detail in [3]. For a minority of websites on which the tool failed, the analysis was done manually with Google Chrome's Developer Tools. Failures were due to the cases where for instance, some of the studied websites, have pop-up elements such as cookie consent banners, that were overlaid with the search functionality, preventing

[2] https://en.wikipedia.org/wiki/List_of_Finnish_municipalities.

our tool from clicking the search button. For each website, the recorded network traffic was analyzed, and the detected third parties and the details on personal data items (did URLs and search terms leak, what kinds of identifying technical details leaked) were written down.

3.2 *Informing the Municipalities*

As the aim of this study is also have societal impact by improving the privacy of the studied municipal websites, we contacted the municipalities with most inadequate online privacy and asked them to consider removing the third-party analytics services and possibly replacing them with better options that store data locally and do not leak users' personal data to third parties. More specifically, we contacted municipalities that were found to leak URL addresses or search terms to Meta/Facebook, Google, or both companies. Although there were several other third parties on the websites as well, these technology giants were seen as the most prominent privacy threats with the capability to identify individual users and build profiles on them for commercial use.

In the sent messages, the chosen municipalities were informed about the fact their websites have been found to leak URLs and/or search terms to third parties. We also explained how severe these leaks can be; sensitive information such as a user looking for a specific medical service may leak to Meta or Google. Furthermore, we informed the municipalities about the Deputy Data Protection Ombudsman's statement on using third-party services in online services provided by authorities and the recommendation that users should be able to visit these websites without data on visited pages ending up in commercial use. After a period of one month, we also analyzed the responses from the municipalities to see how many of them responded and whether they promised to fix the privacy issues on their websites or not.

4 Results

4.1 *Leaked Personal Data*

The data leaked by the municipal websites can be divided into two categories: identifying data and contextual data. *Identifying data* refers to personal data that third parties can use to uniquely identify a website visitor. This can include personally identifiable information (PII) such as IP addresses, device identifiers, names, email addresses, and so on. It can also be a combination of many pieces of technical data that alone do not constitute identifying information, such as operating system, browser version, and screen size. Identifying data can be used to distinguish one individual from another. If this data were to fall into the wrong hands, third parties could use it to build profiles for users and, in certain instances, even share the data with other parties.

Contextual data contains additional details about a specific individual's behavior or situation. On a municipal website, for example, the user may use the search functionality to look for information about a sensitive topic such as drug rehabilitation within a specific region. In this case, the search term is contextual data. This data is not directly identifying on its own, but it becomes much more sensitive when it is combined with identifying personal data. Contextual data provides insights into users' behaviors and preferences, and this information can be exploited—usually commercially—when it is linked to identifying information.

Of the 309 municipal websites studied, 160 (51.8%) included third-party services. Interestingly, all 160 websites with third-party services also exposed users' visited URL addresses. In total, 296 (95.8%) out of the 309 websites featured functional search capabilities. The search term leaked in 48.6% of these websites. This indicates that during our experiments, utilizing the search functionality on a Finnish municipal website led to an almost 50% chance that the search query would be disclosed to a third-party without consent.

Figure 1 shows the third parties receiving visited URL addresses. Google Analytics collected URL addresses on 132 out of 309 (42.7%) municipal websites. When we include YouTube (which on 3 occasions receives the URL through embedded videos), also owned by Google, the number increases to 135 websites (43.7%). These findings underscore the extensive reach of Google in accessing sensitive personal data on public sector websites, even in the absence of user consent. Facebook/Meta comes second with URL leaks on 21 out of 309 (6.9%) websites. Other third-party services only have few occurrences in our data. Categories of these third parties include tracking website visitor behavior (Snoobi, WP Pixel), application performance monitoring and website optimization (Siteimprove, Microsoft Application Insights, New Relic), content sharing widgets and feedback tools (ShareThis, React&Share), social media and networking platforms (Twitter, TikTok, LinkedIn), tools enhancing website search functionality (Site Search 360), AI-powered content recommendations (Rek.ai), digital advertising (Adform), and hosting platform related functionality (Wix).

Furthermore, we have also included the Matomo analytics service as a third-party in the cases where it was hosted in the cloud. Generally, however, Matomo can also be hosted locally so that the website maintainer controls the collected data and it is not shared with any third-party [6, 12]. The cloud-based Matomo is likely to be a safe solution as well, but in principle at least some data could still be collected by the cloud provider.

Figure 2 illustrates the search term leaks to third parties. The situation is quite similar to the case of URL leaks. Google Analytics receives search terms in 122 out of 309 (39.5%) websites, and Facebook/Meta is the runner-up with 18 occurrences (5.8%). The number of services receiving search terms is smaller than the number of those receiving URLs, but a significant portion of the services discussed previously

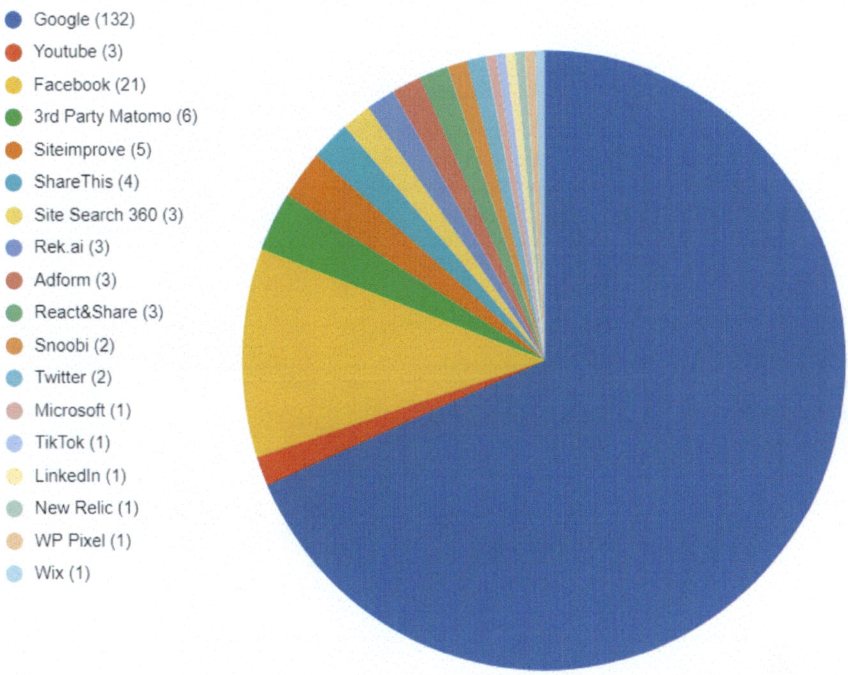

Fig. 1 The third parties receiving URL leaks from the studied municipal websites

still remain. The fact that nearly 40% of the municipal websites studied leak search terms to Google without the user's consent is nothing short of astonishing.

Finally, we also explored whether the municipal websites asked for consent and provided a means to deny the use of cookies and data collection. Consent was requested, and there was an option to refuse cookies on 160 out of 309 websites (51.8%). Conversely, 95 websites (30.7%), a remarkably high number, did not offer the option to refuse cookies but sent the user's personal data traffic to third parties. Lastly, 54 websites (17.5%) neither asked for consent nor contained any third-party services.

4.2 *Responses by Municipalities*

In total, 117 municipalities using Google Analytics or Meta Pixel on their websites were contacted. Additionally, 13 municipalities fixed the privacy flaws even before we contacted them—apparently, the word about our study had spread among the municipalities. We received 53 replies which is 45.3% of the contacted municipalities.

The municipalities we informed about third-party data leaks on their websites responded very positively. They expressed their gratitude for the notification and

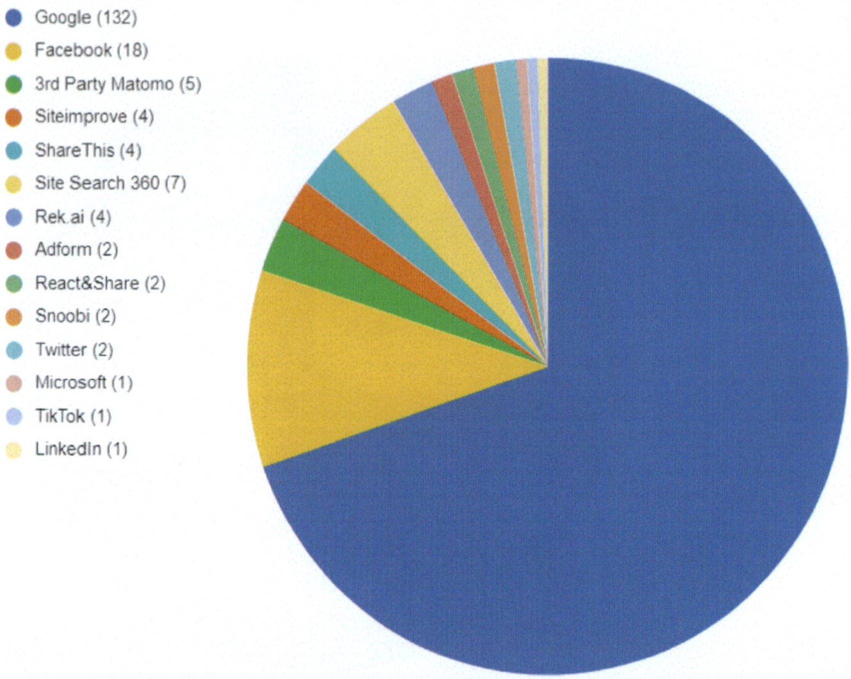

Fig. 2 The third parties receiving search terms leaks from the studied municipal websites

their commitment to addressing the privacy issues. All of the municipalities that responded agreed to take the necessary measures to rectify the privacy issues on their websites.

Many municipalities indicated that they were aware of the challenges associated with using Google Analytics on public sector websites. They expressed their intentions to either remove this service or replace it with an analytics solution that offers improved privacy. Matomo was frequently mentioned as a potential alternative. While many municipalities were eager to enhance the privacy of their websites as part of an upcoming update, some also mentioned a lack of resources as a challenge.

In truth, removing or replacing an analytics service does not demand extensive resources. This was evident from the fact that several small municipalities successfully removed Google Analytics in just a few days. However, even some larger cities claimed that eliminating third-party analytics from their websites could be a substantial undertaking.

While removing an analytics service should be quite easy when done correctly, it is true that changing the used service can of course introduce some complexities in the municipality's own data collection. As one municipality noted: "The collected information contains important statistical data for planning purposes, and we need to make a plan for data extraction before deletion." Deploying the new analytics service also requires some additional planning of data collection and extraction practices.

Although municipalities were willing to improve user privacy by removing the problematic analytics services, many of them failed to do so successfully, and we had to contact them again about this issue. It appears that, despite removing the Google Analytics plugin from their content management systems, Google's tracking scripts often remained in the website's source code. This is why it is important to analyze the network traffic going to third parties to see that no tracking functionality persists.

In some cases, the lack of technical skills was also quite evident from the responses we received. For instance, a reply from one municipality mentioned that "we have deployed cookies on our website to enhance data privacy." This seems quite counter-intuitive, although it could refer to improved implementation of cookie management for better privacy.

Only one municipality was slightly reluctant to implement changes. They argued that because the use of Google Analytics is not outright illegal, and according to their interpretation, there is no strong recommendation from data protection authorities to avoid using Google Analytics, they do not see removing it as a high-priority task right now. Obviously, clearer guidelines from data protection authorities would be beneficial. As the municipality put it: "In this matter, we would have appreciated direct guidance from the relevant authorities."

Finally, 60 out of 117 (51.3%) municipalities had removed Google Analytics or Meta Pixel 3 months after we had contacted them. Some municipalities also indicated that they are going to assess the need for third-party analytics again when the next major update of their website takes place, so more positive changes will probably happen in the future.

5 Discussion

It is evident from our results that the problem posed by the third-party analytics in regards to user privacy is not a small one. When 42.7% of these websites leak data about their users to Google Analytics without the user consent, the situation must be addressed. While there were leakages to other analytics providers too, their scale was so much smaller that it is hardly comparable to Google. On the lighter side, it is reassuring that a large number of the municipalities we contacted about these issues were very willing to make privacy improvements.

From the responses we received from the municipal authorities it is clear that there is strong variance in technological knowledge resources between different municipalities. Some of the municipal administrations we contacted were completely unaware of the issues we have brought to light, while others were already somewhat cognizant of the situation and had, at least allegedly, already discussed possible solutions to it. Then again, many municipalities also were quick to use the excuse of "it is hard to implement these changes." This is a poor excuse, as even several small municipalities which possess very limited resources were able to implement the necessary changes in a matter of days. Removing a web analytics tool from use usually means editing just a few lines of code. It may be that these kinds of objections arise out of the organizational encumbrances—the persons responding to our inquiries might not be themselves knowledgeable or responsible about the privacy matters. Also, the size and complexity of the organization—at least the case of the larger cities—may cause the information to be passed through several levels of administrative bureaucracy before reaching a person who actually can do something about the situation. Such circumstances may entice the people working in these administrations to neglect the issue in favor of "more pressing concerns," and thus come up with these excuses.

Another notable issue that came up in this investigation is the fact that almost one third of the studied municipality websites did not have proper cookie consent banners through which the user could submit consent, but nevertheless these websites leaked the user data to third parties. This kind of behavior is often a direct breach of the GDPR [16], and can potentially lead to legal consequences for these organizations.

To remedy the situation, developers working on public sector websites should conduct a thorough network traffic analysis during the testing phase of the development process, similar to what we have done in this research. This helps to assess whether there is any sensitive personal data being leaked to third parties. Conducting such a technical evaluation requires neither special expertise nor tools developers would not already have at their disposal. Carrying out this examination for a single website does not demand great investment of time either.

In choosing web analytics it would be wise to favor tools which can be deployed locally, for example Matomo, to minimize the potential of leakages. The great irony of the situation is that it has been proven time and again that Google Analytics is a significant reason for personal data leaks [9, 15], both in terms of the amount of data being leaked and due to the nature of this data, yet it is also the most used analytics tool globally. Being the largest analytics service provider, it has also become the go-to choice for most organizations, while at the same time its use both jeopardizes the privacy of the website users and the legal standing of the website proprietors to a great extent.

Awareness about the importance of online privacy among municipal officials and the general public should be increased. This, way the importance of adhering to data protection laws and regulations would also become more apparent. It would also be beneficial for data protection authorities to offer clearer guidelines regarding the use of third-party services on public sector websites.

6 Conclusions

In this study, we have taken a survey on the user privacy aspect of the municipal websites, and found it wanting. In total, 51.8% of all municipalities in Finland leak personal data to third parties, most often to Google Analytics, even when the user has not consented to data collection. Considering that these websites are in actuality part of the Finnish governing infrastructure and as such elementary services provided for citizens, the situation can be considered bleak. This is especially the case, as the Deputy Data Protection Ombudsman has taken the stance that such essential website infrastructure should not deploy tools that collect the user data for commercial purposes.

On the brighter side, a large number of the municipalities we contacted about our findings received our findings well, and committed to improving their data collection practices. Over half (51.3%) of the municipalities we contacted also removed third-party analytics from their websites. While this is reassuring, our findings underscore the issue of a lack of technical expertise within municipal administrations, which is concerning in a technologically advanced society like Finland. As for the future research possibilities, we plan on conducting further reviews on how the privacy situation of the Finnish municipalities evolves, and whether the considerations we have brought forth are addressed. Our findings could also be compared to the other EU countries, for example, to find out whether the situation is similar elsewhere.

Acknowledgements This research has been funded by Academy of Finland project 327397, IDA—Intimacy in Data-Driven Culture.

References

1. Beldad A, de Jong M, Steehouder M (2010) Reading the least read? Indicators of users' intention to consult privacy statements on municipal websites. Gov Inf Quart 27(3):238–244
2. Beldad AD, De Jong M, Steehouder MF (2009) When the bureaucrat promises to safeguard your online privacy: dissecting the contents of privacy statements on Dutch municipal websites. Gov Inf Quart 26(4):559–566
3. Carlsson R, Puhtila P, Rauti S (2023) Towards an automatic tool for detecting third-party data leaks on websites. Accepted to the 10th Workshop on software quality analysis, monitoring (SQAMIA2023)
4. Dias GP, Gomes H, Zúquete A (2016) Privacy policies and practices in Portuguese local e-government. Electron Gov Int J 12(4):301–318
5. Feeney MK, Brown A (2017) Are small cities online? Content, ranking, and variation of us municipal websites. Gov Inf Quart 34(1):62–74
6. Gamalielsson J, Lundell B, Butler S, Brax C, Persson T, Mattsson A, Gustavsson T, Feist J, Lönroth E (2021) Towards open government through open source software for web analytics: the case of Matomo. JeDEM-eJournal eDemocracy Open Gov 13(2):133–153
7. Gomes H, Zúquete A, Dias GP, Marques F (2019) Usage of https by municipal websites in Portugal. In: New knowledge in information systems and technologies, vol 2. Springer, pp 155–164
8. Heino T, Carlsson R, Rauti S, Leppänen V (2022) Assessing discrepancies between network traffic and privacy policies of public sector web services. In: Proceedings of the 17th international conference on availability, reliability and security, pp 1–6
9. Heino T, Rauti S, Carlsson R, Leppänen V (2022) Study of third-party analytics services on university websites. In: International conference on hybrid intelligent systems. Springer, pp 1284–1292
10. Holzer M, Kim ST (2006) Digital governance in municipalities worldwide (2005): a longitudinal assessment of municipal websites throughout the world
11. Kautto T, Henttonen P (2017) Availability and findability of FOI and privacy statements on Finnish municipalities' websites. Tidskriftet Arkiv 8(1)
12. Quintel D, Wilson R (2020) Analytics and privacy. Inf Technol Libr 39(3)

13. Schuele K (2005) Privacy policy statements on municipal websites. J Gov Financ Manage 54(2):20
14. Simelio-Solí N, Ferre-Pavia C, Herrero-Gutierrez FJ (2021) Transparent information and access to citizen participation on municipal websites. Profesional de la Información 30(2)
15. Wambach T, Bräunlich K (2017) The evolution of third-party web tracking. In: Information systems security and privacy: second international conference, ICISSP 2016. Rome, Italy, February 19–21, 2016, Revised Selected Papers 2. Springer, pp 130–147
16. Wesselkamp V, Fouad I, Santos C, Boussad Y, Bielova N, Legout A (2021) In-depth technical and legal analysis of tracking on health related websites with ernie extension. In: Proceedings of the 20th workshop on privacy in the electronic society, pp 151–166
17. Zheutlin AR, Niforatos JD, Sussman JB (2022) Data-tracking on government, non-profit, and commercial health-related websites. J Gen Intern Med 37(5):1315–1317

Evaluating Tabular Data Generation Techniques on the DaFne Platform: Insights from a Predictive Maintenance Case Study on Bridges

Pramod Baddam, Ayse Glass, René Jäkel, Jonathan Jander, Tom Krause, Pamela Kunert, Järg Rainer Noennig, Iryna Okhrin, Mariela Sanchez, and Ulrike Steffens

Abstract In the realm of artificial intelligence (AI) and machine learning (ML), the scarcity of robust and diverse datasets often poses a significant challenge, prompting the need for effective data generation methods. This paper presents an evaluation of tabular data generation techniques on the DaFne platform, centered around a predictive maintenance case study for bridges. The DaFne platform offers a variety of tabular data generation functionalities, including rule-based creation, data fusion (with

P. Baddam (✉) · R. Jäkel · I. Okhrin · M. Sanchez
Center for Information Services and HPC, Tu Dresden, Germany
e-mail: Pramod.Baddam@tu-dresden.de

R. Jäkel
e-mail: rene.jaekel@tu-dresden.de

I. Okhrin
e-mail: Iryna.Okhrin@tu-dresden.de

M. Sanchez
e-mail: Mariela_Rossana.Sanchez_Figueroa@tu-dresden.de

J. Jander · T. Krause · P. Kunert · U. Steffens
HAW Hamburg, Berliner Tor 5, 20099 Hamburg, Germany
e-mail: Jonathan.Jander@haw-hamburg.de

T. Krause
e-mail: Tom.Krause@haw-hamburg.de

P. Kunert
e-mail: Pamela.Kunert@haw-hamburg.de

U. Steffens
e-mail: Ulrike.Steffens@haw-hamburg.de

A. Glass · J. R. Noennig
HafenCity University—Digital City Science, Henning-Voscherau-Platz 1, Hamburg 20457, Germany
e-mail: ayse.glass@hcu-hamburg.de

R. Jäkel · I. Okhrin
Center for Scalable Data Analytics and Artificial Intelligence (ScaDS.AI) Dresden/Leipzig, Dresden University of Technology, Chemnitzer Straße 46b, 2nd floor, 01187 Dresden, Germany

X.-S. Yang et al. (eds.), *Proceedings of Ninth International Congress on Information and Communication Technology*, Lecture Notes in Networks and Systems 1000,
https://doi.org/10.1007/978-981-97-3289-0_49

weather data), and data reproduction. We investigate the utility of these functionalities across different machine learning models for the prediction of bridge conditions. Our analysis includes a descriptive statistical comparison of real and synthetic data. Additionally, we explore the utility of original, weather, and synthetic datasets. We do this through the lens of ML models like MLR, XGBoost, CNN, and GRU, performing a predictive maintenance algorithm on these datasets. Our results indicate that while the inclusion of weather data did not significantly enhance predictive performance, the synthetic dataset shows satisfactory quality. However, the synthetic data's performance is lower than the original data in predictive maintenance tasks, with differences observed in models heavily reliant on sequential data. This research underscores the potential of the DaFne platform in generating high-quality synthetic data. It also highlights areas for future improvement and offers valuable insights for advancing data generation and analysis techniques in predictive maintenance and other AI applications.

Keywords Tabular data generation · Bridge maintenance · Machine learning · Synthetic data · Prediction · Evaluation

1 Introduction

Data generation remains a pivotal challenge in artificial intelligence (AI). For many AI use cases, there is a significant gap in data availability [8], characterized by an absence of required data, insufficient quantity, or a lack of supplementary information [23]. Existing solutions, often conceptual and confined to academic papers, face practical hurdles such as operationalization and hardware requirements.

We introduce the 'DaFne' platform, an online tool for generating and synthesizing tabular data, particularly focused on Smart Cities applications. The platform offers diverse functionalities including rule-based creation, data fusion with weather data, and data reproduction, all tailored for a wide range of data types like numbers, categorical data, timestamps, and geospatial data.

Our research evaluates these functionalities using a predictive maintenance use case for bridges, employing a tabular bridge dataset. We analyze the original dataset for bridge condition prediction and then utilize the DaFne platform to enhance this dataset with new features and generate synthetic data. The efficacy of these datasets is then assessed in the predictive maintenance context.

This study aims to ascertain the robustness, accuracy, and utility of the DaFne platform's data generation functionalities in a real-world scenario. The remainder of this paper is organized as follows: Sect. 2 reviews relevant literature. Section 3 details the DaFne platform's features and functionalities. Section 4 presents a bridge maintenance use case. Section 5 covers practical aspects of handling the bridge dataset, synthetic data generation, and evaluation. It also provides insights into the employed machine learning methods. Section 6 discusses the study's outcomes and their implications. Finally, Sects. 7 and 8 conclude the paper, summarizing key findings and suggesting directions for future research.

2 Related Work

This section explores a range of related studies, starting with research on data generation methods and tools, followed by platforms for ML processes, works on synthetic data evaluation, and finally, studies specific to predictive maintenance for bridges.

In the realm of tabular data synthesis using generative ML models, advancements have led to the development of models like CTGAN and TVAE [49]. These models handle multimodal data distributions and various data types, making them suitable for diverse data synthesis scenarios. CTAB-GAN+ [51] additionally focuses on synthesizing data with privacy considerations. The Synthetic Data Vault (SDV) framework [39] is notable for its implementation of several generative models and evaluation tools.

For rule-based data generation, online services like Mockaroo [35] and Synner [33] provide user-friendly platforms for creating custom tabular datasets. Platforms like MOSTLY AI [36] specialize in data synthesis with an emphasis on maintaining the integrity of the original dataset's features, similar to Gretel.ai [38], which offers detailed comparative analysis tools.

The literature also presents ML platforms tailored for specific domains. Ahmed et al. [2] developed a platform focused on health care, enabling users to create and train ML models. Lee et al. [28] introduced an autonomous ML platform suitable for Smart City data. Tools like Brenningmeijer's Table Evaluator [7] and SDMetrics [44] are critical for evaluating synthetic data, with studies like Hittmeir et al. [20] emphasizing privacy in synthetic data assessment.

In predictive maintenance for bridges, ML models are increasingly used to analyze bridge conditions. Studies employing the NBI dataset, such as those by Kumar et al. [27] and Pengyong Miao [34], demonstrate the impact of dataset size and feature selection on model performance. These works underscore the importance of tailored approaches in predictive maintenance using advanced ML techniques.

3 DaFne Platform

The DaFne platform offers a comprehensive solution for data generation, tailored for a wide user base. It provides intuitive tools for generating various tabular datasets, alongside functionalities for evaluating the generated data, especially when grounded in ML techniques. Services like the generation of a data report allow users to visualize and understand their datasets in terms of distributions and statistical features.

Key functionalities for tabular data generation, depicted in Fig. 1, include:

1. **Rule-Based Data Generation**: Enables users to create custom tabular datasets with specific data types and distributions. It supports generating columns based on defined rules, allowing for the extension of existing datasets.
2. **Data Fusion**: Facilitates the merging of two tabular datasets, such as integrating user data with public datasets like traffic or, as in this case study, weather infor-

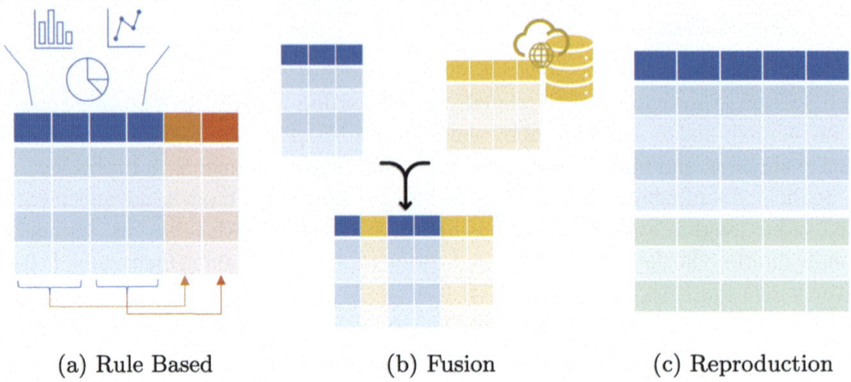

Fig. 1 Data generation functionalities on the platform

mation. This functionality is adaptable, allowing for customization based on user requirements and geospatial elements.

3. **Reproduction**: Synthesizes new rows for an existing dataset, mirroring original attributes and inter-column relationships. The platform includes various evaluation methods to ensure the quality of synthetic data, making it suitable for applications like data augmentation or privacy preservation.

While the DaFne platform also enables the generation of use case-specific data, such as for Smart Cities applications, this paper focuses on its generic functionalities with a bridge maintenance use case as the primary example.

4 Use Case: Predictive Maintenance for Bridges

Transportation systems have great importance for the development of the countries and economic growth [22].

Maintenance is essential to ensure the proper function of smart cities. It is crucial to monitor, track, analyze, diagnose, and control smart city systems. Bridges are an essential part of the infrastructure. There are various approaches and categories of maintenance techniques, from reactive to condition-based [41]. Condition assessment of existing bridges is usually carried out with periodic inspections. In most European countries, the examination is based on visual inspection and damage assessment, and this provides mainly qualitative insights [32]. Very often, professionals do not have enough or suitable data access to make the decisions in a quantitively and data-oriented way [5].

To predict load-carrying capacity, collecting data on existing bridges or reliable synthetic data generated from real conditions is still important. In addition, acoustic, thermal, and ground-penetrating radar inspection techniques may be used in some cases. These physical inspections can have challenges. During bridge inspections, traffic must be controlled, additional staff will be needed for operators and inspectors, access can be difficult, the bridge can be tight and hard to reach, and the operational costs might be high. In Germany, the documentation and inspections are done according to engineering structures on roads and pathways, DIN 1076, or railway bridges RIL 804. Continuous inspection of the bridges is essential to maintain them and ensure their service life.

High-quality synthetic data is a promising alternative to real-world inspection as it mimics the true data. Additionally, the generated tabular datasets can be applied to bridges and other infrastructure or systems that utilize tabular datasets. For example, tunnel systems [45, 46], stormwater and drainage monitoring, management [24], and management of movement in urban districts (mobility) optimize traffic flow and address congestion challenges [47]. Furthermore, these datasets play a pivotal role in hazard prediction at major events, enhancing preparedness and response strategies for various risks and emergencies [50].

5 Experiments

As previously described, the objective of this work is to evaluate the effectiveness of the generic data generation functionalities on the DaFne platform, outlined in Sect. 3, using a predictive maintenance use case, detailed in Sect. 4. This section aims to provide an in-depth explanation of the experiments conducted for this purpose. Section 5.1 presents a detailed description of the bridge dataset used in the experiments. The data generation functionalities of the DaFne platform are then employed to produce the three datasets utilized in the experiments, and this process is elucidated in Sect. 5.2. The evaluation procedure of synthetic data is presented in Sect. 5.3. The algorithms and models used for the predictive maintenance use case, specifically for predicting bridge conditions, will be elaborated in 5.4. The models' evaluation performance methodology is outlined in 5.5.

5.1 Bridge Data

The highway bridge data from the National Bridge Inventory (NBI) database for all the states of the USA for 13 years from 2010 to 2022 was used in this research study. The raw data has more than 130 features, but not all are relevant to prediction. The features selected for this analysis coincide with the prior literature dealing with the NBI data [9, 16, 18, 25, 30, 31, 42]. Later in Sect. 6.1, the descriptive analysis is described in detail and summarized in Tables 1 and 2. According to the Federal Highway Administration, the bridges are scheduled to be inspected once every 24 months. Still, in some cases, due to natural disasters or collisions, bridges require non-scheduled inspections [15]. The dataset contains inconsistency and missing values and had to be cleaned and prepared before analysis. The clean, ready-to-use dataset contains 471123 rows with information on 70898 bridges with only 24 features (12 numerical and 12 categorical).

The target feature, deck condition rating, is labeled from 0 (indicating out of service) to 9 (representing excellent). To address imbalances in label frequencies and enhance their description, the categories {0, 1, 2, 3, 4} are merged into a single one labeled 4, and similar categories {8, 9} are merged into category 8. This pragmatic approach streamlines the labeling structure and facilitates a more coherent analysis of the target feature.

5.2 Datasets: Original, Enrichment, and Synthesis

This section explains how the DaFne platform's functionalities modify, enrich, and synthesize the National Bridge Inventory (NBI) dataset, as outlined in Sect. 3. We created three distinct datasets:

- **Original Dataset**: Utilizing the platform's rule-based service, new attributes like *unique_ID* were added to the existing NBI Bridge dataset during preprocessing. This dataset thus includes the preprocessed NBI data with additional features.
- **Weather Dataset**: The fusion service, particularly the weather fusion component, appended weather data from the MERRA climate data [14] to the bridge dataset. This enriched *weather* dataset now includes twelve more columns of various annual average weather conditions, effectively merging bridge data with corresponding environmental data.
- **Synthetic Dataset**: Employing the reproduction service, a *synthetic* dataset was generated based on the *original* dataset. This new dataset mirrors the bridge data in terms of column distribution and relationships, aiming to serve as a viable substitute for the original data in performance assessments.

These datasets—original, weather, and synthetic—were used to predict bridge conditions in our predictive maintenance case study. The synthetic and original datasets are similar in structure, whereas the weather dataset is an enhanced version of the original, incorporating additional weather data in twelve more columns.

5.3 Evaluation of Synthetic Data

This section introduces certain criteria to explore whether the synthetic data sufficiently replicates real data behavior so that conducting studies on synthetic data can be a viable substitute for real-world investigations.

This study presents a multi-step approach for this evaluation. Initially, a comparison of descriptive statistics of the original and synthetic datasets is recommended, including parameters such as mean, median, and standard deviation (std), tails for numeric features, and shape (the count of distinct categories of a feature), mode, and frequencies of single classes for categorical features. A comparison of the distributions by the Kolmogorov-Smirnov (KS) statistical test [4] for equality of distribution function (goodness of fit tests) has to be considered with caution due to its weaknesses, discussed later. A more robust understanding is presented by a statistical distance between two distributions, served by Kullback-Leibler divergence (KLD) [26]. Its small values can be seen as an indicator of good quality of synthetic data.

Subsequently, the analysis extends to bivariate dependencies between pairs of features in the original and synthetic datasets. Given the heterogeneous nature of the data, comprising both numerical and categorical features, distinct methodologies are employed. Pearson's correlation coefficient [13] is proposed to compare bivariate dependencies among numeric features, while Pearson's contingency coefficient [3] is proposed for assessing dependencies between categorical features. A minor difference in correlation coefficients between the original and synthetic datasets would be an argument for using the synthetic dataset.

Moreover, the assessment extends to the efficacy of the modeling process. This entails evaluating the suitability of synthetic data by comparing their applicability for predictive purposes with original data. The following two sections present the introduction of modeling methods and the evaluation of their quality separately.

5.4 Predictive Maintenance Algorithm

This section provides an overview of all the ML models used in this research study, and a summary of each is presented below. The target variable is categorical, with five categories (4 to 8). Hence, the classification methods are used.

Multinominal Logistic Regression (MLR) is a statistical method used for predicting probabilities where there are three or more categories. The softmax function models the relationship between the influence features and these probabilities and the category with the highest predicted probability is assigned as the predicted outcome [48]. MLR offers the advantage of having an analytical solution for the parameters of the softmax function. However, its effectiveness is contingent on the assumption of symmetric and uncorrelated features, which may not be consistently met in real-world datasets. The next challenge is that it applies to numeric features. So, categorical features must be encoded, increasing the dimension again. This motivates to apply other methods coming from ML.

eXtreme Gradient Boosting (XGBoost) is a ML algorithm rooted in decision trees that operates as an ensemble of numerous small decision trees, and the collective prediction is derived from the majority consensus of individual tree predictions [10]. Notably, these smaller decision trees are grown independently, each fitted to a resampled original data with a distinct feature subset, ensuring their independence, but the resampling weights are dynamically adjusted, amplifying the influence of incorrectly classified observations [10]. Parameter tuning emerges as a crucial part here, serving as a pivotal measure to thwart overfitting.

Convolutional Neural Network (CNN) is a prevalent artificial NN architecture widely employed in computer vision tasks for the analysis of images and videos. At their core, CNNs employ convolutional layers that systematically apply filters to input images, facilitating the automatic learning of hierarchical features and patterns [17, 29]. As information traverses successive convolution, activation, and pooling layers, the acquired features progressively increase in complexity. CNNs showcase remarkable success in predicting time series, [40, 43]. Notably, when configured with a 1D convolutional layer as the initial layer and treating the features as channels, CNNs exhibit proficiency in handling temporal sequences.
Gate Recurrent Units (GRU) is a recurrent NN (RNN) architecture designed to handle long-range dependencies in sequential data and consists of recurrent units that selectively update and reset their internal states based on input information defined by Cho et al. [11]. GRUs introduce two mechanisms: an update gate and a reset gate. The update gate determines the proportion of the previous internal state that should be retained, and the reset gate decides which parts of the previous state should be forgotten, these gating mechanisms enable GRUs to capture and retain relevant information over the longer sequences [12]. The compact design of GRUs makes them computationally efficient compared to other time series Neural Networks.
The deep learning algorithms and their architectures adopted in this research are task-specific. The architecture of the models was optimized through hyperparameter optimization and several experimentations.

5.5 *Models Evaluation*

The aim of the deep learning metrics is to compare the utility of synthetic data and original data for the prediction task. The obtained results give an idea if the generated data is as good as real data or not for predicting the deck conditions ratings.

The selected metrics accuracy, F1 score, precision score, Cohen kappa score and Matthews correlation coefficient (MCC) are described below. Additionally, the training time is computed and considered as part of the evaluation. All the metrics described were selected because they are good performance indicators for multi-class classification tasks such as the predictive maintenance case study described in this research.
Accuracy: stated in [19] is the overall score of how much the model is correctly predicting over the entire dataset. It is computed as the number of corrected classified elements by a model divided by the total number of observations (number of elements in a dataset) [19]. The overall accuracy is computed in percentages from 0 to 100%.
Precision: defined as the portion of an instance of a particular class that is actually correctly predicted by a model (correctly classified) [19]. The overall precision is computed in percentages from 0 to 100%.
F1 Score: defined as the weighted average between precision and recall, where the latest one is the portion of correctly classified instances of a class from the total

number of instances of that specific class [19]. Then, F1 score is computed using a harmonic mean between these two metrics, where the best value of F1 is 1 and the worst is 0 [19]. The overall F1 score is computed in percentages from 0 to 100%.
Kappa Score: defined as the rating values between a model's prediction and the actual classification to determine the independence or dependence among them, it is given by a factor where 0 means the model's prediction is independent of the actual classification and when is 1 model's prediction is totally dependent from actual classification [19]. The overall Cohen Kappa score is computed in percentages from 0 to 100%.
Matthews correlation coefficient (MCC): generally defined as the correlation between observed and predicted classes, for the multi-class case strongly depends on the correctly classified elements of a dataset, where the maximum value is +1 and the minimum value is between −1 and 0 [19]. The overall MCC score is computed in percentages from 0 to 100%.

6 Results

This section explores the evaluation results of the synthetic dataset, as described in Sect. 5.3. Subsequently, it presents the prediction performance for the original dataset, both with and without the inclusion of weather data, and for the synthetic dataset, following Sects. 5.4 and 5.5.

6.1 Evaluation of Synthetic Data

Tables 1 and 2 present a descriptive analysis of numeric and categorical features, respectively, for the original (Panel A) and synthetic (Panel B) data, drawing comparisons between them (Panel C).

First, the numeric features are investigated by referring to Table 1. The means of the original and synthetic data exhibit minor differences if taking into account their standard deviations, reported in columns 'mean' and 'std' in Panels A and B. The two samples t-test (for equality of expectations) confirm that the means are significantly different, as indicated by the nearly zero p-values in column 'p-values t-test' of Panel C, with only two exceptions, namely 'traffic lines on' and 'structure length'. In both original and synthetic datasets, the median is always smaller than the mean, indicating the synthetic data mimics the skewness of the original features to the right. Notably, metrics such as mean, standard deviation (std), and range (min and max) are sensitive to outliers, serving as estimators for location and variation. The comparison of the 95%-quantiles with the range, indicates that some outliers lie extremely far from the rest of the original and synthetic data. The analysis of standard deviations produces more critical outcomes since their ratio, reported in column 'ratio std.' of Panel C, is much smaller than one. The synthetic features are less spread out,

Table 1 Descriptive statistics for numerical features: mean, median, min, max, std. for original data (Panel A) and for synthetic data (Panel B), and comparison of original and synthetic data: absolute difference in mean, the ration between std., percentage of values in original data lying outside the range of synthetic data, p-value of t-test for equality of means, p-values of Kolmogorov-Smirnov (KS) and Anderson tests for equality of distributions, KLD (Panel C)

Panel A: Original

Feature name	Mean	Median	Min	Max	Std	95%-quantile
GPS latitude Degrees						
GPS longitude Degrees						
Traffic lines on	2.21	2.0	1.0	80.0	1.05	4.0
Average daily traffic (adt)	7165.82	572.0	0.0	778093.0	18074.69	35250.0
Degrees skew	9.96	0.0	0.0	99.0	18.44	45.0
Main unit spans	2.91	2.0	0.0	502.0	3.77	7.0
Max span length	17.23	13.7	0.0	2091.8	15.22	37.8
Structure length	57.35	30.5	6.1	12109.1	133.74	166.4
Roadway width	10.09	8.3	0.0	228.0	5.56	20.8
Truck traffic in adt (p_adt)	8.65	8.0	0.0	99.0	8.44	25.0
Age	42.35	39.0	0.0	190.0	25.15	89.0
Years after reconstruction	36.36	32.0	0.0	140.0	22.36	81.0

Panel B: Synthetic

Feature name	Mean	Median	Min	Max	Std	95%-quantile
GPS latitude Degrees						
GPS longitude Degrees						
Traffic lines on	2.21	2.0	1.0	10.0	0.79	4.0
Average daily traffic (adt)	9269.15	6211.0	0	136984	11246.7	30116.04
Degrees skew	0.51	0.0	0.0	99.0	5.67	0.0
Main unit spans	2.47	1.0	0.0	27.0	2.04	7.0
Max span length	16.94	14.0	0.3	135.6	10.16	36.9
Structure length	57.31	34.9	6.1	1959.6	99.99	138.0
Roadway width	9.90	8.3	0.0	55.7	4.55	19.1
Truck traffic in adt (p_adt)	7.66	7.0	0	76	5.56	17.0
Age	44.42	42.0	0	159	20.62	81.0
Years after reconstruction	11.35	3.0	0.0	131.0	16.09	48.0

Panel C: Original versus synthetic

Feature name	Diff. mean	Ratio std.	% outliers	p-value t-test	p-value KS	KLD
GPS latitude Degrees					5e-323	0.010
GPS longitude Degrees					3e-323	0.007
Traffic lines on	0	0.75	0.07	0.58	5e-323	0.002
Average daily traffic (adt)	2103.33	0.62	0.32	0.0	1e-323	0.009
Degrees skew	9.45	0.31	0	0.0	1e-323	0.311
Main unit spans	0.43	0.54	0.27	0.0	2e-323	0.001
Max span length	0.28	0.67	0.07	0.0	5e-323	0.0
Structure length	0.07	0.75	0.05	0.84	1e-323	0.0
Roadway width	0.18	0.82	0.05	0.0	7e-323	0.003
Truck traffic in adt (p_adt)	0.99	0.66	0.02	0.0 0.009		
Age	1.94	0.82	0.02	0.0	2e-323	0.071
Years after reconstruction	25.00	0.72	0.01	0.0	0.0	0.020

Table 2 Categorical features

	Panel A: Original			Panel B: Synthetic		
Feature name	Shape	Mode	Percentage (>95%)	Shape	Mode	Percentage (>95%)
Toll	5	3	(1, 98.6%)	3	3	(1, 99.2%)
Maintenance	25	2	(4, 97.44%)	14	2	(3, 98.16%)
Functional class	12	9	(11, 96.59%)	12	9	(11, 97.6%)
Structure kind	10	5	(6, 97.2%)	9	5	(6, 97.83%)
Structure type	23	2	(6, 95.27%)	19	2	(5, 95.8%)
Deck structure type	10	1	(4, 96.31%)	9	1	(3, 95.98%)
Surface type	11	1	(5, 95.98%)	11	1	(4, 97.77%)
Membrane type	7	0	(2, 95.33%)	6	0	(2, 99.18%)
Deck protection	10	0	(3, 98.66%)	6	0	(3, 99.67%)
Deck condition	10	7	(4, 96.25%)	9	7	(4, 98.3%)
Superstructure condition	10	7	(4, 95.88%)	10	7	(4, 97.56%)
Substructure condition	10	7	(5, 96.74%)	10	7	(4, 96.32%)
Panel C: Original versus synthetic						
Feature name		*p*-value chi-test	KLD	% outliers		
Toll		1.939e-82	0.002	0.03		
Maintenance		0.0	0.018	0.2		
Functional class		1.720e-238	0.012	0		
Structure kind		1.146e-131	0.006	0.01		
Structure type		0.0	0.023	0.02		
Deck structure type		2.364e-96	0.013	0.02		
Surface type		1.462e-134	0.040	0		
Membrane type		2.214e-40	0.027	0.45		
Deck protection		5.894e-39	0.005	0.09		
Deck condition		2.755e-140	0.014	0.01		
Superstructure condition		2.827e-103	0.006	0		
Substructure condition		4.626e-11	0.012	0		

again because of extreme right-hand outliers. Furthermore, the Kolmogorov-Smirnov (KS) goodness of fit test rejects the equivalence of distributions of the original and synthetic datasets, as evidenced by the close-to-zero values in columns '*p* value KS'. The known weakness of KS test is that outliers lead to a rejection of the distribution equality even when the data distributions match. The values of KLD are small. The interpretation of KLD depends on the context of the nature of the data and application. Nevertheless, the values of KLD, reported in Panel C, are generally small enough, indicating a closer match between the original and synthetic data distributions.

Turning to categorical features, Table 2, a reduction in the shape is observed in synthetic data, implying the absence of certain categories present in the original dataset. This discrepancy is further confirmed by the chi-square goodness of fit test, reported in column '*p*-values chi-test' of Panel C. However, column '% outliers' of

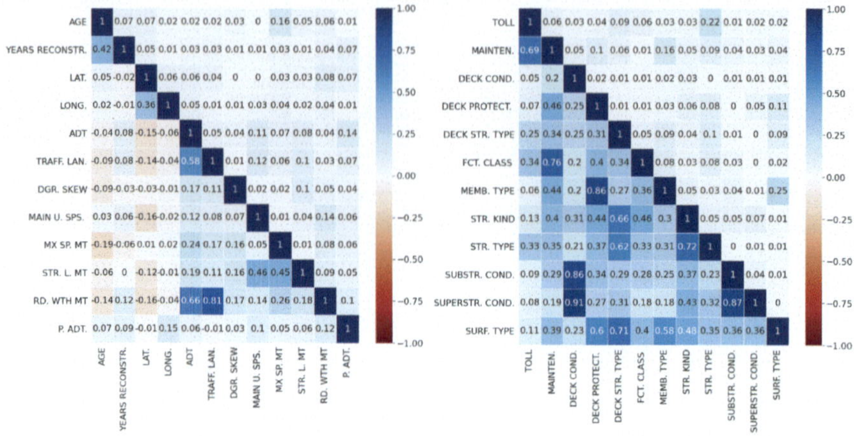

Fig. 2 Pearson's correlation coefficient for numeric features (left) and Pearson's contingency coefficient for categorical features (right)

Panel C presents the proportion of points from the original data that are not covered by the categories of the synthetic data. These percentages consistently are much smaller than 1%. Note that statistical and ML prediction models perform well in regions characterized by dense data and exhibit challenges in handling outliers and observations beyond the data support. Therefore, the reduction in the shapes is not problematic.

Figure 2 illustrates the outcomes of a bivariate dependency analysis conducted on numerical (left) and categorical (right) features, utilizing Pearson's correlation coefficient and Pearson's contingency coefficient, respectively. The lower-left part of the matrices visualizes the coefficients for the original dataset, while the upper-right part displays the differentials between coefficients for the original and synthetic datasets. The color intensity in the visualization corresponds to the magnitude of the coefficients. Notably, certain features exhibit strong dependencies in the original dataset, and the values for the synthetic dataset closely align with those of the original dataset since the differences are small.

In summarizing this section, the comprehensive descriptive analysis has brought attention to certain shortcomings in the synthetic data that can be improved in future research works. Specifically, the synthetic data fails to capture the tails observed in the original data for numerical features. Likewise, the absence of infrequent categories in the synthetic data is noteworthy for categorical features. Nevertheless, it is important to emphasize that, on the whole, the synthetic dataset demonstrates satisfactory quality.

Table 3 Performance measures for all methods. All the values are given in % except the last column training time, measured in minutes

Dataset	Method	Measures					
		Acc	F1	Precision	Kappa	MCC	Train time
Original	MLR	88.4	88.3	88.7	82.9	83.9	3.80
	XGBoost	**90.8**	**90.9**	**91.2**	**86.3**	**86.4**	**2.51**
	CNN	**90.7**	**90.7**	**90.9**	86.0	86.1	141.84
	GRUs	90.1	90.0	90.2	85.2	85.3	180
Weather	MLR	90.8	90.8	91.1	86.2	86.3	4.49
	XGBoost	90.5	90.6	90.9	85.8	85.9	3.70
	CNN	90.5	90.5	90.7	85.6	85.7	168.77
	GRUs	90.8	90.7	90.7	85.8	85.9	300
Synthetic	MLR	82.3	82.2	82.3	73.1	73.2	1.06
	XGBoost	83.3	83.3	83.5	74.8	74.9	0.52
	CNN	83.3	83.2	83.3	74.6	74.7	47.8
	GRUs	72.4	71.9	79.9	58.5	61.0	18.0

The significance of the bold numbers are showing the highest value of all methods

6.2 *Predictive Maintenance*

In the experiments, different models were constructed for the predictive maintenance task such as Multi-Logistic Regression (MLR) [48], Random Forest (RF) [6], K-Nearest Neighbor (KNN) [1], XGBoost [10], 1D Convolution Neural Network (CNN) [29], Long Short-Term Memory (LSTM) [21], Deep Neural Network (DNN) [37], Multi-Layer Perception (MLP) [37], and Gated Recurrent Units (GRUs) [11]. The bridge maintenance dataset spans 12 years, encompassing six inspections per bridge. Beyond the immediate inspection data, historical context is crucial for assessing the current deck condition. While GRU excels in capturing sequential dependencies, other methods require data transformation. For that reason, inspection records from both current and prior years serve as input, except for the target variables, deck, superstructure, and substructure conditions, which exclusively incorporate information from the preceding years. Table 3 highlights the top-performing models, with MLR, XGBoost, CNNs, and GRUs demonstrating the highest performance across real, synthetic, and weather data. Detailed descriptions of these models can be found in Sect. 5.4.

The four models demonstrate strong performance with real data but are less effective with synthetic data. While synthetic data is expected to match rather than exceed the quality of real data, there is still a noticeable gap, with an accuracy difference of about 7% points. This discrepancy can be partly attributed to the synthetic data's inability to capture all data categories. Moreover, the data synthesized by XGBoost and CNN displayed the best overall performance, followed by MLR and GRUs. Furthermore, a few difference in performance was noticed when the models made predictions using additional weather data compared to predictions with the original data.

In the predictive maintenance task evaluation, five metrics were used to assess the utility of data and especially the capabilities of the synthetic data for predicting the bridge's conditions. The description of each metric is mentioned in Sect. 5.5. Also, the results for each metric are given in Table 3, where acc refers to accuracy [19], $F1$ to the F1 score [19], $Kappa$ refers to Cohen Kappa score [19], and MCC to Matthew Correlation Coefficient [19].

7 Conclusion and Discussion

In this research, the evaluation of synthetic data was carried out from two different perspectives. The first one, using descriptive statistics compares real data versus synthetic data using different statistical measures such as median, mean, minimum, maximum, and standard deviation for continuous data, shape and mode for discrete data, and Kullback-Leibler Divergence distance for both types of data. Also, different types of statistical tests as described in Sect. 6 were carried out to evaluate how close is synthetic data to real data. The second perspective addresses the utility of synthetic data through ML models such as MLR, XGBoost, CNN, and GRUs and compares it with the performance of original data and weather data using the same models. The relevance of both perspectives for the study is demonstrated through the analysis of data and the experiment results.

7.1 (Weather) Fusion Service: Weather Data

The inclusion of the weather information in the weather dataset did not bring the expected increase in performance; instead, the performance remained approximately the same as compared to the original dataset. Further analysis revealed that the average weather values added (on a yearly basis) show minimal variation over the years, leading to no significant gain in information for assessing bridge conditions. Consequently, the weather dataset could not achieve better performance compared to the original dataset. The addition of weather data columns resulted in increased computational load without a corresponding increase in informational value. This insight is crucial, as it indicates that for predictive maintenance use cases like bridge condition assessment, yearly average weather values are insufficient. Instead, focusing on extreme values, such as minimum temperatures or the number of extremely hot days, might more precisely aid in predicting bridge conditions.

7.2 Reproduction Service: Synthetic Data

The comparison of real and synthetic data showed that the distributions of the synthetic data do not always align with the original data, mainly due to the absence of outliers in the synthetic data. The tails of these distributions, present in the real data, are notably missing in the synthetic versions. In terms of categorical features, certain categories underrepresented in the original data were not reproduced in the synthetic data.

As anticipated, synthetic data underperformed compared to the original data in the predictive maintenance task. This outcome aligns with the synthesis objective of achieving comparable performance (ML utility), rather than enhancement, allowing synthetic data to act as a viable alternative to original data for purposes like data sharing without breaching privacy regulations. Notably, while models like MLR, XGBoost, and CNN showed a performance drop of around 7% points, the GRU model, particularly reliant on sequential data, exhibited a more significant decline of 18% points. This suggests that the temporal dependencies in the synthetic dataset are not yet adequately represented, highlighting a critical area for future enhancement of the synthesis algorithm.

8 Future Work

The outcomes of this study have paved the way for several promising directions for future research, particularly in the domains of weather data integration and the enhancement of synthetic data quality.

As discussed, the weather data added to the original dataset did not yield an improvement in performance for the predictive maintenance use case under study. To address this, future efforts will focus on expanding the weather fusion service on the DaFne Platform to include a broader range of weather characteristics, particularly extreme values. This enhancement is expected to provide a more relevant and impactful set of data for such use cases. Subsequent validation of these new weather data attributes will be crucial. The re-evaluated datasets will be employed in the predictive maintenance use case to assess their utility and effectiveness in improving model performance.

As shown in the evaluation, there is a vital need to improve the quality of synthetic data. This will involve refining the algorithm used for generating synthetic data, potentially incorporating advanced learning techniques such as Training-by-Sampling for Conditional Models, as proposed by Xu et al. [49]. Such approaches are anticipated to better learn the data distribution of underrepresented categories within features. Additionally, a more nuanced representation of temporal relationships in time-step data is essential. Continual development of the synthesis algorithm is imperative to meet not only the inherent demands for high-quality data generation (in terms of statistical characteristics and sequential relationships) but also to fulfill the broader requirements of the platform: the generic applicability of the algorithm to a diverse array of previously unseen datasets. This endeavor will likely necessitate extensive testing and evaluation of various tabular datasets in conjunction with the synthesis algorithm.

Acknowledgements This work has been developed in the project DaFne (reference number: 01S21059) and is funded by the German Ministry of Education and Research (BMBF).

References

1. Abu Alfeilat HA, Hassanat AB, Lasassmeh O, Tarawneh AS, Alhasanat MB, Eyal Salman HS, Prasath VS (2019) Effects of distance measure choice on k-nearest neighbor classifier performance: a review. Big Data 7(4):221–248
2. Ahmed Z, Mohamed K, Zeeshan S, Dong X (2020) Artificial intelligence with multi-functional machine learning platform development for better healthcare and precision medicine. Database 2020:baaa010
3. Baak M, Koopman R, Snoek H, Klous S (2020) A new correlation coefficient between categorical, ordinal and interval variables with Pearson characteristics. Comput Stat Data Anal 152:107043
4. Berger VW, Zhou Y (2014) Kolmogorov-Smirnov test: overview. Statistics reference online, Wiley statsref
5. Bittencourt T, Futai M, da Conceição Neto A, Ribeiro D (2021) Digital transformation of bridges inspection, monitoring and maintenance processes. In: Bridge maintenance, safety, management, life-cycle sustainability and innovations. CRC Press, pp 11–30
6. Breiman L (2001) Random forests. Mach Learn 45:5–32
7. Brenninkmeijer B, de Vries A, Marchiori E, Hille Y (2019) On the generation and evaluation of tabular data using GANs. Radboud University, PhD dissertation
8. Calafiore A, Palmer G, Comber S, Arribas-Bel D, Singleton A (2021) A geographic data science framework for the functional and contextual analysis of human dynamics within global cities. Comput Environ Urban Syst 85:101539
9. Chang M, Maguire M, Sun Y (2017) Framework for mitigating human bias in selection of explanatory variables for bridge deterioration modeling. J Infrastruct Syst 23(3):04017002. https://doi.org/10.1061/(ASCE)IS.1943-555X.0000352
10. Chen T, Guestrin C (2016) Xgboost: a scalable tree boosting system. In: Proceedings of the 22nd ACM sigkdd international conference on knowledge discovery and data mining, pp 785–794
11. Cho K, van Merriënboer B, Bahdanau D, Bengio Y (2014) On the properties of neural machine translation: encoder–decoder approaches. In: Wu D, Carpuat M, Carreras X, Vecchi EM (eds) Proceedings of SSST-8, eighth workshop on syntax, semantics and structure in statistical translation. Association for Computational Linguistics, Doha, Qatar, pp 103–111. https://doi.org/10.3115/v1/W14-4012, https://aclanthology.org/W14-4012
12. Chung J, Gulcehre C, Cho K, Bengio Y (2014) Empirical evaluation of gated recurrent neural networks on sequence modeling. In: NIPS 2014 workshop on deep learning
13. Cohen I, Huang Y, Chen J, Benesty J (2009) Pearson correlation coefficient. Noise reduction in speech processing, pp 1–4
14. Federal Highway Administration: Long-term pavement performance (ltpp) program. https://infopave.fhwa.dot.gov/Data/DataSelection. Accessed 31 March 2023
15. Federal Highway Administration: National Bridge Inventory. https://www.fhwa.dot.gov/bridge/nbi.cfm. Accessed: 26 Nov 2023

16. Ghonima O, Schumacher T, Unnikrishnan A, Fleischhacker A (2018) Advancing bridge technology, task 10: statistical analysis and modeling of us concrete highway bridge deck performance–internal final report. Civil and Environmental Engineering Faculty Publications and Presentations, vol 443. https://archives.pdx.edu/ds/psu/26267
17. Goodfellow I, Bengio Y, Courville A (2016) Deep learning. MIT Press
18. Goyal R, Whelan MJ, Cavalline TL (2017) Characterising the effect of external factors on deterioration rates of bridge components using multivariate proportional hazards regression. Struct Infrastruct Eng 13(7):894–905. https://doi.org/10.1080/15732479.2016.1217888
19. Grandini M, Bagli E, Visani G (2020) Metrics for multi-class classification: an overview. arXiv preprint arXiv:2008.05756
20. Hittmeir M, Ekelhart A, Mayer R (2019) On the utility of synthetic data: an empirical evaluation on machine learning tasks. In: Proceedings of the 14th international conference on availability, reliability and security, pp 1–6
21. Hochreiter S, Schmidhuber J (1997) Long short-term memory. Neural comput 9(8):1735–1780
22. Ivanova E, Masarova J (2013) Importance of road infrastructure in the economic development and competitiveness. Econ Manage 18(2):263–274
23. Jagatheesaperumal SK, Rahouti M, Ahmad K, Al-Fuqaha A, Guizani M (2022) The duo of artificial intelligence and big data for industry 4.0: applications, techniques, challenges, and future research directions. IEEE Internet Things J 9:12861–12885. https://doi.org/10.1109/jiot.2021.3139827
24. Keung KL, Lee CKM, Ng KKH, Yeung CK (2018) Smart city application and analysis: real-time urban drainage monitoring by IoT sensors: a case study of Hong Kong. In: 2018 IEEE international conference on industrial engineering and engineering management (IEEM). IEEE, pp 521–525. https://doi.org/10.1109/IEEM.2018.8607303
25. Kim YJ, Queiroz LB (2017) Big data for condition evaluation of constructed bridges. Eng Struct 141:217–227. https://doi.org/10.1016/j.engstruct.2017.03.028, https://www.sciencedirect.com/science/article/pii/S0141029617308994
26. Kullback S, Leibler RA (1951) On information and sufficiency. Ann Math Stat 22(1):79–86. https://doi.org/10.1214/aoms/1177729694
27. Kumar A, Singla S, Kumar A, Bansal A, Kaur A (2022) Efficient prediction of bridge conditions using modified convolutional neural network. Wirel Pers Commun 125(1):29–43
28. Lee KM, Yoo J, Kim SW, Lee JH, Hong J (2019) Autonomic machine learning platform. Int J Inf Manage 49:491–501
29. Lin M, Chen Q, Yan S (2013) Network in network. CoRR **abs/1312.4400**, https://api.semanticscholar.org/CorpusID:16636683
30. Liu H, Zhang Y (2020) Bridge condition rating data modeling using deep learning algorithm. Struct Infrastruct Eng 16:10(2):1447–1460. https://doi.org/10.1080/15732479.2020.1712610
31. Madanat SM, Mishalani RG, Ibrahim WHW (1995) Estimation of infrastructure transition probabilities from condition rating data. J Infrastruct Syst 1:120–125
32. Mandić Ivanković A, Skokandić D, Kušter Marić M, Srbić M (2021) Performance-based ranking of existing road bridges. Appl Sci 11(10). https://doi.org/10.3390/app11104398, https://www.mdpi.com/2076-3417/11/10/4398
33. Mannino M, Abouzied A (2020) Synner: generating realistic synthetic data. In: Proceedings of the 2020 ACM SIGMOD international conference on management of data. Association for Computing Machinery, pp 2749–2752
34. Miao P (2021) Prediction-based maintenance of existing bridges using neural network and sensitivity analysis. Adv Civ Eng 2021:1–17
35. Mockaroo LLC: Mockaroo—random data generator and API mocking tool. https://www.mockaroo.com/. Accessed: 19 Nov 2023
36. MOSTLY AI Solutions MP GmbH.: Mostly AI—the synthetic data company. https://mostly.ai/. Accessed: 19 Nov 2023
37. Neagoe VE, Ciotec AD, Cucu GS (2018) Deep convolutional neural networks versus multilayer perceptron for financial prediction. In: 2018 International conference on communications (COMM). IEEE, pp 201–206

38. Noruzman A, Ghani N, Zulkifli N (2021) Gretel. AI: open-source artificial intelligence tool to generate new synthetic data. Malays J Innov Eng Appl Soc Sci (MYJIEAS) 1(01):15–22
39. Patki N, Wedge R, Veeramachaneni K (2016) The synthetic data vault. In: 2016 IEEE international conference on data science and advanced analytics (DSAA). IEEE, pp 399–410 (2016)
40. Purnawansyah P, Haviluddin H, Alfred R, Gaffar AFO (2017) Network traffic time series performance analysis using statistical methods. Knowl Eng Data Sci 1(1):1–7
41. Ran Y, Zhou X, Lin P, Wen Y, Deng R (2019) A survey of predictive maintenance: systems, purposes and approaches. arXiv preprint arXiv:1912.07383
42. Sanders DH, Zhang Y (1994) Bridge deterioration models for states with small bridge inventories. Transp Res Rec
43. Singh J, Tripathi P (2017) Time series forecasting using back propagation neural network with ADE algorithm. Int J Eng Tech Res 7(5):265026
44. The Synthetic Data Vault: Sdmetrics. https://docs.sdv.dev/sdmetrics/. Accessed: 19 Nov 2023
45. Tichý T, Brož J, Bělinová Z, Kouba P (2020) Predictive diagnostics usage for telematic systems maintenance. In: 2020 Smart City symposium Prague (SCSP), pp 1–6. https://api.semanticscholar.org/CorpusID:220472253
46. Tichý T, Brož J, Bělinová Z, Pirník R (2021) Analysis of predictive maintenance for tunnel systems. Sustainability 13(7). https://doi.org/10.3390/su13073977, https://www.mdpi.com/2071-1050/13/7/3977
47. Wang A, Zhang A, Chan EHW, Shi W, Zhou X, Liu Z (2021) A review of human mobility research based on big data and its implication for smart city development. ISPRS Int J Geo-Inf 10(1). https://doi.org/10.3390/ijgi10010013, https://www.mdpi.com/2220-9964/10/1/13
48. Wu Fung S, Tyrväinen S, Ruthotto L, Haber E (2020) Admm-softmax: an admm approach for multinomial logistic regression. ETNA Electron Trans Numer Anal 52:214–229. https://doi.org/10.1553/etna_vol52s214
49. Xu L, Skoularidou M, Cuesta-Infante A, Veeramachaneni K (2019) Modeling tabular data using conditional GAN. In: Wallach H, Larochelle H, Beygelzimer A, d' Alché-Buc F, Fox E, Garnett R (eds) Advances in neural information processing systems, vol 32. Curran Associates, Inc. https://proceedings.neurips.cc/paper/2019/file/254ed7d2de3b23ab10936522dd547b78-Paper.pdf
50. Zhang Y, Geng P, Sivaparthipan C, Muthu BA (2021) Big data and artificial intelligence based early risk warning system of fire hazard for smart cities. Sustain Energy Technol Assess 45:100986. https://doi.org/10.1016/j.seta.2020.100986, https://www.sciencedirect.com/science/article/pii/S2213138820314144
51. Zhao Z, Kunar A, Birke R, Chen LY (2022) CTAB-GAN+: enhancing tabular data synthesis. arXiv preprint arXiv:2204.00401. https://doi.org/10.48550/ARXIV.2204.00401, https://arxiv.org/abs/2204.00401

Cyber Threat Hunting Using Large Language Models

Vinayak Tanksale

Abstract Cyber threat hunting plays a critical role in proactively identifying and mitigating potential cybersecurity threats. The emergence of large language models (LLMs) has increased interest in the application of LLMs in cyber threat hunting. This research paper investigates the utilization of LLMs for cyber threat hunting and provides an overview of their potential benefits, challenges, and future directions. LLMs possess powerful language understanding capabilities and can process vast amounts of data, enabling them to uncover hidden patterns, detect anomalies, and identify vulnerabilities. By incorporating LLMs into workflows, organizations can enhance their ability to detect and respond to emerging threats in real time. The adoption of LLMs in threat hunting raises challenges related to bias and fairness, privacy, and computational efficiency. This paper discusses these challenges and explores potential solutions. The paper outlines future directions, such as improving LLM training with cybersecurity-specific domains and integrating contextual knowledge into LLM-based threat hunting approaches.

Keywords Large language models · Cybersecurity · Machine learning · Cyber threat hunting · Artificial intelligence

1 Introduction

The field of cybersecurity is constantly evolving as threat actors devise increasingly sophisticated techniques to bypass traditional security measures. To combat these evolving threats, cybersecurity professionals are continually exploring new approaches and technologies. One such technology that has gained significant attention and holds immense potential is large language models (LLMs). LLMs are advanced artificial intelligence (AI) models trained on vast amounts of text data,

V. Tanksale (✉)
Ball State University, Muncie, IN 47306, USA
e-mail: vjtanksale@bsu.edu
URL: http://vinayaktanksale.com

X.-S. Yang et al. (eds.), *Proceedings of Ninth International Congress on Information and Communication Technology*, Lecture Notes in Networks and Systems 1000,
https://doi.org/10.1007/978-981-97-3289-0_50

enabling them to understand and generate human-like language. These models have found applications in various domains, including natural language processing, content generation, and now, cyber threat hunting.

Cyber threat hunting is a proactive approach to cybersecurity that focuses on actively searching for and detecting threats that may have bypassed traditional security defenses [2],[?]. It involves analyzing vast amounts of data, identifying anomalies, and investigating potential indicators of compromise (IOCs). Large language models have the potential to enhance cyber threat hunting by augmenting the capabilities of human analysts. LLMs can assist in automating the analysis of security logs, parsing and understanding threat intelligence reports, and identifying patterns or linguistic markers that may indicate malicious activity. By leveraging the power of LLMs, cybersecurity professionals can gain valuable insights, improve the efficiency of their threat hunting efforts, and stay one step ahead of rapidly evolving cyber threats [16].

This research paper aims to explore the intersection of large language models and cyber threat hunting. It investigates the potential applications of LLMs in various stages of the threat hunting lifecycle, including data analysis, anomaly detection, IOC generation, and threat intelligence analysis. Furthermore, it examines the challenges and considerations associated with integrating LLMs into existing cybersecurity frameworks, such as privacy concerns, model biases, and interpretability. By shedding light on the capabilities and limitations of LLMs in the context of cyber threat hunting, this research paper aims to provide valuable insights for cybersecurity professionals seeking to leverage this emerging technology.

By examining the potential of large language models in the context of cyber threat hunting, this research paper contributes to the growing body of knowledge on the application of AI technologies in cybersecurity. Through a comprehensive analysis of the capabilities, limitations, and challenges associated with using LLMs for threat hunting [10][?,?] this research aims to empower cybersecurity professionals with the necessary insights to make informed decisions about integrating LLMs into their existing security frameworks. By harnessing the power of LLMs, cybersecurity professionals can enhance their threat hunting capabilities, improve their defenses against emerging threats, and ultimately safeguard critical assets and data from malicious actors.

2 Overview and Foundations

2.1 Cyber Threat Hunting

Cyber threat hunting refers to the proactive and iterative process of detecting, investigating, and mitigating potential cyber threats that evade traditional security defenses [7]. It involves actively searching for signs of malicious activity within a network environment, identifying and analyzing indicators of compromise, and taking necessary actions to neutralize threats. Cyber threat hunting plays a crucial role in modern cybersecurity, as it enables organizations to detect and respond to threats before they can cause significant damage. By taking a proactive approach, cyber threat hunting helps to reduce dwell time, enhance incident response capabilities, and mitigate the potential impact of cyber attacks [8]. It empowers security teams to stay one step ahead of adversaries and protect critical systems and sensitive information. The importance of cyber threat hunting in the contemporary threat landscape cannot be overstated, as it helps organizations to augment their defensive strategies and maintain a resilient cybersecurity posture.

Cyber threat hunting involves a systematic and proactive approach to detect and mitigate potential threats in an organization's network environment [20]. It consists of several key components that collectively contribute to the effectiveness of the hunting process [1]. First, comprehensive threat intelligence plays a crucial role in providing up-to-date information on the latest attack vectors, known threat actors, and emerging vulnerabilities. Second, advanced data analytics techniques, including machine learning algorithms and behavior analytics, help identify anomalous patterns and indicators of compromise within large datasets. Third, robust network and endpoint visibility tools enable the monitoring and analysis of network traffic, system logs, and user activities, providing valuable insights into potential threats. Additionally, collaborative teamwork and effective communication between cybersecurity professionals and IT operations teams facilitate the sharing of knowledge and expertise, enabling a proactive response to threats. The combination of these key components enhances the organization's ability to proactively detect, investigate, and mitigate cyber threats, strengthening its overall cybersecurity posture.

2.2 Large Language Models

Large language models [23] are trained using vast amounts of textual data, enabling them to understand and generate human-like text. The training process typically involves utilizing deep learning techniques, such as transformer architectures, which have demonstrated exceptional performance in natural language processing tasks. LLMs like OpenAI's GPT-3 employ a transformer architecture consisting of multiple layers of self-attention mechanisms, enabling them to capture long-range dependencies and contextual information within the text. During training, LLMs are exposed to diverse datasets, including books, articles, and web content, allowing them to learn patterns, grammar, and semantics [5]. The training process incorporates techniques like unsupervised learning and self-supervised learning, where LLMs learn to predict missing words or generate coherent text based on context [19]. The resulting trained LLMs can then be fine-tuned for specific applications, such as cyber threat hunting, to leverage their language processing capabilities and contextual understanding.

LLMs have begun to play a significant role in cybersecurity, offering new possibilities for threat detection, analysis, and response [14]. LLMs excel at natural language processing tasks, enabling them to analyze and understand vast amounts of textual data, including security logs, threat intelligence reports, and online discussions. By leveraging their language processing capabilities, LLMs can aid in identifying patterns, detecting anomalies, and uncovering potential cyber threats. They can assist in automated threat hunting, generating security alerts, and even providing contextual information to support incident response activities [21]. LLMs can also contribute to the development of more robust security systems, assisting in the creation of advanced intrusion detection and prevention systems. However, ethical considerations, adversarial attacks, and potential biases within LLMs are challenges that need to be carefully addressed to ensure their responsible and effective integration within cybersecurity practices.

3 LLMs for Cyber Threat Hunting

3.1 Log Analysis and Anomaly Detection

Large language models have emerged as powerful tools for log analysis and anomaly detection in cybersecurity [22],[?]. Logs generated by various systems and applications contain valuable information about system activities, user behaviors, and potential security incidents. LLMs can process and interpret log data, enabling organizations to extract meaningful insights and detect anomalies that may indicate security breaches or suspicious activities. By leveraging their natural language processing capabilities, LLMs can understand log entries, identify patterns, and recognize deviations from normal behavior [6],[?]. They can learn the contextual relationships between log events, aiding in the detection of complex attack scenarios and zero-day exploits. Furthermore, LLMs can assist in automating the analysis of log data, reducing the manual effort required for anomaly detection and providing security analysts with timely alerts.

LLMs can be trained on diverse log datasets, including system logs, network logs, and application logs, to learn the normal behavior of the environment [25]. By comparing incoming log entries to the learned representations, LLMs can identify deviations and anomalies. This approach offers a proactive way to detect potential threats and security incidents. Moreover, LLMs can be combined with other machine learning techniques, such as clustering algorithms or outlier detection methods, to enhance the accuracy of anomaly detection and reduce false positives.

In addition to log analysis, LLMs can also be utilized for log generation. They can generate synthetic log entries based on learned patterns, which can be used to augment existing log datasets for training anomaly detection models. Synthetic logs can introduce new scenarios and attack vectors, helping to improve the resilience of anomaly detection systems and prepare them for emerging threats.

However, the application of LLMs in log analysis and anomaly detection also poses challenges. The sheer volume and velocity of log data require efficient processing and scaling of LLMs. Additionally, ensuring the interpretability and explainability of LLM-based anomaly detection models is crucial for building trust and facilitating effective decision-making by security analysts. Ongoing research is exploring techniques to address these challenges and optimize the integration of LLMs into log analysis and anomaly detection pipelines.

3.2 Generation of Indicators of Compromise (IOCs)

Large Language Models have shown great potential in the generation of Indicators of Compromise [15],[?], which are crucial in detecting and mitigating cyber threats [13]. IOCs are specific artifacts or patterns that indicate the presence of malicious activities or compromises within a system or network. LLMs can leverage their language generation capabilities to automatically generate IOCs based on various contextual factors, such as known attack techniques, threat intelligence reports, or historical data [18]. By understanding the semantics and syntax of IOCs, LLMs can generate accurate and contextually relevant indicators, aiding in the identification and prevention of cyber threats.

The use of LLMs in IOC generation offers several advantages. Firstly, LLMs can generate a wide range of IOCs, including file hashes, IP addresses, domain names, URLs, and behavioral patterns, providing a comprehensive set of indicators for threat detection. Secondly, LLMs can adapt to evolving threat landscapes by learning from large-scale datasets and incorporating the latest attack techniques and trends. This adaptability enables LLMs to generate IOCs that reflect the current threat landscape and anticipate emerging threats. Additionally, LLM-generated IOCs can assist in automated threat hunting and incident response, enabling faster and more efficient detection and mitigation of security incidents.

To ensure the effectiveness and quality of LLM-generated IOCs, careful training and fine-tuning of the models are essential. LLMs can be trained on diverse datasets, including historical attack data, threat intelligence reports, and security blogs, to learn the patterns and characteristics of IOCs. Fine-tuning the models with domain-specific data further enhances their ability to generate accurate and relevant IOCs for specific industries or organizations [4],[?]. Moreover, continuous model evaluation and validation are crucial to identify any biases or limitations in IOC generation and improve the overall performance and reliability of LLM-based IOC generation systems.

However, challenges exist in the utilization of LLMs for IOC generation. The potential generation of false-positive or false-negative IOCs is a concern that needs to be addressed. Adequate validation mechanisms and human oversight are necessary to verify the generated IOCs before operational deployment. Additionally, privacy and security considerations should be taken into account when using LLMs for IOC generation, as the models may inadvertently reveal sensitive information or inadvertently learn and reproduce malicious patterns.

Despite these challenges, LLMs offer a promising avenue for improving the efficiency and effectiveness of IOC generation in cybersecurity. By harnessing the language generation capabilities of LLMs, organizations can enhance their threat detection capabilities and proactively defend against evolving cyber threats.

3.3 Threat Intelligence Analysis

Threat intelligence analysis plays a crucial role in identifying and mitigating cyber threats. With the emergence of large language models (LLMs), such as GPT-3, there is a growing interest in leveraging these models to enhance threat intelligence analysis. LLMs have shown remarkable capabilities in natural language processing, understanding context, and generating human-like text. This opens up new possibilities for extracting valuable insights from vast amounts of unstructured threat intelligence data.

One key area where LLMs can be utilized in threat intelligence analysis is in automating the process of extracting and categorizing threat intelligence information. LLMs can be trained to understand and interpret threat reports, security blogs, forums, and other sources of threat intelligence. They can analyze these texts, identify relevant indicators of compromise (IOCs), extract key information such as malware names, attacker techniques, and affected systems, and categorize the threats based on severity or threat actor attribution. This automation helps in reducing the manual effort required for threat intelligence analysis and enables faster response to emerging threats.

Furthermore, LLMs can assist in identifying patterns and correlations across different threat intelligence sources. By processing a wide range of textual data, LLMs can identify commonalities in attack vectors, TTPs (tactics, techniques, and procedures), and threat actor behaviors. This can enable analysts to detect emerging trends, identify previously unknown threat campaigns, and proactively update defenses. LLMs can also be utilized to identify context-based relationships between threat actors, campaigns, and targeted industries or sectors, providing valuable insights for targeted threat intelligence.

Another potential application of LLMs in threat intelligence analysis is in the generation of high-quality threat reports. LLMs can generate human-like narratives based on the analyzed threat intelligence data, providing detailed and actionable information to security teams and stakeholders. These generated reports can include summaries of threat campaigns, recommended mitigation strategies, and indicators of compromise. By automating the report generation process, LLMs enable analysts to focus on higher-level analysis and decision-making tasks.

Moreover, LLMs can contribute to the improvement of threat intelligence sharing and collaboration. By understanding and interpreting textual descriptions of threat intelligence, LLMs can help in the extraction of relevant information for sharing with trusted partners and communities. LLMs can assist in anonymizing sensitive data, categorizing threat intelligence based on sharing agreements and standards, and facilitating the exchange of actionable intelligence in a timely manner. This enhances the collective defense against cyber threats and promotes a proactive and collaborative approach to cybersecurity.

Several research studies have explored the potential of LLMs in threat intelligence analysis. For example, Kim et al. [15] proposed a framework for generating indicators of compromise (IOCs) based on language models. They demonstrated the effectiveness of their approach in extracting IOCs from threat reports. Additionally, Garg et al. [12] presented a method that utilizes LLMs for threat intelligence analysis and categorization, achieving promising results in identifying threat types and classifying threat severity.

The utilization of large language models (LLMs) in threat intelligence analysis offers significant advantages in automating the extraction, categorization, and correlation of threat intelligence data. LLMs can enhance the speed and accuracy of threat analysis, improve the generation of threat reports, and facilitate information sharing and collaboration among security stakeholders. While challenges related to model bias, interpretability, and data privacy need to be addressed, the potential benefits of incorporating LLMs into threat intelligence analysis are promising, paving the way for more advanced and efficient cybersecurity practices.

4 Evaluation

4.1 Metrics

Metrics and criteria play a significant role in assessing the capabilities and limitations of LLMs in cyber threat hunting. This section explores key metrics and criteria that can be used for evaluating LLMs in threat hunting scenarios. One important metric for evaluating LLMs is the accuracy of threat detection. This metric measures the LLM's ability to correctly identify and classify threats from various sources of data, such as log files, network traffic, and security alerts. Accuracy can be determined by comparing the LLM's output with ground truth labels provided by domain experts or by analyzing the true positive and false positive rates. Additionally, metrics such as precision, recall, and F1 score can provide insights into the LLM's performance in identifying true threats while minimizing false positives.

Another crucial criterion is the interpretability of LLMs. Threat hunters need to understand how LLMs arrive at their predictions to gain insights into the underlying threat landscape. Interpretability metrics assess the ability of LLMs to provide explanations for their decisions, such as highlighting important features, providing contextual information, or generating human-readable explanations. Metrics like explanation fidelity, clarity, and relevance can be employed to evaluate the interpretability of LLMs.

The scalability and efficiency of LLMs are also important considerations in threat hunting. Metrics related to model size, inference time, and resource utilization can gauge the scalability and efficiency of LLMs. These metrics help assess the feasibility of deploying LLMs in real-time threat hunting scenarios, especially when dealing with large volumes of data or time-sensitive investigations. Additionally, criteria such as training time and memory requirements can inform decisions regarding the practicality and resource allocation for LLM deployment.

Robustness and resilience are critical criteria for evaluating LLMs in threat hunting. Robustness refers to the LLM's ability to handle various types of threats, including novel or adversarial examples. Metrics like adversarial robustness, generalization performance, and vulnerability analysis can provide insights into the LLM's resilience against attacks or attempts to deceive the model. Evaluating LLMs under different threat scenarios and perturbations helps assess their robustness in real-world threat hunting environments. Lastly, user satisfaction and usability are subjective but important criteria for evaluating LLMs in threat hunting. User feedback, surveys, and usability testing can be conducted to gather insights on the ease of use, user interface design, and overall satisfaction with the LLM-based threat hunting system. These metrics assess the practicality and user-friendliness of LLMs, considering the perspectives and needs of threat hunters and security analysts.

Several research studies have contributed to evaluating LLMs in the context of threat hunting. For instance, Chen et al. [6] proposed a methodology for evaluating the accuracy and interpretability of LLMs in threat detection tasks. They developed metrics based on interpretability fidelity and semantic consistency to assess LLM performance. Additionally, Nguyen et al. [17] presented a framework for evaluating the scalability and efficiency of LLMs in large-scale log analysis, considering factors such as model size, processing time, and memory usage.

Metrics and criteria for evaluating LLMs in threat hunting encompass accuracy, interpretability, scalability, robustness, and user satisfaction. These metrics provide insights into the LLM's performance, explainability, efficiency, resilience, and usability in real-world threat hunting scenarios. By considering these evaluation aspects, organizations can make informed decisions regarding the deployment and utilization of LLMs in their threat hunting processes, ultimately enhancing their cybersecurity posture.

4.2 Real-Time Analysis

The performance and efficiency of large language models in real-time analysis are critical factors that determine their effectiveness in processing and analyzing data in a timely manner. This section explores the performance and efficiency aspects of LLMs in real-time analysis and their implications for various applications. One key aspect of performance is the speed of LLM inference. Real-time analysis requires LLMs to process and generate predictions within strict time constraints. The inference speed of LLMs can be measured in terms of tokens per second or queries per second. Faster inference enables LLMs to handle high-volume data streams and respond to threats in real time. Techniques such as model optimization, parallel processing, and hardware acceleration can be employed to improve the inference speed of LLMs.

Another important consideration is the memory footprint of LLMs. Efficient memory utilization is crucial for real-time analysis as it directly impacts the scalability and responsiveness of the system. LLMs with large memory requirements may face challenges when processing large-scale datasets or operating in resource-constrained environments. Techniques such as model compression, quantization, and sparsity optimization can help reduce the memory footprint of LLMs without significantly sacrificing performance.

The scalability of LLMs in real-time analysis is a crucial factor, especially in scenarios with rapidly increasing data volumes. Scalability refers to the ability of LLMs to handle larger datasets or higher query rates without compromising performance. Distributed computing frameworks and parallel processing techniques can be employed to scale LLM-based systems horizontally, allowing them to process data in parallel and handle increased workloads effectively.

Energy efficiency is another dimension of efficiency that is gaining importance in real-time analysis. LLMs often require substantial computational resources, which can result in high energy consumption. Optimizing the energy efficiency of LLMs can be achieved through techniques such as model compression, low-power hardware architectures, and intelligent resource allocation. Energy-efficient LLMs enable longer operational times and reduce the environmental impact of real-time analysis systems. Dai et al. [9] proposed a method to improve the inference speed of LLMs by leveraging multi-threading and GPU acceleration. They demonstrated significant speedups in real-time sentiment analysis tasks. Wang et al. [24] introduced a framework for efficient and scalable LLM-based anomaly detection, which utilized parallel computing techniques and achieved high performance in processing large-scale log data.

The performance and efficiency of LLMs in real-time analysis are crucial for their successful deployment in various applications. Improving the inference speed, memory footprint, scalability, and energy efficiency of LLMs enables timely and effective analysis of data streams, enhancing the responsiveness and reliability of real-time analysis systems. By considering these aspects, organizations can select and optimize LLM-based solutions that meet their specific requirements and operational constraints.

4.3 Bias and Fairness

Assessing model bias and fairness in threat detection using large language models is crucial to ensure equitable and unbiased outcomes in cybersecurity systems. LLMs have gained popularity due to their ability to process and analyze vast amounts of textual data, including threat intelligence reports and online content. However, like any machine learning model, LLMs can be susceptible to bias, and it is essential to evaluate their fairness to avoid discriminatory outcomes.

One approach to assessing model bias and fairness in LLMs is to examine the training data. Biases present in the training data can be reflected in the model's outputs. Therefore, it is important to ensure that the training data is diverse, representative, and free from biases. Furthermore, methods such as debiasing techniques and data augmentation can be applied to reduce biases and promote fairness during the training phase of LLMs.

Several metrics and evaluation techniques can be employed to assess model bias and fairness in LLMs. For example, the disparate impact ratio measures the ratio of positive predictions for different demographic groups. The equal opportunity difference quantifies the difference in true positive rates across different groups. Other metrics such as the false positive and false negative rates can also be used to evaluate fairness. Additionally, fairness-aware fine-tuning methods can be applied to mitigate biases and enhance fairness in LLMs.

Research has been conducted on assessing model bias and fairness in LLMs. Bolukbasi et al. [3] proposed a method to learn gender-neutral word embeddings, aiming to mitigate gender biases in natural language processing tasks. Gonen and Goldberg [11] introduced a method for quantifying and mitigating biases in word embeddings. These studies provide insights and techniques for addressing biases in LLMs and promoting fairness in threat detection applications.

To assess model bias and fairness in LLMs, it is crucial to consider the social and ethical implications. Transparency in model development, documentation, and decision-making processes is vital to ensure accountability and enable external auditing. Additionally, involving diverse stakeholders and experts from various domains can help identify and address potential biases in LLMs, fostering a collaborative and inclusive approach toward fairness.

Assessing model bias and fairness in threat detection using LLMs is essential for promoting equitable and unbiased outcomes. By applying appropriate evaluation metrics, considering diverse training data, and employing fairness-aware techniques, biases in LLMs can be identified and mitigated. The continuous efforts toward assessing and enhancing fairness in LLMs contribute to the development of more reliable and unbiased threat detection systems, promoting fairness and equality in cybersecurity.

5 Future Directions and Challenges

One future direction for large language models in cyber threat hunting is the integration of explainability and interpretability techniques. As LLMs become more sophisticated, understanding the reasoning behind their predictions becomes crucial. Research, such as the work by Ribeiro et al. [21], on explaining the predictions of complex models can be leveraged to provide insights into the decision-making process of LLMs and improve their transparency.

Enhancing the robustness and resilience of LLMs to adversarial attacks is an ongoing challenge. Adversarial attacks aim to manipulate LLMs' behavior by introducing subtle changes to input data. Research, such as the work by Mateusz et al. [14], on poisoning attacks and countermeasures can inform the development of defense mechanisms to mitigate the impact of such attacks on LLMs used in cyber threat hunting.

Scaling LLMs to handle the increasing volume and complexity of cybersecurity data is a pressing future direction. Efficient architectures and distributed computing techniques, such as those explored by Bonawitz et al. [4] in federated learning, can enable LLMs to process and analyze massive datasets while maintaining privacy and security. Addressing ethical considerations and biases in LLMs is crucial for their responsible deployment in cyber threat hunting.

Collaborative efforts and shared datasets are essential for advancing LLMs in cyber threat hunting. Incorporating real-time analysis capabilities into LLMs for proactive threat hunting is an exciting future direction. Real-time monitoring and analysis can enable faster response to emerging threats. Metrics and criteria specific to LLMs can be devised to assess their performance, accuracy, and resilience against attacks.

The use of LLMs in threat hunting raises important privacy concerns that need to be addressed. LLMs are trained on large amounts of data, which may include sensitive and private information. The models have the potential to learn and retain information about individuals and organizations, leading to privacy risks if not handled properly. It is crucial to establish robust privacy protection mechanisms when utilizing LLMs in threat hunting scenarios. To address privacy concerns, researchers and practitioners need to explore methods for privacy-preserving LLM training and deployment.

While LLMs have shown tremendous potential in cyber threat hunting, future research should focus on enhancing interpretability, addressing bias and fairness, improving scalability and efficiency, developing evaluation methodologies, integrating LLMs with other cybersecurity domains, and ensuring ethical use. These advancements will pave the way for more effective and responsible utilization of LLMs in cyber threat hunting. Techniques for data anonymization and differential privacy can be applied to ensure that sensitive information is not directly exposed during the training and deployment of LLMs. It is crucial to strike a balance between the utility of LLMs in threat hunting and the protection of individuals' privacy rights.

6 Conclusion

Large language models have emerged as powerful tools in the field of cyber threat hunting. Their ability to process and analyze vast amounts of text data, combined with their language understanding capabilities, makes them valuable assets in detecting and mitigating cyber threats. LLMs offer several advantages, including their capacity to uncover hidden patterns, identify indicators of compromise, and aid in anomaly detection.

However, the adoption of LLMs in cyber threat hunting also presents challenges and considerations. These include the need for interpretability and explainability of

LLM predictions, addressing model bias and fairness, ensuring privacy protection, and dealing with the trade-off between model performance and computational efficiency. To fully harness the potential of LLMs, researchers and practitioners should continue to explore and develop innovative approaches in these areas.

Moreover, future research should focus on improving the training process of LLMs with respect to cybersecurity-specific domains, as well as integrating domain-specific knowledge and context into their learning. Collaboration between cybersecurity experts, data scientists, and language model developers is crucial to enhance the effectiveness and applicability of LLMs in threat hunting.

Despite the challenges and considerations, LLMs hold significant promise for the future of cyber threat hunting. Their ability to continuously learn and adapt to evolving threats, combined with the advancements in interpretability and privacy-preserving techniques, can greatly enhance organizations' cybersecurity posture. By leveraging LLMs effectively, organizations can gain valuable insights, proactively detect and respond to threats, and ultimately strengthen their defense against cyber-attacks.

The integration of large language models into cyber threat hunting represents an exciting and rapidly evolving field. With continued research, development, and collaboration, LLMs have the potential to revolutionize the way we approach cybersecurity, enabling more proactive and effective threat detection and response.

References

1. Baccas T, Gharakheili Habib H, Armita S (2021) Cyber threat hunting: a systematic review. Comput Secur 106:102295
2. Bandyopadhyay D, Joshi A, Dutt R, Chakraborty A (2019) Cyber threat hunting: an overview. Int J Netw Secur Appl 11(6):1–20
3. Bolukbasi T, Chang K-W, Zou JY, Saligrama V, Kalai A (2016) Man is to computer programmer as woman is to homemaker? debiasing word embeddings. In: Advances in neural information processing systems, pp 4349–4357
4. Bonawitz K, Ivanov V, Kreuter B, Marcedone A, Brendan McMahan H, Patel S, Yarmolinsky A (2019) Towards federated learning at scale: system design. *arXiv preprint* arXiv:1902.01046
5. Brown TB, Mann B, Ryder N, Subbiah M, Kaplan J, Dhariwal P, Amodei D (2020) Language models are few-shot learners. *arXiv preprint* arXiv:2005.14165
6. Chen Z, Bai Y, Zhou L, Wu Q (2020) Logbert+: pretrained natural language understanding for log anomaly detection. *arXiv preprint* arXiv:2010.13002
7. Cole E, Ring M (2014) Hiding in plain sight: cyber threat hunting. IEEE Secur Privacy 12(4):74–77
8. Crawford J (2019) Cyber threat hunting and incident response: defending organizations against advanced threats. IT Professional 21(5):24–31
9. Dai Z, Yang Z, Yang Y, Carbonell J, Le QV, Salakhutdinov R (2020) Transformer-xl: attentive language models beyond a fixed-length context. In: Proceedings of the 57th annual meeting of the association for computational linguistics, pp 2978–2988
10. Garg A, Saha SB, Poria S (2022) Large language models: a review. ACM Trans Asian Low-Resour Lang Inf Process (TALLIP) 21(1):1–32
11. Gonen H, Goldberg Y (2019) Lipstick on a pig: debiasing methods cover up systematic gender biases in word embeddings but do not remove them. *arXiv preprint* arXiv:1903.03862

12. Gwak J, Park J, Lee H, Kim J (2020) Towards automated threat intelligence analysis using large-scale language models. In: Proceedings of the 2020 IEEE international conference on big data (big data)
13. Islam R, Mahbub S, Ahmed I (2020) Generation of cyber threat intelligence indicators of compromise using LSTM and GAN. IEEE Access 8:96888–96898
14. Jagielski M, Oprea A, Biggio B, Liu C, Nita-Rotaru C, Li B (2019) Manipulating machine learning: poisoning attacks and countermeasures for regression learning. ACM Trans Intell Syst Technol 10(3):1–27
15. Kim Y, Yun H, Lee Y, Kim J, Kim D (2021) Indicators of compromise generation based on language models. J Inf Sci Eng 37(5):1163–1177
16. Montenegro V, Ortiz M et al (2018) Hunting cyber criminals: a hacker's guide to online intelligence gathering tools and techniques. Syngress
17. Nguyen DC, Tran TT, Phung D (2021) Sc-transformer: scalable log analysis with pretrained language models. *arXiv preprint* arXiv:2103.07535
18. Panda A, Ghosh S, Dasgupta D (2021) Automated indicator generation using transformer based language models for cybersecurity operations. In: 2021 IEEE symposium series on computational intelligence (SSCI)
19. Radford A, Wu J, Child R, Luan D, Amodei D, Sutskever I (2019) Language models are unsupervised multitask learners. OpenAI Blog 1(8)
20. Ramachandran P (2017) Cyber threat hunting: a proactive approach to advanced threats. IEEE Secur Privacy 15(5):46–53
21. Ribeiro MT, Singh S, Guestrin C (2016) "Why should i trust you?" explaining the predictions of any classifier. In: Proceedings of the 22nd ACM SIGKDD international conference on knowledge discovery and data mining. ACM, pp 1135–1144
22. Sun J, Zhu Y, Zhu W, Liu S, Tan Y (2021) Logbert: transformer-based log anomaly detection. IEEE Trans Ind Inform 17(10):6771–6780
23. Vaswani A, Shazeer N, Parmar N, Uszkoreit J, Jones L, Gomez AN, Polosukhin I(2017) Attention is all you need. In: Advances in neural information processing systems, pp 5998–6008 (2017) Attention is all you need. In: Advances in neural information processing systems, pp 5998–6008
24. Wang K, Zheng S, Liu Z, Zhou L, Zhang X (2021) Gluoncv and gluonnlp: deep learning in computer vision and natural language processing. J Mach Learn Res 22(137):1–7
25. Yang M, Zheng S, Wang K, Zhang X, Zhou L (2021) Deeplog: anomaly detection and diagnosis from system logs through deep learning. ACM Trans Internet Technol 21(4):1–26

Transforming Patient Experience in Underserved Areas with Innovative Voice-Based Healthcare Solutions

Minliang Xia, Bo Huang, Yan Yan, Wenjun Zhang, and Wenjun Lin

Abstract This study presents a novel voice-driven online appointment system aimed at improving healthcare access in rural and indigenous communities. Utilizing existing telephone infrastructure, the system transcends the limitations of traditional chatbots by leveraging Large Language Models (LLMs) and advanced voice recognition technologies like Whisper. This approach enables the transformation of conventional phone calls into a streamlined digital booking process. The system's integration with current booking processes in rural healthcare facilities and facilitates a smooth transition from voice-based to digital scheduling, providing an intuitive and efficient user experience. This innovation addresses critical healthcare accessibility challenges, notably reducing appointment booking barriers and enhancing the overall patient experience. The implementation of this technology in underserved areas demonstrates a significant advancement in patient-centered care, highlighting the role of LLMs in harmonizing traditional and digital healthcare practices. The paper provides a comprehensive overview of the system's development, explores the challenges that may encounter, and discusses the significant potential of this approach in shaping future healthcare technologies and influencing policy decisions in healthcare accessibility.

M. Xia
Mechanical Engineering, University of Saskatchewan, Saskatoon, Saskatchewan, Canada
e-mail: svi089@mail.usask.ca

B. Huang · Y. Yan
School of Computer Science, University of Guelph, Geulph, Ontario, Canada
e-mail: bhuang06@uoguelph.ca

Y. Yan
e-mail: yyan15@uoguelph.ca

W. Zhang (✉)
Biomedical Engineering, University of Saskatchewan, Saskatoon, Saskatchewan, Canada
e-mail: chris.zhang@usask.ca

W. Lin (✉)
Digital Healthcare Innovation Lab, School of Computer Science and Technology, Algoma University, Brampton, Ontario, Canada
e-mail: randy.lin@algomau.ca

X.-S. Yang et al. (eds.), *Proceedings of Ninth International Congress on Information and Communication Technology*, Lecture Notes in Networks and Systems 1000,
https://doi.org/10.1007/978-981-97-3289-0_51

Keywords Healthcare accessibility · Rural and indigenous communities · Voice-driven technology · Large language models · Digital healthcare innovation · Online appointment system · Patient-centered care · Telecommunications in healthcare · Digital booking systems

1 Introduction

The healthcare sector has recently seen a significant shift toward digitalization, with online appointment systems becoming increasingly prevalent. However, these systems often fail to address the specific needs of rural and indigenous communities, where inconsistent Internet access and socio-economic factors pose barriers. Our research focuses on developing a voice-driven online appointment system that leverages existing telephone infrastructure to enhance accessibility in these communities.

Traditional phone-based healthcare appointments, often the norm in remote areas, face challenges such as long wait times and limited availability [7, 10]. On the other hand, online booking systems, efficient in their operation, depend heavily on reliable Internet connectivity, which is not always a given in rural settings [2]. This disparity highlights the need for a solution that combines digital convenience with the simplicity of traditional communication.

Conventional voice chatbots, a potential solution for booking appointments, generally require resources beyond the reach of smaller healthcare facilities, especially in rural areas [16]. Our project addresses this gap by integrating advanced voice recognition technologies and Large Language Models (LLMs) like ChatGPT 3.5 and Whisper. This system transforms spoken language into digital data, facilitating seamless integration with both online and manual booking systems. This innovative approach not only makes appointment scheduling more intuitive but also overcomes the limitations of Internet dependency.

This paper aims to demonstrate how our system can democratize healthcare access, particularly in rural and indigenous communities. It is structured to first review relevant literature, emphasizing the importance of our approach in bridging the digital divide in healthcare. We then detail the development and integration of the Voice-Driven Appointment System Integration (VDAI), including the use of LLMs. The subsequent sections present our experimental findings on the system's functionalities and discuss their implications for enhancing healthcare access and patient experience. The paper concludes by summarizing our key findings and discussing their potential impact on healthcare innovation and policy.

2 Literature Review

In this section, we explore pivotal areas including the digital transformation in healthcare, challenges specific to rural and indigenous healthcare access, and the role of LLMs and voice recognition technologies in healthcare. This literature review lays the foundational context for our project, which aims to enhance healthcare accessibility through a novel voice-driven online appointment system.

The digital transformation in healthcare is marked by the integration of technologies such as machine learning and the Internet of Medical Things (IoMT). Sun et al. [20] review these systems, emphasizing their role in personalizing healthcare and telemedicine. Concurrently, Alazab et al. [1] and Lin et al. [12] discuss the emergence of digital twins in healthcare, while Lakhan et al. [9] and Tian et al. [22] focus on optimizing the efficiency and security of these digital systems. The legal and ethical aspects, including privacy and informed consent, are examined by Jain [6] and Borrelli et al. [3]. Olanrewaju et al. [18] and Lin et al. [11] analyze the application of these technologies during public health crises like the COVID-19 pandemic.

Studies by Kuupiel et al. [8] and Benevenuto et al. [2] highlight the challenges in healthcare accessibility in rural areas, focusing on resource availability and geographical barriers. Mangundu et al. [15] and Tha et al. [17] emphasize societal and infrastructural barriers in healthcare access. Proposed solutions for improving accessibility in rural China and Poland are discussed by Lu et al. [13] and Lechowski et al. [10].

The evolving integration of LLMs and voice recognition in healthcare is explored in studies by Cascella et al. [4] and Yeung et al. [23], examining their capabilities and limitations in medical contexts. Dash et al. [5] and Thirunavukarasu et al. [21] explore the practical application of LLMs in clinical settings, while Macaluso [14] discusses their use in infection prevention in healthcare. The work of Yang et al. [24] on the Talk2Care system demonstrates the utility of combining LLMs with voice technologies to enhance healthcare accessibility.

These studies collectively highlight the transformative potential of digital technologies, especially LLMs and voice recognition, in the healthcare sector, pointing toward specialized applications in medical settings. Their relevance to our project is evident in demonstrating the potential of tailoring LLMs to meet specific healthcare communication and accessibility needs.

3 Methodology: Voice-Driven Appointment System Integration

The VDAI methodology is designed to enhance the appointment booking process in healthcare settings, particularly focusing on rural and indigenous communities. This methodology is structured to integrate seamlessly with existing systems, using cloud-based telephony services and advanced algorithms for voice-to-text transformation and information extraction. The key phases of the VDAI methodology include initial set-up, voice-to-text conversion, identification of time preferences and patient identity, and integration with online booking platforms.

3.1 Initial Set-Up

The initial set-up involves configuring a cloud-based telephony service as the primary interface for receiving patient calls. These services are adept at handling high call volumes and provide features such as call recording, which is essential for the subsequent voice-to-text transformation process. The cloud-based nature of these services ensures that the system remains accessible and functional even in areas with limited on-site technical infrastructure. The choice of a cloud-based solution for the initial set-up of VDAI methodology aligns with the project's goal of providing a low-cost, low-maintenance, and scalable solution to improve healthcare accessibility. This approach ensures that the system is not only efficient and reliable but also adaptable to the specific needs of rural and indigenous communities.

3.2 Voice to Text Conversion Process

The voice-to-text transformation is an essential component of the VDAI methodology, crucial for transforming patient interactions into actionable data. This process utilizes deep learning algorithms found in advanced speech recognition technologies like Whisper [19].

In rural or remote areas, varying levels of background noise and call quality can significantly challenge speech recognition accuracy. The system employs a dual approach to address this. Firstly, it uses traditional voice enhancing techniques such as noise reduction filters, echo cancelation, and automatic gain control to improve the clarity of spoken words. This preprocessing step is vital for enhancing the accuracy of the speech recognition software.

Secondly, the system uses LLMs for clarification in scenarios with high noise levels. When traditional techniques are insufficient, LLMs interactively clarify key information by asking targeted, context-aware questions to confirm crucial details for appointment booking. This novel approach merges advanced speech recognition with LLMs' conversational capabilities, ensuring accurate patient information capture, even in challenging audio conditions typical in rural settings.

Additionally, to address the varying accents in different rural areas, especially for local indigenous languages, data collection on local accents and fine-tuning Whisper is considered for more accurate results.

3.3 Identification of Time Preferences and Patient Identity

The process of identifying time preferences and patient identity in our system is organized into a three-stage approach using two AI agents: "Directive Coordinator" and "Interactive Extractor". This structure effectively leverages LLMs for efficient information gathering.

Stage 1: Directive Setting by the Directive Coordinator The Directive Coordinator establishes foundational rules and guidelines for the appointment scheduling process. It conveys instructions to the Interactive Extractor on critical patient information, including time availabilities, necessary patient details (contact information, identifiers), and appointment constraints such as available slots and healthcare services offered.

Stage 2: Interactive Extraction by the Interactive Extractor In this stage, the Interactive Extractor directly interacts with patients, applying guidelines from the Directive Coordinator. Utilizing LLM capabilities, it identifies and records time preferences and patient details through natural language interactions, ensuring an intuitive process for patients.

Stage 3: Confirmation, Error Identification, and Clarification Loop This stage involves the interaction between the Directive Coordinator and the Interactive Extractor, focusing on confirming collected information and identifying errors. The Directive Coordinator reviews extracted data against initial instructions, identifying any inaccuracies or missing elements. It communicates specific concerns back to the Interactive Extractor for re-engagement with patients to resolve issues or gather additional information. This loop continues until all data is accurate and complete, ensuring reliability in the appointment scheduling process.

Employing the Directive Coordinator and Interactive Extractor in this three-stage process guarantees precise collection of appointment-related information. This methodology leverages LLMs' strengths in directive understanding and patient interaction, providing a sophisticated approach to appointment scheduling that caters to the unique needs of rural and indigenous communities.

3.4 Integration with Online Booking Platforms

The integration strategy of the VDAI system is designed to be adaptable and efficient, catering to different types of existing booking systems within healthcare facilities. This strategy encompasses three distinct cases:

Integration with Database-Driven Systems via API: In this scenario, the VDAI system utilizes API integration to directly communicate with the healthcare facility's database-driven booking system. After processing the appointment details, the system automatically updates the booking platform, ensuring accurate and efficient scheduling.

Booking via Existing Online Pages Using LLM: For facilities with online booking interfaces that do not offer API integration, the VDAI system employs LLM to navigate and interact with the booking page. The system identifies and fills out necessary fields, such as patient details and appointment times, directly on the webpage, mimicking the process a human user would undertake.

Adaptation for Manual Booking Systems with Cloud-Based Solutions: Recognizing that some facilities may still rely on manual methods like Excel spreadsheets for scheduling, the VDAI system adapts to integrate with cloud-based spreadsheet solutions like Google Sheets. This allows for real-time updates and parallel operation of manual and automated booking processes.

3.5 Information Collection and Privacy

To ensure patient privacy and security, VDAI will only collect necessary information to ensure the security such as the patient ID, phone number, and appointment time. The Patient ID is a short, unique identifier with two letters and two numbers, such as AA11. This ensures that the patient's name will not be disclosed and also resolves the issue of inaccuracies in collecting uncommon names.

3.6 Preventing Double Booking

To prevent double booking in the VDAI system, we employ several protocols alongside real-time synchronization through API or cloud-based solutions:

Time-Block Reservation System: The system temporarily reserves the chosen time slot as soon as an appointment is initiated. This reservation prevents the slot from being booked by others, allowing time to finalize the booking and update the system. **Confirmation Callbacks**: Post-booking, a confirmation callback is sent to the patient. This step confirms the appointment's accuracy and informs the patient. In cases where the patient uses a landline that cannot receive callbacks, the system prompts them to remember their patient number for later confirmation or modification via phone call.

4 Experiments

The experimentation phase of the VDAI methodology aimed to evaluate the system's performance and gather feedback from end users. This phase was structured to assess various aspects of the system, including its accuracy, efficiency, user experience, and the challenges encountered during its implementation.

Experiment 1: System Accuracy and Efficiency

In our assessment of the VDAI using the Whisper voice recognition technology, we conducted a series of tests to evaluate the system's accuracy in transcribing key information under different noise conditions. Over five rounds of testing, we used consistent voice inputs in both street and quiet environments to understand how background noise affected the system's performance.

The test included two specific statements crucial for appointment scheduling: a phone number and time preferences for an appointment. These were chosen to reflect typical information required in healthcare appointment settings. The accuracy was measured based on the correct transcription of these details and the impact of any errors on the comprehension of the content.

Throughout the five rounds, we noticed varying levels of accuracy, with the system's performance notably fluctuating in the presence of background noise. While Whisper demonstrated proficiency in voice-to-text conversion, the accuracy was compromised in noisier environments. This underlines the importance of incorporating traditional voice enhancing techniques to mitigate background noise. Additionally, the integration of LLMs to interactively ask for clarifications when the noise level is highly emerged as a crucial requirement. Such interactive clarification would allow the system to confirm and rectify key information directly with the patient, ensuring accuracy in appointment scheduling.

Experiment 2: Feasibility and Efficiency of ChatGPT in Scheduling

In this segment of our study, we focused on evaluating ChatGPT 3.5's ability to simulate a hospital's telephone appointment service, specifically its efficiency in scheduling and managing patient information. We created eight virtual patient profiles, each with distinct availability scenarios, and engaged ChatGPT 3.5 in a simulated booking process. The system's performance was scored based on how accurately it collected and scheduled appointments according to patients' details and the hospital's available time slots.

The scoring system was devised to assess the correct identification of critical elements like names, times, addresses, phone numbers, and reasons for visits, with a total possible score of five. The patients' availability fell into three categories:

- Those available at any time within the hospital's specified slots.
- Those whose availability partially overlapped with the hospital's time slots.
- Those available only outside the hospital's designated appointment times.

During the testing, ChatGPT 3.5 demonstrated varying degrees of success. In scenarios where patients had open availability (matching the hospital's schedule), the system performed well, accurately aligning patient appointments within the hospital's available times. However, challenges arose in more complex scenarios. For patients with partial or no alignment with the hospital's schedule, ChatGPT 3.5 tended to prioritize patients' most recently mentioned availability, often ignoring the hospital's time constraints. This led to several instances where appointments were inaccurately scheduled outside of the hospital's available slots.

While ChatGPT 3.5 was adept at gathering patient information, its ability to negotiate between patient preferences and hospital constraints was limited. Integrating a Directive Coordinator to establish scheduling rules and an Interactive Extractor to interact with patients could potentially bridge this gap. This combined approach would enable the system to consider both sets of requirements, ensuring more accurate and efficient scheduling in line with hospital operations.

This experiment highlights the limitations of using a standalone LLM for complex appointment scheduling tasks in healthcare settings. It underscores the necessity of a collaborative and nuanced AI system capable of handling the intricacies of real-world appointment scheduling.

Experiment 3: Error Handling and Information Update Test

In this part of our experimentation, we aimed to test the error handling capabilities of ChatGPT 3.5, particularly its proficiency in updating and correcting patient information during the appointment scheduling process. The test involved simulating scenarios where patients either changed their appointment details or corrected previously provided information. This setup was designed to reflect real-world situations where patients might need to update their appointment times or personal details.

The results revealed mixed effectiveness in ChatGPT 3.5's ability to handle these updates. ChatGPT 3.5, while adept at processing individual requests, showed limitations in maintaining the accuracy of information across multiple sequential updates. This is a crucial aspect in healthcare settings, where the accuracy of patient data and appointment details is paramount.

The experiment underscores the need for a more robust system capable of handling dynamic changes and corrections. Integrating a sophisticated error handling mechanism, potentially through the two AI agent model with a Directive Coordinator and an Interactive Extractor, could significantly enhance the system's capability to manage such complexities. This would ensure that updates and corrections are processed accurately and efficiently, maintaining the reliability and trustworthiness of the appointment scheduling system.

5 Discussion

The experimentation phase with ChatGPT 3.5 provided valuable insights into the capabilities and limitations of utilizing AI for healthcare appointment scheduling. One of the key challenges highlighted in the experiments was ChatGPT 3.5's difficulty in balancing patient availability with the hospital's scheduling constraints. While it efficiently collected patient data, it often prioritized the most recent patient availability information, neglecting the hospital's time slots. This issue is particularly critical in healthcare settings, where scheduling efficiency can significantly impact service delivery and patient satisfaction. It underscores the necessity for a system that can intelligently balance and negotiate between patient preferences and hospital operational requirements.

The error handling tests revealed another significant challenge: maintaining the integrity of patient data through sequential updates and modifications. ChatGPT 3.5's performance varied when handling changes in appointment details, especially when multiple modifications were requested in succession. In a real-world scenario, this limitation could lead to scheduling errors, causing inconvenience to both patients and healthcare providers. Therefore, a robust system that can accurately track and manage sequential data changes is crucial.

The experiments suggest a need for a more comprehensive AI solution, like the proposed two AI agent model in our methodology. This model, comprising a Directive Coordinator and an Interactive Extractor, could potentially address the identified challenges more effectively. The Directive Coordinator would establish and enforce scheduling rules, ensuring that appointments align with hospital availability. In contrast, the Interactive Extractor would interact with patients, accommodating their preferences while adhering to these rules. Such a collaborative system could offer a more sophisticated approach to appointment scheduling, balancing efficiency with the flexibility required in dynamic healthcare settings.

6 Conclusion

In conclusion, this research has highlighted the significant potential of the VDAI in enhancing healthcare accessibility, particularly in rural and indigenous communities. Our experiments with ChatGPT 3.5 demonstrated that while existing LLMs offer a solid foundation, they require substantial augmentation to effectively handle the complexities of healthcare scheduling. This need is particularly pronounced in scenarios involving intricate scheduling preferences and error handling.

The VDAI system, with its innovative integration of advanced voice-to-text technology and a dual AI agent approach, represents a significant step forward in addressing these challenges. Our findings underscore the necessity of a sophisticated system like VDAI, which combines the strengths of LLMs with specialized mechanisms for managing patient information and appointment scheduling nuances.

This research contributes to the evolving narrative of digital transformation in healthcare, underscoring the importance of developing technology solutions that are not just technologically advanced but also finely attuned to the specific needs of healthcare delivery in diverse settings. The insights gained from this project are instrumental in guiding future developments in AI-driven healthcare systems, emphasizing the need for solutions that are inclusive, patient-centered, and adaptable to the unique challenges faced by different communities.

Acknowledgements The authors wish to express their sincere gratitude to the NSERC CREATE Fund (565429-2022) and Algoma University Research Fund for their generous support of this work. Their commitment to fostering innovative research has been invaluable to the success of this project.

References

1. Alazab M, Khan LU, Koppu S, Ramu SP, Iyapparaja M, Boobalan P, Baker T, Maddikunta PKR, Gadekallu TR, Aljuhani A (2022) Digital twins for healthcare 4.0-recent advances, architecture, and open challenges. IEEE Consum Electron Mag
2. Benevenuto RG, Azevedo ICC, Caulfield B (2019) Assessing the spatial burden in health care accessibility of low-income families in rural Northeast Brazil. J Transp Health 14:100595
3. Borrelli N, Grimaldi N, Papaccioli G, Fusco F, Palma M, Sarubbi B (2023) Telemedicine in adult congenital heart disease: usefulness of digital health technology in the assistance of critical patients. Int J Environ Res Public Health 20(10):5775
4. Cascella M, Montomoli J, Bellini V, Bignami E (2023) Evaluating the feasibility of chatgpt in healthcare: an analysis of multiple clinical and research scenarios. J Med Syst 47(1):33
5. Dash D, Thapa R, Banda JM, Swaminathan A, Cheatham M, Kashyap M, Kotecha N, Chen JH, Gombar S, Downing L et al (2023) Evaluation of gpt-3.5 and gpt-4 for supporting real-world information needs in healthcare delivery. arXiv preprint arXiv:2304.13714
6. Jain D (2023) Regulation of digital healthcare in India: ethical and legal challenges. Healthcare MDPI 11:911
7. Koebisch SH, Rix J, Holmes MM (2020) Recruitment and retention of healthcare professionals in rural Canada: a systematic review. Can J Rural Med 25(2):67–78
8. Kuupiel D, Tlou B, Bawontuo V, Mashamba-Thompson TP (2019) Accessibility of pregnancy-related point-of-care diagnostic tests for maternal healthcare in rural primary healthcare facilities in Northern Ghana: a cross-sectional survey. Heliyon 5(2)
9. Lakhan A, Mohammed MA, Nedoma J, Martinek R, Tiwari P, Vidyarthi A, Alkhayyat A, Wang W (2022) Federated-learning based privacy preservation and fraud-enabled blockchain iomt system for healthcare. IEEE J Biomed Health Inform 27(2):664–672
10. Lechowski Ł, Jasion A (2021) Spatial accessibility of primary health care in rural areas in Poland. Int J Environ Res Public Health 18(17):9282
11. Lin W, Babyn P, Yan Y, Zhang W (2023) A novel scheduling method for reduction of both waiting time and travel time of patients to visit health care units in the case of mobile communication. In: Enterprise information systems, p 2188124
12. Lin W, Babyn P, Yan Y, Zhang W (2023) Optimizing healthcare resource allocation through digital twins: a multi-objective approach for efficiency, equity, and resilience. In: 3rd Digital twin international conference, pp 1–4. Taylor & Francis Group
13. Lu R, Li Y, Zheng Z, Yan Z (2021) Exploring factors associated with self-management compliance among rural elders with diabetes. INQUIRY J Health Care Organ Provision Financ 58:00469580211012491
14. Macaluso Jr JN (2023) Hospital, catheter, peritoneal dialysis acquired infections: visible light as a new solution to reduce risk and incidence. Cureus 15(8)
15. Mangundu M, Roets L, Janse van Rensburg E (2023) A strategic action plan to enhance the accessibility of healthcare in rural areas of Zimbabwe. Global Public Health 18(1):2234982
16. Nagarhalli TP, Vaze V, Rana N (2020) A review of current trends in the development of chatbot systems. In: 2020 6th International conference on advanced computing and communication systems (ICACCS). IEEE, pp 706–710
17. Naing OT, Shoesmith WD, Tan CB, Ibrahim MY, Hussein SS (2020) Geographic accessibility of healthcare services and health seeking behaviours of rural communities in Kudat and pitas areas of Sabah. Borneo Epidemiol J 1(1):46–54
18. Olanrewaju Z, Faisal M, Randell R (2023) A scoping review of digital technologies used in monitoring COVID-19. Stud Health Technol Inform 309:310–311
19. Radford A, Kim JW, Xu T, Brockman G, McLeavey C, Sutskever I (2023) Robust speech recognition via large-scale weak supervision. In: International conference on machine learning. PMLR, pp 28492–28518
20. Sun T, He X, Li Z (2023) Digital twin in healthcare: recent updates and challenges. Digit Health 9:20552076221149652

21. Thirunavukarasu AJ, Ting DSJ, Elangovan K, Gutierrez L, Tan TF, Ting DSW (2023) Large language models in medicine. Nat Med 29(8):1930–1940
22. Tian Y, Wang S, Xiong J, Bi R, Zhou Z, Bhuiyan MZA (2023) Robust and privacy-preserving decentralized deep federated learning training: focusing on digital healthcare applications. IEEE/ACM Trans Comput Biol Bioinform
23. Yang K, Ji S, Zhang T, Xie Q, Ananiadou S (2023) On the evaluations of chatgpt and emotion-enhanced prompting for mental health analysis. arXiv preprint arXiv:2304.03347
24. Yang Z, Xu X, Yao B, Zhang S, Rogers E, Intille S, Shara N, Wang D et al (2023) Talk2care: facilitating asynchronous patient-provider communication with large-language-model. arXiv preprint arXiv:2309.09357

Fortifying Machine Learning-Powered Intrusion Detection: A Defense Strategy Against Adversarial Black-Box Attacks

Medha Pujari and Weiqing Sun

Abstract Intrusion detection systems (IDSs) powered by machine learning (ML) techniques have greatly advanced in detecting security threats. However, with advancements come new challenges and vulnerabilities. As most of the ML-based IDSs prioritize performance, their robustness and security are in question, which becomes an advantage for bad actors to exploit the vulnerabilities and target IDS models. It is a serious concern in the cybersecurity domain. This paper emphasizes the impact of adversarial machine learning (AML)—a well-known vulnerability of ML techniques—and proposes a defense mechanism to improve the robustness of ML-powered IDSs to adversarial black-box attacks. The primary motive of this research is to defend the defenders of cybersecurity, in particular, the IDSs. The paper provides a background of the elements involved in the experimentation, presents the methodology followed for the experiments, and analyzes the evaluation results obtained. Further, it discusses the directions and possibilities for future endeavors.

Keywords Intrusion detection systems · Machine learning · Adversarial machine learning · Black-box attacks · Defense mechanism · Research dataset

1 Introduction

Intrusion detection systems (IDSs) have been one of the vital defenders in the domain of cybersecurity. The introduction of machine learning (ML) techniques to improve the abilities of IDSs has produced promising results in expanding the detection poten-

M. Pujari (✉)
Western New England University, Springfield, Massachusetts 01119, USA
e-mail: medha.pujari@wne.edu
URL: https://wne.edu/directory/medha-pujari.cfm

W. Sun
University of Toledo, Toledo, Ohio 43606, USA
e-mail: weiqing.sun@utoledo.edu
URL: https://www.utoledo.edu/engineering/engineering-technology/cset/sun.html

X.-S. Yang et al. (eds.), *Proceedings of Ninth International Congress on Information and Communication Technology*, Lecture Notes in Networks and Systems 1000,
https://doi.org/10.1007/978-981-97-3289-0_52

tial of these devices thereby increasing the scope of network security, in general [10, 13, 21, 31, 35]. However, ML has not succeeded so much in cybersecurity as it did in other domains such as image processing [33] and text recognition because of the complexity of the domain and its ever-evolving attack behaviors [12, 13, 20]. Further, with every advancement come new challenges, risks, and vulnerabilities, and it is essential to proactively study them and develop mechanisms to address them effectively. In a domain such as cybersecurity, even a small vulnerability can cause a huge disruption, therefore, it is necessary to make network defenders more robust and reliable. It is necessary for the defenders to be capable of defending themselves too.

Adversarial machine learning (AML) is a vulnerability of ML models which is being widely investigated in the research community lately [11, 25, 26, 28, 32]. It is a concept using which a bad actor can deceive an ML model by introducing subtle noise into data in a way to make the target model render incorrect predictions. The noise, often called perturbation, is carefully crafted to achieve the predictions that the bad actor aims for while ensuring that the data, after modification, looks as similar as possible to its original form (before modification). White-box and black-box attacks are two prominent categories of AML algorithms. A white-box attack algorithm has almost complete knowledge about the target model, including details like the data used by the model, training parameters, and so on. A black-box attack algorithm is one which has almost zero (or very limited) knowledge of the target model. There is another category, called gray-box attacks, wherein the algorithms have partial knowledge about their target models. While white-box attacks tend to be more powerful than black-box attacks, they are not so practical in nature because it is usually not easy (or realistic) for an attacker to have so much information about a target model. White-box attacks are more applicable to research and experimentation settings wherein all the information required by an adversarial algorithm can be provided. The footprint of AML research has been relatively wider and deeper in domains involving tasks such as image processing, and image-based classification [1, 34]. Although its research is gaining momentum in the areas of security, there is a need for more investigation into the impacts of AML algorithms on various ML-powered components/devices.

Shahad Alahmed et al. [3] presented their work with a motive of mitigating the impact of black-box attacks on ML-based IDSs using zeroth order optimization (ZOO) algorithm which is also a black-box algorithm. ZOO is a type of optimization algorithm that estimates the gradients (or derivatives) using just the function evaluations instead of using explicit gradient or derivative values of the objective function. Their paper uses two feature reduction techniques to boost the performance of the target model. The defense mechanism proposed through the current work achieved significant resistance to the adversarial attack even without incorporating any feature reduction techniques. Gabriel Amaizu et al. [8], in their paper on investigating intrusion detection datasets evaluated the performance of IDS models when trained with three datasets - NSL-KDD, UNSW-NB15, and CSE-CIC-IDS2018. The accuracy they obtained for CSE-CIC-IDS2018 through their experiments was 76.47%. The performance scores produced by RF and DT in the current work are higher than the scores presented based on their experiments. Islam Debicha et al. [15] investigate the transferability property of adversarial data and examine how it succeeds in defeating multiple black-box anomaly-based network IDSs. Furhter, the authors implement an ensemble IDS by combining five ML-based IDS models, to detect and reject the deceptive samples. Their experiments are based on NSL-KDD, an old research dataset. Yasmeen Alslman et al. [7] proposes a defense mechanism using a denoiser auto-encoder (DAE). Their experiments are based on two datasets—SNMP-MIB and UNSW-NB15. They validate the defense using four white-box attacks and a black-box attack known as Simultaneous perturbation stochastic approximation (SPSA) attack.

As significant as it is to explore and formulate more ways to break a model, it is essential to develop mechanisms to counter them. Ultimately, only a system that is prepared to tackle adversarial activities can be considered fit to be deployed in a real-time security-critical environment. The following are the contributions of this paper to the literature.

- In-Depth Analysis of Genetic Algorithm (GA): The paper conducts a comprehensive analysis of the impact of an efficient evolutionary algorithm (EA), specifically the Genetic Algorithm (GA). While existing research extensively explores various evolutionary algorithms, particularly in domains like image processing [4], this study aims to contribute to the ongoing research progress in Adversarial Machine Learning (AML). By focusing on the GA, the authors provide insights into its effectiveness and applicability in the context of AML, extending the understanding of evolutionary algorithms in cybersecurity.
- Innovative Defense Strategy: The paper proposes a defense mechanism designed to mitigate the impact of adversarial data on Machine Learning (ML)-based Intrusion Detection Systems (IDS) models. In doing so, the authors emphasize the critical importance of defending the defenders in the realm of cybersecurity. The proposed mechanism addresses the challenges posed by adversarial attacks on ML-based IDS models, contributing a practical solution to enhance the robustness and reliability of cybersecurity systems.
- The combination of the evolutionary algorithm and a modern benchmark research dataset such as CSE-CIC-IDS2018 is a novelty of this study in addition to the strategy for defense.

Nature or bio-inspired algorithms draw inspiration from biological phenomena to solve complex problems. They have unique abilities in terms of adaptability and evolution, which makes them a potential avenue for research in a complex domain such as cybersecurity. It is the main motive behind conducting this research with focus on GA. The authors believe that bio-inspired approaches hold promising solutions in tackling real-world security threats, and intend to continue research in this avenue.

The remainder of this paper is organized as follows. Section 2 provides background on various elements involved in this work. Section 3 discusses recent works related to this research available in the literature. Section 4 covers the design and architecture of the idea proposed in this paper. Section 5 provides the experimentation methodology followed. Section 6 presents the evaluation results and analyzes them. Section 7 concludes the paper and discusses the directions for future work.

2 Related Work

There are relatively more works proposing approaches that can deceive ML-based models, some of which can be very invasive [14] because they have complete knowledge of the target models. However, with complete knowledge of the target model, an adversary can choose to craft much more disruptive attacks rather than relying on adding subtle adversarial perturbations to the data [9].

Yiran Zhu et al. [38] use generative adversarial networks (GAN) to create adversarial samples and use them to examine the robustness of ML-based anomaly detection algorithms. Further, the authors point that the feature reduction techniques implemented in ML-based anomaly detectors have a vulnerability and put forth a defense mechanism to cover it up.

Shuang Zhao et al. [37] also use a GAN-based architecture to generate an adversarial attack on black-box IDS. The paper by Olakunle Ibitoye et al. [19] presents a comparison of performance between forward neural networks (FNNs) and self-normalizing neural networks (SNNs) in intrusion detection tasks in the context of IoT, and also their robustness to adversarial inputs. Their analysis suggests that the impact of adversarial inputs is more on the FNN model than on SNN.

Hossein Gharaee et al. [17] introduce an anomaly-based Intrusion Detection System (IDS) that combines Genetic Algorithm (GA) and Support Vector Machine (SVM) with a novel feature selection approach. The proposed model employs a GA-based feature selection method with an innovative fitness function, aiming to reduce data dimensionality, enhance true-positive detection, and concurrently diminish false positive detection. Their experiments are based on KDD CUP 99 and UNSW-NB15 datasets, one of which is a very old dataset.

Mohammad Sazzadul Hoque et al. [18] use GA to build an IDS to improve its detection capabilities and evaluate their experiments using KDD99 dataset.

There are surveys [5, 6, 23] that discuss and analyze recently proposed adversarial attack techniques and defenses in the literature. Some researchers [29] focus on improving the robustness of features by identifying vulnerable features using certain feature robustness metrics. While there are many works that perform feature reduction techniques to enhance fake data detection capabilities as part of defense mechanisms, it is crucial to evaluate how such defenses perform in non-adversarial settings given the feature reduction. This is because feature reduction may bring down the performance of ML-powered IDSs in non-adversarial settings while rendering outstanding performance in adversarial environments.

3 Background

This section briefly introduces the elements involved in the experimentation method, such as the type of adversarial algorithm used, the dataset, and classification algorithms.

3.1 Adversarial Machine Learning: Black-Box Attacks

Unlike a white-box attack, an adversarial black-box attack does not need any information about its target model to implement the attack. Broadly classified, a black-box attack algorithm can either be transferability-based [22], or query-based [24]. This work uses a query-based attack for evaluation.

Query-Based Attacks The core idea of a query-based black-box attack is to query the target ML model with input data and observe the output, without having any knowledge of the internal workings or parameters of the model. The algorithm uses the input-output behavior of the target to craft malicious input data that can exploit vulnerabilities of the target, such as biases or other weaknesses.

Query-based attacks can be implemented in many ways, but a common approach is to use an optimization algorithm, such as a genetic algorithm, or a gradient-based optimization, to generate input data that maximizes a certain objective, such as causing the model to misclassify a specific input, or to exhibit unexpected behavior. Such data is usually referred to as adversarial or deceptive data.

Based on the type of output a query-based attack algorithm is expecting, it can be classified as decision-based or score-based. A decision-based algorithm depends on the output label provided by the target model for a given input, to craft deceptive data. A score-based algorithm relies on the confidence scores or probabilities associated with each output.

In a practical setting, an adversary typically has a limited number of queries they can make to the model, meaning they can only make a certain number of input-output observations. This makes the attack more challenging because the adversary has to carefully choose which inputs to use for each query in order to maximize the chances of crafting a successful adversarial example.

Genetic Algorithm An EA, in general, is an optimization-based algorithm that is inspired by natural selection. Natural selection is a natural phenomenon that drives the process of evolution, where organisms with advantageous characteristics are more likely to survive and reproduce to pass on those characteristics to their offspring.

GA starts by selecting a set of initial inputs (or population) that are fed into the target model to observe its output. The attacker then applies genetic operators such as mutation, crossover, and selection to these inputs to generate a new set of candidate adversarial examples. The candidate examples are then evaluated by feeding them to the target model to see if they are successful in fooling it. The best-performing candidate examples are then selected and used to create the next generation of inputs, and the process is repeated until a possibly optimal adversarial example is generated. Evolutionary algorithms, in general, can generate adversarial examples that are highly tailored to the specific model and data being used. However, they can also be computationally expensive and time-consuming.

The following are the generalized steps involved in an evolutionary algorithm:

- Generate initial population: Create a population of N individuals with randomly assigned values for the decision variables.

- Evaluate each individual: A fitness value of each individual is determined by querying the target model with the individual and observing the output provided by the model. It can be the classification accuracy of the model on each individual.
- Select high-performing individuals: Select individuals with high fitness values.
- Recombination: Combine the decision variables of the selected individuals to create new individuals.
- Mutation: Randomly perturb the decision variables of some individuals.
- Evaluation of new individuals: Evaluate the fitness of the new individuals.
- Replacement: Replace the least fit individuals in the population with the new individuals.

The steps of evaluation, recombination, mutation, and replacement are repeated until a pre-determined stopping criterion is met. The stopping criterion can be based on various factors such as reaching a certain number of generations, achieving a satisfactory fitness level, or exceeding a set computation time.

3.2 Dataset

There are certain characteristics that determine the quality of an IDS dataset, some of which are—the framework used for generating the traffic scenarios; diversity of attack scenarios captured; anonymity; the variety of protocols included; capturing complete network interactions; configurations in the network; feature set; labeled data samples; heterogeneity; and metadata [16, 30]. Considering the above characteristics, the CSE-CIC-IDS2018 dataset was chosen for this effort.

3.3 Classification Algorithms

Random forest (RF) algorithm is used for attack data generation. The target black-box model is trained and evaluated using RF, Decision Tree (DT), and Support Vector Machine (SVM). The reason behind choosing the RF classifier is the efficiency and robustness it has projected in the literature, which is also observed in the evaluation results of the experiments presented by Medha Pujari et al. [27] and Giovanni Apruzzese et al. [9].

4 Design and Architecture

The defense framework used for this experiment uses a generative adversarial network (GAN) with additional adversarial training as a reinforcement. Figure 1 outlines the higher-level idea of this work.

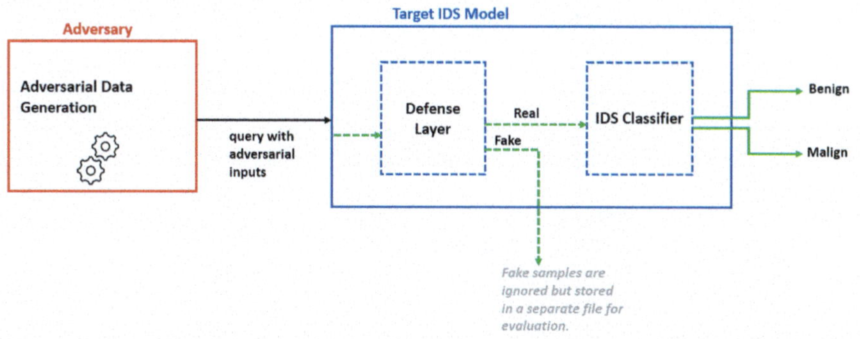

Fig. 1 Outline of the overall architecture

The adversary is actively involved in crafting adversarial samples while also interacting with the target black-box model with inputs. GA uses a local model for data generation along with the responses provided by the target. The proposed design for the target model has two core components—the defense layer and the IDS classifier. Every incoming data input passes through GAN which filters the inputs that it considers *fake/deceptive*. Inputs that are considered *real* are forwarded to the IDS classifier for final classification. The IDS determines whether the data coming through it is benign or malign. During the evaluation, the filtered fake inputs are stored separately to assess the performance of the defense layer. More details on the internal design and workflow are discussed in Sect. 5.

4.1 Strategy for the Detection of Adversarial Inputs

A query-based attack is usually built in a way that it takes the attack algorithm several queries to create a successful adversarial sample. The key is that each of the several queries is designed in a way that the sample is very close to the original input in a data distribution. An adversarial sample can only be designed if all or most of the iterative queries are answered by the target model. If a model can identify and not respond to the queries that are part of an attack, the adversary will not be able to craft an adversarial input, at least not in a feasible amount of time. This is the thought process driving the development of this defense.

5 Experimentation

This section provides details of the methodology followed for this experiment in multiple modules.

5.1 Data Preprocessing

The original dataset is preprocessed before providing it to the ML model, either for training or validation purposes, to improve the reliability of its data interpretation. The following steps are performed for preprocessing in this work.

- Removal of records with NaN and Infinity values.
- Deletion of the Timestamp column from the dataset.
- One-Hot encoding to convert nominal values to numerical data.
- Separation of data into training (70%) and test (30%) sets.
- Separation of features and labels from training and test sets.
- Min-Max normalization to all features in the train and test sets to scale the data between 0 and 1.

5.2 Train the IDS

The IDS classifier is trained well with the original dataset using the three learning algorithms specified in Sect. 2.3, as shown in Fig. 2. During this training, the IDS classifier becomes efficient in classifying benign and malicious data. Table 1 provides the hyperparameters of the classification algorithms during model training.

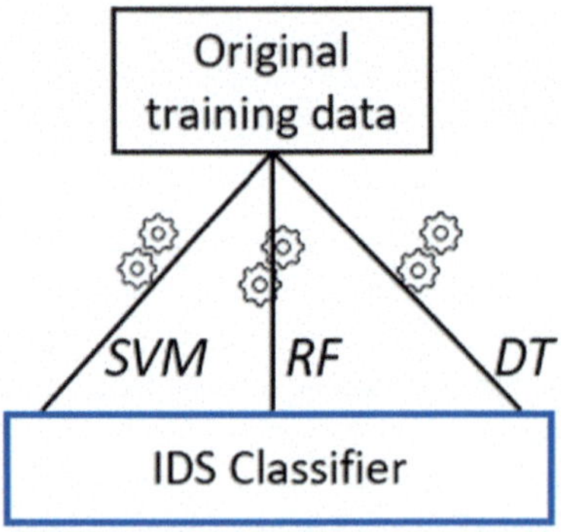

Fig. 2 Training the IDS classifier with original data

Table 1 Hyperparameters of the classification algorithms used for training the IDS classifier

Classifier	Hyperparameters
Decision tree	criterion = gini, max_depth = 12
Random forest	n_estimators = 200, criterion = gini, max_depth = 20
Support vector machine	C = 1, random_state = 42, loss = hinge

5.3 Adversarial Data Generation

Adversarial data generation involves a series of steps. The very first step is to pre-process the dataset as described in Sect. 5.1. A substitute classification model is used for the adversarial data generation and its partial evaluation to reduce the number of queries made to the target model. The substitute model is trained with the preprocessed training data using the RF classification algorithm. The core process of the attack starts with the creation of an initial population. A random population is created based on the data distribution of the dataset used. Each data point, hereafter referred to as an individual, in the population looks similar to a record in the dataset. Each individual is evaluated using a fitness function to determine how well it performs against the target model. This evaluation is achieved by passing each individual to the substitute RF model as a test input, and retrieving its accuracy, F1-score, recall, and AUC values. The higher the values are, the more the fitness of an individual is, which thereby means more misclassified adversarial examples. Each batch of individuals, i.e., the population, is passed to the genetic algorithm wherein the individuals are combined with one another, mutated, and new ones are generated. The fitness values of new ones are measured and the individuals with the least fitness values in the previous population are replaced with high-performing individuals from the new batch. For this experiment, the attack took 20 iterations, meaning, 20 generations of individuals were created to arrive at one best individual. Each of such best individuals can be considered an adversarial perturbation to be added to the original data. Adding the perturbation to the original data provides adversarial data. The resultant adversarial data is clipped to make sure it is present within the valid range of the original dataset. In this case, the adversarial data is clipped to be between 0 and 1.
Querying the Target Model with Adversarial Data Once a batch of adversarial data is ready, it is divided into training and test sets for this experiment. The algorithm queries the target model with the adversarial test set and observes the outputs provided by the model. The target model here represents only the IDS classifier without a defense layer. Therefore, the GA interacts directly with the IDS classifier in this module of the experiment, and the performance scores of the IDS classifier are presented later in Sect. 6 of this paper. Figures 3 and 4 show an overview of the process described in this section.

The hyperparameters of the RF algorithm used for the local substitute model during the adversarial data generation are provided in Table 2.

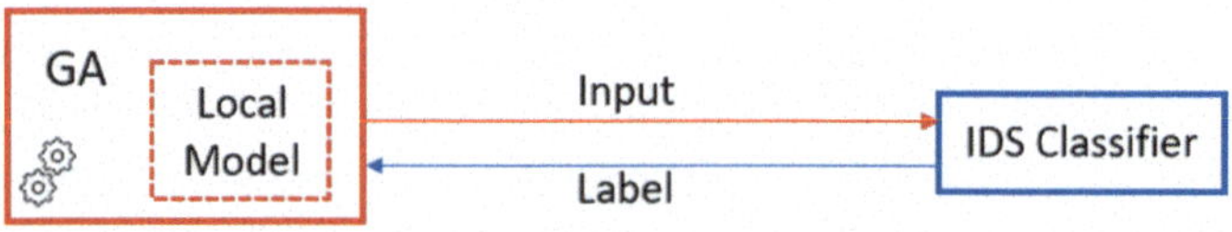

Fig. 3 Adversarial data is generated by interacting with the target black-box model and using the local model

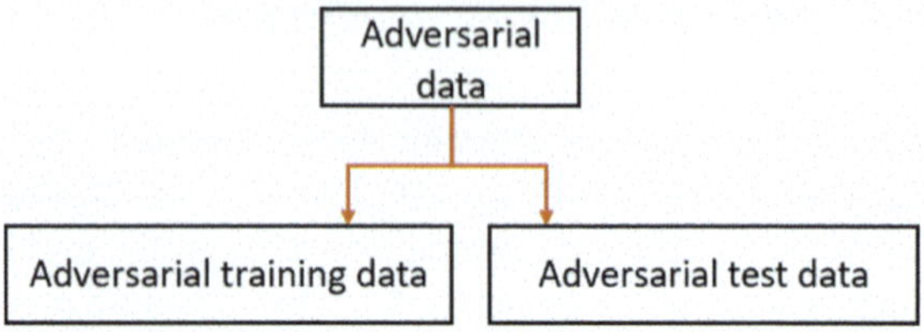

Fig. 4 Adversarial data is separated into training and test sets

Table 2 The hyperparameters for random forest algorithm used during adversarial data generation.

Parameter	Value
n_estimators	100
criterion	gini
max_depth	10
random_state	0

5.4 Training the Defense Layer

This module is for training the components present in the defense layer. As mentioned earlier in this paper, the proposed defense framework uses GAN. A GAN has two neural networks (NN) namely, a generator and a discriminator. The role of a generator is primarily in the training phase of a GAN. Its purpose is to create fake data that resembles the given original dataset, in this case, the CSE-CIC-IDS2018, by taking a random noise vector as input. The purpose of a discriminator is to distinguish between the fake data created by the generator and the original data. During the training phase, the generator network is trained to craft more and more realistic samples to increase the error rates of the discriminator; while the discriminator is trained to correctly classify the fake samples of the generator and the real data from the dataset. During this training, the discriminator sends feedback to the generator about how close its fake samples are to the original data, which is used by the generator to adjust its parameters and create more realistic data in the next iteration. Meanwhile, the discriminator also keeps adjusting its parameters in every iteration to reduce its error rates in detecting fake samples.

An additional step introduced in the proposed defense process is augmented adversarial training. The discriminator, in addition to being trained with the original training set and the fake data coming from the generator, is also trained with the adversarial training set that was separated after adversarial data generation as mentioned in Sect. 5.3. The reason behind performing the augmented training is to provide additional knowledge of adversarial inputs to the discriminator and make it more prepared for the validation phases. Such additional knowledge coming from the adversarial data generated by a powerful algorithm can promote the discriminator's ability to distinguish between real and fake data. It exposes the discriminator to not only fake data generated by GAN, but also a good number of samples generated by a sophisticated algorithm. Therefore, GAN and augmented training together bolster the detection capabilities of the defense layer, thereby, improving the robustness of IDS. Figure 5

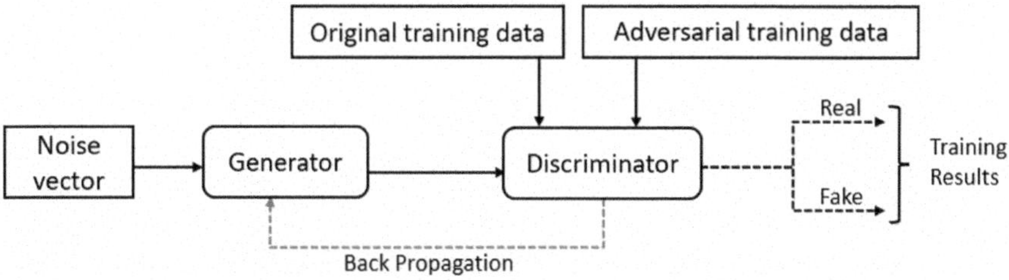

Fig. 5 Preparation of defense layer by training the GAN

Table 3 Architecture of GAN's generator network

Layer	Number of neurons	Activation
Dense_1	256	LeakyReLU
Dense_2	512	LeakyReLU
Dense_3	1024	LeakyReLU
Dense_4	2048	LeakyReLU
Output	79	tanh

Table 4 Architecture of GAN's discriminator network

Layer	Number of neurons	Activation
Dense_1	2048	LeakyReLU
Dense_2	1024	LeakyReLU
Dense_3	512	LeakyReLU
Dense_4	256	LeakyReLU
Output	1	sigmoid

outlines the GAN training. Tables 3 and 4 summarize the network configurations of the generator and discriminator respectively.

5.5 *Evaluate the Target Model with Defense*

In this module, the defense framework is plugged into the target model and the performance of the target model is assessed by providing the adversarial test inputs. In this scenario, the IDS classifier processes only the data that is forwarded to it by the discriminator of the defense layer. Meaning, the adversarial samples that are correctly classified by the discriminator are not acted upon, therefore, the adversarial algorithm does not receive any response from the target model.

6 Results and Discussion

This section presents evaluation results from the modules of the experiment discussed in Sect. 5.

Table 5 summarizes the performance scores of the IDS classifier when tested with original test data and later with adversarial test data. The scores presented here are the average values of five rounds of this experiment. Under adversarial settings, there is a significant drop in the performance scores overall, with values of almost all the metrics dropping by at least 50%. Table 6 summarizes the evaluation results of the target model under adversarial settings before and after the incorporation of the defense layer. The results show a promising improvement in the resistance of the target system to adversarial data, enabling it to perform almost as efficiently as it did in the absence of adversarial data.

Figures 6, 7, 8 and 9 show the changes observed in various performance metrics of the IDS classifier under the following three scenarios—when tested with original test data; when tested with a combination of original and adversarial test data in the absence of defense; when tested with a combination of original and adversarial test data in the presence of defense.

Analyzing the performance from the perspective of classification algorithms, scores were the highest and the percentage drop in the evaluation metrics overall was the least when the IDS classifier was trained with the RF algorithm. After integrating the defense layer into the IDS model, overall performance metrics, including accuracy, improved. The model demonstrated enhanced accuracy, correctly predicting more inputs. Given the dataset imbalance, additional metrics such as F1-score, recall, and AUC were considered. Recall, indicating the true-positive prediction rate, increased with the defense, improving the model's ability to classify different data classes accurately. AUC scores also showed improvement, highlighting the model's

Table 5 Comparison of performance of the IDS classifier in normal and adversarial settings

Classifier	Metric	Normal	Adversarial
RF	Accuracy	0.95	0.39
	F1-Score	0.91	0.37
	Recall	0.91	0.36
	AUC	1	0.5
DT	Accuracy	0.87	0.26
	F1-Score	0.91	0.29
	Recall	0.90	0.25
	AUC	1	0.47
SVM	Accuracy	0.72	0.23
	F1-Score	0.77	0.28
	Recall	0.91	0.30
	AUC	1	0.49

Table 6 Comparison of the performance in adversarial settings without and with defense

Classifier	Metric	Performance in adversarial settings without defense	Performance in adversarial settings with defense
RF	Accuracy	0.39	0.84
	F1-Score	0.37	0.82
	Recall	0.36	0.83
	AUC	0.5	0.89
DT	Accuracy	0.26	0.79
	F1-Score	0.29	0.82
	Recall	0.25	0.84
	AUC	0.47	0.90
SVM	Accuracy	0.23	0.61
	F1-Score	0.28	0.65
	Recall	0.30	0.87
	AUC	0.49	0.92

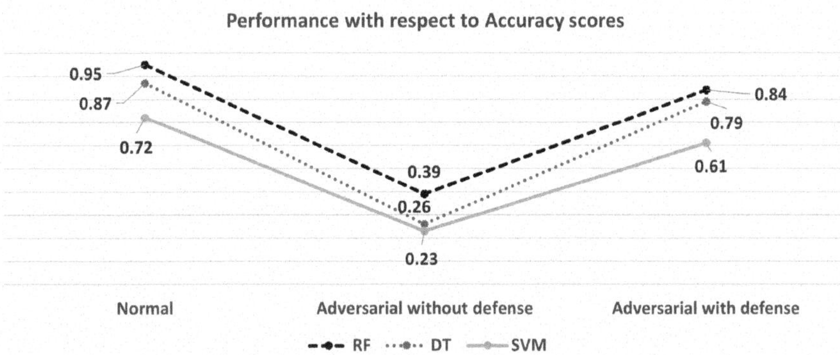

Fig. 6 Changes in the performance of target model with respect to accuracy scores

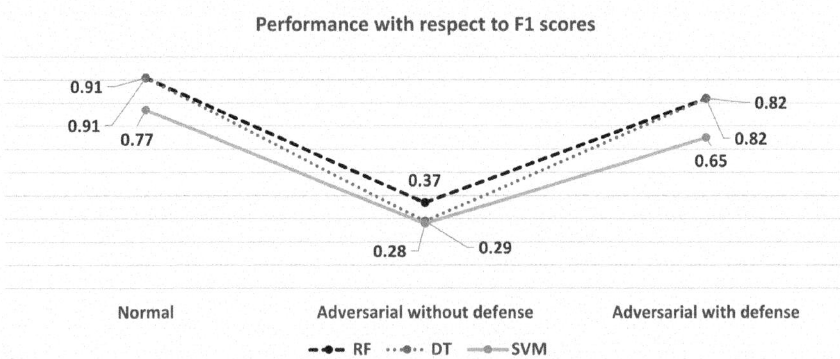

Fig. 7 Changes in the performance of target model with respect to F1-scores

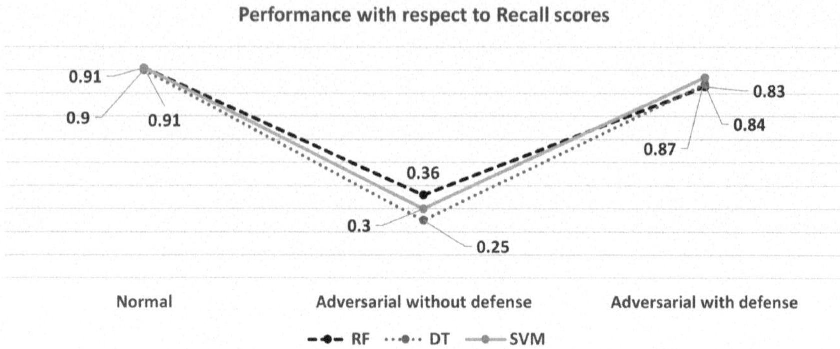

Fig. 8 Changes in the performance of target model with respect to recall scores

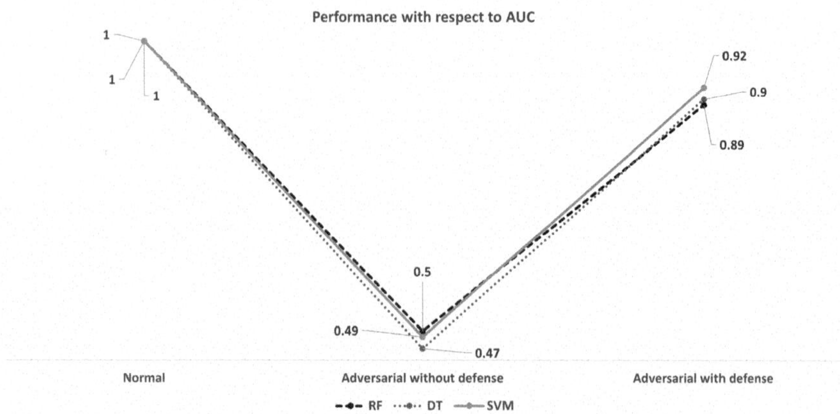

Fig. 9 Changes in the performance of target model with respect to AUC scores

enhanced capability to distinguish between normal and malign data. The defense mechanism exhibited promising enhancements in F1-scores, providing a comprehensive evaluation considering false-positives and false-negatives. Based on the evaluation of the defense layer's performance, on average, 87% of the adversarial samples were correctly classified as fake and filtered out. The performance of the defense layer is promising and may be further improved through our future work in this direction. The defense relies on a computationally intensive GA algorithm to train the defense layer, increasing time and resource consumption. Balancing resource usage and performance is a continual trade-off.

The work presented by Muhammad Usama et al. [36] discusses creating a GAN-based adversarial attack and proposes a GAN-based defense to thwart their attack. However, the evaluation of their work was performed using KDD99 dataset [2]. Based on the evaluation results of their GAN-based attack and defense mechanisms, the approach of this work gives a much better performance considering the following factors—(a) the normal/baseline scores of the target model produced in this work are much higher than those from their experiment; (b) the drop in the scores from the attack performed in the current work is far greater than that presented in their results; (c) the final scores after incorporating the current defense proposed are, for almost all the metrics, higher.

7 Conclusions and Future Work

This work proposes a defensive approach to mitigate the influence of powerful adversarial black-box attacks on ML-based IDSs. An evolutionary attack algorithm known as the genetic algorithm is implemented to generate adversarial data and to analyze the performance of the IDS black-box model under two scenarios—in the absence of the proposed defense; and in the presence of the defense. The defense proposed here follows a GAN-based approach combined with additional adversarial training wherein the discriminator of GAN is trained with a subset of the adversarial data generated using the genetic algorithm. The experiment results are presented summarizing the performance of the model under normal and adversarial settings, followed by the evaluation results with the proposed defense in place.

For the future work, the defense shall be evaluated using other sophisticated black-box attacks. Another idea is to use an IDS model that is deployed in a real-time environment, for example, a publicly accessible cloud-based IDS model to compare its resistance to black-box attacks with the model (equipped with defense) evaluated in this work. The work also proceeds in the direction of enhancing the detection capacity with the proposed defense framework and thereby making it harder to penetrate. There are considerations to deploy the model in a real-time cybersecurity environment once some of the above-mentioned milestones are accomplished.

Disclosure of Interest The authors have no competing interests.

References

1. Agarwal A, Singh R, Vatsa M, Ratha N (2020) Image transformation-based defense against adversarial perturbation on deep learning models. IEEE Trans Dependable Secure Comput 18(5):2106–2121. https://doi.org/10.1109/TDSC.2020.3027183
2. Al-Daweri MS, Zainol Ariffin KA, Abdullah S, Md. Senan MFE (2020) An analysis of the kdd99 and unsw-nb15 datasets for the intrusion detection system. Symmetry 12(10):1666. https://doi.org/10.3390/sym12101666
3. Alahmed S, Alasad Q, Hammood MM, Yuan JS, Alawad M (2022) Mitigation of black-box attacks on intrusion detection systems-based ml. Computers 11(7):115. https://doi.org/10.3390/computers11070115
4. Alarsan FI, Younes M (2021) Best selection of generative adversarial networks hyper-parameters using genetic algorithm. SN Comput Sci 2(4):283. https://doi.org/10.1007/s42979-021-00689-3

5. Alatwi HA, Aldweesh A (2021) Adversarial black-box attacks against network intrusion detection systems: a survey. In: 2021 IEEE world AI IoT congress (AIIoT). IEEE, pp 34–40. https://doi.org/10.1109/AIIOT52608.2021.9454214
6. Alotaibi A, Rassam MA (2023) Adversarial machine learning attacks against intrusion detection systems: a survey on strategies and defense. Future Internet 15(2):62. https://doi.org/10.3390/fi15020062
7. Alslman Y (2023) A robust SNMP-MIB intrusion detection system against adversarial attacks. Arab J Sci Eng. https://doi.org/10.1007/s13369-023-08315-5
8. Amaizu GC, Nwakanma CI, Lee JM, Kim DS (2020) Investigating network intrusion detection datasets using machine learning. In: 2020 International conference on information and communication technology convergence (ICTC). IEEE, pp 1325–1328. https://doi.org/10.1109/ictc49870.2020.9289329
9. Apruzzese G, Andreolini M, Colajanni M, Marchetti M (2020) Hardening random forest cyber detectors against adversarial attacks. IEEE Trans Emerg Top Comput Intell 4(4):427–439. https://doi.org/10.1109/tetci.2019.2961157
10. Apruzzese G, Colajanni M, Ferretti L, Guido A, Marchetti M (2018) On the effectiveness of machine and deep learning for cyber security. In: 2018 10th international conference on cyber conflict (CyCon). IEEE, pp 371–390. https://doi.org/10.23919/CYCON.2018.8405026
11. Biggio B, Corona I, Maiorca D, Nelson B, Šrndić N, Laskov P, Giacinto G, Roli F (2013) Evasion attacks against machine learning at test time. In: Machine learning and knowledge discovery in databases: European conference, ECML PKDD 2013, Prague, Czech Republic, September 23–27, 2013, Proceedings, Part III 13. Springer, pp 387–402. https://doi.org/10.1007/978-3-642-40994-3_25
12. Blanzieri E, Bryl A (2008) A survey of learning-based techniques of email spam filtering. Artif Intell Rev 29:63–92. https://doi.org/10.1007/s10462-009-9109-6
13. Buczak AL, Guven E (2015) A survey of data mining and machine learning methods for cyber security intrusion detection. IEEE Commun Surv Tutorials 18(2):1153–1176. https://doi.org/10.1109/comst.2015.2494502
14. Carlini N, Wagner D (2017) Towards evaluating the robustness of neural networks. In: 2017 IEEE symposium on security and privacy (sp). IEEE, pp 39–57. https://doi.org/10.1109/sp.2017.49
15. Debicha I (2021) Detect and reject for transferability of black-box adversarial attacks against network intrusion detection systems. In: Advances in cyber security. Springer Singapore, pp 329–339. https://doi.org/10.1007/978-981-16-8059-5_20
16. Dwibedi S, Pujari M, Sun W (2020) A comparative study on contemporary intrusion detection datasets for machine learning research. In: 2020 IEEE international conference on intelligence and security informatics (ISI). IEEE, pp 1–6. https://doi.org/10.1109/ISI49825.2020.9280519
17. Gharaee H, Hosseinvand H (2016) A new feature selection ids based on genetic algorithm and SVM. In: 2016 8th International symposium on telecommunications (IST). IEEE, pp 139–144. https://doi.org/10.1109/ISTEL.2016.7881798
18. Hoque MS, Mukit MA, Bikas MAN (2012) An implementation of intrusion detection system using genetic algorithm. arXiv preprint arXiv:1204.1336. https://doi.org/10.48550/arXiv.1204.1336
19. Ibitoye O, Shafiq O, Matrawy A (2019) Analyzing adversarial attacks against deep learning for intrusion detection in IoT networks. In: 2019 IEEE global communications conference (GLOBECOM). IEEE, pp 1–6. https://doi.org/10.1109/GLOBECOM38437.2019.9014337
20. Kettani H, Wainwright P (2019) On the top threats to cyber systems. In: 2019 IEEE 2nd international conference on information and computer technologies (ICICT). IEEE, pp 175–179. https://doi.org/10.1109/INFOCT.2019.8711324
21. Kim J, Kim J, Thu HLT, Kim H (2016) Long short term memory recurrent neural network classifier for intrusion detection. In: 2016 international conference on platform technology and service (PlatCon). IEEE, pp 1–5. https://doi.org/10.1109/PlatCon.2016.7456805
22. Liu Y, Chen X, Liu C, Song D (2016) Delving into transferable adversarial examples and black-box attacks. arXiv preprint arXiv:1611.02770. https://doi.org/10.48550/arXiv.1611.02770

23. McCarthy A, Ghadafi E, Andriotis P, Legg P (2022) Functionality-preserving adversarial machine learning for robust classification in cybersecurity and intrusion detection domains: a survey. J Cybersecur Privacy 2(1):154–190. https://doi.org/10.3390/jcp2010010
24. Papernot N, McDaniel P, Goodfellow I, Jha S, Celik ZB, Swami A (2017) Practical black-box attacks against machine learning. In: Proceedings of the 2017 ACM on Asia conference on computer and communications security, pp 506–519. https://doi.org/10.1145/3052973.3053009
25. Papernot N, McDaniel P, Jha S, Fredrikson M, Celik ZB, Swami A (2016) The limitations of deep learning in adversarial settings. In: 2016 IEEE European symposium on security and privacy (EuroS&P). IEEE, pp 372–387. https://doi.org/10.1109/EuroSP.2016.36
26. Papernot N, McDaniel P, Sinha A, Wellman MP (2018) Sok: security and privacy in machine learning. In: 2018 IEEE European symposium on security and privacy (EuroS&P). IEEE, pp 399–414. https://doi.org/10.1109/EuroSP.2018.00035
27. Pujari M, Cherukuri BP, Javaid AY, Sun W (2022) An approach to improve the robustness of machine learning based intrusion detection system models against the Carlini-Wagner attack. In: 2022 IEEE international conference on cyber security and resilience (CSR). IEEE, pp 62–67. https://doi.org/10.1109/CSR54599.2022.9850306
28. Pujari M, Pacheco Y, Cherukuri B, Sun W (2022) A comparative study on the impact of adversarial machine learning attacks on contemporary intrusion detection datasets. SN Comput Sci 3(5):412. https://doi.org/10.1007/s42979-022-01321-8
29. Schneider M, Aspinall D, Bastian ND (2021) Evaluating model robustness to adversarial samples in network intrusion detection. In: 2021 IEEE international conference on big data (big data). IEEE, pp 3343–3352. https://doi.org/10.1109/bigdata52589.2021.9671580
30. Sharafaldin I, Gharib A, Lashkari AH, Ghorbani AA (2018) Towards a reliable intrusion detection benchmark dataset. Softw Netw 2018(1):177–200. https://doi.org/10.13052/jsn2445-9739.2017.009
31. Sommer R, Paxson V (2010) Outside the closed world: on using machine learning for network intrusion detection. In: 2010 IEEE symposium on security and privacy. IEEE, pp 305–316. https://doi.org/10.1109/sp.2010.25
32. Su J, Vargas DV, Sakurai K (2019) One pixel attack for fooling deep neural networks. IEEE Trans Evolut Comput 23(5):828–841. https://doi.org/10.1109/TEVC.2019.2890858
33. Subedi R, Shrestha P, Pujari M, Chou EY (2022) Vehicle classification, rumble strips detection, and mapping using artificial intelligence. In: International conference on transportation and development 2022, pp 46–56. https://doi.org/10.1061/9780784484319.005
34. Tian S, Yang G, Cai Y (2018) Detecting adversarial examples through image transformation. In: Proceedings of the AAAI conference on artificial intelligence, vol 32, p 8. https://doi.org/10.1609/aaai.v32i1.11828
35. Torres P, Catania C, Garcia S, Garino CG (2016) An analysis of recurrent neural networks for botnet detection behavior. In: 2016 IEEE biennial congress of Argentina (ARGENCON). IEEE, pp 1–. https://doi.org/10.1109/ARGENCON.2016.7585247
36. Usama M, Asim M, Latif S, Qadir J et al (2019) Generative adversarial networks for launching and thwarting adversarial attacks on network intrusion detection systems. In: 2019 15th International wireless communications and mobile computing conference (IWCMC). IEEE, pp 78–83. https://doi.org/10.1109/iwcmc.2019.8766353
37. Zhao S, Li J, Wang J, Zhang Z, Zhu L, Zhang Y (2021) Attackgan: adversarial attack against black-box ids using generative adversarial networks. Proc Comput Sci 187:128–133. https://doi.org/10.1016/j.procs.2021.04.118
38. Zhu Y, Cui L, Ding Z, Li L, Liu Y, Hao Z (2022) Black box attack and network intrusion detection using machine learning for malicious traffic. Comput Secur 123:102922. https://doi.org/10.1016/j.cose.2022.102922

Author Index

X.-S. Yang et al. (eds.), *Proceedings of Ninth International Congress on Information and Communication Technology*, Lecture Notes in Networks and Systems 1000,
https://doi.org/10.1007/978-981-97-3289-0

The manufacturer's authorised representative in the EU is Springer
Nature Customer Service Centre GmbH, Europaplatz 3, 69115 Heidelberg,
Germany. If you have any concerns regarding our products, please
contact ProductSafety@springernature.com

Printed and bound by CPI Group (UK) Ltd, Croydon, CR0 4YY
30/09/2026
02225232-0014